# A Textbook of Physical Chemistry

## Volume III

**A Textbook of Physical Chemistry**

| | | |
|---|---|---|
| *Volume I* | : | States of Matter and Ions in Solution |
| *Volume II* | : | Thermodynamics and Chemical Equilibrium |
| *Volume III* | : | Applications of Thermodynamics |
| *Volume IV* | : | Quantum Chemistry and Molecular Spectroscopy |
| *Volume V* | : | Dynamics of Chemical Reactions, Statistical Thermodynamics Macromolecules, and Irreversible Processes |
| *Volume VI* | : | Computational Aspects in Physical Chemistry |

# A Textbook of Physical Chemistry

## Volume III
## (SI Units)

## Applications of Thermodynamics

**Fourth Edition**

**K L KAPOOR**

Former Associate Professor
Hindu College
University of Delhi
New Delhi

**McGraw Hill Education (India) Private Limited**
NEW DELHI

*McGraw Hill Education Offices*

**New Delhi** New York St Louis San Francisco Auckland Bogotá Caracas
Kuala Lumpur Lisbon London Madrid Mexico City Milan Montreal
San Juan Santiago Singapore Sydney Tokyo Toronto

**McGraw Hill Education (India) Private Limited**

Published by McGraw Hill Education (India) Private Limited,
P-24, Green Park Extension, New Delhi 110 016.

**A Textbook of Physical Chemistry, Vol III**



**Print Edition**
ISBN 13: 9789339218928
ISBN 10: 9339218922

**E-book Edition**
ISBN (13): 978-93-392-0428-0
ISBN (10): 93-392-0428-X

Managing Director: *Kaushik Bellani*

Head—Higher Education (Publishing and Marketing): *Vibha Mahajan*

Senior Publishing Manager (SEM & Tech. Ed.): *Shalini Jha*
Associate Sponsoring Editor: *Smruti Snigdha*
Senior Editorial Researcher: *Amiya Mahapatra*
Senior Development Editor: *Renu Upadhyay*
Manager—Production Systems: *Satinder S Baveja*
Assistant Manager—Editorial Services : *Sohini Mukherjee*

Assistant General Manager (Marketing)—Higher Education: *Vijay Sarathi*
Senior Graphic Designer—Cover: *Meenu Raghav*
General Manager—Production: *Rajender P Ghansela*
Manager—Production: *Reji Kumar*



Typeset at Script Makers, 19, A1-B, DDA Market, Paschim Vihar, New Delhi 110 063, and text printed at

*To the Memory of*
*My Parents*

# Preface

In recent years, the teaching curriculum of Physical Chemistry in many Indian universities has been restructured with a greater emphasis on a theoretical and conceptual methodology and the applications of the underlying basic concepts and principles. This shift in the emphasis, as I have observed, has unduly frightened undergraduates whose performance in Physical Chemistry has been otherwise generally far from satisfactory. This poor performance is partly because of the non-availability of a comprehensive textbook which also lays adequate stress on the logical deduction and solution of numericals and related problems. Naturally, the students find themselves unduly constrained when they are forced to refer to various books to collect the necessary reading material.

It is primarily to help these students that I have ventured to present a textbook which provides a systematic and comprehensive coverage of the theory as well as of the illustration of the applications thereof.

The present volumes grew out of more than a decade of classroom teaching through lecture notes and assignments prepared for my students of BSc (General) and BSc (Honours). The schematic structure of the book is assigned to cover the major topics of Physical Chemistry in six different volumes. **Volume I** discusses the states of matter and ions in solutions. It comprises five chapters on the gaseous state, physical properties of liquids, solid state, ionic equilibria and conductance. **Volume II** describes the basic principles of thermodynamics and chemical equilibrium in seven chapters, viz., introduction and mathematical background, zeroth and first laws of thermodynamics, thermochemistry, second law of thermodynamics, criteria for equilibrium and $A$ and $G$ functions, systems of variable composition, and thermodynamics of chemical reactions. **Volume III** seeks to present the applications of thermodynamics to the equilibria between phases, colligative properties, phase rule, solutions, phase diagrams of one-, two- and three-component systems, and electrochemical cells. **Volume IV** deals with quantum chemistry, molecular spectroscopy and applications of molecular symmetry. It focuses on atomic structure, chemical bonding, electrical and magnetic properties, molecular spectroscopy and applications of molecular symmetry. **Volume V** covers dynamics of chemical reactions, statistical and irreversible thermodynamics, and macromolecules in six chapters, viz., adsorption, chemical kinetics, photochemistry, statistical thermodynamics, macromolecules and introduction to irreversible processes. **Volume VI** describes computational aspects in physical chemistry in three chapters, viz., synopsis of commonly used statements in BASIC language, list of programs, and projects.

The study of Physical Chemistry is incomplete if students confine themselves to the ambit of theoretical discussions of the subject. They must grasp the practical significance of the basic theory in all its ramifications and develop a clear perspective to appreciate various problems and how they can be solved.

It is here that these volumes merit mention. Apart from having a lucid style and simplicity of expression, each has a wealth of carefully selected examples and solved illustrations. Further, three types of problems with different objectives in view are listed at the end of each chapter: (1) Revisionary Problems, (2) Try Yourself Problems, and (3) Numerical Problems. Under *Revisionary Problems*, only those problems pertaining to the text are included which should afford an opportunity to the students in self-evaluation. In *Try Yourself Problems*, the problems related to the text but not highlighted therein are provided. Such problems will help students extend their knowledge of the chapter to closely related problems. Finally, unsolved *Numerical Problems* are pieced together for students to practice.

Though the volumes are written on the basis of the syllabi prescribed for undergraduate courses of the University of Delhi, they will also prove useful to students of other universities, since the content of physical chemistry remains the same everywhere. In general, the SI units (*Systeme International d'unite's*), along with some of the common non-SI units such as atm, mmHg, etc., have been used in the books.

## Salient Features

- Comprehensive coverage to applications of thermodynamics to the equilibria between phases, colligative properties, phase rule, solutions, phase diagrams of one-, two-and three-component systems and electrochemical cells
- Emphasis given to applications and principles
- Explanation of equations in the form of solved problems and numericals
- IUPAC recommendations and SI units have been adopted throughout
- Rich and illustrious pedagogy

## Acknowledgements

I wish to acknowledge my greatest indebtedness to my teacher, late Prof. R P Mitra, who instilled in me the spirit of scientific inquiry. I also record my sense of appreciation to my students and colleagues at Hindu College, University of Delhi, for their comments, constructive criticism and valuable suggestions towards improvement of the book. I am grateful to late Dr Mohan Katyal (St. Stephen's College), and late Prof. V R Shastri (Ujjain University) for the numerous suggestions in improving the book. I would like to thank Sh. M M Jain, Hans Raj College, for his encouragement during the course of publication of the book.

I wish to extend my appreciation to the students and teachers of Delhi University for the constructive suggestions in bringing out this edition of the book. I also wish to thank my children, Saurabh-Urvashi and Surabhi-Jugnu, for many useful suggestions in improving the presentation of the book.

Finally, my special thanks go to my wife, Pratima, for her encouragement, patience and understanding.

## Feedback Request

The author takes the entire responsibility for any error or ambiguity, in fact or opinion, that may have found its way into this book. Comments and criticism from readers will, therefore, be highly appreciated and incorporated in subsequent editions.

**K L Kapoor**

## Publisher's Note

McGraw-Hill Education (India) invites suggestions and comments from you, all of which can be sent to *info.india@mheducation.com* (kindly mention the title and author name in the subject line).
Piracy-related issues may also be reported.

# Contents

# 1 Equilibrium Between Phases

## 1.1 THERMODYNAMIC CRITERION OF PHASE EQUILIBRIA

**An Example of a Substance Distributed Between Two Phases**

Consider a system in which a pure substance B is present in two phases in equilibrium with each other. Let $\mu_{B(\alpha)}$ and $\mu_{B(\beta)}$ denote the chemical potentials of the substance in phase α and phase β, respectively. The Gibbs function of the system is given by

$$G = f(T, p, n_{B(\alpha)}, n_{B(\beta)}) \qquad (1.1.1)^{\dagger}$$

where $n_{B(\alpha)}$ and $n_{B(\beta)}$ are the amounts of the substance B in phases α and β, respectively. The change in Gibbs function of the system as a result of changes in $T$, $p$, $n_{B(\alpha)}$ and $n_{B(\beta)}$ is given by

$$dG = \left(\frac{\partial G}{\partial T}\right)_{p,\, n_{B(\alpha)},\, n_{B(\beta)}} dT + \left(\frac{\partial G}{\partial p}\right)_{T,\, n_{B(\alpha)},\, n_{B(\beta)}} dp + \left(\frac{\partial G}{\partial n_{B(\alpha)}}\right)_{T,\, p,\, n_{B(\beta)}} dn_{B(\alpha)} + \left(\frac{\partial G}{\partial n_{B(\beta)}}\right)_{T,\, p,\, n_{B(\alpha)}} dn_{B(\beta)}$$

or

$$dG = \left(\frac{\partial G}{\partial T}\right)_{p,\, n_{B(\alpha)},\, n_{B(\beta)}} dT + \left(\frac{\partial G}{\partial p}\right)_{T,\, n_{B(\alpha)},\, n_{B(\beta)}} dp + \mu_{B(\alpha)}\, dn_{B(\alpha)} + \mu_{B(\beta)}\, dn_{B(\beta)} \qquad (1.1.2)$$

If the amount $dn_B$ of the substance is transferred from phase α to phase β at constant temperature and pressure, then from Eq. (1.1.2), we have

$$\begin{aligned} dG &= \mu_{B(\alpha)}\,(-dn_B) + \mu_{B(\beta)}(dn_B) \\ &= (\mu_{B(\beta)} - \mu_{B(\alpha)})\, dn_B \end{aligned} \qquad (1.1.3)$$

If the above transfer takes place reversibly, then

$$dG = 0$$

---

†Throughout this book, phase is written within brackets drawn immediately after the substance.

and Eq. (1.1.3) reduces to

$$(\mu_{B(\beta)} - \mu_{B(\alpha)})\, dn_B = 0 \tag{1.1.4}$$

Since $dn_B \neq 0$, it follows that

$$\mu_{B(\alpha)} = \mu_{B(\beta)} \tag{1.1.5}$$

that is, if a substance is in equilibrium between two phases, its chemical potential must have the same value in both the phases.

**Generalized Treatment**

The above criterion of a substance present in two phases in equilibrium may be generalized for a system containing more than one phase and each phase containing more than one substance. Let us consider a closed system at a given temperature and pressure containing the amounts $n_1$, $n_2$, $n_3$, ... of components 1, 2, 3, ..., respectively. The differential of the Gibbs function of each phase at constant temperature and pressure may be written as

$$\begin{aligned}
dG_{(\alpha)} &= \mu_{1(\alpha)}\, dn_{1(\alpha)} + \mu_{2(\alpha)}\, dn_{2(\alpha)} + \mu_{3(\alpha)}\, dn_{3(\alpha)} + \cdots \\
dG_{(\beta)} &= \mu_{1(\beta)}\, dn_{1(\beta)} + \mu_{2(\beta)}\, dn_{2(\beta)} + \mu_{3(\beta)}\, dn_{3(\beta)} + \cdots \\
dG_{(\gamma)} &= \mu_{1(\gamma)}\, dn_{1(\gamma)} + \mu_{2(\gamma)}\, dn_{2(\gamma)} + \mu_{3(\gamma)}\, dn_{3(\gamma)} + \cdots \\
&\cdots\cdots\cdots\cdots\cdots\cdots
\end{aligned} \tag{1.1.6}$$

The total change of Gibbs function is given by

$$dG = dG_{(\alpha)} + dG_{(\beta)} + dG_{(\gamma)} + \cdots = 0 \tag{1.1.7}$$

(This is zero because of equilibrium between the phases.)

Since the system over all is a closed system, the sum of changes of the substances in various phases must be zero. Thus, we have

$$\begin{aligned}
&dn_{1(\alpha)} + dn_{1(\beta)} + dn_{1(\gamma)} + \cdots = 0 \\
&dn_{2(\alpha)} + dn_{2(\beta)} + dn_{2(\gamma)} + \cdots = 0 \\
&dn_{3(\alpha)} + dn_{3(\beta)} + dn_{3(\gamma)} + \cdots = 0 \\
&\cdots\cdots\cdots\cdots\cdots\cdots
\end{aligned} \tag{1.1.8}$$

Equation (1.1.8) can be rewritten as

$$\begin{aligned}
dn_{1(\alpha)} &= -(dn_{1(\beta)} + dn_{1(\gamma)} + \cdots) \\
dn_{2(\alpha)} &= -(dn_{2(\beta)} + dn_{2(\gamma)} + \cdots) \\
dn_{3(\alpha)} &= -(dn_{3(\beta)} + dn_{3(\gamma)} + \cdots) \\
&\cdots\cdots\cdots\cdots\cdots\cdots
\end{aligned}$$

Substituting these in Eq. (1.1.7) and rearranging, we get

$$\begin{aligned}
&(\mu_{1(\beta)} - \mu_{1(\alpha)})\, dn_{1(\beta)} + (\mu_{1(\gamma)} - \mu_{1(\alpha)})\, dn_{1(\gamma)} + \cdots \\
&+ (\mu_{2(\beta)} - \mu_{2(\alpha)})\, dn_{2(\beta)} + (\mu_{2(\gamma)} - \mu_{2(\alpha)})\, dn_{2(\gamma)} + \cdots \\
&+ (\mu_{3(\beta)} - \mu_{3(\alpha)})\, dn_{3(\beta)} + (\mu_{3(\gamma)} - \mu_{3(\alpha)})\, dn_{1(\gamma)} + \cdots = 0
\end{aligned}$$

Since all d*n*s are independent variables, we must have

$$\mu_{1(\alpha)} = \mu_{1(\beta)} = \mu_{1(\gamma)} = \cdots$$
$$\mu_{2(\alpha)} = \mu_{2(\beta)} = \mu_{2(\gamma)} = \cdots$$
$$\mu_{3(\alpha)} = \mu_{3(\beta)} = \mu_{3(\gamma)} = \cdots$$
$$\cdots\cdots\cdots \qquad (1.1.9)$$

that is, if the system is in equilibrium, the chemical potential of any one component must have the same value in each phase in which it exists.

## 1.2 CHEMICAL POTENTIAL VERSUS TEMPERATURE GRAPHS OF A PURE SUBSTANCE

**Factor Determining the Slope of Graph**

For a pure substance, we have

$$\mathrm{d}\mu = -S_\mathrm{m}\,\mathrm{d}T + V_\mathrm{m}\,\mathrm{d}p \qquad (1.2.1)$$

where $S_\mathrm{m}$ and $V_\mathrm{m}$ are its molar entropy and volume, respectively. The expression for the slope of $\mu$ versus $T$ curve at a constant pressure as obtained from Eq. (1.2.1) is

$$\left(\frac{\partial \mu}{\partial T}\right)_p = -S_\mathrm{m} \qquad (1.2.2)$$

Now $S_\mathrm{m}$ of a given substance in any phase, i.e. solid(s), liquid(1) or gas(g), is always positive with the result that the slope of $\mu$ versus $T$ curve at a constant pressure is always negative

For the three phases of a given substance, we have

$$\left(\frac{\partial \mu_\mathrm{s}}{\partial T}\right)_p = -S_\mathrm{m,s}; \quad \left(\frac{\partial \mu_\mathrm{l}}{\partial T}\right)_p = -S_\mathrm{m,l}; \quad \left(\frac{\partial \mu_\mathrm{g}}{\mathrm{d}T}\right)_p = -S_\mathrm{m,g}$$

At any temperature, since

$$S_\mathrm{m,s} < S_\mathrm{m,l} << S_\mathrm{m,g}$$

it follows that

$$-S_\mathrm{m,s} > -S_\mathrm{m,l} >> -S_\mathrm{m,g} \qquad (1.2.3)$$

In other words, we have:

- The slope of $\mu$ versus $T$ curve for a substance in the solid phase has a small negative value as $S_\mathrm{m,s}$ has a small value.
- The slope of $\mu$ versus $T$ curve for the liquid phase is slightly more negative than that of the solid phase.
- The slope of $\mu$ versus $T$ curve for the gaseous phase has a large negative value as $S_\mathrm{m,g}$ is very much larger than that of the liquid.

**Diagrammatic Representation**

The above characteristics are shown in Fig. 1.2.1.

**Conclusion**

It will be seen from Fig. 1.2.1 that the curve for the solid and liquid phases intersect each other at temperature $T_\mathrm{m}$. At this temperature, the chemical potential

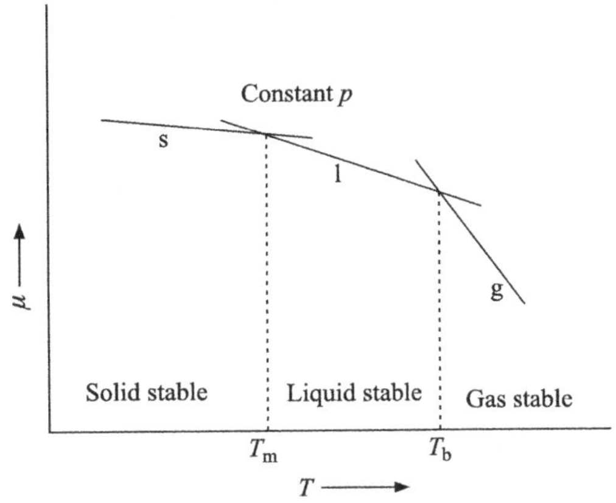


**Fig. 1.2.1 Chemical potential versus temperature graphs of solid, liquid and gaseous phases of a pure substance**

of the substance is same in both the phases and, consequently, it represents the solid-liquid equilibrium. The temperature $T_m$ is the melting point (m.pt.) of the solid. Similarly, the curves for the liquid and gaseous phases intersect each other at temperature $T_b$ where $\mu_l = \mu_g$. At this point, the liquid and gaseous phases coexist in equilibrium with each other. The temperature $T_b$ is the boiling point (b.pt.) of the liquid.

The state of the system at any temperature can be obtained from Fig. 1.2.1. Thus, we have:

(i) $T < T_m$ : $\mu_s$ has the lowest value and thus the solid phase is stable.
(ii) $T = T_m$ : $\mu_s = \mu_l$; solid and liquid phases are in equilibrium.
(iii) $T_b > T > T_m$ : $\mu_l$ has the lowest value and thus the liquid phase is stable.
(iv) $T = T_b$ : $\mu_l = \mu_g$; liquid and gaseous phases are in equilibrium.
(v) $T > T_b$ : $\mu_g$ has the lowest value and thus the gaseous phase is stable.

## 1.3 EFFECT OF PRESSURE ON THE CHEMICAL POTENTIAL VERSUS TEMPERATURE GRAPHS

**Factor Determining the Slope of Graph**

It can be seen from Eq. (1.2.1) that at constant temperature

$$\left(\frac{\partial \mu}{\partial p}\right)_T = V_m \tag{1.3.1}$$

Since $V_m$ is always positive, it follows that d$\mu$ has the same sign as that of d$p$, i.e. d$\mu$ is positive if d$p$ is positive and it is negative if d$p$ is negative.

If the pressure is changed by an infinitesimal amount d$p$ at constant temperature, the change in the chemical potential is given by

$$d\mu = V_m\, dp \tag{1.3.2}$$

Since for most substances

$$V_{m,g} >> V_{m,l} > V_{m,s}$$

it follows that the order of change in chemical potential is

$$|d\mu_g| >> |d\mu_l| > |d\mu_s| \tag{1.3.3}$$

**Diagrammatic Representation**

Figure 1.3.1 shows qualitatively the effects which are produced on the chemical potential (shown by dotted lines) of a substance in the three phases when pressure of the system is reduced. The decrease at a given temperature is smallest for the solid phase and largest for the gaseous phase (Eq. 1.3.3). For example, at a temperature $T$, the decrease in chemical potential for solid is from a to a′, for liquid from b to b′ and that for gas from c to c′. The decrease c to c′ is much larger than from a to a′ and b to b′.

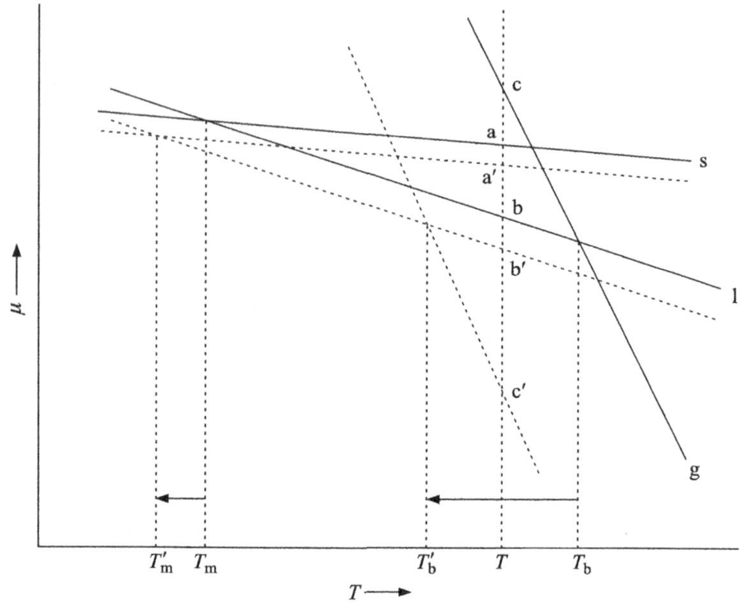


**Fig. 1.3.1 Effect of pressure on the chemical potential versus temperature plots of a substance having $V_{m,\,l} > V_{m,\,s}$**

**Conclusion**

The following conclusions my be drawn from Fig. 1.3.1.

The melting point as well as the boiling point shifts to a lower value as the pressure is decreased; the shift is relatively larger for boiling point.

**Condition for Sublimation of a Substance**

The temperature difference between $T_m$ and $T_b$ decreases with decrease in pressure. This indicates that the temperature range over which the liquid phase can exist is decreased. If the pressure is decreased to a sufficiently low value, it may happen that the liquid is not formed, the solid directly passes over to the gaseous form. This happens when the boiling point of the liquid falls below the melting point of the solid as shown in Fig. 1.3.2.

The temperature $T_s$ is the sublimation temperature and is found to be very sensitive to changes in pressure.

From Fig. 1.3.2, it may be concluded that if:

(i) $T < T_s$ : $\mu_s$ has the lowest value and thus the solid phase is stable,
(ii) $T = T_s$ : $\mu_s = \mu_g$, both solid and gaseous phases coexist in equilibrium,
(iii) $T > T_s$ : $\mu_g$ has the lowest value and thus the gaseous phase is stable.

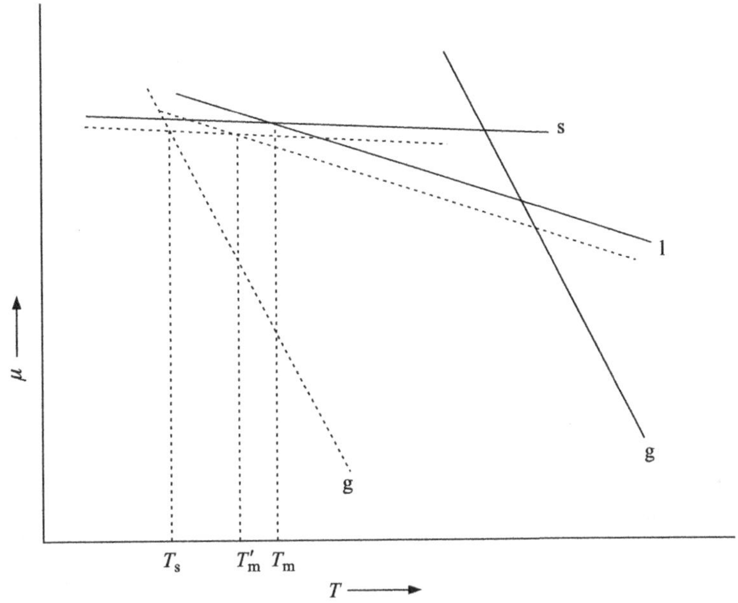


**Fig. 1.3.2 Effect of a large decrease in pressure on $\mu$ versus $T$ plots**

The effects which are produced when the pressure is increased are exactly opposite to those described above. There is an increase in melting point as well as boiling point of the substance; the increase in boiling point is relatively larger, and this enhances the stability of the liquid phase.

**Effect on a Substance Involving a Decrease in Volume on Melting**

The effect of pressure on the melting and boiling points of a substance which shows a decrease in volume on melting can likewise be discussed. In this case, the decrease of chemical potential of the solid will be larger than that of the liquid as shown in Fig. 1.3.3. It is obvious from Fig. 1.3.3 that there occurs an increase instead of a decrease in the melting point of the solid as the pressure is decreased. The increase in pressure will consequently decrease the melting point of such type of substances. Examples include water, bismuth and antimony.

## 1.4 CLAPEYRON EQUATION

**Derivation of Clapeyron Equation**

Let us consider a system in which a pure substance B is present in two phases $\alpha$ and $\beta$. The two phases may be either solid and liquid, or solid and vapour, or liquid and vapour. At equilibrium, we will have

$$\mu_{B(\alpha)} = \mu_{B(\beta)} \tag{1.4.1}$$

Now let the temperature of the system be changed by an infinitesimal amount. When the equilibrium is re-established, the pressure also undergoes a change by an infinitesimal amount. Under these conditions, the chemical potentials $\mu_{B(\alpha)}$ and $\mu_{B(\beta)}$ also change by infinitesimal amounts $d\mu_{B(\alpha)}$ and $d\mu_{B(\beta)}$, respectively. Since the system is again in equilibrium, if follows that

$$\mu_{B(\alpha)} + d\mu_{B(\alpha)} = \mu_{B(\beta)} + d\mu_{B(\beta)} \tag{1.4.2}$$

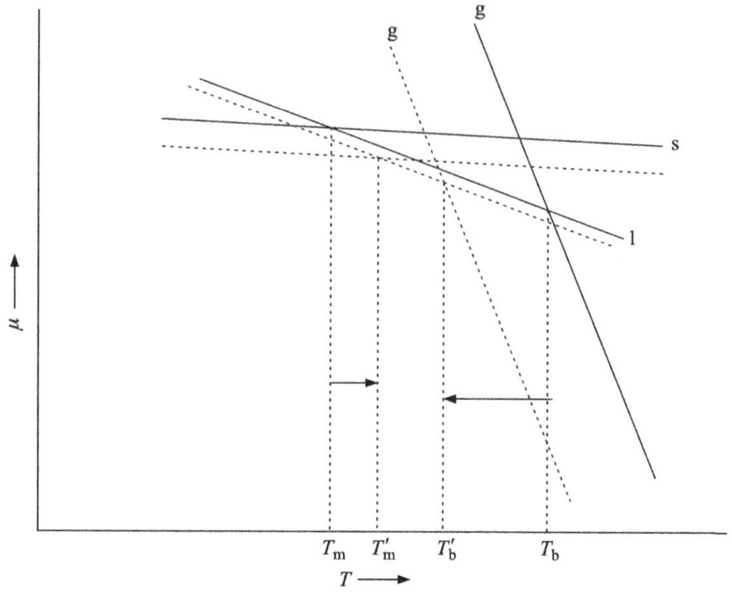


**Fig. 1.3.3 Effect of pressure on chemical potential versus temperature plot of a substance having $V_{m,l} < V_{m,s}$**

Making use of Eq. (1.4.1), this reduces to

$$d\mu_{B(\alpha)} = d\mu_{B(\beta)} \tag{1.4.3}$$

Writing d$\mu$s in the explicit form in terms of d$T$ and d$p$, we get

$$\begin{aligned} d\mu_{B(\alpha)} &= -S_{m,B(\alpha)}\,dT + V_{m,B(\alpha)}\,dp \\ d\mu_{B(\beta)} &= -S_{m,B(\beta)}\,dT + V_{m,B(\beta)}\,dp \end{aligned} \tag{1.4.4}$$

where $S_{m,B(\alpha)}$ and $V_{m,B(\alpha)}$ are molar entropy and molar volume of the substance B in the phase $\alpha$, respectively, and $S_{m,B(\beta)}$ and $V_{m,B(\beta)}$ are those in the phase $\beta$, Substituting Eq. (1.4.4) in Eq. (1.4.3), we get

$$-S_{m,B(\alpha)}\,dT + V_{m,B(\alpha)}\,dp = -S_{m,B(\beta)}\,dT + V_{m,B(\beta)}\,dp$$

Rearranging, we get

$$\frac{dp}{dT} = \frac{S_{m,B(\beta)} - S_{m,B(\alpha)}}{V_{m,B(\beta)} - V_{m,B(\alpha)}} = \frac{\Delta_{trs}S_{m,B}}{\Delta_{trs}V_{m,B}} \tag{1.4.5}$$

where $\Delta_{trs}S_{m,B}$ and $\Delta_{trs}V_{m,B}$ are the respective changes in entropy and volume of the system when 1 mol of pure substance B is transferred from the phase $\alpha$ to the phase $\beta$.

Equation (1.4.5) is known as the *Clapeyron equation.*

## 1.5 APPLICATION OF CLAPEYRON EQUATION

In this section, we consider the application of Clapeyron equation to the two-phase equilibria involving solid and liquid, solid and vapour, liquid and vapour, and solid and solid.

**Solid-Liquid Equilibrium**

If one mole of the substance B is transformed from the solid phase to the liquid phase, then we have

$$\Delta_{trs}S_{m,B} = S_{m,B(l)} - S_{m,B(s)} = \Delta_{fus}S_{m,B} \tag{1.5.1}$$

$$\Delta_{trs}V_{m,B} = V_{m,B(l)} - V_{m,B(s)} = \Delta_{fus}V_{m,B} \tag{1.5.2}$$

At the equilibrium temperature, the transformation of the substance from solid to liquid is reversible, hence

$$\Delta_{fus}S_{m,B} = \Delta_{fus}H_{m,B}/T \tag{1.5.3}$$

Substituting Eq. (1.5.3) in Eq. (1.4.5), we get

$$\frac{dp}{dT} = \frac{\Delta_{fus}S_{m,B}}{\Delta_{fus}V_{m,B}} = \frac{\Delta_{fus}H_{m,B}}{T\,\Delta_{fus}V_{m,B}} \tag{1.5.4}$$

The transformation of solid to liquid is always an endothermic process and hence $\Delta_{fus}H_m$ is a positive quantity. The term $\Delta_{fus}V_m$ may be positive or negative depending upon which one (whether solid or liquid) is more dense. For most substances, $\rho_s$ is greater than $\rho_l$ and hence $\Delta_{fus}V_m$ is positive. For substances such as water, bismuth and antimony, $\Delta_{fus}V_m$ is negative as the solid phase is less dense than the liquid phase.

**Magnitude of d*p*/d*T* for Fusion Process**

The ordinary magnitudes of the quantities $\Delta_{fus}S_m$ and $\Delta_{fus}V_m$ are

$$\Delta_{fus}S_m = 8 \text{ to } 25 \text{ J K}^{-1}\text{ mol}^{-1} \quad\text{and}\quad \Delta_{fus}V_m = \pm\ (1 \text{ to } 10)\text{ cm}^3\text{ mol}^{-1}$$

Taking the typical values of

$$\Delta_{fus}S_m = 20 \text{ J K}^{-1}\text{ mol}^{-1} \quad\text{and}\quad \Delta_{fus}V_m = \pm\ 5\text{ cm}^3\text{ mol}^{-1}$$

we get

$$\frac{dp}{dT} = \frac{20\text{ J K}^{-1}\text{ mol}^{-1}}{\pm\ 5\text{ cm}^3\text{ mol}^{-1}} = \pm\,4\text{ J K}^{-1}\text{ cm}^{-3} = \pm\,4(\text{MPa cm}^3)\text{ K}^{-1}\text{ cm}^{-3}$$

$$= \pm\,4\text{ MPa K}^{-1} = \pm\,4\times10^3\text{ kPa K}^{-1} = \pm\,4\times10^3\left(\frac{1}{101.325}\text{ atm}\right)\text{K}^{-1}$$

$$= \pm\,39.5\text{ atm K}^{-1} \tag{1.5.5}$$

That is, the slope of pressure versus temperature graph for solid-liquid equilibrium has a large value. The graph is almost vertical. It is slightly tilted to the left (negative slope) if $\Delta_{fus}V_m$ is negative (Fig. 1.5.1).

**Diagrammatic Representation of Pressure Variation with Temperature**

The line of Fig. 1.5.1 represents equilibrium temperatures of solid-liquid equilibrium at various pressures, i.e. along the line solid and liquid phases coexist in a state of equilibrium with each other. A point lying anywhere on the left of the line corresponds to temperature below the melting point of the substance, and thus in this region only the solid form exists. A point on the right of the line corresponds to temperature above the melting point, and hence in this region only the liquid form exists.

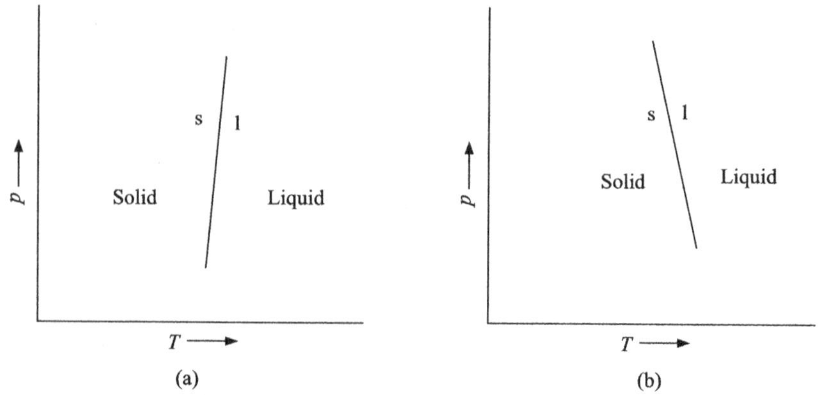


**Fig. 1.5.1 Plot of $p$ versus $T$ for solid-liquid equilibrium, when (a) $\Delta_{fus}V_m$ = +ve, and (b) $\Delta_{fus}V_m$ = –ve**

Now inverting Eq. (1.5.5), we get

$$\frac{dT}{dp} = \pm \frac{1}{39.5} \text{ K atm}^{-1} = \pm\, 0.025 \text{ K atm}^{-1}$$

Thus, change in the melting point of a solid with variation of external pressure is very small. The plus sign is meant for those substances which show an increase in volume on melting whereas the negative sign is meant for those substances where a decrease in volume on melting is observed. For most substances, $\rho_s > \rho_l$, and thus an increase in the melting point is observed with increase in external pressure. In case of substances for which volume decreases on melting (i.e. $\rho_s < \rho_l$), a decrease in the melting point occurs when the external pressure increases. Examples include ice, bismuth and antimony.

## Integrated Form of Clapeyron Equation

The Clapeyron equation (Eq. 1.5.4) can be written as

$$dp = \frac{\Delta_{fus}H_m}{\Delta_{fus}V_m}\frac{dT}{T} \tag{1.5.6}$$

Integrating Eq. (1.5.6), we get

$$\int_{p_1}^{p_2} dp = \int_{T_m}^{T'_m} \frac{\Delta_{fus}H_m}{\Delta_{fus}V_m}\frac{dT}{T}$$

where $T_m$ and $T'_m$ are the melting points at pressures $p_1$ and $p_2$, respectively. If $\Delta_{fus}H_m$ and $\Delta_{fus}V_m$ are considered independent of temperature and pressure, the above equation becomes

$$p_2 - p_1 = \frac{\Delta_{fus}H_m}{\Delta_{fus}V_m} \ln\left(\frac{T'_m}{T_m}\right)$$

Since the difference $T'_m - T_m$ is usually small, the logarithm term can be approximated to

$$\ln\left(\frac{T'_m}{T_m}\right) = \ln\left(\frac{T_m + T'_m - T_m}{T_m}\right) = \ln\left(1 + \frac{T'_m - T_m}{T_m}\right)$$

$$\simeq \frac{T'_m - T_m}{T_m} = \frac{\Delta T}{T_m}$$

Thus, we have

$$p_2 - p_1 = \Delta p = \frac{\Delta_{\text{fus}} H_{\text{m}}}{\Delta_{\text{fus}} V_{\text{m}}} \frac{\Delta T}{T_{\text{m}}} \tag{1.5.7}$$

Equation (1.5.7) describes variation in the melting point of a substance with the change in external pressure. The form of Eq. (1.5.7) is identical to that of Eq. (1.5.6) except that the infinitesimal changes d$p$ and d$T$ are replaced by finite changes $\Delta p$ and $\Delta T$, respectively.

**Conclusion**

It may be pointed out once again here that for a given increase in external pressure (i.e. $\Delta p$ positive), the sign of $\Delta T$ depends on the sign of $\Delta_{\text{fus}} V_{\text{m}}$. Thus, we will have

$\Delta T$ positive if $\Delta_{\text{fus}} V_{\text{m}}$ is positive

$\Delta T$ negative if $\Delta_{\text{fus}} V_{\text{m}}$ is negative.

---

***Example 1.5.1***

The melting point of mercury is 234.5 K at 1.0132 5 bar pressure and it increases 5.033 × $10^{-3}$ K per bar increase in pressure. The densities of solid and liquid mercury are 14.19 and 13.70 g $\text{cm}^{-3}$, respectively, (a) Determine the molar enthalpy of fusion, (b) Calculate the pressure required to raise the melting point to 273 K.

***Solution***

(**a**) Since $\Delta p = (\Delta_{\text{fus}} H_{\text{m}} / \Delta_{\text{fus}} V_{\text{m}})\,(\Delta T / T_{\text{m}})$, we have

$$\Delta_{\text{fus}} H_{\text{m}} = \frac{(\Delta_{\text{fus}} V_{\text{m}})\, T_{\text{m}}}{(\Delta T / \Delta p)}$$

where $\Delta_{\text{fus}} V_{\text{m}} = V_{\text{m, l}} - V_{\text{m, s}} = M\left(\frac{1}{\rho_{\text{l}}} - \frac{1}{\rho_{\text{s}}}\right)$

From the given data, we get

$$\Delta_{\text{fus}} V_{\text{m}} = (200.59 \text{ g mol}^{-1})\left(\frac{1}{13.70 \text{ g cm}^{-3}} - \frac{1}{14.19 \text{ g cm}^{-3}}\right)$$

$$= 0.51 \text{ cm}^3 \text{ mol}^{-1} = 0.51 \times 10^{-3} \text{ dm}^3 \text{ mol}^{-1}$$

Hence, $$\Delta_{\text{fus}} H_{\text{m}} = \frac{(\Delta_{\text{fus}} V_{\text{m}}) T_{\text{m}}}{(\Delta T / \Delta p)} = \frac{(0.51 \times 10^{-3} \text{ dm}^3 \text{ mol}^{-1})(234.5 \text{ K})}{(5.033 \times 10^{-3} \text{ K bar}^{-1})}$$

$$= 23.30 \text{ dm}^3 \text{ bar mol}^{-1}$$

$$= (23.30 \text{ dm}^3 \text{ bar mol}^{-1})\left(\frac{100 \text{ kPa}}{1 \text{ bar}}\right) = 23.30 \times 10^2 \text{ J mol}^{-1}$$

$$= 2.33 \text{ kJ mol}^{-1}$$

(**b**) Substituting the given data in the expression

$$\Delta p = \frac{\Delta_{\text{fus}} H_{\text{m}}}{\Delta_{\text{fus}} V_{\text{m}}} \frac{\Delta T}{T_{\text{m}}}$$

we get $$\Delta p = \frac{(23.30 \text{ dm}^3 \text{ bar mol}^{-1})(38.5 \text{ K})}{(0.51 \times 10^{-3} \text{ dm}^3 \text{ mol}^{-1})(234.5 \text{ K})}$$

$$= 7\,501 \text{ bar}$$

---

**Liquid-Vapour Equilibrium**

For the liquid-vapour equilibrium of the substance B, we have

$$\Delta_{vap}S_{m,B} = S_{m,B(g)} - S_{m,B(l)} = \frac{\Delta_{vap}H_m}{T} \tag{1.5.8}$$

$$\Delta_{vap}V_{m,B} = V_{m,B(g)} - V_{m,B(l)}$$

where $\Delta_{vap}S_{m,B}$ and $\Delta_{vap}V_{m,B}$ are the respective changes in entropy and volume during the vaporization of one mole of liquid B at its boiling point $T$. Since both $\Delta_{vap}S_m$ and $\Delta_{vap}V_m$ are positive, the quantity $dp/dT$ is always positive, i.e. the slope of $p$ versus $T$ is always positive.

**A Specific Illustration**

For most substances, $\Delta_{vap}S_m$ 90 $\simeq$ J K$^{-1}$ mol$^{-1}$ (Trouton's rule) and $\Delta_{vap}V_m = 24.0$ dm$^3$ mol$^{-1}$, the quantity $dp/dT$ has a value of the order of 0.037 atm K$^{-1}$ as shown in the following.

$$\frac{dp}{dT} = \frac{90\ \text{J K}^{-1}\ \text{mol}^{-1}}{24.0\ \text{dm}^3\ \text{mol}^{-1}} = 3.75\ \text{kPa K}^{-1}$$

$$= (3.75\ \text{kPa K}^{-1})\left(\frac{1\ \text{atm}}{101.325\ \text{kPa}}\right) = 0.037\ \text{atm K}^{-1}$$

This value is much smaller than that of the solid-liquid equilibrium.

**Solid-Vapour Equilibrium**

For the solid-vapour equilibrium of the substance B, we have

$$\Delta_{sub}S_{m,B} = S_{m,B(g)} - S_{m,B(s)} = \frac{\Delta_{sub}H_m}{T};\ T \text{ is the sublimation temperature}$$

$$\Delta_{sub}V_{m,B} = V_{m,B(g)} - V_{m,B(s)}$$

where the subscript 'sub' stands for the sublimation process. Since both $\Delta_{sub}S_m$ and $\Delta_{sub}V_m$ are positive, the value of $dp/dT$ is also positive.

**Important Conclusion**

At the triple point, where the three curves solid-liquid, solid-vapour and liquid vapour meet one another (Fig. 1.5.2), the slope of the solid-vapour curve is steeper than that of the liquid-vapour curve. This can be seen from the following analysis.

At the triple point

$$\Delta_{sub}H_m = \Delta_{fus}H_m + \Delta_{vap}H_m$$

and the slopes of the two curves are

$$\left(\frac{dp}{dT}\right)_{l\rightleftharpoons v} = \frac{\Delta_{vap}H_m}{T\,\Delta_{vap}V_m} \quad \text{and} \quad \left(\frac{dp}{dT}\right)_{s\rightleftharpoons v} = \frac{\Delta_{sub}H_m}{T\,\Delta_{sub}V_m}$$

Since $\Delta_{sub}H_m > \Delta_{vap}H_m$, it follows that

$$\left(\frac{dp}{dT}\right)_{s\rightleftharpoons v} > \left(\frac{dp}{dT}\right)_{l\rightleftharpoons v} \tag{1.5.9}$$

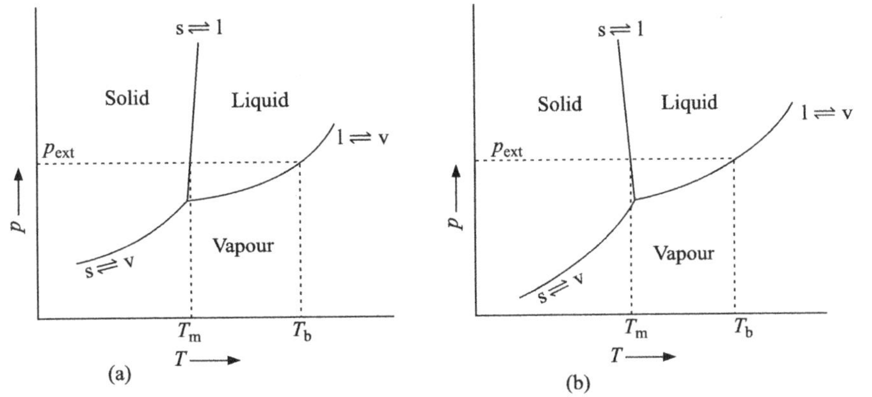


**Fig. 1.5.2 Phase diagrams (a) carbon dioxide where $V_{m,1} > V_{m,s}$ and (b) water where $V_{m,1} < V_{m,s}$**

## Depiction of Phase Diagram

The three curves representing s ⇋ l, l ⇋ v and s ⇋ v are shown together in Fig. 1.5.2. This diagram, known as the *phase diagram*, shows the region of stability of different phases. It also depicts at a glance properties of the substance such as melting point, boiling point, transition points, triple points, etc. The melting point, boiling point and transition temperature (if any) are represented by the intersection points of a line drawn horizontally from the given external pressure with those of s ⇋ l, l ⇋ v and s ⇋ s curves, respectively. The state of the system at various temperatures for a given pressure (say, 1 atm) can also be predicted from the phase diagram. For example, in Fig. 1.5.2, we have

| | | |
|---|---|---|
| $T < T_m$ | ; | only solid phase |
| $T = T_m$ | ; | solid and liquid phases in equilibrium |
| $T_b > T > T_m$ | ; | only liquid phase |
| $T = T_b$ | ; | liquid and vapour phases in equilibrium |
| $T > T_b$ | ; | only vapour phase. |

The l ⇋ v curve extends only up to a limit of critical pressure and temperature, because above these conditions, it is not possible to distinguish the vapour phase from the liquid phase.

We will show in Chapter 3 that the triple point of a given substance is observed at definite values of temperature and pressure. At this point, all the three phases are at equilibrium. For water, the triple point is observed at 4.58 Torr and 0.009 8 °C and that for $CO_2$ at 5.11 atm and –56.6 °C.

## Integrated Form of Clapeyron Equation (Clausius-Clapeyron Equation)

The Clapeyron equation is

$$\frac{dp}{dT} = \frac{\Delta_{trs}S_{m,B}}{\Delta_{trs}V_{m,B}} \qquad \text{(Eq. 1.4.5)}$$

At the transformation temperature $T$, the above expression becomes

$$\frac{dp}{dT} = \frac{\Delta_{trs}H_{m,B}}{T\,\Delta_{trs}V_{m,B}} \qquad (1.5.10)$$

where $\Delta_{trs}H_{m,B}$ is either the molar enthalpy of vaporization of the liquid or the molar enthalpy of sublimation of the solid and $\Delta_{tras}V_{m,B}$ is given by

$$\Delta_{tras}V_{m,B} = V_{m,B(g)} - V_{m,B(s\text{ or }l)} \qquad (1.5.11)$$

**Clausius Simplification**

Clausius simplifies Eq. (1.5.11) by introducing the following two approximations.

1. $V_{m,B(g)} - V_{m,B(s\ or\ l)} \simeq V_{m,B(g)}$; since $V_{m,B(s\ or\ l)} << V_{m,B(g)}$.
2. If the gaseous phase is assumed to be ideal, the molar volume of the gaseous phase may be replaced by $RT/p$.

With these approximations, the Clapeyron equation becomes

$$\frac{dp}{dT} = \frac{\Delta_{trs}H_m}{RT^2}p \quad \text{or} \quad \frac{dp}{p} = \frac{\Delta_{trs}H_m}{R}\frac{dT}{T^2} \tag{1.5.12}$$

Equation (1.5.12) is known as the Clausius-Clapeyron equation. It relates the vapour pressure of a condensed phase (solid or liquid) with its molar enthalpy of phase transformation (sublimation or vaporization) and the temperature of transformation.

**Integrated Form of Clausius-Clapeyron Equation**

Integration of Eq. (1.5.12) gives

$$\int_{p_1}^{p_2}\frac{dp}{p} = \int_{T_1}^{T_2}\frac{\Delta_{trs}H_m}{R}\frac{dT}{T^2}$$

where $p_1$ and $p_2$ are the vapour pressures of the condensed phase (solid or liquid) at temperatures $T_1$ and $T_2$, respectively. If $\Delta_{trs}H_m$ is assumed to be independent of temperature, we get

$$\ln\left(\frac{p_2}{p_1}\right) = -\frac{\Delta_{trs}H_m}{R}\left(\frac{1}{T_2} - \frac{1}{T_1}\right) \tag{1.5.13}$$

If $p_1 = 1$ bar, then $T_1$ represents the normal sublimation (or boiling) point of the solid (or liquid) phase. Representing this by $T^*$, we get

$$\ln\left(\frac{p_2}{1\ \text{bar}}\right) = -\frac{\Delta_{trs}H_m}{R}\left(\frac{1}{T_2} - \frac{1}{T^*}\right)$$

Dropping the subscript 2, we get

$$\ln\left(\frac{p}{1\ \text{bar}}\right) = -\frac{\Delta_{trs}H_m}{R}\left(\frac{1}{T} - \frac{1}{T^*}\right) \tag{1.5.14}$$

**Diagrammatic Depiction of Eq. (1.5.14)**

If we plot ln ($p$/bar) (or log $p$/bar) versus $1/T$, we get a linear curve with slope equal to $-\Delta_{trs}H_m/R$ (or $-\Delta_{trs}H_m/2.303R$) as shown in Fig. 1.5.3. The intercept at $1/T = 0$ yields the value of $\Delta_{trs}H_m/RT^*$ (or $\Delta_{trs}H_m/2.303RT^*$). From the determined values of slope and intercept, we can determine both $\Delta_{trs}H_m$ and $T^*$ of the substance under study.

**Open Integration of Eq. (1.5.12)**

Dividing the numerator and denominator of left side of Eq. (1.5.12) by the unit vapour pressure (represented by $p^\circ$), we get

$$\frac{(1/p^\circ)\,dp}{(p/p^\circ)} = \frac{\Delta_{trs}H_m}{R}\frac{dT}{T^2}$$

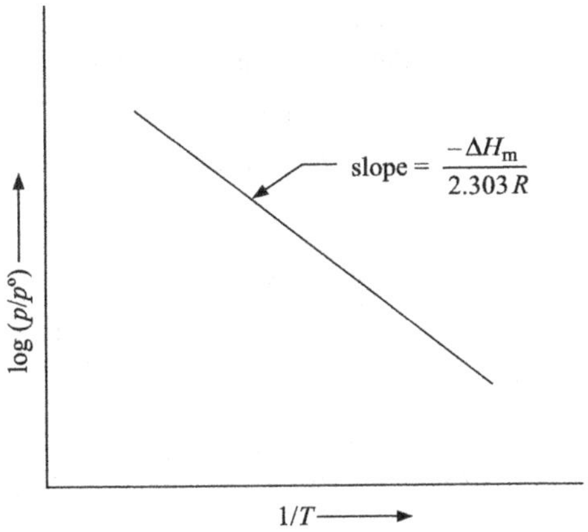


**Fig. 1.5.3 Plot of log $(p/p^\circ)$ versus 1/T for the condensed phase-vapour equilibria**

The open integration of this expression gives

$$\ln\left(\frac{p}{p^\circ}\right) = -\frac{\Delta_{trs}H_m}{R}\frac{1}{T} + I \tag{1.5.15}$$

where $I$ is a integration constant. Its values may be determined if the value of vapour pressure $p$ at some known temperature $T$ is known. In terms of common logarithm, Eq. (1.5.15) becomes

$$\log\left(\frac{p}{p^\circ}\right) = -\frac{\Delta_{trs}H_m}{20.303\,R}\frac{1}{T} + I' \tag{1.5.16}$$

Note that the value of slope of Eq. (1.5.16) is independent of the unit of $p^\circ$.

---

***Example 1.5.2***

At 373.15 K and 1 bar, the specific volume of water vapour is 1 696 $cm^3\ g^{-1}$ and the value of $dp/dT$ is 0.036 2 bar $K^{-1}$. Calculate $\Delta_{vap}H_m$.

***Solution***

The given data are

$$T = 373.15\ K$$

$$V_g = 1\,696\ cm^3\ g^{-1} \equiv (1\,696 \times 18)\ cm^3\ mol^{-1} = 1.696 \times 18\ dm^3\ mol^{-1}$$

$$p = 1\ bar$$

$$\frac{dp}{dT} = 0.036\ 2\ bar\ K^{-1}$$

From the expression

$$\frac{dp}{dT} = \frac{\Delta_{vap}H_m}{T\,\Delta_{vap}V_m} \simeq \frac{\Delta_{vap}H_m}{TV_{m,v}}$$

we get
$$\Delta_{vap}H_m = \left(\frac{dp}{dT}\right)TV_{m,v}$$

$$= (0.036\ 2\ bar\ K^{-1})\,(373.15\ K)\,(1.696 \times 18\ dm^3\ mol^{-1})$$

$$= 412.37\ bar\ dm^3\ mol^{-1} \equiv 41\ 237\ kPa\ dm^3\ mol^{-1}$$

$$= 41\ 237\ J\ mol^{-1}$$

---

***Example 1.5.3***

The vapour pressure of toluene is 0.078 8 bar at 313.75 K and 0.398 bar at 353.15 K. Calculate the molar enthalpy of vaporization.

***Solution***

The given data are

$$T_1 = 313.75\ \text{K} \qquad T_2 = 353.15\ \text{K}$$

$$p_1 = 0.078\,8\ \text{bar} \qquad p_2 = 0.398\ \text{bar}$$

Substituting these values in the expression

$$\ln \frac{p_2}{p_1} = -\frac{\Delta_{\text{vap}}H_{\text{m}}}{R}\left(\frac{1}{T_2} - \frac{1}{T_1}\right)$$

we get

$$\ln\left(\frac{0.398\ \text{bar}}{0.078\,8\ \text{bar}}\right) = -\frac{\Delta_{\text{vap}}H_{\text{m}}}{(8.314\ \text{J K}^{-1}\ \text{mol}^{-1})}\left(\frac{1}{353.15\ \text{K}} - \frac{1}{313.75\ \text{K}}\right)$$

Thus

$$\Delta_{\text{vap}}H_{\text{m}} = \frac{(8.314\ \text{J K}^{-1}\ \text{mol}^{-1})\,(353.15\ \text{K})\,(313.75\ \text{K})}{(353.15\ \text{K} - 313.75\ \text{K})}\ln\left(\frac{0.398}{0.078\,8}\right)$$

$$= 37\,866\ \text{J mol}^{-1}$$

---

***Example 1.5.4***

A certain liquid has a boiling point of 338.15 K at 1.013 25 bar pressure. Using Trouton's rule (a) estimate the vapour pressure at 325.15 K, (b) estimate the boiling point at a pressure of 0.267 bar, and (c) obtain an equation for the vapour pressure in bar of this liquid as a function of temperature.

***Solution***

(**a**) The given data are

$$T_1 = 338.15\ \text{K} \qquad p_1 = 1.013\,25\ \text{bar}$$

$$T_2 = 325.15\ \text{K} \qquad p_2 = ?$$

$$\Delta_{\text{vap}}H_{\text{m}} \simeq 10.5\,RT_1 \qquad \text{(Trouton's rule)}$$

Substituting the given data in the expression

$$\ln \frac{p_2}{p_1} = -\frac{\Delta_{\text{vap}}H_{\text{m}}}{R}\left(\frac{1}{T_2} - \frac{1}{T_1}\right)$$

we get

$$\ln \frac{p_2}{p_1} = -(10.5\,T_1)\left(\frac{1}{T_2} - \frac{1}{T_1}\right) = -(10.5)\left(\frac{T_1}{T_2} - 1\right)$$

i.e.

$$\ln\left(\frac{p_2}{1.013\,25\ \text{bar}}\right) = -(10.5)\left(\frac{338.15\ \text{K}}{325.15\ \text{K}} - 1\right)$$

$$= -0.42$$

or

$$p_2 = (\text{e}^{-0.42})\,(1.013\,25\ \text{bar}) = (0.657)\,(1.013\,25\ \text{bar})$$

$$= 0.666\ \text{bar}$$

(**b**) In this case, we have

$$\ln\left(\frac{0.267\ \text{bar}}{1.013\,25\ \text{bar}}\right) = -(10.5)\left(\frac{338.15\ \text{K}}{T_2} - 1\right)$$

$$-1.334 = -(10.5)\left(\frac{338.15\ \text{K}}{T_2} - 1\right)$$

or $$\frac{338.15\ \text{K}}{T_2} = \frac{1.334}{10.5} + 1 = \frac{11.834}{10.5}$$

or $$T_2 = \left(\frac{10.5}{11.834}\right)(338.15\ \text{K}) = 300.00\ \text{K}$$

(c) Substituting the values of $T_1$ and $p_1$ in the expression

$$\ln\frac{p_2}{p_1} = -(10.5)\left(\frac{T_1}{T_2} - 1\right)$$

we get $$\ln\left(\frac{p_2}{1.013\ 25\ \text{bar}}\right) = -(10.5)\left(\frac{338.15\ \text{K}}{T_2} - 1\right)$$

or $$\ln\left(\frac{p_2}{\text{bar}}\right) = \ln(1.013\ 25) - \frac{3\ 550.58\ \text{K}}{T_2} + 10.5$$

$$= 0.013\ 2 - \frac{3\ 550.58\ \text{K}}{T_2} + 10.5 = 10.51 - \frac{3\ 550.58\ \text{K}}{T_2}$$

or $$\log\left(\frac{p_2}{\text{bar}}\right) = 4.564 - \frac{1\ 541.72\ \text{K}}{T_2}$$

---

***Example 1.5.5***

The vapour pressure of a liquid which obeys Trouton's rule rises by 52 Torr between temperatures 1 K below and 1 K above the normal boiling point. Determine the value of the normal boiling point and the molar enthalpy of vaporization of the liquid.

***Solution***

The given data are

$$\mathrm{d}p = 52\ \text{Torr} \qquad \mathrm{d}T = 2\ \text{K}$$

$$\frac{\Delta_{\text{vap}}H_{\text{m}}}{RT_{\text{b}}} = 10.5 \quad \text{i.e.} \quad \Delta_{\text{vap}}H_{\text{m}} = (10.5)\ RT_{\text{b}} \qquad \text{(Trouton's rule)}$$

The Clapeyron equation becomes

$$\frac{\mathrm{d}p}{\mathrm{d}T} = \frac{\Delta_{\text{vap}}H_{\text{m}}}{T_{\text{b}}\,\Delta_{\text{vap}}V_{\text{m}}} \simeq \frac{(10.5)\ R}{V_{\text{m, v}}} = \frac{(10.5)\ R}{(RT_{\text{b}}/p)} = \frac{(10.5)\ p}{T_{\text{b}}}$$

This gives

$$T_{\text{b}} = \frac{(10.5)\ p}{\mathrm{d}p/\mathrm{d}T} = \frac{(10.5)\ (760\ \text{Torr})}{(52\ \text{Torr})/(2\ \text{K})} = 306.9\ \text{K}$$

Also $$\Delta_{\text{vap}}H_{\text{m}} = (10.5)\ RT_{\text{b}} = (10.5)\ (8.314\ \text{J K}^{-1}\ \text{mol}^{-1})\ (306.9\ \text{K})$$

$$= 26\ 791\ \text{J mol}^{-1}$$

---

***Example 1.5.6***

If the enthalpy of vaporization of water is assumed to be constant at 2 255 kJ $\text{g}^{-1}$, calculate the temperature at which water will boil under a pressure of 77.0 cmHg, the boiling point of water being 373.15 K at 76.0 cmHg. The specific volume of water vapour at 373.15 K and 76 cmHg is 1 644 $\text{cm}^3\ \text{g}^{-1}$ and that of liquid water is 1 $\text{cm}^3\ \text{g}^{-1}$.

***Solution***

The given data are

$$T_1 = 373.15\ \text{K} \qquad p_1 = 76.0\ \text{cmHg}$$
$$T_2 = ? \qquad p_2 = 77.0\ \text{cmHg}$$

$$\Delta_{\text{vap}}H_{\text{m}} = (2\,255\ \text{J g}^{-1})(18\ \text{g mol}^{-1}) = 40\,590\ \text{J mol}^{-1}$$

$$V_{\text{m,v}} = (1\,664 \times 18)\ \text{cm}^3\ \text{mol}^{-1} = 29\,952\ \text{cm}^3\ \text{mol}^{-1}$$
$$= 29.952\ \text{dm}^3\ \text{mol}^{-1}$$
$$V_{\text{m,l}} = 1 \times 18\ \text{cm}^3\ \text{mol}^{-1} = 0.018\ \text{dm}^3\ \text{mol}^{-1}$$

Using the expression

$$\ln\left(\frac{p_2}{p_1}\right) = -\frac{\Delta_{\text{vap}}H_{\text{m}}}{R}\left(\frac{1}{T_2} - \frac{1}{T_1}\right)$$

we get $$2.303 \log\left(\frac{77.0\ \text{cmHg}}{76.0\ \text{cmHg}}\right) = -\frac{(40\,590\ \text{J mol}^{-1})}{(8.314\ \text{J K}^{-1}\ \text{mol}^{-1})}\left(\frac{1}{T_2} - \frac{1}{373.15\ \text{K}}\right)$$

$$0.013\,07 = -(4\,882.1\ \text{K})\left(\frac{1}{T_2} - \frac{1}{373.15\ \text{K}}\right)$$

$$\frac{1}{T_2} = -\frac{0.013\,07}{4\,882.1\ \text{K}} + \frac{1}{373.15\ \text{K}} = (-\ 2.677 \times 10^{-6} + 2.680 \times 10^{-3})\ \text{K}^{-1}$$
$$= 2.677 \times 10^{-3}\ \text{K}^{-1}$$

$$T_2 = \frac{1}{(2.677 \times 10^{-3}\ \text{K}^{-1})} = 373.5\ \text{K}$$

---

***Example 1.5.7***

The mean enthalpy of vaporization of water in the temperature range between 363.15 K and 373.15 K is 2 268 J $g^{-1}$. Calculate the vapour pressure of water at 363.15 K, its value at 373.15 K is being 76.0 cmHg.

***Solution***

The given data are

$$T_1 = 373.15\ \text{K} \qquad p_1 = 76.0\ \text{cmHg}$$
$$T_2 = 363.15\ \text{K} \qquad p_2 = ?$$
$$\Delta_{\text{vap}}H_{\text{m}} = 2\,268 \times 18\ \text{J mol}^{-1} = 40\,824\ \text{J mol}^{-1}$$

Using the expression

$$\log\left(\frac{p_2}{p_1}\right) = -\frac{\Delta_{\text{vap}}H_{\text{m}}}{2.303\,R}\left(\frac{1}{T_2} - \frac{1}{T_1}\right)$$

we get $$\log\left(\frac{p_2}{76.0\ \text{cmHg}}\right) = -\frac{40\,824\ \text{J mol}^{-1}}{(2.303)(8.314\ \text{J K}^{-1}\ \text{mol}^{-1})}\left(\frac{1}{363.15\ \text{K}} - \frac{1}{373.15\ \text{K}}\right)$$
$$= -\frac{(40\,824)(10)}{(2.303)(8.314)(363.15)(373.15)} = -\ 0.157\,3$$

$$\frac{p_2}{76.0\ \text{cmHg}} = 0.696 \quad \text{i.e.} \quad p_2 = (0.696)(76.0\ \text{cmHg}) = 52.9\ \text{cmHg}$$

---

***Example 1.5.8***

Iodine boils at 456.15 K and the vapour pressure at 389.65 K is 100 Torr. The enthalpy of fusion is 15.65 kJ mol$^{-1}$ and the vapour pressure of the solid is 1 Torr at 311.85 K. Calculate the triple point temperature and pressure.

***Solution***

The given data are

***Liquid-Vapour Equilibrium***

$$T_1 = 456.15 \text{ K} \qquad p_1 = 760 \text{ Torr}$$

$$T_2 = 389.65 \text{ K} \qquad p_2 = 100 \text{ Torr}$$

***Solid-Liquid Equilibrium***

$$\Delta_{\text{fus}}H_{\text{m}} = 15.65 \text{ kJ mol}^{-1}$$

***Solid-Vapour Equilibrium***

$$T = 311.85 \text{ K} \qquad p = 1 \text{ Torr}$$

***Calculation of $\Delta_{\text{vap}}H_{\text{m}}$:*** Using the Clausius-Clapeyron equation

$$\log\left(\frac{p_2}{p_1}\right) = -\frac{\Delta_{\text{vap}}H_{\text{m}}}{2.303\,R}\left(\frac{1}{T_2} - \frac{1}{T_1}\right)$$

we get

$$\log\left(\frac{100}{760}\right) = -\frac{\Delta_{\text{vap}}H_{\text{m}}}{(2.303)\,(8.314 \text{ J K}^{-1} \text{ mol}^{-1})}\left[\frac{1}{389.65 \text{ K}} - \frac{1}{456.15 \text{ K}}\right]$$

This gives

$$\Delta_{\text{vap}}H_{\text{m}} = \frac{(2.303)\,(8.314 \text{ J K}^{-1} \text{ mol}^{-1})\,(389.65 \text{ K})\,(456.15 \text{ K})}{(456.15 \text{ K} - 389.65 \text{ K})}\log\left(\frac{760}{100}\right)$$

$$= 45\,076 \text{ J mol}^{-1}$$

***Value of $\Delta_{\text{sub}}H_{\text{m}}$:*** We have

$$\Delta_{\text{sub}}H_{\text{m}} = \Delta_{\text{fus}}H_{\text{m}} + \Delta_{\text{vap}}H_{\text{m}} = 15.65 \times 10^3 \text{ J mol}^{-1} + 45\,076 \text{ J mol}^{-1}$$

$$= 60\,726 \text{ J mol}^{-1}$$

***Calculation of Triple Point Temperature and Triple Point Pressure***

Let $p$ and $T$ be the pressure and temperature at the triple point, respectively. At this point, the vapour pressure of liquid will be the same as that of solid. Hence, applying Clapeyron equation at the triple point to liquid-vapour and solid-vapour equilibria, we get

*Liquid-Vapour Equilibrium*

$$\log\left(\frac{p}{100 \text{ Torr}}\right) = -\left(\frac{(45\,076 \text{ J mol}^{-1})}{(2.303)\,(8.314 \text{ J K}^{-1} \text{ mol}^{-1})}\right)\left(\frac{1}{T} - \frac{1}{389.65 \text{ K}}\right) \qquad (1)$$

$$= -(2\,354.2 \text{ K})\left(\frac{1}{T} - \frac{1}{389.65 \text{ K}}\right)$$

*Solid-Vapour Equilibrium*

$$\log\left(\frac{p}{1 \text{ Torr}}\right) = -\left(\frac{(60\,726 \text{ J mol}^{-1})}{(2.303)\,(8.314 \text{ J K}^{-1} \text{ mol}^{-1})}\right)\left(\frac{1}{T} - \frac{1}{311.85 \text{ K}}\right) \qquad (2)$$

$$= -(3\,171.5 \text{ K})\left(\frac{1}{T} - \frac{1}{311.85 \text{ K}}\right)$$

Subtracting Eq. (1) from Eq. (2), we get

$$\log(100) = -\frac{3\,171.5\ \text{K}}{T} + \frac{3\,171.5}{311.85} + \frac{2\,354.5\ \text{K}}{T} - \frac{2\,354.2}{389.65}$$

$$2.00 = \frac{1}{T}(-\ 3\,171.5\ \text{K} + 2\,354.2\ \text{K}) + (10.170 - 6.042)$$

$$T = \frac{-817.3\ \text{K}}{-2.128} = 384.07\ \text{K}$$

Substituting the value of $T$ in Eq. (1), we get

$$\log\left(\frac{p}{100\ \text{Torr}}\right) = -(2\,354.2\ \text{K})\left(\frac{1}{384.07\ \text{K}} - \frac{1}{389.65\ \text{K}}\right)$$

$$= -\,0.0878$$

$$p = (0.817)\,(100\ \text{Torr}) = 81.7\ \text{Torr}$$

---

***Example 1.5.9***

The vapour pressures of solid and liquid white phosphorus are given by the expressions

$$\log\left(\frac{p_2}{\text{atm}}\right) = -\frac{(2\,875\ \text{K})}{T} + 5.36 \quad \text{and} \quad \log\left(\frac{p_1}{\text{atm}}\right) = -\frac{(2\,740\ \text{K})}{T} + 4.95$$

respectively. Calculate (a) the temperature and pressure of the triple point of phosphorus, and (b) the molar enthalpy and molar entropy of fusion of phosphorus at the triple point.

***Solution***

(**a**) At triple point, the vapour pressures of solid and liquid will be identical. Equating the expressions of vapour pressure, we have

$$\log\left(\frac{p_2}{\text{atm}}\right) = \log\left(\frac{p_1}{\text{atm}}\right)$$

or
$$-\frac{(2\,875\ \text{K})}{T} + 5.36 = -\frac{(2\,740\ \text{K})}{T} + 4.95$$

or
$$\frac{(2\,875\ \text{K})}{T} - \frac{(2\,740\ \text{K})}{T} = 5.36 - 4.95$$

or
$$\frac{(135\ \text{K})}{T} = 0.41 \quad \text{or} \quad T = \frac{(135\ \text{K})}{0.41} = 329.3\ \text{K}$$

The vapour pressure at the triple point can be obtained by substituting this temperature in either of the vapour pressure expressions. Thus, we have

$$\log\left(\frac{p_2}{\text{atm}}\right) = -\frac{(2\,875\ \text{K})}{(329.3\ \text{K})} + 5.36 = -\,8.732 + 5.36 = -\,3.372$$

Hence, $p_2 = 4.246 \times 10^{-4}$ atm

(**b**) The expression of variation of vapour pressure with temperature is given by

$$\ln\left(\frac{p}{\text{atm}}\right) = -\frac{\Delta H_{\text{m}}}{RT} + I \quad \text{i.e.} \quad \log\left(\frac{p}{\text{atm}}\right) = -\frac{\Delta H_{\text{m}}}{2.303\,RT} + I'$$

Comparing this expression with the given expressions of vapour pressure, we get

$$\frac{\Delta_{sub}H_m}{2.303\,R} = 2\,875\ \text{K} \quad \text{and} \quad \frac{\Delta_{vap}H_m}{2.303\,R} = 2\,740\ \text{K}$$

Thus $\quad \Delta_{sub}H_m = (2.303)\,(8.314\ \text{J K}^{-1}\ \text{mol}^{-1})\,(2\,875\ \text{K}) = 55\,048\ \text{J mol}^{-1}$

$\quad \Delta_{vap}H_m = (2.303)\,(8.314\ \text{J K}^{-1}\ \text{mol}^{-1})\,(2\,740\ \text{K}) = 52\,463\ \text{J mol}^{-1}$

Hence $\quad \Delta_{fus}H_m = \Delta_{sub}H_m - \Delta_{vap}H_m = (55\,048 - 52\,463)\ \text{J mol}^{-1}$

$\quad = 2\,585\ \text{J mol}^{-1}$

Since $\Delta G = 0$ for the fusion at equilibrium temperature, therefore

$$\Delta_{fus}S_m = \frac{\Delta_{fus}H_m}{T} = \frac{2\,585\ \text{J mol}^{-1}}{329.3\ \text{K}} = 7.85\ \text{J K}^{-1}\ \text{mol}^{-1}$$

***Example 1.5.10***

The vapour pressure of a liquid changes with temperature according to the expression

$$\log\left(\frac{p}{\text{mmHg}}\right) = 11.36 - \frac{273\ \text{K}}{T}$$

Calculate the enthalpy of vaporization per mole of the liquid.

***Solution***

The variation of vapour pressure with temperature is given by

$$\ln\left(\frac{p}{p^\circ}\right) = -\frac{\Delta_{vap}H_m}{RT} + I \quad \text{or} \quad \log\left(\frac{p}{p^\circ}\right) = -\frac{\Delta_{vap}H_m}{2.303\,RT} + I'$$

Comparing this with the given expression, we get

$$\frac{\Delta_{vap}H_m}{2.303\,R} = 273\ \text{K}$$

or $\quad \Delta_{vap}H_m = (2.303\,R)\,(273\ \text{K}) = (2.303 \times 8.314\ \text{J K}^{-1}\ \text{mol}^{-1})\,(273\ \text{K})$

$\quad = 5\,227\ \text{J mol}^{-1}$

## Solid-Solid Equilibrium

This type of equilibrium is observed if a substance exists in more than one allotropic form. The typical example is that of sulphur, which exists in two allotropic forms. At ordinary temperatures and under 1 atm pressure, its rhombic form is stable. However, it passes over to the monoclinic form at 95.4 °C if heated slowly. Above 95.4 °C, the monoclinic form is stable, until at 119 °C, where it melts to give liquid sulphur. We can have the following equilibria in the sulphur system:

S(rhombic) $\rightleftharpoons$ S(monoclinic)
S(monoclinic) $\rightleftharpoons$ S(liquid)
S(liquid) $\rightleftharpoons$ S(vapour)
S(rhombic) $\rightleftharpoons$ S(vapour)
S(monoclinic) $\rightleftharpoons$ S(vapour)

Besides the above equilibria, we can also have

$$\text{S(rhombic)} \rightleftharpoons \text{S(liquid)}$$

This represents the metastable equilibrium and is observed only when the heating of rhombic sulphur is done quickly. For 1 atm external pressure, the above metastable equilibrium is observed at 114 °C.

The slope of the curve representing the allotropic transformation can be obtained from the Clapeyron equation

$$\frac{dp}{dT} = \frac{\Delta_{trs} S_m}{\Delta_{trs} V_m} = \frac{\Delta_{trs} H_m}{T\,\Delta_{trs} V_m}$$

The slope is usually quite large because of the small change in the molar volume of transformation.

## Phase Diagram of Sulphur

The phase diagram for the sulphur system is shown in Fig. 1.5.4. The metastable equilibria are shown by dotted lines.

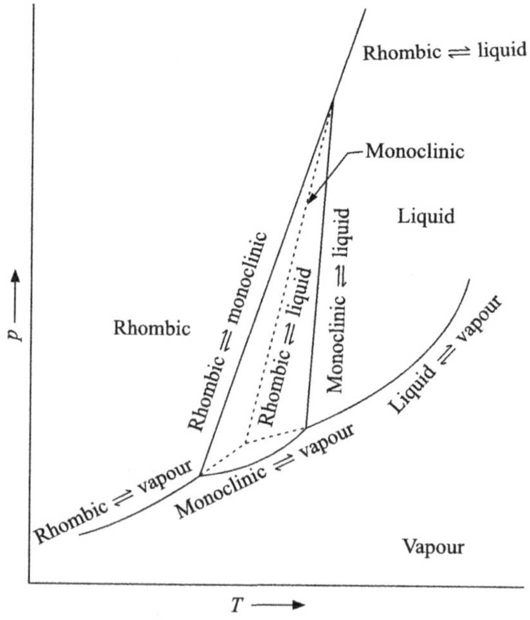


**Fig. 1.5.4 Schematic Phase diagram of sulphur**

There are three stable and one metastable triple points:

(i) at 95.4 °C $\mu_{rhomb} = \mu_{mono} = \mu_v$

(ii) at 119 °C $\mu_{mono} = \mu_l = \mu_v$

(iii) at 151 °C $\mu_{rhomb} = \mu_{mono} = \mu_l$

(iv) at 114 °C $\mu_{rhomb} = \mu_l = \mu_v$

***Example 1.5.11***

The density of diamond is 3.52 g $cm^{-3}$, while that of graphite is 2.25 g $cm^{-3}$. At 298 K, the free energy of formation of diamond from graphite is 2.866 kJ $mol^{-1}$. At 298 K, what pressure must be applied to bring diamond and graphite into equilibrium?

***Solution***

The given data are

$$\text{C(graphite)} \rightarrow \text{C(diamond)} \qquad \Delta_r G = 2\ 866\ \text{J mol}^{-1}$$
$$\rho\,(\text{graphite}) = 2.25\ \text{g cm}^{-3}$$
$$\rho(\text{diamond}) = 3.52\ \text{g cm}^{-3}$$

Since at equilibrium $\Delta_r G$ of the reaction is zero, it is obvious that we would have to decrease $\Delta_r G$ of the reaction by 2 866 J $mol^{-1}$ by increasing the external pressure. We start with the relation

$$\left(\frac{\partial(\Delta_r G)}{\mathrm{d}p}\right)_T = \Delta_{\text{trs}} V_{\text{m}}$$

or
$$\mathrm{d}(\Delta_r G) = (\Delta_{\text{trs}} V_{\text{m}})\,\mathrm{d}p \qquad (T\ \text{constant})$$
$$= M_{\text{carbon}}\left(\frac{1}{\rho_{\text{diamond}}} - \frac{1}{\rho_{\text{graphite}}}\right)(p_2 - p_1)$$

Substituting the given data, we get

$$(-2\ 866\ \text{J mol}^{-1}) = (12\ \text{g mol}^{-1})\left(\frac{1}{3.52\ \text{g cm}^{-3}} - \frac{1}{2.25\ \text{g cm}^{-3}}\right)(p_2 - 1\ \text{bar})$$

or
$$(p_2 - 1\ \text{bar}) = \frac{-2\ 866\ \text{J mol}^{-1}}{(3.409 - 5.333)\ (\text{cm}^3\ \text{mol}^{-1})} = 1\ 489.6\ \text{J cm}^{-3}$$

Now, since 1 J = 1 kPa $dm^3$ = $10^{-2}$ bar $dm^3$ = 10 bar $cm^3$, we get

$$p_2 - 1\ \text{bar} = 14\ 896\ \text{bar}$$

or
$$p_2 = 14\ 897\ \text{bar}$$

***Example 1.5.12***

The density of ice II is 1.21 g $cm^{-3}$, that of ice III is 1.10 g $cm^{-3}$. Calculate $\Delta_{\text{trs}}H_{\text{m}}$ for the conversion of 1 mol of ice II into ice III, if these two phases are in equilibrium at 3 260 atm and 298.15 K and at 2 450 atm and 242.15 K.

***Solution***

The Clapeyron equation is

$$\frac{\mathrm{d}p}{\mathrm{d}T} = \frac{\Delta_{\text{trs}}S_{\text{m}}}{\Delta_{\text{trs}}V_{\text{m}}} = \frac{\Delta_{\text{trs}}H_{\text{m}}}{T\,\Delta_{\text{trs}}V_{\text{m}}}$$

or
$$\mathrm{d}p = \frac{\Delta_{\text{trs}}H_{\text{m}}}{T\,\Delta_{\text{trs}}V_{\text{m}}}\mathrm{d}T$$

For a finite change, we have

$$\Delta p = \frac{\Delta_{\text{trs}}H_{\text{m}}}{T\,\Delta_{\text{trs}}V_{\text{m}}}\Delta T$$

or
$$\Delta_{\text{trs}}H_{\text{m}} = \frac{(\Delta p)\,T\,\Delta_{\text{trs}}V_{\text{m}}}{\Delta T} = \frac{(p_2 - p_1)T_1\,[V_{\text{m, ice(III)}} - V_{\text{m, ice(II)}}]}{(T_2 - T_1)}$$

Substituting the given data, we get

$$\Delta_{\text{trs}} H_{\text{m}} = \frac{(3\,260\ \text{atm} - 2\,450\ \text{atm})\,(242.15\ \text{K})\left(\dfrac{18\ \text{g mol}^{-1}}{1.1\ \text{g cm}^{-3}} - \dfrac{18\ \text{g mol}^{-1}}{1.21\ \text{g cm}^{-3}}\right)}{(6\ \text{K})}$$

$$= \frac{(810\ \text{atm})\,(242.15\ \text{K})\,(1.488\ \text{cm}^3\ \text{mol}^{-1})}{(6\ \text{K})}$$

$$= 48\,643\ \text{atm cm}^3\ \text{mol}^{-1} = 48.643\ \text{atm dm}^3\ \text{mol}^{-1}$$

$$= (48.643\ \text{atm dm}^3\ \text{mol}^{-1})\left(\frac{8.314\ \text{J}}{0.082\ \text{atm dm}^3}\right)$$

$$= 4\,932\ \text{J mol}^{-1}$$

***Example 1.5.13***

The specific volume of monoclinic sulphur which is stable above the transition temperature is greater than that of rhombic sulphur by $0.0126\ \text{cm}^3\ \text{g}^{-1}$. The transition point at one atm pressure is 368.65 K and it increases at the rate of $0.035\ \text{K atm}^{-1}$. Calculate the molar enthalpy of transition.

***Solution***

The given data are

$$p = 1\ \text{atm}$$

$$T_{\text{i}} = 368.65\ \text{K}$$

$$\Delta_{\text{trs}} V_{\text{m}} = V_{\text{m, monoclinic}} - V_{\text{m, rhombic}} = 0.012\,6\ \text{cm}^3\ \text{g}^{-1}$$

$$= (0.012\,6 \times 32)\ \text{cm}^3\ \text{mol}^{-1} = 0.012\,6 \times 32 \times 10^{-3}\ \text{dm}^3\ \text{mol}^{-1}$$

$$= 4.032 \times 10^{-4}\ \text{dm}^3\ \text{mol}^{-1}$$

$$\text{d}T/\text{d}p = 0.035\ \text{K atm}^{-1} \quad \text{i.e.} \quad \text{d}p/\text{d}T = 28.57\ \text{atm K}^{-1}$$

Using the Clapeyron equation $\text{d}p/\text{d}T = \Delta_{\text{trs}}H_{\text{m}}/(T\ \Delta_{\text{trs}}V_{\text{m}})$, we get

$$\Delta_{\text{trs}} H_{\text{m}} = T_{\text{i}}\ \Delta_{\text{trs}} V_{\text{m}}\ (\text{d}p/\text{d}T)$$

$$= (368.65\ \text{K})\,(4.032 \times 10^{-4}\ \text{dm}^3\ \text{mol}^{-1})\,(28.57\ \text{atm K}^{-1})$$

$$= 4.247\ \text{dm}^3\ \text{atm mol}^{-1} = 4.247\ \text{dm}^3\ (101.325\ \text{kPa})\ \text{mol}^{-1}$$

$$= 430.3\ \text{J mol}^{-1}$$

## 1.6 FIRST- AND SECOND-ORDER PHASE TRANSITIONS

### First-Order Phase Transition

A given phase transition of a substance may be classified into first-order or second-order depending upon the following characteristics.

A first-order phase transition of a substance is accompanied with changes in the values of enthalpy and volume of the substance, i.e. $\Delta_{\text{trs}}H$ and $\Delta_{\text{trs}}V$ are nonzero. Examples include familiar phase transitions such as solid to liquid, solid to vapour, and liquid to vapour.

For a substance B, we have

$$\left(\frac{\partial \mu_{\text{B}}}{\partial p}\right)_T = V_{\text{m,B}} \quad \text{and} \quad \left(\frac{\partial \mu_{\text{B}}}{\partial T}\right)_p = -\,S_{\text{m,B}}$$

If the substance B undergoes phase transition $\alpha \rightarrow \beta$, we will have

$$\Delta_{trs} V_{m,B} = \left(\frac{\partial \Delta \mu_B}{\partial p}\right)_T = \left(\frac{\partial \mu_{B(\beta)}}{\partial p}\right)_T - \left(\frac{\partial \mu_{B(\alpha)}}{\partial p}\right)_T \neq 0 \qquad (1.6.1)$$

and $$\Delta_{trs} H_{m,B} = T\,\Delta_{trs} S_{m,B} = -T\left(\frac{\partial \Delta \mu_B}{\partial T}\right)_p$$

$$= -T\left\{\left(\frac{\partial \mu_{B(\beta)}}{\partial T}\right)_p - \left(\frac{\partial \mu_{B(\alpha)}}{\partial T}\right)_p\right\} \neq 0 \qquad (1.6.2)$$

Equations (1.6.1) and (1.6.2) imply that the derivatives $(\partial \mu_B/\partial p)_T$ and $(\partial \mu_B/\partial T)_p$ have different values on either side of the transition or in other words, these derivatives are discontinuous.

**Main Characteristics of First-Order Transitions**

Thus the main characteristics of first-order transitions are:

(1) *The chemical potential is a continuous function of temperature.*
(2) *The derivatives of chemical potential are discontinuous.*

These characteristics are illustrated in Fig. 1.6.1.

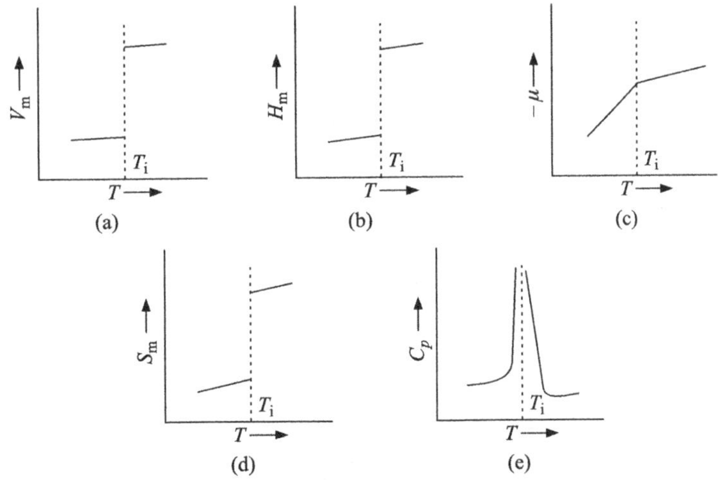


**Fig. 1.6.1 Typical variations of**
**(a) $V_m$ versus $T$,**
**(b) $H_m$ versus $T$,**
**(c) $-\mu$ versus $T$,**
**(d) $S_m$ versus $T$, and**
**(e) $C_p$ versus $T$ of first-order transition**

The fact that $(\partial \mu_B/\partial T)_p$ is discontinuous at the transition point leads to another important characteristic that the first-order transition is accompanied by an infinite heat capacity at the transition point. This may be understood from the following analysis.

The heat capacity of the substance B is given by the relation

$$C_{p,B} = \left(\frac{\partial H_B}{\partial T}\right)_p$$

Now since the transition takes place at constant temperature, the heat supplied to the system is used in driving the transition (that is why $H$ is discontinuous) rather than in raising the temperature of the system. Thus at the transition point, we have

$$(\partial H_B)_p = \text{positive}$$
$$(\partial T)_p = 0$$

Hence $$C_{p,B} = \frac{(\partial H_B)_p}{(\partial T)_p} = \frac{\text{positive}}{0} = \infty$$

The variation of $C_p$ of the system with temperature is also illustrated in Fig. 1.6.1.

**Second-Order Phase Transition**

In a second-order phase transition of a substance B, the chemical potential and its derivatives $(\partial \mu_B/\partial T)_p$ and $(\partial \mu_B/\partial p)_T$ are continuous functions of temperature whereas the second derivatives are discontinuous.

The fact that the first derivatives are continuous implies that changes in the values of enthalpy and volume during transition are zero. In other words, the entropy, the enthalpy and the volume of the system do not change when the transition occurs.

Although the entropy of the substance changes in a continuous manner with temperature, yet the rate of change in disorder (i.e. $dS_B$) alters at the phase transition point. Hence, the first derivative $(\partial S_{m,B}/\partial T)_p$ is a discontinuous function of temperature. Since

$$-\left(\frac{\partial^2 \mu_B}{\partial T^2}\right)_p = \left(\frac{\partial S_{m,B}}{\partial T}\right)_p = \frac{C_{p,m,B}}{T}$$

it follows that the second derivative of $\mu_B$ and $C_{p,m,B}$ are discontinuous functions of temperature.
Similarly, from the relations

$$-\left(\frac{\partial^2 \mu_B}{\partial p^2}\right)_p = -\left(\frac{\partial V_{m,B}}{\partial p}\right)_T = \kappa_T V_{m,B} \qquad (\kappa_T \text{ is isothermal compressibility})$$

$$\frac{\partial^2 \mu_B}{\partial T\, \partial p} = \left(\frac{\partial V_{m,B}}{\partial T}\right)_p = \alpha V_{m,B} \qquad (\alpha \text{ is cubic expansion coefficient})$$

and the fact that the rates of variation of $V_m$ with $T$ and $p$ alter at the phase transition, indicate that the second derivatives $(\partial^2 \mu_B/\partial p^2)$ and $(\partial^2 \mu_B/\partial T\, \partial p)$ are discontinuous.

**Diagrammatic Depiction of Characteristics**

The various characteristics of second-order transition discussed above are illustrated in Fig. 1.6.2. The heat capacity versus temperature curve often has the shape of Greek λ, and because of this reason the second-order transition is frequently referred to as a λ-transition.

Examples of second-order transitions include order-disorder transitions in alloys and the fluid-superfluid transition in helium (line AB in Fig. 1.6.3).

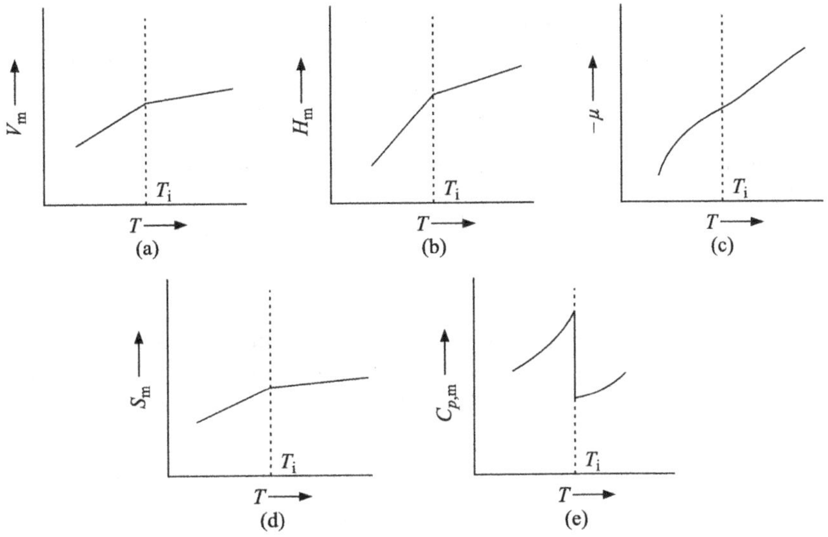


**Fig. 1.6.2 Typical variations of (a) $V_m$ versus $T$, (b) $H_m$ versus $T$, (c) $-\mu$ versus $T$, (d) $S_m$ versus $T$, and (e) $C_{p,m}$ versus $T$ of second-order transition**

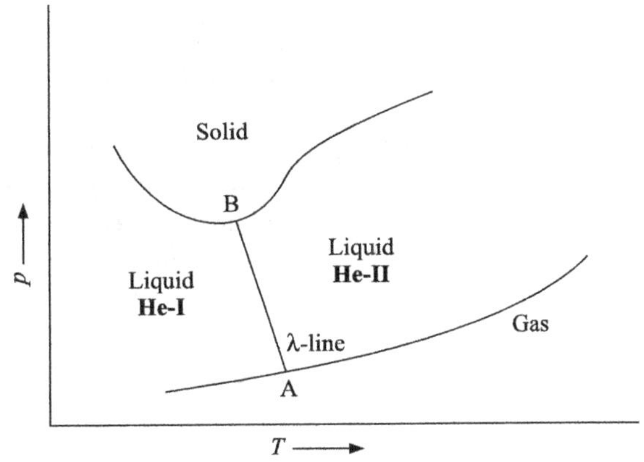


**Fig. 1.6.3 λ-transition of helium. Solid helium is never in equilibrium with the gas phase**

## 1.7 EFFECT OF PRESSURE ON THE VAPOUR PRESSURE OF A LIQUID

**Qualitative Prediction**

The variation in chemical potential of a liquid B (or solid B) with change in the external pressure is given by

$$\left(\frac{\partial \mu_{B(l)}}{\partial p}\right)_p = V_{m,B(l)} \qquad \text{(Eq. 1.3.1)}$$

Since $V_{m,B(l)}$ is positive, it follows that the chemical potential (or the escaping tendency) of the liquid B is increased on increasing the external pressure.

For a liquid B in equilibrium with its vapour, we have

$$\mu_{B(l)} = \mu_{B(v)}$$

Moreover, the chemical potential of vapours of B is related to its pressure by the relation

$$\mu_{B(v)} = \mu^{\circ}_{B(v)} + RT \ln (p/p^{\circ})$$

where $p^\circ$ represents the standard unit pressure. From the above expressions, it follows that the increased value of $\mu_{B(l)}$ (or $\mu_{B(v)}$) due to increase in the external pressure is associated with increase in the vapour pressure of the liquid B(or solid B).

**Derivation of Expression**

To determine the actual relation between the vapour pressure of the liquid and the total external pressure on the liquid phase, we proceed as follows. The thermodynamic condition of equilibrium is

$$\mu_{B(v)}(T, p) = \mu_{B(l)}(T, P) \tag{1.7.1}$$

Equation (1.7.1) implies that at constant temperature, vapour pressure $p$ is a function of the external pressure $P$. To determine the relation between these, we differentiate Eq. (1.7.1) with respect to $P$, such that

$$\left(\frac{\partial \mu_{B(v)}}{\partial p}\right)_T \left(\frac{\partial p}{\partial P}\right)_T = \left(\frac{\partial \mu_{B(l)}}{\partial P}\right)_T$$

or
$$\left(\frac{\partial p}{\partial P}\right)_T = \frac{(\partial \mu_{B(l)}/\partial p)_T}{(\partial \mu_{B(v)}/\partial p)_T} = \frac{V_{m,\,B(l)}}{V_{m,\,B(v)}} \tag{1.7.2}$$

where $V_{m,\,B(v)}$ is the molar volume of vapour at pressure $p$. Since both $V_{m,\,B(l)}$ and $V_{m,\,B(v)}$ are positive, it follows that the derivative $(\partial p/\partial P)_T$ is positive and hence vapour pressure $p$ increases with increase in external pressure $P$. However, the rate of increase is small as $V_{m,\,B(l)} << V_{m,\,B(v)}$.

If the vapour is assumed to be ideal, we have

$$\left(\frac{\partial p}{\partial P}\right)_T = \frac{V_{m,B(l)}}{RT/p}$$

or
$$RT\frac{dp}{p} = V_{m,B(l)}\,dP \qquad (T \text{ constant}) \tag{1.7.3}$$

On integrating the above expression, we get

$$\int_{p_1}^{p_2} RT\frac{dp}{p} = \int_{P_1}^{P_2} V_{m,B(l)}\,dP$$

If $V_{m,\,B(l)}$ is assumed to be independent of pressure, we get

$$RT\ln\frac{p_2}{p_1} = V_{m,B(l)}(P_2 - P_1) \tag{1.7.4}$$

where $p_1$ and $p_2$ are the vapour pressures of the liquid under the external pressures $P_1$ and $P_2$, respectively. If one of the external pressures on the liquid is its own vapour pressure $p^*$, then Eq. (1.7.4) reduces to

$$RT\ln\frac{p_2}{p^*} = V_{m,\,B(l)}(P_2 - p^*) \tag{1.7.5}$$

Expressions for the variation of vapour pressure of a solid phase with the external pressure can be derived similarly and is given by

$$RT \ln \frac{p_2}{p_1} = V_{\mathrm{m,\,B(s)}}(P_2 - P_1) \tag{1.7.6}$$

---

***Example 1.7.1***

What is the vapour pressure of ice at 263.15 K due to the pressure of air at atmospheric pressure? The density of ice is 0.920 g $\mathrm{cm^{-3}}$ and its vapour pressure at 263.15 K is $2.6 \times 10^{-3}$ bar.

***Solution***

The given data are

$$T = 263.15\ \mathrm{K} \qquad p^* = 2.6 \times 10\ ^{3}\ \mathrm{bar}$$

$$P_2 = 1.01\,325\ \mathrm{bar} \qquad p_2 = ?$$

$$\rho(\mathrm{ice}) = 0.920\ \mathrm{g\ cm}\ ^{3}$$

The molar volume of solid ice is

$$V_{\mathrm{m,\,s}} = \frac{M}{\rho} = \frac{18.0\ \mathrm{g\ mol^{-1}}}{0.920\ \mathrm{g\ cm^{-3}}} = 19.57\ \mathrm{cm^3\ mol^{-1}} = 19.57 \times 10^{-3}\ \mathrm{dm^3\ mol^{-1}}$$

Substituting the given data in the expression

$$RT \ln \frac{p_2}{p^*} = V_{\mathrm{m,s}}(P_2 - p^*)$$

we get $\quad (0.083\,14\ \mathrm{dm^3\ bar\ K^{-1}\ mol^{-1}})\,(263.15\ \mathrm{K}) \ln\left(\dfrac{p_2}{2.6 \times 10^{-3}\ \mathrm{bar}}\right)$

$$= (19.57 \times 10^{-3}\ \mathrm{dm^3\ mol^{-1}})\,(1.01325\ \mathrm{bar} - 2.6 \times 10^{-3}\ \mathrm{bar})$$

$$\ln\left(\frac{p_2}{2.6 \times 10^{-3}\ \mathrm{bar}}\right) = \frac{(19.57 \times 10^{-3})\,(1.010\,65)}{(0.083\,14)\,(263.15)} = 9.0 \times 10^{-4}$$

$$\frac{p_2}{(2.6 \times 10^{-3}\ \mathrm{bar})} = 1.000\,9$$

i.e. $\quad p_2 = (1.000\,9)\,(2.6 \times 10^{-3}\ \mathrm{bar}) = 2.602 \times 10^{-3}\ \mathrm{bar}$

---

***Example 1.7.2***

Vapour pressure of water at 303.15 K is 0.042 4 bar when exposed to an insoluble gas at 1.013 25 bar pressure. What will be the vapour pressure if the gas pressure were increased to 101.325 bar? Density of water is 0.996 g $\mathrm{cm^{-3}}$ and it may be assumed to be incompressible.

***Solution***

The given data are

$$P_1 = 1.013\,25\ \mathrm{bar} \qquad p_1 = 0.042\,4\ \mathrm{bar}$$

$$P_2 = 101.325\ \mathrm{bar} \qquad p_2 = ?$$

$$\rho_{\mathrm{water}} = 0.996\ \mathrm{g\ cm^{-3}}$$

The molar volume or liquid water is

$$V_{\mathrm{m,l}} = \frac{M}{\rho} = \frac{18\ \mathrm{g\ mol^{-1}}}{0.996\ \mathrm{g\ cm^{-3}}} = 18.07\ \mathrm{cm^3\ mol^{-1}} = 18.07 \times 10^{-3}\ \mathrm{dm^3\ mol^{-1}}$$

Substituting the given data in the expression

$$RT \ln \frac{p_2}{p_1} = V_{m,water}(P_2 - P_1)$$

we get $(0.083\,14 \text{ dm}^3 \text{ bar K}^{-1} \text{ mol}^{-1})(303.15 \text{ K}) \ln \left(\frac{p_2}{0.042\,4 \text{ bar}}\right)$

$$= (18.07 \times 10^{-3} \text{ dm}^3 \text{ mol}^{-1})(101.325 \text{ bar} - 1.013\,25 \text{ bar})$$

or $$\ln \left(\frac{p_2}{0.042\,4 \text{ bar}}\right) = \frac{(18.07 \times 10^{-3})(100.312)}{(0.083\,14)(303.15)} = 0.071\,92$$

or $$\frac{p_2}{0.042\,4 \text{ bar}} = 1.075 \quad \text{or} \quad p_2 = (1.075)(0.042\,4 \text{ bar}) = 0.045\,6 \text{ bar}$$

## REVISIONARY PROBLEMS

1.1 Show that the chemical potential of a substance present in more than one phase in equilibrium with each other has the same value in all the phases.

1.2 (a) Show that the slope of $\mu$ versus $T$ curve for a pure substance is equal to $-S_m$ and hence show that this slope is always negative.

(b) Show that the slope of $\mu$ versus $T$ curve of a pure substance B follows the order

$$\left(\frac{\partial \mu_{B(s)}}{\partial T}\right)_p > \left(\frac{\partial \mu_{B(l)}}{\partial T}\right)_p >> \left(\frac{\partial \mu_{B(v)}}{\partial T}\right)_p$$

and hence draw a schematic diagram of $\mu$ versus $T$ curves for the above-mentioned three phases. Locate the melting point and the boiling point of the substance on the diagram.

(c) What is the effect of pressure on the $\mu$ versus $T$ curves? Show that the shift in the curves due to the change in pressure follows the order

$$d\mu_{B(v)} >> d\mu_{B(l)} > d\mu_{B(s)}$$

and hence show the shift in the melting point and the boiling point with increase in the external pressure.

(d) Show qualitatively that on increasing the external pressure the melting point of a substance increases when melting results in an increase in volume, but a decrease is observed when melting results in a decrease in volume.

1.3 (a) Derive the Clapeyron equation

$$\frac{dp}{dT} = \frac{\Delta_{trs}S_m}{\Delta_{trs}V_m}$$

as applicable to the phase transformation $\alpha \rightarrow \beta$, where $\Delta_{trs}S_m$ and $\Delta_{trs}V_m$ are the respective changes in molar entropy and molar volume accompanying the phase transformation.

(b) Show that:

(i) The slope $dp/dT$ of solid to liquid transformation has a very high positive (or negative) value.

(ii) At the triple point, the slope $dp/dT$ of solid to vapour transformation is steeper than that of liquid to vapour transformation.

(c) On what factors does the sign of the slope $dp/dT$ of solid to liquid transformation depend?

(d) Show that the integrated form of Clapeyron equation as applicable to solid-solid and solid-liquid transformations is

$$\frac{\Delta p}{\Delta T} = \frac{1}{T_{\text{trs}}} \frac{\Delta_{\text{trs}} H_{\text{m}}}{\Delta_{\text{trs}} V_{\text{m}}}$$

1.4 (a) Discuss the Clausius simplifications to the Clapeyron equation and hence derive the Clausius-Clapeyron equation as applicable to the condensed phase-gas equilibrium. Does this relation hold good for the solid-liquid equilibrium?
(b) Draw a schematic plot of log $(p/p^\circ)$ versus $1/T$. What will be its slope and intercept? Do the values of slope and intercept depend upon the units in which pressure is expressed?
(c) Is the relation

$$\Delta_{\text{sub}} H_{\text{m}} = \Delta_{\text{fus}} H_{\text{m}} + \Delta_{\text{vap}} H_{\text{m}}$$

for a substance true at all conditions of temperature and pressure?
(d) Show qualitatively that the shift in the boiling point due to the given pressure change is relatively larger than that of the melting point.
(e) Under what circumstances can the solid pass directly to the gaseous form without the formation of liquid state?

1.5 What do you understand by the first-order and second-order phase transitions? What are their main characteristics?

1.6 Derive the relation

$$RT \ln \frac{p}{p^*} = V_{\text{m, cond.phase}} (P - p^*)$$

where $p$ is the vapour pressure under a pressure of $P$ and $p^*$ is the vapour pressure when the condensed phase and vapour are under the same pressure. From this relation, show that the vapour pressure of the condensed phase increases on increasing the total external pressure.

1.7 (a) For the liquid-vapour transformation, derive the following form of Clausius-Clapeyron equation:

$$\ln\left(\frac{p}{p^*}\right) = -\frac{\Delta_{\text{vap}} H_{\text{m}}}{R}\left(\frac{1}{T} - \frac{1}{T^*}\right)$$

where $T^*$ is the normal boiling of the liquid where $p^* = 1$ atm; $p$ is the vapour pressure at temperature $T$ and $\Delta_{\text{vap}} H_{\text{m}}$ is the temperature-independent molar enthalpy change of the transformation.
(b) Draw a schematic plot of log $(p/\text{atm})$ versus $1/T$. What do the slope and intercept of such a graph represent? How are the slope and intercept affected if $p$ is expressed in cmHg?

## TRY YOURSELF PROBLEMS

1.1 Trouton's rule says that the molar entropy of vaporization of liquids is about 87.86 J $K^{-1}$ $mol^{-1}$ at the normal boiling point $T^*$. From this show that the boiling point at pressure $p$ can be corrected to 760 mmHg by the relation

$$\Delta T_{\text{b}} = 0.000\,125\, T^* (760 - p/\text{mmHg}).$$

$$\left\{\text{Hint. } \frac{\Delta p}{\Delta T} = \frac{\Delta_{\text{trs}} S_{\text{m}}}{\Delta_{\text{trs}} V_{\text{m}}} \simeq \frac{\Delta_{\text{trs}} S_{\text{m}}}{V_{\text{m}}} = \frac{\Delta_{\text{trs}} S_{\text{m}}}{RT/p} = \frac{\Delta_{\text{trs}} S_{\text{m}}}{T^*}\left(\frac{760 \text{ mmHg}}{R}\right)\right\}$$

1.2 (a) In deriving Eq. (1.5.13), it has been assumed that $\Delta H_m$ of the system is independent of temperature. Suppose $\Delta H_m$ depends upon the temperature and is given by

$$\Delta H_m = \Delta H_0 + (\Delta a)T + \left(\frac{\Delta b}{2}\right)T^2 + \frac{(\Delta c)T^3}{3} + \cdots$$

where $\Delta H_0$, $(\Delta a)$, $(\Delta b)$, . . . , etc., are constants. Show that with this expression of $\Delta H_m$, the Clausius-Clapeyron equation is given by

$$R \ln\left(\frac{p_2}{p_1}\right) = \frac{\Delta H_0 \, (T_2 - T_1)}{T_2 T_1} + (\Delta a) \ln \frac{T_2}{T_1} + \frac{(\Delta b)}{2}(T_2 - T_1) + \frac{\Delta c}{6}(T_2^2 - T_1^2) + \cdots$$

(b) Show that with the expression of $\Delta H_m$ given in part (a), the open integration of Clausius-Clapeyron equation yields

$$R \ln\left(\frac{p}{p^\circ}\right) = -\frac{\Delta H_0}{T} + \Delta a \ln (T/\text{K}) + \frac{\Delta b}{2}T + \frac{\Delta c}{6}T^2 + \cdots + I$$

where $I$ is the constant of integration.

1.3 For the condensed phase-vapour equilibrium, it is known that $K_{eq} = p$, where $p$ is the pressure of the vapour. Show that on substituting this relation in the Clausius-Clapeyron equation, we get

$$\ln \frac{K_2}{K_1} = \frac{\Delta H_m}{R}\left[\frac{1}{T_1} - \frac{1}{T_2}\right]$$

1.4 The vapour pressure of a particular liquid varies with temperature as follows:

$$\log\left(\frac{p}{\text{atm}}\right) = -\frac{(4\,849 \text{ K})}{T} - 14.70 \log\left(\frac{T}{\text{K}}\right) + 50.24$$

How does the enthalpy of vaporization vary with temperature?

1.5 The melting points of mercury at external pressures of 10 atm and 3 540 atm are 234.3 K and 257.2 K, respectively. Predict whether the transformation solid Hg $\rightarrow$ liquid Hg involves an increase or a decrease of volume.

(Ans. Increase)

1.6 (a) A solid substance directly passes to the gaseous form at STP conditions. State the conditions under which the liquid form of the substance can be obtained.
(b) For a given substance, $\rho_s = \rho_l$. Predict the nature of solid to liquid equilibrium line in its phase diagram.

1.7 (a) The rate of change of chemical potential with temperature [i.e. $(\partial\mu/\partial T)_p$] is an important quantity in the discussion of phase transitions. What is the difference in slope on either side of (i) the freezing point of water, and (ii) the normal boiling point of water? Given: $\Delta_{vap}H_m = 40.59$ kJ mol$^{-1}$ and $\Delta_{fus}H_m = 6.008$ kJ mol$^{-1}$.

(Ans. 22 J K$^{-1}$ mol$^{-1}$, 109 J K$^{-1}$ mol$^{-1}$)

(b) The gradient $(\partial\mu/\partial p)_T$ is also an important quantity in order to decide the nature of transition. What is the difference of this quantity on either side of the two transitions referred to in part (a)? Given: $\rho$(ice) = 0.917 g cm$^{-3}$, $\rho$(liquid water) = 1.0 g cm$^{-3}$ at 0 °C and $\rho$(liquid water) = 0.958 g cm$^{-3}$ and $\rho$(steam) = 0.598 kg m$^{-3}$ at 100 °C.

1.8 Combine the barometric formula for the dependence of pressure on altitude with the Clausius-Clapeyron equation, and predict how the boiling point of a liquid depends on altitude.

$$\left[\text{Ans. } \frac{1}{T} = \frac{1}{T^*} + \left(\frac{Mgh}{T^*\Delta_{\text{vap}}H_{\text{m}}}\right)\right]$$

1.9 Why is the following relation not applicable for graphite-diamond transformation?

$$\ln \frac{p_2}{p_1} = \frac{\Delta_{\text{trs}}H_{\text{m}}}{R}\left(\frac{1}{T_1} - \frac{1}{T_2}\right)$$

1.10 How do the slopes of $G_{\text{m}}$ versus $p$ curves for a liquid and its vapour vary as the temperature of the liquid is increased towards its critical temperature?

## NUMERICAL PROBLEMS

1.1 The melting point of benzene increases by about 0.03 K/atm. If the specific gravities of solid and liquid benzene are 1.02 and 0.89 at the melting point of 278.65 K at 1 atm, what is the molar enthalpy of fusion of benzene? (Ans. 10.52 kJ $\text{mol}^{-1}$)

1.2 For uranium hexafluoride the vapour pressures for the solid and liquid are given by

$$\log(p_s/\text{Torr}) = 10.648 - 2\,559.5/(T/\text{K})$$
$$\log(p_l/\text{Torr}) = 7.540 - 1\,511.3/(T/\text{K})$$

Calculate (i) the temperature and pressure of the triple point and (ii) the molar enthalpy of fusion of $UF_6$.

(Ans. 337.2 K, 1 122 Torr, 20.08 kJ $\text{mol}^{-1}$)

1.3 What is the boiling point of water on a mountain where the barometric reading is 500 Torr? The enthalpy of vaporization of water may be taken as 2.259 kJ $\text{g}^{-1}$.

(Ans. 362 K)

1.4 The densities of α- and β-sulphur are 2.00 and 1.95 g $\text{cm}^{-3}$, respectively, at their transition temperature 369 K. The transition temperature changes by 0.036 K for every atmosphere rise in pressure. Determine the molar enthalpy of transition.

(Ans. 426.4 J $\text{mol}^{-1}$)

1.5 A particular liquid obeys both Clausius-Clapeyron equation and Trouton's rule. With this information calculate the vapour pressure of the liquid in atm, at a temperature equal to one-third of the normal boiling point.

(Ans. $6.4 \times 10^{-10}$ atm)

1.6 $NH_4NO_3$ undergoes a transition from one solid modification to another at 398.65 K and 1 atm, and at 408.15 K and 1 000 atm pressure. The form stable at higher temperatures has an average volume of 0.0126 $\text{cm}^3\ \text{g}^{-1}$ greater than the other modification over the pressure range studied. From these data, calculate the molar enthalpy of transition. (Ans. 4.284 kJ $\text{mol}^{-1}$)

1.7 AgI exists in two forms, α and β, which are in equilibrium at 419.65 K at 1 atm pressure. For the change of α to β, $\Delta_{\text{trs}}H_{\text{m}} = 6.40$ kJ $\text{mol}^{-1}$, while $\Delta_{\text{trs}}V_{\text{m}} = -2.2\ \text{cm}^3\ \text{mol}^{-1}$. Find the pressure at which the transition temperature will be 418.15 K.

(Ans. 67.37 atm)

1.8 Propene has the following vapour pressures:

| $T$/K | 150 | 200 | 250 | 300 |
|---|---|---|---|---|
| $p$/Torr | 3.82 | 198.0 | 2 074 | 10 040 |

From these data calculate (a) the molar enthalpy of vaporization, (b) the normal boiling point, and (c) the vapour pressure at 225 K

(Ans. (a) 19.54 kJ $mol^{-1}$, (b) 741 Torr)

1.9 (a) The boiling point of $n$-butyl chloride is 77.96 °C. Without using any other data, determine the boiling point at vapour pressure 650 Torr. (Ans. 72.8 °C)

(b) The variation of vapour pressure of $n$-butyl chloride is also given by the empirical equation

$$\log\left(\frac{p}{\text{Torr}}\right) = -\frac{(1763\ \text{K})}{T} + 7.912$$

Calculate the molar enthalpy of vaporization from this equation.

(Ans. 33.76 kJ $mol^{-1}$)

1.10 The vapour pressure of a given liquid varies with temperature according to the following expression

$$\log\left(\frac{p}{\text{Torr}}\right) = -\frac{(4\,849.3\ \text{K})}{T} - 14.701\log\left(\frac{T}{\text{K}}\right) + 53.118\left(\frac{T}{\text{K}}\right)$$

Determine the molar enthalpy of vaporization at 298 K and at 423 K.

(Ans. 39.23 kJ $mol^{-1}$, 79.04 kJ $mol^{-1}$)

1.11 Find out the difference in the freezing points of ice at 1 atm pressure and under its own vapour pressure. The vapour pressure of ice at 273 K is 4.579 mmHg. The densities of ice and water at 273 K are 1.000 and 0.916 8 g $cm^{-3}$, respectively, and the enthalpy of fusion of ice is 6.009 kJ $mol^{-1}$. (Ans. 0.007 5 K)

1.12 From the following data at 298 K, determine (i) the free energy change of transition from graphite to diamond at 1 atm and 298 K, (ii) direction of spontaneous process of graphite to diamond conversion, (iii) the pressure at which the two forms would be in equilibrium at 298 K and at 1 273 K. Assume densities to be independent of pressure.

| Form | $\Delta_c H_m$ / (kJ $mol^{-1}$) | $S_m$/J $K^{-1}$ $mol^{-1}$ | *Density* / (g $cm^{-3}$) |
|---|---|---|---|
| diamond | −395.32 | 2.439 | 3.513 |
| graphite | −393.42 | 5.694 | 2.260 |

# 2 Colligative Properties

## 2.1 SOLUTION

**Definition**

A solution may be defined as a homogeneous mixture of a single phase containing one or more of the chemical species dispersed on a molecular scale. The characteristics of any one section of the homogeneous solution will be completely identical to those of any other section of the solution. Depending upon the total constituents present in the solution, it is called the binary solution (two constituents), ternary solution (three constituents), quarternary solution (four constituents), and so on.

**Classification of Solute and Solvent**

By convention, the constituent present in the largest amount is called the *solvent* and the constituents present in relatively small amounts are called the *solutes*. However, this classification is completely arbitrary, since it is not possible to classify the two constituents of a binary solution into solute and solvent when both of them are present in equal amounts. Any one of the two constituents may be classified as the solute or the solvent. Thus this classification is a matter of convenience only.

**Colligative Properties**

In this chapter, we shall consider only those properties which result because of the dissolution of a nonvolatile solute in a volatile liquid solvent. These are, the relative lowering of vapour pressure, elevation of boiling point, depression of freezing point and the osmotic pressure. These are collectively known as the *colligative*[†] properties as they depend only on the number of species present in the solution and not on their nature. Before considering these properties, it is worth reviewing the units in which the composition of the solution can be expressed.

## 2.2 METHODS OF EXPRESSING CONCENTRATION OF A SOLUTION

**Introduction**

There are many ways of expressing the concentration of a solution which are used for different purposes. Some of these are:

- Mass percentage.
- Volume percentage.
- Mass of a solute in a definite mass of solvent.
- Mass of a solute per definite mass of solution.

---

[†]Colligative, a latin word; *co,* together; *ligare,* to bind.

- Molarity; amount (in moles) of solute per $dm^3$ of solution.
- Normality; amount (in equivalents) of solute per $dm^3$ of solution.
- Molality; amount (in moles) of solute per kg of solvent.
- Amount (or mole) fraction; amount (in moles) of a given substance divided by the total amount (in moles) of all constituents of a solution.

Since the density of a solution varies with temperature, presumably due to the variation of volume, it is obvious that the concentration $c_2$, which is expressed as mol $dm^{-3}$ of the solution, varies with temperature. On the other hand, molality and amount fraction are independent of temperature.

---

***Example 2.2.1*** Calculate the amount fraction, molal and molar concentrations of a solution containing 20per cent acetic acid by mass in water, if the density of the solution at 20 °C is 1.026 g $cm^{-3}$.

***Solution*** In 100 g of the solution, we have

Mass of acetic acid, $m_2 = 20$ g

Mass of water, $m_1 = 80$ g

Molar mass of acetic acid, $M_2 = 60 \text{ g mol}^{-1}$

$$\text{Amount of acetic acid, } n_2 = \frac{m_2}{M_2} = \frac{20 \text{ g}}{60 \text{ g mol}^{-1}} = 0.333 \text{ mol}$$

$$\text{Amount of water, } n_1 = \frac{m_1}{M_1} = \frac{80 \text{ g}}{18 \text{ g mol}^{-1}} = 4.444 \text{ mol}$$

$$\text{Amount fraction of acetic acid, } x_2 = \frac{n_2}{n_2 + n_1} = \frac{0.333 \text{ mol}}{0.333 \text{ mol} + 4.444 \text{ mol}}$$

$$= \frac{0.333}{4.777} = 0.069\,7$$

$$\text{Molality of acetic acid solution, } m = \frac{n_2}{m_1} = \frac{0.333 \text{ mol}}{(80/1\,000) \text{ kg}} = 4.163 \text{ mol kg}^{-1}$$

Volume of the solution containing 20 per cent acetic acid by mass is

$$V = \frac{\text{mass}}{\text{density}} = \frac{100 \text{ g}}{1.026 \text{ g cm}^{-3}} = 97.45 \text{ cm}^3$$

$$\text{Molarity of acetic acid solution, } M = \frac{n_2}{V} = \frac{0.333 \text{ mol}}{(97.45/1\,000) \text{ dm}^3} = 3.417 \text{ mol dm}^{-3}$$

---

Of the various units, molality, molarity and amount fraction are commonly employed to express the composition of a solution. Quite often, we require the interconversion of these concentration units.† These can be carried out as shown in the following.

---

†Throughout this chapter, solvent is represented by the subscript 1, solute is represented by the subscript 2. To avoid any confusion between the symbol $m$ used simultaneously for mass and molality, we use subscripted $m$ for mass whereas unsubscripted $m$ for molality of the solution. To be specific, $m_1$ and $m_2$ represent masses of solvent and solute, respectively, whereas $m$ represents molality of the solution.

As per IUPAC recommendations, the symbol $b$ may be used for molality to avoid the confusion of notation.

**Amount Fraction into Molarity**

By definition, we have

$$x_1 = \frac{n_1}{n_1 + n_2}; \quad x_2 = \frac{n_2}{n_1 + n_2} \quad \text{and} \quad M = \frac{n_2}{V}$$

We will have

$$M = \frac{n_2}{V} = \frac{n_2}{m/\rho} = \frac{n_2}{(n_1 M_1 + n_2 M_2)/\rho}$$

Dividing numerator and denominator by $n_1 + n_2$ we get

$$M = \frac{x_2 \rho}{x_1 M_1 + x_2 M_2} \tag{2.2.1}$$

For a dilute solution $x_2 M_2 << x_1 M_1$, $x_1 \rightarrow 1$ and $\rho_{\text{solution}} \rightarrow \rho_{\text{solvent}}$. Under these approrimations, Eq. (2.2.1) may be written as

$$M \simeq \frac{x_2 \rho}{M_1} \quad \text{i.e.} \quad M \propto x_2 \tag{2.2.2}$$

that is, molarity of the solution is directly proportional to the amount fraction of solute in the solution.

---

***Example 2.2.2*** The amount fraction of methanol in aqueous solution is 0.02 and its density is 0.994 g $mL^{-1}$. Determine the molarity of the solution.

***Solution*** Let there be 1 mol of total amount of solvent and solute in the solution. We will have

Amount of solute, $n_2 = 0.02$ mol

Amount of solvent, $n_1 = 0.98$ mol

Mass of solution, $m = \text{mass of (solvent + solute)} = n_1 M_1 + n_2 M_2$
$= (0.98 \times 18 + 0.02 \times 32)\ \text{g} = 18.28\ \text{g}$

Volume of solution, $V = \frac{m}{\rho} = \frac{18.28\ \text{g}}{0.994\ \text{g cm}^{-3}} = 18.39\ \text{cm}^3$

Finally, the molarity of solution would be

$$M = \frac{n_2}{V} = \frac{0.02\ \text{mol}}{18.39 \times 10^{-3}\ \text{dm}^3} = 1.088\ \text{mol dm}^{-3} \qquad (V \text{ expressed in dm}^3)$$

**Alternatively,** using the conversion formula directly, we get

$$M = \frac{x_2 \rho}{x_1 M_1 + x_2 M_2} = \frac{(0.02)(0.994\ \text{g cm}^{-3})}{(0.98)(18\ \text{g mol}^{-1}) + (0.02)(32\ \text{g mol}^{-1})}$$

$= 1.088 \times 10^{-3}$ mol cm$^{-3}$ $= 1.088 \times 10^{-3}$ mol $(10^{-1}$ dm$)^{-3}$

$= 1.088$ mol dm$^{-3}$

---

## Molarity into Amount Fraction

The expression relating molarity ($M$) with amount fractions ($x_1$ and $x_2$) is

$$M = \frac{x_2 \rho}{x_1 M_1 + x_2 M_2}$$

where $M_1$ and $M_2$ are molar masses of solvent and solute, respectively.

Since $x_2 = 1 - x_1$, we get

$$M = \frac{x_2 \rho}{(1 - x_2)M_1 + x_2 M_2}$$

which on rearranging gives

$$x_2 = \frac{MM_1}{\rho + M(M_1 - M_2)} \qquad (2.2.3)$$

---

***Example 2.2.3***

The molarity of $NH_3$ in aqueous solution is 11.8 mol dm$^{-3}$. Calculate the amount fraction of $NH_3$ in solution. The density of solution is 0.916 g cm$^{-3}$.

***Solution***

Let the volume of solution be 1 dm$^3$. We will have

Amount of $NH_3$, $n_2 = 11.8$ mol

Volume of solution, $V = 1$ dm$^3$

Mass of 1 dm$^3$ solution, $m = V\rho = (10^3\ \text{cm}^3)(0.916\ \text{g cm}^{-3}) = 916\ \text{g}$

Mass of $NH_3$ in 1 dm$^3$ solution, $m_2 = n_2 M_2 = (11.8\ \text{mol})(17\ \text{g mol}^{-1})$
$= 200.6\ \text{g}$

Mass of solvent in 1 dm$^3$ solution, $m_1$ = mass of (solution – solute)
$= 916\ \text{g} - 200.6\ \text{g} = 715.4\ \text{g}$

Amount of solvent in 1 dm$^3$ solution, $n_1 = \dfrac{m_1}{M_1} = \dfrac{715.4\ \text{g}}{18\ \text{g mol}^{-1}} = 39.74\ \text{mol}$

Amount fraction of $NH_3$ in solution, $x_2 = \dfrac{n_2}{n_1 + n_2} = \dfrac{11.8}{39.74 + 11.8} = 0.229$

**Alternatively,** using the conversion formula, we get

$$x_2 = \frac{MM_1}{\rho + M(M_1 - M_2)}$$

$$= \frac{(11.8\ \text{mol dm}^{-3})(18\ \text{g mol}^{-1})}{(0.916 \times 10^3\ \text{g dm}^{-3}) + (11.8\ \text{mol dm}^{-3})\{(18 - 17)\ \text{g mol}^{-1}\}}$$

$$= \frac{212.4}{916 + 11.8} = \frac{212.4}{927.8} = 0.229$$

---

**Amount Fraction into Molality**

By definition

$$x_1 = \frac{n_1}{n_1 + n_2}; \quad x_2 = \frac{x_2}{n_1 + n_2}; \quad m = \frac{n_2}{m_1}$$

We will have

$$m = \frac{n_2}{m_1} = \frac{n_2}{n_1 M_1}$$

Dividing numerator and denominator by $n_1 + n_2$, we get

$$m = \frac{x_2}{x_1 M_1} \qquad (2.2.4)$$

For a dilute solution $x_1 \to 1$. With this approximation, Eq. (2.2.4) may be written as

$$m \simeq \frac{x_2}{M_1} \quad \text{or} \quad m \propto x_2 \qquad (2.2.5)$$

that is, molality of solution is directly proportional to the amount fraction of solute in the solution.

---

*Example 2.2.4*

The amount fraction of methanol in aqueous solution is 0.02. Determine the molality of methanol in solution.

*Solution*

Let there be 1 mol of total amount of solvent and solute. Then

Amount of solute, $n_2 = 0.02$ mol

Amount of solvent, $n_1 = 0.98$ mol

Mass of solvent, $m_1 = n_1 M_1 = (0.98 \text{ mol})(18 \text{ g mol}^{-1}) = 17.64 \text{ g}$

Finally, the molality of solution is

$$m = \frac{n_2}{m_1} = \frac{0.02 \text{ mol}}{17.64 \times 10^{-3} \text{ kg}} = 1.134 \text{ mol kg}^{-1} \qquad (m_1 \text{ expressed in kg})$$

**Alternatively,** using the conversion expression directly, we get

$$m = \frac{x_2}{x_1 M_1} = \frac{0.02}{(0.98)(18 \text{ g mol}^{-1})}$$
$$= 1.134 \times 10^{-3} \text{ mol g}^{-1} = 1.134 \times 10^{-3} \text{ mol } (10^{-3} \text{ kg})^{-1}$$
$$= 1.134 \text{ mol kg}^{-1}$$

---

**Molality into Amount Fraction**

The expression relating amount fractions ($x_1$ and $x_2$) and molality ($m$) is

$$m = \frac{x_2}{x_1 M_1} \qquad (M_1 \text{ is the molar mass of solvent})$$

Replacing $x_1 = 1 - x_2$ and rearranging, we get

$$x_2 = \frac{m M_1}{1 + m M_1} \qquad (2.2.6)$$

***Example 2.2.5***

An aqueous solution is 1.133 molal in methanol. Determine the amount fractions of methanol and water.

***Solution***

Let the mass of solvent be 1 kg. We will have

$$\text{Amount of solute, } n_2 = 1.133 \text{ mol}$$

$$\text{Mass of solvent, } m_1 = 1 \text{ kg} = 1\,000 \text{ g}$$

$$\text{Amount of solvent, } n_1 = \frac{m_1}{M_1} = \frac{1\,000 \text{ g}}{18 \text{ g mol}^{-1}} = 55.56 \text{ mol}$$

Amount fraction of solute (i.e. methanol) is

$$x_2 = \frac{n_2}{n_1 + n_2} = \frac{1.133}{55.56 + 1.133} = 0.02$$

Amount fraction of solvent (i.e. water) is

$$x_1 = 1 - x_2 = 1 - 0.02 = 0.98$$

**Alternatively,** using the formula directly, we get

$$x_2 = \frac{mM_1}{1 + mM_1} = \frac{(1.133 \text{ mol kg}^{-1})(18 \times 10^{-3} \text{ kg mol}^{-1})}{1 + (1.133 \text{ mol kg}^{-1})(18 \times 10^{-3} \text{ kg mol}^{-1})}$$

$$= \frac{0.020\,394}{1 + 0.020\,394} = 0.02$$

$$x_1 = 1 - x_2 = 1 - 0.02 = 0.98$$

**Molarity into Molality**

Since molarity is given as $M = n_2/V$, we assume that

$$\text{Amount of solute} = n_2$$

$$\text{Volume of solution} = V$$

$$\text{Mass of solute, } m_2 = n_2M_2 \qquad (M_2 \text{ is molar mass of solute})$$

If $\rho$ is the density of solution, then

$$\text{Mass of solution} = V\rho$$

Now Mass of solvent, $m_1$ = mass of (solution – solute) = $V\rho - n_2M_2$

Finally, molality of solution is

$$\text{molality} = \frac{\text{amount of solute}}{\text{mass of solvent expressed in kg}}$$

i.e.

$$m = \frac{n_2}{V\rho - n_2M_2}$$

Dividing numerator and denominator of the right side by $V$, we get

$$m = \frac{(n_2/V)}{\rho - (n_2/V)M_2} = \frac{M}{\rho - MM_2} \qquad (2.2.7)$$

---

***Example 2.2.6*** The density of 1.35 M solution of acetic acid is 1.012 g $cm^{-3}$. Calculate its molality.

***Solution*** Let the volume of solution be 1 $dm^3$. We have

Amount of solute, $n_2 = 1.35$ mol

Volume of solution, $V = 1\ dm^3$

Mass of solute, $m_2 = n_2M_2 = (1.35\ mol)\ (60\ g\ mol^{-1}) = 81\ g$

Mass of solution $= V\rho = (10^3\ cm^3)\ (1.012\ g\ cm^{-3}) = 1012\ g$

Mass of solvent, $m_1$ = Mass of (solution – solute) = (1 012 – 81) g

= 931 g = 0.931 kg

Finally, the molality of solution is

$$m = \frac{n_2}{m_1} = \frac{1.35\ \text{mol}}{0.931\ \text{kg}} = 1.45\ \text{mol kg}^{-1} \qquad (m_1 \text{ expressed in kg})$$

**Alternatively,** using the conversion formula directly, we get

$$m = \frac{M}{\rho - MM_2} = \frac{(1.35\ \text{mol dm}^{-3})}{(1.012\ \text{g cm}^{-3}) - (1.35\ \text{mol dm}^{-3})\ (60\ \text{g mol}^{-1})}$$

$$= \frac{1.35\ \text{mol dm}^{-3}}{\{1.012\ (10^{-3}\ \text{kg})\ (10^{-1}\ \text{dm})^{-3}\} - (1.35\ \text{mol dm}^{-3})\ \{60\ (10^{-3}\ \text{kg})\ \text{mol}^{-1}\}}$$

$$= \frac{1.35}{1.012 - 1.35 \times 60 \times 10^{-3}}\ \text{mol kg}^{-1} = 1.45\ \text{mol kg}^{-1}$$

---

## Molality into Molarity

The expression relating molality ($m$) and molarity ($M$) is

$$m = \frac{M}{\rho - MM_2} \qquad (\rho \text{ is density of solution and } M_2 \text{ is molar mass of solute})$$

Rearranging this expression, we get

$$M = \frac{m\rho}{1 + mM_2} \tag{2.2.8}$$

---

***Example 2.2.7*** The density of 1.450 molal solution of acetic acid is 1.012 g $cm^{-3}$. Calculate its molarity.

***Solution*** Let the mass of solvent be 1 kg. We have

Amount of solute, $n_2 = 1.450$ mol

Mass of solvent, $m_1 = 1$ kg

Mass of solute, $m_2 = n_2M_2 = (1.450\ mol)\ (60\ g\ mol^{-1}) = 87.0\ g$

Mass of solution, $m_{soln}$ = mass of (solvent + solute)

= 1 000 g + 87.0 g = 1 087 g

$$\text{Volume of solution, } V = \frac{m_{\text{soln}}}{\rho} = \frac{1\ 087\ \text{g}}{1.012\ \text{g cm}^{-3}} = 1\ 074.1\ \text{cm}^3$$

Finally, molarity of solution is

$$M = \frac{n_2}{V} = \frac{1.450\ \text{mol}}{1.074\ 1\ \text{dm}^3} = 1.35\ \text{mol dm}^{-3} \qquad (V\ \text{expressed in dm}^3)$$

**Alternatively,** using the conversion formula directly, we get

$$\begin{aligned} M &= \frac{m\rho}{1 + mM_2} = \frac{(1.450\ \text{mol kg}^{-1})(1.012\ \text{g cm}^{-3})}{1 + (1.450\ \text{mol kg}^{-1})(60\ \text{g mol}^{-1})} \\ &= \frac{(1.450\ \text{mol kg}^{-1})\{1.012(10^{-3}\ \text{kg})(10^{-1}\ \text{dm})^{-3}\}}{1 + (1.450\ \text{mol kg}^{-1})\{60(10^{-3}\ \text{kg})\ \text{mol}^{-1}\}} \\ &= \frac{1.450 \times 1.012}{1 + 1.450 \times 60 \times 10^{-3}}\ \text{mol dm}^{-3} \\ &= 1.35\ \text{mol dm}^{-3} \end{aligned}$$

## 2.3 LOWERING OF VAPOUR PRESSURE — EXPERIMENTAL OBSERVATIONS

**Experimental Fact**

It is observed that when a nonvolatile solute is dissolved in a volatile solvent, vapour pressure of the latter is lowered.

**Qualitative Explanation**

If a liquid is placed in an evacuated vessel, there occurs migration of solvent molecules from the liquid surface to the empty space above and vice versa. The system attains an equilibrium when the rates of these two migrations become equal. At this stage, the pressure recorded at the space above the liquid is known as the *vapour pressure* of the liquid. In the presence of a nonvolatile solute, the number of solvent molecules available for migration from surface of solution relative to those present in the pure solvent is decreased. As a consequence of this, the vapour pressure of the solution is lesser than that of the pure solvent.

**Experimental Observation of Variation in the Vapour Pressure**

Let $p^*$ be the vapour pressure of the pure solvent and $p$ be that of a solution. Experimentally, it is found that the decrease in vapour pressure $p^* - p$ depends upon the concentration of the solute in the solution. Figure 2.3.1 shows the vapour pressure of the solvent when plotted against the amount fraction of the solute in the solution. It can be seen from the figure that the vapour pressure of the solution approaches the dashed line connecting $p^*(x_2 = 0)$ to zero $(x_2 = 1)$ in the lower concentration region ($x_2$ near zero). At higher concentration region, the observed pressure is found to lie below the dashed line—the deviation becoming more as $x_2$ increases.†

The above experimental observation at $x_2 \rightarrow 0$ can be summarized as follows.

The vapour pressure of the solution approaches that of the dashed line as the concentration decreases. In the lower concentration range ($x_2 \rightarrow 0$), the

†The case discussed here is the negative deviation from the straight-dashed line. Examples with positive deviation, i.e. the actual vapour pressure line lies above the dashed line are also known. These lines show the characteristic of approaching dashed line as $x_2 \rightarrow 0$. The dashed-line plot is exhibited by the so-called ideal solutions.

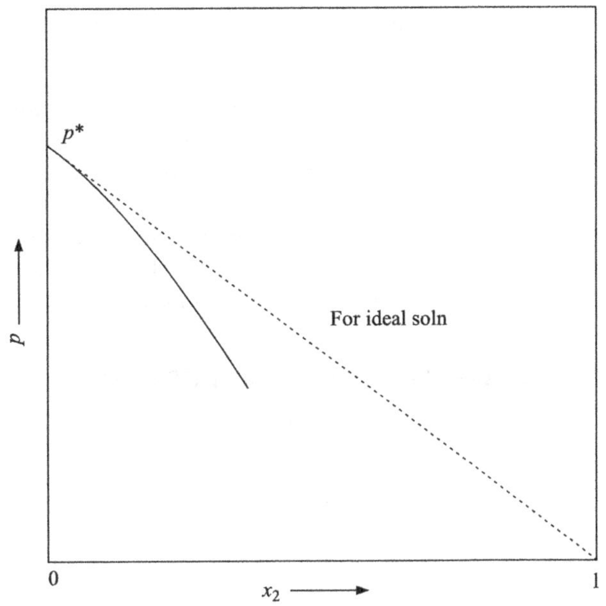


**Fig. 2.3.1 Variation in vapour pressure of a solution with the amount fraction of solute in the solution**

dashed line is tangent to the vapour pressure curve. In other words, the dashed line represents the behaviour of a solution when its concentration is very small.

## Postulating an Ideal Solution

The above behaviour is exactly parallel to that observed in case of real gases. It is found that the real gases deviate from the experimental gas laws such as Boyle's law, Charles law, etc. The deviation from these laws becomes less and less as the pressure decreases and it is only in the very low pressure region as $p \rightarrow 0$, the real gases follow the experimental gas laws. There we have imagined an ideal gas, which is supposed to follow these gas laws under all conditions of temperature and pressure. The real gas behaves like an ideal gas only in the low pressure region and exactly as an ideal gas only when $p \rightarrow 0$. Thus the characteristics of an ideal gas can be obtained from the characteristics of a real gas by extrapolation to $p = 0$. In a similar way, we can define an *ideal solution* which is supposed to follow the dashed line for variation of vapour pressure of the solution with the concentration of the solute. Since for real solutions, this dashed line is approached only when the concentration of the solution is small (i.e. in dilute solution), it, therefore, follows that only the dilute solution for which $x_2 \rightarrow 0$ behaves like an ideal solution. The characteristics of an ideal solution can be obtained by extrapolating the characteristics of a real solution to $x_2 = 0$.

## Raoult's Law

The dashed line can be represented by an equation of straight line of the type

$$p = b + mx_2 \tag{2.3.1}$$

with the characteristics of

$$p = p^* \text{ at } x_2 = 0$$

and $\quad p = 0 \text{ at } x_2 = 1.$

With these characteristics, Eq. (2.3.1) modifies to

$$p = (1 - x_2)p^* = x_1 p^* \qquad (2.3.2)$$

that is, *the vapour pressure of the solution containing a nonvolatile solute is equal to the vapour pressure of the pure solvent multiplied by the amount fraction of the solvent in the solution.* This law, which is a limiting law, was discovered by P. M. Raoult and is known as the *Raoult's law.*[†] It is followed by an ideal solution over the entire range of concentrations, whereas for a real solution it is followed only when the concentration of solute approaches zero (i.e. ideally dilute solution).[‡]

## 2.4 RELATIVE LOWERING OF VAPOUR PRESSURE

**Definition**

According to Raoult's law, the lowering of vapour pressure of a solvent is given by

$$p^* - p = p^* - x_1 p^* = p^*(1 - x_1) = p^* x_2$$

Thus
$$\frac{p^* - p}{p^*} = x_2 \qquad (2.4.1)$$

The expression $(p^* - p)/p^*$ is known as the relative lowering of vapour pressure and according to Eq. (2.4.1), this is equal to the amount fraction of the solute in the solution.

---

***Example 2.4.1***

Exactly 100 g water contains 1.0 g urea and 2.0 g sucrose at 298 K. The vapour pressure of water at 298 K is 0.031 67 bar. Predict the vapour pressure of the solution.

***Solution***

Amount fraction of solvent

$$= \frac{\text{Amount of solvent}}{\text{Amount of (solvent + solute)}}$$

$$= \frac{(100\ \text{g}/18\ \text{g mol}^{-1})}{(100\ \text{g}/18\ \text{g mol}^{-1}) + (1\ \text{g}/60\ \text{g mol}^{-1}) + (2\ \text{g}/342\ \text{g mol}^{-1})}$$

$$= \frac{5.556}{5.577}$$

According to Raoult's law

$$p_1 = p_1^* x_1 = (0.031\,67\ \text{bar})\,(5.556/5.577) = 0.031\,55\ \text{bar}$$

---

**A Colligative Property**

The amount fraction $x_2$ of solute is given by

$$x_2 = \frac{n_2}{n_1 + n_2} = \frac{n_2 N_A}{n_1 N_A + n_2 N_A}$$

$$= \frac{\text{Number of molecules of solute}}{\text{Total number of molecules of solute and solvent}}$$

---

[†]See Section 4.3 for the kinetic-molecular treatment of Raoult's law.

[‡]Raoult's law applicable to solvent in a nonideal solution may be written as $p = a_1 p^*$, where $a_1$ is the activity of solvent in the solution. See also, Section 4.14.

Since $x_2$ involves only the number of solute molecules relative to all the molecules present and not on their individual nature, it is obvious that the relative lowering of vapour pressure is a colligative property.

**Determination of Molar Mass of the Solute**

By definition,

$$\frac{p_1^* - p_1}{p_1^*} = x_2 \qquad \text{(Eq. 2.4.1)}$$

where $x_2$, the amount fraction of the solute, is given by

$$x_2 = \frac{m_2/M_2}{(m_1/M_1) + (m_2/M_2)} \qquad (2.4.2)$$

where $m_2$ is the mass of solute (molar mass $M_2$) that is dissolved in the solvent (molar mass $M_1$) of mass $m_1$. Substituting Eq. (2.4.2) in Eq. (2.4.1), we get

$$\frac{p_1^* - p_1}{p_1^*} = \frac{m_2/M_2}{(m_1/M_1) + (m_2/M_2)} \qquad (2.4.3)$$

Equation (2.4.3) can be employed to determine the molar mass of a solute if all other quantities are known. However, a simplified expression can be used if the solution is dilute. In this case since $m_2/M_2 << m_1/M_1$, we have

$$\frac{p_1^* - p_1}{p_1^*} \simeq \frac{m_2/M_2}{m_1/M_1}$$

i.e.

$$M_2 = \left(\frac{m_2}{m_1} M_1\right)\left(\frac{p_1^*}{p_1^* - p_1}\right) \qquad (2.4.4)$$

---

*Example 2.4.2*

Ten grams of a solute was dissolved in 80 g of acetone at 303 K. The vapour pressure of the solution was found to be 271 Torr. Calculate the molar mass of the solute. Given: Vapour pressure of pure acetone at 303 K is 283 Torr.

*Solution*

The given data are

$$m_1 = 10\text{ g}, \quad m_2 = 80\text{ g}, \quad p_1 = 271\text{ Torr}, \quad p_1^* = 283\text{ Torr},$$
$$M_1 = 58\text{ g mol}^{-1}, \quad M_2 = ?$$

Substituting the data in the expression

$$\frac{p_1^* - p_1}{p_1^*} = \frac{m_2/M_2}{(m_1/M_1) + (m_2/M_2)}$$

we get

$$\frac{283\text{ Torr} - 271\text{ Torr}}{283\text{ Torr}} = \frac{(10\text{ g}/M_2)}{(80\text{ g}/58\text{ g mol}^{-1}) + (10\text{ g}/M_2)}$$

$$0.042\,4 = \frac{(10\text{ g}/M_2)}{(1.379\text{ mol}) + (10\text{ g}/M_2)}$$

This gives $M_2 = \dfrac{(10\text{ g})\,(1 - 0.042\,4)}{(0.042\,4)\,(1.379\text{ mol})} = 163.8\text{ g mol}^{-1}$

---

***Example 2.4.3***

When 2 g of a nonvolatile hydrocarbon containing 94.4 per cent carbon is dissolved in 100 g benzene, the vapour pressure of benzene at 20 °C is lowered from 0.099 54 bar to 0.098 67 bar. Calculate the molecular formula of the hydrocarbon.

***Solution***

The given data are

$$m_1 = 100 \text{ g}, \quad m_2 = 2 \text{ g}, \quad p_1^* = 0.099\,54 \text{ bar}, \quad p_1 = 0.098\,67 \text{ bar}$$

$$M_1 = 78 \text{ mol}^{-1}, \quad M_2 = ?$$

Substituting the given data in the expression

$$\frac{p_1^* - p_1}{p_1^*} = x_2 = \frac{(m_2/M_2)}{(m_1/M_1) + (m_2/M_2)}$$

we get $\dfrac{0.099\,54 \text{ bar} - 0.098\,67 \text{ bar}}{0.099\,54 \text{ bar}} = \dfrac{(2 \text{ g}/M_2)}{(100 \text{ g}/78 \text{ g mol}^{-1}) + (2 \text{ g}/M_2)}$

$$0.008\,740 = \frac{(2 \text{ g}/M_2)}{(1.282 \text{ mol}) + (2 \text{ g}/M_2)}$$

This gives $M_2 = \dfrac{(2 \text{ g})(1 - 0.008\,740)}{(0.008\,740)(1.282 \text{ mol})} = 176.9 \text{ g mol}^{-1}$

Given mass ratio is $m_C : m_H :: 94.4 : 5.6$

Thus, atomic ratio is

$$N_C : N_H :: \frac{94.4}{12} : \frac{5.6}{1} \Rightarrow 7.87 : 5.6 \Rightarrow 1.4 : 1 \Rightarrow 7 : 5$$

Hence, Empirical formula : $C_7H_5$
Empirical molar mass : $(12 \times 7 + 5 \times 1) \text{ g mol}^{-1} = 89 \text{ g mol}^{-1}$
Number of units of $C_7H_5$ present in the given molecule

$$n = \frac{\text{Molar mass}}{\text{Empirical molar mass}} = \frac{176.9 \text{ g mol}^{-1}}{89 \text{ g mol}^{-1}} \simeq 2$$

Thus, Molecular formula: $C_{14}H_{10}$
The substance is anthracene.

## 2.5 CHEMICAL POTENTIALS OF SOLUTE AND SOLVENT IN AN IDEAL LIQUID SOLUTION

**Chemical Potential of Solvent**

For the equilibrium

Solvent molecules in solution $\rightleftharpoons$ Solvent molecules in vapour

the thermodynamic condition of equilibrium is

$$\mu_{1(\text{sol})} = \mu_{1(\text{v})}^* \tag{2.5.1}$$

If the vapour is assumed to be an ideal gas, the chemical potential of vapour in terms of its pressure is given by

$$\mu_{1(\text{v})}^* = \mu_{1(\text{v})}^\circ + RT \ln (p/p^\circ) \tag{2.5.2}$$

where $\mu^{\circ}_{1(v)}$ is the standard chemical potential of the solvent vapour and $p$ is the vapour pressure of the solution. Substituting Eq. (2.5.2) in Eq. (2.5.1), we have

$$\mu_{1(sol)} = \mu^{\circ}_{1(v)} + RT \ln (p/p^{\circ}) \tag{2.5.3}$$

Using Raoult's law ($p = x_1 p_1^*$), we get

$$\begin{aligned} \mu_{1(sol)} &= \mu^{\circ}_{1(v)} + RT \ln (x_1 p_1^*/p^{\circ}) \\ &= \mu^{\circ}_{1(v)} + RT \ln (p_1^*/p^{\circ}) + RT \ln x_1 \end{aligned} \tag{2.5.4}$$

If in the above expression $x_1 = 1$ (i.e. pure liquid solvent), we will have

$$\mu^*_{1(l)} = \mu^{\circ}_{1(v)} + RT \ln (p_1^*/p^{\circ}) \tag{2.5.5}$$

With this, Eq. (2.5.4) becomes

$$\mu_{1(sol)} = \mu^*_{1(l)} + RT \ln x_1 \tag{2.5.6}$$

The form of Eq. (2.5.6) is exactly the same as that of an ideal gas in a mixture of gases.

**Chemical Potential of Solute**

The expression of chemical potential for the solute in a binary ideal solution can be obtained by making use of the Gibbs-Duhem equation:

$$n_1\, d\mu_{1(sol)} + n_2\, d\mu_{2(sol)} = 0 \qquad \text{(constant } T \text{ and } p) \tag{2.5.7}$$

Rearranging Eq. (2.5.7), we get

$$d\mu_{2(sol)} = -\left(\frac{n_1}{n_2}\right) d\mu_{1(sol)} \tag{2.5.8}$$

Since $n_1/n_2 = x_1/x_2$, we have

$$d\mu_{2(sol)} = -\left(\frac{x_1}{x_2}\right) d\mu_{1(sol)} \tag{2.5.9}$$

For the solvent, we have

$$\mu_{1(sol)} = \mu^*_{1(l)} + RT \ln x_1 \qquad \text{(Eq. 2.5.6)}$$

Therefore, at constant $T$ and $p$, we have

$$d\mu_{1(sol)} = \frac{RT}{x_1} dx_1 \tag{2.5.10}$$

Substituting this relation in Eq. (2.5.9), we have

$$\begin{aligned} d\mu_{2(sol)} &= -\left(\frac{x_1}{x_2}\right)\left(\frac{RT}{x_1}\right) dx_1 \\ &= -RT \frac{dx_1}{x_2} \end{aligned} \tag{2.5.11}$$

Now, since $x_1 + x_2 = 1$, therefore

$$\mathrm{d}x_1 + \mathrm{d}x_2 = 0 \quad \text{or} \quad \mathrm{d}x_1 = -\,\mathrm{d}x_2$$

With this, Eq. (2.5.11) becomes

$$\mathrm{d}\mu_{2(\mathrm{sol})} = RT\frac{\mathrm{d}x_2}{x_2} \tag{2.5.12}$$

Equation (2.5.12) is similar in form as Eq. (2.5.10). Integrating Eq. (2.5.12), we have

$$\mu_{2(\mathrm{sol})} = RT \ln x_2 + I \tag{2.5.13}$$

where $I$ is the constant of integration. For pure liquid solute $x_2 = 1$. Thus $I = \mu^*_{2(\mathrm{l})}$ and Eq. (2.5.13) becomes

$$\mu_{2(\mathrm{sol})} = \mu^*_{2(\mathrm{l})} + RT \ln x_2 \tag{2.5.14}$$

a form similar to that of the solvent.

**Conclusion**

Thus, we conclude that the expression

$$\mu_{i(\mathrm{sol})} = \mu^*_{i(\mathrm{l})} + RT \ln x_i \tag{2.5.15}$$

is applicable to both the constituents of an ideal solution. Since $x_1$ or $x_2$ is less than one, it is obvious that

$$\mu_{i(\mathrm{sol})} < \mu^*_{i(\mathrm{l})} \tag{2.5.16}$$

## 2.6 ORIGIN OF COLLIGATIVE PROPERTIES

**Qualitative Explanation**

The depression of freezing point and the elevation of boiling point when a nonvolatile solute is dissolved in a volatile solvent can be explained qualitatively by plotting $\mu$ versus $T$ curves. For a pure substance in solid, liquid and vapour phases, the variations of $\mu^*$ with $T$ are shown by solid lines in Fig. 2.6.1. The point of intersection of the $\mu^*_\mathrm{s}$ line with the $\mu^*_\mathrm{l}$ line and that of the $\mu^*_\mathrm{l}$ line with the $\mu^*_\mathrm{v}$ line represent, respectively, the freezing point and the boiling point of the solvent.

Consider now a solution formed by the dissolution of a nonvolatile solute in the above solvent. At the freezing point, we have the separation of the solid solvent from the solution and thus an equilibrium of the type

$$\text{Solvent in solution} \rightleftharpoons \text{Solid solvent}$$

exists at the freezing point of the solution. Similarly at the boiling point of the solution, the equilibrium of the type

$$\text{Solvent in solution} \rightleftharpoons \text{Vapour solvent}$$

exists. Since the solute is nonvolatile, the vapour will not contain any solute molecules.

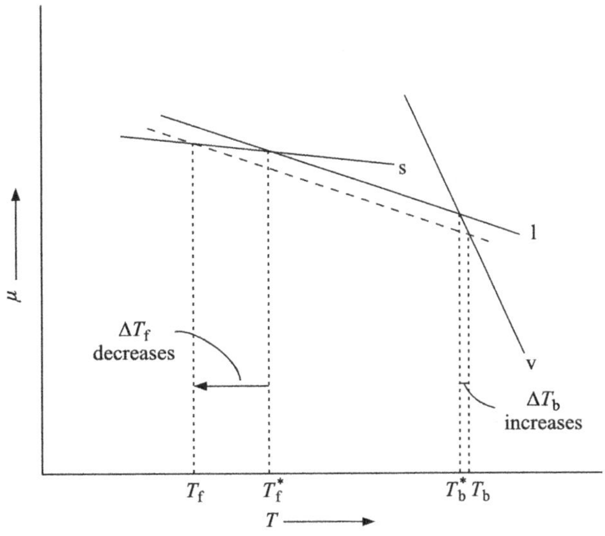


**Fig. 2.6.1 The elevation of boiling point and the depression of freezing point on the basis of $\mu$ versus $T$ curves**

Since the solid which separates at the freezing point of the solution and the vapour which appears at the boiling point are of pure solvent, the variations of chemical potentials for these two phases with temperature remain unaltered and thus are represented by the same solid and vapour lines in Fig. 2.6.1. However, the chemical potential of the liquid solvent will be decreased by an amount $RT \ln x_1$ (Eq. 2.5.6). This decrease is shown by the broken line in Fig. 2.6.1. This line intersects $\mu^*_{\text{solid}}$ and $\mu^*_{\text{vapour}}$ curves at points $T_f$ and $T_b$ and are, therefore, the respective freezing and boiling points of the solutions. It can be seen from Fig. 2.6.1 that:

- The boiling point of the solution is higher than that of the solvent ($T_b > T_b^*$) and thus there occurs an elevation of the boiling point.
- The freezing point of the solution is smaller than that of the solvent ($T_f < T_f^*$) and thus there occurs a depression of the freezing point.
- The depression of freezing point for a solution of a given concentration is greater in magnitude than the corresponding elevation of the boiling point.

**Display on the Phase Diagram**

The elevation of boiling point and the depression of freezing point can also be illustrated on the phase diagram. Let the solvent chosen be water. Its phase diagram is represented by the solid lines in Fig. 2.6.2.

Now if a nonvolatile solute is added to it, the vapour pressure is lowered at every temperature. The new liquid-vapour equilibrium curve will lie below the corresponding liquid-vapour curve of the pure solvent. The new solid-liquid equilibrium curve can be drawn from the point where the new liquid-vapour curve meets the solid-vapour curve of the pure solvent. The freezing point and the boiling point can be obtained from the intersections of a line drawn horizontally from the given external pressure to the solid-liquid and liquid-vapour curves, respectively. From Fig. 2.6.2, it is obvious that the boiling point is elevated, whereas the freezing point is depressed on adding a nonvolatile solute in a volatile solvent.

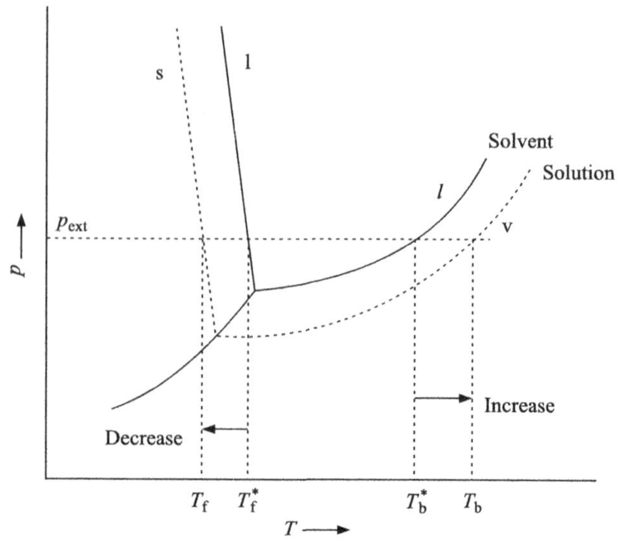


**Fig. 2.6.2 The elevation of the boiling point and the depression of freezing point on the basis of phase diagram**

## Explanation of Osmotic Pressure

The third colligative property, the osmotic pressure, arises because of the equilibrium between the solvent in solution and the pure liquid solvent. Let the given solution be separated from the pure solvent by a semipermeable membrane (Fig. 2.6.3) which allows only the flow of solvent molecules from one side to the other. Since

$$\mu_{1(sol)} < \mu^*_{1(l)}$$

therefore, the situation shown in Fig. 2.6.3 will not be at equilibrium. The solvent will flow from the region of high chemical potential (pure solvent) to the region of low chemical potential (solution). This can, however, be stopped either by raising the chemical potential of the solvent in the solution by increasing the external pressure on the solution or by lowering the chemical potential of the pure solvent by decreasing the external pressure on the pure solvent till the chemical potential of the solvent becomes the same on both sides of the membrane. This excessive pressure on the solution side or the lowering of external pressure on the solvent side which is just sufficient to equate the chemical potential of solvent on both sides is known as the *osmotic pressure* of the solution.

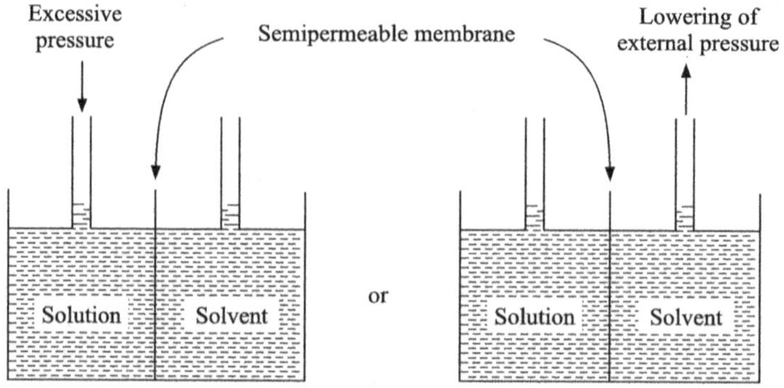


**Fig. 2.6.3 Osmotic pressure of the solution**

## 2.7 ELEVATION OF BOILING POINT

**Definition of Boiling Point**

A liquid starts boiling when its vapour pressure becomes equal to the external pressure. The temperature at which this occurs is known as the boiling point of the liquid. If the external pressure is 1 atm, this temperature is known as the normal boiling point of the liquid. When a nonvolatile solute is added to a liquid solvent, the boiling point of the resultant solution is increased.

**Qualitative Explanation of Elevation of Boiling Point**

The vapour pressure of the solution is comparatively smaller than that of the pure solvent (Raoult's law). Consequently, the temperature at which the vapour pressure of solution becomes equal to the external pressure, will be greater than that of pure solvent, thus raising its boiling point. Since the decrease in vapour pressure is directly proportional to the amount fraction of the solute in the solution, it is, therefore, expected that the corresponding increase in the boiling point also depends on the amount fraction of the solute in the solution.

**Thermodynamic Expression of Boiling Point of a Solution**

The quantitative relation between the elevation of boiling point and the concentration of the solution can be derived thermodynamically. Consider a solution which is in equilibrium with the vapour of the pure solvent. Basically, we have the following equilibrium at the boiling point of the solution.

$$\text{Solvent in solution} \rightleftharpoons \text{Vapour of pure solvent}$$

The thermodynamic condition of equilibrium is

$$\mu_{1(\text{sol})}(T_\text{b}, p) = \mu^*_{1(\text{v})}(T_\text{b}, p) \tag{2.7.1}$$

where $T_\text{b}$ is the boiling point of the solution. At constant pressure, this temperature is a function of $x_1$, the amount fraction of the solvent in solution. For an ideal solution, we have

$$\mu_{1(\text{sol})} = \mu^*_{1(\text{l})} + RT_\text{b} \ln x_1 \qquad \text{(Eq. 2.5.6)}$$

Substituting the above relation in Eq. (2.7.1) and rearranging, we have

$$\ln x_1 = \frac{\mu^*_{1(\text{v})} - \mu^*_{1(\text{l})}}{RT_\text{b}} \tag{2.7.2}$$

or

$$\ln x_1 = \frac{\Delta_{\text{vap}}\mu^*_1}{RT_\text{b}} \tag{2.7.3}$$

where $\Delta_{\text{vap}}\mu^*_1 = \mu^*_{1(\text{v})} - \mu^*_{1(\text{l})}$ and is the molar free energy of evaporation of the pure solvent at temperature $T_\text{b}$ and pressure $p$. Equation (2.7.3) is the required relation between the composition of the solution and the boiling point.

**Alternative Expression of Boiling Point**

As stated above $T_\text{b} = f(x_1)$. The actual relation between $T_\text{b}$ and $x_1$ can be determined by differentiating Eq. (2.7.3) with respect to $x_1$ at constant $p$, such that

$$\frac{1}{x_1} = \frac{1}{R}\left\{\frac{\partial(\Delta_{\text{vap}}\mu^*_1/T_\text{b})}{\partial T_\text{b}}\right\}_p \left(\frac{\partial T_\text{b}}{\partial x_1}\right)_p$$

Using the Gibbs-Helmholtz equation

$$\left\{\frac{\partial(\Delta\mu/T)}{\partial T}\right\}_p = -\frac{\Delta H_m}{T^2}$$

we get $$\frac{1}{x_1} = -\frac{\Delta_{vap}H_{1,m}}{RT_b^2}\left(\frac{\partial T_b}{\partial x_1}\right)_p \quad (2.7.4)$$

or $$\frac{dx_1}{x_1} = -\frac{\Delta_{vap}H_{1,m}}{RT_b^2}dT_b \qquad (p \text{ constant}) \quad (2.7.5)$$

where $\Delta_{vap}H_{1,m}$ is the molar enthalpy of vaporization of the pure solvent at temperature $T_b$.

The relation between boiling point ($T_b$) and amount fraction of solvent ($x_1$) can be obtained by integrating Eq. (2.7.5). Thus, we have

$$\int_1^{x_1}\frac{dx_1}{x_1} = -\int_{T_b^*}^{T_b}\frac{\Delta_{vap}H_{1,m}}{R}\frac{dT_b}{T_b^2} \quad (2.7.6)$$

where $T_b$ and $T_b^*$ are the boiling points of solution and pure solvent, respectively. If $\Delta_{vap}H_{1,m}$ is assumed to be independent of over the temperature range $T_b$ and $T_b^*$, Eq. (2.7.6) on integrating gives

$$\ln x_1 = -\frac{\Delta_{vap}H_{1,m}}{R}\left(-\frac{1}{T_b}+\frac{1}{T_b^*}\right) \quad (2.7.7)$$

i.e. $$\frac{1}{T_b} = \frac{1}{T_b^*} + \frac{R\ln x_1}{\Delta_{vap}H_{1,m}} \quad (2.7.8)$$

Equation (2.7.8) can be used to compute the boiling point of a solution in which the amount fraction of solvent is $x_1$, $T_b^*$ and $\Delta_{vap}H_{1,m}$ are the boiling point and the molar enthalpy of vaporization of the pure solvent, respectively.

**Boiling Point Related to Molality of Solution**

By definition, we have

$$x_1 = \frac{n_1}{n_1+n_2} = \frac{m_1/M_1}{(m_1/M_1)+n_2} = \frac{1}{1+(n_2/m_1)M_1} = \frac{1}{1+mM_1}$$

where $m$ is the molality of the solution.

Hence, $$\ln x_1 = \ln\left(\frac{1}{1+mM_1}\right) = -\ln(1+mM_1)$$

Differentiating this expression, we get

$$\frac{dx_1}{x_1} = -\frac{1}{1+mM_1}M_1\,dm$$

Substituting this in Eq. (2.7.5), we get

$$-\frac{1}{1+mM_1}M_1\,dm = -\frac{\Delta_{vap}H_{1,m}}{RT_b^2}\,dT_b$$

or

$$dT_b = \left(\frac{RT_b^2}{\Delta_{vap}H_{1,m}}\right)\left(\frac{M_1}{1+mM_1}\right)dm \qquad (2.7.9)$$

**Definition of Ebullioscopic Constant**

For a dilute solution, as $m \to 0$, $T_b \to T_b^*$. Thus, the limiting value of the derivative $\partial T_b/\partial m$ at zero molality is

$$\left(\frac{\partial T_b}{\partial m}\right)_{m\to 0} = \frac{RM_1T_b^{*2}}{\Delta_{vap}H_{1,m}} \qquad (2.7.10)$$

From Eq. (2.7.10), it follows that the limiting value of $(\partial T_b/\partial m)$ depends only on the characteristics of the solvent. The derivative $(\partial T_b/\partial m)_{m\to 0}$ is known as the **boiling point elevation constant** or **the ebullioscopic constant** and is represented by the symbol $K_b$.

Equation (2.7.9) may be written as

$$dT_b = K_b\,dm \qquad (p\text{ constant, very dilute solution}) \qquad (2.7.11)$$

For a finite value of d$m$, we have

$$\Delta T_b = K_b\,m \qquad (2.7.12)^\dagger$$

**Unit of $K_b$**

The unit of $K_b$ may be worked out as follows.

$$\text{unit of } K_b = \frac{\text{unit of } \Delta T_b}{\text{unit of } m} = \frac{\text{K}}{\text{mol kg}^{-1}} = \text{K kg mol}^{-1}$$

**Value of $K_b$ for Water**

The numerical value of $K_b$ for a given solvent can be determined from its $T_b^*$ and $\Delta_{vap}H_m$ values. For example, for water we have

$$K_b = \frac{M_1\,RT_b^{*2}}{\Delta_{vap}H_m} = \frac{(18\times10^{-3}\text{ kg mol}^{-1})(8.314\text{ J K}^{-1}\text{ mol}^{-1})(373\text{ K})^2}{(40\,658\text{ J mol}^{-1})}$$

$$= 0.51\text{ K kg mol}^{-1}$$

**Values of $K_b$ for Some Solvents**

The values of $K_b$ for other commonly used solvents are given in Table 2.7.1.

---

†Equation (2.7.12) may be derived directly from Eq. (2.7.8) by invoking the following approximations.

$$\ln x_1 = -\ln(1 + mM_1) \simeq -mM_1$$

$$T_b\,T_b^* \simeq T_b^{*2}$$

**Table 2.7.1** Boiling-Point Elevation Constants

| *Solvent* | *Molar mass* / g mol$^{-1}$ | b.pt./°C | $K_b$/(K kg mol$^{-1}$) |
|---|---|---|---|
| Water | 18.0 | 100 | 0.51 |
| Methyl alcohol | 32.0 | 64.7 | 0.86 |
| Ethyl alcohol | 46.1 | 78.5 | 1.23 |
| Acetone | 58.1 | 56.1 | 1.71 |
| Benzene | 78.1 | 80.2 | 2.53 |
| Cyclohexane | 84.2 | 81.3 | 2.79 |
| Chloroform | 119.5 | 61.3 | 3.63 |
| Carbon tetrachloride | 154.0 | 76.8 | 5.03 |

***Problem 2.7.1***

The expression $\Delta T_b = K_b m$ for the boiling point elevation can also be obtained by applying the Clausius-Clapeyron equation to the solution-vapour equilibrium curves. Derive the relation by drawing the necessary diagram.

***Solution***

The variations in vapour pressure of the solvent and that of the solution with temperature are given by the respective solvent-vapour and solution-vapour curves of the phase diagram (Fig. 2.7.1).

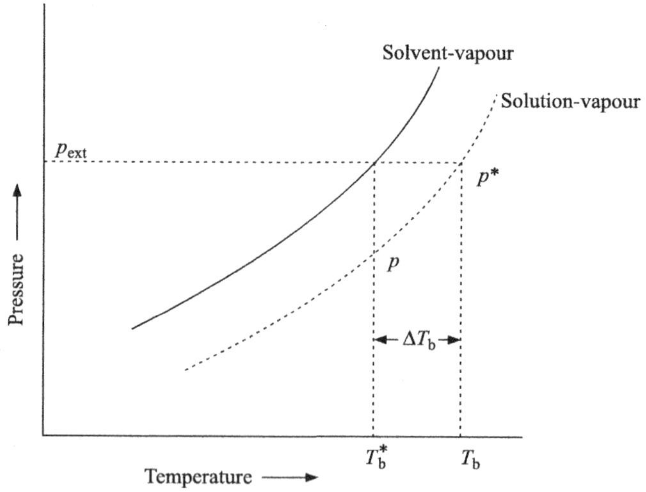


**Fig. 2.7.1 Clausisus-Clapeyron equation as applied to the solution-vapour curve**

For a given value of the external pressure $p_{ext}$, the pure solvent will boil at temperature $T_b^*$ and the solution at temperature $T_b$. Let $p_{ext}$ be equal to $p^*$, the vapour pressure of pure solvent. Let $p$ be the vapour pressure of the solution at temperature $T_b^*$.

Applying the Clausius-Clapeyron equation to the solution-vapour equilibrium for two values of $p$, $T_b^*$ and $p^*$, $T_b$, we have

$$\ln \frac{p^*}{p} = \frac{\Delta_{vap}H_{1,m}}{R}\left(\frac{1}{T_b^*} - \frac{1}{T_b}\right) = \frac{\Delta_{vap}H_{1,m}}{R}\frac{\Delta T_b}{T_b^* T_b} \tag{2.7.13}$$

where $\Delta T_b = T_b - T_b^*$, the elevation of boiling point. For a dilute solution, Eq. (2.7.13) may be simplified under the following approximations.

(1) $T_b$ will be very near to $T_b^*$, so that the term $T_b T_b^*$ can be replaced by $T_b^{*2}$, i.e.

$$T_b^* T_b \simeq T_b^{*2} \tag{2.7.14}$$

(2) The term ln $p^*/p$ can be written as

$$\ln \frac{p^*}{p} = -\ln \frac{p}{p^*} = -\ln \left( \frac{p^* - p^* + p}{p^*} \right) = -\ln \left( 1 - \frac{p^* - p}{p^*} \right)$$

The term $(p^* - p)/p^*$, the relative lowering of vapour pressure, is very small in comparison to 1, thus the term $-\ln(1 - (p^* - p)/p^*)$ can be approximated to

$$-\ln \left( 1 - \frac{p^* - p}{p^*} \right) \simeq - \left( - \frac{p^* - p}{p^*} \right) = \frac{p^* - p}{p^*}$$

Hence $\quad \ln \dfrac{p^*}{p} = \dfrac{p^* - p}{p^*}$

(3) According to Raoult's law, we have

$$\frac{p^* - p}{p^*} = x_2 = \frac{n_2}{n_1 + n_2}$$

For a dilute solution $n_2 \ll n_1$, and, therefore, $n_2$ can be neglected in comparison to $n_1$. Thus, we have

$$\ln \frac{p^*}{p} \simeq \frac{p^* - p}{p^*} \simeq \frac{n_2}{n_1} = \frac{n_2}{m_1/M_1} \tag{2.7.15}$$

Substituting Eqs (2.7.14) and (2.7.15) in Eq. (2.7.13), we get

$$\frac{n_2}{m_1/M_1} = \frac{\Delta_{vap}H_{1,m}}{R} \frac{\Delta T_b}{T_b^{*2}}$$

Rearranging this, we get

$$\Delta T_b = \left( \frac{M_1 R T_b^{*2}}{\Delta_{vap}H_{1,m}} \right) \left( \frac{n_2}{m_1} \right) = K_b\, m$$

---

***Problem 2.7.2***

According to Raoult's law $p_1 = x_1 p_1^*$ or $\ln(p_1/p^\circ) = \ln x_1 + \ln(p_1^*/p^\circ)$. If $p_1 = p_{ext}$ = constant, then

$$\frac{d \ln (p_1/p^\circ)}{dT_b} = 0 = \frac{d \ln x_1}{dT_b} + \frac{d \ln (p_1^*/p^\circ)}{dT_b}$$

Now if the Clapeyron equation

$$\frac{d \ln (p_1/p^\circ)}{dT} = \frac{\Delta_{vap}H_m}{RT^2}$$

is introduced in the previous expression, show that it leads to the expression $\Delta T_b = K_b m$ for a dilute solution.

***Solution***

Substituting the Clapeyron equation $\dfrac{d \ln (p_1^*/p^\circ)}{dT_b^*} = \dfrac{\Delta_{vap}H_{1,m}}{RT_b^{*2}}$ in the expression

$$\frac{\mathrm{d}\ln x_1}{\mathrm{d}T_\mathrm{b}^*} = -\frac{\mathrm{d}\ln (p_1^*/p_1^\circ)}{\mathrm{d}T_\mathrm{b}^*},$$

we get $$\frac{\mathrm{d}\ln x_1}{\mathrm{d}T_\mathrm{b}^*} = -\frac{\Delta_\mathrm{vap}H_{1,\mathrm{m}}}{RT_\mathrm{b}^{*2}} \quad \text{or} \quad \frac{\mathrm{d}\ln (1-x_2)}{\mathrm{d}T_\mathrm{b}^*} = -\frac{\Delta_\mathrm{vap}H_{1,\mathrm{m}}}{RT_\mathrm{b}^{*2}}$$

Integrating within the limits, we get

$$\int_0^{x_2} \mathrm{d}\ln (1-x_2) = -\int_{T_\mathrm{b}^*}^{T_\mathrm{b}} \frac{\Delta_\mathrm{vap}H_{1,\mathrm{m}}}{R}\frac{\mathrm{d}T_\mathrm{b}^*}{T_\mathrm{b}^{*2}}$$

i.e. $$\ln (1-x_2) = -\frac{\Delta_\mathrm{vap}H_{1,\mathrm{m}}}{R}\left(-\frac{1}{T_\mathrm{b}} + \frac{1}{T_\mathrm{b}^*}\right)$$

For a dilute solution, we will have

$$\ln (1-x_2) \simeq -x_2 = -\frac{n_2}{n_1+n_2} \simeq -\frac{n_2}{n_1} = -\frac{n_2}{m_1}M_1 = -mM_1$$

$$T_\mathrm{b} \simeq T_\mathrm{b}^* \text{ so that } T_\mathrm{b}T_\mathrm{b}^* \simeq T_\mathrm{b}^{*2}$$

With these, the previous expression becomes

$$-mM_1 = -\frac{\Delta_\mathrm{vap}H_{1,\mathrm{m}}}{R}\frac{\Delta T_\mathrm{b}}{T_\mathrm{b}^{*2}}$$

or $$\Delta T_\mathrm{b} = \left(\frac{M_1RT_\mathrm{b}^{*2}}{\Delta_\mathrm{vap}H_{1,\mathrm{m}}}\right)m = K_\mathrm{b}m$$

---

***Example 2.7.1*** A solution containing 0.512 6 g of naphthalene (molar mass 128.2 g mol$^{-1}$) in 50 g $CCl_4$, yields a boiling point elevation of 0.402 K, while a solution of 0.621 6 g of an unknown solute in the same mass of the solvent gives a boiling point elevation of 0.647 K. Find $K_\mathrm{b}$ of the solvent the molar mass of the unknown solute.

***Solution*** Mass of dissolved naphthalene = 0.512 6 g
Mass of solvent = 50 g

Molality of the solution, $$m = \frac{n_2}{m_1} = \frac{m_2/M_2}{m_1} = \frac{(0.512\,6\ \mathrm{g}/128.2\ \mathrm{g\ mol^{-1}})}{(0.050\ \mathrm{kg})}$$

$$= 0.079\,97\ \mathrm{mol\ kg^{-1}}$$

From the relation, $\Delta T_\mathrm{b} = K_\mathrm{b}m$, we get

$$K_\mathrm{b} = \frac{\Delta T_\mathrm{b}}{m} = \frac{0.402\ \mathrm{K}}{0.079\,97\ \mathrm{mol\ kg^{-1}}} = 5.027\ \mathrm{K\ kg\ mol^{-1}}$$

Let $M_2$ be the molar mass of the unknown substance. The molality of solution containing unknown compound is

$$m = \frac{m_2/M_2}{m_1} = \frac{(0.621\,6\ \mathrm{g})/M_2}{0.050\ \mathrm{kg}} = \frac{12.432}{M_2}\ \mathrm{g\ kg^{-1}}$$

Substituting the above molality along with $\Delta T_b$ = 0.647 K in the relation $\Delta T_b = K_b m$, we get

$$0.647 \text{ K} = (5.027 \text{ K mol}^{-1} \text{ kg})\left(\frac{12.432}{M_2} \text{g kg}^{-1}\right)$$

or $$M_2 = \frac{5.027 \times 12.432}{0.647} \text{ g mol}^{-1} = 96.59 \text{ g mol}^{-1}$$

---

***Example 2.7.2*** A complex compound of unknown molar mass is known to dissolve in benzene without association or dissociation. A solution of this compound in benzene has a vapour pressure of 100 Torr at 300.15 K. Pure benzene boils at 353.15 K at 760 Torr pressure; its entropy of vaporization is 87.03 J $K^{-1}$ $mol^{-1}$. What would be the amount fraction of the complex compound in this solution and the boiling point of the solution?

***Solution*** ***Vapour Pressure of the Pure Solvent at* 300.15 K** The given data are

$$p_2 = 760 \text{ Torr} \qquad T_2 = 353.15 \text{ K}$$
$$p_1 = ? \qquad T_1 = 300.15 \text{ K}$$

The Clausius-Clapeyron equation is

$$\ln \frac{p_2}{p_1} = \frac{\Delta_{vap}H_m}{R}\left(\frac{1}{T_1} - \frac{1}{T_2}\right) = \left(\frac{\Delta_{vap}H_m}{R}\right)\left(\frac{\Delta T_b}{T_1 T_2}\right)$$

But $\Delta_{vap}H_m = T_2\,\Delta_{vap}S_m$, therefore

$$\log \frac{p_2}{p_1} = \left(\frac{\Delta_{vap}S_m}{2.303\,R}\right)\left(\frac{\Delta T_b}{T_1}\right)$$

Substituting the given data, we get

$$\log\left(\frac{760 \text{ Torr}}{p_1}\right) = \left[\frac{(87.03 \text{ J K}^{-1} \text{ mol}^{-1})}{(2.303)\,(8.314 \text{ J K}^{-1} \text{ mol}^{-1})}\right]\left[\frac{(53 \text{ K})}{(300.15 \text{ K})}\right]$$
$$= 0.802\,6$$

or $$\frac{760 \text{ Torr}}{p_1} = 6.35 \quad \text{or} \quad p_1 = \frac{760 \text{ Torr}}{6.35} = 119.7 \text{ Torr}$$

***Amount Fraction of the Solute*** According to the relative lowering of vapour pressure, we get

$$x_2 = \frac{p^* - p}{p^*} = \frac{119.7 \text{ Torr} - 100 \text{ Torr}}{119.7 \text{ Torr}} = 0.164\,6$$

***Boiling Point of the Solution*** Substituting the data in the expression

$$\frac{1}{T_b} = \frac{1}{T_b^*} + \frac{R \ln x_1}{\Delta_{vap}H_{1,m}} \quad \text{i.e.} \quad \frac{1}{T_b} = \frac{1}{T_b^*} + \frac{R \ln(1 - x_2)}{T_b^* \Delta_{vap}S_{1,m}}$$

we get $$\frac{1}{T_b} = \frac{1}{353\,15 \text{ K}} + \frac{(8.314 \text{ J K}^{-1} \text{ mol}^{-1}) \ln(1 - 0.164\,6)}{(353.15 \text{ K})\,(87.03 \text{ J K}^{-1} \text{ mol}^{-1})}$$

$$= 0.002\,832 \text{ K}^{-1} - 0.000\,048\,6 \text{ K}^{-1} = 0.002\,783 \text{ K}^{-1}$$

$$T_b = 359.3 \text{ K}$$

---

***Example 2.7.3***

For a solution of 3.795 g sulphur in 100 g $CS_2$ the boiling point was 319.81 K. For pure $CS_2$, the boiling point is 319.45 K and the enthalpy of vaporization is 351.87 J $g^{-1}$. What is the molar mass and formula of sulphur in $CS_2$?

***Solution***

The given data are

$$m_2 = 3.795\ \text{g} \qquad m_1 = 100\ \text{g} \qquad M_{CS_2} = 76\ \text{g mol}^{-1}$$

$$T_b = 319.81\ \text{K} \qquad T_b^* = 319.45\ \text{K}$$

$$\Delta_{vap}H_{1,m} = 351.87 \times 76\ \text{J mol}^{-1}$$

***Determination of*** $K_b(CS_2)$

$$K_b = \frac{M_1 R T_b^{*2}}{\Delta_{vap}H_{1,m}} = \frac{(76 \times 10^{-3}\ \text{kg mol}^{-1})\,(8.314\ \text{J K}^{-1}\ \text{mol}^{-1})(319.45\ \text{K})^2}{(351.87 \times 76\ \text{J mol}^{-1})}$$

$$= 2.41\ \text{K kg mol}^{-1}$$

***Molality of the given solution***

$$m = \frac{\Delta T_b}{K_b} = \frac{T_b - T_b^*}{K_b} = \frac{(319.81\ \text{K} - 319.45\ \text{K})}{(2.41\ \text{K kg mol}^{-1})} = 0.149\ \text{mol kg}^{-1}$$

***Molar mass of solute*** Since $m = (m_2/M_2)/m_1$, we get

$$M_2 = \frac{m_2}{m m_1} = \frac{(3.795\ \text{g})}{(0.149\ \text{mol kg}^{-1})\,(100\ \text{g})} = 0.255\ \text{kg mol}^{-1}$$

$$= 255\ \text{g mol}^{-1}$$

Number or sulphur atoms in 1 molecule of sulphur $= \dfrac{(255\ \text{g mol}^{-1})}{(32\ \text{g mol}^{-1})} \simeq 8$

Hence, Molecular formula = $S_8$

## 2.8 DEPRESSION OF FREEZING POINT

**Thermodynamic Expression of Freezing Point of a Solution**

We have seen in Section 2.6 that when a nonvolatile solute is dissolved in a volatile solvent, the freezing point of the resultant solution is lowered relative to that of the pure solvent. The quantitative relation between the depression of freezing point and the concentration of the solution can be derived thermodynamically following the sequence of steps adopted in the derivation of expression for the elevation of boiling point.

Basically, at the freezing point of the solution, we have the equilibrium

$$\text{Solvent in solution} \rightleftharpoons \text{Solid solvent}$$

The condition of equilibrium is

$$\mu_{1(sol)}(T_f, p) = \mu_{1(s)}^*(T_f, p) \qquad (2.8.1)$$

where $T_f$ is the freezing point of the solution. At a constant pressure, this temperature will be a function of $x_1$, the amount fraction of the solvent in solution. If the solution is ideal, then

$$\mu_{1(\text{sol})} = \mu^*_{1(\text{l})} + RT_\text{f} \ln x_1$$

Substituting the above relation in Eq. (2.8.1) and rearranging, we have

$$\ln x_1 = -\frac{\mu^*_{1(\text{l})} - \mu^*_{1(\text{s})}}{RT_\text{f}} \tag{2.8.2}$$

or

$$\ln x_1 = -\frac{\Delta_\text{fus}\mu^*_1}{RT_\text{f}} \tag{2.8.3}$$

where $\Delta_\text{fus}\mu^*_1$ is the molar free energy of fusion of the pure solvent at temperature $T_\text{f}$ and pressure $p$.

**Alternative Expression of Freezing Point**

In order to derive the relation governing the freezing point of a solution with its composition, we differentiate Eq. (2.8.3) with respect to $x_1$ at constant pressure, such that

$$\frac{1}{x_1} = -\frac{1}{R}\left\{\frac{\partial(\Delta_\text{fus}\mu^*_1/T_\text{f})}{\partial T_\text{f}}\right\}_p \left(\frac{\partial T_\text{f}}{\partial x_1}\right)_p$$

Using the Gibbs-Helmholtz equation

$$\left\{\frac{\partial(\Delta\mu/T)}{\partial T}\right\}_p = -\frac{\Delta H_\text{m}}{T^2}$$

we get

$$\frac{1}{x_1} = \frac{\Delta_\text{fus}H_{1,\text{m}}}{RT_\text{f}^2}\left(\frac{\partial T_\text{f}}{\partial x_1}\right)_p \tag{2.8.4}$$

or

$$\frac{\text{d}x_1}{x_1} = \frac{\Delta_\text{fus}H_{1,\text{m}}}{RT_\text{f}^2}\,\text{d}T_\text{f} \qquad (p \text{ constant}) \tag{2.8.5}$$

where $\Delta_\text{fus}H_{1,\text{m}}$ is the molar enthalpy of fusion of the pure solvent at temperature $T_\text{f}$.

The relation between freezing point ($T_\text{f}$) and amount fraction of solvent ($x_1$) can be obtained by integrating Eq. (2.8.5). Thus, we have

$$\int_1^{x_1}\frac{\text{d}x_1}{x_1} = \int_{T_\text{f}^*}^{T_\text{f}}\frac{\Delta_\text{fus}H_{1,\text{m}}}{R}\frac{\text{d}T_\text{f}}{T_\text{f}^2} \tag{2.8.6}$$

where $T_\text{f}$ and $T_\text{f}^*$ are the freezing points of solution and pure solvent, respectively. If $\Delta_\text{fus}H_{1,\text{m}}$ is assumed to be independent over the temperature range $T_\text{f}$ and $T_\text{f}^*$, Eq. (2.8.6) on integrating gives

$$\ln x_1 = \frac{\Delta_\text{fus}H_{1,\text{m}}}{R}\left(-\frac{1}{T_\text{f}} + \frac{1}{T_\text{f}^*}\right) \tag{2.8.7}$$

i.e.

$$\frac{1}{T_\text{f}} = \frac{1}{T_\text{f}^*} - \frac{R\ln x_1}{\Delta_\text{fus}H_{1,\text{m}}} \tag{2.8.8}$$

Equation (2.8.8) can be used to compute the freezing point of a solution in which the amount fraction of solvent is $x_1$; $T_f^*$ and $\Delta_{fus}H_{1,m}$ are the freezing point and the molar enthalpy of fusion of the pure solvent, respectively.

**Freezing Point Related to Molality of Solution**

By definition, we have

$$x_1 = \frac{n_1}{n_1 + n_2} = \frac{m_1/M_1}{(m_1/M_1) + n_2} = \frac{1}{1 + (n_2/m_1)M_1} = \frac{1}{1 + mM_1}$$

where $m$ is the molality of the solution.

Hence, $$\ln x_1 = \ln\left(\frac{1}{1 + mM_1}\right) = -\ln(1 + mM_1)$$

Differentiating this expression, we get

$$\frac{dx_1}{x_1} = -\frac{1}{1 + mM_1} M_1 \, dm$$

Substituting this in Eq. (2.8.5), we get

$$-\frac{1}{1 + mM_1} M_1 \, dm = \frac{\Delta_{fus}H_{1,m}}{RT_f^2} dT_f$$

or $$dT_f = -\left(\frac{RT_f^2}{\Delta_{fus}H_{1,m}}\right)\left(\frac{M_1}{1 + mM_1}\right) dm \qquad (2.8.9)$$

**Definition of Cryscopic Constant**

For a dilute solution, as $m \to 0$, $T_f \to T_f^*$. Thus, the limiting value of the derivative $\partial T_f/\partial m$ at zero molality is

$$\left(\frac{\partial T_f}{\partial m}\right)_{m \to 0} = \frac{RM_1T_f^{*2}}{\Delta_{fus}H_{1,m}} \qquad (2.8.10)$$

From Eq. (2.8.10), it follows that the limiting value of $(\partial T_f/\partial m)$ depends only on the characteristics of the solvent. The derivative $(\partial T_f/\partial m)_{m \to 0}$ is known as the **freezing point depression constant** or the **cryscopic constant** of the solvent and is represented by the symbol $K_f$.

Equation (2.8.9) may be written as

$$-dT_f = K_f \, dm \qquad (p \text{ constant, very dilute solution}) \qquad (2.8.11)$$

For a finite value of $dm$, we have

$$-\Delta T_f = K_f m \qquad (2.8.12)^\dagger$$

---

†Equation (2.8.12) may be derived directly from Eq. (2.8.8) by invoking the following approximations.

$$\ln x_1 = -\ln(1 + mM_1) = -mM_1$$

$$T_f T_f^* \simeq T_f^{*2}$$

**Unit of $K_f$**

The unit of $K_f$ may be worked out as follows.

$$\text{unit of } K_f = \frac{\text{unit of } \Delta T_f}{\text{unit of } m} = \frac{\text{K}}{\text{mol kg}^{-1}} = \text{K kg mol}^{-1}$$

**Value of $K_f$ for Water**

The numerical value of $K_f$ for a given solvent can be determined from its $T_f^*$ and $\Delta_{fus}H_{1,m}$ values. For example, for water, we have

$$K_f = \frac{M_1 R T_f^{*2}}{\Delta_{fus}H_{1,m}} = \frac{(18 \times 10^{-3}\ \text{kg mol}^{-1})(8.314\ \text{J K}^{-1}\ \text{mol}^{-1})(273\ \text{K})^2}{(6\,009.5\ \text{J mol}^{-1})}$$

$$= 1.86\ \text{K kg mol}^{-1}$$

**Values of $K_f$ for Some Solvents**

The values of $K_f$ for other commonly used solvents are given in Table 2.8.1.

**Table 2.8.1** Freezing Point Depression Constants

| *Solvent* | *Molar mass* / g mol$^{-1}$ | *m. pt.*/°C | $K_f$ / K kg mol$^{-1}$ |
|---|---|---|---|
| Water | 18.0 | 0 | 1.86 |
| Acetic acid | 60.0 | 16.6 | 3.57 |
| Benzene | 78.1 | 5.45 | 5.07 |
| Cyclohexane | 84.2 | 6.5 | 20.0 |
| Dioxane | 88.1 | 11.7 | 4.71 |
| Naphthalene | 128.3 | 80.1 | 6.98 |
| Camphor | 152.2 | 178.4 | 37.7 |

***Problem 2.8.1***

The expression $-\Delta T_f = K_f m$ for the freezing point depression can also be obtained from the shift in triple point of a solvent on forming a solution[†] of the substance and by applying the Clausius-Clapeyron equation to the solid-vapour and the solution-vapour equilibria. Derive the relation by drawing the necessary diagram.

***Solution***

The solution-vapour curve will lie below that of the solvent-vapour curve and thus it intersects the solid-vapour curve at a temperature (triple temperature) smaller than the corresponding intersection temperature for the solvent-vapour curve. This is shown in Fig. 2.8.1. There occurs a depression in the triple point of the substance when a nonvolatile solute is dissolved in it.

Applying the Clausius-Clapeyron equation to the solution-vapour equilibrium for the two values of $p_2$, $T$ and $p_1$, $T_f^*$, we have

$$\ln \frac{p_1}{p_2} = \frac{\Delta_{vap}H_{1,m}}{R}\left(\frac{1}{T} - \frac{1}{T_f^*}\right) = \frac{\Delta_{vap}H_{1,m}}{R}\,\frac{(-\Delta T_f)}{T\,T_f^*} \tag{2.8.13}$$

where $\quad \Delta T_f = T - T_f^*$

[†]The effect of pressure on the equilibrium temperature of solid to liquid equilibrium is usually very small and thus the triple point may be taken as the freezing point. Take, for example, the triple point of water system, it lies at 0.0075 °C and 4.6 mmHg. The normal freezing point of water is 0 °C (at 1 atm).

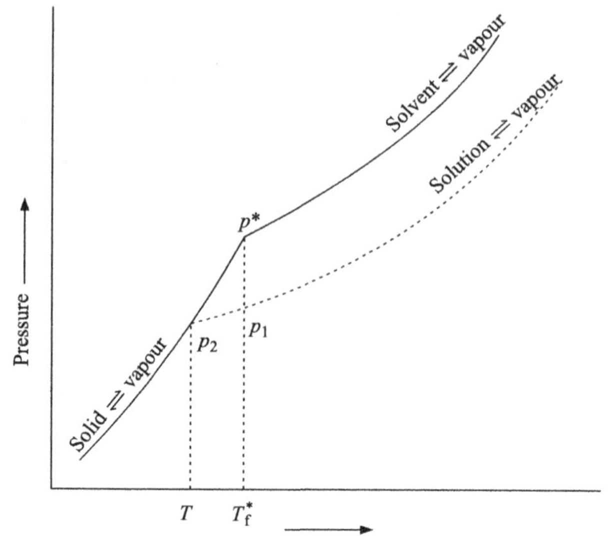


**Fig. 2.8.1 Clausius-Clapeyron equation as applied to the depression of triple point**

Applying now the Clausius-Clapeyron equation to the solid-vapour equilibrium for the two values of $p_2$, $T$ and $p^*$, $T_f^*$, we have

$$\ln \frac{p^*}{p_2} = \frac{\Delta_{sub}H_{1,m}}{R}\left(\frac{1}{T} - \frac{1}{T_f^*}\right) = \frac{\Delta_{sub}H_{1,m}}{R}\left(\frac{-\Delta T_f}{TT_f^*}\right) \tag{2.8.14}$$

Subtracting Eq. (2.8.13) from Eq. (2.8.14), we get

$$\ln \frac{p^*}{p_2} - \ln \frac{p_1}{p_2} = \frac{\Delta_{sub}H_{1,m}}{R}\left(\frac{-\Delta T_f}{TT_f^*}\right) - \frac{\Delta_{vap}H_{1,m}}{R}\left(\frac{-\Delta T_f}{TT_f^*}\right)$$

or $$\ln \frac{p^*}{p_1} = \frac{-\Delta T_f}{TT_f^*R}(\Delta_{sub}H_{1,m} - \Delta_{vap}H_{1,m}) = \frac{-\Delta T_f}{TT_f^*R}(\Delta_{fus}H_{1,m}) \tag{2.8.15}$$

Equation (2.8.15) can be simplified under the following approximations.

(i) $T$ will be very near $T_f^*$, so that the term $T\,T_f^*$ may be replaced by $T_f^{*2}$, i.e.

$$TT_f^* \simeq T_f^{*2} \tag{2.8.16}$$

(ii) The term $\ln p^*/p_1$ can be written as

$$\ln \frac{p^*}{p_1} = -\ln \frac{p_1}{p^*} = -\ln\left(\frac{p^* - p^* + p_1}{p^*}\right) = -\ln\left(1 - \frac{p^* - p_1}{p^*}\right)$$

Since the term $(p^* - p_1)/p^*$, the relative lowering of vapour pressure at temperature $T_f^*$, is very small in comparison to 1, the term $-\ln(1 - (p^* - p_1)/p^*)$ can be approximated as

$$-\ln\left(1 - \frac{p^* - p_1}{p^*}\right) = -\left(-\frac{p^* - p_1}{p^*}\right) = \frac{p^* - p_1}{p^*}$$

(iii) According to Raoult's law

$$\frac{p^* - p_1}{p^*} = x_2 = \frac{n_2}{n_1 + n_2}$$

For a dilute solution $n_2 \ll n_1$, and therefore $n_2$ can be neglected in comparison to $n_1$. Thus, we have

$$\ln \frac{p^*}{p_1} \simeq \frac{p^* - p_1}{p^*} \simeq \frac{n_2}{n_1} = \frac{m_2/M_2}{m_1/M_1} \tag{2.8.17}$$

Substituting Eqs (2.8.16) and (2.8.17) in Eq. (2.8.15), we get

$$\frac{m_2/M_2}{m_1/M_1} = \frac{\Delta_{\text{fus}}H_{1,\text{m}}}{R} \frac{(-\Delta T_{\text{f}})}{T_{\text{f}}^{*2}}$$

Rearranging this, we get

$$(-\Delta T_{\text{f}}) = \frac{M_1 R T_{\text{f}}^{*2}}{\Delta_{\text{fus}}H_{1,\text{m}}} \left( \frac{m_2/M_2}{m_1} \right) = K_{\text{f}} m \tag{2.8.18}$$

---

***Example 2.8.1*** A sample of camphor used in the Rast method for the determination molar mass had melting point of 451.55 K. The melting point of a solution containing 0.522 g camphor and 0.038 6 g of an unknown compound was 433.85 K. If the unknown compound was a hydrocarbon containing 7.7% H, find its molecular formula. The molar mass of camphor is 152.2 g mol$^{-1}$ and its $\Delta_{\text{fus}}H_{1,\text{m}}$ = 6.844 kJ mol$^{-1}$.

***Solution*** The given data are

$$m_1 = 0.522\ \text{g}, \quad m_2 = 0.038\,6\ \text{g}, \quad T_{\text{f}} = 433.85\ \text{K}, \quad T_{\text{f}}^* = 451.55\ \text{K},$$
$$M_1 = 152.2\ \text{g mol}^{-1}, \quad \Delta_{\text{fus}}H_{1,\text{m}} = 6.844\ \text{kJ mol}^{-1}$$

***Calculation of $K_f$***

$$K_{\text{f}} = \frac{M_1 R T_{\text{f}}^{*2}}{\Delta_{\text{fus}}H_{1,\text{m}}} = \frac{(152.2 \times 10^{-3}\ \text{kg mol}^{-1})\,(8.314\ \text{J K}^{-1}\ \text{mol}^{-1})(451.55\ \text{K})^2}{(6.844 \times 10^3\ \text{J mol}^{-1})}$$
$$= 37.7\ \text{K kg mol}^{-1}$$

***Calculation of molality of solution***

$$m = \frac{-\Delta T_{\text{f}}}{K_{\text{f}}} = \frac{-(T_{\text{f}} - T_{\text{f}}^*)}{K_{\text{f}}} = \frac{-(433.85\ \text{K} - 451.55\ \text{K})}{(37.7\ \text{K kg mol}^{-1})} = 0.47\ \text{mol kg}^{-1}$$

***Calculation of molar mass of solute*** Since $m = (m_2/M_2)/m_1$, we get

$$M_2 = \frac{m_2}{m m_1} = \frac{(0.038\,6\ \text{g})}{(0.47\ \text{mol kg}^{-1})(0.522\ \text{g})} = 0.157\ \text{kg mol}^{-1}$$

***Determination of molecular formula***

The given mass ratio is $m_C : m_H :: 92.3 : 7.7$

The atomic ratio is $N_C : N_H :: \dfrac{92.3}{12} : \dfrac{7.7}{1} \Rightarrow 7.7 : 7.7 \Rightarrow 1 : 1$

Hence, Empirical formula : CH

Empirical molar mass = 13 g mol$^{-1}$

$$\text{Number of units of CH} = \frac{\text{molar mass}}{\text{empirical molar mass}} = \frac{157\ \text{g}}{13\ \text{g}} \simeq 12$$

Hence, Molecular formula = $C_{12}H_{12}$

***Example 2.8.2***

A solution comprising 0.1 mol of naphthalene and 0.9 mol of benzene is cooled until some solid benzene freezes out. The solution is then decanted off from the solid, and warmed to 353 K, where its vapour pressure is found to be 670 Torr. The freezing and normal boiling points of benzene are 278.5 K and 353 K, respectively, and its enthalpy of fusion is 10.67 kJ mol$^{-1}$. Calculate the temperature to which the solution was cooled originally and the amount of benzene that must have frozen out. Assume conditions of ideal solution.

***Solution***

The given data are:

$$n_1 = 0.9\ \text{mol}, \quad n_2 = 0.1\ \text{mol}, \quad T_f^* = 278.5\ \text{K}, \quad T_b^* = 353\ \text{K},$$

$$p^* = 760\ \text{Torr}, \quad p = 670\ \text{Torr}, \quad \Delta_{fus}H_{1,m} = 10.67\ \text{kJ mol}^{-1}$$

From the relative lowering of vapour pressure, we obtain the amount fraction of the solute (i.e. naphthalene).

$$x_2 = \frac{p^* - p}{p^*} = \frac{760\ \text{Torr} - 670\ \text{Torr}}{760\ \text{Torr}} = 0.118\,5$$

Since $x_2 = n_2/(n_1 + n_2)$, we get

$$n_1 + n_2 = \frac{n_2}{x_2} = \frac{0.1\ \text{mol}}{0.118\,5} = 0.844\ \text{mol}$$

Since $n_2 = 0.1$ mol, we get

$$n_1 = 0.844\ \text{mol} - n_2 = 0.844\ \text{mol} - 0.1\ \text{mol} = 0.744\ \text{mol}$$

Hence, the amount of benzene frozen out = 0.9 mol – 0.744 mol = 0.156 mol
The freezing point depression constant of benzene is

$$K_f = \frac{M_1 R T_f^{*2}}{\Delta_{fus}H_{1,m}} = \frac{(0.078\ \text{kg mol}^{-1})(8.314\ \text{J K}^{-1}\ \text{mol}^{-1})(278.5\ \text{K})^2}{(10\,670\ \text{J mol}^{-1})}$$

$$= 4.714\ \text{K kg mol}^{-1}$$

Molality of the solution is

$$m = \frac{n_2}{m_1} = \frac{n_2}{n_1 M_1} = \frac{(0.1\ \text{mol})}{(0.744\ \text{mol})(0.078\ \text{kg mol}^{-1})} = 1.723\ \text{mol kg}^{-1}$$

Finally $-\Delta T_f = K_f m = (4.714\ \text{K kg mol}^{-1})\,(1.723\ \text{mol kg}^{-1}) = 8.12\ \text{K}$

$$T_f = T_f^* + \Delta T_f = 278.5\ \text{K} - 8.12\ \text{K} = 270.38\ \text{K}$$

## 2.9 OSMOSIS AND OSMOTIC PRESSURE

**Phenomenon of Osmosis**

We have seen in Section 2.6 that a solution when separated from the pure solvent with the help of a semipermeable membrane, the flow of solvent molecules from the pure solvent side to the solution side takes place. This flow of solvent molecules from a region of higher concentration (i.e. pure solvent) to the region of lower concentration (i.e. solution) of solvent molecules is known as the phenomenon of osmosis.

**Osmotic Pressure**

As a result of osmosis, level of the solution in the tube rises, whereas that of the pure solvent falls (Fig. 2.9.1). After some time an equilibrium is reached where the level of the solution in the tube does not rise further. At this stage, the excessive hydrostatic pressure created on the solution side exactly balances the tendency of the solvent to pass through the membrane as shown in Fig. 2.9.1. This excessive hydrostatic pressure is known as the osmotic pressure of the solution and is represented by the symbol $\Pi$.

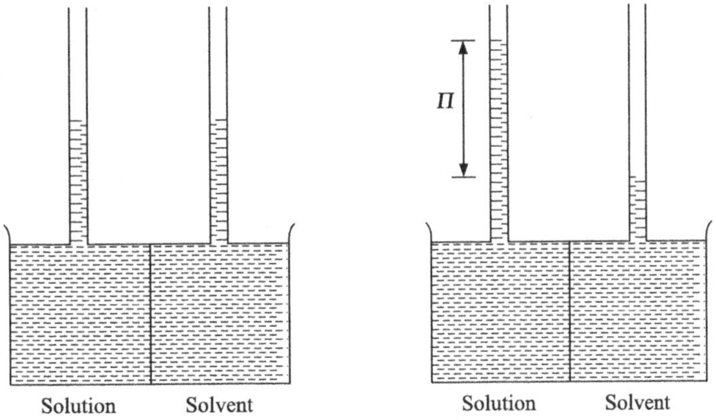


**Fig. 2.9.1 The phenomenon of osmosis and generation of osmotic pressure**

**Alternative way of Checking Osmosis**

The phenomenon of osmosis can also be checked by increasing the external pressure on the solution by an amount equivalent to the hydrostatic pressure $\Pi$ as shown in Fig. 2.9.2.[†]

Alternatively, this can also be checked by reducing the external pressure on the solvent side by an amount equivalent to this hydrostatic pressure.

**Thermodynamic Derivation of Osmotic Pressure**

The osmotic pressure of a solution depends on its concentration—larger the concentration larger the osmotic pressure. The expression relating the osmotic pressure with concentration of the solution can be derived thermodynamically.

Initially, there exists a difference of chemical potential of the solvent on either sides of the membrane and thus the solvent flows from the pure solvent side to the solution side. This flow of solvent to the solution side generates an

---

[†]If the applied excessive pressure becomes larger than the osmotic pressure, the reverse of osmosis is observed, i.e. solvent molecules will flow from the solution to the solvent side. This is due to larger increase of chemical potential than the required value.

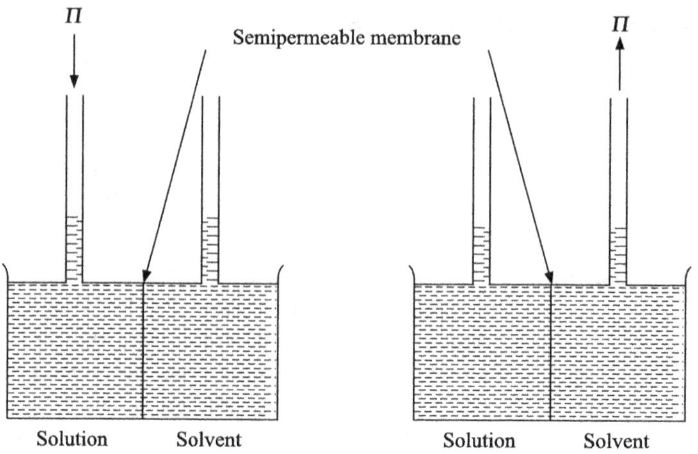


**Fig. 2.9.2 An alternative way of stopping the phenomenon of osmosis**

excessive hydrostatic pressure on the solution side, which in turn increases the chemical potential of the solvent in solution. Consequently, the difference of chemical potential of solvent on either sides is decreased. After some time, enough solvent has flown into the solution side so that the difference of chemical potential is altogether eliminated and thus osmosis stops. At this stage, the system is in equilibrium and no more solvent flows from the solvent to the solution. Basically, in the present case we have equilibrium of the type

$$\text{Solvent in solution} \rightleftharpoons \text{Pure solvent}$$

The thermodynamic condition of equilibrium is

$$\mu_{1(\text{sol})}(T, p + \Pi) = \mu^*_{1(1)}(T, p) \tag{2.9.1}$$

where $\Pi$ is the excessive hydrostatic pressure. At constant temperature, $\Pi$ is a function of $x_1$, the amount fraction of the solvent in solution. If the solution is assumed to be ideal, we have

$$\mu_{1(\text{sol})}(T, p + \Pi) = \mu^*_{1(1)}(T, p + \Pi) + RT \ln x_1 \tag{2.9.2}$$

Substituting Eq. (2.9.2) in Eq. (2.9.1), we get

$$\mu^*_{1(1)}(T, p + \Pi) + RT \ln x_1 = \mu^*_{1(1)}(T, p)$$

or

$$\mu^*_{1(1)}(T, p + \Pi) - \mu^*_{1(1)}(T, p) = - RT \ln x_1 \tag{2.9.3}$$

The left side of Eq. (2.9.3) represents the difference in chemical potentials of the pure solvent under the two different pressures $p + \Pi$ and $p$. This difference may be evaluated by employing the relation

$$\left(\frac{\partial \mu^*_{1(1)}}{\partial p}\right)_T = V^*_{1,\text{m}} \tag{2.9.4}$$

where $V^*_{1,\text{m}}$ is the molar volume of the pure solvent. Rewriting Eq. (2.9.4), we have

$$d\mu^*_{1(1)} = V^*_{1,m}\, dp \qquad (T \text{ constant})$$

Integrating the above relation within the limits $p$ and $p + \Pi$, we get

$$\int_p^{p+\Pi} d\mu^*_{1(1)} = \int_p^{p+\Pi} V^*_{1,m}\, dp \qquad (T \text{ constant})$$

or
$$\mu^*_{1(1)}(T, p+\Pi) - \mu^*_{1(1)}(T, p) = \int_p^{p+\Pi} V^*_{1,m}\, dp$$

If the solvent is incompressible, $V^*_{1,m}$ is independent of pressure and can be removed from the integral. Thus, we have

$$\mu^*_{1(1)}(T, p+\Pi) - \mu^*_{1(1)}(T, p) = V^*_{1,m}\Pi \qquad (2.9.5)$$

Substituting Eq. (2.9.5) in Eq. (2.9.3), we get

$$V^*_{1,m}\Pi = -RT \ln x_1 \qquad (2.9.6)$$

**Expression of Osmotic Pressure for a Dilute Solution (van't Hoff Equation)**

For a dilute solution, Eq. (2.9.6) can be simplified under the following approximations.

(i) Since the solution is dilute, $x_2 << 1$, the term $\ln x_1$ can be approximated as

$$\ln x_1 = \ln(1 - x_2) \simeq -x_2 \qquad (2.9.7)$$

(ii) The amount fraction of the solute is given as

$$x_2 = \frac{n_2}{n_1 + n_2} \simeq \frac{n_2}{n_1} \qquad (\text{since } n_2 << n_1) \quad (2.9.8)$$

(iii) For a solution, the total volume $V$ can be written in terms of the partial molar volumes following the additivity rule

$$V = n_1V_1 + n_2V_2$$

where $V_1$ is the partial molar volume of the solvent and $V_2$ is the corresponding volume of the solute. For a dilute solution, we assume that:

(a) Partial molar volume of the solvent in the solution is the same as that of the pure solvent, i.e. $V_1 = V^*_{1,m}$.

(b) The factor $n_2V_2 << n_1V_1$ since $n_2 << n_1$ and can, therefore, be neglected. Thus we have

$$V \simeq n_1V_1 = n_1V^*_{1,m} \qquad (2.9.9)$$

Substituting Eqs (2.9.7)-(2.9.9) in Eq. (2.9.6), we have

$$\frac{V}{n_1}\Pi = -RT(-x_2) = RTx_2$$

or
$$\frac{V}{n_1}\Pi = RT\frac{n_2}{n_1}$$

or $$V\Pi = n_2 RT$$

or $$\Pi = \left(\frac{n_2}{V}\right) RT = cRT \tag{2.9.10}$$

where $c$ is the molar concentration of solute in the solution. Equation (2.9.10) is known as the **van't Hoff equation** for osmotic pressure of the solution.

**Parallelism with Ideal Gas Equation**

The form of van't Hoff equation is exactly identical to that of the ideal gas equation except that $\Pi$ replaces the gas pressure $p$. This striking resemblance suggests that the solute molecules are dispersed in the solvent the way the gas molecules are dispersed in the empty space. Thus the solute is analogous to the gas molecules and the solvent is analogous to the empty space between the gas molecules.

---

***Problem 2.9.1***

(a) Show that the difference in chemical potential of the solvent on either side of semipermeable membrane shown in Fig. 2.9.1 is given by

$$\mu^*_{1(1)}(T, p^*) - \mu_{1(\text{sol})}(T, p) = - RT \ln x_1$$

(b) The osmotic pressure of a solution may be defined as the excessive pressure which needs to be applied on the solution side so as to prevent osmosis. Using this fact, derive the relation

$$V_1 \Pi = - RT \ln x_1$$

where $V_1$ is the partial molar volume of solvent in solution.
(c) The osmotic pressure of a solution may be defined as the pressure which needs to be lowered on the solvent side so as to prevent osmosis. Using this fact, show that

$$V^*_{1,\text{m}} \Pi = - RT \ln x_1$$

***Solution***

(**a**) Let $p^*$ and $p$ be the vapour pressures of pure solvent and solution, respectively. In Fig. 2.9.1, we have the following two equilibria.

Solvent in solution $\rightleftharpoons$ Solvent vapour at pressure $p$

Pure solvent $\rightleftharpoons$ Solvent vapour at pressure $p^*$

The thermodynamic relations for these two equilibria are:

$$\mu_{1(\text{sol})}(T, p) = \mu^*_{1(\text{v})}(T, p) = \mu^\circ_{1(\text{v})} + RT \ln (p/p^\circ)$$

$$\mu^*_1(T, p^*) = \mu^*_{1(\text{v})}(T, p^*) = \mu^\circ_{1(\text{v})} + RT \ln (p^*/p^\circ)$$

Hence $$\mu^*_1(T, p^*) - \mu_{1(\text{sol})}(T, p) = RT \ln (p^*/p^\circ) - RT \ln (p/p^\circ) = RT \ln \frac{p^*}{p}$$

For an ideal solution $p = p^* x_1$. With this, the above expression becomes

$$\mu^*_1(T, p^*) - \mu_{1(\text{sol})}(T, p) = RT \ln (1/x_1) = - RT \ln x_1$$

(**b**) By applying the pressure $\Pi$ on the solution side, the osmosis is prevented. At this stage, we will have

$$\mu^*_1(T, p^*) = \mu_{1(\text{sol})}(T, p + \Pi)$$

Substituting this expression in the expression derived in part (a), we get

$$\mu_{1(\text{sol})}(T, p + \Pi) - \mu_{1(\text{sol})}(T, p) = -RT \ln x_1 \tag{1}$$

Now $$\left(\frac{\partial \mu_{1(\text{sol})}}{\partial p}\right)_T = V_1$$

where $V_1$ is the partial molar volume of solvent in the solution. Rewriting and integrating within the limits $p$ and $p + \Pi$, we get

$$\int_p^{p+\Pi} \mathrm{d}\mu_{1(\text{sol})} = \int_p^{p+\Pi} V_1 \, \mathrm{d}p$$

Assuming $V_1$ to be independent of pressure, we get

$$\mu_{1(\text{sol})}(T, p + \Pi) - \mu_{1(\text{sol})}(T, p) = V_1(p + \Pi - p)$$
$$= V_1 \Pi$$

Hence Eq. (1) becomes

$$V_1 \Pi = -RT \ln x_1$$

(**c**) By lowering the pressure on the solvent side by an amount $\Pi$, the phenomenon of osmosis is prevented. At this stage, we will have

$$\mu_1^*(T, p^* - \Pi) = \mu_{1(\text{sol})}(T, p)$$

Substituting this in the expression derived in part (a), we get

$$\mu_1^*(T, p^*) - \mu_1^*(T, p^* - \Pi) = -RT \ln x_1 \tag{2}$$

Now $$\left(\frac{\partial \mu_1^*}{\partial p}\right)_T = V_{1,\text{m}}^*$$

or $$\mathrm{d}\mu_1^* = V_{1,\text{m}}^* \, \mathrm{d}p; \qquad (T \text{ constant})$$

Thus $$\int_{p^*-\Pi}^{p^*} \mathrm{d}\mu_1^* = \int_{p^*-\Pi}^{p^*} V_{1,\text{m}}^* \, \mathrm{d}p; \qquad (T \text{ constant})$$

Assuming $V_{1,\text{m}}^*$ to be independent of pressure, we have

$$\mu_1^*(T, p^*) - \mu_1^*(T, p^* - \Pi) = V_{1,\text{m}}^*\{p^* - (p^* - \Pi)\} = V_{1,\text{m}}^* \Pi$$

Hence Eq. (2) becomes

$$V_{1,\text{m}}^* \Pi = -RT \ln x_1$$

---

**Determination of Molar Mass**

The principal application of osmotic pressure measurement is in the determination of the molar mass of a substance which is either slightly soluble or has a very high molar mass such as proteins, polymers of various types and colloids. This is due to the fact that even a very small concentration of the solution produces fairly large magnitude of the osmotic pressure. In the laboratory, the concentrations usually employed are of the order of 0.001 mol dm$^{-3}$. At concentration of 0.001 mol dm$^{-3}$, the magnitude of osmotic pressure at 298 K is

$$\begin{aligned}\Pi = cRT &= (0.001\ \text{mol dm}^{-3})(8.314\ \text{J K}^{-1}\ \text{mol}^{-1})(298\ \text{K})\\ &= 2.4776\ \text{J dm}^{-3} = 2\,477.6\ \text{J m}^{-3} = 2\,477.6\ \text{Pa}\\ &= 0.0245\ \text{atm} = 18.58\ \text{Torr}\end{aligned}$$

At this concentration, the values of other colligative properties such as the boiling point elevation and the freezing point depression are too small to be determined experimentally.

The van't Hoff equation is expected to be valid only at infinite dilution or to a very dilute solution where the solute-solute interaction is negligible. This, however, might not be applicable to a substance of high molar mass at concentrations which are usually employed for experimental measurements. The method of extrapolation can be employed for such type of substances. The van't Hoff equation can be written as

$$\Pi = \frac{n_2}{V}RT = \frac{(m_2/M_2)}{V}RT = \left(\frac{m_2}{V}\right)\left(\frac{RT}{M_2}\right) = \rho\left(\frac{RT}{M_2}\right)$$

i.e.

$$\frac{\Pi}{\rho} = \frac{RT}{M_2}$$

where $\rho$ is the density of the solution. The limiting value of $\Pi/\rho$ at infinite dilution can be determined by plotting $\Pi/\rho$ versus $\rho$ and determining the intercept when $\rho \rightarrow 0$. This limiting value of $(\Pi/\rho)_{\rho\rightarrow 0}$ will be equal to $RT/M_2$.

Thus

$$M_2 = \frac{RT}{(\Pi/\rho)_{\rho\rightarrow 0}}$$

### Number and Mass Average Molar Masses

The molar mass determined from any of the colligative properties is the number average molar mass, designated as $\bar{M}_n$ and is given by

$$\bar{M}_n = \frac{N_1M_1 + N_2M_2 + \cdots}{N_1 + N_2 + \cdots} = \frac{\sum_i N_iM_i}{\sum_i N_i}$$

where $N_i$ is the number of molecules having molar mass $M_i$. The above average gives equal weightage to large and small molecules and simply counts the number of molecules. It is thus independent of the individual characteristics of the molecules. For example, all the polymer molecules in any sample do not contain the same number of units in every chain, and, therefore, there will be a range of molecular masses in the polymer sample. The molar mass determined with the help of colligative property will be the number average molar mass.

A second average molar mass is the mass average molar mass, $\bar{M}_m$, and is defined as

$$\bar{M}_m = \frac{N_1M_1^2 + N_2M_2^2 + \cdots}{N_1M_1 + N_2M_2 + \cdots} = \frac{\sum_i N_i\,M_i^2}{\sum_i N_iM_i}$$

This average gives more weightage to the heavier molecules as compared to the number average molar mass. This mass can be determined from the property which depends upon the individual characteristics of the molecules, such as the light-scattering measurements.

***Example 2.9.1***

The osmotic pressure values of various concentrations of polyisobutylene in two different solvents at 298 K are given below. Calculate the molar mass of polyisobutylene by the method of extrapolation to the infinite dilution.

| | | | | | | |
|---|---|---|---|---|---|---|
| $\rho$/g cm$^{-3}$ | 0.020 0 | 0.015 0 | 0.010 0 | 0.007 5 | 0.005 0 | 0.002 5 |
| $\Pi$/ atm in benzene | 0.002 08 | 0.001 52 | 0.000 99 | | 0.000 49 | |
| in cyclohexane | 0.011 7 | 0.006 6 | 0.00 30 | 0.001 73 | 0.000 9 | 0.000 35 |

***Solution***

The values of $(\Pi/\rho)/(\text{atm g}^{-1}\ \text{cm}^3)$ are as follows.

| | | | | | | |
|---|---|---|---|---|---|---|
| $\rho$/g cm$^{-3}$ | 0.020 0 | 0.015 0 | 0.010 0 | 0.007 5 | 0.005 0 | 0.002 5 |
| $\Pi/\rho$ in benzene | 0.104 | 0.101 | 0.099 | | 0.098 | |
| in cyclohexane | 0.585 | 0.440 | 0.300 | 0.230 | 0.18 | 0.140 |

The plots of $\Pi/\rho$ versus $\rho$ are shown in Fig. 2.9.3.

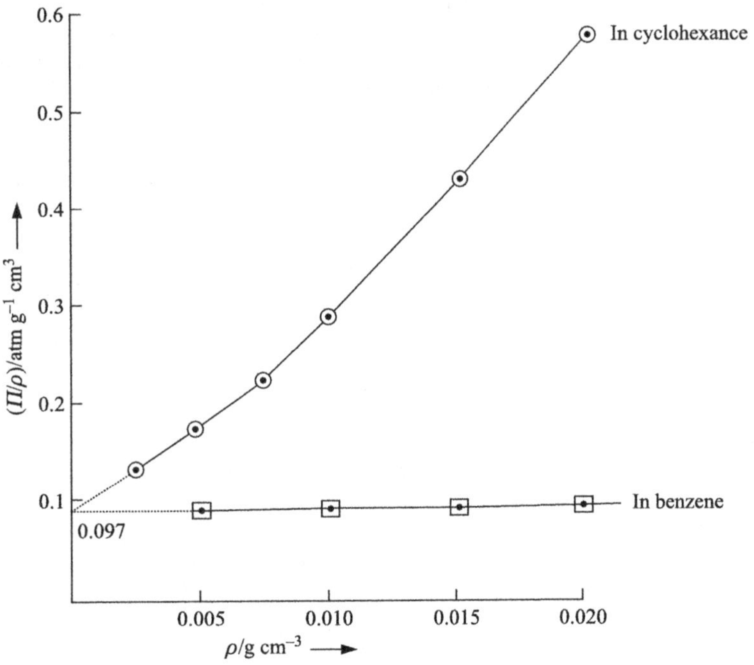


**Fig. 2.9.3 Plot of $\Pi/\rho$ versus $\rho$ for polyisobutylene in benzene and cyclohexane**

Thus Molar mass of polyisobutylene $= \dfrac{RT}{(\Pi/\rho)_{\rho\to 0}} = \dfrac{(82.0\ \text{cm}^3\ \text{atm K}^{-1}\ \text{mol}^{-1})(298\ \text{K})}{(0.097\ \text{atm g}^{-1}\ \text{cm}^3)}$

$\simeq 250\ 000\ \text{g mol}^{-1}$

***Example 2.9.2***

One gram of serum albumin dissolved in 1 $dm^3$ of water gives a solution height of 3.9 mm when separated from pure water by a semipermeable membrane at 25 °C. What is the molar mass of serum albumin? Density of the solution may be taken as 1.0 g $cm^{-3}$.

***Solution***

Osmotic pressure of the solution is

$$\Pi = h\rho g = (3.9 \times 10^{-3}\,\text{m})\,(10^3\ \text{kg m}^{-3})\,(9.81\ \text{m s}^{-2})$$

$$= 38.26\ \text{N m}^{-2}$$

Now $$M_2 = \frac{m_2RT}{\Pi V} = \frac{(10^{-3}\ \text{kg})\,(8.314\ \text{J K}^{-1}\ \text{mol}^{-1})\,(298\ \text{K})}{(38.26\ \text{N m}^{-2})\,(10^{-3}\ \text{m}^3)}$$

$$= 64.78\ \text{kg mol}^{-1}$$

---

***Example 2.9.3***

The osmotic pressure is measured between water and a solution containing 1 g of glucose $C_6H_{12}O_6$ and 1 g of sucrose $C_{12}H_{22}O_{11}$ in 1 000 g of water, the temperature being maintained at 25 °C, and $\rho$ (soln) = 1 g $cm^{-3}$.

(a) Calculate osmotic pressure of the solution.
(b) If this pressure were measured and if the nature of solutes were not known, what molar mass would have been calculated?
(c) The measurement of the osmotic pressure (and other colligative properties) gives an average molar mass of the solute. What kind of average (i.e. number or mass) is this? Verify the answer.

***Solution***

(**a**) Total amount of solutes $= \left(\dfrac{1\ \text{g}}{180\ \text{g mol}^{-1}} + \dfrac{1\ \text{g}}{342\ \text{g mol}^{-1}}\right) = 0.008\,479\ \text{mol}$

Since $\Pi V = n_2RT$, we have

$$\Pi = \frac{n_2RT}{V} = \frac{(0.008\,479\ \text{mol})\,(8.314\ \text{J K}^{-1}\ \text{mol}^{-1})\,(298\ \text{K})}{(10^{-3}\ \text{m}^3)}$$

$$= 2.101 \times 10^4\ \text{J m}^{-3} \equiv 2.101 \times 10^4\ \text{N m}^{-2}$$

(**b**) Solution will contain a total of 2 g of the solute. Let the molar mass of solute be $M$.

Thus $$\Pi V = nRT = \frac{m}{M}RT$$

or $$M = \frac{mRT}{\Pi V} = \frac{(2 \times 10^{-3}\ \text{kg})(8.314\ \text{J K}^{-1}\ \text{mol}^{-1})(298\ \text{K})}{(2.101 \times 10^4\ \text{N m}^{-2})(10^{-3}\ \text{m}^3)}$$

$$= 0.236\ \text{kg mol}^{-1}$$

(**c**) It is the number average molar mass as may be seen from the following calculations.

$$\overline{M}_n = \frac{N_1M_1 + N_2M_2}{N_1 + N_2} = \frac{n_1M_1 + n_2M_2}{n_1 + n_2} = \frac{m_1 + m_2}{n_1 + n_2}$$

$$= \frac{10^{-3}\ \text{kg} + 10^{-3}\ \text{kg}}{0.008\,479\ \text{mol}} = 0.236\ \text{kg mol}^{-1}$$

---

## 2.10 RELATIONS BETWEEN DIFFERENT COLLIGATIVE PROPERTIES

The different colligative properties can be correlated with each other through the following expressions.

$$\frac{p^* - p}{p^*} = x_2 \qquad \text{(Eq. 2.4.1)}$$

$$\ln x_1 = -\frac{\Delta_{\text{vap}} H_{1,\text{m}}}{R}\left(\frac{1}{T_\text{b}^*} - \frac{1}{T_\text{b}}\right) \qquad \text{(Eq. 2.7.7)}$$

$$\ln x_1 = \frac{\Delta_{\text{fus}} H_{1,\text{m}}}{R}\left(\frac{1}{T_\text{f}^*} - \frac{1}{T_\text{f}}\right) \qquad \text{(Eq. 2.8.7)}$$

and $\quad V\Pi = -RT \ln x_1 \qquad$ (Eq. 2.9.6)

where $V = V_{1,\text{m}}^*$. For dilute solutions, Eqs (2.7.7), (2.8.7) and (2.9.6) may be approximated as shown in the following.

(i) Replacing $x_1$ by $1 - x_2$ in Eq. (2.7.7), we get

$$\ln(1 - x_2) = -\frac{\Delta_{\text{vap}} H_{1,\text{m}}}{R}\left(\frac{T_\text{b} - T_\text{b}^*}{T_\text{b} T_\text{b}^*}\right)$$

Since $x_2 << 1$ and $T_\text{b} \simeq T_\text{b}^*$, we may write

$$x_2 \simeq \frac{\Delta_{\text{vap}} H_{1,\text{m}}}{R}\left(\frac{\Delta T_\text{b}}{T_\text{b}^{*2}}\right) \qquad (2.10.1)$$

(ii) Replacing $x_1$ by $1 - x_2$ in Eq. (2.8.7), we get

$$\ln(1 - x_2) = \frac{\Delta_{\text{fus}} H_{1,\text{m}}}{R}\left(\frac{T_\text{f} - T_\text{f}^*}{T_\text{f} T_\text{f}^*}\right)$$

Since $x_2 << 1$ and $T_\text{f} \simeq T_\text{f}^*$, we may write

$$x_2 = \frac{\Delta_{\text{fus}} H_{1,\text{m}}}{R}\left(\frac{-\Delta T_\text{f}}{T_\text{f}^{*2}}\right) \qquad (2.10.2)$$

(iii) Replacing $x_1$ by $1 - x_2$ in Eq. (2.9.6), we get

$$V\Pi = -RT \ln(1 - x_2)$$

Since $x_2 << 1$, we may write

$$V\Pi \simeq RT\, x_2 \qquad (2.10.3)$$

## Relations Of Relative Lowering Of Vapour Pressure With Other Colligative Properties

**With Elevation of Boiling Point**

Comparing Eq. (2.4.1) with Eq. (2.10.1), we get

$$\frac{p^* - p}{p^*} = x_2 = \frac{\Delta_{\text{vap}} H_{1,\text{m}}}{R}\left(\frac{\Delta T_{\text{b}}}{T_{\text{b}}^{*2}}\right) \tag{2.10.4}$$

**With Depression of Freezing Point**

Comparing Eq. (2.4.1) with Eq. (2.10.2), we get

$$\frac{p^* - p}{p^*} = x_2 = \frac{\Delta_{\text{fus}} H_{1,\text{m}}}{R}\left(\frac{-\Delta T_{\text{f}}}{T_{\text{f}}^{*2}}\right) \tag{2.10.5}$$

**With Osmotic Pressure**

Comparing Eq. (2.4.1) with Eq. (2.10.3), we get

$$\frac{p^* - p}{p^*} = x_2 = \frac{V\Pi}{RT} \tag{2.10.6}$$

## Relations Of Elevation Of Boiling Point With Other Colligative Properties

**With Depression of Freezing Point**

Comparing Eq. (2.10.1) with Eq. (2.10.2), we get

$$\frac{\Delta_{\text{vap}} H_{1,\text{m}}}{R}\left(\frac{\Delta T_{\text{b}}}{T_{\text{b}}^{*2}}\right) = x_2 = \frac{\Delta_{\text{fus}} H_{1,\text{m}}}{R}\left(\frac{-\Delta T_{\text{f}}}{T_{\text{f}}^{*2}}\right) \tag{2.10.7}$$

or

$$\frac{\Delta T_{\text{b}}}{(-\Delta T_{\text{f}})} = \frac{\Delta_{\text{fus}} H_{1,\text{m}}}{T_{\text{f}}^{*2}} \frac{T_{\text{b}}^{*2}}{\Delta_{\text{vap}} H_{1,\text{m}}} \tag{2.10.8}$$

**With Osmotic Pressure**

Comparing Eq. (2.10.1) with Eq. (2.10.3), we get

$$\frac{\Delta_{\text{vap}} H_{1,\text{m}}}{R}\left(\frac{\Delta T_{\text{b}}}{T_{\text{b}}^{*2}}\right) = x_2 = \frac{\Pi V}{RT}$$

or

$$\Pi = \left(\frac{T\,\Delta_{\text{vap}} H_{1,\text{m}}}{V T_{\text{b}}^{*2}}\right)\Delta T_{\text{b}} \tag{2.10.9}$$

or

$$\Delta T_{\text{b}} = \left(\frac{V T_{\text{b}}^{*2}}{T\,\Delta_{\text{vap}} H_{1,\text{m}}}\right)\Pi \tag{2.10.10}$$

## Relations Of Depression Of Freezing Point With Osmotic Pressure

Comparing Eq. (2.10.2) with Eq. (2.10.3), we get

$$\frac{\Delta_{\text{fus}} H_{1,\text{m}}}{R}\left(\frac{-\Delta T_{\text{f}}}{T_{\text{f}}^{*2}}\right) = x_2 = \frac{\Pi V}{RT} \tag{2.10.11}$$

or $$\Pi = \left(\frac{T\,\Delta_{\rm fus}H_{1,\rm m}}{VT_{\rm f}^{*2}}\right)(-\Delta T_{\rm f}) \tag{2.10.12}$$

or $$-\Delta T_{\rm f} = \left(\frac{VT_{\rm f}^{*2}}{T\,\Delta_{\rm fus}H_{1,\rm m}}\right)\Pi \tag{2.10.13}$$

---

***Example 2.10.1***

The osmotic pressure of an aqueous solution of sucrose at 303 K is 2.47 atm; the molar volume of water at this temperature is 18.10 cm$^3$. Calculate the elevation of boiling point of this solution. Given: $\Delta_{\rm vap}H_1$ = 539 cal g$^{-1}$.

***Solution***

The given data are

$$\Pi = 2.47\ \text{atm} = 2.47 \times 101\,325\ \text{N m}^{-2} = 250\,273\ \text{N m}^{-2}$$

$$\Delta_{\rm vap}H_{1,\rm m} = (539 \times 18\ \text{cal mol}^{-1})\left(\frac{8.314\ \text{J}}{1.987\ \text{cal}}\right) = 40\,595\ \text{J mol}^{-1}$$

$$T = 303\ \text{K} \qquad T_{\rm b}^{*} = 373\ \text{K} \qquad V_{1,\rm m}^{*} = 18.1 \times 10^{-6}\ \text{m}^3\ \text{mol}^{-1}$$

Substituting the data in the expression

$$\Delta T_{\rm b} = \frac{\Pi V_{1,\rm m}^{*}\, T_{\rm b}^{*2}}{T\,\Delta_{\rm vap}H_{1,\rm m}}$$

we get $$\Delta T_{\rm b} = \frac{(250\,273\ \text{N m}^{-2})\,(18.1 \times 10^{-6}\ \text{m}^3\ \text{mol}^{-1})\,(373\ \text{K})^2}{(303\ \text{K})\,(40\,595\ \text{J mol}^{-1})}$$

$$= 0.051\ \text{K}$$

---

***Example 2.10.2***

The addition of 3 g of a substance to 100 g of $CCl_4$ raises the boiling point of $CCl_4$ by 0.60 K, $K_{\rm b}$ = 5.03 K mol$^{-1}$ kg. Calculate the freezing point depression ($K_{\rm f}$ = 31.8 K kg mol$^{-1}$); the relative vapour pressure lowering; the osmotic pressure of the solution at 298 K, and the molar mass of the substance. The density of $CCl_4$ is 1.59 g cm$^{-3}$.

***Solution***

The given data are

$$m_1 = 100\ \text{g}, \quad m_2 = 3\ \text{g}, \quad \Delta T_{\rm b} = 0.60\ \text{K}, \quad \rho_1(\text{CCl}_4) = 1.59\ \text{g cm}^{-3},$$
$$K_{\rm b} = 5.03\ \text{K kg mol}^{-1}, \quad K_{\rm f} = 31.8\ \text{K kg mol}^{-1}, \quad M_1(\text{CCl}_4) = 154\ \text{g mol}^{-1}$$

Since $\Delta T_{\rm b} = K_{\rm b}m$, we get

$$m = \frac{\Delta T_{\rm b}}{K_{\rm b}} = \frac{0.60\ \text{K}}{(5.03\ \text{K kg mol}^{-1})} = 0.12\ \text{mol kg}^{-1}$$

Since $m = n_2/m_1 = (m_2/M_2)/m_1$, we get

$$M_2 = \frac{m_2}{m\,m_1} = \frac{(3\ \text{g})}{(0.12\ \text{mol kg}^{-1})\,(100\ \text{g})} = 0.25\ \text{kg mol}^{-1}$$

Now $-\Delta T_\mathrm{f} = K_\mathrm{f} m = (31.8 \text{ K kg mol}^{-1})(0.12 \text{ mol kg}^{-1}) = 3.82 \text{ K}$

$$\frac{\Delta p}{p^*} = x_2 = \frac{n_2}{n_1 + n_2} = \frac{(m_2/M_2)}{(m_1/M_1) + (m_2/M_2)}$$

$$= \frac{(0.003 \text{ kg}/0.25 \text{ kg mol}^{-1})}{(0.1 \text{ kg}/0.154 \text{ kg mol}^{-1}) + (0.003 \text{ kg}/0.25 \text{ kg mol}^{-1})} = 0.018$$

$$V^*_{1,\mathrm{m}} = \frac{m_1}{\rho_1} = \frac{0.1 \text{ kg}}{1.59 \text{ kg dm}^{-3}} = 0.062\,9 \text{ dm}^3$$

$$\Pi = \frac{n_2 RT}{V^*_{1,\mathrm{m}}} = \frac{(3 \text{ g}/250 \text{ g mol}^{-1})(8.314 \text{ J K}^{-1} \text{ mol}^{-1})(298 \text{ K})}{(0.062\,9 \text{ dm}^3)}$$

$$= 472.67 \text{ kPa} \equiv 4.66 \text{ atm}$$

## 2.11 COLLIGATIVE PROPERTIES OF STRONG AND WEAK ELECTROLYTES

**Colligative Properties of a Strong Electrolyte**

The colligative properties of an electrolytic solution of a given concentration are found to be larger than those of a nonelectrolytic solution of the same concentration. This is due to the fact that an electrolyte in solution dissociates to give positive and negative charged ions in such a manner that the sum of the charges on positive ions is equal to the sum of the charges on negative ions (condition of electroneutrality). The net result is that the total number of species irrespective of their nature is increased. Since colligative properties depend only on the number of species and not on their nature, it is obvious that the colligative effects produced by an electrolytic solution will be larger than those produced by a nonelectrolytic solution of the same concentration. The colligative properties of a dilute solution of a strong electrolyte which undergoes complete dissociation are found to be approximately an integral multiple of the corresponding values of a nonelectrolytic solution of the same concentration. The value of the integral number is equal to the total number of ions that are produced on dissociation of a molecule of the strong electrolyte.

**Colligative Properties of a Weak Electrolyte**

The colligative properties produced by a solution of a weak electrolyte of known concentration are found to lie between the corresponding properties produced by the same concentration of a nonelectrolytic solution and of a strong electrolytic solution; the stoichiometry of the strong electrolyte is the same as that of the weak electrolyte. This indicates that the weak electrolyte is partially dissociated in the solution.

**Van't Hoff Factor**

The colligative properties of an electrolytic solution can be expressed in terms of those of a nonelectrolytic solution by use of the van't Hoff factor *i*. It is defined as the ratio of colligative effect produced by a given concentration of an electrolytic solution and of a nonelectrolytic solution of the same concentration, i.e.

$$i = \frac{\text{Colligative effect produced by a given concentration of an electrolytic solution}}{\text{Colligative effect produced by the same concentration of a nonelectrolytic solution}} \tag{2.11.1}$$

Thus we may write

$$i = \frac{-\Delta T_\mathrm{f}}{(-\Delta T_\mathrm{f})_0} = \frac{\Delta T_\mathrm{b}}{(\Delta T_\mathrm{b})_0} = \frac{\Pi}{(\Pi)_0} = \frac{\Delta p}{(\Delta p)_0} \qquad (2.11.2)$$

where the quantities without subscript refer to electrolytic solution and those with subscript refer to nonelectrolytic solution of the same concentration. Writing Eq. (2.11.2) more explicitly, we have

$$-\Delta T_\mathrm{f} = i(-\Delta T_\mathrm{f})_0 = i\,K_\mathrm{f}\,m$$
$$\Delta T_\mathrm{b} = i\,(\Delta T_\mathrm{b})_0 = i\,K_\mathrm{b}\,m$$
$$\Pi = i(\Pi)_0 = i\,c_2 RT \qquad (2.11.3)$$

The value of $i$ must be calculated from the experimental data for each electrolyte at various concentrations.

## Variation of van't Hoff Factor with Dilution

**For a Strong Electrolyte**

It is found that for a solution of strong electrolyte, the value of $i$ increases on dilution and approaches the limit of an integral number equal to the ions produced by a molecule of the electrolyte. A few examples are given below.

| *Substance* | *Limiting value of i* |
|---|---|
| (1) HCl, $HNO_3$, $NH_4Cl$, $CuSO_4$ | |
| NaCl | 2 |
| (2) $H_2SO_4$, $CoCl_2$, $K_2SO_4$ | 3 |
| (3) $K_3[Fe(CN)_6]$ | 4 |
| (4) $K_4[Fe(CN)_6]$ | 5 |

**For a Weak Electrolyte**

The variation of $i$ for a solution of a weak electrolyte is more wide in nature, i.e. it increases more rapidly with dilution and attains the limiting value only at infinite dilution.

**Explanation for Weak Electrolytes**

The variation of van't Hoff factor $i$ for a weak electrolyte can be explained on the basis of Arrhenius theory of electrolytic dissociation. According to this theory, a weak electrolyte is in equilibrium with its ions. Thus, for $A_xB_y$ we have

$$A_xB_y \rightleftharpoons xA^{y+} + yB^{x-} \qquad (2.11.4)$$

Let $\alpha$ be the degree of dissociation, then the molalities of various species at equilibrium will be

$$m(A_xB_y) = m(1-\alpha); \quad m(A^{y+}) = m(x\alpha); \quad m(B^{x-}) = m(y\alpha)$$

where $m$ is the original molality of the electrolyte.

The total molality $m_\mathrm{t}$ of substances of all types present in the solution is

$$m_\mathrm{t} = m(1-\alpha) + m(x\alpha) + m(y\alpha) = m[1 + \alpha(x + y - 1)] \qquad (2.11.5)$$
$$= m[1 + \alpha(\nu - 1)]$$

where $\nu$ is the total number of ions that are produced when an electrolytic molecule completely dissociates in the solution and is equal to $(x + y)$.

The value of $\alpha$ increases on dilution (Ostwald dilution law) and thus the molality of the solution increases on dilution. Consequently, the colligative effects also increase on dilution. Taking, for example, the depression of freezing point, we have

$$-\Delta T_f = K_f m_t = K_f m[1 + \alpha(\nu - 1)]$$

Solving for $\alpha$, we obtain

$$\alpha = \frac{(-\Delta T_f / K_f m) - 1}{\nu - 1} = \frac{(-\Delta T_f) - K_f m}{(\nu - 1)K_f m} \tag{2.11.6}$$

The degree of dissociation $\alpha$ can also be written in terms of van't Hoff factor $i$. Since for an electrolytic solution

$$-\Delta T_f = iK_f m = K_f m_t$$

Hence $\quad i\,m = m_t = m[1 + \alpha(\nu - 1)]$

or $\quad i = 1 + \alpha(\nu - 1)$

Thus
$$\alpha = \frac{i - 1}{\nu - 1} \tag{2.11.7}$$

In fact, Eq. (2.11.7) is applicable to any of the colligative properties and can equally be derived from other colligative properties. This equation provides a method to calculate the degree of dissociation from the van't Hoff factor $i$ for a given weak electrolyte.

**Explanation for Strong Electrolytes**

The variation of colligative effects of a solution of a strong electrolyte on dilution is not due to the increase in number of species on dilution since it is known that strong electrolyte is already present in the completely ionized form. The variation, however, can be explained on the basis of Debye-Hückel theory of interionic-attraction according to which each ion in the solution is surrounded on an average with more of the oppositely charged ions than like charges. In other words, there exists an ionic atmosphere around each and every ion whose net charge is opposite to that of the central ion. The properties of the electrolyte are determined by the interaction of the central ion with its atmosphere. For a given solvent and at a fixed temperature, the interaction between the ions depend only on the charges of the ions and their concentrations and not at all on their nature.

The interaction of the ions can be represented by a quantity, called the *ionic strength* of the solution. It is defined as

$$\mu = \frac{1}{2}\sum_i c_i z_i^2 \tag{2.11.8}$$

where $c_i$ is the concentration of the $i$th ions in mol per dm$^3$ and $z_i$ is its charge number. The summation is to be carried over all types of ions present in the solution.

Debye-Hückel showed that the variation of van't Hoff factor on dilution is due to the variation in ionic strength of the solution on dilution. The expression relating these two is

$$i = \nu(1 - 0.375\, z_+ |z_-| \sqrt{\mu/\mu^\circ}) \tag{2.11.9}$$

where $\nu$ is the number of ions yielded by a given molecule of electrolyte.

The agreement between the calculated values of $i$ and the experimental values is good for very dilute solutions up to molarity equal to 0.01 mol dm$^{-3}$. The deviations increase rapidly when molarity is greater than 0.01 mol dm$^{-3}$ and are larger for electrolytes with higher values of $z_+|z_-|$.

---

***Example 2.11.1*** A solution containing 0.122 kg of benzoic acid in 1 kg of benzene (b. pt. 353 K) boils at 354.5 K. Determine the apparent molar mass of benzoic acid (which dimerizes) in the solution and the degree of dimerization. Given: $\Delta_{vap}H$(benzene) = 394.57 J g$^{-1}$.

***Solution*** The given data are

$$m_1 = 1\ \text{kg}, \quad m_2 = 0.122\ \text{kg}, \quad T_b^* = 353\ \text{K}, \quad T_b = 354.5\ \text{K},$$

$$M_1 = 78\ \text{g mol}^{-1}, \quad \Delta_{vap}H_{1,m} = 394.57 \times 78\ \text{J mol}^{-1}$$

The value of $K_b$(benzene) is

$$K_b = \frac{M_1 R T_b^{*2}}{\Delta_{vap}H_{1,m}} = \frac{(0.078\ \text{kg mol}^{-1})(8.314\ \text{J K}^{-1}\ \text{mol}^{-1})(353\ \text{K})^2}{(394.57 \times 78\ \text{J mol}^{-1})}$$

$$= 2.626\ \text{K kg mol}^{-1}$$

The molality of the solution is

$$m = \frac{\Delta T_b}{K_b} = \frac{T_b - T_b^*}{K_b} = \frac{(354.5\ \text{K} - 353\ \text{K})}{(2.626\ \text{K kg mol}^{-1})} = 0.57\ \text{mol kg}^{-1}$$

Since $m = n_2/m_1 = (m_2/M_2)/m_1$, we get

$$M_2 = \frac{m_2}{m m_1} = \frac{0.122\ \text{kg}}{(0.57\ \text{mol kg}^{-1})(1\ \text{kg})} = 0.214\ \text{kg mol}^{-1}$$

Let $\alpha$ be the degree of dimerization of benzoic acid. Since the solution contains 1 mol of benzoic acid, we will have

$$\underset{(1-\alpha)\ \text{mol}}{2C_6H_5COOH} \rightleftharpoons \underset{(\alpha/2)\ \text{mol}}{(C_6H_5COOH)_2}$$

Total amount of solutes = $\{(1 - \alpha) + \alpha/2)\}$ mol = $(1 - \alpha/2)$ mol

The number average molar mass of solutes in solution is given by

$$\bar{M}_n = \frac{N_1 M_1 + N_2 M_2}{N_1 + N_2} = \frac{n_1 M_1 + n_2 M_2}{n_1 + n_2} = \frac{m}{n_1 + n_2} = \frac{0.122\ \text{kg}}{(1 - \alpha/2)\ \text{mol}}$$

Equating this to 0.214 kg mol$^{-1}$, we get

$$\frac{0.122}{1 - \alpha/2} = 0.214 \quad \text{or} \quad \alpha = 2(1 - 0.122/0.214) = 0.86$$

---

***Example 2.11.2*** A solution containing 0.011 kg of barium nitrate in 0.1 kg of water boils at 100.46 °C. Calculate the degree of ionization of the salt. $K_b$(water) = 0.52 K kg mol$^{-1}$.

***Solution*** The given data are

$$m_1 = 0.1\ \text{kg}, \quad m_2 = 0.011\ \text{kg}, \quad T_b = 100.46\ ^\circ\text{C}, \quad T_b^* = 100\ ^\circ\text{C},$$

$$M_2 = M_{Ba(NO_3)_2} = 261.3\ \text{g mol}^{-1}$$

The molality of the solution is

$$m = \frac{n_2}{m_1} = \frac{m_2/M_2}{m_1} = \frac{(0.011\ \text{kg}/0.261\,3\ \text{kg mol}^{-1})}{(0.1\ \text{kg})} = 0.421\ \text{mol kg}^{-1}$$

The value of $\Delta T_b$ assuming the solute to be nonelectrolyte is

$$(\Delta T_b)_0 = K_b m = (0.52\ \text{K kg mol}^{-1})\,(0.421\ \text{mol kg}^{-1}) = 0.22\ \text{K}$$

The value of van't Hoff factor is

$$i = \frac{\Delta T_b}{(\Delta T_b)_0} = \frac{0.46\ \text{K}}{0.22\ \text{K}} = 2.09$$

Finally, the degree of dissociation of the electrolyte is

$$\alpha = \frac{i-1}{\nu-1} = \frac{2.09-1}{3-1} = \frac{1.09}{2} = 0.55$$

***Example 2.11.3***

When 3.24 g of mercuric nitrate $Hg(NO_3)_2$ is dissolved in 1 kg of water, the freezing point of the solution is found to be – 0.055 8 °C. When 10.84 g of mercuric chloride $HgCl_2$ is dissolved in 1 kg of water, the freezing point of the solution is – 0.074 4 °C. $K_f$ = 1.86 mol$^{-1}$ K kg. Will either of these dissociate into ions in an aqueous solution?

***Solution***

Molar mass of $Hg(NO_3)_2$ = 324.6 g mol$^{-1}$

Molality of $Hg(NO_3)_2$ solution, $m = \frac{n_2}{m_1} = \frac{m_2/M_2}{m_1} = \frac{(3.24\ \text{g}/324.6\ \text{g mol}^{-1})}{(1\ \text{kg})}$

$$= 0.01\ \text{mol kg}^{-1}$$

We will have

$$(-\Delta T_f)_0 = K_f m = (1.86\ \text{K kg mol}^{-1})(0.01\ \text{mol kg}^{-1}) = 0.018\,6\ \text{K}$$

$$i = \frac{(-\Delta T_f)}{(-\Delta T_f)_0} = \frac{0.055\,8\ \text{K}}{0.018\,6\ \text{K}} = 3$$

$$\alpha = \frac{i-1}{\nu-1} = \frac{3-1}{3-1} = 1$$

that is, the degree of dissociation of $Hg(NO_3)_2$ is one and thus the salt is completely ionized in the solution.

For mercuric chloride, we have

Molar mass of $HgCl_2$ = 271.5 g mol$^{-1}$

Molality of $HgCl_2$ solution, $m = \frac{m_2/M_2}{m_1} = \frac{(10.84\ \text{g}/271.5\ \text{g mol}^{-1})}{(1\ \text{kg})}$

$$= 0.04\ \text{mol kg}^{-1}$$

$$(-\Delta T_f)_0 = K_f m = (1.86\ \text{K kg mol}^{-1})(0.04\ \text{mol kg}^{-1}) = 0.074\ 4\ \text{K}$$

Since $(\Delta T_f)_0 = \Delta T_f$, the salt $HgCl_2$ does not ionize in the solution.

***Example 2.11.4***

The freezing point of a solution of acetic acid (amount fraction = 0.02) in benzene is 277.4 K. Acetic acid exists partly as a dimer

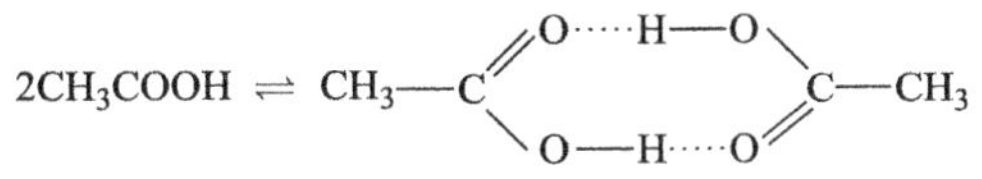

Calculate the equilibrium constant for the dimerization. The melting point of pure benzene is 278.4 K, its molar enthalpy of fusion is 10.042 kJ mol$^{-1}$.

***Solution***

The value of $K_f$(benzene) is

$$K_f = \frac{M_1 R T_f^{*2}}{\Delta_{fus} H_{1,m}} = \frac{(78 \times 10^{-3}\ \text{kg})\ (8.134\ \text{J K}^{-1}\ \text{mol}^{-1})\ (278.4\ \text{K})^2}{(10\ 042 \times 10^3\ \text{J mol}^{-1})}$$

$$= 5.0\ \text{K kg mol}^{-1}$$

Molality of the solution is

$$m = \frac{(-\Delta T_f)}{K_f} = \frac{(1.0\ \text{K})}{(5.0\ \text{K kg mol}^{-1})} = 0.2\ \text{mol kg}^{-1}$$

Given amount fraction of acetic acid = 0.02

For such a dilute solution, we may write

$$x_2 = \frac{n_2}{n_1 + n_2} \simeq \frac{n_2}{n_1} = \frac{n_2}{(m_1/M_1)} = (n_2/m_1)M_1 = mM_1$$

Hence $\quad m = \dfrac{x_2}{M_1} = \dfrac{0.02}{78 \times 10^{-3}\ \text{kg mol}^{-1}} = 0.257\ \text{mol kg}^{-1}$

Thus, we have

Given: $\quad m_m + 2m_d = 0.257\ \text{mol kg}^{-1}$

Found: $\quad m_m + m_d = 0.2\ \text{mol kg}^{-1}$

where the subscripts m and d stand for monomer and dimer, respectively. Solving for $m_m$ and $m_d$, we get

$$m_m = 0.143\ \text{mol kg}^{-1} \quad \text{and} \quad m_d = 0.057\ \text{mol kg}^{-1}$$

The equilibrium constant for the dimerization of acetic acid is

$$K_{eq}^{\circ} = \frac{(m_d/m^{\circ})}{(m_m/m^{\circ})^2} = \frac{0.057}{(0.143)^2} = 2.79$$

---

***Example 2.11.5***

The freezing point of an aqueous solution of KCN containing 0.189 mol kg$^{-1}$ was – 0.704 °C. On adding 0.095 mol of $Hg(CN)_2$, the freezing point of the solution became – 0.530 °C. Assuming that the complex is formed according to the equation

$$Hg(CN)_2 + xCN^- \rightarrow Hg(CN)_{x+2}^{x-}$$

find the formula of the complex.

***Solution***

Molality of the solution containing only KCN is

$$m = \frac{(-\Delta T_f)}{K_f} = \frac{(0.704\ \text{K})}{(1.86\ \text{K kg mol}^{-1})} = 0.379\ \text{mol kg}^{-1}$$

This is just double of the given molality (= 0.189 mol $kg^{-1}$) of KCN, indicating complete dissociation of KCN. Molality of the solution after the formation of the complex

$$m = \frac{(-\Delta T_f)}{K_f} = \frac{(0.530\ \text{K})}{(1.86\ \text{K kg mol}^{-1})} = 0.285\ \text{mol kg}^{-1}$$

If it be assumed that the whole of $Hg(CN)_2$ is converted into complex, then the amounts of various species in 1 kg of solvent after the formation of the complex will be

$$n(\text{K}^+) = 0.189\ \text{mol}, \quad n(\text{CN}^-) = (0.189 - x)\ \text{mol}$$

$$n(\text{Hg(CN)}_{x+2}^{x-}) = 0.095\ \text{mol}$$

Total amount of species in 1 kg solvent becomes

$$n_{\text{total}} = [0.189 + (0.189 - x) + 0.095]\ \text{mol} = (0.473 - x)\ \text{mol}$$

Equating this to 0.285 mol, we get

$$(0.473 - x)\ \text{mol} = 0.285\ \text{mol}$$

i.e. $\quad x = (0.473 - 0.285) = 0.188$

$$\text{Number of CN}^- \text{ units combined} = \frac{0.188\ \text{mol}}{0.095\ \text{mol}} = 2$$

Thus, the formula of the complex is $Hg(CN)_4^{2-}$.

***Example 2.11.6***

The vapour pressure of a solution containing 6.69 g of $Mg(NO_3)_2$ dissolved in 100 g of water is 747 Torr at 373 K. Calculate the degree of dissociation of the salt in the solution.

***Solution***

Molar mass of $Mg(NO_3)_2$ = 148.31 g $mol^{-1}$

Amount fraction of the solute in the solution is

$$x_2 = \frac{n_2}{n_1 + n_2} = \frac{(m_2/M_2)}{(m_1/M_1) + (m_2/M_2)} = \frac{(6.69/148.31)}{(100/18) + (6.69/148.3)}$$

$$= \frac{0.045}{5.556 + 0.045} = 8.03 \times 10^{-3}$$

The depression in the vapour pressure is

$$(\Delta p)_0 = x_2 p^* = (8.03 \times 10^{-3})(760\ \text{Torr}) = 6.10\ \text{Torr}$$

The value of van't Hoff factor is

$$i = \frac{\Delta p}{(\Delta p)_0} = \frac{(760 - 747)\ \text{Torr}}{(6.10\ \text{Torr})} = 2.13$$

The degree of dissociation of the salt is

$$\alpha = \frac{i - 1}{\nu - 1} = \frac{2.13 - 1}{3 - 1} = \frac{1.13}{2} = 0.565$$

***Example 2.11.7***

A solution is 0.5 molar in $MgSO_4$, 0.1 molar in $AlCl_3$ and 0.2 molar in $(NH_4)_2SO_4$. What is the total ionic strength?

***Solution***

The ionic strength of the solution is given by

$$\mu = \frac{1}{2}\sum_i c_i z_i^2$$

where $c_i$ is the molar concentration of the *i*th ion and $z_i$ is its charge number. Thus

$$\mu = \frac{1}{2}[(c_{Mg^{2+}})(z_{Mg^{2+}})^2 + (c_{SO_4^{2-}})(z_{SO_4^{2-}})^2 + (c_{Al^{3+}})(z_{Al^{3+}})^2 + (c_{Cl^-})(z_{Cl^-})^2 + (c_{NH_4^+})(z_{NH_4^+})^2]$$

Substituting the given data, we get

$$\mu/\text{mol dm}^{-3} = \frac{1}{2}[(0.5)(2)^2 + (0.7)(-2)^2 + (0.1)(3)^2 + (0.3)(-1)^2 + (0.4)(1)^2]$$
$$= \frac{1}{2}[2.0 + 2.8 + 0.9 + 0.3 + 0.4] = \frac{1}{2}(6.4)$$
$$= 3.2$$

---

***Example 2.11.8***

A 0.2 molar aqueous solution of KCl freezes at – 0.680 °C. (a) Calculate the van't Hoff factor *i* and the apparent degree of dissociation of KCl. (b) Using the Debye-Hückel equation, calculate the value of the van't Hoff factor and compare it with the value obtained in part (a), (c) Calculate the expected values of $\Delta T_b$ and $\Pi$ at 0 °C using the value of *i* obtained in part (a). Assume molality and molarity have the same value.

***Solution***

(**a**) We have

$$(-\Delta T_f)_0 = K_f m = (1.86\ \text{K kg mol}^{-1})(0.2\ \text{mol kg}^{-1}) = 3.72\ \text{K}$$

$$i = \frac{(-\Delta T_f)}{(-\Delta T_f)_0} = \frac{0.680\ \text{K}}{0.372\ \text{K}} \simeq 1.83$$

The apparent degree of dissociation of KCl is

$$\alpha = \frac{i-1}{\nu-1} = \frac{1.83-1}{2-1} = 0.83$$

(**b**) The ionic strength of the solution is given by

$$\mu = \frac{1}{2}\sum_i c_i z_i^2$$

Substituting the given data, we get

$$\mu/\text{mol dm}^{-3} = \frac{1}{2}[(0.2)(1)^2 + (0.2)(-1)^2] = 0.2$$

According to Debye-Hückel theory, we get

$$i = \nu(1 - 0.375\, z_+ \,|\, z_- \,|\, \sqrt{\mu/\mu^\circ})$$

$$= 2[1 - (0.375)(1)(1)\sqrt{0.2}] = 2\,(1 - 0.168)$$

$$= 1.644$$

(c) $$\Delta T_b = i\,(\Delta T_b)_0 = i\,K_b m$$

$$= (1.83)(0.52 \text{ K kg mol}^{-1})(0.2 \text{ mol kg}^{-1})$$

$$= 0.19 \text{ K}$$

$$\Pi = i(\Pi)_0 = icRT$$

$$= (1.83)(0.2 \times 10^3 \text{ mol m}^{-3})(8.314 \text{ J K}^{-1} \text{ mol}^{-1})\,(273 \text{ K})$$

$$= 8.307 \times 10^5 \text{ Pa} \equiv (8.307 \times 10^5 \text{ Pa})\left(\frac{1 \text{ atm}}{1.01325 \text{ Pa}}\right) = 8.2 \text{ atm}$$

---

## 2.12 SOLUBILITY OF A SOLUTE IN AN IDEAL SOLUTION

**Thermodynamic Derivation of Ideal Law of Solubility**

At saturation point, where no more solute dissolves in a solution, there exists an equilibrium between the solute in solution and the pure solid solute, i.e.

$$\text{Solute in solution} \rightleftharpoons \text{Pure solid solute}$$

The thermodynamic condition of equilibrium is

$$\mu_{2(\text{sol})}(T, p, x_2) = \mu^*_{2(\text{s})}(T, p) \tag{2.12.1}$$

where $x_2$ is the amount fraction of solute in the saturated solution. If the solution is ideal, then

$$\mu_{2(\text{sol})}(T, p, x_2) = \mu^*_{2(\text{l})}(T, p) + RT \ln x_2 \tag{2.12.2}$$

Substituting Eq. (2.12.2) in Eq. (2.12.1) and rearranging, we have

$$\mu^*_{2(\text{l})} + RT \ln x_2 = \mu^*_{2(\text{s})}$$

or $$\ln x_2 = \frac{\mu^*_{2(\text{s})} - \mu^*_{2(\text{l})}}{RT} = -\frac{\Delta_{\text{fus}}\mu^*_2}{RT} \tag{2.12.3}$$

According to Eq. (2.12.3), the amount fraction of solute in the saturated solution is a function of the temperature of the solution. In order to determine the functionality, we differentiate Eq. (2.12.3) with respect to $x_2$, keeping $p$ constant. Thus, we have

$$\frac{1}{x_2} = -\frac{1}{R}\left\{\frac{\partial(\Delta_{\text{fus}}\mu^*_2/T)}{\partial T}\right\}_p \left\{\frac{\partial T}{\partial x_2}\right\}_p$$

Using Gibbs-Helmholtz equation, we get

$$\frac{1}{x_2} = \frac{\Delta_{\text{fus}}H_{2,\text{m}}}{RT^2}\left(\frac{\partial T}{\partial x_2}\right)_p$$

or $$\frac{dx_2}{x_2} = \frac{\Delta_{fus}H_{2,m}}{R}\frac{dT}{T^2} \qquad (p \text{ constant}) \qquad (2.12.4)$$

To determine the variation of $x_2$ with $T$, we integrate the above expression to give

$$\int_1^{x_2} \frac{dx}{x_2} = \int_{T_f^*}^{T} \frac{\Delta_{fus}H_{2,m}}{R}\frac{dT}{T^2} \qquad (2.12.5)$$

where $T_f^*$ is the freezing point of the pure solute ($x_2 = 1$) and $x_2$ is the solubility of the solute at temperature $T$ of the solution.

Assuming $\Delta_{fus}H_{2,m}$ to be independent of temperature, Eq. (2.12.5) yields

$$\ln x_2 = -\frac{\Delta_{fus}H_{2,m}}{R}\left(\frac{1}{T} - \frac{1}{T_f^*}\right) \qquad (2.12.6)$$

Equation (2.12.6) is an expression of the *ideal law of solubility*. According to this law, the solubility of a substance depends only on the characteristics of the solute through $\Delta_{fus}H_{2,m}$ and $T_f^*$ and is independent of the nature of the solvent.

**Factors Determining Solubility**

The following dependence on the solubility of the solute may be concluded.

- A comparatively lesser value of $T_f^*$ (i.e. low freezing point of the solute) makes the right side of Eq. (2.12.6) more positive indicating larger value of $x_2$ (i.e. more is the solubility).
- A comparatively lesser value of $\Delta_{fus}H_{2,m}$ makes the right side of Eq. (2.12.6) less negative indicating larger value of $x_2$ (i.e. more is the solubility).
- Increasing the value of $T$ also makes the right side of Eq. (2.12.6) lesser negative indicating larger value of $x_2$ (i.e. more is the solubility).

**Diagrammatic Representation of Solubility**

Equation (2.12.6) can be written as

$$\ln x_2 = \frac{\Delta_{fus}S_{2,m}}{R}\left(1 - \frac{T_f^*}{T}\right) \qquad (2.12.7)$$

where $\Delta_{fus}S_{2,m}$ is the molar entropy of fusion of solute. Figure 2.12.1 shows the variation of solubility, $x$, as a function of temperature for two substances with the same entropy of fusion but different freezing points. Larger the value $T_f^*$, lesser the solubility of the substance.

**Limitation of Eq. (2.12.6)**

In deriving Eq. (2.12.6), we have assumed that $\Delta_{fus}H_{2,m}$ remains constant within the temperature range $T_f^*$ and $T$. This is true only if the temperature of the solution is not far away from the melting point of the solute. If the temperature of the solution is very far away from $T_f^*$, then the solubility as calculated from Eq. (2.12.6) is no longer accurate.

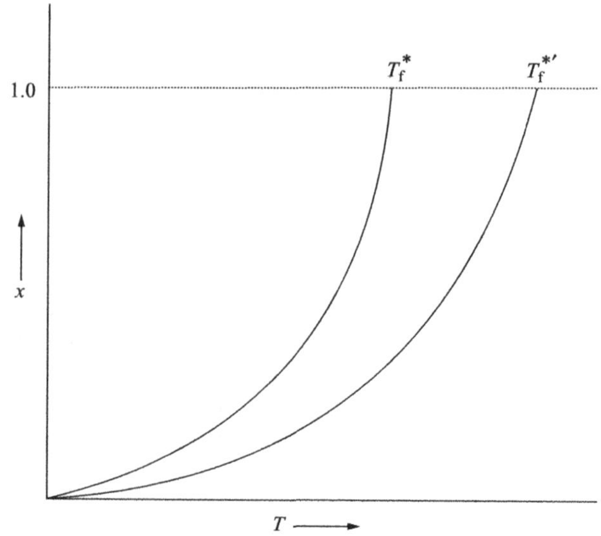


**Fig. 2.12.1 Variation of solubility with temperature**

## Correlating Solubilities at Two Temperatures

If $x_1$ and $x_2$ are the solubilities of solute at temperatures $T_1$ and $T_2$, respectively, we can write

$$\ln x_1 = -\frac{\Delta_{\text{fus}} H_{2,\text{m}}}{R}\left(\frac{1}{T_1} - \frac{1}{T_f^*}\right) \tag{2.12.8}$$

$$\ln x_2 = -\frac{\Delta_{\text{fus}} H_{2,\text{m}}}{R}\left(\frac{1}{T_2} - \frac{1}{T_f^*}\right) \tag{2.12.9}$$

Subtracting Eq. (2.12.8) from Eq. (2.12.9), we get

$$\ln \frac{x_2}{x_1} = -\frac{\Delta_{\text{fus}} H_{2,\text{m}}}{R}\left(\frac{1}{T_2} - \frac{1}{T_1}\right) \tag{2.12.10}$$

Equation (2.12.10) relates the solubility of the solute in a solvent with temperature of the solution.

---

***Example 2.12.1***

The molar enthalpy of fusion of naphthalene is 10.0 kJ mol$^{-1}$ at 298.15 K and its normal melting point is 353.35 K. Calculate the solubility of naphthalene at 298.15 K in a liquid with which it forms an ideal solution.

***Solution***

The given data are

$\Delta_{\text{fus}} H_{2,\text{m}} = 10.0\ \text{kJ mol}^{-1}$, $\quad T = 298.15\ \text{K}$, $\quad T_f^* = 353.35\ \text{K}$

Substituting the given data in the expression

$$\ln x_2 = -\frac{\Delta_{\text{fus}} H_{2,\text{m}}}{R}\left(\frac{1}{T} - \frac{1}{T_f^*}\right)$$

we get $$\ln x_2 = -\frac{(10^4\ \text{J mol}^{-1})}{(8.314\ \text{J K}^{-1}\ \text{mol}^{-1})}\left(\frac{1}{298.15\ \text{K}} - \frac{1}{353.35\ \text{K}}\right) = -0.273\,6$$

$$x_2 = 0.532\,6$$

---

***Example 2.12.2***

Given the following data, compute the molar enthalpy of fusion of sucrose assuming ideal solution.

Saturation solubility at 298 K = 6.2 mol per kg of solvent

Freezing point of solute = 473 K.

***Solution***

The given data are

$$T = 298\ \text{K}, \qquad T_f^* = 473\ \text{K}$$

$$x_2 = \frac{n_2}{n_1 + n_2} = \frac{(6.2\ \text{mol})}{(1\ 000/18)\ \text{mol} + (6.2\ \text{mol})} = 0.100\ 4$$

Substituting the data in the expression

$$\ln x_2 = -\frac{\Delta_{\text{fus}}H_{2,\text{m}}}{R}\left(\frac{1}{T} - \frac{1}{T_f^*}\right)$$

we get

$$\ln 0.100\ 4 = -\frac{\Delta_{\text{fus}}H_{2,\text{m}}}{(8.314\ \text{J K}^{-1}\ \text{mol}^{-1})}\left(\frac{1}{298\ \text{K}} - \frac{1}{473\ \text{K}}\right)$$

This gives

$$\Delta_{\text{fus}}H_{2,\text{m}} = -\frac{(\ln 0.100\ 4)(8.314\ \text{J K}^{-1}\ \text{mol}^{-1})(298\ \text{K})(473\ \text{K})}{(473\ \text{K} - 298\ \text{K})}$$

$$= 15\ 422\ \text{J mol}^{-1}$$

## REVISIONARY PROBLEMS

2.1 What do you understand by the terms (a) solution, (b) solute molecules in a solution, and (c) colligative properties?

2.2 (a) Explain the following concentration units.

(1) Mass %. (2) Volume %. (3) Molarity. (4) Molality. (5) Normality. (6) Amount fraction.

(b) How will you convert:

(1) Amount fraction into molarity and vice versa?

(2) Amount fraction into molality and vice versa?

(3) Molarity into molality and vice versa?

(c) Explain which of the concentration units given in part (a) vary with temperature.

2.3 Derive the following conversion expressions of concentration units.

| | |
|---|---|
| Amount fraction into molarity | $M = x_2\,\rho/(x_1M_1 + x_2M_2)$ |
| Molarity into amount fraction | $x_2 = MM_1/\{\rho + M(M_1 - M_2)\}$ |
| Amount fraction into molality | $m = x_2/(x_1M_1)$ |
| Molality into amount fraction | $x_2 = mM_1/(1 + mM_1)$ |
| Molarity into molality | $m = M/(\rho - MM_2)$ |
| Molality into molarity | $M = m\rho/(1 + mM_2)$ |

Show that for a very dilute solution, molality of the solution is virtually identical with its molarity.

2.4 (a) Describe the experimental behaviour regarding the variation of vapour pressure when a nonvolatile solute is added to a volatile solvent.
(b) What is the characteristic of the curve of vapour pressure versus amount fraction of solute $x_2$ as $x_2 \to 0$?
(c) What do you understand by an ideal solution? Explain, why a real solution deviates from the ideal behaviour as $x_2$ is increased.
(d) State Raoult's law as applicable to ideal solutions. Under what conditions does a real solution follow Raoult's law?

2.5 Show that the relative lowering of vapour pressure of a solvent is equal to the amount fraction of the nonvolatile solute and hence show that this relative lowering is a colligative property. Discuss how the relative lowering of vapour pressure of a dilute solution can be utilized in determining the molar mass of the solute.

2.6 Show that the chemical potential for the solute and solvent in an ideal solution is given by

$$\mu_{i(\text{sol})} = \mu_i^* + RT \ln x_i$$

2.7 Show that the origin of colligative properties such as elevation of boiling point, depression of freezing point and osmotic pressure lies in the relation

$$\mu_{1(\text{sol})} = \mu_{1(\text{l})}^* + RT \ln x_1$$

where $\mu_{1(\text{sol})}$ is the chemical potential of the solvent in the solution, $\mu_{1(\text{l})}^*$ is that of pure liquid solvent and $x_1$ is the amount fraction of the solvent in the solution.

2.8 Draw chemical potential versus temperature diagrams for a solvent in solid, liquid and vapour phases. What changes in the above diagrams occur when a nonvolatile solute is added to the solvent?
Explain with the help of these diagrams that there occurs (a) an elevation in boiling point and (b) a depression in freezing point of the solvent when a nonvolatile solute is added to the solvent. Show also that the depression in freezing point is larger than the elevation in boiling point.

2.9 (a) Derive thermodynamically the relation

$$\frac{1}{T_\text{b}} = \frac{1}{T_\text{b}^*} + \frac{RT \ln x_1}{\Delta_\text{vap} H_{1,\text{m}}}$$

where $T_\text{b}$ is the boiling point of the solution in which the amount fraction of the solvent is $x_1$ and $T_\text{b}^*$ is the boiling point of the pure solvent.
(b) Starting from the relation

$$\frac{\text{d}x_1}{x_1} = -\frac{\Delta_\text{vap} H_{1,\text{m}}}{RT_\text{b}^2} \text{d}T_\text{b}$$

derive the relation

$$\text{d}T_\text{b} = K_\text{b} \, \text{d}m \qquad \text{i.e.} \qquad \Delta T_\text{b} = K_\text{b} m$$

as applicable to dilute solutions. Explain the meaning of the term $K_\text{b}$ and the factors on which its value depends. What is the unit of $K_\text{b}$?

2.10 (a) Derive thermodynamically the relation

$$\frac{\text{d}x_1}{x_1} = \frac{\Delta_\text{fus} H_{1,\text{m}}}{R} \frac{\text{d}T_\text{f}}{T_\text{f}^2}$$

and its integrated form,

$$\frac{1}{T_\text{f}} = \frac{1}{T_\text{f}^*} - \frac{R \ln x_1}{\Delta_\text{fus} H_{1,\text{m}}}$$

where $T_f$ is the freezing point of the solution, $T_f^*$ is that of the pure solvent and $x_1$ is the amount fraction of the solvent in the solution.
(b) Starting from the relation

$$\frac{dx_1}{x_1} = \frac{\Delta_{fus}H_{1,m}}{R}\frac{dT_f}{T_f^2}$$

derive the relation

$$-\,dT_f = K_f dm \qquad \text{i.e.} \qquad -\,\Delta T_f = K_f m$$

On what factors does the value of $K_f$ depend? What is its unit?

2.11 (a) What do you understand by the osmotic pressure of the solution? Derive thermodynamically the relation

$$\Pi V = -\,RT \ln x_1$$

where $\Pi$ is the osmotic pressure, $V$ is the molar volume of the pure solvent and $x_1$ is the amount fraction of the solvent in solution.
(b) Show that the above relation can be simplified to

$$\Pi = cRT$$

where $c$ is the molar concentration of the solute in the solution. Discuss the types of approximations made in deriving the above relation.
(c) Explain why the osmotic pressure measurement is preferred for the determination of molar mass of polymers, proteins and colloids over the other colligative properties. Discuss whether the molar mass obtained is the number average molar mass or the mass average molar mass.

2.12 Derive the following relations connecting the different colligative properties.

$$\frac{p^* - p}{p^*} = \frac{\Delta_{vap}H_{1,m}}{R}\left(\frac{\Delta T_b}{T_b^{*2}}\right) = \frac{\Delta_{fus}H_{1,m}}{R}\left(\frac{-\,\Delta T_f}{T_f^{*2}}\right) = \frac{\Pi V}{RT}$$

$$\frac{\Delta T_b}{-\,\Delta T_f} = \frac{\Delta_{fus}H_{1,m}}{T_f^{*2}}\,\frac{T_b^{*2}}{\Delta_{vap}H_{1,m}}$$

$$\Delta T_b = \frac{\Pi V T_b^{*2}}{T\,\Delta_{vap}H_{1,m}} \qquad \text{and} \qquad -\,\Delta T_f = \frac{\Pi V T_f^{*2}}{T\,\Delta_{fus}H_{1,m}}$$

where $V = V_{1,m}^*$.

2.13 (a) Discuss why the colligative properties of a solution containing either a strong electrolyte or a weak electrolyte are larger than those of a solution containing nonelectrolyte of the same concentration. What effect does the dilution have on the colligative properties of an electrolytic solution?
(b) What is the van't Hoff factor $i$? Show that this is given as

$$i = \frac{-\,\Delta T_f}{(-\,\Delta T_f)_0} = \frac{\Delta T_b}{(\Delta T_b)_0} = \frac{\Pi}{(\Pi)_0} = \frac{\Delta p}{(\Delta p)_0}$$

where the quantities without subscript refer to the electrolytic solution and those with subscript refer to the nonelectrolytic solution of the same concentration.

(c) Discuss how the Arrhenius theory of electrolytic dissociation for a weak electrolyte and the Debye-Hückel theory of interionic attraction for a strong electrolyte explain the variation of colligative properties on dilution? Derive the relation

$$\alpha = \frac{i-1}{\nu-1}$$

where $\alpha$ is the degree of dissociation, $i$ is the van't Hoff factor and $\nu$ is the total number of ions that are produced on complete dissociation of the electrolytic molecule.

2.14 (a) Show that the ideal solubility expression at temperature $T$ is given by

$$\ln x_2 = -\frac{\Delta_{\text{fus}}H_{2,\text{m}}}{R}\left[\frac{1}{T}-\frac{1}{T_{\text{f}}^*}\right]$$

where $x_2$ is the amount fraction of the solute in the saturated solution, $\Delta_{\text{fus}}H_{2,\text{m}}$ is the molar enthalpy of fusion of the pure solute and $T_{\text{f}}^*$ is the freezing point of the pure solute. Discuss the approximations made in deriving the above relation.

(b) On what factors does the ideal solubility of the solute depend?

## TRY YOURSELF PROBLEMS

2.1 In the Rast method of determining molar mass, camphor is used as the solvent. Outline the merits of choosing camphor as the solvent. How will you determine the value of $K_{\text{f}}$ for camphor theoretically?

2.2 Osmotic pressure measurements are preferred over the other colligative properties to determine the molar mass of high polymers. Give reasons.

2.3 The accurate determination of molar mass is done by the method of extrapolation where the plot of $(\Pi/c)$ versus $c$ is extrapolated to zero concentration. Justify this procedure.

2.4 Justify the statements:

(a) The classification of solute and solvent in a solution is a matter of convenience only.

(b) Molality of a solution is independent of temperature of the solution.

(c) The expressions $-\Delta T_{\text{f}} = K_{\text{f}}\, m$, $\Delta T_{\text{b}} = K_{\text{b}}\, m$ and $\Pi = cRT$ are applicable only for dilute solutions.

(d) For solvents obeying Trouton's rule, the values of $K_{\text{b}}$ and $K_{\text{f}}$ are directly proportional to the factor $(MT^*)$.

(e) The solubility as calculated from the ideal solubility law shows large deviation from the observed value if the temperature of interest is far below the melting point of the solid.

(f) Colligative properties of a given electrolytic solution are larger than the corresponding properties produced by the same concentration of a nonelectrolytic solution.

2.5 Discuss what types of graphs do you expect for the following. What will be their slopes and intercepts?

(a) $\ln x_1$ versus $1/T_{\text{b}}$. (b) $\ln x_1$ versus $1/T_{\text{f}}$. (c) $\ln x_2$ versus $1/T$, where $x_2$ is the amount fraction of solute in the saturated ideal solution.

2.6 (a) If a system contains particles of the same mass, it is called the *monodisperse* system, whereas if it contains particles of different masses, it is called the *polydisperse* system. Show that for a monodisperse system, the number average and mass average molar masses have the same value whereas for a polydisperse system, the mass average is greater than the number average.

(b) Name at least one property which depends on each of the number average molar mass and mass average molar mass.
(c) Determine the number average and the mass average of a system containing the following particles.

5 particles of $m = 1$ g

3 particles of $m = 2$ g

3 particles of $m = 3$ g

1 particle of $m = 4$ g

(Ans. (b) any colligative property and light scattering, (c) 2.0 g and 2.5 g)

2.7 (a) In deriving Eq. (2.12.6), it has been assumed that $\Delta_{\text{fus}}H_{2,\text{m}}$ is independent of the temperature. Suppose $\Delta_{\text{fus}}H_{2,\text{m}}$ depends upon the temperature and is given by

$$\Delta H = \Delta H_0 + (\Delta a)T + \left(\frac{\Delta b}{2}\right)T^2 + \left(\frac{\Delta c}{3}\right)T^3 + \cdots$$

where $\Delta H_0$, $\Delta a$, $\Delta b$, $\Delta c, \ldots$, etc., are constants. Taking the above temperature dependence, show that the solubility expression is given by

$$\ln x_2 = -\frac{\Delta H_0}{RT} + \frac{\Delta a}{R}\ln (T/\text{K}) + \frac{\Delta b}{2R}T + \frac{\Delta c}{3R}\frac{T^2}{2} + \cdots + K$$

where $K$ is the constant of integration.
(b) Predict for which substance (1 or 2) having following characteristics, the ideal solubility is larger, and give reasons:
(i) Same $\Delta_{\text{fus}}H_{\text{m}}$ but $T_1^* > T_2^*$.
(ii) Same $T^*$ but $\Delta_{\text{fus}}H_{1,\text{m}} > \Delta_{\text{fus}}H_{2,\text{m}}$

2.8 Given below is a cell.

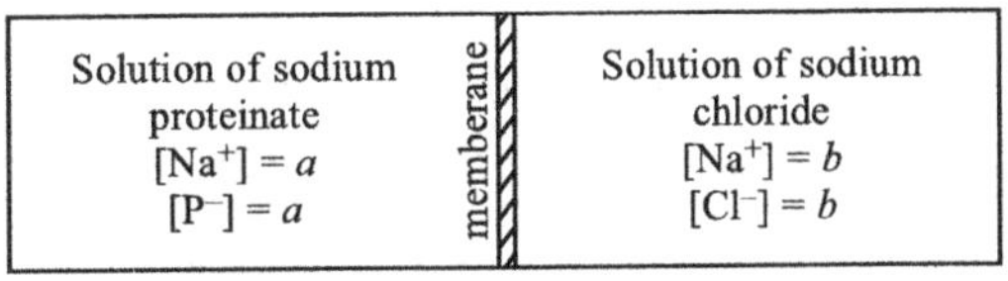


The membrane allows the diffusion of small ions and not of large ions such as $P^-$. Such a system will not be in equilibrium as the $Cl^-$ ions can diffuse from right side to left side. This diffusion will also be accompanied by the diffusion of $Na^+$ ions in order to maintain the electroneutrality. Equilibrium will be achieved when the chemical potential of NaCl is same on each side of the membrane, i.e.

$$(\mu_{\text{NaCl}})_\text{L} = (\mu_{\text{NaCl}})_\text{R}$$

For dilute solutions, we can write

$$\mu^*_{\text{NaCl}} + RT\ln (m_{\pm,\text{NaCl}})^2_\text{L} = \mu^*_{\text{NaCl}} + RT\ln (m_{\pm,\text{NaCl}})^2_\text{R}$$

or $$(m_{\pm,\text{NaCl}})_\text{L} = (m_{\pm,\text{NaCl}})_\text{R}$$

or $$(m_{\text{Na}^+})_\text{L}\,(m_{\text{Cl}^-})_\text{L} = (m_{\text{Na}^+})_\text{R}\,(m_{\text{Cl}^-})_\text{R}$$

If $x$ is the concentration of NaCl that has diffused from right side to left side, show that it is equal to

$$x = \frac{b^2}{a + 2b}$$

The above equilibrium is known as *Donnan membrane equilibrium.*

2.9 Given below is a cell

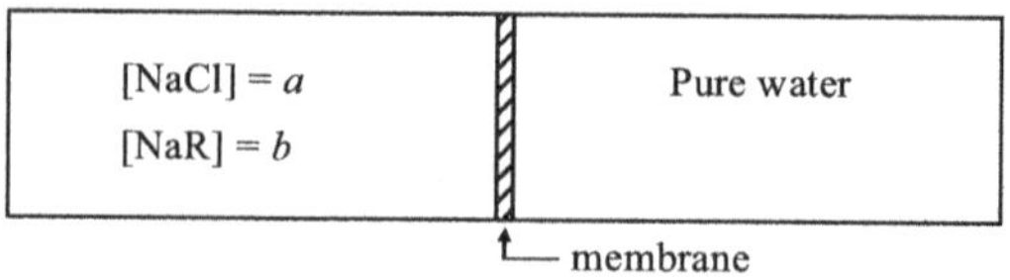


The membrane allows the passage of small ions $Na^+$ and $Cl^-$ and not large ion such as $R^-$. If $x$ is the concentration of NaCl that has diffused from left to right side, show that

$$x = \frac{a(a+b)}{2a+b}$$

2.10 Given below is a cell.

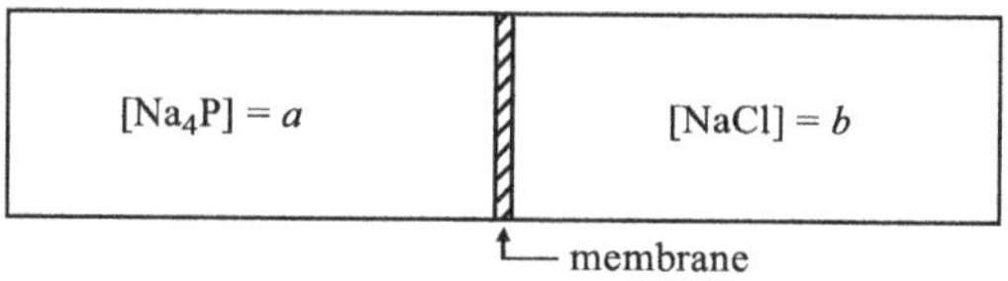


The membrane allows the passage of small ions $Na^+$ and $Cl^-$ and not large ion such as $P^{4-}$. If $b > a$ and if $x$ is the concentration of NaCl that has diffused from right to left side, show that

$$x = \frac{b^2}{4a+2b}$$

2.11 Given below is a cell.

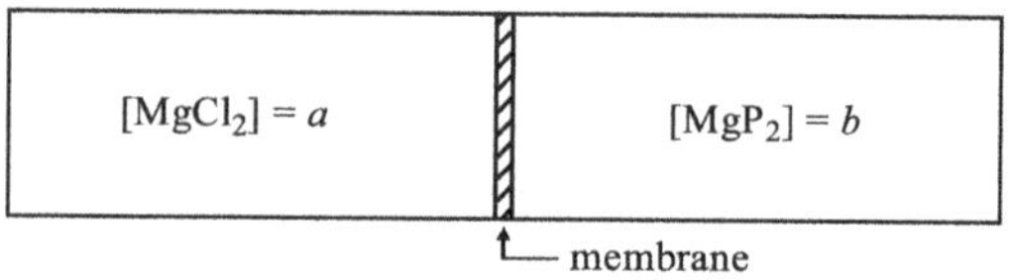


The membrane allows the passage of small ion $Mg^{2+}$ and $Cl^-$ and not large ion such as $P^-$. If $a > b$ and if $x$ is the concentration of $MgCl_2$ that has diffused from left to right, show that the value $x$ is obtainable from the expression.

$$2x^3 + x^2(b - 3a) + 3a^2x - a^3 = 0$$

(Hint: $[Mg^{2+}]_L\ [Cl^-]_L^2 = [Mg^{2+}]_R\ [Cl^-]_L^2$)

2.12 Why does increase in pressure reduce cooking time?

2.13 Will an egg cook faster if boiled in pure water or in water to which salt has been added? Give reason.

2.14 The phenomenon of osmosis can be stopped by either increasing the external pressure on the solution side or by lowering the external pressure on the solvent side. Explain.

2.15 Show that the relative lowering in the freezing $-\Delta T_f/T_f^*$, when a nonvolatile solute is dissolved in a volatile solvent, is given by $RT_f^* x_2/\Delta_{fus}H_{1,m}$, where $x_2$ is the amount fraction of solute in the solution. Does the freezing point remain constant as more and more of the solvent crystallizes out?

## NUMERICAL PROBLEMS

**Concentration Units**

2.1 An aqueous solution of HCl has 2.0% by mass of HCl and a density of 1.016 g $cm^{-3}$. Calculate the amount fraction of solute, molarity and molality of the solution.
(Ans. 0.01, 0.557 mol $dm^{-3}$, 0.559 mol $kg^{-1}$)

**Vapour Pressure**

2.2 In 804 g of water, 30 g sucrose ($C_{12}H_{22}O_{11}$) and 30 g glucose ($C_6H_{12}O_6$) were dissolved. What is the vapour pressure of this solution at 303 K if the vapour pressure of water is 31.82 mmHg? (Ans 31.64 mmHg)

2.3 The relative lowering of vapour pressure can be determined conveniently by the Ostwald and Walker's dynamic method. In this method, a fixed volume of dry air is first passed through bulbs containing the solution and then through the bulbs containing the pure solvent, the losses in masses of these two sets of bulbs are related to $p_{sol}$ and $\Delta p$ as follows.

$$\Delta m_1 \propto p_{sol}$$

$$\Delta m_2 \propto p^* - p_{sol}$$

Hence $$\frac{\Delta p}{p^*} = \frac{p^* - p_{sol}}{p^*} = \frac{\Delta m_2}{\Delta m_1 + \Delta m_2}$$

Using the above relation, determine the molar mass of solute from the following data:
$\Delta m_1$ = 1.949 g, $\Delta m_2$ = 0.087 g, the solution contains 26.66 g of solute in 200 g of water.
(Ans. 56.15 g $mol^{-1}$)

**Elevation of Boiling Points**

2.4 The boiling point of ethyl alcohol is 78.4 °C and its $\Delta_{vap}H_m$ = 39.42 kJ $mol^{-1}$. Calculate the boiling point elevation constant for ethyl alcohol.
(Ans. 1.19 K kg $mol^{-1}$)

2.5 Benzene and cyclohexane follow Trouton's rule. If their boiling points are 80.2 °C and 81.4 °C, respectively, calculate the values of the boiling point elevation constants.
(Ans. 2.625 K kg $mol^{-1}$, 2.833 K kg $mol^{-1}$)

2.6 A 10% solution of cane sugar has undergone partial inversion according to the reaction

sucrose + water $\rightarrow$ glucose + fructose

The boiling point of the solution is 100.27 °C: (a) What is the average molar mass of the dissolved material? (b) What fraction of the sugar has inverted?
(Ans. (a) 209.9 g $mol^{-1}$, (b) 0.63)

**Depression of Freezing Point**

2.7 (a) The molar mass and the melting point of camphor are 152.2 g $mol^{-1}$ and 178.4 °C, respectively. The cryscopic constant of camphor is 37.7 K kg $mol^{-1}$. Calculate $\Delta_{fus}H_{1,m}$ for camphor. (Ans. 6.844 kJ $mol^{-1}$)
(b) A mixture which contains 0.52 g of camphor and 0.045 g of an organic solute freezes at 157.2 °C. The solute contains 93.46% of C and 6.54% of H by mass. Determine the molecular formula of the compound. (Ans. $C_{12}H_{10}$)

2.8 Pure benzene freezes at 5.4 °C and a solution of 0.223 g of phenylacetic acid ($C_6H_5CH_2COOH$) in 4.4 g of benzene freezes at 4.47 °C. The molar enthalpy of fusion of benzene is 9.89 kJ $mol^{-1}$. Calculate the relative molar mass of phenylacetic acid and comment on the form of the acid in the solution.
(Ans. $M$ = 273 g $mol^{-1}$, dimerizes in solution)

2.9 (a) Assuming $\Delta_{fus}H_{1,m}$ to be independent of temperature. Show that the integrated expression of Eq. (2.8.9) is given by

$$\frac{1}{T_f} = \frac{R \ln (1 + mM_1)}{\Delta_{fus}H_{1,m}} + \frac{1}{T_f^*}$$

Calculate $T_f$ for aqueous solutions having

$m$/mol kg$^{-1}$ 0.1, 0.2, · · ·, 0.8, 0.9.

Given: $\Delta_{fus}H_{1,m}$ = 6009.5 J mol$^{-1}$.

(b) Calculate $T_f$ by using the expression, $-\Delta T_f = K_f m$ (an expression derived for dilute solutions in which $m \to 0$) for the aqueous solutions mentioned in part (a). Given: $K_f = M_1RT_f^{*2}/\Delta_{fus}H_{1,m}$.

(c) Determine graphically or otherwise the maximum molality of aqueous solution for which $T_f$'s calculated in parts (a) and (b) do not differ more than 0.01 K

(Ans. 0.59 mol kg$^{-1}$)

**Osmotic Pressure**

2.10 The average osmotic pressure of human blood is 7.7 atm at 40 °C. (a) What should be the total concentration of various solutes in the blood? (b) Assuming this concentration to be essentially the same as the molality, find the freezing point of blood.

(Ans. – 0.56 °C)

2.11 Consider the following arrangement, in which a solution containing 20 g of hemoglobin in 1 dm$^3$ of the solution is placed in the right compartment and pure water is placed in the left compartment. At equilibrium, the height of water in the right-side column is 77.8 mm in excess of that in the left-side column. What is the molar mass of hemoglobin? The temperature of the system is maintained at 298 K.

(Ans. 65 040 g mol$^{-1}$)

2.12 The chemical potential of pure water is −285.84 kg mol$^{-1}$. What is the chemical potential of water in a solution containing 0.1 mol of solute per 120 g of water at 298.15 K and 1 atm total pressure? Assume ideal behaviour.

(Ans. – 285.88 kJ mol$^{-1}$)

**Correlation between Colligative Properties**

2.13 The freezing point of a solution of 0.138 g of a nonvolatile solute in 15.2 g benzene is 0.53 K less than the freezing point of pure benzene. What are the molar mass of the substance and the normal boiling point of solution?

(Ans. 88 g mol$^{-1}$, 80.4 °C)

2.14 An aqueous solution of cane sugar ($M$ = 342 g mol$^{-1}$) has an osmotic pressure of 1.5 × 101.3 kPa at 18 °C. What will be the vapour pressure of this solution at 40 °C? If 100 g of this solution is cooled to – 3.0 °C, what mass of ice will separate out? Vapour pressure of water at 40 °C = 55.324 mmHg.

(Ans. 55.26 mmHg, 94 g).

2.15 Blood freezes at – 0.56 °C and a solution of 3.0 g of urea ($M$ = 60 g mol$^{-1}$) in 250 g of water freezes at – 0.37 °C. What is the concentration of blood solution and its osmotic pressure at 37 °C?

(Ans. 0.3 mol dm$^{-3}$, 750 kPa)

2.16 The vapour pressure of a solution of urea is 736.2 mmHg at 100 °C. What is the osmotic pressure of this solution at 15 °C? At what temperature would the solution begin to freeze? $K_f$ for water = 1.86 K kg mol$^{-1}$.

(Ans. 4.17 kPa, –3.2 °C.)

2.17 Two grams of a complex of high molar mass compound ($M$ = 12 400 g mol$^{-1}$) is dissolved in 100 g of water at 25 °C. Calculate $\Delta T_f$, $\Delta T_b$, $\Delta p$ and $\Pi$ of the solution at 25 °C. On the basis of values obtained suggest the most appropriate method for determining the molar mass.

(Ans. – $\Delta T_f = 3 \times 10^{-3}$ °C, $\Pi$ = 29 mmHg, $\Delta T_b = 8 \times 10^{-4}$ °C, $\Delta p = 6.9 \times 10^{-4}$ mmHg)

**Colligative Properties of Electrolytic Solution**

2.18 A 0.4 molal solution of $K_2SO_4$ freezes at – 1.52 °C. Assuming that van't Hoff factor $i$ is constant with temperature, calculate the vapour pressure at 25 °C and the normal boiling point of the solution. Given: $p_1^*$(298 K) = 0.031 67 bar.

(Ans. 0.0312 bar, 100.42 °C)

2.19 An aqueous solution contains 5% by mass of urea, 1.0% by mass of KCl and 10% by mass of glucose. Calculate its freezing point. Assume complete ionization for KCl.

(Ans. – 4.29 °C)

2.20 When phenol ($C_6H_5OH$) is dissolved in a particular solvent, it is partially dimerized. If 2.58 g of phenol is dissolved in 100 g of the solvent, the freezing point of solvent ($K_f$ = 14.1 K mol$^{-1}$ kg) is depressed by 2.37 K. Find how much of the phenol is dimerized in solution. (Ans. 77.55%)

2.21 (a) An involatile solute A dimerizes in a solvent:

$$2A \rightleftharpoons A_2$$

show that $$K_{eq} = \frac{K_b(K_b m - \Delta T_b)}{(2\,\Delta T_b - K_b m)^2}$$

where $m$ is the molality of the solute in the solvent and $K_b$ is the boiling point elevation constant of the solvent.

(b) A 0.103 3 molal solution of A in the above solvent produces $\Delta T_b$ = (0.070 4 mol kg$^{-1}$) $K_b$. Calculate the value of $K_{eq}$. (Ans. 23.4 mol$^{-1}$ kg)

2.22 125 g benzoic acid (molar mass = 121 g mol$^{-1}$) when dissolved in 100 cm$^3$ of benzene produces osmotic pressure of 1.73 atm at 300 K. Benzoic acid is known to form dimer in benzene. Calculate (i) the percentage of benzoic acid in the associated state and (ii) the equilibrium constant of the dimerization reaction.

(Ans. 63.93%, 23.78 mol$^{-1}$ dm$^3$)

2.23 (a) The freezing point of 0.01 molal $CaCl_2$ is – 0.051 12 °C. What freezing point is predicted by the Debye-Hückel theory? What is the osmotic pressure of 0.01 molal $CaCl_2$ at 0 °C?

(b) Using the Debye-Hückel equation, calculate the values of $i$ at 0 °C for 0.000 5 molar aqueous solutions of HCl, $BaCl_2$, $H_2SO_4$, and $La(NO_3)_3$.

**Ideal Solubility**

2.24 The solubility values of picric acid in benzene are:

| | | | | | | |
|---|---|---|---|---|---|---|
| $\theta_c$ / °C | 5 | 10 | 15 | 20 | 25 | 35 |
| g/100 g benzene | 3.70 | 5.37 | 7.29 | 9.56 | 12.66 | 21.38. |

Calculate the freezing point and molar enthalpy of fusion of picric acid.

**Donnan-Membrane Equilibrium**

2.25 If in Q 2.11 of Try Yourself Problem, $a$ = 0.003 M and $b$ = 0.001 M, Show that $x = 1.3628 \times 10^{-3}$ M.

(Hint: By the method of successive approximation, determine the value of $x$ using the expression $x = [\{x^2(3a - b) - 3a^2x + a^3\}/2]^{1/3}$

# 3 Phase Rule

## 3.1 INTRODUCTION AND DEFINITIONS

The various conditions of equilibrium between phases, such as the number of phases, the number of components and the degrees of freedom (or the variance), can be correlated with one another with the help of a general rule, known as the *phase rule*. This rule was deduced theoretically by J.W. Gibbs. Before proceeding with the deduction of the rule, it is necessary to define and explain the terms involved, viz., phase, component and degree of freedom.

**Phase**

*A phase is defined as any homogeneous and physically distinct part of a system which is separated from other parts of the system by definite bounding surfaces.* The term homogeneous means that the system has identical physical properties and chemical composition throughout the whole of the system.

In general, we have:

***For gaseous system*** Only one phase is possible since gases are completely miscible with one another in all proportions.

***For liquid system*** The number of phases is equal to the number of layers present in the system. For completely miscible liquids, the number of phases is equal to one.

***For solid system*** In general, every solid constitutes a single phase except when a solid solution is formed. Thus the number of phases in the solid system is equal to the number of solids present. In solid solution, the number of phases is equal to one. Each polymorphic form and allotropic modification constitutes a separate phase.

**Number of Components**

*It is the smallest number of independent chemical constituents by means of which the composition of each and every phase can be expressed.* The independent chemical constituent is the one whose concentration can be varied independent of other constituents of the system. Some typical examples are given below.

***One-component system*** The water system is a typical example of one-component system. We can have the following equilibria between the different phases of water:

$$\text{solid water (ice)} \rightleftharpoons \text{liquid water}$$

$$\text{solid water} \rightleftharpoons \text{water vapour}$$

$$\text{liquid water} \rightleftharpoons \text{water vapour}$$

$$\text{solid water} \rightleftharpoons \text{liquid water} \rightleftharpoons \text{water vapour}$$

The composition of any one phase (solid, liquid or vapour) can be expressed in terms of the single constituent water and hence the number of components is one. The variable could be either $H_2$ or $O_2$ since specification of one fixes the other through the formula $H_2O$. However, this choice is never considered since $H_2$ or $O_2$ does not exist as the individual constituent in the above equilibrium.

***Two-component system*** The typical example of a two-component system is the salt-water system such as sodium sulphate-water, copper sulphate-water, etc. In $Na_2SO_4$-$H_2O$ system, the various phases that may occur are $Na_2SO_4$, $Na_2SO_4 \cdot 7H_2O$, $Na_2SO_4 \cdot 10H_2O$, an aqueous solution of $Na_2SO_4$, water present as ice, liquid and vapour. The composition of any of the above phases can be stated in terms of the two constituents, namely, sodium sulphate and water:

| | | |
|---|---|---|
| $Na_2SO_4$ | : | $Na_2SO_4 + 0H_2O$ |
| $Na_2SO_4 \cdot 7H_2O$ | : | $Na_2SO_4 + 7H_2O$ |
| $Na_2SO_4 \cdot 10H_2O$ | : | $Na_2SO_4 + 10H_2O$ |
| $Na_2SO_4(aq)$ | : | $Na_2SO_4 + x\ H_2O$ |
| $H_2O(s)$, $H_2O(l)$, $H_2O(g)$ | : | $0Na_2SO_4 + H_2O$ |

It can be seen that the composition of some of the phases can be represented by only one constituent while the others necessarily require two constituents. Since two constituents are the smallest number by which the compositions of all the phases can be defined, the number of components in sodium sulphate-water system is two.

Another example is the equilibrium between $CaCO_3(s)$, $CaO(s)$ and $CO_2(g)$ as represented by the equation

$$CaCO_3(s) \rightleftharpoons CaO(s) + CO_2(g)$$

Here the number of phases are three, viz., solid $CaCO_3$, solid CaO and gaseous $CO_2$. The number of components in the above system is two since two species out of the three are sufficient to express the compositions of all the three phases. It is immaterial which of the two are selected for this purpose. For example:

(a) If CaO and $CO_2$ are chosen, then

| | | |
|---|---|---|
| $CaCO_3(s)$ | : | $CaO + CO_2$ |
| $CaO(s)$ | : | $CaO + 0CO_2$ |
| $CO_2(g)$ | : | $0CaO + CO_2$ |

(b) If $CaCO_3$ and CaO are chosen, then

| | | |
|---|---|---|
| $CaCO_3(s)$ | : | $CaCO_3 + 0CaO$ |
| $CaO(s)$ | : | $0CaCO_3 + CaO$ |
| $CO_2(g)$ | : | $CaCO_3 - CaO$ |

and so on.

Thus it is the number and not the nature that is important for determining the components of the system.

Another example, which makes a clear distinction between the one component system and the two-component system, is the dissociation of ammonium chloride:

$$NH_4Cl(s) \rightleftharpoons NH_3(g) + HCl(g)$$

This system has two phases, viz., solid $NH_4Cl$ and gaseous phase containing ammonia and hydrogen chloride. If the dissociation is carried out in vacuum, then the number of components of the system is one. This is due to the fact that in the gaseous phase both HCl and $NH_3$ are always present in equal amounts and thus its composition can be represented by $NH_4Cl$ ($= x\ NH_3 + x\ HCl$). But if the dissociation is carried out in the presence of arbitrary amounts of $NH_3$ or/ and HCl, then the number of components of the system becomes two:

| | | |
|---|---|---|
| Solid $NH_4Cl$ | : | $NH_3 + HCl$ |
| Gaseous phase | : | $x\ NH_3 + y\ HCl$ |

**Degree of Freedom**

*The degree of freedom or variance of the system is the minimum number of independent variables such as temperature, pressure and concentration, that must be ascertained so that a given system in equilibrium is completely defined.*

For an illustration, we cite an example of water system. If only one phase is present (solid, liquid or gas), then we have to describe values of at least two parameters, namely, temperature and pressure, in order to define the system. For example, we can have liquid water under different conditions of temperature and pressure. Thus the water system has two degrees of freedom or it is bivariant. Now consider two phases in equilibrium, for example, solid and liquid water, i.e.

$$\text{solid water} \rightleftharpoons \text{liquid water}$$

It is well known that the temperature at which a solid melts depends upon the external pressure. Thus if the value of one of the variables is fixed, the other will automatically have a definite value. The usual melting point of 0 °C is referred to one atm external pressure. In order to describe the above equilibrium completely, we need to specify only one variable—either temperature or pressure; hence the system has one degree of freedom, i.e. it is an univariant system.

***Alternatively****, The degree of freedom of the system may be defined as the number of factors, such as temperature, pressure and concentration, which can be varied independently without altering the number of phases.* For a single phase system, both temperature and pressure can be varied independently of each other and thus the system is bivariant. For two phases in equilibrium, only one variable can be varied as the other one will automatically have a fixed value and thus the system is univariant.

Consider now the situation at the triple point of water where all the three phases are in equilibrium with each other. This equilibrium is achieved at a temperature of 0.0098 °C and a pressure of 4.58 mmHg. Any change in the values of these variables, converts the three-phase system into either two-phase or one-phase system. Thus, none of the variables can be changed so long as we want to have all the three phases in equilibrium. Thus the system has zero degree of freedom, or it is nonvariant (or invariant).

## 3.2 DERIVATION OF PHASE RULE

Phase rule is a general rule which is applicable to all types of reactive and nonreactive systems. In a nonreactive system, we simply have the distribution of various components in different phases without any complications such as the chemical reaction between the components. We derive the rule, first, for a nonreactive system and then show how the same rule can be used for a reactive system.

**Nonreactive System** Consider a heterogeneous system of $P$ phases at equilibrium containing in all $C$ components. Let us start with the assumption that all the $C$ components are present in all $P$ phases. The system at equilibrium can be completely described if we know the values of the variables listed in Table 3.2.1.

**Table 3.2.1** Number of Variables to be Known to Define a Given System Completely

| *Variables* | *Number* |
|---|---|
| (i) Temperature of the system | 1 |
| (ii) Pressure of the system | 1 |
| (iii) Concentration of each and every component in all the $P$ phases. For each phase, we will have to specify the values of $C$ concentration terms and thus for $P$ phases, we will have to specify in all $PC$ values | $PC$ |
| Total number of variables that need to be specified | $PC + 2$ |

Values of these variables can be obtained by solving the equations which are applicable when the system is at equilibrium. There are two types of equations which are available (Table 3.2.2).

Mathematically, we know that the number of variables that can be obtained from a set of equations is equal to the number of equations. Thus if there are as many equations as there are variables, then the temperature, pressure and composition of the whole system in equilibrium can be determined. Such a system, as stated earlier, is the *nonvariant system.* If the number of variables exceeds the number of equations by one, then the equilibrium of the system cannot be determined until one of the variables is arbitrarily chosen. Such a system is called *monovariant* (*or univariant*) and has one degree of freedom. Thus we see that the excess of variables over equations, which is called the variance $F$ of the system, is given as

$$\text{Variance} = \begin{pmatrix}\text{Total number of variables}\\ \text{that need to be specifed}\end{pmatrix} - \begin{pmatrix}\text{Total number of equations}\\ \text{that are available}\end{pmatrix}$$

Thus, from Tables 3.2.1 and 3.2.2, we have

$$F = (PC + 2) - \{P + C(P - 1)\}$$

**Table 3.2.2** Types and Number of Equations which are Available

| *Equations* | *Number* |
|---|---|
| (i) Condition of sum of the amount fractions in any phase being equal to one, $x_{1(1)} + x_{2(1)} + \cdots + x_{c(1)} = 1$; $x_{1(2)} + x_{2(2)} + \cdots + x_{c(2)} = 1$; ......; $x_{1(P)} + x_{2(p)} + \cdots + x_{c(P)} = 1$. There will be as many equations as the number of phases. Thus their total number will be $P$. | $P$ |
| (ii) Thermodynamic condition of phase equilibria: According to this, the species will be distributed in such a manner that value of free energy of the system at equilibrium is minimum. The condition for this is that the chemical potential of any component will have the same value in all the $P$ phases, i.e. $\mu_{1(1)} = \mu_{1(2)} = \cdots = \mu_{1(P)}$; $\mu_{2(1)} = \mu_{2(2)} = \cdots = \mu_{2(P)}$; ... ... ... ... ...; $\mu_{c(1)} = \mu_{c(2)} = \cdots = \mu_{c(P)}$. For each component, we will have $(P - 1)$ equations and thus for $C$ components, the number of equations will be $C(P - 1)$ | $C(P - 1)$ |
| Total number of equations that are available | $P + C(P - 1)$ |

or $$F = C - P + 2$$

or $$F + P = C + 2 \tag{3.2.1}$$

Equation (3.2.1) is the phase rule which connects the number of phases and components with the variance of the system.

**Verification of the Phase Rule**

Let us verify the rule by calculating the variance of a one-component system:

*One phase only* $P = 1,\quad F = C - P + 2 = 1 - 1 + 2 = 2$

*Two phases in equilibrium* $P = 2,\quad F = C - P + 2 = 1 - 2 + 2 = 1$

*Three phases in equilibrium* $P = 3,\quad F = C - P + 2 = 1 - 3 + 2 = 0$

Thus, we see that the calculated degrees of freedom agree with experimentally observed values of the one-component water system.

**Comment on the Phase Rule**

The condition that all the components must be present in every phase, which has been used above in deriving the phase rule, is not altogether essential. The rule is applicable even if any component is missing in any of the phases. Say, for example, the component number 1 is missing from the phase number 1. We can once again derive the phase rule by following the procedure described earlier. Tables 3.2.3 and 3.2.4 describe the number of variables and equations for the present case.

**Table 3.2.3** Number of Variables to be Known to Define a given System Completely

| *Variables* | *Number* |
|---|---|
| (i) Temperature | 1 |
| (ii) Pressure | 1 |
| (iii) Concentration of components in different phases: For the first phase, $(C-1)$ concentration terms and for the remaining $(P-1)$ phases, $C$ terms for each phase | $(C-1)+C(P-1)$ |
| Total number of variables that need to be specified | $(C-1)+C(P-1)+2$ $= CP + 1$ |

**Table 3.2.4** Types and Number of Equations Available

| *Equations* | *Number* |
|---|---|
| (i) Condition of amount fraction: | |
| $x_{2(1)} + x_{3(1)} + \cdots + x_{c(1)} = 1$ | |
| $x_{1(2)} + x_{2(2)} + \cdots + x_{c(2)} = 1$ | |
| ......................................... | |
| $x_{1(P)} + x_{2(P)} + \cdots + x_{c(P)} = 1$ | |
| There will again be in all $P$ equations | $P$ |
| (ii) Condition of thermodynamic equilibrium: | |
| For the component 1 | |
| $\mu_{1(2)} = \mu_{1(3)} = \cdots = \mu_{1(P)}$ | $(P-2)$ |
| For other components | |
| $\mu_{2(1)} = \mu_{2(2)} = \cdots = \mu_{2(P)}$ | |
| ......................................... | |
| $\mu_{c(1)} = \mu_{c(2)} = \cdots = \mu_{c(P)}$ | $(C-1)(P-1)$ |
| There will be $(P-2)$ equations for the component number 1, one less as compared to others as it is absent in the phase number 1 | |
| Total number of equations that are available | $P + (P-2) + (C-1)(P-1)$ $= \{P + C(P-1)\} - 1$ |

From Tables 3.2.3 and 3.2.4, we have

$$\text{Variance of the system} = \begin{pmatrix}\text{Total number}\\ \text{of variables}\end{pmatrix} - \begin{pmatrix}\text{Total number of}\\ \text{equations}\end{pmatrix}$$

$$F = CP + 1 - [P + C(P-1) - 1]$$

$$F = C - P + 2$$

The above equation is the same as Eq. (3.2.1). Thus we see that the effect of one of the components missing is that both the number of variables and the number of equations have been reduced by one. Thus the difference is the same and the phase rule remains unchanged

## Reactive System

Let us consider a heterogeneous system composed of arbitrary amounts of $C$ constituents. For the sake of simplicity we assume that four of the constituents are chemically active, capable of undergoing the reaction

$$\nu_1 A_1 + \nu_2 A_2 \rightleftharpoons \nu_3 A_3 + \nu_4 A_4 \qquad \text{; (All } \nu\text{s are positive)}$$

As usual, we can write down the number of variables that are needed to describe the system completely and the number of equations that are available at equilibrium (Tables 3.2.5 and 3.2.6).

**Table 3.2.5** Number of Variables to be Stated to Define a Given System Completely

| *Variables* | *Number* |
|---|---|
| (i) Temperature | 1 |
| (ii) Pressure | 1 |
| (iii) Concentration of constituents in different phases | $CP$ |
| Total number of variables | $CP + 2$ |

**Table 3.2.6** Types and Number of Equations which are Available

| *Equations* | *Number* |
|---|---|
| (i) Condition of amount fraction:<br>$x_{1(1)} + x_{2(1)} + \cdots + x_{c(1)} = 1$<br>$x_{1(2)} + x_{2(2)} + \cdots + x_{c(2)} = 1$<br>..........................................<br>$x_{1(P)} + x_{2(P)} + \cdots + x_{c(P)} = 1$ | $P$ |
| (ii) Condition of thermodynamic equilibrium between phases:<br>$\mu_{1(1)} = \mu_{1(2)} = \cdots = \mu_{1(P)}$<br>$\mu_{2(1)} = \mu_{2(2)} = \cdots = \mu_{2(P)}$<br>..........................................<br>$\mu_{c(1)} = \mu_{c(2)} = \cdots = \mu_{c(P)}$ | $C(P - 1)$ |
| (iii) Besides above two conditions, there is another condition which has to be satisfied for the reactive system, that is, the thermodynamic condition of reaction equilibrium. At equilibrium, the value of reaction potential, $\Delta_r\tilde{G}$ is zero. Thus<br>$0 = \Delta_r\tilde{G} = \sum_j \nu_j\mu_j \qquad \text{where } \nu_j \begin{cases} \text{positive for products} \\ \text{negative for reactants} \end{cases}$<br>For the given reaction, we have<br>$0 = \nu_3\mu_3 + \nu_4\mu_4 - \nu_1\mu_1 - \nu_2\mu_2$<br>or $\nu_1\mu_1 + \nu_2\mu_2 = \nu_3\mu_3 + \nu_4\mu_4$<br>Thus, we get one more equation | 1 |
| Total number of equations that are available | $P + C(P - 1) + 1$ |

From Tables 3.2.5 and 3.2.6, we have

$$\text{Variance} = \text{Number of variables} - \text{Number of equations}$$

$$F = CP + 2 - [P + C(P - 1) + 1\}$$

$$F = (C - 1) - P + 2$$

Similarly, if two independent reactions are possible, then we get

$$F = (C - 2) - P + 2$$

Generalizing, we have

$$F = (C - r) - P + 2$$

where $r$ is the number of independent reactions that are taking place in the system.

### Example Involving Chemical Equilibrium with Additional Restricting Conditions

Up to now, we have considered three kinds of equations amongst the variables $T$, $P$, and the $x$s. These are: (a) equations of phase equilibria; (b) equations of reaction equilibria; and (c) equations of the type $\sum_i x_i = 1$. Sometimes a chemical reaction takes place in such a manner that additional equations expressing further restriction upon the $x$s are to be satisfied. One such example, as cited earlier, is the decomposition of $NH_4Cl(s)$ in vacuum:

$$NH_4Cl(s) \rightleftharpoons NH_3(g) + HCl(g)$$

In the gaseous phase, the additional restriction which always exists is that

$$x_{NH_3} = x_{HCl}$$

This constitutes a fourth types of equation, which should also be included wherever it is applicable.

Another example of additional restricting equation among the $x$s is provided by the phenomenon of dissociation in solution. Suppose that we have a system in which a salt is dissolved in a solvent. Suppose that the salt dissociates according to the scheme given below:

$$AB \rightarrow A^+ + B^-$$

All the ions remain in the solution and no precipitate is formed. Consequently, we have the additional restricting equation

$$x_{A^+} = x_{B^-}$$

If multi-dissociation takes place as shown below.

$$A \rightarrow B^+ + C^-$$

$$C^- \rightleftharpoons D^+ + E^{2-}$$

$$E^{2-} \rightleftharpoons F^+ + G^{3-}$$

then three independent restricting equations among the *xs* are available. These are

$$x_{B^+} = x_{G^{3-}} + x_{E^{2-}} + x_{C^-}$$

$$x_{D^+} = x_{G^{3-}} + x_{E^{2-}}$$

and $$x_{F^+} = x_{G^{3-}}$$

Adding these equations, we get the dependent equation

$$x_{B^+} + x_{D^+} + x_{F^+} = 3x_{G^{3-}} + 2x_{E^{2-}} + x_{C^-}$$

which expresses the condition of electrical neutrality.

**Final Form of the Phase Rule**

If, in general, a reactive system has $r$ independent reactions and $Z$ independent restrictive conditions, then the number of equations that are available at equilibrium is:

$$\begin{aligned}\begin{pmatrix}\text{Total number}\\ \text{of equations}\end{pmatrix} &= \begin{pmatrix}\text{Equations due to the}\\ \text{condition of amount fractions}\end{pmatrix} + \begin{pmatrix}\text{Equations due to the}\\ \text{phase equilibria}\end{pmatrix}\\ &+ \begin{pmatrix}\text{Equations due to the}\\ \text{chemical reactions}\end{pmatrix} + \begin{pmatrix}\text{Equations due to}\\ \text{restricting conditions}\end{pmatrix}\\ &= P + C(P-1) + r + Z\end{aligned}$$

Thus the variance is

$$F = (CP + 2) - \{P + C(P-1) + r + Z\}$$

or $$F = (C - r - Z\} - P + 2$$

or $$F = C' - P + 2$$

where $C' = C - r - Z$ and is known as the number of components of the system. It is thus equal to the total number of constituents present in the system less the number of independent chemical reactions and the number of independent restricting equations. With this definition, we have put the phase rule in the form that is applicable for a nonreactive system.

---

***Problem 3.2.1***

Show explicitly without using the phase rule that the one-component system has (1) $F = 2$ when $P = 1$, (2) $F = 1$ when $P = 2$, and (3) $F = 0$ when $P = 3$.

***Solution***

For a one-component system, we usually state the values of two variables, namely, temperature and pressure, in order to define it completely. The concentration term is not stated as it has a constant value for a given component. For example, water has a concentration of 55.56 mol $dm^{-3}$ (= 1000 g $dm^{-3}$/18 g $mol^{-1}$ at $t$ = 4 °C and $p$ = 1 atm) when present in the liquid state and in the gaseous form (assumed to be ideal) it is

$$(n/V) = p/RT$$

Now if we have only one phase, then two variables $T$ and $p$ will have to be stated and thus its variance is two. If the number of phases at equilibrium are two, then we have one relation

$$\mu_{(1)}(T, p) = \mu_{(2)}(T, p)$$

where $\mu_{(1)}$ and $\mu_{(2)}$ are the chemical potentials in phases 1 and 2, respectively. Thus, we have one relation connecting $T$ and $p$ and we have seen earlier that it is Clapeyron equation. Hence if one variable is stated, then the other one can be calculated from the above relation and thus the variance is only one.

For three phases in equilibrium, two equations are available:

$$\mu_{(1)} = \mu_{(2)} \quad \text{and} \quad \mu_{(2)} = \mu_{(3)}$$

Thus the values of the two variables, namely, temperature and pressure, can be obtained from the above relations and therefore the system is nonvariant.

## 3.3 SOME TYPICAL EXAMPLES TO COMPUTE THE NUMBER OF COMPONENTS

We have earlier defined the number of components as the minimum number of independent chemical constituents required to express the composition of each and every phase of a system. This definition is completely in agreement with the definitions given below for the reactive and the nonreactive systems.

For nonreactive system: Number of components = Number of constituents

For reactive system: Number of components = (Number of constituents) – (Number of chemical reactions) – (Number of restricting conditions)

Let us illustrate the above consistency with the help of a few typical examples.

### Sodium Sulphate-Water System

As stated earlier, it is a two-component system. It can be shown that so long as no precipitate is formed by virtue of a reaction between salt and water, the number of components is always two no matter what else is assumed to take place in the solution. We consider below four typical assumptions regarding the nature of the system.

***Neglecting all dissociations*** There are two constituents, no chemical reaction, and no restricting equation. Hence

$$C' = C - r - Z = 2 - 0 - 0 = 2$$

***Complete dissociation of the salt*** The total number of constituents present in the solution are three: $Na^+$, $SO_4^{2-}$ and $H_2O$. Since there is one restricting condition:,

$$x_{Na^+} = 2x_{SO_4^{2-}}$$

we have

$$C' = C - r - Z = 3 - 0 - 1 = 2$$

***Partial dissociation of the salt*** The dissociation may be represented as

$$Na_2SO_4 \rightleftharpoons 2Na^+ + SO_4^{2-}$$

Now the system possesses four constituents, namely, $Na_2SO_4$, $Na^+$, $SO_4^{2-}$ and $H_2O$. There is one reaction and one restricting condition:

$$x_{Na^+} = 2x_{SO_4^{2-}}$$

Thus, we have

$$C' = C - r - Z = 4 - 1 - 1 = 2$$

***Dissociation of water also*** We have the following two equilibrium reactions:

$$Na_2SO_4 \rightleftharpoons 2Na^+ + SO_4^{2-}$$

$$H_2O \rightleftharpoons H^+ + OH^-$$

The different constituents are $Na_2SO_4$, $Na^+$, $SO_4^{2-}$, $H_2O$, $H^+$ and $OH^-$, a total of six. There are two reactions and two restricting conditions:

$$x_{Na^+} = 2x_{SO_4^{2-}} \quad \text{and} \quad x_{H^+} = x_{OH^-}$$

Thus $\quad C' = C - r - Z = 6 - 2 - 2 = 2$

### Sodium Dihydrogen Phosphate-Water System

It can be shown that this system is a two-component system irrespective of whatever chemical changes may be considered in the solution except that no precipitate is formed.

***Neglecting all dissociations*** There are two constituents, no chemical reaction and no restricting equation. Hence

$$C' = C - r - Z = 2 - 0 - 0 = 2$$

***Single dissociation of the salt*** We may represent the dissociation as

$$NaH_2PO_4 \rightleftharpoons Na^+ + H_2PO_4^-$$

There are four constituents ($NaH_2PO_4$, $H_2O$, $Na^+$ and $H_2PO_4^-$), one chemical reaction and one restricting equation:

$$x_{Na^+} = x_{H_2PO_4^-}$$

Hence $\quad C' = C - r - Z = 4 - 1 - 1 = 2$

***Multiple dissociation of the salt*** The multistep dissociation may be represented as

$$NaH_2PO_4 \rightleftharpoons Na^+ + H_2PO_4^-$$

$$H_2PO_4^- \rightleftharpoons H^+ + HPO_4^{2-}$$

$$HPO_4^{2-} \rightleftharpoons H^+ + PO_4^{3-}$$

There are seven constituents, $NaH_2PO_4$, $Na^+$, $H_2PO_4^-$, $H^+$, $HPO_4^{2-}$, $PO_4^{3-}$ and $H_2O$.

Number of chemical reactions = 3
Number of independent restricting equations = 2:

$$x_{Na^+} = x_{PO_4^{3-}} + x_{HPO_4^{2-}} + x_{H_2PO_4^-} \quad (3.3.1)$$

$$x_{H^+} = 2x_{PO_4^{3-}} + x_{HPO_4^{2-}} \quad (3.3.2)$$

(**Note:** One may consider the charge balance expression

$$x_{Na^+} + x_{H^+} = x_{H_2PO_4^-} + 2x_{HPO_4^{2-}} + 3x_{PO_4^{3-}}$$

Equation (3.3.2) may be obtained by subtracting Eq. (3.3.1) from the above expression).

Thus $\quad C' = C - r - Z = 7 - 3 - 2 = 2$

***Dissociation of water also*** Besides the dissociation given in part (iii), we also have

$$H_2O \rightleftharpoons H^+ + OH^-$$

There are total eight constituents, $NaH_2PO_4$, $Na^+$, $H_2PO_4^-$, $H^+$, $HPO_4^{2-}$, $PO_4^{3-}$, $OH^-$ and $H_2O$.

Number of chemical reactions = 4
Number of restricting equations = 2:

$$x_{Na^+} = x_{PO_4^{3-}} + x_{HPO_4^{2-}} + x_{H_2PO_4^-}$$

$$x_{H^+} = 2x_{PO_4^{3-}} + x_{HPO_4^{2-}} + x_{OH^-}$$

Thus $\quad C' = C - r - Z = 8 - 4 - 2 = 2$

**Aluminium Chloride-Water System**

In this case, $AlCl_3$ combines with water according to the equation:

$$AlCl_3 + 3H_2O = Al(OH)_3 + 3HCl$$

Some of the $Al(OH)_3$ is precipitated out. The dissociation of the various species will be as follows:

$$AlCl_3 \rightleftharpoons Al^{3+} + 3Cl^-; \qquad Al(OH)_3 \rightleftharpoons Al^{3+} + 3OH^-$$

$$H_2O \rightleftharpoons H^+ + OH^-; \qquad HCl \rightleftharpoons H^+ + Cl^-$$

Total number of constituents = 8:

$$AlCl_3,\ H_2O,\ Al(OH)_3,\ HCl,\ Al^{3+},\ Cl^-,\ H^+ \text{ and } OH^-$$

Number of chemical reactions = 4

Since some of the $Al(OH)_3$ has precipitated out, there is only one restricting equation, namely, that expressing the electrical neutrality of the solution

$$3x_{Al^{3+}} + x_{H^+} = x_{OH^-} + x_{Cl^-}$$

Thus the number of components = 8 – 1 – 4 = 3.

If we assume complete dissociation of $AlCl_3$ and that of HCl, we will have

$$C = 6,\ r = 2 \text{ and } Z = 1$$

Thus $\quad C' = C - Z - r = 6 - 1 - 2 = 3$

---

***Example 3.3.1*** Show that an aqueous system containing $K^+$, $Na^+$ and $Cl^-$ is a three-component system whereas $K^+$, $Na^+$, $Cl^-$ and $Br^-$ is a four-component system. What are the number of components if the salts are present in equal amounts?

***Solution***

(**a**) $K^+$, $Na^+$, $Cl^-$ *and* $H_2O$ *system*: In this case, we have

Number of reactions, $r = 3$:

$$H_2O \rightleftharpoons H^+ + OH^-$$

$$KCl \rightleftharpoons K^+ + Cl^-$$

$$NaCl \rightleftharpoons Na^+ + Cl^-$$

Number of constituents, $C = 8$:

$H_2O$, KCl, NaCl, $H^+$, $OH^-$, $K^+$, $Cl^-$ and $Na^+$

Number of restricting equations, $Z = 2$:

*Electrical neutrality* $\quad x_{Na^+} + x_{K^+} + x_{H^+} = x_{Cl^-} + x_{OH^-}$

Water *dissociation* $\quad x_{H^+} = x_{OH^-}$

Hence, Number of components, $C' = C - r - Z = 8 - 3 - 2 = 3$

If the salts are present in equal amounts, then one more restricting equation exists, i.e.

$$x_{Na^+} = x_{K^+}$$

Hence, here

Number of components, $C' = C - r - Z = 8 - 3 - 3 = 2$

**Alternatively,** leaving KCl and NaCl on the basis that they are strong electrolytes, we have

Number of constituents, $C = 6$:

$H_2O$, $H^+$, $OH^-$, $Na^+$, $K^+$, $Cl^-$

Number of reactions, $r = 1$:

$$H_2O \rightleftharpoons H^+ + OH^-$$

Number of restricting equations, $Z = 2$:

$$x_{H^+} = x_{OH^-} \quad \text{and} \quad x_{Na^+} + x_{K^+} = x_{Cl^-}$$

Hence, Number of components, $C' = C - r - Z = 6 - 1 - 2 = 3$

If the salts are present in equal amounts, then one more restricting equation exists, i.e.

$$x_{Na^+} = x_{K^+}$$

Hence, here

Number of components, $C' = C - r - Z = 6 - 1 - 3 = 2$

(**b**) $K^+$, $Na^+$, $Cl^-$, $Br^-$ *and* $H_2O$ *system*: In this case, we have

Number of reactions, $r = 5$;

$$H_2O \rightleftharpoons H^+ + OH^- \qquad NaCl \rightleftharpoons Na^+ + Cl^-$$

$$KCl \rightleftharpoons K^+ + Cl^- \qquad KBr \rightleftharpoons K^+ + Br^-$$

$$NaBr \rightleftharpoons Na^+ + Br^-$$

Number of constituents, $C = 11$

$H_2O$, $H^+$, $OH^-$, KCl, $K^+$, $Cl^-$, NaBr, $Na^+$, $Br^-$, NaCl and KBr

Number of restricting equations, $Z = 2$:

*Electrical neutrality* $x_{Na^+} + x_{K^+} + x_{H^+} = x_{Cl^-} + x_{Br^-} + x_{OH^-}$

*Water dissociation* $x_{H^+} = x_{OH^-}$

Hence, Number of components, $C' = C - r - Z = 11 - 5 - 2 = 4$

If the salts are present in equal amounts, then we have three independent restricting equations:

$$x_{Na^+} = x_{K^+}; \qquad x_{H^+} = x_{OH^-}; \qquad \text{and} \qquad x_{Cl^-} = x_{Br^-}$$

Hence, Number of components, $C' = C - r - Z = 11 - 5 - 3 = 3$

**Alternatively,** leaving KCl, NaBr, NaCl and KBr as they are strong electrolytes, we have:

Number of constituents, $C = 7$:

$H_2O$, $H^+$, $OH^-$, $Na^+$, $K^+$, $Cl^-$ and $Br^-$

Number of reactions, $r = 1$:

$$H_2O \rightleftharpoons H^+ + OH^-$$

Number of restricting conditions, $Z = 2$:

$$x_{H^+} = x_{OH^-} \quad \text{and} \quad x_{Na^+} + x_{K^+} = x_{Cl^-} + x_{Br^-}$$

Hence, Number of components, $C' = C - r - Z = 7 - 1 - 2 = 4$

If the salts are present in equal amounts, then we have three independent restricting equations:

$$x_{Na^+} = x_{K^+}; \quad x_{Cl^-} = x_{Br^-} \quad \text{and} \quad x_{H^+} = x_{OH^-}$$

Hence, Number of components, $C' = C - r - Z = 7 - 1 - 3 = 3$

---

***Example 3.3.2*** Show that $NH_4Cl(s) - NH_3(g) - HCl(g)$ system in which $p_{NH_3} = p_{HCl}$ is a one-component system whereas when $p_{NH_3} \neq p_{HCl}$ is a two-component system.

***Solution*** (**a**) $NH_4Cl(s) - NH_3(g) - HCl(g)$ *when* $p_{NH_3} = p_{HCl}$. The condition of $p_{NH_3} = p_{HCl}$ would arise only when the gases are obtained by the sublimation of $NH_4Cl(s)$. Thus, we have

Number of reactions, $r = 1$ :

$$NH_4Cl(s) \rightleftharpoons NH_3(g) + HCl(g)$$

Number of constituents, $C = 3$:

$$NH_4Cl(s),\ NH_3(g) \text{ and } HCl(g)$$

Number of restricting equations, $Z = 1$:

$$p_{NH_3} = p_{HCl}$$

Hence, Number of components, $C' = C - Z - r = 3 - 1 - 1 = 1$

(**b**) $NH_4Cl(s) - NH_3(g) - HCl(g)$ *when* $p_{NH_3} \neq p_{HCl}$: Here, we have

Number of reactions, $r = 1$
Number of constituents, $C = 3$
Number of restricting equations, $Z = 0$

Hence, Number of components, $C' = C - Z - r = 3 - 0 - 1 = 2$

---

***Example 3.3.3*** Determine the number of components in a system containing $NH_4Cl(s)$, $NH_4^+(aq)$, $Cl^-(aq)$, $H_2O(1)$, $H_3O^+(aq)$, $H_2O(g)$, $NH_3(g)$, $OH^-(aq)$, and $NH_4OH(aq)$.

***Solution*** We have

Number of constituents, $C = 9$

Number of equilibrium reactions, $r = 5$:

$$NH_4Cl(s) + aq \rightleftharpoons NH_4^+(aq) + Cl^-(aq)$$

$$NH_4^+(aq) + 2H_2O \rightleftharpoons NH_4OH(aq) + H_3O^+(aq)$$

$$NH_3(g) + H_2O(l) \rightleftharpoons NH_4OH(aq)$$

$$2H_2O(1) \rightleftharpoons H_3O^+(aq) + OH^-(aq)$$

$$H_2O(l) \rightleftharpoons H_2O(g)$$

Number of restricting conditions, $Z = 1$:

$$x_{NH_4^+} + x_{H_3O^+} = x_{Cl^-} + x_{OH^-} \qquad \text{(condition of electroneutrality)}$$

Hence, Number of components, $C' = C - r - Z = 9 - 5 - 1 = 3$

***Example 3.3.4***

Consider a homogeneous mixture of four ideal gases capable of undergoing the reaction

$$\nu_1 A_1 + \nu_2 A_2 \rightleftharpoons \nu_3 A_3 + \nu_4 A_4$$

Determine the components if we start with (a) arbitrary amounts of $A_1$ and $A_2$ only, (b) arbitrary amounts of all the four gases, and (c) $\nu_1$ moles of $A_1$ and $\nu_2$ moles of $A_2$ only.

***Solution***

**(a)** Number of constituents, $C = 4$:

$$A_1,\ A_2,\ A_3 \text{ and } A_4$$

Number of reactions, $r = 1$
Number of restrictions, $Z = 1$:

$$[A_3] : [A_4] :: \nu_3 : \nu_4$$

Hence, Number of components, $C' = C - r - Z = 4 - 1 - 1 = 2$

**(b)** Number of constituents, $C = 4$
Number of reactions, $r = 1$
Number of restrictions, $Z = 0$
Number of components, $C' = C - r - Z = 4 - 1 - 0 = 3$

**(c)** Number of constituents, $C = 4$
Number of reactions, $r = 1$
Number of restrictions, $Z = 2$

$$[A_1] : [A_2] :: \nu_1 : \nu_2 \text{ and } [A_3] : [A_4] :: \nu_3 : \nu_4$$

Hence, Number of components, $C' = C - r - Z = 4 - 1 - 2 = 1$

## REVISIONARY PROBLEMS

3.1 (a) Explain with suitable examples the terms involved in the phase rule

$$F + P = C + 2$$

(b) What do you understand by the reactive and nonreactive systems?

3.2 (a) Derive the expression of phase rule as applicable to a nonreactive system.
(b) Prove that the expression of phase rule remains unchanged even if one of the components is missing in some of the phases present at equilibrium.

3.3 (a) What is the difference between the number of constituents and the number of components of a given system?
(b) Show that for the reactive systems, the form of the phase rule remains unaltered provided we define the number of components as

$$C' = C - r - Z$$

where $C$ is the number of constituents present, $r$ is the number of independent reactions, and $Z$ is the number of independent restricting equations.

## TRY YOURSELF PROBLEMS

3.1 Show explicitly without using the phase rule that the two-component system has (1) $F = 3$ if $P = 1$; (2) $F = 2$ if $P = 2$; (3) $F = 1$ if $P = 3$; and (4) $F = 0$ if $P = 4$.

3.2 Show that it is not possible to have more than three phases at equilibrium for a one-component system.

3.3 Blue copper sulphate crystals decompose and release their water of hydration when heated. How many phases and components are present in an otherwise empty heated vessel?

3.4 The absence of a few components in some phases does not alter the phase rule relation, $F + P - C + 2$. Derive the above conclusion by taking the following system of three components distributed over five phases:

$$\underset{\text{Phase 1}}{A + B} \rightleftharpoons \underset{\text{Phase 2}}{B + C} \rightleftharpoons \underset{\text{Phase 3}}{A + B + C} \rightleftharpoons \underset{\text{Phase 4}}{A} \rightleftharpoons \underset{\text{Phase 5}}{A + C}$$

3.5 How many degrees of freedom, number of phases and number of components are present in each of the following systems:

(a) Solid carbon in equilibrium with gaseous CO, $CO_2$ and $O_2$ at 100 °C.

(b) Solid $CaCO_3$ is heated in a sealed tube.

(c) $I_2$ dispersed between liquid water and $CCl_4$ at 1 atm pressure with no solid $I_2$ present.

(d) $CuSO_4 \cdot 5H_2O(s) \rightleftharpoons CuSO_4 \cdot 3H_2O(s) + 2H_2O(g)$

(e) $Fe(s) + H_2O(g) \rightleftharpoons FeO(s) + H_2(g)$

(f) Liquid and its vapour in equilibrium with each other.

(g) An aqueous solution of NaCl and $Na_2SO_4$.

(h) An aqueous solution of KCl and $Na_2SO_4$.

(i) $CH_3COONH_4(s)$, $CH_3COO^-(aq)$, $NH_4^+(aq)$, $H_3O^+(aq)$, $OH^-(aq)$, $CH_3COOH(aq)$, $H_2O(1)$, $H_2O(g)$, taking hydrolysis into account.

(j) An azeotrope in a binary system.

(k) Water vapour heated to 2000 K so that $H_2O$, $H_2$, $O_2$, OH, O and H are present at equilibrium.

(l) An arbitrary mixture of $H_2$ and $O_2$ has been heated to 2000 K, so that $H_2O$, $H_2$, $O_2$, OH, H and O are present at equilibrium.

(m) A pure substance at its critical point.

(n) An aqueous solution containing LiCl, NaCl, KC1 and CsCl.

3.6 Suppose that a one-component system exhibits a gas phase, a liquid phase and three solid modifications. How many one-, two-, three-, and four-phase, equilibria are theoretically possible?

3.7 Compute the number of components in the following systems:

(i) A solution containing $Na^+$, $Cl^-$, $Ag^+$, $NO_3^-$, AgCl(s) and $H_2O$.

(ii) A solution containing $H^+$, $OH^-$, $Na^+$, $Cl^-$, $Ag^+$, $NO_3^-$, AgCl(s) and $H_2O$.

(iii) A solution containing $H_2O$, $Na^+$, $Cl^-$, $K^+$, $NO_3^-$, $NH_4^+$, $NH_3$, $H^+$ and $OH^-$.

(iv) An aqueous solution containing $H_3PO_4$, $H_2PO_4$, $HPO_4^-$, $PO_4^{3-}$, $Na^+$, and $H^+$ at 1 atm pressure.

(v) $NH_4Cl(s)$, $NH_4^+(aq)$, $Cl^-(aq)$, $H_2O(1)$, $H_3O^+(aq)$, $H_2O(g)$, $NH_3(g)$, $OH^-(aq)$, and $NH_4OH(aq)$.

(Ans. (i) 4, (ii) 4, (iii) 6, (iv) 3, (v) 3)

# 4 Solutions

## 4.1 IDEAL SOLUBILITY OF GASES IN LIQUIDS

**Application of the Phase Rule**

Solubility of a gas in a liquid depends on temperature and pressure. Therefore, both these variables should be stated while expressing the solubility of a gas in a liquid. This conclusion can also be arrived at with the help of the phase rule. In the present case, we have

Number of components, $C = 2$ (Gas and liquid)
Number of phases, $P = 2$ (Liquid phase and gaseous phase)

Thus Degrees of freedom of the system, $F = C + P - 2 = 2 + 2 - 2 = 2$

that is, the system is bivariant and values of two parameters, namely, temperature and pressure, must be stated in order to define the system completely.

**Thermodynamic Derivation of Ideal Law of Solubility of Gases**

At the saturation point, where no more gas dissolves in a solution, there exists an equilibrium between the dissolved gas and the undissolved gas, i.e.

Dissolved gas in solution $\rightleftharpoons$ Undissolved gas

The thermodynamic condition of equilibrium is

$$\mu_{2(\text{sol})}(T, p, x_2) = \mu^*_{2(\text{g})}(T, p) \tag{4.1.1}$$

where $x_2$ is the amount fraction of the gas in the saturated solution. If the solution is assumed to be ideal, then

$$\mu_{2(\text{soln})} = \mu^*_{2(\text{l})} + RT \ln x_2 \qquad (4.1.2)^{\dagger}$$

Substituting Eq. (4.1.2) in Eq. (4.1.1), we get

$$\mu^*_{2(\text{l})} + RT \ln x_2 = \mu^*_{2(\text{g})}$$

or

$$\ln x_2 = \frac{\mu^*_{2(\text{g})} - \mu^*_{2(\text{l})}}{RT} = \frac{\Delta_{\text{vap}}\mu^*_2}{RT} \tag{4.1.3}$$

---

†If the solution is not ideal but an ideally dilute solution, then $\mu^*_{2(\text{l})}$ is replaced by $\mu_{2(\text{l, hs})}$ where $\mu_{2(\text{l, hs})}$ represents the chemical potential of liquid solute in a hypothetical state. See, section 4.6 for detail. In applying Eq. (4.1.2), the temperature of the gas is below its critical temperature so that the gas exists in the liquid form in the solution.

According to Eq. (4.1.3), the amount fraction of solute in a saturated solution is a function of temperature of the solution. To determine this functionality, we differentiate Eq. (4.1.3) with respect to $x_2$, keeping $p$ constant. Thus, we have

$$\frac{1}{x_2} = \frac{1}{R}\left\{\frac{\partial(\Delta_{\text{vap}}\mu_2^*/T)}{\partial T}\right\}_p \left\{\frac{\partial T}{\partial x_2}\right\}_p$$

Using Gibbs-Helmhlotz equation, we get

$$\frac{1}{x_2} = -\frac{\Delta_{\text{vap}}H_{2,\text{m}}}{RT^2}\left(\frac{\partial T}{\partial x_2}\right)_p$$

or

$$\frac{\mathrm{d}x_2}{x_2} = -\frac{\Delta_{\text{vap}}H_{2,\text{m}}}{R}\frac{\mathrm{d}T}{T^2} \qquad (p \text{ constant}) \qquad (4.1.4)$$

To determine the variation of $x_2$ with $T$, we integrate the above expression to give

$$\int_1^{x_2}\frac{\mathrm{d}x_2}{x_2} = -\int_{T_\text{b}^*}^{T}\frac{\Delta_{\text{vap}}H_{2,\text{m}}}{R}\frac{\mathrm{d}T}{T^2} \qquad (4.1.5)$$

where $T_\text{b}^*$ is the boiling point of the pure liquid solute ($x_2 = 1$) and $x_2$ is the solubility of the solute at temperature $T$ of the solution.

Assuming $\Delta_{\text{vap}}H_{2,\text{m}}$ to be independent of temperature, Eq. (4.1.5) yields

$$\ln x_2 = \frac{\Delta_{\text{vap}}H_{2,\text{m}}}{R}\left(\frac{1}{T} - \frac{1}{T_\text{b}^*}\right) \qquad (4.1.6)$$

Equation (4.1.6) is an expression of the *ideal law of solubility*. According to this law, the solubility of a substance depends only on the characteristics of the gas through $\Delta_{\text{vap}}H_{2,\text{m}}$ and $T_\text{b}^*$ and is independent of the nature of the solvent.

**Factors Determining Ideal Solubility**

The following dependence on the solubility of the gases may be concluded.

- A comparatively lesser value of $T_\text{b}^*$ makes the right side of Eq. (4.1.6) more negative indicating the lesser value of $x_2$ (i.e. lesser is the solubility). For example, hydrogen and helium are much less soluble as compared to carbon dioxide ammonia and sulphur dioxide.
- A comparatively larger value of $T_\text{b}^*$ makes the right side of Eq. (4.1.6) lesser negative indicating the larger value of $x_2$ (i.e. larger is the solubility).
- A comparatively lesser value of $\Delta_{\text{vap}}H_{2,\text{m}}$ makes the right side of Eq. (4.1.6) lesser positive indicating the lesser value of $x_2$ (i.e. lesser is the solubility).
- Increasing the value of $T$ makes the right side lesser positive indicating the lesser value of $x_2$ (i.e. lesser is the solubility).

**Correlating Solubilities at Two Temperatures**

If $x_2$ and $x_2'$ are the solubilities of gases at temperatures $T_1$ and $T_2$, respectively, we may write

$$\ln x_2 = \frac{\Delta_{vap} H_{2,m}}{R}\left(\frac{1}{T_1} - \frac{1}{T_b^*}\right) \tag{4.1.7}$$

$$\ln x_2' = \frac{\Delta_{vap} H_{2,m}}{R}\left(\frac{1}{T_2} - \frac{1}{T_b^*}\right) \tag{4.1.8}$$

Subtracting Eq. (4.1.7) from Eq. (4.1.8), we get

$$\ln \frac{x_2'}{x_2} = \frac{\Delta_{vap} H_{2,m}}{R}\left(\frac{1}{T_2} - \frac{1}{T_1}\right) \tag{4.1.9}$$

Equation (4.1.9) relates the solubility of a gas in a solvent at two temperatures of the solution.

**Exception to Ideal Solubility**

It may be point out here that Eq. (4.1.6) represents ideal gas solubility and is independent of the nature of the solvent. Solubility of a gas which can interact with the solvent is very much enhanced. That is why gases like ammonia, carbon dioxide, sulphur dioxide and hydrochloric acid are very easily soluble in water. The interaction reactions are:

$$NH_3 + H_2O \rightleftharpoons NH_4OH \rightleftharpoons NH_4^+ + OH^-$$
$$CO_2 + H_2O \rightleftharpoons H_2CO_3 \rightleftharpoons HCO_3^- + H^+$$
$$SO_2 + H_2O \rightleftharpoons H_2SO_3 \rightleftharpoons HSO_3^- + H^+$$
$$HCl + H_2O \rightleftharpoons H_3O^+ + Cl^-$$

## Coefficients for Expressing Solubility of a Gas in a Liquid

Solubilities of gases can be expressed either in terms of *absorption coefficient* or the *coefficient of solubility.*

**Absorption Coefficient**

The absorption coefficient was suggested by R. Bunsen and is known as *Bunsen absorption coefficient.* At a given temperature, this coefficient is defined as *the volume of gas, reduced to STP* (0 °C and 1 atm pressure), *that has been dissolved by unit volume of solvent under a partial pressure of* 1 atm *of the gas.* If $\upsilon_0$ is the volume of gas dissolved, reduced to STP, by the volume $V$ of the solvent under the partial pressure $p$ of the gas, the absorption coefficient is given by

$$\alpha = \frac{\upsilon_0/p}{V} \tag{4.1.10}$$

The amount of gas dissolved by unit volume of solvent under the partial pressure $p$ is

$$c = \frac{(\upsilon_0/\upsilon_m)}{Vp} = \frac{\alpha}{\upsilon_m} \tag{4.1.11}$$

where $\upsilon_m$ is the molar volume of the gas at STP.

**Coefficient of Solubility**

The second unit of *coefficient of solubility* was suggested by W. Ostwald and is defined as *the volume of gas measured under given conditions of temperature and pressure, that has been dissolved in a unit volume of the solvent.* If $\upsilon$ is the volume of gas dissolved by the volume $V$ of the solvent, then the coefficient of solubility is

$$\beta = \frac{\upsilon}{V} \tag{4.1.12}$$

**Relation between the Two Coefficients**

Assuming ideal behaviour for gas, we can write

$$\frac{p\upsilon}{nRT} = \frac{(1\ \text{atm})\upsilon_0}{nR(273\ \text{K})}$$

Thus $$\upsilon = \frac{\upsilon_0 T}{(273\ \text{K})(p/\text{atm})} \tag{4.1.13}$$

Substituting Eq. (4.1.13) in Eq. (4.1.12), we get

$$\beta = \frac{\upsilon}{V} = \frac{\upsilon_0 T}{(273\ \text{K})(p/\text{atm})} \cdot \frac{1}{V} = \left(\frac{\upsilon_0}{Vp}\right)\frac{T(1\ \text{atm})}{(273\ \text{K})}$$

or $$\beta = \alpha \frac{T(1\ \text{atm})}{(273\ \text{K})} \tag{4.1.14}$$

## 4.2 EFFECT OF PRESSURE AND TEMPERATURE ON THE SOLUBILITY OF GASES

**Effect of Pressure – Henry's Law**

The solubility of a gas in a liquid increases with the increase in the pressure of the gas. The quantitative relation connecting the solubility with pressure was stated by W. Henry and is known as *Henry's law.* The law states that *at a given temperature the mass of dissolved gas in a given volume of solvent is proportional to the pressure of the gas with which it is in equilibrium.*[†] If $m$ is the mass of the gas dissolved by unit volume of solvent at equilibrium pressure $p$, then according to Henry's law, we have

$$m \propto p$$

or $$m = kp \tag{4.2.1}$$

where $k$ is the constant of proportionality. Thus, if a graph of the solubility and the external equilibrium pressure is plotted, a straight line would be observed (Fig. 4.2.1).

***Example 4.2.1***

At 273.15 K, 500 $cm^3$ of water dissolves 15.03 $cm^3$ of $CH_4$(STP) under a partial pressure of methane of 1 atm. If Henry's law holds, what pressure is required to dissolve 0.001 mol methane in 300 $cm^3$ water?

***Solution***

Amount of methane dissolved in 500 $cm^3$ of water under a partial pressure of 1 atm of methane is

[†]For the solubility of a gas from a mixture of gases, the equilibrium pressure is the partial pressure of the gas.

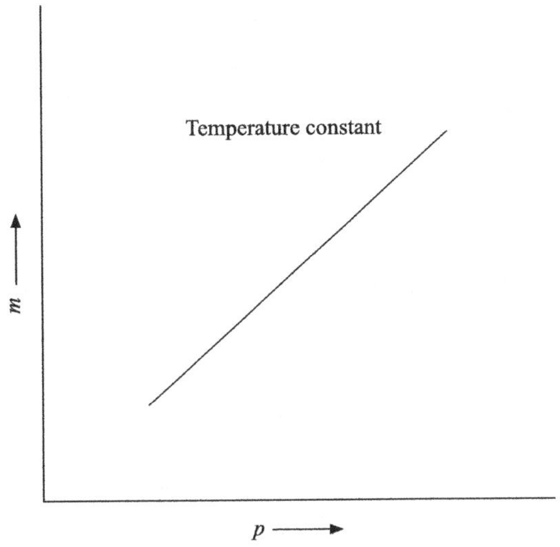


**Fig. 4.2.1 Variation of solubility with pressure**

$$n = \frac{v_0}{V_m} = \frac{15.03\ \text{cm}^3}{22\,414\ \text{cm}^3\ \text{mol}^{-1}} = 6.706 \times 10^{-4}\ \text{mol}$$

Amount of methane dissolved per unit volume of water is

$$n' = \frac{6.706 \times 10^{-4}\ \text{mol}}{500\ \text{cm}^3} = 1.341 \times 10^{-6}\ \text{mol cm}^{-3}$$

Amount of methane to be dissolved per unit volume of water is

$$n'' = \frac{0.001\ \text{mol}}{300\ \text{cm}^3} = 3.333 \times 10^{-6}\ \text{mol cm}^{-3}$$

Since mass of gas $\propto$ amount of gas, we can write Henry's law as

$$n'' = k'p'' \quad \text{and} \quad n' = k'p'$$

Hence,
$$p'' = \left(\frac{n''}{n'}\right)p' = \left(\frac{3.33 \times 10^{-6}\ \text{mol cm}^{-3}}{1.341 \times 10^{-6}\ \text{mol cm}^{-3}}\right)(1\ \text{atm})$$
$$= 2.485\ \text{atm}$$

---

## A Consequence of Henry's Law

A direct consequence of Henry's law is that the volume of an ideal gas, measured at the experimental pressure, that is dissolved in a given volume of the solvent has a constant value. This can be proved as follows.

If $v$ is the dissolved volume of the gas, measured at the experimental pressure $p$, in a given volume of the solvent at temperature $T$, then according to ideal gas law, we have

$$pv = nRT = \frac{m}{M}RT$$

or
$$v = \left(\frac{m}{p}\right)\frac{RT}{M}$$

Substituting the value of $(m/p)$ from Eq. (4.2.1), we have

$$v = k\frac{RT}{M}$$

Since the right side of the above equation is constant for a given gas ($M$ constant) at a given temperature ($T$ constant), it follows that the volume of the gas dissolved in a given volume of the solvent has a constant value.

**Thermodynamic Derivation of Henry's Law**

The thermodynamic condition of equilibrium between dissolved and undissolved gases is

$$\mu_{2(\text{sol})}(T, p, x_2) = \mu^*_{2(\text{g})}(T, p) \tag{4.2.2}$$

If the solution is considered to be ideally dilute, we have

$$\mu_{2(\text{sol})} = \mu^*_{2(1,\,\text{hs})} + RT \ln x_2$$

where the subscript 2(1, hs) represents hypothetical state of pure solute in the liquid phase.[†] If the gas is assumed to be ideal, then

$$\mu^*_{2(\text{g})} = \mu^\circ_{2(\text{g})} + RT \ln (p/p^\circ)$$

With these, Eq. (4.2.2) becomes

$$\mu^*_{2(1,\,\text{hs})} + RT \ln x_2 = \mu^\circ_{2(\text{g})} + RT \ln (p/p^\circ)$$

or
$$\ln \frac{x_2}{(p/p^\circ)} = \frac{\mu^\circ_{2(\text{g})} - \mu^*_{2(1,\,\text{hs})}}{RT} \tag{4.2.3}$$

At a constant temperature, the right side of Eq. (4.2.3) is a constant. Hence

$$\ln \frac{x_2}{(p/p^\circ)} = \text{constant}$$

or
$$x_2 = (\text{constant})\ p \tag{4.2.4}$$

For a dilute solution, we can write

$$x_2 = \frac{n_2}{n_1 + n_2} \simeq \frac{n_2}{n_1} = \frac{m_2/M_2}{n_1} = \left(\frac{1}{M_2 n_1}\right) m_2$$

i.e.
$$x_2 \propto m_2$$

With this, Eq. (4.2.4) becomes

$$m_2 = kp \qquad \text{(Eq. 4.2.1)}$$

which is Henry's law.

---

[†]See Section 4.6 for the hypothetical state of liquid solute.

**Applicability of Henry's Law**

It will be shown that Henry's law is a special case of more generalized laws such as the distribution law and Raoult's law. Since these laws are applicable only for ideal systems, it is, therefore, expected that Henry's law is also applicable only to ideal systems.

A solution of gas in liquid will show ideal behaviour, provided the solubility of the gas in the liquid is low (i.e. dilute solution) and that the gas itself shows ideal behaviour. The real gases are known to behave ideally under the conditions of high temperature and low pressure, but show deviations at low temperature and high pressure. The deviations are pronounced for easily liquefiable gases. Thus, it is expected that Henry's law will be applicable for low soluble gases at high temperatures and low pressures. The deviations from Henry's law are expected to be large at low temperatures and high pressures. These facts are in agreement with the experimental results.

**Henry's Law and the Distribution Law**

It can be shown that Henry's law is a special case of the distribution law.† According to Henry's law, we have

$$\frac{m}{p} = k \tag{4.2.5}$$

where $m$ is the mass of the gas dissolved by a unit volume of the solvent, $p$ is the equilibrium pressure of the gas, and $k$ is constant whose value depends upon the temperature of the system.

The term $m$ in Eq. (4.2.5) is the concentration term of the dissolved gas expressed in mass of the gas dissolved per unit volume of solvent. It is proportional to the concentration expressed as amount of gas dissolved per unit volume of solvent or per unit volume of solution, if it is assumed that no change of volume occurs when the gas dissolves in the solvent. Similarly, the pressure $p$ of the gas will be proportional to the concentration of the gas expressed as amount of gas per unit volume of the gaseous phase. Thus Eq. (4.2.5) can be written as

$$\frac{\text{Concentration of gas in liquid phase}}{\text{Concentration of gas in gaseous phase}} = \text{Constant} \tag{4.2.6}$$

The value of constant depends only on temperature and not on pressure of the gas.

Equation (4.2.6) is, in fact, the statement of the distribution law. It may be stated that the distribution law is applicable only for common species in both the phases. Thus a gas which undergoes a change in a solvent, such as change in molecular complexity, compound formation between solute and solvent, ionization, etc., will not follow the conventional definition of Henry's law as given by Eq. (4.2.6). However, if concentration of the gas in the liquid phase which exists in the same form as that of the gaseous phase, is determined and substituted in Eq. (4.2.6), the ratio is found to be independent of the pressure of the gas. Thus

---

†See Section 4.19

Henry's law is still applicable provided concentration of the gas in the liquid phase, which exists in the same form as that of gaseous phase, is taken into consideration. One example is the dissolution of ammonia in water, where the following reaction takes place:

$$NH_3 + H_2O \rightleftharpoons NH_4^+ + OH^-$$

If $c_l$ and $c_g$ represent the total concentrations of ammonia in the liquid and gaseous phases, respectively, then $c_l/c_g$ is not constant. If $\alpha$ is the fraction of the dissolved ammonia which has undergone reaction with water, then the ratio $c_l(1 - \alpha)/c_g$ is found to have a constant value irrespective of the total amount of $NH_3$ that has been dissolved.

**Effect of Temperature on Solubility of Gases in Water**

The solubility of gases in water is generally exothermic in nature. In accordance to Le-Chatelier principle, their solubility decreases with increase in temperature. If Henry's law is interpreted as

$$K = \frac{\text{Concentration of gas in liquid phase}}{\text{Concentration of gas in gaseous phase}} \qquad \text{(Eq. 4.2.6)}$$

then the effect of temperture on $K$ may be expressed by the van't Hoff equation:

$$\frac{\mathrm{d}\ln K}{\mathrm{d}T} = \frac{\Delta_r H_m^\circ}{RT^2} \qquad (4.2.7)$$

where $\Delta_r H_m^\circ$ is the differential enthalpy of solution for 1 mol of the gas in the saturated solution at temperature $T$. If $\Delta_r H_m^\circ$ is assumbed to be independent of temperature, the integrated form of the above expression is

$$\ln \frac{K_2}{K_1} = -\frac{\Delta_r H_m}{R}\left(\frac{1}{T_2} - \frac{1}{T_1}\right) \qquad (4.2.8)$$

For the constant external pressure of the gas, we find from Eq. (4.2.6) that

$$K \propto (c_{gas})_{liquid}$$

Hence, Eq. (4.2.8) may be written as

$$\ln \frac{c_2}{c_1} = -\frac{\Delta_r H_m}{R}\left(\frac{1}{T_2} - \frac{1}{T_1}\right) \qquad (4.2.9)^\dagger$$

where $c_1$ and $c_2$ are the solubilites of the gas at tempeturares $T_1$ and $T_2$, respectively.

Since Bunsen absorption coefficient ($\alpha$), is directly proportional to the concentration is gas in the solution, Eq. (4.2.9) may also be expressed as

---

†Comparing Eqs (4.1.9) and (4.2.9) we find that $\Delta_r H_m = -\Delta_{vap} H_{2,m}$ for an ideal solution.

$$\ln \frac{\alpha_2}{\alpha_1} = -\frac{\Delta_r H_m}{R}\left(\frac{1}{T_2} - \frac{1}{T_1}\right) \tag{4.2.10}$$

Accoridng to Eqs (4.2.9) and (4.2.10), it follows that the graph between $\ln c_2$ (or $\ln \alpha$) and $1/T$ is linear with the slope equal to $(-\Delta_r H_m/R)$.

## 4.3 HENRY'S LAW AND RAOULT'S LAW

**Henry's Law in Amount Fraction Form**

Equation (4.2.4) is, in fact, Henry's law expressed in amount fraction form. This, however, may be obtained from the conventional Henry's law expressed in the mass of the gas dissolved in the given volume of solution. If $m_2$ is the mass of the gas that is dissolved at the gaseous pressure $p_2$, then according to Henry's law, we have

$$m_2 = kp_2 \qquad \text{(Eq. 4.2.1)}$$

Dividing throughout by $m_1$, the mass of the solvent, we have

$$\frac{m_2}{m_1} = \left(\frac{k}{m_1}\right)p_2 = k'p_2$$

Dividing the two masses by the respective molar masses, we have

$$\frac{m_2/M_2}{m_1/M_1} = \left(k'\frac{M_1}{M_2}\right)p_2 = k''p_2$$

or

$$\frac{n_2}{n_1} = k''p_2 \tag{4.3.1}$$

The amount fraction of the gas in the above solution is given by

$$x_2 = \frac{n_2}{n_1 + n_2}$$

For a dilute solution as is the case for gases in liquids, $n_2 << n_1$, and thus the amount fraction can be approximated as

$$x_2 \simeq \frac{n_2}{n_1}$$

Substituting this in Eq. (4.3.1), we get

$$x_2 = k''p_2 \tag{4.3.2}$$

Thus for a dilute solution Henry's law can also be stated in terms of amount fraction of gaseous solute in the saturated solution. According to this, *the solubility of a gas expressed in amount fraction is directly proportional to the pressure of the gas.*

**Henry's Law Constant**

The solubility of a gas may be considered from the following two points of view:

(i) $x_2$ may be regarded as the solubility of the gas under pressure $p_2$.

(ii) $p_2$ may be taken as the vapour pressure of a volatile solute when its amount fraction in the dilute solution is $x_2$, the vapour pressure of a solute in a solution is proportional to its amount fraction. Mathematically, we have

$$p_2 = \frac{1}{k''} x_2 = k_H x_2 \tag{4.3.3}$$

where $k_H$ is equal to $1/k''$ and is known as *Henry's law constant.* It has the unit of pressure.

The values of Henry's law constant for some gases in water and benzene at 25 °C are shown in Table 4.3.1.

**Table 4.3.1** Henry's Law constant for some Gases at 25 °C

| | $k_H$/kbar | | | $k_H$/kbar | |
|---|---|---|---|---|---|
| Gas | Water | Benzene | Gas | Water | Benzene |
| $H_2$ | 71.7 | 3.93 | CO | 58.8 | 1.52 |
| $N_2$ | 86.4 | 2.27 | $CH_4$ | 40.4 | 0.49 |
| $O_2$ | 44.1 | 1.24 | $C_2H_6$ | 30.3 | 0.068 |

From Eq. (4.3.3), it follows that the gases will have the same value of $k_H x_2$ provided the gaseous pressure $p_2$ has the same value. From this it follows that the larger the value of $k_H$, the smaller the solubility (in terms of $x_2$) of the gas.

**Correspondence between Henry's and Raoult's Laws**

In a special case where Henry's law is applicable over the entire range of concentrations starting from $x_2 = 0$ (infinitely dilute solution, i.e. pure solvent) to $x_2 = 1$ (pure liquid solute), the constant $k''$ is given as

$$k'' = \frac{x_2}{p_2} = \frac{1}{p_2^*} \quad \text{and thus} \quad k_H = \frac{1}{k''} = p_2^*$$

where $p_2^*$ is the vapour pressure of pure liquid solute ($x_2 = 1$). With this Eq. (4.3.3) can be written as

$$x_2 = \frac{p_2}{p_2^*} \quad \text{or} \quad p_2 = x_2 p_2^* \tag{4.3.4}$$

Equation (4.3.4) is the statement of Raoult's law when applied to a volatile solute. Hence it may be concluded that Raoult's law is a special case of Henry's law; **all systems which obey Raoult's law must satisfy Henry's law, but the reverse is not true unless Henry's law is applicable over the entire range of concentrations.**

**Comment on Eq. (4.3.4)**

Equation (4.3.4) is useful in computing the solubility (in terms of amount fraction $x_2$) of a gas in a solvent at a temperature lower than its critical temperature, provided the solution behaves in an ideal manner. Since $x_2$ and $p_2^*$ have inverse dependence, a lower value of $p_2^*$ results into higher value of $x_2$ (i.e. solubility) and vice versa. A gas which is difficult to liquefy has a low boiling point. Consequently, it has a larger vapour pressure and thus has a lower solubility in a solvent. Easily liquefiable gases have higher boiling points, lower vapour pressures and thus exhibit larger solubility.

Equation (4.3.4) is also useful in estimating the solubility of a gas at a temperature higher than its critical temperature provided the hypothetical vapour pressure of the gas in the liquid form at the said temperature is available. This vapour pressure may be estimated by the extrapolation method or by the use of Clapeyron-Clausisus equation

$$\ln \frac{p_2}{p_1} = -\frac{\Delta_{\text{vap}} H_{\text{m}}}{R}\left(\frac{1}{T_2} - \frac{1}{T_1}\right)$$

provided $\Delta_{\text{vap}}H_{\text{m}}$ may be assumed to be independent of temperatures. For example, the hypothetical vapour pressure of liquid methane (critical temperature – 95.5 °C) at 20 °C is estimated to be 310 atm. Its solubility at 1 atm gaseous pressure in any solvent would be

$$x_2 = \frac{p_2}{p_2^*} = \frac{1\ \text{atm}}{310\ \text{atm}} = 0.003\,2$$

which is very near to its solubility in *n*-hexane.

The solubility of gas in amount fraction is theoretically independent of the nature of the solvent. Any deviation may be attributed to departure from the ideal behaviour.

---

***Example 4.3.1***

(a) For a solution of acetone in chloroform, it is found that the value of Henry's law constant is 150 Torr when the solution is at a temperature of 308 K. Calculate the value of vapour pressure of acetone when the amount fraction is 0.12.
(b) Assuming that Henry's law is applicable over sufficient range of composition to make the calculation valid, compute the composition at which Henry's law pressure of chloroform is equal to Henry's law pressure of acetone at 308 K. (Given: Henry's law constant for chloroform 175 Torr.)

***Solution***

(**a**) Henry's law is given as $p = k_{\text{H}}\, x$

Substituting the given values of $k_{\text{H}}$ and $x$, we get

$$p = (150\ \text{Torr})\ (0.12) = 18.0\ \text{Torr}$$

(**b**) Here $p$ = 18.0 Torr and $k_H$ = 175.0 Torr.

Hence $$x = \frac{p}{k_H} = \frac{18.0 \text{ Torr}}{175 \text{ Torr}} = 0.103$$

***Example 4.3.2***

Henry's law constants as defined by the relation $k_H = p_A/x_A$ for oxygen and nitrogen dissolved in water at 273 K are $2.53 \times 10^9$ Pa and $5.47 \times 10^9$ Pa, respectively. A sample of water at a temperature just above 273 K was equilibrated with air (20% oxygen and 80% nitrogen) at 1 atm. The dissolved gas was separated from a sample of this water and then dried. (a) Determine the composition of this gas. (b) Calculate the freezing point of the saturated water. Given: $K_f$ = 1.86 K kg mol$^{-1}$.

***Solution***

(**a**) Partial pressure of oxygen, $p_{O_2}$ = 0.2 atm = 20 265 Pa

Partial pressure of nitrogen, $p_{N_2}$ = 0.8 atm = 81 060 Pa

Amount fraction of dissolved oxygen in water

$$= \frac{p_{O_2}}{k_H(O_2)} = \frac{20\,265 \text{ Pa}}{2.53 \times 10^9 \text{ Pa}} = 8.01 \times 10^{-6}$$

Amount fraction of dissolved nitrogen in water

$$= \frac{p_{N_2}}{k_H(N_2)} = \frac{81\,060 \text{ Pa}}{5.47 \times 10^9 \text{ Pa}} = 1.48 \times 10^{-5}$$

$$\frac{\text{Amount of dissolved oxygen}}{\text{Amount of dissolved nitrogen}} = \frac{8.01 \times 10^{-6}}{1.48 \times 10^{-5}} = 0.541$$

Hence $$\text{Amount per cent of } O_2 = \frac{0.541}{1.541} \times 100 = 35.11$$

$$\text{Amount per cent of } N_2 = 100 - 35.11 = 64.89$$

(**b**) Total amount fraction of dissolved gas = $8.01 \times 10^{-6} + 1.48 \times 10^{-5} = 2.281 \times 10^{-5}$
For such a small value of amount fraction, we may write

$$n_{\text{solute}} = 2.281 \times 10^{-5} \text{ mol} \quad \text{and} \quad n_{\text{water}} = 1 \text{ mol}$$

Hence, molality of the solution is

$$m = \frac{n_{\text{solute}}}{m_{\text{solvent}}} = \frac{(2.28 \times 10^{-5} \text{ mol})}{(18 \times 10^{-3} \text{ kg})} = 1.267 \times 10^{-3} \text{ mol kg}^{-1}$$

The depression in freezing point of water is

$$-\Delta T_f = K_f m = (1.86 \text{ K kg mol}^{-1})\,(1.267 \times 10^{-3} \text{ mol kg}^{-1})$$
$$= 0.002\,36 \text{ K}$$

The freezing point of the saturated water is – 0.002 36 °C.

## 4.4 RAOULT'S LAW FOR SOLVENT AND HENRY'S LAW FOR SOLUTE

**Application of Duhem-Margules Equation**

It can be shown that if Henry's law is applicable to the volatile solute, then Raoult's law is applicable to the volatile solvent. We start with the Duhem-Margules equation (Section 4.15), such that

$$\frac{\mathrm{d}\log(p_1/p^\circ)}{\mathrm{d}\log x_1} = \frac{\mathrm{d}\log(p_2/p^\circ)}{\mathrm{d}\log x_2} \tag{4.4.1}$$

Henry's law in the amount fraction form as applicable to the solute is

$$p_2 = k_\mathrm{H} x_2 \qquad \text{(Eq. 4.3.3)}$$

Taking logarithm, we have

$$\log(p_2/p^\circ) = \log(k_\mathrm{H}/p^\circ) + \log x_2 \quad \text{or} \quad \mathrm{d}\log(p_2/p^\circ) = \mathrm{d}\log x_2$$

i.e.

$$\frac{\mathrm{d}\log(p_2/p^\circ)}{\mathrm{d}\log x_2} = 1 \tag{4.4.2}$$

Hence, from Eq. (4.4.1), we have

$$\frac{\mathrm{d}\log(p_1/p^\circ)}{\mathrm{d}\log x_1} = 1 \tag{4.4.3}$$

which on integration gives

$$\log(p_1/p^\circ) = \log x_1 + \text{constant}$$

$$p_1/p^\circ = K' x_1$$

or

$$p_1 = K x_1 \tag{4.4.4}$$

For the pure solvent $x_1 = 1$ and $p_1 = p_1^*$ where $p_1^*$ is the vapour pressure of the pure solvent. Thus

$$K = p_1^*$$

With the above identity, Eq. (4.4.4) becomes

$$p_1 = x_1 p_1^* \tag{4.4.5}$$

Equation (4.4.5) is obtained provided Eq. (4.4.3) is applicable to the solvent. In deriving Eq. (4.4.3), we have used Henry's law in the amount fraction form as applicable to the solute (Eq. 4.3.3). Thus, it is obvious that Eq. (4.4.5) holds good for the solvent as long as Eq. (4.3.3) is obeyed by the solute. If Eq. (4.3.3) is applicable to solute only in the limited range of low concentration (in dilute solutions where $x_2$ is small), then Eq. (4.4.5) will be applicable to solvent over the same range of concentration. When Henry's law is not obeyed by the solute, Raoult's law is also not obeyed by the solvent.

**Illustration**

To make the above discussion more clear, let us take a specific example of a solution of liquid A in liquid B. If the partial pressure of B is plotted against its amount fraction in the solution, we get a curve as shown in Fig. 4.4.1a. The curve as expected from Raoult's law can be obtained by connecting the zero point at the left ($x_B = 0$) with the point representing $p_B^*$ at the right ($x_B = 1$). It can be seen from Fig. 4.4.1a that Raoult's law curve meets the partial vapour pressure curve tangentially in the region where $x_B \rightarrow 1$. Thus, we see that Raoult's law is applicable only in the limited range of concentrations where B is present in excess amounts and is thus acting as a solvent. Now if a line tangent to the curve of the observed partial vapour pressure at very low concentration of B is drawn, we get the graph of Henry's law. The extrapolation of this line to the right axis where $x_B = 1$ gives the hypothetical vapour pressure which would be observed if Henry's law be applicable over the entire range of concentration, that is, up to $x_B = 1$.

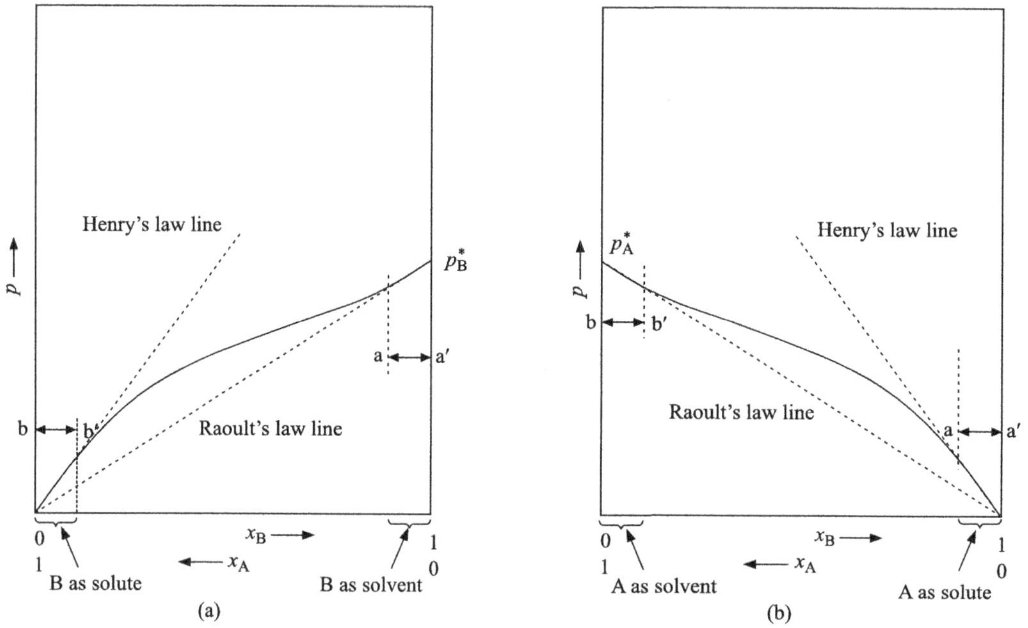


**Fig. 4.4.1 Schematic graphs indicating applicability of Henry's law to the solute and Raoult's law to the solvent. (a) For liquid B and (b) for liquid A.**

From Eq. (4.3.3), it follows that at $x_B = 1$, we have

$$p_B = k_H$$

Thus, Henry's law constant has a value of the vapour pressure which pure B would have if the vapour pressure obeyed Henry's law instead of Raoult's law. Note that according to Raoult's law, the constant $K_H$ would have been equal to $p_B^*$. In other words, Henry's law constant for nonideal system is different from Raoult's law constant.

Now, if a graph between the partial vapour pressure of liquid A and its amount fraction in the solution is drawn, we would get a plot as shown in Fig. 4.4.1b. The associated Raoult's law curve and Henry's law curve are also shown in Fig. 4.4.1b. It may be seen that Raoult's law curve meets the partial vapour pressure curve tangentially in the region $x_A \rightarrow 1$ indicating that Raoult's law is applicable to the solvent. On the other hand, Henry's law curve meets the partial vapour pressure curve in the region $x_A \rightarrow 0$ indicating that Henry's law is applicable to solute.

Figure 4.4.1 also illustrates the fact that Raoult's law is applicable to the solvent over the same range where Henry's law is applicable to the solute. For example, if the applicable range of Henry's law for liquid A (solute) were from 0 to 8% of A (shown by the line aa′ in Fig. 4.4.1b), then the applicable range for B (solvent) will be from 92 to 100% of B as shown by the line aa′ in Fig. 4.4.1a. On the other side, if Henry's law applicable range for B were from 0 to 5% of B (shown by the line bb′ in Fig. 4.4.la), then Raoult's law applicable range for liquid A will be from 95 to 100% of A as shown by the lines bb′ in Fig. 4.4.1b.

***Example 4.4.1***

The total vapour pressure of a 4 mol % solution of a gas B in water at 293 K is 50.0 Torr; the vapour pressure of pure water is 17.0 Torr at this temperature. Applying Henry's and Raoult's laws, calculate the two partial pressures and the total vapour pressure for a solution of 5 mol % B.

***Solution***

The given data are

$$p^*_{\text{water}} = 17.0 \text{ Torr}$$

$$p_{\text{total}} \text{ (4 mol \% solution)} = p_B + p_{\text{water}} = 50.0 \text{ Torr}$$

$$x_B = 0.04 \quad \text{and} \quad x_{\text{water}} = 0.96$$

Now according to Raoult's law, we have

$$p_{\text{water}} = x_{\text{water}} p^*_{\text{water}} = 0.96 \times 17.0 \text{ Torr} = 16.3 \text{ Torr}$$

Thus $$p_B = p_{\text{total}} - p_{\text{water}} = 50.0 \text{ Torr} - 16.3 \text{ Torr} = 33.7 \text{ Torr}$$

Now Henry's law constant for the gas B is

$$k_H(B) = \frac{p_B}{x_{NH_3}} = \frac{33.7 \text{ Torr}}{0.04} = 842 \text{ Torr}$$

Hence for 5 mol % solution, we have

$$p_{NH_3} = k_H(B)\, x_{NH_3} = (842 \text{ Torr})\,(0.05) = 42.1 \text{ Torr}$$

$$p_{\text{water}} = p^*_{\text{water}}\, x_{\text{water}} = (17 \text{ Torr})\,(0.95) = 16.1 \text{ Torr}$$

Thus $$p_{\text{total}} \text{ (5 mol \% solution)} = p_B + p_{\text{water}} = 42.1 \text{ Torr} + 16.1 \text{ Torr} = 58.2 \text{ Torr}$$

## 4.5 DERIVATIONS OF RAOULT'S AND HENRY'S LAWS FROM KINETIC-MOLECULAR THEORY

Raoult's and Henry's laws may be derived from the view-point of kinetic-molecular theory as follows.

**Raoult's Law**

For a pure liquid solvent in equilibrium with its vapour, we may write

$$r_{1,\,\mathrm{cond}} = r_{1,\,\mathrm{evap}} \tag{4.5.1}$$

where $r_{1,\,\mathrm{cond}}$ and $r_{1,\,\mathrm{evap}}$ are the rates of condensation of solvent molecules from vapour phase and of evaporation of solvent molecules from pure liquid solvent, respectively. The rate of condensation may be taken as proportional to the number of collisions per second of vapour molecules with the liquid's surface and this number is proportional to pressure $p_1^*$. Thus, we have

$$r_{1,\,\mathrm{cond}} = r_{1,\,\mathrm{evap}} = bp_1^* \tag{4.5.2}$$

where $b$ is the constant of proportionality.

In an ideal binary liquid solution, the vapour pressure of solvent is $p_1$. Hence

$$r_{1,\,\mathrm{cond}} = bp_1 \tag{4.5.3}$$

Since the sizes of solute and solvent molecules in an ideal solution are identical, the number of molecules of solvent in unit surface area of the solution will be $x_1$ times the number of solvent molecules in unit surface area of pure solvent, where $x_1$ is the amount fraction of solvent in solution. Consequently, the rate of evaporation of molecules from the surface of solution will also be $x_1$ times the rate of evaporation from the pure solvent's surface. Hence

$$r_{1,\,\mathrm{evap}} = x_1(bp_1^*) \tag{4.5.4}$$

Equating Eqs (4.5.3) and (4.5.4), we get

$$p_1 = x_1 p_1^*$$

which is Raoult's law.

**Henry's Law**

For a pure liquid solute in equilibrium with its vapour, we may write

$$r_{2,\,\mathrm{cond}} = r_{2,\,\mathrm{evap}} \tag{4.5.5}$$

where $r_{2,\,\mathrm{cond}}$ and $r_{2,\,\mathrm{evap}}$ are the rates of condensation of solute molecules from vapour phase and of evaporation of solute molecules from pure liquid solute, respectively. As in the case of pure solvent, we write

$$r_{2,\,\mathrm{cond}} = r_{2,\,\mathrm{evap}} = bp_2^* \tag{4.5.6}$$

In an ideally dilute solution, the rate of condensation of solute molecules is proportional to the partial vapour pressure $p_2$. But because of the fact that in the liquid phase solute interacts only with solvent molecules, the proportionality constant in the present case will be different from that appearing in Eq. (4.5.6). Thus we write

$$r_{2,\,\mathrm{cond}} = b'p_2 \tag{4.5.7}$$

For the rate of evaporation of solute molecules from an ideally dilute solution, we consider the following two facts:

(i) *Size of solute and solvent molecules are different* If $x_2$ is the amount fraction of solute in solution, then the number of solute molecules present in unit surface area of solution will be reduced by a factor $ax_2$ compared with those present in unit surface area of pure solute, where the constant $a$ takes into account the different sizes of solute and solvent molecules.

(ii) *Solute molecule interacts with solvent molecules only* Because of this interaction, the rate of evaporation of solute molecules will not be simply equal to $ax_2$ times but $c(ax_2)$ times the rate of evaporation from pure liquid solute, where the constant $c$ takes into account the different interactions in pure solute and in ideally dilute solution. Thus, we write

$$r_{2,\,\text{evap}} = \{c(ax_2)\}\{bp_2^*\} \tag{4.5.8}$$

Equating Eqs (4.5.7) and (4.5.8), we have

$$b'p_2 = \{c(ax)\}\{bp_2^*\} \quad \text{or} \quad p_2 = \left(\frac{c\,a\,b\,p_2^*}{b'}\right)x_2$$

i.e. $$p_2 = k_H x_2$$

which is Henry's law.

## 4.6 IDEAL AND IDEALLY DILUTE SOLUTIONS

**Ideal Solution**

*An ideal solution is the one whose components follow Raoult's law in the entire range of composition.* Each component also follows Henry's law where the Henry's law constant is equal to the vapour pressure of the pure component (see Section 4.15).

**Ideally Dilute Solution**

A real solution in the limit $x_1 \rightarrow 1$ and $x_2 \rightarrow 0$ is known as *ideally dilute solution.* In this solution, each solute molecule is surrounded by solvent molecules. The separation between solute molecules is so large that they do not interact with one another. *Raoult's law for solvent* and *Henry's law* for *solute are the defining equations for an ideally dilute solution.* Thus, the vapour pressures of solvent and solute of an ideally dilute solution are given by the expressions:

*Raoult's law for solvent* $$p_1 = x_1\, p_1^* \tag{4.6.1a}$$

*Henry's law for solute* $$p_2 = x_2 k_H \tag{4.6.1b}$$

**Expressions of Chemical Potentials**

Equations (4.6.1) can be used to derive the expressions of chemical potential for solvent and solute in an ideally dilute solution. From the thermodynamic criterion of equilibrium, we write

$$\mu_{1(\text{sol})} = \mu_{1(\text{v})} \tag{4.6.2a}$$

$$\mu_{2(\text{sol})} = \mu_{2(\text{v})} \tag{4.6.2b}$$

For the vapour phase, we have

$$\mu = \mu^* + RT \ln (p/p^\circ)$$

With the above expression, Eqs (4.6.2a) and (4.6.2b) become

$$\mu_{1(\text{sol})} = \mu^*_{1(\text{v})} + RT \ln(p_1/p^\circ)$$

$$\mu_{2(\text{sol})} = \mu^*_{2(\text{v})} + RT \ln(p_2/p^\circ)$$

which on making use of Eqs (4.6.1a) and (4.6.1b) become

$$\mu_{1(\text{sol})} = \{\mu^*_{1(\text{v})} + RT \ln(p^*_1/p^\circ)\} + RT \ln x_1 \tag{4.6.3a}$$

$$\mu_{2(\text{sol})} = \{\mu^*_{2(\text{v})} + RT \ln(k_\text{H}/p^\circ)\} + RT \ln x_2 \tag{4.6.3b}$$

or

$$\mu_{1(\text{sol})} = \mu^*_{1(\text{l})} + RT \ln x_1 \tag{4.6.4a}$$

$$\mu_{2(\text{sol})} = \mu^*_{2(\text{l, hs})} + RT \ln x_2 \tag{4.6.4b}$$

where the subscript 'hs' stands for the hypothetical state. From Eq. (4.6.4a), we conclude that the standard state of solvent in ideally dilute solution is the pure liquid solvent. For solute, the standard state is the hypothetical state of solute where its vapour pressure is $k_\text{H}$. Its value is determined by extrapolating the experimental vapour pressure of solute in the region $x_2 \rightarrow 0$ to a value of $x_2 \rightarrow 1$. This hypothetical state has the properties which the pure solute would have if its limiting low-concentration properties in solution were to be retained in a pure substance. The actual 'properties of the pure solute will be different from those of the hypothetical state, since in the pure form, the solute molecule is surrounded by another solute molecules whereas in the hypothetical state solute molecule is surrounded by solvent molecules only. From Eq. (4.6.4b), we may define the chemical potential for the hypothetical state as

$$\mu^*_{2(\text{l, hs})} = \lim_{x_2 \rightarrow 1} (\mu_{2(\text{sol})} - RT \ln x_2) \tag{4.6.5}$$

## 4.7 VARIATION OF HENRY'S LAW CONSTANT WITH TEMPERATURE

From Eqs (4.6.3b) and (4.6.4b), we may write

$$\mu^*_{2(\text{l, hs})} = \mu^*_{2(\text{v})} + RT \ln(k_\text{H}/p^\circ)$$

or

$$\ln(k_\text{H}/p^\circ) = \frac{\mu^*_{2(\text{l, hs})} - \mu^*_{2(\text{v})}}{RT} = -\frac{\Delta_\text{vap}\mu^*_{2(\text{l, hs})}}{RT} \tag{4.7.1}$$

According to Eq. (4.7.1), the Henry's law constant is a function of temperature of the solution. To determine functionality, we differentiate Eq. (4.7.1) with respect to temperature, i.e.

$$\left(\frac{\partial \ln(k_\text{H}/p^\circ)}{\partial T}\right)_p = -\frac{1}{R}\left(\frac{\partial(\Delta_\text{vap}\mu^*_{2(\text{l, hs})}/T}{\partial T}\right)_p \tag{4.7.2}$$

Using Gibbs-Helmhlotz equation, we get

$$\left(\frac{\partial \ln (k_H/p^\circ)}{\partial T}\right)_p = \frac{\Delta_{vap}H^*_{m,2(1,hs)}}{RT^2} \tag{4.7.3}$$

To determine the variation of $k_H$ with $T$, we integrate the above expression to give

$$\int_{T_1}^{T_2} d(\ln k_H/p^\circ) = \int_{T_1}^{T_2} \frac{\Delta_{vap}H^*_{m,2(1,hs)}}{R}\frac{dT}{T^2} \tag{4.7.4}$$

$$\ln \frac{k'_H}{k_H} = \frac{\Delta_{vap}H^*_{m,2(1,hs)}}{R}\left(\frac{1}{T_1} - \frac{1}{T_2}\right) \tag{4.7.5}$$

where $k'_H$ and $k_H$ are Henry's law constants at temperatures $T_2$ and $T_1$, respectively.

Since $p_2 = k_H x_2$, the variation of $x_2$ with temperature at constant $p_2$ will given by

$$\ln \frac{x_2}{x'_2} = \frac{\Delta_{vap}H^*_{m,2(1,hs)}}{R}\left(\frac{1}{T_1} - \frac{1}{T_2}\right) \tag{4.7.6}$$

which is the same as given by Eq. (4.1.9).

From Eq. (4.7.5), it follows that $k_H$ increases with increase in temperature. Since $p_2 = k_H x_2$, it may be concluded that

(i) $p_2$ increases with increases in $T$ for a constant $x_2$

(ii) $x_2$ decreases with increase in $T$ for a constant $p_2$.

## 4.8 THERMODYNAMICS OF IDEAL SOLUTIONS OF LIQUID IN LIQUID

**Definition of an Ideal Solution**

When a liquid is completely miscible with another liquid, a homogeneous solution consisting of a single phase is formed. If such a solution is placed in a closed evacuated vessel, the total pressure exerted by the vapour, after the system has attained equilibrium, will be equal to the sum of partial pressures of its constituents. *A solution is said to be an ideal solution if its constituents follow Raoult's law under all conditions of compositions,* i.e. the partial pressure of each and every constituent is given by

$$p_i = x_i\, p_1^* \tag{4.8.1}$$

where $p_i$ is the partial pressure of the constituent $i$ whose amount fraction in the solution is $x_i$ and $p_i^*$ is the corresponding vapour pressure of the pure constituent.

**Thermodynamics of an Ideal Solution**

The various constituents of an ideal solution follow the relation

$$\mu_{i(\text{soln})} = \mu^*_{i(1)} + RT \ln x_i \qquad \text{(Eq. 2.5.15)}$$

where $\mu_{i(\text{sol})}$ is the chemical potential of $i$th constituent of the solution, $\mu^*_{i(1)}$ is that of the pure liquid constituent $i$ and $x_i$ is the amount fraction of the constituent in the solution. The changes in the thermodynamic functions when an ideal solution is formed by mixing pure components can be readily calculated as described below.

***Change of Gibbs free-energy*** We have

$$\Delta_{\text{mix}}G = G_{\text{final}} - G_{\text{initial}} = \sum_i n_i\mu_{i(\text{sol})} - \sum_i n_i\mu^*_{i(1)}$$

$$= \sum_i n_i(\mu_{i(\text{sol})} - \mu^*_{i(1)})$$

Making use of Eq. (2.5.15), we get

$$\Delta_{\text{mix}}G = \sum_i n_iRT \ln x_i = n_{\text{total}}RT \sum_i x_i \ln x_i \tag{4.8.2}$$

where $n_{\text{total}}$ is the total amount of all the constituents present in the solution.

***Change of entropy on mixing*** The entropy of mixing can be obtained from the relation

$$\left\{\frac{\partial(\Delta_{\text{mix}}G)}{\partial T}\right\}_{p,n_j\text{s}} = -\,\Delta_{\text{mix}}S$$

Substituting $\Delta_{\text{mix}}G$ from Eq. (4.8.2) and carrying out the desired differentiation, we get

$$\Delta_{\text{mix}}S = -\left\{\frac{\partial(n_{\text{total}}RT \sum_i x_i \ln x_i)}{\partial T}\right\}_{p,n_j\text{s}} = -\,n_{\text{total}}R\sum_i x_i \ln x_i \tag{4.8.3}$$

***Change of enthalpy on mixing*** This can be determined from the relation

$$\Delta_{\text{mix}}G = \Delta_{\text{mix}}H - T\,\Delta_{\text{mix}}S \quad \text{i.e.} \quad \Delta_{\text{mix}}H = \Delta_{\text{mix}}G + T\,\Delta_{\text{mix}}S$$

Substituting $\Delta_{\text{mix}}G$ and $\Delta_{\text{mix}}S$ from Eq. (4.8.2) and (4.8.3), we get

$$\Delta_{\text{mix}}H = n_{\text{total}}RT \sum_i x_i \ln x_i - T\,n_{\text{total}}\,R\sum_i x_i \ln x_i = 0 \tag{4.8.4}$$

*Thus in the formation of an ideal solution neither heating nor cooling is observed.*

***Change of volume on mixing*** This can be determined from the relation

$$\Delta_{\text{mix}}V = \left(\frac{\partial(\Delta_{\text{mix}}G)}{\partial p}\right)_{T,n_j\text{s}}$$

Substituting $\Delta_{\text{mix}}G$ from Eq. (4.8.2), we get

$$\Delta_{\text{mix}}V = \left(\frac{\partial(\sum_i n_{\text{total}}\;RT\,x_i \ln x_i)}{\partial p}\right)_{T,n_j\text{s}} = 0 \tag{4.8.5}$$

as $\Delta_{\text{mix}}G$ is independent of pressure. Equation (4.8.5) implies that the *total volume of an ideal solution will be equal to the sum of individual volumes of its constituents.*

**Molecular Interpretation of $\Delta_{mix}H = 0$ and $\Delta_{mix}V = 0$**

The fact that $\Delta_{mix}H$ is zero for the formation of an ideal solution implies that at the molecular level, the cohesive forces between the molecules of its constituents must be completely identical in nature, since only then it is expected that on mixing, the environmental forces of each remain almost unchanged. In other words, the two liquids A and B are expected to form a binary ideal solution provided the forces of attraction between A ··· A molecules are identical with those of B ··· B molecules and thus are also identical with those between A ··· B molecules which exist in the solution. Thus the molecule A in solution will not be knowing whether it is attracting a like molecule A or an unlike molecule B and vice versa. Besides the identical forces between all pairs of molecules in an ideal solution, the different molecules must also have equal volumes, since only then we would expect no volume change on mixing its components.

The partial molar volume of a constituent in an ideal solution is equal to the molar volume of the constituent when present in the pure form.

**Examples of Ideal Binary Liquid Solutions**

The above conditions are, however, met only in few cases and thus limited examples of ideal binary liquid solutions are known. Some of the known examples which form nearly ideal solutions are:

(i) Ethylene dibromide and propylene dibromide at 358 K
(ii) Ethyl bromide and ethyl iodide at 303 K
(iii) *n*-Butyl chloride and *n*-butyl bromide at 323 K
(iv) *n*-Hexane and *n*-heptane at 303 K.

The temperatures mentioned along with the above liquid pairs are those at which measurements have been reported.

---

***Example 4.8.1***

Assuming ideal behaviour, determine $\Delta_{mix}G_m$, $\Delta_{mix}G$, $\Delta_{mix}S_m$ and $\Delta_{mix}S$ when 270.0 g of sugar with a molar mass of 358 g mol$^{-1}$ is dissolved in 1.0 kg of water at 298 K.

***Solution***

$$\text{Amount of sugar} = \frac{270.0\ \text{g}}{358\ \text{g mol}^{-1}} = 0.754\ \text{mol}$$

$$\text{Amount of water} = \frac{1\,000\ \text{g}}{18\ \text{g mol}^{-1}} = 55.556\ \text{mol}$$

Total amount, $n_{total}$ = 0.754 mol + 55.556 mol = 56.310 mol

$$\text{Amount fraction of sugar, } x_s = \frac{0.754\ \text{mol}}{56.310\ \text{mol}} = 0.013\,4$$

$$\text{Amount fraction of water, } x_w = \frac{55.556\ \text{mol}}{56.310\ \text{mol}} = 0.986\,6$$

Substituting the data in

$$\Delta_{mix}G_m = x_s\ RT\ \ln x_s + x_w\ RT\ \ln x_w$$

we get

$$\begin{aligned}\Delta_{mix}G_m &= (0.013\,4)\,(8.314\ \text{J K}^{-1}\ \text{mol}^{-1})\,(298\ \text{K})\,(\ln 0.013\,4) \\ &\quad + (0.986\,6)\,(8.314\ \text{J K}^{-1}\ \text{mol}^{-1})\,(298\ \text{K})\,(\ln 0.986\,6) \\ &= -143.173\ \text{J mol}^{-1} - 32.976\ \text{mol}^{-1} \\ &= -176.149\ \text{J mol}^{-1}\end{aligned}$$

Now $\Delta_{\text{mix}}G = n_{\text{total}}\,\Delta_{\text{mix}}G_{\text{in}} = (56.310\text{ mol})\,(-176.149\text{ J mol}^{-1})$

$= -9\,919.0\text{ J}$

$$\Delta_{\text{mix}}S_{\text{m}} = -\frac{\Delta_{\text{mix}}G_{\text{m}}}{T} = -\frac{(-176.149\text{ J mol}^{-1})}{298\text{ K}} = 0.591\text{ J mol}^{-1}\text{ K}^{-1}$$

$$\Delta_{\text{mix}}S = -\frac{\Delta_{\text{mix}}G}{T} = -\frac{(-9\,919.0\text{ J})}{298\text{ K}} = 33.285\text{ J K}^{-1}$$

---

***Example 4.8.2***

Assuming ideal behaviour, determine $\Delta_{\text{mix}}G$ and $\Delta_{\text{mix}}S$ when a solution of 1 mol of liquid A in 9 mol of liquid B is mixed with 10 mol of liquid B at 300 K.

***Solution***

We consider the following three processes.

(i) 9 mol of liquid B + 1 mol of liquid A $\rightarrow$ (solution)$_1$; $\Delta_{\text{mix}}G_1$
(ii) 10 mol of liquid B + (solution)$_1$ $\rightarrow$ (solution)$_2$; $\Delta_{\text{mix}}G_2$
(iii) 19 mol of liquid B + 1 mol of liquid A $\rightarrow$ (solution)$_2$; $\Delta_{\text{mix}}G_3$

Obviously, we have

$$\Delta_{\text{mix}}G_3 = \Delta_{\text{mix}}G_1 + \Delta_{\text{mix}}G_2 \quad \text{or} \quad \Delta_{\text{mix}}G_2 = \Delta_{\text{mix}}G_3 - \Delta_{\text{mix}}G_1$$

Similarly $\Delta_{\text{mix}}S_2 = \Delta_{\text{mix}}S_3 - \Delta_{\text{mix}}S_1$

Now $\Delta_{\text{mix}}G_1 = n_{\text{total}}\,RT\sum_i x_i \ln x_i$

$$= (10\text{ mol})\,(8.314\text{ J K}^{-1}\text{ mol}^{-1})\,(300\text{ K})\left(\frac{1}{10}\ln\frac{1}{10} + \frac{9}{10}\ln\frac{9}{10}\right)$$

$= -8\,108.2\text{ J}$

$$\Delta_{\text{mix}}S = -\frac{\Delta_{\text{mix}}G_1}{T} = -\frac{(-8112\text{ J})}{(300\text{ K})} = 27.03\text{ J K}^{-1}$$

$$\Delta_{\text{mix}}G_3 = n_{\text{total}}\,RT\sum_i x_i \ln x_i$$

$$= (20\text{ mol})\,(8.314\text{ J K}^{-1}\text{ mol}^{-1})\,(300\text{ K})\left(\frac{1}{20}\ln\frac{1}{20} + \frac{19}{20}\ln\frac{19}{20}\right)$$

$= -9\,902.7\text{ J}$

$$\Delta_{\text{mix}}S_3 = -\frac{\Delta_{\text{mix}}G_3}{T} = -\frac{(-9\,902.7\text{ J})}{(300\text{ K})} = 33.01\text{ J K}^{-1}$$

Hence $\Delta_{\text{mix}}G_2 = \Delta_{\text{mix}}G_3 - \Delta_{\text{mix}}G_1 = (-9\,902.7\text{ J mol}^{-1}) - (-8\,108.2\text{ J mol}^{-1})$

$= -1\,794.5\text{ J mol}^{-1}$

$\Delta_{\text{mix}}S_2 = \Delta_{\text{mix}}S_3 - \Delta_{\text{mix}}S_1 = (33.01\text{ J K}^{-1}) - (27.03\text{ J K}^{-1})$

$= 5.98\text{ J K}^{-1}$

---

## 4.9 VAPOUR PRESSURE OF AN IDEAL BINARY LIQUID SOLUTION

**Expression of Vapour Pressure**

Since both the constituents of an ideal binary liquid system follow Raoult's law over the entire range of composition, the partial pressures exerted by vapours of these constituents over the solution will be given by

$$p_A = x_A p_A^* \tag{4.9.1}$$

$$p_B = x_B p_B^* \tag{4.9.2}$$

where $x_A$ and $x_B$ are the amount fractions of the two constituents in the liquid phase and $p_A^*$ and $p_B^*$ are the respective vapour pressures of the pure constituents.

The total pressure over the solution will be sum of these partial pressures. Thus, we have

$$p = p_A + p_B = x_A p_A^* + x_B p_B^*$$

Since in a binary solution, $x_A + x_B = 1$ or $x_B = 1 - x_A$, therefore, we have

$$p = x_A p_A^* + (1 - x_A) p_B^*$$

$$= p_B^* + (p_A^* - p_B^*) x_A \tag{4.9.3}$$

Thus, the total vapour pressure over the solution is a linear function of $x_A$ with intercept equal to $p_B^*$ and slope equal to $p_A^* - p_B^*$. The slope can be positive or negative depending upon the relative magnitudes of $p_A^*$ and $p_B^*$. If the constituent A is more volatile, then $p_A^* > p_B^*$ and the term $p_A^* - p_B^*$ is positive. For such a case, the total pressure $p$ increases with increase in amount fraction $x_A$ of the more volatile constituent. If the constituent A is less volatile, then $p_A^* < p_B^*$. In this case, the term $p_A^* - p_B^*$ is negative and there occurs a decrease in vapour pressure with increase in amount fraction $x_A$ of the less volatile constituent.

Equations (4.9.1), (4.9.2) and (4.9.3) have been shown schematically in Fig. 4.9.1 with $p_A^* > p_B^*$.

**Composition of the Vapour Phase**

Composition of the vapour phase can be determined with the help of Dalton's law of partial pressures. Thus, we have

$$y_A = \frac{p_A}{p} = \frac{x_A p_A^*}{p_B^* + (p_A^* - p_B^*) x_A} \tag{4.9.4}$$

and $$y_B = \frac{p_B}{p} = 1 - y_A = \frac{x_B p_B^*}{p_A^* + (p_B^* - p_A^*) x_B} \tag{4.9.5}$$

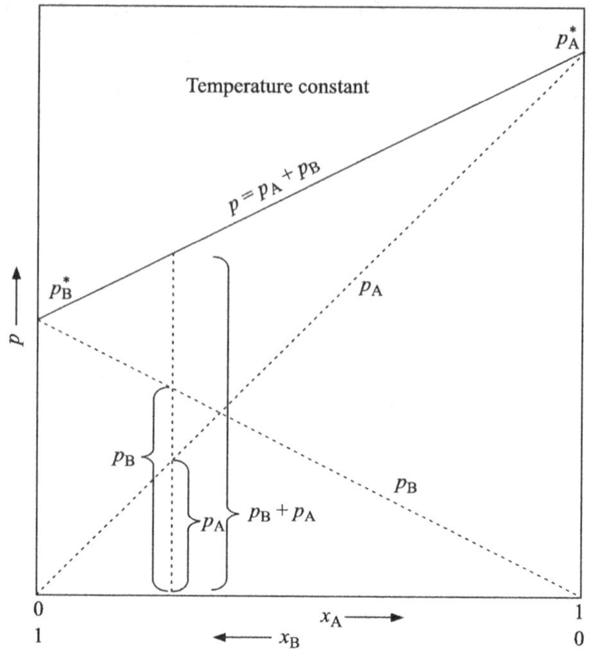


**Fig. 4.9.1 Plots of $p_A$, $p_B$ and $p$ against the amount fraction $x_A$ or $x_B$ in the liquid state**

## Vapour Pressure Related to Vapour Composition

To express the total vapour pressure in terms of the composition of vapour phase, we rearrange Eq. (4.9.4) as

$$x_A = \frac{y_A\, p_B^*}{p_A^* + (p_B^* - p_A^*)y_A} \tag{4.9.6}$$

and then substitute $x_A$ in Eq. (4.9.3). Thus, we have

$$p = p_B^* + (p_A^* - p_B^*)x_A$$

$$= p_B^* + (p_A^* - p_B^*)\frac{y_A p_B^*}{p_A^* + (p_B^* - p_A^*)y_A}$$

i.e.

$$p = \frac{p_B^* p_A^*}{p_A^* + (p_B^* - p_A^*)y_A} \tag{4.9.7}$$

The plot of $p$ versus $y_A$ for a solution in which $p_A^* > p_B^*$ is shown in Fig. 4.9.2.

Equation (4.9.7) can be written in a linear form by taking its reciprocal. Thus, we have

$$\frac{1}{p} = \frac{p_A^* + (p_B^* - p_A^*)y_A}{p_B^* p_A^*} = \frac{1}{p_B^*} + \left(\frac{1}{p_A^*} - \frac{1}{p_B^*}\right)y_A \tag{4.9.8}$$

According to Eq. (4.9.8), the reciprocal of the total pressure varies linearly with $y_A$, the amount fraction of the constituent A in the vapour phase. This variation is shown in Fig. 4.9.3.

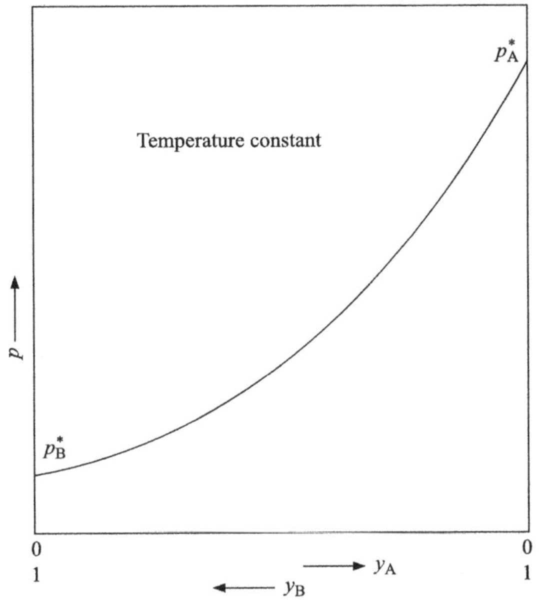


**Fig. 4.9.2 Plot of $p$ against $y_A$, the composition of A in vapour phase**

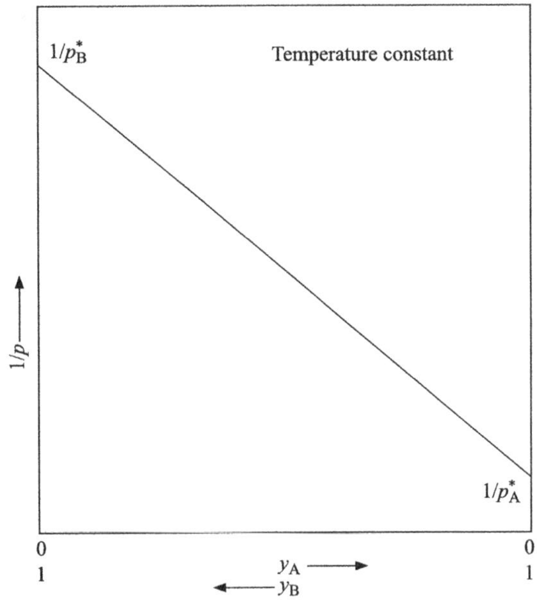


**Fig. 4.9.3 Plot of $1/p$ versus $y_A$, the composition of A in vapour phase**

***Example 4.9.1***

Suppose that the vapour over an ideal solution contains the amounts $n_1$ of 1 and $n_2$ of 2 and occupies a volume $V$ under the pressure $p = p_1 + p_2$. If we define $V_{1,\,m}^* = RT/p_1^*$ and $V_{2,\,m}^* = RT/p_2^*$, then show that Raoult's law implies that $V = n_1 V_{1,\,m}^* + n_2 V_{2,\,m}^*$.

***Solution***

The pressure $p$ over a solution as calculated by using Raoult's law is given by

$$\frac{1}{p} = \frac{1}{p_2^*} + \left(\frac{1}{p_1^*} - \frac{1}{p_2^*}\right) y_1 \qquad \text{(Eq. 4.9.8)}$$

Replacing $p$s in terms of $V$s we get

$$\frac{V}{(n_1 + n_2)RT} = \frac{V_{2,\,m}^*}{RT} + \left(\frac{V_{1,\,m}^*}{RT} - \frac{V_{2,\,m}^*}{RT}\right)\left(\frac{n_1}{n_1 + n_2}\right)$$

or $$V = (n_1 + n_2)V_{2,\,m}^* + n_1(V_{1,\,m}^* - V_{2,\,m}^*)$$

or $$V = n_1 V_{1,\,m}^* + n_2 V_{2,\,m}^*$$

***Example 4.9.2***

The composition of vapour over a binary ideal solution is determined by the composition of the liquid. If $x_A$ and $y_A$ are the amount fractions of A in the liquid and vapour, respectively, find the value of $x_A$ for which $y_A - x_A$ has a maximum. What is the value of the pressure at this composition?

***Solution***

We have $y_A = \dfrac{x_A p_A^*}{p_B^* + (p_A^* - p_B^*)x_A}$

Subtracting $x_A$ from both sides, we get

$$y_A - x_A = \frac{x_A p_A^*}{p_B^* + (p_A^* - p_B^*)x_A} - x_A$$

Differentiating with respect to $x_A$, we get

$$\frac{d(y_A - x_A)}{dx_A} = \frac{p_A^*}{p_B^* + (p_A^* - p_B^*)x_A} - \frac{x_A p_A^*(p_A^* - p_B^*)}{\{p_B^* + (p_A^* - p_B^*)x_A\}^2} - 1$$

The value of $x_A$ at which $y_A - x_A$ has a maximum value can be obtained by setting the above differential equal to zero. Thus, we have

$$\frac{p_A^*}{p_B^* + (p_A^* - p_B^*)x_A} - \frac{x_A p_A^*(p_A^* - p_B^*)}{\{p_B^* + (p_A^* - p_B^*)x_A\}^2} - 1 = 0$$

Solving for $x_A$, we get

$$x_A = \frac{\sqrt{p_A^* p_B^*} - p_B^*}{p_A^* - p_B^*}$$

The value of $p$ at this composition is

$$p = x_A p_A^* + x_B p_B^* = p_B^* + (p_A^* - p_B^*)x_A$$

$$= p_B^* + (p_A^* - p_B^*)\left(\frac{\sqrt{p_A^* p_B^*} - p_B^*}{p_A^* - p_B^*}\right) = \sqrt{p_A^* p_B^*}$$

---

***Example 4.9.3***

Two liquids A and B form an ideal solution at temperature $T$. When the total vapour pressure above the solution is 600 Torr, the amount fraction of A in the vapour phase is 0.35 and in the liquid phase is 0.70. What are the vapour pressures of pure A and pure B at temperature $T$?

***Solution***

The given data are

$$x_A = 0.70 \quad y_A = 0.35 \quad p = 600 \text{ Torr} \quad p_A^* = ? \quad p_B^* = ?$$

Now $\quad y_A = \frac{p_A}{p} = \frac{x_A p_A^*}{p}$

Therefore $\quad p_A^* = \frac{y_A p}{x_A} = \frac{(0.35)(600 \text{ Torr})}{(0.70)} = 300 \text{ Torr}$

Similarly, $\quad p_B^* = \frac{y_B p}{x_B} = \frac{(0.65)(600 \text{ Torr})}{(0.30)} = 1\,300 \text{ Torr}$

---

***Example 4.9.4***

At 323 K, the vapour pressures of pure *n*-hexane and *n*-heptane are 54.4 and 18.8 kN $m^{-2}$, respectively. Assuming ideality, determine:

(a) The amount fraction of *n*-hexane in a solution in equilibrium with a vapour in which the amount fraction of *n*-hexane is 0.85.

(b) The molar entropy change when this solution is formed from pure components at 323 K.

***Solution***

The given data are

$p_A^* = p_{n\text{-hexane}}^* = 54.4 \text{ kN m}^{-2}$ $\quad$ $p_B^* = p_{n\text{-heptane}}^* = 18.8 \text{ kN m}^{-2}$

$x_A = x_{n\text{-hexane}} = ?$ $\quad$ $y_A = y_{n\text{-hexane}} = 0.85$

$\Delta S_{\text{mix}} = ?$

**(a)** Now $\quad y_A = \dfrac{p_A}{p} = \dfrac{x_A p_A^*}{p_B^* + (p_A^* - p_B^*)x_A}$

Therefore $\quad x_A = \dfrac{y_A p_B^*}{p_A^* - (p_A^* - p_B^*)y_A}$

Substituting the values, we get

$$x_A = \frac{0.85 \times 18.8 \text{ kN m}^{-2}}{54.4 \text{ kN m}^{-2} - (54.4 \text{ kN m}^{-2} - 18.8 \text{ kN m}^{-2}) \times 0.85} = 0.662$$

**(b)** Now $\quad \Delta_{\text{mix}} S/R = -x_A \ln x_A - x_B \ln x_B$

Substituting the values of $x_A$ and $x_B$, we get

$$\Delta_{\text{mix}} S/R = -0.662 \times 2.303 \times \log(0.662) - 0.338 \times 2.303 \times \log(0.338)$$
$$= 0.273\,1 + 0.366\,7 = 0.639\,8$$

Hence $\quad \Delta S_{\text{mix}} = (0.639\,8)(8.314 \text{ J K}^{-1} \text{ mol}^{-1}) = 5.319 \text{ J K}^{-1} \text{ mol}^{-1}$

---

***Example 4.9.5***

At 303 K, the vapour pressure of pure toluene and pure benzene are 36.7 and 118.2 Torr, respectively, and the two liquids form a nearly ideal solution, (a) For a solution containing 50.0 mass % of toluene, calculate the total vapour pressure and the amount fraction of each compound in the vapour phase. (b) Determine the composition of a solution of benzene and toluene that will boil at 303 K at a pressure of 50.0 Torr.

***Solution***

The given data are

$p_{\text{toluene}}^* = p_t^* = 36.7 \text{ Torr}$ $\quad$ $p_{\text{benzene}}^* = p_b^* = 118.2 \text{ Torr}$

**(a)** From the given data, we have

$$x_t = \frac{n_t}{n_t + n_b} = \frac{(50 \text{ g}/92 \text{ g mol}^{-1})}{(50 \text{ g}/92 \text{ g mol}^{-1}) + (50 \text{ g}/78 \text{ g mol}^{-1})} = 0.459$$

$$p_t = x_t p_t^* = 0.459 \times 36.7 \text{ Torr} = 16.85 \text{ Torr}$$

$$p_b = x_b p_b^* = (1 - 0.459) \times 118.2 \text{ Torr} = 63.95 \text{ Torr}$$

$$p = p_t + p_b = 16.85 \text{ Torr} + 63.95 \text{ Torr} = 80.80 \text{ Torr}$$

Now $\quad y_t = \dfrac{p_t}{p} = \dfrac{16.85 \text{ Torr}}{80.80 \text{ Torr}} = 0.21 \quad$ and $\quad y_b = 1 - y_t = 0.79$

(**b**) We have

$$p = p_t + p_b = x_t p_t^* + x_b p_b^* = x_t p_t^* + (1 - x_t) p_b^* = p_b^* + (p_t^* - p_b^*) x_t$$

Thus $$x_t = \frac{p - p_b^*}{p_t^* - p_b^*}$$

Substituting the given data, we get

$$x_t = \frac{50.0 \text{ Torr} - 118.2 \text{ Torr}}{36.7 \text{ Torr} - 118.2 \text{ Torr}} = 0.84$$

$$x_b = 1 - x_t = 0.16$$

---

## 4.10 VAPOUR PRESSURE-COMPOSITION DIAGRAM OF A BINARY LIQUID SOLUTION

### Application of the Phase Rule

Since the number of components in a binary mixture are two, according to the phase rule, we have

$$F = C - P + 2 = 4 - P \qquad (4.10.1)$$

The variance of the system will depend on the number of phases present in the system.

***System consisting of only one Phase*** If the system has only one phase (liquid or gaseous), then it is trivariant. This means that the values of three variables will have to be stated in order to define the state of the system completely. These are temperature, pressure and composition. If one of them is held constant, say temperature, then the values of other two variables, namely, pressure and composition, will have to be stated in order to describe the system completely. The state of such a system can be represented by a point in a two-dimensional plot of pressure versus composition at a given temperature. In Fig. 4.9.1, we have plotted the total vapour pressure of a binary liquid mixture against the amount fraction of the constituent A (or constituent B as $x_A + x_B = 1$) in the liquid phase. Since the total pressure is an equilibrium vapour pressure, it is obvious that the system would consist of only liquid phase if it is subjected to a pressure greater than its equilibrium vapour pressure. Hence a point anywhere above the total vapour pressure line in Fig. 4.9.1 represents the system in the liquid phase. Similarly, a point anywhere below the total pressure line in Fig. 4.9.2 represents the system in the vapour phase.

***System consisting of two phases*** If the system has two phases in equilibrium, namely, liquid and vapour, then the variance of the system is two. This means that the values of only two variables will have to be stated in order to describe the system completely. Any two of the three variables, temperature, pressure and composition, can be stated. The third one automatically will have a definite value. For example, if temperature and composition are stated, then the vapour pressure of the system will have a definite value. The straight line of Fig. 4.9.1 and the curved line of Fig. 4.9.2 represent the two phases, liquid and vapour, in equilibrium with each other.

## Depiction of the Diagram

The two states, namely, liquid and vapour, can be shown in one diagram when the plots of Figs 4.9.1 and 4.9.2 are drawn together in one graph. This is shown in Fig. 4.10.1. The upper curve is called the *liquidus curve* since this represents the line above which only liquid phase exists. A point anywhere on this line represents the two phases, liquid and vapour, in equilibrium with each other. Similarly, the lower curve is called the *vaporous curve* since it represents the curve below which only vapour exists. Along this line also the two phases, liquid and vapour, are in equilibrium with each other.

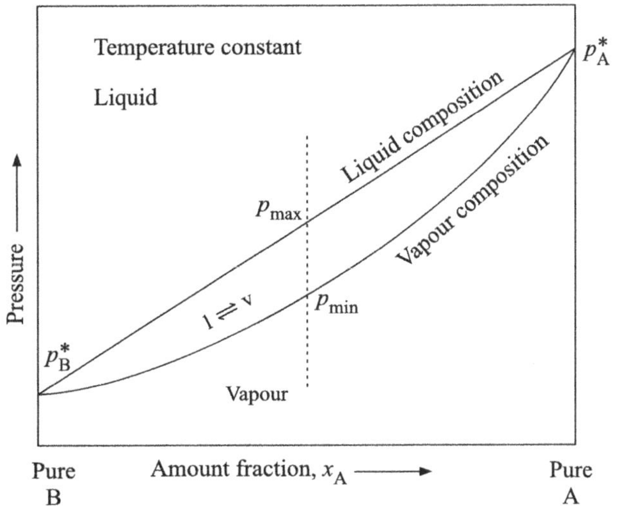


**Fig. 4.10.1 Vapour pressure-composition diagram**

A point anywhere in the region between the two curves represents the system in which both liquid and vapour coexist in equilibrium with each other. It is for this reason, this region is known as the *liquid-vapour region*. For a given composition, the points on the liquidus and vaporous curves represent, respectively, the maximum and the minimum pressures within which the two phases can exist in equilibrium with each other. The above facts can be summarized as follows:

(i) If $p_{system} > p_{max}$; only liquid phase exists.
(ii) If $p_{system} < p_{min}$; only vapour phase exists.
(iii) For $p_{max} > p_{system} > p_{min}$; the liquid and vapour phases are in equilibrium with each other.

## Tie Line

As indicated above, the region between the liquidus and vaporous curves of Fig. 4.10.1 represents a region where both the phases exist in equilibrium with each other. For this region, $P = 2$ and thus $F = 2$. Since the temperature is fixed, one other variable (any one of $p$, $x_A$ and $y_A$) is sufficient to describe the system completely. For instance, if $p$ is chosen, then the amount fractions of the component A in the liquid and vapour phases are represented by the intersection points of a horizontal line, known as the *tie line*, drawn from the given $p$ with those of liquidus and vaporous curves, respectively. If $x_A$ is chosen as the describing variable, then the intersection of a vertical line from $x_A$ with the liquidus curve yields the value of $p$ and a horizontal line from $p$ cutting the vaporous curve gives the vapour composition $y_A$.

## The Lever Rule

A point within the liquid and vapour curves represents two phases, liquid and vapour, in equilibrium with each other. For the point a shown in Fig. 4.10.2, the amount fraction of the constituent A in the liquid phase is given by the point b (i.e. $x_A$) and that in the vapour phase is given by the point c (i.e. $y_A$).

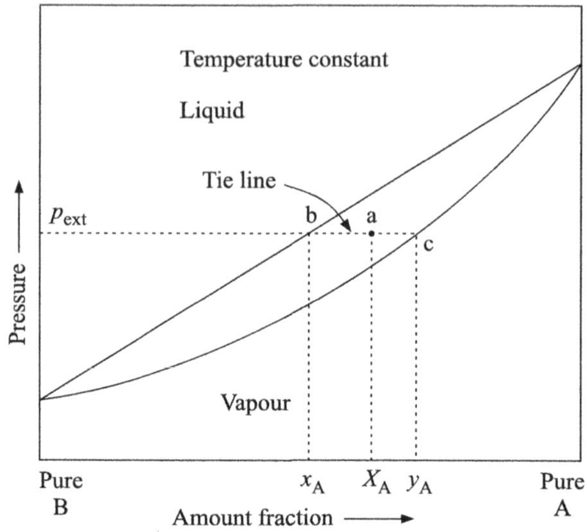


**Fig. 4.10.2 The lever rule**

***Qualitative introduction of the lever rule*** In fact, any point a on the tie line bc represents the same compositions of liquid and vapour phases, namely, $x_A$ and $y_A$, respectively. The only difference that exists from point to point is the relative amounts of the two phases. If a coincides with b then the vapour phase has just started forming with amount fraction $y_A$ and if it coincides with c, then the last drop of liquid phase with amount fraction $x_A$ is just going to be converted into the vapour phase. As we move from b to c, more and more of the liquid phase is converted into the vapour phase, keeping the amount fractions of the two phases equal to $x_A$ and $y_A$, respectively. Thus if the point a is nearer to b, the system will consist of a larger amount of liquid and a relatively smaller amount of vapour; but if it is nearer to c, then the amount of liquid present is relatively small as compared with the amount of vapour present. As a moves from b to c, more and more of the vapour is formed, thus the distance ba represents the amount of vapour formed. Similarly as a moves from c to b, more and more of the liquid is formed and thus the distance ac represents the amount of liquid formed. Taking these together, we have

distance ab $\propto$ amount of the vapour formed
distance ac $\propto$ amount of the liquid formed

Taking the ratio, we have

$$\frac{\text{ab}}{\text{ac}} = \frac{\text{Amount of the vapour formed}}{\text{Amount of the liquid formed}} \tag{4.10.2}$$

Equation (4.10.2) is knows as the lever rule.

***Quantitative derivation of the lever rule*** The amount fraction $X_A$ corresponding to the point a represents the amount fraction of the component A in the entire system consisting of liquid and vapour phases, i.e.

$$X_A = \frac{n_{A(l)} + n_{A(v)}}{n_{A(l)} + n_{A(v)} + n_{B(l)} + n_{B(v)}} \quad (4.10.3)$$

From Fig. 4.10.2, we have

$$(ab) = X_A - x_A = X_A - \frac{n_{A(l)}}{n_{A(l)} + n_{B(l)}} \quad (4.10.4)$$

$$(ac) = y_A - X_A = \frac{n_{A(v)}}{n_{A(v)} + n_{B(v)}} - X_A \quad (4.10.5)$$

Multiplying Eq. (4.10.4) by $n_{A(l)} + n_{B(l)}$ and Eq. (4.10.5) by $n_{A(v)} + n_{B(v)}$, we have

$$(n_{A(l)} + n_{B(l)})\,(ab) = (n_{A(l)} + n_{B(l)})X_A - n_{A(l)} \quad (4.10.6)$$

$$(n_{A(v)} + n_{B(v)})\,(ac) = n_{A(v)} - (n_{A(v)} + n_{B(v)})X_A \quad (4.10.7)$$

Subtracting Eq. (4.10.7) from Eq. (4.10.6), we get

$$(n_{A(l)} + n_{B(l)})\,(ab) - (n_{A(v)} + n_{B(v)})(ac)$$
$$= X_A(n_{A(l)} + n_{B(l)} + n_{A(v)} + n_{B(v)}) - (n_{A(l)} + n_{A(v)})$$

which on making use of Eq. (4.10.3) becomes

$$(n_{A(l)} + n_{B(l)})\,(ab) - (n_{A(v)} + n_{B(v)}(ac) = 0$$

or $$\left(\frac{ab}{ac}\right) = \frac{n_{A(v)} + n_{B(v)}}{n_{A(l)} + n_{B(l)}}$$

i.e. $$\left(\frac{ab}{ac}\right) = \frac{\text{Amount in the vapour phase}}{\text{Amount in the liquid phase}} \quad (4.10.8)$$

In words, the ratio of the amount of vapour to the amount of liquid is given by the ratio of lengths of the line segments connecting a to b and a to c. Thus if a lies very close to b, then ab is very small and, therefore, $n_v << n_l$, i.e. the system consists mainly of liquid phase. If a lies very close to c, then $n_v >> n_l$ and thus the system consists mainly of vapour phase.

***Example 4.10.1***

At 353 K, the vapour pressures of pure ethylene bromide and propylene bromide are 22.93 and 16.93 kN m$^{-2}$, respectively, and these compounds form a nearly ideal solution. 3 mol of ethylene bromide and 2 mol of propylene bromide are equilibrated at 353 K and a total pressure of 20.4 kN m$^{-2}$. (a) What is the composition of the liquid phase? (b) What amount of each compound is present in the vapour phase?

***Solution***

The given data are

$$p_A^* = p^*_{\substack{\text{ethylene}\\ \text{bromide}}} = 22.93 \text{ kN m}^{-2} \qquad p_B^* = p^*_{\substack{\text{propylene}\\ \text{bromide}}} = 16.93 \text{ kN m}^{-2}$$

$$n_A = 3 \text{ mol} \qquad n_B = 2 \text{ mol} \qquad p = 20.4 \text{ kN m}^{-2}$$

(**a**) We have

$$p = p_A + p_B = x_A p_A^* + x_B p_B^* = p_B^* + (p_A^* - p_B^*) x_A$$

Therefore $x_A = \dfrac{p - p_B^*}{p_A^* - p_B^*}$

Substituting the given data, we have

$$x_A = \frac{20.4 \text{ kN m}^{-2} - 16.93 \text{ kN m}^{-2}}{22.93 \text{ kN m}^{-2} - 16.93 \text{ kN m}^{-2}} = 0.578$$

$$x_B = 1 - 0.578 = 0.422$$

(**b**) Now $y_A = \dfrac{p_A}{p} = \dfrac{x_A p_A^*}{p} = \dfrac{0.578 \times 22.93 \text{ kN m}^{-2}}{20.4 \text{ kN m}^{-2}} = 0.649\,7$

Let $n_A$ and $n_B$ be the amounts of vaporized ethylene bromide and propylene bromide, respectively, when $p = 20.4 \text{ kN m}^{-2}$. Hence we have

$$y_A = \frac{n_A}{n_A + n_B} = 0.649\,7$$

$$x_A = \frac{3 \text{ mol} - n_A}{(3 \text{ mol} - n_A) + (2 \text{ mol} - n_B)} = \frac{3 \text{ mol} - n_A}{5 \text{ mol} - (n_A + n_B)}$$

$$= \frac{3 \text{ mol} - n_A}{5 \text{ mol} - n_A/0.649\,7} = 0.578$$

or $$(3 \text{ mol} - n_A) = 0.578\left(5 \text{ mol} - \frac{n_A}{0.649\,7}\right)$$

Therefore $n_A = \dfrac{3 \text{ mol} - 0.578 \times 5 \text{ mol}}{1 - 0.578/0.649\,7} = 0.996\,7 \text{ mol}$

Since $\dfrac{n_A}{n_A + n_B} = 0.649\,7,$ we, therefore, have

$$n_B = \frac{n_A}{0.649\,7} - n_A = n_A\left(\frac{1}{0.649\,7} - 1\right) = \frac{(0.996\,7 \text{ mol})(0.350\,3)}{(0.649\,7)}$$

$$= 0.537\,4 \text{ mol}$$

***Example 4.10.2*** For the preceding problem, use the lever rule to obtain the amount of each compound in the vapour phase.

***Solution*** Now, we know

$$\frac{\text{Amount in the liquid phase}}{\text{Amount in the vapour phase}} = \frac{y_A - X_A}{X_A - x_A}$$

where $X_A$ is the amount fraction of ethylene bromide in the overall system. Substituting the values of $x_A$, $y_A$ and $X_A$, we have

$$\frac{\text{Amount in the liquid phase}}{\text{Amount in the vapour phase}} = \frac{0.649\,7 - (3/5)}{(3/5) - 0.578} = \frac{0.049\,7}{0.022}$$
$$= 2.259$$

Adding one on both sides and taking inverse, we get

$$\frac{\text{Amount in the vapour phase}}{\text{Amount in the (liquid + vapour) phase}} = \frac{1}{1 + 2.259}$$

$$\text{Amount in the vapour phase} = \frac{1}{(1 + 2.259)}(5\ \text{mol})$$

Therefore $$n_A = y_A\left(\frac{5\ \text{mol}}{3.259}\right) = (0.649\,7)\left(\frac{5\ \text{mol}}{3.259}\right) = 0.996\,8\ \text{mol}$$

$$n_B = y_B\left(\frac{5\ \text{mol}}{3.259}\right) = (0.350\,3)\left(\frac{5\ \text{mol}}{3.259}\right) = 0.537\,4\ \text{mol}$$

## 4.11 ISOTHERMAL FRACTIONAL DISTILLATION OF AN IDEAL BINARY SOLUTION

**Underlying Principle** Isothermal distillation of an ideal binary solution is based on the fact that whenever a liquid phase is in equilibrium with the vapour phase, the latter is relatively richer in the more volatile component. This can be proved from Eq. (4.9.4). We have

$$y_A = \frac{x_A p_A^*}{x_A p_A^* + x_B p_B^*} \quad \text{or} \quad \frac{y_A}{x_A} = \frac{1}{x_A + x_B(p_B^*/p_A^*)} \tag{4.11.1}$$

If the constituent A is more volatile than the constituent B, i.e. $p_A^* > p_B^*$, then $p_B^*/p_A^*$ will be less than one and so also the denominator of Eq. (4.11.1), since $x_A + x_B$ can only be equal to one. The result of this is that the right side of Eq. (4.11.1) is greater than one and hence $y_A > x_A$, i.e. the amount fraction of the component A, the more volatile one, in the vapour phase is larger than that present in the liquid phase.

**Effects of Isothermal Decrease in Pressure** Before we discuss the procedure that is employed for isothermal distillation, we will discuss the effects which are produced when the pressure of the system is reduced isothermally.

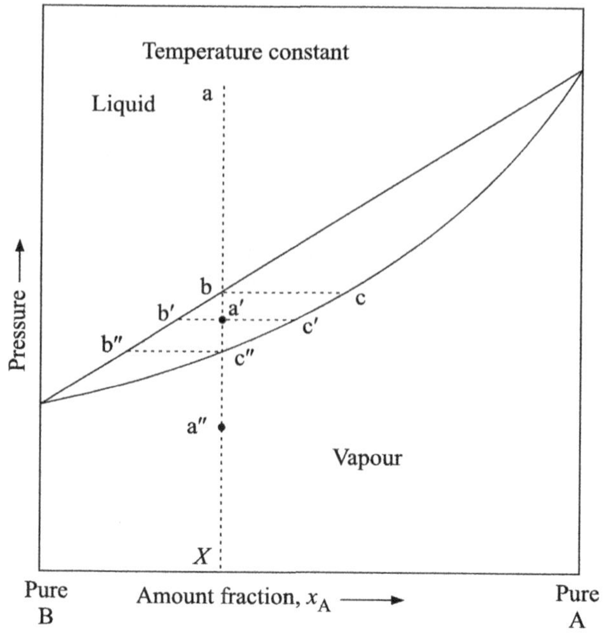


**Fig. 4.11.1 Effect of reducing pressure at constant temperature on the ideal binary mixture**

Let us start with the liquid phase as shown by the point a in Fig. 4.11.1. Let the pressure of the system be reduced, while its temperature remains constant. The effects to be observed are summarized as follows.

- The state of the system will move along the vertical line aa′a″.
- The system will remain of one phase (i.e. liquid) till the pressure corresponding to the point b is reached.
- At the state b of the system, the vapour phase just starts forming. Its composition will correspond to the point c. The obtained vapour phase is more rich in the component A (which is more volatile) as compared to the liquid phase b.
- Removal of more of the component A in the vapour phase makes the liquid phase more rich in the component B with the result that the composition of liquid phase moves along bb′.
- If the vaporization is to be continued, the pressure of the system must be lowered. The overall state of the system will still move along the same vertical line as shown, for example, by the point a′.
- At the point a′, the compositions of the liquid and vapour phases are given by the points b′ and c′, respectively. The relative amounts of the two phases as given by the lever rule are

$$\frac{\text{Amount in liquid phase}}{\text{Amount in vapour phase}} = \frac{a'c'}{a'b'}$$

- Continued reduction of pressure brings the state point to c″ where only a negligible trace of liquid b″ is left behind and the vapour phase has the composition $X$ which is the same as that of starting liquid (represented by the point a).

- As the pressure is further reduced, the two-phase system is converted into one-phase system consisting of the vapour phase.

**A Specific Example**

To illustrate the above changes, we take up a specific example of a binary liquid solution with $p_A^* = 200$ Torr and $p_B^* = 100$ Torr at a temperature $T$. Let us have, for example, a solution of 1 mol of constituent A and 2 mol of constituent B.

- The total vapour pressure of the system would be

$$p = x_A p_A^* + x_B p_B^* = (1/3)\,(200\text{ Torr}) + (2/3)\,(100\text{ Torr})$$
$$= 133.33\text{ Torr}$$

If the pressure on the system is higher than this value, no vapour phase can exist and the system consists of only liquid phase. This may be represented by the point a in Fig. 4.11.1.

- Let the pressure on the system is lowered to a value just less than 133.33 Torr (point b). At this pressure, a bubble of vapour will appear whose composition would be

$$y_A = \frac{p_A}{p} = \frac{x_A p_A^*}{p} = \frac{(1/3)\,(200\text{ Torr})}{(133.33\text{ Torr})} = 0.50$$

The state representing this vapour phase is represented by the point c in Fig. 4.11.1.

- As the pressure on the system is gradually lowered, more and more of liquid evaporates and the vapour phase becomes more rich in the component B.

As the last drop of the liquid evaporates, the composition of the vapour phase approaches the composition of the entire system and thus $y_A = (1/3)$ mol. The composition of last trace of liquid to be evaporated can be calculated by using the expression

$$y_A = \frac{p_A}{p} = \frac{x_A p_A^*}{x_A p_A^* + x_B p_B^*} = \frac{x_A p_A^*}{x_A p_A^* + (1 - x_A) p_B^*}$$

Rearranging this expression, we get

$$x_A = \frac{p_B^* y_A}{p_A^* + (p_B^* - p_A^*) y_A}$$

Substituting the values, we get

$$x_A = \frac{(100\text{ Torr})(1/3)}{(200\text{ Torr}) + (100\text{ Torr} - 200\text{ Torr})(1/3)} = 0.20$$

- The total pressure of the system at point b″ is

$$p = x_A p_A^* + x_B p_B^* = (0.20)(200\text{ Torr}) + (0.80)(100\text{ Torr}) = 120\text{ Torr}$$

- The system will consist of two phases between the pressure range 120 Torr – 133.33 Torr. The composition of liquid and vapour phases at any pressure can be computed using the following expressions.

(i) $p = x_A p_A^* + (1 - x_A)p_B^* \quad \Rightarrow \quad x_A = \dfrac{p - p_B^*}{p_A^* - p_B^*}$

(ii) $y_A = \dfrac{p_A}{p} = \dfrac{x_A p_A^*}{p} = \dfrac{x_A p_A^*}{p_B^* + (p_A^* - p_B^*)x_A}$

For example, if $p$ = 125 Torr, then

$$x_A = \frac{125\ \text{Torr} - 100\ \text{Torr}}{200\ \text{Torr} - 100\ \text{Torr}} = 0.25$$

$$y_A = \frac{x_A p_A^*}{p} = \frac{(0.25)(200\ \text{Torr})}{(125\ \text{Torr})} = 0.40$$

- The individual amounts in liquid and vapour phases can be determined by using the level rule.

$$\frac{\text{Amount in liquid phase}}{\text{Amount in vapour phase}} = \frac{\text{a}'\text{c}'}{\text{a}'\text{b}'} = \frac{y_A - X}{X - x_A} = \frac{0.40 - (1/3)}{(1/3) - 0.25} = 0.80$$

Adding one to both sides and taking the inverse, we get

$$\frac{\text{Amount in vapour phase}}{\text{Amount in liquid and vapour phases}} = \frac{1}{1.80}$$

Thus Amount in vapour phase = (1/1.80) (3.0 mol) = 1.67 mol
Amount in liquid phase = 3.0 mol – 1.67 mol = 1.33 mol

Now Amount of A in liquid phase = $x_A n_l$ = (0.25) (1.33 mol) = 0.33 mol
Amount of B in liquid phase = 1.33 mol – 0.33 mol = 1.00 mol
Amount of A in vapour phase = $y_A n_v$ = (0.40)(1.67 mol) = 0.67 mol
Amount of B in vapour phase = 1.67 mol – 0.67 mol = 1.00 mol

---

***Example 4.11.1***

(a) A liquid mixture of benzene and toluene is composed of 1 mol of benzene and 1 mol of toluene. If the pressure over the mixture at 300 K is reduced, at what pressure does the first vapour form? Given: $p_t^* = 4.274\ \text{kN m}^{-2}$ and $p_b^* = 13.734\ \text{kN m}^{-2}$, where the subscripts t and b stand for toluene and benzene, respectively.
(b) What is the composition of the first trace of vapour formed?
(c) If the pressure is reduced further, at what pressure does the last trace of liquid disappear?
(d) What is the composition of the last trace of liquid?
(e) What will be the pressure, the composition of the liquid and the composition of the

vapour, when 1 mol of the mixture has been vaporized?

***Solution***

(**a**) The first vapour will be formed when the external pressure becomes equal to the vapour pressure of the system. Thus, the required pressure is

$$p = x_t p_t^* + x_b p_b^*$$

$$= \frac{1}{2}(4.274 \text{ kN m}^{-2}) + \frac{1}{2}(13.734 \text{ kN m}^{-2}) = 9.004 \text{ kN m}^{-2}$$

(**b**) Composition of the first vapour can be calculated using the equation

$$y_t = \frac{p_t}{p} = \frac{x_t p_t^*}{p} = \frac{(0.5)(4.274 \text{ kN m}^{-2})}{9.004 \text{ kN m}^{-2}} = 0.237\ 3$$

$$y_b = 1 - y_t = 0.762\ 7$$

(**c**) The last trace of liquid will disappear when the composition of the vapour phase becomes

$$y_b = \frac{1}{2} \quad \text{and} \quad y_t = \frac{1}{2} \qquad \text{(i.e. practically whole of the liquid has vaporized)}$$

The pressure at which this occurs can be calculated from the expression

$$p = \frac{p_b^* p_t^*}{p_t^* + (p_b^* - p_t^*) y_t} \qquad \text{(Eq. 4.9.7)}$$

$$= \frac{(13.734 \text{ kN m}^{-2})(4.274 \text{ kN m}^{-2})}{(4.274 \text{ kN m}^{-2}) + (13.734 \text{ kN m}^{-2} - 4.274 \text{ kN m}^{-2})(0.5)}$$

$$= 6.519 \text{ kN m}^{-2}$$

(**d**) Composition of the last trace of the liquid can be calculated using the expression

$$x_t = \frac{y_t p}{p_t^*} = \frac{(0.5)(6.519 \text{ kN m}^{-2})}{4.274 \text{ kN m}^{-2}} = 0.762\ 6$$

Therefore $x_b = 1 - x_t = 0.237\ 4$

(**e**) Let $n_b$ be the amount of benzene present in 1 mol of the mixture that has been vaporized.

Thus $\quad y_b = n_b/1 \text{ mol}$

The amount fraction of benzene in the remaining liquid phase will be

$$x_b = \frac{1 \text{ mol} - n_b}{1 \text{ mol}}$$

Now from the expression $p = p_t^* + (p_b^* - p_t^*) x_b$, we get

$$x_b = \frac{p - p_t^*}{p_b^* - p_t^*} \qquad (1)$$

Also

$$y_b = \frac{p_b}{p} = \frac{x_b p_b^*}{p} \qquad (2)$$

Substituting the values of $x_b$ and $y_b$ in Eqs (1) and (2), we have

$$\frac{1\text{ mol} - n_b}{1\text{ mol}} = \frac{p - p_t^*}{p_b^* - p_t^*} \tag{3}$$

and
$$n_b = \frac{(1\text{ mol} - n_b)\, p_b^*}{p} \tag{4}$$

Rearranging Eq. (4) for $n_b$, we have

$$n_b = \frac{(1\text{ mol})\, p_b^*}{p + p_b^*} \tag{5}$$

Substituting this in Eq. (3), we get

$$1 - \frac{p_b^*}{p + p_b^*} = \frac{p - p_t^*}{p_b^* - p_t^*} \tag{6}$$

Rearranging the above equation, we have

$$p = \sqrt{p_t^*\, p_b^*}$$

Substituting the given values of $p_t^*$ and $p_b^*$, we get

$$p = \{(4.274\text{ kN m}^{-2})\,(13.734\text{ kN m}^{-2})\}^{1/2} = 7.662\text{ kN m}^{-2}$$

Substituting $p$ in Eq. (3), we get

$$\frac{1\text{ mol} - n_b}{(1\text{ mol})} = \frac{(7.662\text{ kN m}^{-2}) - (4.274\text{ kN m}^{-2})}{(13.734\text{ kN m}^{-2}) - (4.274\text{ kN m}^{-2})} = \frac{3.388}{9.460}$$

or
$$n_b = \left(1 - \frac{3.388}{9.460}\right)\text{mol} = \frac{6.072}{9.480}\text{ mol} = 0.642\text{ mol}$$

Thus $\quad y_b = n_b/\text{mol} = 0.642 \quad$ and $\quad y_t = 1 - y_b = 0.358$

---

***Example 4.11.2*** The vapour pressures of two pure liquids, A and B, that form an ideal solution are 300 and 800 Torr, respectively, at temperature $T$. A mixture of the vapours of A and B for which the amount fraction of A is 0.25 is slowly compressed at temperature $T$. Calculate (a) the composition of the first drop of the condensate, (b) the total pressure when this drop is formed, (c) the composition of the solution whose normal boiling point is $T$. (d) the pressure when only the last bubble of vapour remains, and (e) the composition of the last bubble.

***Solution*** The given data are

$$p_A^* = 300\text{ Torr} \qquad p_B^* = 800\text{ Torr} \qquad y_A = 0.25$$

(**a**) We have $y_A = \dfrac{p_A}{p} = \dfrac{x_A p_A^*}{x_A p_A^* + x_B p_B^*} = \dfrac{x_A p_A^*}{p_B^* + (p_A^* - p_B^*)x_A}$

Therefore $x_A = \dfrac{y_A p_B^*}{p_A^* - (p_A^* - p_B^*) y_A}$

Substituting the given data, we have

$$x_A = \frac{(0.25)\,(800\ \text{Torr})}{300\ \text{Torr} - (300\ \text{Torr} - 800\ \text{Torr})\,(0.25)} = 0.470\,6$$

(**b**) We have

$$p = x_A p_A^* + x_B p_B^* = x_A p_A^* + (1 - x_A) p_B^*$$

Substituting the given data, we get

$$\begin{aligned} p &= 0.470\,6 \times 300\ \text{Torr} + 0.529\,4 \times 800\ \text{Torr} \\ &= 141.18\ \text{Torr} + 423.52\ \text{Torr} = 564.7\ \text{Torr} \end{aligned}$$

(**c**) For the normal boiling point, we have

$$p = 760\ \text{Torr} = x_A p_A^* + (1 - x_A)\, p_B^*$$

Therefore $x_A = \dfrac{760\ \text{Torr} - p_B^*}{p_A^* - p_B^*}$

Substituting the given data, we have

$$x_A = \frac{760\ \text{Torr} - 800\ \text{Torr}}{300\ \text{Torr} - 800\ \text{Torr}} = 0.08$$

Thus $\quad x_B = 1 - 0.08 = 0.92$

(**d**) Here, the amount fraction of A in the liquid phase will be the same as that of A in the original vapour phase, i.e. $x_A = 0.25$

Hence $\quad p = x_A p_A^* + x_B p_B^* = 0.25 \times 300\ \text{Torr} + 0.75 \times 800\ \text{Torr}$

$$= 75\ \text{Torr} + 600\ \text{Torr} = 675\ \text{Torr}$$

(**e**) In this case, we have

$$y_A = \frac{x_A p_A^*}{p} = \frac{0.25 \times 300\ \text{Torr}}{675\ \text{Torr}} = 0.111$$

$$y_B = 1 - y_A = 0.889$$

---

***Example 4.11.3***

The vapour pressure of two pure liquids A and B which form an ideal solution are 300 and 800 Torr, respectively, at temperature *T*. A liquid solution of A and B for which the amount fraction of A is 0.60 is contained in a cylinder closed by a piston on which the pressure can be varied. The solution is slowly vaporized at temperature *T* by decreasing the applied pressure, starting with a pressure of about 1 atm. Calculate (a) the pressure at which the first bubble of vapour is formed, (b) the composition of the vapour in this bubble, (c) the composition of the last droplet, and (d) the pressure when only this last droplet of liquid remains.

***Solution***

The given data are

$$p_A^* = 300 \text{ Torr} \qquad p_B^* = 800 \text{ Torr} \qquad x_A = 0.60$$

(**a**) *The pressure at which first bubble of vapour is formed*

We have

$$p = x_A p_A^* + x_B p_B^* = 0.6 \times 300 \text{ Torr} + 0.4 \times 800 \text{ Torr}$$
$$= 180 \text{ Torr} + 320 \text{ Torr} = 500 \text{ Torr}$$

(**b**) *Amount fractions of components in the first bubble of vapour*

We have

$$y_A = \frac{p_A}{p} = \frac{x_A p_A^*}{500 \text{ Torr}} = \frac{0.60 \times 300 \text{ Torr}}{500 \text{ Torr}} = 0.36$$

Thus $y_B = 1 - 0.36 = 0.64$

(**c**) *Amount fractions of components in the last drop of liquid*

Here we will have $y_A = 0.60$ (as most of the liquid has vaporized).

Now $$y_A = \frac{p_A}{p} = \frac{x_A p_A^*}{x_A p_A^* + x_B p_B^*} = \frac{x_A p_A^*}{p_B^* + (p_A^* - p_B^*)x_A}$$

Substituting the given data, we get

$$0.60 = \frac{x_A (300 \text{ Torr})}{800 \text{ Torr} + (300 \text{ Torr} - 800 \text{ Torr})x_A}$$

Solving for $x_A$, we have

$$x_A = 0.80$$

Thus $x_B = 1 - x_A = 0.20$

(**d**) *Pressure when the last droplet of liquid remains*

We have $$p = x_A p_A^* + x_B p_B^* = 0.80 \times 300 \text{ Torr} + 0.2 \times 800 \text{ Torr}$$
$$= 240 \text{ Torr} + 160 \text{ Torr} = 400 \text{ Torr}$$

---

**Procedure of Isothermal Fractional Distillation**

Suppose the vapour which is formed over the solution is removed and is condensed separately. This new liquid which is formed will be richer in the more volatile constituent. Now, let again the vapour which is formed over this new solution be removed and recondensed. the new liquid which is formed will be still richer in the more volatile constituent. Let this procedure of removing the vapour from the newer solution and condensing them be repeated several times. Ultimately, a stage will come when the vapour consists only of the more volatile constituent and the liquid that of the less volatile constituent. Thus a separation of constituents has been achieved.

The above method is useful for those mixtures which would decompose if distilled by ordinary methods. However, it is very inconvenient and is used only when the other method of separation is not available.

**Konowaloff's Rule**

From Eqs (4.11.1) and (4.9.3), we have drawn the following two conclusions.

(i) If $p_A^* > p_B^*$, the amount fraction of the component A in the vapour phase is larger than in the liquid phase. In other words, the vapour phase is richer in the component A than the liquid phase.

(ii) If $p_A^* > p_B^*$, the component A is more volatile than the component B. In a system where component A is more volatile, the total vapour pressure of the system increases with increase in the amount fraction $x_A$ of the more volatile component. In other words, if we add more of the component A in the liquid phase, the total vapour pressure of the system is increased.

We can combine the above two conclusions in a statement given below.
*The vapour phase is richer in the component whose addition to the liquid mixture results in an increase in total vapour pressure.*

The above statement is, in fact, the *Konowaloff's rule,* which we shall derive more rigorously in Section 4.15.

## 4.12 TEMPERATURE-COMPOSITION DIAGRAM OF A BINARY LIQUID SOLUTION

**Qualitative Description**

We have already seen that if the two phases are in equilibrium in a binary system, then its variance is two. Thus fixing the values of any of the two variables amongst temperature, pressure and composition, completely defines the system. The third variable automatically has a fixed value. For example, if temperature and composition are chosen, then the equilibrium pressure of the system has a fixed value. Mathematically, this can be stated as

$$p = f(\text{temperature, composition}) \tag{4.12.1}$$

If the temperature is held constant, then $p$ depends only on the composition of the mixture. Thus

$$p = f(\text{composition}) \qquad (T \text{ constant}) \tag{4.12.2}$$

Likewise, if pressure and composition be chosen as the independent variables, then the temperature becomes the dependent variable, i.e.

$$T = f(\text{pressure, composition}) \tag{4.12.3}$$

If the pressure is kept constant for a series of solutions of different compositions, then

$$T = f(\text{composition}) \qquad (p \text{ constant}) \tag{4.12.4}$$

that is, temperature at which the given system has the equilibrium pressure $p$ will depend on the composition of the system.

Equations (4.12.2) and (4.12.4) may be understood by taking an example of liquid-vapour equilibrium of an ideal binary liquid system. From Raoult's law, the total vapour pressure $p$ of the system is given by

$$p = p_A + p_B = x_A p_A^* + x_B p_B^* = x_A p_A^* + (1 - x_A) p_B^*$$

$$= p_B^* + (p_A^* - p_B^*) x_A \qquad (4.12.5)$$

Thus $p$ depends on $x_A$, the composition of the solution and on the values of $p_A^*$ and $p_B^*$, the vapour pressures of the pure constituents. Since $p_A^*$ and $p_B^*$ depend only on temperature, it follows that $p$ is a function of temperature and composition. Now if the temperature is kept constant (i.e. $p_A^*$ and $p_B^*$ are constant), then $p$ depends only on the composition of the solution. Figure 4.9.1 shows the dependence of $p$ on composition of an ideal binary liquid system at a constant temperature. Now suppose we want to keep the equilibrium pressure $p$ to a constant value. For a system, this given value of $p$ can be obtained by varying the temperature of the system. This time $p_A^*$ and $p_B^*$ are being changed so that the value of $p$ as calculated from Eq. (4.12.5) becomes equal to the given value of the pressure. The temperature at which this equality is obtained will depend upon the composition of the system.

**Dependence of Boiling Point of a Solution on its Composition**

The boiling point of a liquid (or a solution) is the temperature at which the vapour pressure of the liquid (or the total vapour pressure of the solution) becomes equal to that of the external pressure. If the external pressure is 1 atm, the boiling point is known as the normal boiling point. A liquid which has a higher vapour pressure will have a lower boiling point since less heating will be required to attain vapour pressure equal to the external pressure. Similarly, a liquid with low vapour pressure will have a high boiling point. The boiling point of a solution will have a definite relation with the composition of the solution. The relation connecting these two may be determined with the help of Clausius-Clapeyron equation.

Let $T_A^*$ and $T_B^*$ be the normal boiling points of constituents A and B, respectively. Let $T$ be the normal boiling point of a solution of known composition and let $p_A^*$ and $p_B^*$ be the vapour pressures of pure constituents at temperature $T$. Applying Clausius-Clapeyron equation separately to both the constituents, we have

$$\ln\left(\frac{p_A^*}{1 \text{ atm}}\right) = \frac{\Delta_{vap}H_{m,A}}{R}\left(\frac{1}{T_A^*} - \frac{1}{T}\right)$$

or $$\frac{p_A^*}{\text{atm}} = \exp\left\{\frac{\Delta_{vap}H_{m,A}}{R}\left(\frac{1}{T_A^*} - \frac{1}{T}\right)\right\} \qquad (4.12.6a)$$

$$\ln\left(\frac{p_B^*}{1 \text{ atm}}\right) = \frac{\Delta_{vap}H_{m,B}}{R}\left(\frac{1}{T_B^*} - \frac{1}{T}\right)$$

or $$\frac{p_B^*}{\text{atm}} = \exp\left\{\frac{\Delta_{vap}H_{m,B}}{R}\left(\frac{1}{T_B^*} - \frac{1}{T}\right)\right\} \qquad (4.12.6b)$$

Now the total vapour pressure of the solution of known composition is

$$p = x_A p_A^* + x_B p_B^*$$

and this, by definition, is equal to 1 atm (the external pressure). Thus, we have

$$1 = x_A\left(\frac{p_A^*}{\text{atm}}\right) + x_B\left(\frac{p_B^*}{\text{atm}}\right)$$
$$= x_A \exp\left\{\frac{\Delta_{vap}H_{m,A}}{R}\left(\frac{1}{T_A^*} - \frac{1}{T}\right)\right\} + x_B \exp\left\{\frac{\Delta_{vap}H_{m,B}}{R}\left(\frac{1}{T_B^*} - \frac{1}{T}\right)\right\}$$
$$= x_A \exp\left\{\frac{\Delta_{vap}S_{m,A}}{R}\left(1 - \frac{T_A^*}{T}\right)\right\} + x_B \exp\left\{\frac{\Delta_{vap}S_{m,B}}{R}\left(1 - \frac{T_B^*}{T}\right)\right\} \quad (4.12.7)$$

where $\Delta_{vap}S_{A,m}$ and $\Delta_{vap}S_{B,m}$ are the molar entropy changes for the liquid to vapour conversion for the components A and B, respectively. Assuming that the liquids follow Trouton's rule ($\Delta_{vap}S_m = \Delta_{vap}S_{m,A} = \Delta_{vap}S_{m,B} = 88$ J K$^{-1}$ mol$^{-1}$), we have

$$1 = x_A \exp\left\{\frac{\Delta_{vap}S_m}{R}\left(1 - \frac{T_A^*}{T_A}\right)\right\} + x_B \exp\left\{\frac{\Delta_{vap}S_m}{R}\left(1 - \frac{T_B^*}{T}\right)\right\}$$

or
$$\exp\left(-\frac{\Delta_{vap}S_m}{R}\right) = x_A \exp\left\{-\frac{\Delta_{vap}S_m}{R}\left(\frac{T_A^*}{T}\right)\right\} + x_B \exp\left\{-\frac{\Delta_{vap}S_m}{R}\left(\frac{T_B^*}{T}\right)\right\} \quad (4.12.8)$$

Thus, knowing $T_A^*$, $T_B^*$ and $x_A$ (or $x_B$, since $x_A + x_B = 1$), we can solve the above equation for $T$, the boiling point of the solution.

**Composition of Vapour at a Given Temperature**

Composition of vapour at temperature $T$ can be determined as usual from the relations

$$y_A = \frac{x_A(p_A^*/p^\circ)}{x_A(p_A^*/p^\circ) + x_B(p_B^*/p^\circ)} = \frac{x_A \exp\left\{\frac{\Delta_{vap}S_m}{R}\left(1 - \frac{T_A^*}{T}\right)\right\}}{1}$$

and
$$y_B = \frac{x_B(p_B^*/p^\circ)}{x_A(p_A^*/p^\circ) + x_B(p_B^*/p^\circ)} = \frac{x_B \exp\left\{\frac{\Delta_{vap}S_m}{R}\left(1 - \frac{T_B^*}{T}\right)\right\}}{1} \quad (4.12.9)$$

where $p^\circ$ is the unit pressure.

**Depiction of the Diagram**

Figure 4.12.1 shows the graph of boiling point versus composition for the ideal solution corresponding to Fig. 4.10.1.

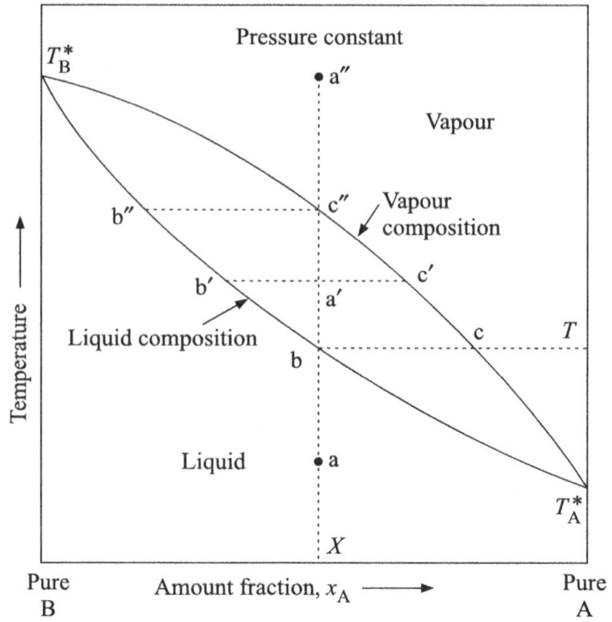


**Fig. 4.12.1 Graph of boiling point versus amount fraction corresponding to Fig. 4.10.1**

## Qualitative Features of the Diagram

- Neither the liquidus nor the vapourous curve is a straight line. This follows immediately from the forms of Eqs (4.12.8) and (4.12.9).
- The lower region of the figure represents the liquid phase as this phase is stable at lower temperatures. Thus, the lower curved line represents the liquidus curve. Similarly, it can be concluded that upper region of the figure represents the vapour phase as this phase is stable at higher temperatures. Thus, the upper curved line represents the vaporous curve.
- The normal boiling point $T_B^*$ of the constituent B is higher than that of the constituent A, i.e. $T_A^*$. This follows immediately from the fact that the vapour pressure of the constituent A is higher than that of the constituent B.
- The central region between the two curved lines represents the liquid-vapour equilibrium.

## Effects Produced on Heating a System

Now we can examine the sequence of events that takes place when a given liquid mixture under a constant pressure is heated from a lower temperature to a high temperature. Let us start with a system represented by the point a in Fig. 4.12.1.

- On increasing the temperature at constant pressure, the state of the system will move along the vertical line aa′a″.
- The system will remain of one phase till the temperature $T$ is reached where the liquid just starts boiling with the vapour composition corresponding to the point c.
- The vapour phase is richer than the liquid in the component A, the lower boiling component. Removal of more of the component A in the vapour phase makes the solution more rich in the second component with the result that the composition of the solution moves along bb′.

- If the boiling is to be continued, the temperature of the system must be increased; there occurs an increase in the boiling point of the solution.
- The overall state of the system at any temperature will be represented by a point on the vertical line aa′a″. For example, at point a′, the compositions of liquid and vapour phases are represented by the points b′ and c′, respectively. The relative amounts of these two phases as given by the lever rule are

$$\frac{\text{Amount in liquid phase}}{\text{Amount in vapour phase}} = \frac{\text{a}'\text{c}'}{\text{a}'\text{b}'}$$

- The increase in temperature will eventually bring the system to the point c″ where only a last trace of liquid of composition b″ is left and the vapour phase has the composition $X$.
- If the temperature of the system is slightly increased the liquid phase disappears completely and the system becomes of one phase comprising of the vapour phase.

---

***Example 4.12.1*** The boiling points of pure benzene and toluene are 353.25 K and 383.75 K under 1 atm pressure. By employing appropriate equations, determine the compositions of the liquid mixture and its vapour when the former just starts boiling at 368.15 K. Assume that Trouton's law is applicable to both the constituents.

***Solution*** We know from Eq. (4.12.8) that

$$\exp\left(-\frac{\Delta_{\text{vap}}S_{\text{m}}}{R}\right) = x_{\text{b}}\exp\left\{-\frac{\Delta_{\text{vap}}S_{\text{m}}}{R}\left(\frac{T_{\text{b}}^*}{T}\right)\right\} + x_{\text{t}}\exp\left\{-\frac{\Delta_{\text{vap}}S_{\text{m}}}{R}\left(\frac{T_{\text{t}}^*}{T}\right)\right\} \quad (1)$$

where, according to Trouton's law, $\Delta_{\text{vap}}S_{\text{m}}/R = 10.6$

Now $T_{\text{b}}^* = T_{\text{benzene}}^* = 353.25\ \text{K}$ $\quad T_{\text{t}}^* = T_{\text{toluene}}^* = 383.75\ \text{K}$

$T = 368.15\ \text{K}$ $\quad x_{\text{t}} = 1 - x_{\text{b}}$

Substituting these in Eq. (1), we get

$$\exp(-10.6) = x_{\text{b}}\exp\left\{-10.6\left(\frac{353.25}{368.15}\right)\right\} + (1 - x_{\text{b}})\exp\left\{-10.6\left(\frac{383.75}{368.15}\right)\right\}$$

$$2.576 \times 10^{-5} = x_{\text{b}}(3.908 \times 10^{-5}) + (1 - x_{\text{b}})(1.629 \times 10^{-5})$$

or
$$x_{\text{b}} = \frac{2.576\times10^{-5} - 1.629\times10^{-5}}{3.908\times10^{-5} - 1.629\times10^{-5}} = 0.415$$

$$x_{\text{t}} = 1 - x_{\text{b}} = 1 - 0.415 = 0.585$$

Composition of the issuing vapour can be calculated using the expression

$$y_{\text{b}} = x_{\text{b}}\exp\left\{\frac{\Delta S_{\text{m}}}{R}\left(1 - \frac{T_{\text{b}}^*}{T}\right)\right\} \quad \text{(Eq. 4.12.9)}$$

Substituting the data, we get

$$y_{\text{b}} = 0.415\times\exp\left\{10.6\left(1 - \frac{353.25}{368.15}\right)\right\} = 0.415\times\exp(0.429)$$

$$= 0.637$$

Thus $y_{\text{t}} = 1 - y_{\text{b}} = 0.363$

---

## 4.13 ISOBARIC FRACTIONAL DISTILLATION OF AN IDEAL BINARY SOLUTION

**Underlying Principle** The isobaric process of distillation in which the external pressure is kept constant instead of temperature is more convenient and is often employed for the separation of constituents of a binary liquid system. The principle of separation of constituents using the isobaric fractional distillation may be explained with the help of temperature-composition diagram shown in Fig. 4.13.1.

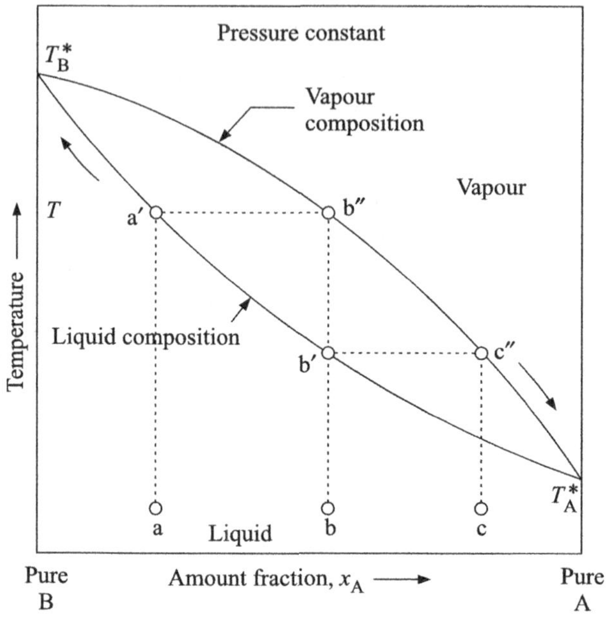


**Fig. 4.13.1 Principle of isobaric distillation**

- Let the starting composition of the solution to be distilled be represented by the point a. Let the temperature of the solution be raised till it starts boiling. At this stage, the vapour pressure of the solution is equal to the external pressure, which is kept constant throughout the distillation process. Usually the atmospheric pressure of 1 atm is employed for this purpose. The vapour which appeared at the boiling point $T$ is represented by b″. Since the latter is richer in the constituent A (the more volatile constituent), the residual liquid will become richer in the constituent B and will boil at a slightly higher temperature.
- Let the vapour formed at $T$ be removed and condensed separately to yield distillate b. Let this new distillate be heated till it starts boiling, the vapour which now emerges is represented by c″ and is still richer in the constituent A.
- If the above sequence of collecting the vapour, condensing them to give a new distillate and heating the new distillate to its boiling point is repeated several times, the vapour will continue to contain more and more of the constituent A and ultimately a stage would be reached where it would contain only this constituent.

The residual liquid at any stage can be mixed with the previous residual liquid and can be treated in the same way. The residual liquid continues to contain lesser and lesser of the more volatile constituent and thus more and more of the lesser volatile constituent. When the process is repeated several times, a stage would be reached when the residual liquid would contain only the lesser volatile constituent.

Thus, we see that by carrying out the above fractional distillation process, it is possible to separate the two constituents of a binary liquid mixture; vapour containing more volatile constituent and the liquid containing less volatile constituent.

**Theoretical Plate**

Each horizontal line in the zig-zag steps shown in Fig. 4.13.1 is known as the *theoretical plate*. For example, in going from liquid of composition a to c, we have two theoretical plates.

---

***Example 4.13.1***

The vapour pressures of two liquids A and B that form an ideal solution are given by the equations:

$$\log(p_A/\text{Torr}) = -\frac{5\,100\text{ K}}{T} + 16.24$$

$$\log(p_B/\text{Torr}) = -\frac{4\,530\text{ K}}{T} + 13.38$$

Draw the temperature-composition phase diagram for solutions of A and B at a total pressure of 1 atm. How many theoretical plates in the fractionating column will be present in going from liquid composition $x_A = 0.03$ to $x_A = 0.96$?

***Solution***

***Boiling points of the two liquids*** These can be obtained by substituting $p_A = p_B =$ 760 Torr in the given expressions

$$\log(p_A/\text{Torr}) = -\frac{5\,100\text{ K}}{T} + 16.24$$

$$\log(p_B/\text{Torr}) = -\frac{4\,530\text{ K}}{T} + 13.38$$

Thus

$$T_A^* = \frac{5\,100\text{ K}}{16.24 - \log(p_A/\text{Torr})} = \frac{5\,100\text{ K}}{16.24 - \log(760)} = \frac{5\,100\text{ K}}{16.24 - 2.880\,8} = 381.7\text{ K}$$

$$T_B^* = \frac{4\,530\text{ K}}{13.38 - \log(p_B/\text{Torr})} = \frac{4\,530\text{ K}}{13.38 - \log(760)} = \frac{4\,530\text{ K}}{13.38 - 2.880\,8} = 413.4\text{ K}$$

***Compositions of the solution and vapour at different temperatures*** For a given temperature, $p_A^*$ and $p_B^*$ can be obtained from the given relations

$$\log(p_A^*/\text{Torr}) = -\frac{5\,100\text{ K}}{T} + 16.24 \qquad (1)$$

$$\log(p_B^*/\text{Torr}) = -\frac{4\,530\text{ K}}{T} + 13.38 \qquad (2)$$

Let $x_A$ be the amount fraction of liquid A in the solution. The amount fraction of liquid B will be equal to $(1 - x_A)$. Since the two liquids form an ideal solution, the partial vapour pressures of the two liquids can be obtained by using Raoult's law, i.e.

$$p_A = x_A p_A^*$$

$$p_B = x_B p_B^* = (1 - x_A) p_B^*$$

Since the total pressure is 760 Torr, therefore

$$p = p_A + p_B = x_A p_A^* + (1 - x_A) p_B^* = 760 \text{ Torr}$$

or
$$x_A = \frac{760 \text{ Torr} - p_B^*}{p_A^* - p_B^*} \qquad (3)$$

The composition of constituent A in the vapour phase is given by

$$y_A = \frac{p_A}{p} = \frac{x_A p_A^*}{760 \text{ Torr}} \qquad (4)$$

The values of $p_A^*$, $p_B^*$, $x_A$ and $y_A$ as calculated from Eqs (1) to (4) at different temperatures are given below.

| $T$/K | $p_A^*$/Torr | $p_B^*$/Torr | $x_A = \frac{760 \text{ Torr} - p_B^*}{p_A^* - p_B^*}$ | $y_A = \frac{x_A p_A^*}{760 \text{ Torr}}$ |
|---|---|---|---|---|
| 385 | 1 000 | 39.81 | 0.750 | 0.986 |
| 390 | 1 445 | 58.88 | 0.506 | 0.962 |
| 400 | 3 090 | 114.8 | 0.217 | 0.882 |
| 410 | 6 310 | 213.8 | 0.089 | 0.743 |
| 420 | 12 590 | 398.1 | 0.030 | 0.491 |
| 425 | 17 380 | 573.0 | 0.029 | 0.255 |

On plotting temperature versus composition, we get a graph as shown in Fig. 4.13.2. The number of theoretical plates in a column in going from $x_A = 0.03$ (point $b_3$) to $x_A = 0.96$ (point $c_1$) are two since there are two horizontal lines present in the zig-zag steps.

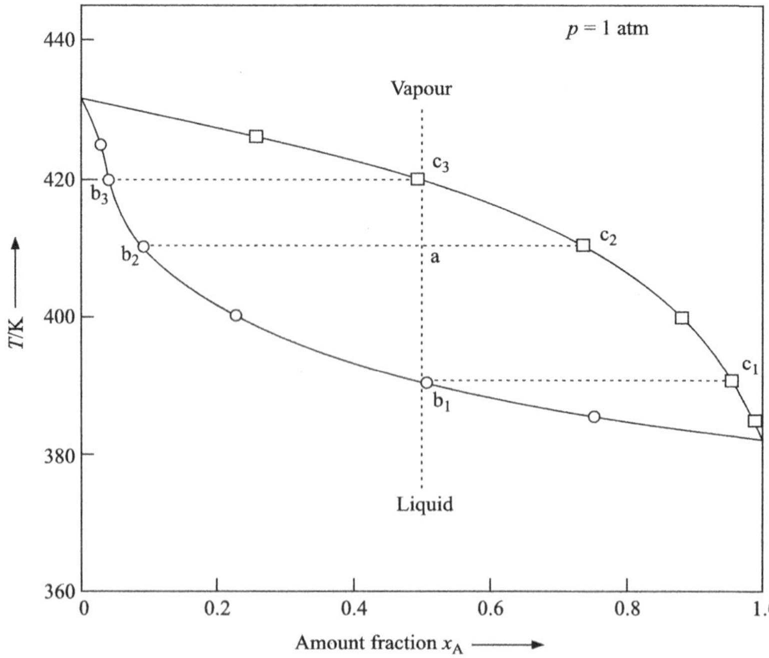


**Fig. 4.13.2 The required temperature versus composition plot**

***Example 4.13.2***

A liquid mixture containing 50 mol per cent of liquid A in Example 4.13.1 was heated gradually. Answer the following from the phase diagram thus obtained:

(i) What is the normal boiling point of the mixture?
(ii) What is the composition of the vapour that appears at the normal boiling point?
(iii) Indicate whether the boiling point of the remaining liquid mixture increases or decreases.
(iv) What are the compositions of liquid and vapour at temperature 410 K? What are their relative amounts?
(v) What is the composition of the last drop of the liquid to be vaporized?
(vi) If the given mixture of 50 mol per cent of liquid A is distilled without removing the distillate till the temperature of residue is 410 K, what will be the composition of the resultant distillate?

***Solution***

**(i)** Normal boiling point of the mixture = 391 K
**(ii)** Composition of vapour issuing at 391 K (near the point $c_1$);

$$y_B = 0.045 \qquad \text{and} \qquad y_A = 0.955$$

**(iii)** Boiling point of the remaining liquid mixture increases.
**(iv)** Compositions at 410 K are

Composition of liquid (point $b_2$): $x_A = 0.09$, $x_B = 0.91$

Composition of vapour (point $c_2$): $y_A = 0.74$, $y_B = 0.26$

$$\frac{\text{Amount of liquid mixture}}{\text{Amount of vapour}} = \frac{ac_2}{b_2a} = \frac{y_A - 0.5}{0.5 - x_A} = \frac{0.24}{0.41}$$

$$\text{Amount \% of vapour} = \frac{0.41}{0.24 + 0.41} \times 100 = 63.08$$

$$\text{Amount \% of liquid} = 100 - 63.08 = 36.92$$

**(v)** Composition of the last drop of the liquid to be vaporized (near the point $b_3$);

$$x_A = 0.035 \qquad \text{and} \qquad x_B = 0.965$$

**(vi)** The composition of the resultant distillate will be the mean of the initial composition of vapour (which appears when the composition of liquid mixture is 50 mol per cent in A) and the final composition of vapour (when the temperature of the residue is 410 K). The two compositions are:

$$(y_A)_{\text{initial}} = 0.955 \qquad \text{and} \qquad (y_A)_{\text{final}} = 0.743$$

Hence, the amount fraction of A in the resultant distillate will be

$$\frac{0.955 + 0.743}{2} = 0.743 \simeq 0.85$$

## The Fractionating Column

The process of fractional distillation is extremely tedious and involves more time and labour as the separation is carried out in batches and in a discontinuous manner. However, these difficulties can be overcome by employing a fractionating column, which essentially carries the distillation in a continuous manner. Figure 4.13.3 displays one such column commonly employed in industry. This is known as the *bubble-cap column.* It consists of a long tube carrying a large number of bubble-cap plates and is attached to a boiler at the bottom and to a condenser at the top. Each plate can hold a thin layer of liquid and has an overflow mechanism through which the excess liquid can pass to the plate just below it. It also has many bubble-caps through which the vapour passes upward after bubbling through the liquid. There is a temperature gradient along the length of the column, the top being cooler than the bottom. The various plates are thus situated at different temperatures and also hold the liquid at that temperature.

**Principle Underlying Fractionating Column** The principle of bubble-cap column can be illustrated very nicely with the help of temperature-composition diagram (Fig. 4.13.4).

Let the liquid be boiled at the bottom, say at temperature $T_0$. The vapour issuing has composition $v_0$. When this vapour is passed through the first plate, it is cooled to temperature $T_1$ and thus its state is moved to the point a. At this state, some of the vapour condenses to form liquid of composition $l_1$ and the

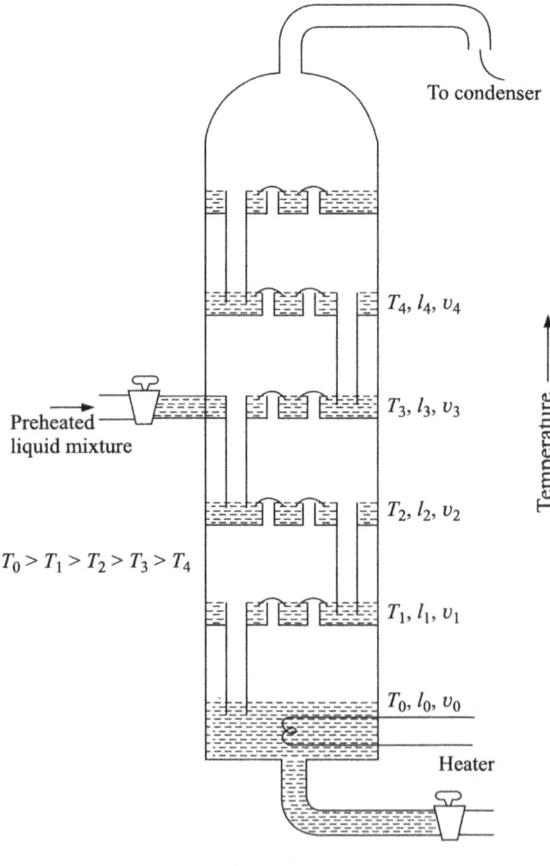


Fig. 4.13.3 Bubble-cap distilling column

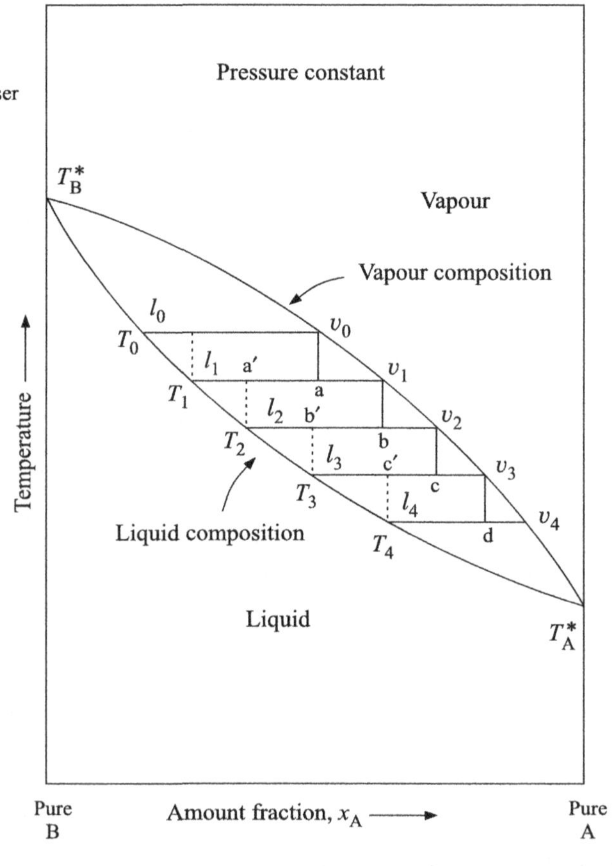


Fig. 4.13.4 Scheme of redistribution of constituents in the distilling column

remaining vapour has composition $v_1$. The liquid formed contains more of the less volatile constituent. Next, the vapour of composition $v_1$ is passed through the second plate whose temperature is $T_2(T_2 < T_1)$. Here the vapour is cooled to $T_2$ and thus the state of system is moved from $v_1$ to b. At this state, again part of the vapour is condensed to give liquid of composition $l_2$ and the remaining vapour has composition $v_2$. Now the vapour has become more enriched in the more volatile constituent. This happens at every plate of the column as is also shown in Fig. 4.13.4. As the vapour moves up the column, it is being cooled; this cooling condenses the less volatile component preferentially, so that the vapour becomes increasingly enriched in the more volatile component as it passes upward from one plate to another.

Similarly, as the liquid flows down, its temperature is increased and again there is a redistribution of the constituents. For example, the liquid of composition $l_3$ has flown from the plate 3 to plate 2. The liquid has been heated from $T_3$ to $T_2$ and thus the state of the system has moved from $l_3$ to b′. Thus part of the liquid vaporizes to yield vapour of composition $v_2$ containing more of the more volatile constituent. The resultant liquid has a composition of $l_2$ and thus contains more of the less volatile constituent. This happens at every plate of the column as shown in Fig. 4.13.4. As the liquid moves down the column, it is being heated; this heating vaporizes the more volatile constituent preferentially, so that the liquid becomes increasing enriched in the less volatile component as it moves downward from one plate to another.

If sufficient number of plates are used, it is possible to separate the two constituents of a binary liquid mixture; the more volatile constituent in the vapour form is collected from the top of the distilling column and that of lesser volatile constituent from the bottom of the column. The vapour from the top of the column is fed into a condenser where it is liquefied. Part of this liquid is drawn off and the rest is returned to the column in order to maintain the stock of essentially pure distillate on the upper plates. In order to make the separation continuous, the preheated liquid mixture is introduced somewhere within the column as shown in Fig. 4.13.3.

## 4.14 NONIDEAL SOLUTIONS OF LIQUID IN LIQUID

**Nonideal Solutions** It was pointed out earlier that an ideal binary liquid solution is formed only when two liquids having more or less identical intermolecular forces of attraction are mixed together, so that the environment around any particular molecule of either of the two liquids is identical to that existing in the liquid mixture. This condition is, however, not satisfied for most liquid pairs and thus they form nonideal solutions. These solutions show deviations from Raoult's law. The nature and the extent of deviation depends upon the types of liquids that are being mixed up. Two types of deviations are observed as describe below.

**Positive Deviation** If the forces of attraction between unlike molecules (i.e. between A and B) are weaker than those between like molecules (i.e. between either A and A or B and B), the solution shows positive deviation from Raoult's law. In this case, the vapour pressure curves of the constituents and the mixture lie above those of ideal lines (Fig. 4.14.1).

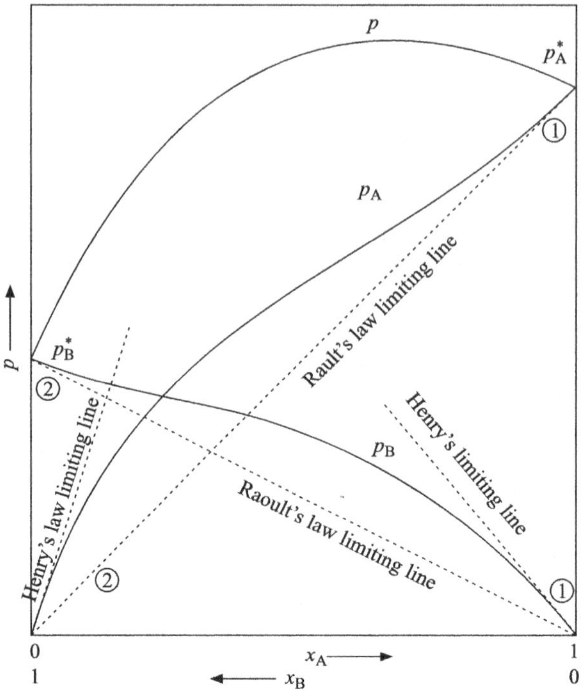


(1) Region (labelled as 1) of concentration where Raoult's law is applicable to the component A (solvent) and Henry's law to the component B (solute)
(2) Region (labelled as 2) of concentration where Raoult's law is applicable to B (solvent) and Henry's law to A (solute)

**Fig. 4.14.1 Pressure-composition curves of a typical nonideal solution exhibiting positive deviation from ideality**

The positive deviation arises from the fact that the tendency of molecules of each kind to escape from the solution to vapour is greater in nonideal solution than that prevailing in an ideal solution. The extent of positive deviation depends upon many factors. Some of them are: (i) difference in polarity of the molecules; (ii) difference in the length of hydrocarbon chain or analogous grouping of the molecules; (iii) difference in the intermolecular forces of attraction; and (iv) association of either of the constituent in the liquid state. The deviation is larger if the two liquids differ greatly in respect to the above factors in which cases a maximum in the plot of total vapour pressure versus composition of solution is observed.

## Examples of Positive Deviation

Examples of solutions showing positive deviation are numerous. In fact, most liquid pairs fall into this category. Some of these are:

(i) Carbon tetrachloride and heptane at 323 K: This solution shows small deviations from Raoult's law as both the constituents are nonpolar and involve low forces of attraction.

(ii) Ethyl ether and acetone at 293 K and 303 K: Deviations here are bit larger as the components differ appreciably in the intermolecular attraction forces.

(iii) Heptane and ethyl alcohol at 323 K: This solution shows very large deviations as the two liquids differ appreciably in all the above four factors, namely, polarity, chain length, intermolecular forces of attraction and the association of alcohol in the liquid state.

## Negative Deviation

If, on the other hand, the forces of attraction between unlike molecules (i.e. between A and B) are stronger than those between like molecules (i.e. between either A and A or B and B), then the solution shows negative deviation from Raoult's law. Here the vapour pressure curves of the constituents and the mixture lie below those of the ideal lines. In this case, the tendency of molecules of each kind to escape from the solution into the vapour is weak relative to that found in an ideal solution. This is primarily due to the stronger interactions between the unlike molecules. Figure 4.14.2 exhibits pressure-composition curves of a typical solution exhibiting negative deviation from ideality. The deviation is larger if the two constituents form partial compound in the solution and in such cases a minimum in the plot of total vapour pressure versus composition of solution is observed.

**Examples of Negative Deviation**

Examples of solutions showing negative deviation are not very common. A few examples are:

(i) Pyridine and formic acid (or acetic or propanoic acid): One of the constituents is basic and the other one is acidic.

(ii) Mixture of halomethane (e.g., chloroform) with an oxygen or nitrogen compound (e.g., a ketone, ether, ester or amine): In this case, there occurs a partial association between the molecules through the hydrogen bonding.

(iii) An aqueous solution of a strong volatile acid such as halogen acids, nitric acid and perchloro acids: In this case, non-volatile ions are formed with the interaction of ions of the acid with water.

**Ideally Dilute Solution**

There is one common feature in Figs 4.14.1 and 4.14.2 that the partial pressure curves of each component approaches the ideal ones asymptotically as the amount fraction of the component approaches one. This is to be expected since at this stage, the solution is very dilute and hence behaves ideally. This dilute solution is known as *ideally dilute solution*. Thus, we can say that Raoult's law is a limiting law for the solvent (major component as $x_A \rightarrow 1$ or $x_B \rightarrow 1$) in very dilute solutions. Mathematically, it may be expressed as

$$\lim_{x_A \to 1} \left( \frac{p_A}{x_A} \right) = p_A^* \quad \text{and} \quad \lim_{x_B \to 1} \left( \frac{p_B}{x_B} \right) = p_B^* \tag{4.14.1}$$

It was shown earlier that if Raoult's law is applicable to the major component (solvent) over a certain region of concentration, then Henry's law is followed by the minor component (solute) over the same region of concentration. It is thus expected that the partial vapour pressure curve of each of the two components should also approach linearity asymptotically, as the amount fraction of the component approaches zero. In fact, it is found to be so and thus the linearity observed at $x \rightarrow 0$ represents the limiting behaviour of Henry's law as applicable to the solute. Mathematically, the linearity may be expressed as

$$\lim_{x_A \to 0} \left( \frac{p_A}{x_A} \right) = k_H(A) \quad \text{and} \quad \lim_{x_B \to 0} \left( \frac{p_B}{x_B} \right) = k_H(B) \tag{4.14.2}$$

where $k_H(A)$ is a constant for a given solute (A) in a particular solvent (B) and is not equal to $p_A^*$. Likewise, $k_H(B)$ is also constant and will not be equal to $p_B^*$.

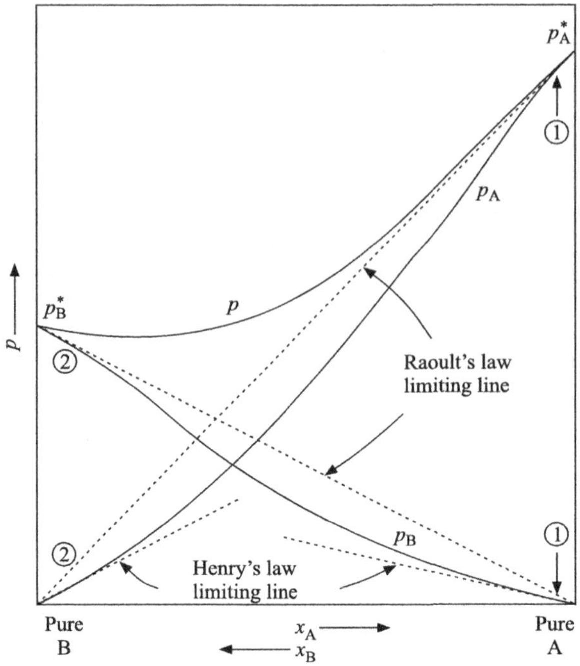


(1) Region (labelled as 1) of concentration over which Raoult's law is applicable to A and Henry's law to B
(2) Region (labelled as 2) of concentration onver which Raoult's law is applicable to B and Henry's law to A.

**Fig. 4.14.2 Pressure-composition diagram of a typical system exhibiting negative deviation**

In general, the slope of the limiting Henry's law line is greater than that of the limiting Raoult's law line in case of positive deviation from ideality and is smaller in case of negative deviation from ideality.

## Expression of Raoult's Law for a Nonideal Solution

For an ideal solution, Raoult's law is expressed as

$$p_i = x_i p_i^* \tag{4.14.3}$$

For the *i*th constituent in a nonideal solution, Raoult's law may be expressed as

$$p_i = a_i p_i^* \tag{4.14.4}$$

where $a_i$ is the activity of *i*th constituent in the solution.† In terms of amount fraction, we have

$$p_i = (x_i \gamma_i)\, p_i^* \tag{4.14.5}$$

where $\gamma_i$ is the activity coefficient of *i*th constituent in the solution. From Eq. (4.14.5), it follows that

$$a_i = x_i \gamma_i = \frac{p_i}{p_i^*} \tag{4.14.6a}$$

or

$$\gamma_i = \frac{p_i}{x_i p_i^*} = \frac{(p_i)_{\text{real}}}{(p_i)_{\text{ideal}}} \tag{4.14.6b}$$

†See Annexure I at the end of Chapter 8 for the concept of activity.

For a solution exhibiting positive deviation from ideality, we have

$$(p_i)_{\text{real}} > (p_i)_{\text{ideal}} \quad \text{i.e.} \quad \gamma_i > 1 \tag{4.14.7}$$

and for a solution exhibiting negative deviation

$$(p_i)_{\text{real}} < (p_i)_{\text{ideal}} \quad \text{i.e.} \quad \gamma_i < 1 \tag{4.14.8}$$

From Figs 4.14.1 and 4.14.2, it is obvious that the activity coefficient $\gamma_i$ will vary as the amount fraction of $i$th constituent in the solution is varied. But there is one common feature that $\gamma_i \rightarrow 1$ as $x_i \rightarrow 1$, since $(p_i)_{\text{real}}$ approaches $(p_i)_{\text{ideal}}$ asymptotically as the amount fraction of the constituent approaches one. The typical variations of activity coefficient with mole fraction for positive and negative deviations are of the type shown in Fig. 4.14.2.

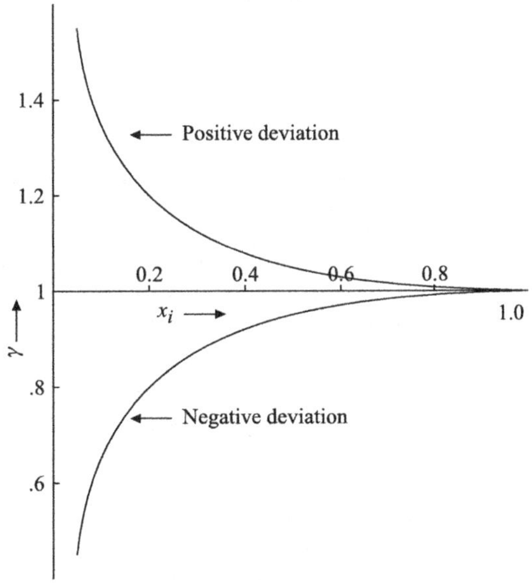


**Fig. 4.14.3 Typical variations of activity coefficient with amount fraction**

---

***Example 4.14.1***

Starting from the fact that for a binary solution exhibiting negative deviation from Raoult's law, solute-solvent interactions are strong compared to solute-solute and solvent-solvent interactions, show that for such a solution

$$\Delta_{\text{mix}}H < \Delta_{\text{mix}}H(\text{ideal}); \qquad \Delta_{\text{mix}}S < \Delta_{\text{mix}}S(\text{ideal})$$

$$\Delta_{\text{mix}}G < 0; \qquad p_{\text{A}} < p_{\text{A}}(\text{ideal})$$

***Solution***

Since solute-solvent interactions are stronger than solute-solute interactions (B · · · B) and solvent-solvent interactions (A · · · A), we have

$$\Delta H_{\text{A}\cdots\text{B}} > \Delta H_{\text{A}\cdots\text{A}} \quad \text{and} \quad \Delta H_{\text{A}\cdots\text{B}} > \Delta H_{\text{B}\cdots\text{B}} \tag{1}$$

where $\Delta H_{\text{A}\cdots\text{A}}$, $\Delta H_{\text{B}\cdots\text{B}}$ and $\Delta H_{\text{A}\cdots\text{B}}$ represent the enthalpies required to separate A · · · A in the solvent, B · · · B in the solute and A · · · B in the solution, respectively. For an ideal solution, we have

$$\Delta H_{\text{A}\cdots\text{A}}(\text{ideal}) = \Delta H_{\text{B}\cdots\text{B}}(\text{ideal}) = \Delta H_{\text{A}\cdots\text{B}}(\text{ideal}) \tag{2}$$

In the solution process, we will be breaking A · · · A and B · · · B interactions and will be generating 2(A · · · B) interactions, i.e.

$$\Delta_{mix}H = \Delta H_{A\cdots A} + \Delta H_{B\cdots B} + 2(-\Delta H_{A\cdots B}) \tag{3}$$

For an ideal solution, From Eq. (2), we get

$$\Delta_{mix}H(\text{ideal}) = 0$$

For a nonideal solution which satisfies Eq. (1), we get

$$\Delta_{mix}H = -\text{ ve} \quad \text{i.e.} \quad \Delta_{mix}H < \Delta_{mix}H(\text{ideal}) \tag{4}$$

that is, heat will be released on mixing solute and solvent to form a solution. $\Delta_{mix}S$ of the solution process will always be positive as the solute and solvent molecules are more disordered in solution than in the pure liquids. For an ideal solution, $\Delta_{mix}S$ will have a maximum value as there exists no interaction terms. If A · · · B interactions are strong, there will be some ordering in the solution, which make $\Delta_{mix}S$ smaller than $\Delta_{mix}S$(ideal), i.e.

$$\Delta_{mix}S < \Delta_{mix}S(\text{ideal}) \tag{5}$$

$\Delta_{mix}G$ will be given by

$$\Delta_{mix}G = \Delta_{mix}H - T\,\Delta_{mix}S$$

Now $\Delta_{mix}H$ is negative; tending to make $\Delta_{mix}G$ more negative than for an ideal solution. But $T\,\Delta_{mix}S$ is less positive than $T\,\Delta_{mix}S$(ideal), which tends to make $-\,T\,\Delta_{mix}S$ and $\Delta_{mix}G$ less negative. Though the final $\Delta_{mix}G$ is negative, but its value relative to the ideal solution depends on the relative magnitudes of $\Delta_{mix}H$ and $T\,\Delta_{mix}S$.

Finally, because of A · · · B interactions, A will not be as free to vaporize as in the ideal solution. Thus, we will have

$$p_A < p_A(\text{ideal}) \tag{6}$$

## 4.15 THE DUHEM-MARGULES EQUATION

**Thermodynamic Derivation**

A quantitative relation between the partial vapour pressures of the two constituents of a binary liquid system and their corresponding amount fractions in liquid phase can be derived thermodynamically. We start with the Gibbs-Duhem equation for a system of two components A and B, such that

$$n_A d\mu_A + n_B d\mu_B = 0$$

or

$$n_A d\mu_A = -n_B\, d\mu_B$$

Dividing both sides by $n_A + n_B$, we have

$$\frac{n_A}{n_A + n_B} d\mu_A = -\frac{n_B}{n_A + n_B} d\mu_B \tag{4.15.1}$$

or

$$x_A d\mu_A = -x_B d\mu_B$$

where $x_A$ and $x_B$ are the amount fractions of components A and B, respectively. Dividing throughout by $dx_A$, we get

$$x_A \frac{d\mu_A}{dx_A} = -x_B \frac{d\mu_B}{dx_A} \tag{4.15.2}$$

Since $x_A + x_B = 1$, therefore

$$dx_A + dx_B = 0$$

or $$\mathrm{d}x_A = -\,\mathrm{d}x_B$$

Substituting the above relation in Eq. (4.15.2), we get

$$x_A \frac{\mathrm{d}\mu_A}{\mathrm{d}x_A} = x_B \frac{\mathrm{d}\mu_B}{\mathrm{d}x_B} \tag{4.15.3}$$

The chemical potential of any constituent of a liquid mixture is given by

$$\mu_{i(\mathrm{sol})} = \mu_i^* + RT \ln(p/p^\circ) \tag{4.15.4}$$

where $p$ is the partial vapour pressure of that constituent. In deriving the above expression, it is assumed that the vapour behaves as an ideal gas. For the component A, the above equation on differentiating with respect to $x_A$ at constant temperature and total pressure gives

$$\frac{\mathrm{d}\mu_A}{\mathrm{d}x_A} = RT \frac{\mathrm{d}\ln(p_A/p^\circ)}{\mathrm{d}x_A} \tag{4.15.5}$$

Similarly, for the component B, differentiation with respect to $x_B$ gives

$$\frac{\mathrm{d}\mu_B}{\mathrm{d}x_B} = RT \frac{\mathrm{d}\ln(p_B/p^\circ)}{\mathrm{d}x_B} \tag{4.15.6}$$

Multiplying Eq. (4.15.5) by $x_A$ and Eq. (4.15.6) by $x_B$ and equating the resultant expressions (Eq. 4.15.3), we get

$$x_A \frac{\mathrm{d}\ln(p_A/p^\circ)}{\mathrm{d}x_A} = x_B \frac{\mathrm{d}\ln(p_B/p^\circ)}{\mathrm{d}x_B}$$

or $$\frac{\mathrm{d}\ln(p_A/p^\circ)}{\mathrm{d}\ln x_A} = \frac{\mathrm{d}\ln(p_B/p^\circ)}{\mathrm{d}\ln x_B} \tag{4.15.7}$$

Equation (4.15.7) is known as the *Duhem-Margules equation.* It relates the partial pressures of the two constituents with their corresponding amount fractions. Equation (4.15.7) is applicable to both ideal and nonideal liquid mixtures, since in its derivation nowhere the nature of the liquid mixture is involved. The only assumption made was that the vapour behaves as an ideal gas.

## Application of Duhem-Margules Equation

One of the relations for the partial pressure of the component A in a binary liquid solution is

$$p_A = p_A^* x_A \exp(\alpha\, x_B^2) \tag{4.15.8a}$$

and that for the component B is

$$p_B = p_B^* x_B \exp(\alpha\, x_A^2) \tag{4.15.8b}$$

where $p_A^*$ and $p_B^*$ are the vapour pressures of pure liquids A and B, respectively. The symbol $\alpha$ is a constant which has the same value for both the components. For ideal solutions $\alpha = 0$; for nonideal solutions $\alpha$ is positive for solutions exhibiting positive deviation from ideality, and negative for solutions exhibiting negative deviation from ideality. Since $\alpha$ is same for both the components, it follows that both components behave in an identical manner. Thus, we have:

(i) If one component behaves in an ideal manner, the other one will also behave ideally.

(ii) If one component shows positive (or negative) deviation from ideality so also must the other.

The above conclusions can be derived directly from Eq. (4.15.7) as follows.

***When One component behaves ideally*** Let this component be A. Then, according to Raoult's law, we have

$$p_A = p_A^* x_A$$

or $$\ln(p_A/p^\circ) = \ln(p_A^*/p^\circ) + \ln x_A$$

Hence $$\mathrm{d}\ln(p_A/p^\circ) = \mathrm{d}\ln x_A$$

or $$\frac{\mathrm{d}\ln(p_A/p^\circ)}{\mathrm{d}\ln x_A} = 1$$

Thus, according to the Duhem-Margules relation (Eq. 4.15.7), we must also have

$$\frac{\mathrm{d}\ln(p_B/p^\circ)}{\mathrm{d}\ln x_B} = 1 \qquad \text{or} \qquad \mathrm{d}\ln(p_B/p^\circ) = \mathrm{d}\ln x_B$$

Integration of the above expression gives

$$\int \mathrm{d}\ln(p_B/p^\circ) = \int \mathrm{d}\ln x_B \qquad \text{or} \qquad \ln(p_B/p^\circ) = \ln x_B + I$$

where $I$ is the constant of integration. Its value may be determined from the fact that at $x_B = 1$, $p_B = p_B^*$ and hence

$$I = \ln(p_B^*/p^\circ)$$

Thus $$\ln(p_B/p^\circ) = \ln x_B + \ln(p_B^*/p^\circ) = \ln(x_B p_B^*/p^\circ)$$

or $$p_B = x_B\, p_B^*$$

that is, the second component also follows Raoult's law and thus behaves in an ideal manner.

***When one component exhibits positive deviation*** Let this component be A. Obviously, we have

$$p_{A(\text{real})} > p_{A(\text{ideal})} = p_A^* x_A$$

or $$\ln(p_{A(\text{real})}/p^\circ) > \ln(p_{A(\text{ideal})}/p^\circ) = \ln(p_A^*/p^\circ) + \ln x_A$$

Therefore

$$\frac{\mathrm{d}\ln(p_{A(\text{real})}/p^\circ)}{\mathrm{d}x_A} > \frac{\mathrm{d}\ln(p_{A(\text{ideal})}/p^\circ)}{\mathrm{d}x_A} = \frac{1}{x_A}$$

or $$x_A \frac{d \ln (p_{A(real)}/p^\circ)}{dx_A} > \frac{x_A \, d \ln (p_{A(ideal)}/p^\circ)}{dx_A} = 1$$

i.e. $$\frac{d \ln (p_{A(real)}/p^\circ)}{d \ln x_A} > 1$$

Thus, according to the Duhem-Margules equation, we must also have

$$\frac{d \ln (p_{B(real)}/p^\circ)}{d \ln x_B} > 1$$

After integration, we will have

$$p_{B(real)} > p_{B(ideal)} = p_B^* x_B$$

that is, the second component also exhibits positive deviation from ideality.

Arguing in a similar manner, it can be shown that if one component exhibits negative deviation, the other will also do so.

**Konowaloff's Rule**

A general conclusion regarding the composition of the vapour phase in equilibrium with any particular solution (ideal or nonideal) can be obtained theoretically through the Duhem-Margules equation.

From Eq. (4.15.7), we have

$$\frac{d (\ln p_A/p^\circ)}{d \ln x_A} = \frac{d (\ln p_B/p^\circ)}{d \ln x_B} \quad \text{or} \quad \frac{x_A}{p_A}\frac{dp_A}{dx_A} = \frac{x_B}{p_B}\frac{dp_B}{dx_B} \qquad (4.15.9)$$

Since $x_A + x_B = 1$ therefore $dx_A + dx_B = 0$

or $dx_B = - dx_A$

Substituting the above expression in Eq. (4.15.9), we get

$$\frac{x_A}{p_A}\frac{dp_A}{dx_A} + \frac{x_B}{p_B}\frac{dp_B}{dx_A} = 0 \qquad (4.15.10)$$

Now $p = p_A + p_B$, therefore $$\frac{dp}{dx_A} = \frac{dp_A}{dx_A} + \frac{dp_B}{dx_A} \qquad (4.15.11)$$

Substituting $dp_A/dx_A$ from Eq. (4.15.10) in the above equation, we get

$$\frac{dp}{dx_A} = - \frac{x_B}{p_B}\frac{dp_B}{dx_A}\frac{p_A}{x_A} + \frac{dp_B}{dx_A} = \frac{dp_B}{dx_A}\left[1 - \frac{x_B p_A}{x_A p_B}\right] \qquad (4.15.12)$$

From Eq. (4.8.1), it follows that the addition of either of the components causes an increase in the corresponding partial vapour pressure (i.e. $dp_A/dx_A$ and $dp_B/dx_B$ are positive). Hence, the derivative $dp_B/dx_A$, which is equal to $- dp_B/dx_B$, must be negative. The change in the total vapour pressure with the addition of $dx_A$ (i.e. the derivative $dp/dx_A$ in Eq. 4.15.12) can be either positive or negative depending upon the sign of the expression within brackets in Eq. (4.15.12). Thus, we have two alternatives as given below.

(i) $x_B p_A > x_A p_B$: In this case $dp/dx_A$ is positive. The condition

$$x_B p_A > x_A p_B$$

can be written as

$$\frac{p_A}{p_B} > \frac{x_A}{x_B}$$

Using Dalton's law of partial pressures, we have

$$\frac{y_A}{y_B} > \frac{x_A}{x_B} \tag{4.15.13}$$

that is, the ratio of amount fractions of A and B in the vapour phase is greater than the corresponding ratio in the liquid phase. This means that the vapour is relatively richer in A than is the liquid from which it vaporizes. Thus, the vapour is richer in the component (here A) whose addition to the liquid mixture results in an increase in total vapour pressure (i.e. $dp/dx_A$ is positive).

(ii) $x_B p_A < x_A p_B$: In this case, $dp/dx_A$ is negative or $dp/dx_B$ is positive. The condition

$$x_B p_A < x_A p_B$$

can be written as

$$\frac{p_A}{p_B} < \frac{x_A}{x_B}$$

Employing Dalton's law of partial pressures, we get

$$\frac{y_A}{y_B} < \frac{x_A}{x_B} \quad \text{or} \quad \frac{y_B}{y_A} > \frac{x_B}{x_A} \tag{4.15.14}$$

that is, the ratio of amount fractions of B and A in the vapour phase is greater than the corresponding ratio in the liquid phase. This means that the vapour phase is relatively richer in B than is the liquid from which it vaporizes. Thus, the vapour is richer in the component (here B) whose addition to the liquid mixture results in an increase in total vapour pressure (i.e. $dp/dx_B$ is positive).

**Statement of Konowaloff's Rule**

Hence we have a general conclusion that:

*The vapour phase is richer in the component whose addition to the liquid mixture results in an increase in total vapour pressure, or alternatively, the liquid phase is richer in the component whose addition to the liquid mixture results in a decrease in total vapour pressure.*

This rule was stated empirically by D.P. Konowaloff on the basis of his systematical measurements of total vapour pressure of homogeneous liquid systems and is, therefore, known as *Konowaloff's rule*.

**Alternative Statement of Konowaloff's Rule**

Since the boiling point of a solution is inversely related to its vapour pressure (higher vapour pressure means lower boiling point and *vice versa*), we can state Konowaloff's rule in terms of boiling point of the mixture:

> *The vapour phase is richer in the component whose addition to the liquid mixture causes a decrease in its boiling point, or alternatively, the liquid phase is richer in the component whose addition to the liquid mixture causes an increase in its boiling point.*

## 4.16 TEMPERATURE-COMPOSITION DIAGRAMS OF NONIDEAL SOLUTIONS

Based on the extent of deviations, a nonideal solution can be classified into the following two categories.

**A Little Deviation from Ideality**

In this case the vapour pressure versus composition curve and the boiling point versus composition curve are very similar to those of an ideal solution (Figs. 4.10.1 and 4.12.1). The total vapour pressure of the liquid mixture lies between those of the pure components. It increases or decreases regularly with the composition of the mixture. Consequently, boiling point of the liquid mixture also lies in between those of the pure components and it increases or decreases regularly with the composition.

**Very Large Deviation from Ideality**

In this category, the solution exhibits either maximum or minimum in the pressure versus composition curve. Since boiling point is inversely related to vapour pressure, it is expected that the solution will exhibit either a minimum or a maximum in the boiling point curve at a concentration where maximum or minimum in the total vapour pressure is observed. It is also expected that the general appearance of temperature versus composition curve will have one to one inverse correspondence with that of the vapour pressure versus composition curve.

**An Important Comment**

The temperature versus composition curve for a given solution can be drawn in accordance with Konowaloff's rule. Before doing so, we consider the composition of the vapour phase relative to that of the liquid phase at a point where maximum or minimum in the vapour pressure versus composition curve is observed. At this point, $\mathrm{d}p/\mathrm{d}x_A$ is zero, and hence from Eq. (4.15.12), we have either $\mathrm{d}p_B/\mathrm{d}x_A$ equal to zero or $x_B p_A$ is equal to $x_A p_B$. The former will not be true as it means that the partial pressure would remain constant in spite of a change in composition of the solution. The only alternative is that

$$x_B p_A = x_A p_B \quad \text{or} \quad \frac{p_A}{p_B} = \frac{x_A}{x_B} \quad \text{or} \quad \frac{y_A}{y_B} = \frac{x_A}{x_B} \tag{4.16.1}$$

Thus, the ratio of amount fractions of A and B in the vapour phase is the same as that in the liquid phase. We conclude that the composition of the vapour phase is the same as that of the liquid with which it is in equilibrium. It is, therefore, expected that the vaporous curve and the liquidus curve in a temperature versus composition diagram meet each other at maximum or minimum point, so that both the phases have same composition at this point.

**Depiction of Diagrams**

Keeping in mind the above fact and that the vapour phase is stable at higher temperatures, we can readily draw temperature versus composition diagrams for nonideal solutions as described in the following.

***Solution exhibiting a maximum in the total vapour pressure curve corresponding to Fig. 4.14.1*** It will exhibit a minimum in the boiling point and thus the temperature versus composition curves will have an appearance as given in Fig. 4.16.1.

***Solution exhibiting a minimum in the total vapour pressure curve corresponding to Fig. 4.14.2*** It will exhibit a maximum in the boiling point and thus the temperature versus composition curves will have an appearance as given in Fig. 4.16.2.

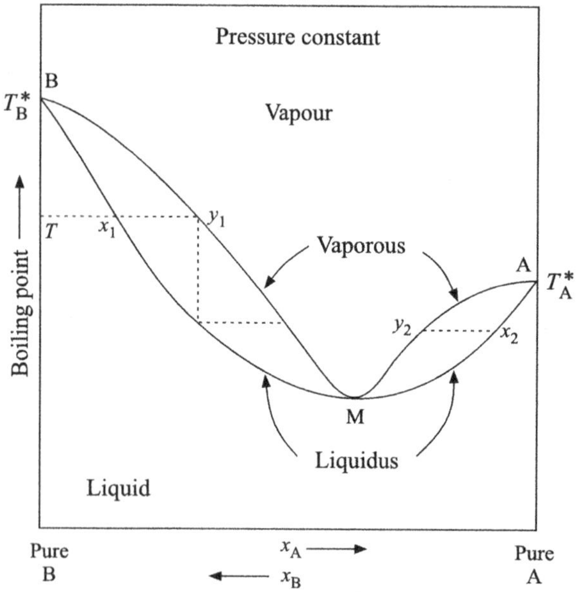


**Fig. 4.16.1 Temperature-composition diagram for a system exhibiting a minimum in boiling point**

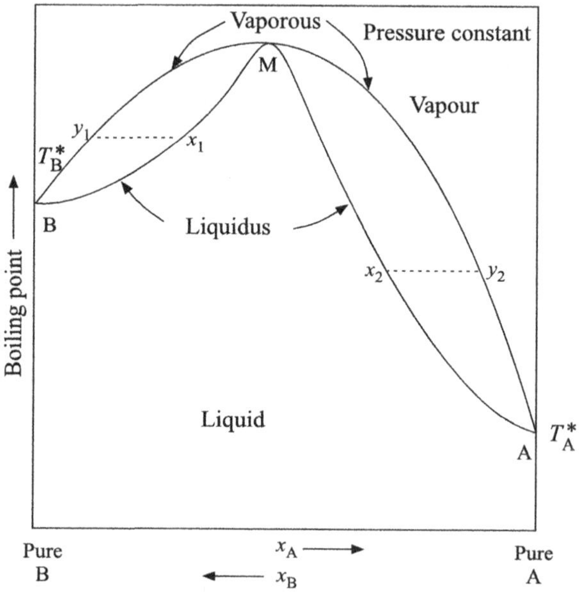


**Fig. 4.16.2 Temperature-composition diagram for a system exhibiting a maximum in boiling point**

## FRACTIONAL DISTILLATION OF NONIDEAL SOLUTIONS

We now consider the results that are obtained when a nonideal liquid mixture is subjected to fractional distillation.

**A Little Deviation from Ideality**

Since the temperature-composition diagram for this type of solution is very similar to that of an ideal solution, it is obvious that the given liquid mixture on fractionating will yield the pure constituents. During distillation, the liquid phase becomes richer in the lesser volatile component, the boiling point of which rises gradually, and the vapour phase becomes richer in the more volatile component. Thus using a fractionating column, the more volatile constituent can be removed from the top of the fractionating column and the lesser volatile one from the bottom of the column.

**Very Large Deviations from Ideality**

Fractional distillation of nonideal solutions exhibiting minimum and maximum boiling points yield, in general, only one of the constituents in the pure form, depending upon the composition of the liquid mixture, as discussed below.

***Solutions exhibiting a minimum in the boiling point*** Consider the distillation of a liquid mixture of composition, say $x_1$, lying between the points B and M as shown in Fig. 4.16.1. The liquid mixture will start boiling at temperature $T$ and the vapour which appeared on boiling will have composition $y_1$. Since the vapour phase carries relatively more of the constituent A, it is obvious that the remaining liquid will contain more of the constituent B and will thus boil at a higher temperature. As the distillation proceeds, composition of the liquid phase will move along $x_1$B, i.e. towards the pure B, and that of the vapour phase will move along $y_1$M, i.e. towards the point M. At the point M, the vapour phase and the liquid phase have identical compositions. Thus, condensing the vapour and distillating the resultant solution will not yield any further separation of the constituents. The liquid at this point would boil at a constant temperature, as if it consisted of a single constituent. Thus, we see that the fractional distillation of liquid mixture of composition $x_1$ will yield vapour of composition corresponding to the point M and the pure liquid B. The former can be collected from the top of the fractionating column and the latter from the bottom of the column.

A similar conclusion is reached when the composition of the liquid mixture lies between A and M (say $x_2$) as shown in Fig. 4.16.1. In this case, as the distillation proceeds, the vapour phase will contain more and more of B and ultimately acquire the composition corresponding to the point M and the liquid phase will contain more and more of A and ultimately yield pure A. Thus, the fractional distillation of this liquid mixture will yield vapour of composition corresponding to the point M, which can be collected from the top of the fractionating column and the liquid of pure constituent A, which can be collected from the bottom of the column.

Thus, we see that the fractional distillation of a solution exhibiting a minimum in the boiling point yields:

***Distillate*** A constant boiling liquid mixture of composition corresponding to the point M.

***Residue*** Pure A (and not B at all) if the composition of the liquid mixture lies between A and M or pure B (and not A at all) if the composition lies between B and M.

***Solutions exhibiting a maximum in the boiling point*** Such a solution can be analyzed in a similar manner. Consider, for example, a liquid mixture of composition $x_1$ (or $x_2$) as shown in Fig. 4.16.2. When this liquid mixture starts boiling, the vapour has a composition $y_1$ (or $y_2$). Since the vapour is relatively richer in the constituent B (or A), the remaining liquid becomes richer in the constituent A (or B). Thus, during distillation, composition of the liquid phase moves along $x_1$M (or $x_2$M), the boiling point of which rises gradually and that of the vapour phase moves along $y_1$B (or $y_2$A). Hence the end products of fractional distillation of a liquid mixture are:

***Distillate*** Pure A (and not B at all) if the composition of the liquid mixture lies between A and M or pure B (and not A at all) if the composition lies between B and M.

***Residue*** A constant boiling liquid mixture of composition corresponding to the point M.

**Azeotropic Mixture**

We have seen that a liquid mixture of composition M corresponding to the minimum or maximum boiling point, boils at a constant temperature and can be distilled without any change of composition. Such a liquid mixture is known as *azeotropic mixture* or simply *azeotrope* (Greek: to boil unchanged). For a given pair of liquids and at a given pressure, the composition of the azeotropic mixture and the corresponding boiling point are quite definite. Their values vary as the external pressure is changed. This fact supports that the solution at the point M is really a mixture of the two liquids and not a liquid containing only one component formed as a result of definite union of the two liquids. Thus the compound formation at the point M is excluded. Some examples of azeotropic mixtures are given in Table 4.16.1.

**Table 4.16.1** Some Examples of Azeotropic Mixtures. The boiling Points (B.Pt.; $T$/K = $\theta_c$/°C + 273) Referred to are at 1 atm External Pressure

| *Type* | *Components* | | *Azeotrope* | |
|---|---|---|---|---|
| | A | B | mass % A | b.pt./K |
| Minimum | Water (373 K) | Ethyl alcohol (351.3 K) | 4 | 351.2 |
| | Water (373 K) | Ethyl acetate (352.6 K) | 11.3 | 346.4 |
| | Carbon tetrachloride (349.8 K) | Methyl alcohol (337.7 K) | 79.49 | 328.7 |
| | Chloroform (334.2 K) | Methyl alcohol (337.7 K) | 87.4 | 326.4 |
| Maximum | Water (373 K) | Nitric acid (359.7 K) | 68 | 393.5 |
| | Water (373 K) | Hydrochloric acid (353 K) | 20.22 | 381.6 |
| | Chloroform (334.2 K) | Acetrone (329.1 K) | 78.5 | 337.4 |
| | Phenol (455.2 K) | Aniline (457.4 K) | 42 | 459.2 |

The variation of the composition and boiling point of azeotropic mixture of HCl-water and pyridine-formic acid solutions with pressure is given in Table 4.16.2. Since the composition of azeotrope varies with pressure, it indicates that the azeotrope consists of pure mixture and not a single compound.

## 4.17 KONOWALOFF'S RULE — REVISITED

**Thermodynamic Derivation**

An alternative proof of Konowaloff's rule can be derived from the quantitative relation of shift in the boiling point of the solvent when a volatile solute is added to it. We will derive the relation on the assumption that both liquid mixture and vapour behave ideally. This assumption will be applicable only when the solution is very dilute.

**Table 4.16.2** Variation of Composition and the Boiling Point of the Azeotropic Mixture with the External Pressure

| HCl–$H_2O$ | | | Pyridine + Formic acid | | |
|---|---|---|---|---|---|
| $p$/Torr | mass % HCl | b.pt./°C | $p$/Torr | mol % | b.pt./°C |
| 500 | 20.916 | 97.578 | 760 | 58.4 | 138.25 |
| 700 | 20.360 | 106.424 | 570 | 58.9 | 129.4 |
| 760 | 20.22 | 108.584 | 380 | 59.5 | 117.5 |
| 800 | 20.155 | 110.007 | 190 | 60.2 | 99.5 |
| | | | 120 | 60.8 | 87.10 |

The condition of phase equilibrium between the liquid mixture and the vapour for one of the components (say solvent) is

$$\mu_{1(1)} = \mu_{1(v)}$$

where the subscripts l and v within the brackets stand for liquid and vapour, respectively.

If it is assumed that both liquid and vapour mixtures behave ideally, then we have

$$\mu^*_{1(1)} + RT \ln x_{1(1)} = \mu^*_{1(v)} + RT \ln y_{1(v)}$$

Rearranging the above expression, we have

$$R(\ln x_{1(1)} - \ln y_{1(v)}) = \frac{\mu^*_{1(v)} - \mu^*_{1(1)}}{T} = \frac{\Delta_{vap}G^*_{1,m}}{T} \tag{4.17.1}$$

where $\Delta_{vap}G^*_{1,m}$ is the molar free energy of vaporization of the pure solvent at temperature $T$ and pressure $p$. For $p = 1$ atm, we have

$$\Delta_{vap}G^{\circ}_{1,m} = \Delta_{vap}H^{\circ}_{1,m} - T\,\Delta_{vap}S^{\circ}_{1,m}$$

and at the normal boiling point $T^*$ of the pure solvent

$$\Delta_{vap}S^{\circ}_{1,m} = \frac{\Delta_{vap}H^{\circ}_{1,m}}{T^*}$$

Therefore, we have

$$\Delta_{vap}G^{\circ}_{1,m} = \Delta_{vap}H^{\circ}_{1,m} - T\frac{\Delta_{vap}H^{\circ}_{1,m}}{T^*} = \Delta_{vap}H^{\circ}_{1,m}\left(1 - \frac{T}{T^*}\right)$$

Substituting the above expression in Eq. (4.17.1), we get

$$R(\ln x_{1(1)} - \ln y_{1(v)}) = \frac{\Delta_{vap}H^{\circ}_{1,m}}{T}\left(1 - \frac{T}{T^*}\right) = \frac{\Delta_{vap}H^{\circ}_{1,m}}{T}\frac{(-\Delta T_b)}{T^*} \tag{4.17.2}$$

where $\Delta T_b = T - T^*$.

Equation (4.17.2) can be simplified under the following approximations.

(i) For a dilute solution, $T$ will be very near to $T^*$, so that the term $TT^*$ may be replaced by $T^{*2}$.

(ii) For a dilute solution, we have

$$\ln x_{1(1)} = \ln (1 - x_{2(1)}) \simeq - x_{2(1)}$$

$$\ln y_{1(v)} = \ln (1 - y_{2(v)}) \simeq - y_{2(v)}$$

With these approximations, Eq. (4.17.2) reduces to

$$R(- x_{2(1)} + y_{2(v)}) = \frac{\Delta_{vap} H^{\circ}_{1,m} (- \Delta T_b)}{T^{*2}}$$

or

$$- R\, x_{2(1)} \left(1 - \frac{y_{2(v)}}{x_{2(1)}}\right) = \frac{\Delta_{vap} H^{\circ}_{1,m} (- \Delta T_b)}{T^{*2}}$$

or

$$\Delta T_b = \frac{RT^{*2}}{\Delta_{vap} H^{\circ}_{1,m}} x_{2(1)} \left(1 - \frac{y_{2(v)}}{x_{2(1)}}\right) \tag{4.17.3}$$

Equation (4.17.3) can be written in terms of molality of the liquid solution using the approximation:

$$x_{2(1)} = \frac{n_{2(1)}}{n_{1(1)} + n_{2(1)}} \simeq \frac{n_{2(1)}}{n_{1(1)}} = \frac{n_{2(1)}}{m_{1(1)}/M_1} = \left(\frac{n_{2(1)}}{m_{1(1)}}\right) M_1 = mM_1$$

Thus, Eq. (4.17.3) modifies to

$$\Delta T_b = \frac{RT^{*2}\, M_1}{\Delta_{vap} H^{\circ}_{1,m}} m \left(1 - \frac{y_{2(v)}}{x_{2(1)}}\right)$$

or

$$\Delta T_b = K_b m (1 - K) \tag{4.17.4}$$

where $K$ is the distribution coefficient of the solute between the vapour phase and the liquid phase and is given by

$$K = \frac{\text{Amount fraction of the solute in vapour phase}}{\text{Amount fraction of the solute in liquid phase}} \tag{4.17.5}$$

If in Eq. (4.17.4), $K = 0$ (i.e. the solute is nonvolatile), it reduces to $\Delta T_b = K_b m$, an expression which we derived earlier in Section 2.7 for the elevation of boiling point of the solvent when a nonvolatile solute is dissolved in it.

**Application of Eq. (4.17.4)**

$\Delta T_b$ in Eq. (4.17.4) is equal to $T - T^*$ and can have either positive or negative value depending upon the value of $K$. Hence, we will have the following two cases.

(i) $\Delta T_b$ *Positive Provided* $K < 1$ (i.e. $y_{2(v)} < x_{2(1)}$)

In this case, $T > T^*$ and an elevation in the boiling point of the solvent occurs. The presence of $(1 - K)$ factor in Eq. (4.17.4) indicates that the numerical value of $\Delta T_b$ is smaller than the corresponding value expected for a solution containing a nonvolatile solute.

(ii) $\Delta T_b$ *Negative Provided* $K > 1$ (i.e. $y_{2(v)} > x_{2(l)}$)

In this case, $T < T^*$ and a depression in the boiling point of the solvent occurs.

**Agreement with Konowaloff's Rule**

The above two conclusions are, in fact, the statements of Konowaloff 's rule as may be seen from the following analysis.

According to Konowaloff's rule, the vapour phase is richer in the component whose addition to the liquid mixture causes a decrease in its boiling point. This is, in fact, the second conclusion which we have derived above. According to this rule, the depression of boiling point takes place only when $K > 1$, i.e. the vapour contains relatively more of the solute than the liquid and thus the addition of such a solute in the solvent (or solution) will cause a decrease in the boiling point of the latter.

Alternatively, the liquid phase is richer in the component whose addition to the liquid mixture causes an increase in its boiling point. This is, in fact, the first conclusion which we have derived above. According to this rule the elevation of boiling point takes place only when $K < 1$, i.e. the vapour contains relatively lesser of the solute than the liquid (or the liquid contains relatively more of the solute than the vapour) and thus the addition of such a solute in the solvent (or the solution) will cause an increase in the boiling point of the latter.

Now the nature of curves depicted in Figs 4.16.1 and 4.16.2 may be understood qualitatively. Nonideal solutions exhibiting maximum boiling point are observed whenever molecular attractions between unlike molecules are greater than those between like molecules. In such cases, the escaping tendencies of molecules of both the components are weakened with the result that the amount fraction of the solute in the vapour phase is smaller than the corresponding amount fraction in the liquid phase. Consequently, $K_{\text{A in B}}$ (i.e $y_{A(v)}/x_{A(l)}$, where A is solute and B is solvent) and $K_{\text{B in A}}$ (i.e. $y_{B(v)}/x_{B(l)}$, where B is solute and A is solvent) are less than one and thus there occurs an elevation in the boiling point when either B is added to A or A is added to B. Hence the boiling point versus composition curves exhibit upward trends on either end of the diagram and thus a maximum at a point where the liquidus and vaporous fragments meet each other is observed (Fig. 4.16.2).

On the other hand, solutions exhibiting minimum boiling point are observed whenever molecular attractions between unlike molecules are smaller than those between like molecules. This results in the increase in the escaping tendencies of molecules of both the components with the result that the amount fraction of the solute in the vapour phase is greater than the corresponding amount fraction in the liquid phase. Consequently, $K_{\text{A in B}}$ and $K_{\text{B in A}}$ are greater than one and thus there occurs a depression in the boiling point when either B is added to A or A is added to B (Fig. 4.16.1). Hence the boiling point versus composition curves exhibit downward trends on either end and thus a minimum at a point where the liquidus and vaporous fragments meet each other is observed.

## 4.18 PARTIALLY MISCIBLE LIQUIDS

### General Considerations

Certain pairs of liquids are miscible with each other in limited range of concentrations and are known as partial miscible liquids. For example, if a small quantity of phenol, ether or aniline is added to water at room temperature, and the mixture is shaken, it is observed that the organic compound dissolves completely in water. If the addition of the organic compound is continued, a stage is reached where the compound is no more soluble and thus two solutions are formed. These are: (i) a saturated solution of the organic compound in water, and (ii) a saturated solution of water in the organic compound.

These two solutions are in equilibrium with each other and are known as *conjugate solutions*. The compositions of these solutions are fixed and are independent of the relative amounts of the two solutions. Thus, if more of organic compound is added, it is dispersed into these two solutions in such a way that the compositions of both the solutions remain unaltered, the only thing that vary is the relative amounts of the two solutions. As more and more of organic compound is added, it continues to disperse in these two solutions and the net effect is that the amount of the solution of organic compound in water is gradually decreased, whereas that of water in organic compound is gradually increased. Ultimately, a stage is reached where the solution of organic compound in water is present in an infinitesimal amount. At this stage, the addition of a very small amount of organic compound converts the system once again to a single solution, but this time to a saturated solution of water in organic compound. The addition of more organic compound simply makes the solution more and more unsaturated with respect to water.

The above sequence of formation of a single solution followed by two solutions and once again into a single solution is observed only in case of those liquids which differ very widely from each other and show large positive deviations from Raoult's law.

### Theoretical Consideration

The formation of partial miscible liquid mixtures has a sound theoretical background. The extent of positive deviation from Raoult's law is proportional to the value of $\alpha$ in Eq. (4.15.8), the deviation increases as the value of $\alpha$ increases. For a given pair of liquids, the value of $\alpha$ can be increased by lowering the temperature of the system: the deviations being larger at lower temperature and smaller at higher temperatures. In general, the partial vapour pressure curve for each component undergoes the changes of types I, II, III and IV shown in Fig. 4.18.1, as the temperature of system is lowered (or as $\alpha$ is increased). Curve I corresponds to a higher temperature where lesser deviations are observed and curve IV corresponds to a lower temperature where the deviations are larger. Special interest attaches to curves of types III and IV, where instead of a smooth curve, a curly shaped curve is obtained. These curves have maximum and minimum in the curly portion. The composition of solution at maximum or minimum can be obtained by differentiating Eq. (4.15.8) with respect to $x$ and setting the resultant equation to zero.

Writing Eq. (4.15.8) without subscripts, we have

$$p = p^{*}x \exp\{\alpha(1 - x)^2\}$$

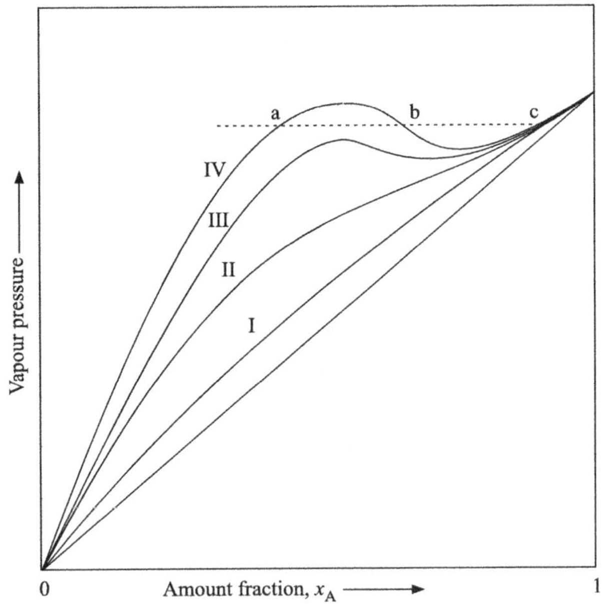


**Fig. 4.18.1 Vapour pressure-composition diagram for partial miscible liquids**

Differentiating this with respect $x$, we get

$$\frac{dp}{dx} = p^* \exp\{\alpha(1-x)^2\} + p^*x\,\{-2\alpha(1-x)\}\exp\{\alpha(1-x)^2\}$$

or
$$\frac{dp}{dx} = p^* \exp\{\alpha(1-x)^2\}\,\{1-2\alpha x(1-x)\}$$

Equating the above expression to zero, we get

$$1 - 2\alpha x(1-x) = 0$$

or
$$2\alpha x^2 - 2\alpha x + 1 = 0 \qquad (4.18.1)$$

Solving for $x$, we have

$$x = \frac{2\alpha - \sqrt{4\alpha^2 - 8\alpha}}{4\alpha} \qquad (4.18.2)$$

The roots of Eq. (4.18.2) are real if $\alpha > 2$ and approaches each other as $\alpha \to 2$. At $\alpha = 2$, the two roots are identical and for $\alpha < 2$, they are imaginary.

It can be seen from the curly curve that there are three solutions of different concentrations which have the same vapour pressure. For example, on curve IV, the three solutions are represented by the points a, b and c. This situation, however, is not consistent with the phase rule, according to which, we have

$$F = C - P + 2 = 4 - P \qquad (\text{as } C = 2)$$

or
$$P = 4 - F$$

If the state of the system is to be defined completely by only one degree of freedom (i.e. temperature alone), then we can have at the most three phases in equilibrium with each other. Of these three phases, one is the vapour phase thus, we are left with only two liquid phases and, therefore, we cannot have more than two liquid phases in equilibrium with each other. In fact, the curly shape abc is not found experimentally and one observes a flat portion between a and c, the liquid phase b having no real existence. The vapour pressure remains constant between a and c.

Thus, we see that for a binary liquid system exhibiting large deviations from Raoult's law, there occurs limited solubility of one liquid into another and hence two saturated solutions are formed over a certain range of composition. Beyond this range, the two liquids are completely miscible with each other and thus there exists a miscible gap in the system.

**Temperature-Composition Diagram**

The compositions of the two conjugate solutions in the miscible gap are quite definite for a given temperature and are independent of the relative amounts of the two solutions. This fact is consistent with the phase rule. For $P = 2$ and $C = 2$, we have

$$F = C - P + 2 = 2$$

One of these two degrees of freedom represents the external pressure and has a fixed value. Hence stating the value of one other parameter, namely, temperature or composition of either of the conjugate solutions, defines the given system completely. If, for example, temperature is stated then compositions of the two solutions are fixed. If a graph between compositions of the two solutions and the corresponding temperature is plotted, we obtain a curve (known as solubility curve) of the type shown in Fig. 4.18.2.

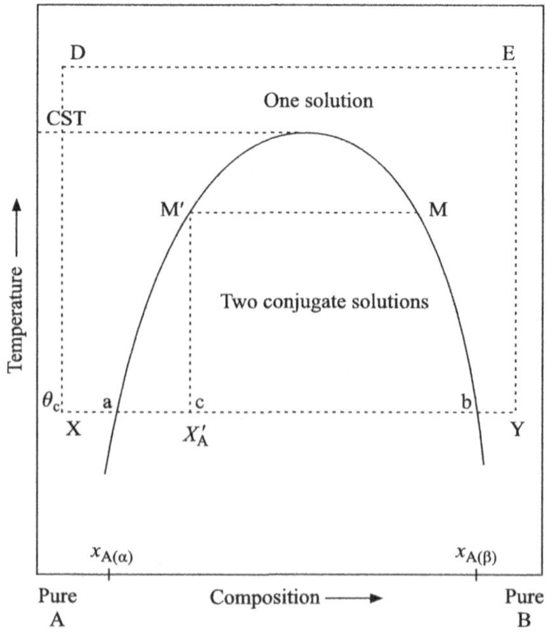


**Fig. 4.18.2 Temperature-composition diagram for partial miscible liquids**

**Critical Solution Temperature**

The compositions of the two solutions become closer as temperature is raised. This is due to the increase in mutual solubilities of the two components as the temperature is raised. At a certain temperature, the compositions of the two solutions become identical and both the solutions merge into each other, thereby forming a single solution. This temperature is known as the *consolute temperature* or the *critical solution temperature* (CST). When the temperature is greater than CST, the two liquids are miscible in all proportions and thus always form a single solution. In fact, a point outside the curve of Fig. 4.18.2 represents a single solution, whereas that within the curve represents two conjugate solutions in equilibrium with each other.

The compositions of the two conjugate solutions at a given temperature are given by the two points on the solubility curve. Consider, for example, the point c of composition $X'_A$ within the curve at temperature $\theta_c$. At this point, the system consists of two solutions α and β (represented by the points a and b, respectively) of compositions $x_{A(\alpha)}$ and $x_{A(\beta)}$. In fact, the point c anywhere on the horizontal line ab represents two solutions of compositions $x_{A(\alpha)}$ and $x_{A(\beta)}$. The only thing that varies from point to point is the relative amounts of the two solutions, which can be calculated by using the lever rule as derived below.

**Lever Rule**

Let the amounts of species A and B at the point c be $n_A$ and $n_B$, respectively. Now at the point c the system contains two conjugate solutions α and β of compositions $x_{A(\alpha)}$ and $x_{A(\beta)}$, respectively. Of the total amount of A, let the amounts present in these two solutions be $n_{A(\alpha)}$ and $n_{A(\beta)}$. Similarly, let $n_{B(\alpha)}$ and $n_{B(\beta)}$ be the amounts of B present in these two solutions. Obviously,

$$X'_A = \frac{n_A}{n_A + n_B} = \frac{n_{A(\alpha)} + n_{A(\beta)}}{n_{A(\alpha)} + n_{A(\beta)} + n_{B(\alpha)} + n_{B(\beta)}} \tag{4.18.3}$$

Now $$ac = X'_A - x_{A(\alpha)} = X'_A - \frac{n_{A(\alpha)}}{n_{A(\alpha)} + n_{B(\alpha)}}$$

or $$(n_{A(\alpha)} + n_{B(\alpha)})\,(ac) = (n_{A(\alpha)} + n_{B(\alpha)})\,X'_A - n_{A(\alpha)} \tag{4.18.4}$$

and $$bc = x_{A(\beta)} - X'_A = \frac{n_{A(\beta)}}{n_{A(\beta)} + n_{B(\beta)}} - X'_A$$

or $$\left(n_{A(\beta)} + n_{B(\beta)}\right)(bc) = n_{A(\beta)} - (n_{A(\beta)} + n_{B(\beta)})\,X'_A \tag{4.18.5}$$

Subtracting Eq. (4.18.5) from Eq. (4.18.4), we get

$$(n_{A(\alpha)} + n_{B(\alpha)})(ac) - (n_{A(\beta)} + n_{B(\beta)})(bc)$$
$$= (n_{A(\alpha)} + n_{B(\alpha)} + n_{A(\beta)} + n_{B(\beta)})\,X'_A - (n_{A(\alpha)} + n_{A(\beta)}) \tag{4.18.6}$$

In view of Eq. (4.18.3), the right side of Eq. (4.18.6) is zero. Hence

$$\frac{\text{ac}}{\text{bc}} = \frac{n_{\text{A}(\beta)} + n_{\text{B}(\beta)}}{n_{\text{A}(\alpha)} + n_{\text{B}(\alpha)}}$$

$$= \frac{\text{Total amount of substances in solution } \beta}{\text{Total amount of substances in solution } \alpha} \qquad (4.18.7)$$

Equation (4.18.7) is the required lever rule.

**Application of the Phase Rule**

We consider below the application of the phase rule to the following three typical cases.

***A point outside the solubility curve*** Here we have

$$P = 1 \quad \text{and} \quad C = 2$$

Therefore

$$F = C - P + 2 = 3$$

Since the pressure is kept constant (usually 1 atm), the values of two variables, namely temperature and composition of the solution, should be stated in order to define the system completely.

***A point inside the solubility curve*** Here we have

$$P = 2 \quad \text{and} \quad C = 2$$

Therefore

$$F = C - P + 2 = 2$$

Once again, since pressure is constant, we would have to state the value of one variable (usually temperature) in order to define the system completely. As stated earlier, the compositions of the two solutions will have fixed values and will be represented by the two points on the solubility curve corresponding to the given temperature.

***At critical solution temperature*** At this point, we have one restricted condition of identical compositions of the two solutions. Thus, we have

$$F = (C - r) - P + 2 = (2 - 1) - 2 + 2 = 1$$

Since $p$ is kept constant, the system is invariant, i.e. the critical solution temperature and the corresponding composition on the curve have fixed values for a given value of pressure (usually 1 atm is chosen for this purpose).

**Examples**

A few examples of pairs of liquids showing partial miscibility along with their critical solution temperatures and the corresponding compositions are given in Table 4.18.1.

**Effects of Changing the Composition**

Let us have a system represented by a point X in Fig. 4.18.2. This represents an unsaturated solution of B in A. Now if more of B is added, it continues to dissolve till the point a is reached. At this point, the single solution just splits into two; a saturated solution of B in A of composition $x_{\text{A}(\alpha)}$, which is presents

**Table 4.18.1** Examples of Pairs of Liquids showing Partial Miscibility

| *System* | CST/°C | *Composition* |
|---|---|---|
| Water-Phenol | 65.9 | 66 mass % of water |
| Aniline-Hexane | 59.6 | 48 mass % of aniline |
| Methyl alcohol-Cyclohexane | 49.1 | 29 mass % of methyl alcohol |
| Methyl alcohol-Carbon disulphide | 40.5 | 20 mass % of methyl alcohol |

in the large amount and a saturated solution A in B of composition $x_{A(\beta)}$, which has just started forming. Now as more of B is added, system continues to possess the above two solutions till the point b is reached. The only change that is observed during this interval is that the relative amounts of the two solutions vary; the amount of the solution of B in A gradually decreases whereas that of A in B gradually increases. The relative ratio at any stage can be determined with the help of the lever rule. At point b, the solution of B in A is just going to disappear, and, therefore, is present in a very small amount. Further addition of B gives rise once again to a single solution, but this time an unsaturated solution of A in B (point Y). We can also go from the point X to Y without going through the formation of two solutions. This can be done as follows:

(a) Raise the temperature of the solution represented by the point X to a point above the critical solution temperature.
(b) Add more of B till the desired composition of A in B, corresponding to the point Y, is obtained.
(c) Cool the resultant solution till the point Y is reached.

The above procedure is also shown in Fig. 4.18.2.

### Effect of Changing the Temperature

Let us have a system respresented by the point c (Fig. 4.18.2). This represents two solutions of B in A and A in B with compositions $x_{A(\alpha)}$ and $x_{A(\beta)}$, respectively. If the temperature of the system is increased, the state of the system is moved vertically, the composition of the solution of A in B changes along bM and that of B in A along aM′. The relative amounts of the two solutions also vary; the amount of B in A is gradually increased whereas that of A in B is gradually decreased. The system continues to have two solutions till the point M′ is reached where the solution of A in B is present in vanishing amount. In fact, the system of two solutions becomes a single saturated solution of B in A. The temperature at which this happens is known as the *mutual solubility temperature* (MST). When the temperature is further increased, we get an unsaturated solution of B in A. Obviously, critical solution temperature is the maximum of all mutual solubility temperatures.

### Lower Consolute Temperature

There are some binary liquid mixtures which show increase in the mutual solubilities as the temperature is decreased. Such systems, therefore, are expected to have temperature-composition curve just reverse of the curve of Fig. 4.18.2. This is shown in Fig. 4.18.3. The system now has a lower consolute temperature and thus liquids are completely miscible with each other in all respects so long as the temperature is below the lower consolute temperature. Systems showing the

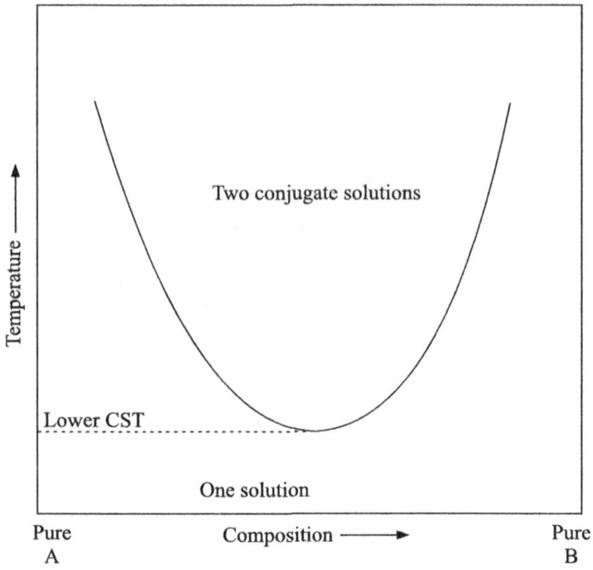


**Fig. 4.18.3 The lower consolute temperature**

lower CST almost invariably consist of either a hydroxy-compound and an amine or a hydroxy-compound and a keto grouping (or an ether grouping). It is, therefore, expected that the two components partially associate through the hydrogen bonding

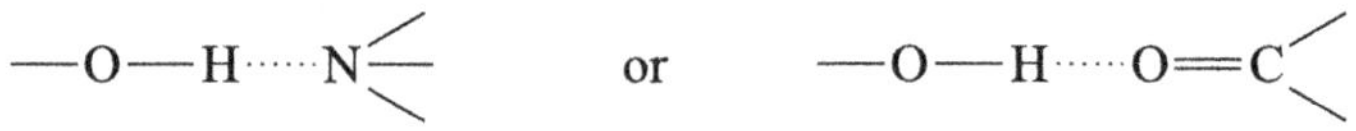


The extent of such association depends upon the temperature of the system and it decreases as the temperature is raised. This probably is the reason why their mutual solubilities are decreased as the temperature is increased.

A few examples of pairs of liquids showing lower consolute temperature are given in Table 4.18.2.

**Table 4.18.2** Examples of Systems showing Lower CST

| *System* | *Lower CST*/°C |
|---|---|
| Triethylamine-water | 18.5 |
| $\gamma$-Collidine-water | 6 |
| Glycerol-*m*-toluidine | 6.7 |

## Lower and Upper Consolute Temperatures

It is found that the mutual solubilities do not continue to decrease as the temperature is increased. After a certain temperature the solutions behave in a normal manner and the solubilities again start increasing with rise in temperature It is, therefore, expected that such solutions will also show the upper critical solution temperature. The solubility curve for such systems consists of a closed curve and is shown in Fig. 4.18.4. At any point within the closed loop, two liquid phases are present, while that lying outside the loop represents a single solution. The first case of this type to be established was the nicotine-water system, the

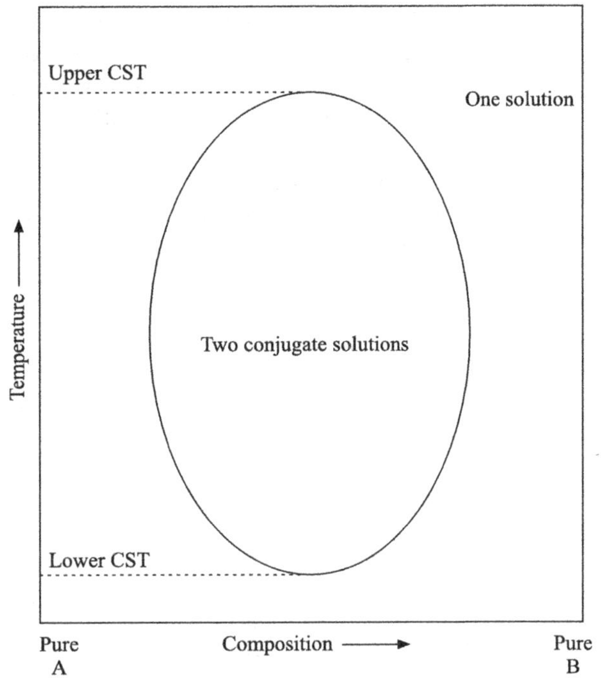


**Fig. 4.18.4 Temperature-composition curve of a system having both upper and lower CST**

upper CST being at 208 °C and the lower one at 60.8 °C. Thus nicotine and water are completely miscible in each other at temperatures greater than 208 °C and less than 60.8 °C.

The upper critical solution temperatures for most liquids pairs are very high and are realized only under high pressures. In many cases, it has not been possible to attain this upper CST experimentally as this lies above the respective critical temperatures of the liquids with the effect that the system volatalizes much before CST is reached.

In few cases such as the ethyl acetate-water mixture and the ether-water mixture, it has not been possible to attain experimentally both the upper and the lower critical solution temperatures. In the above two systems, theoretically one would expect the lower critical solution temperatures since the solubilities of ethyl alcohol and ether in water decrease with increasing temperature. However, the critical solution temperatures for both the above systems lie below the freezing point of water. Thus water freezes completely before these temperatures are reached. Similarly, the upper CST for chloroform-water mixture has not been realized experimentally; here the upper CST lies above the critical temperature of chloroform. Thus, it is obvious that one would not observe experimentally the critical solution temperature when (i) the upper CST lies above the critical temperature of either of the two liquids, and (ii) the lower CST lies below the freezing point of either of the two liquids.

## Effect of Pressure on CST

It was shown earlier with the help of phase rule that the CST of partially miscible liquids has a definite value for a given value of the external pressure. Thus, it is expected that the CST will vary as the external pressure is varied. Since the mutual solubilities of the two components invariably increase with increase in external pressure, it is, therefore, expected that the upper CST will decrease whereas the lower CST will increase as the external pressure is increased. This is shown in Fig. 4.18.5. In fact, it is found to be so experimentally. The area of the loop thus becomes smaller and smaller with increase in pressure, ultimately, a stage is reached when the area is just reduced to a point. At this stage, the two liquids become miscible with each other in all proportions and thus form a single liquid solution.

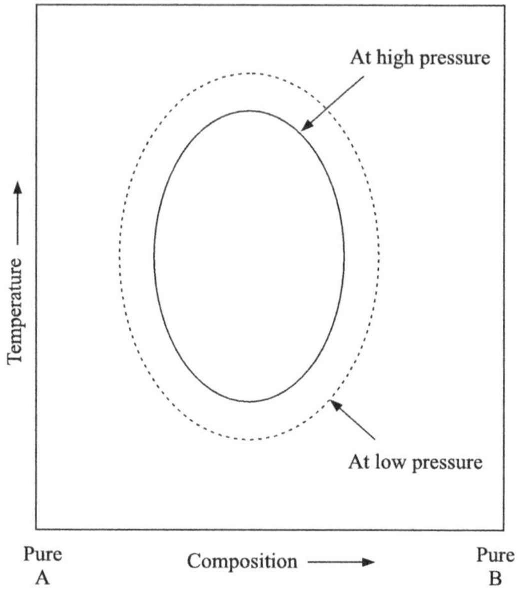


**Fig. 4.18.5 Effect of pressure on CST; dotted curve at low pressure, solid curve at high pressure**

## Effect on Impurities on CST

The impurities have a marked effect on the critical solution temperature, the nature of the effect depends whether the added substance is soluble in one liquid or in both the liquids.

If the substance is soluble only in one liquid, it will effect the solubility of the other liquid in this liquid. In general, this solubility is decreased, i.e. the mutual solubilities of the two liquids are decreased and, consequently, the CST is raised as shown by the dotted curve I in Fig. 4.18.6. For example, the addition of 0.1 mol of KCl per $dm^3$ of water raises the CST of the phenol-water mixture by about 8 K whereas the addition of 0.1 mol of naphthalene per $dm^3$ of phenol raises the temperature nearly by 20 K.

If the added substance is soluble in both the liquids, it increases the mutual solubilities of the two liquids and, consequently, the CST is lowered as shown by the dotted curve II in Fig. 4.18.6. This probably is due to the fact that the added substance will distribute itself between the two conjugate liquid mixtures in a definite manner and is being continuously exchanged between them. Thus,

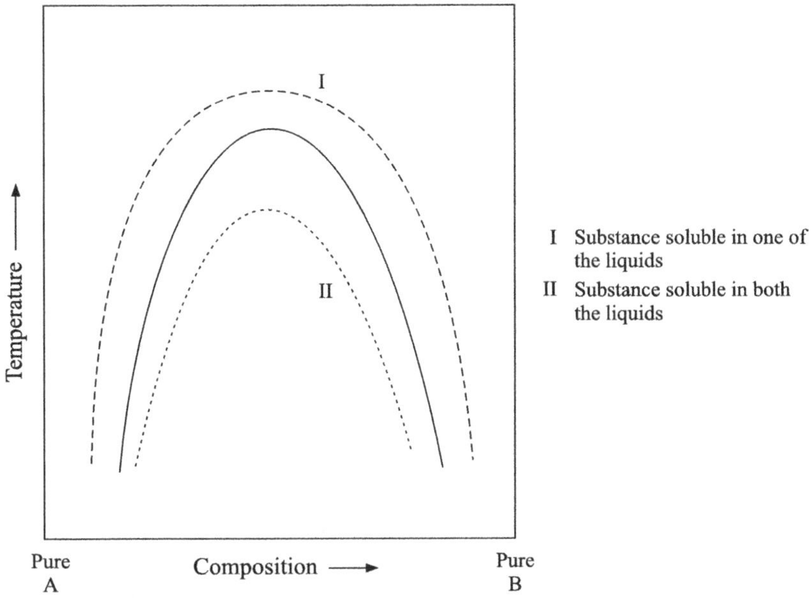


**Fig. 4.18.6 Effect of impurities on CST**

a stage of dynamic equilibrium exists. This to and fro movement of molecules of the added substance acts as a cementing force between the two conjugate liquid mixtures and hence brings about more solubility of the two liquids. For example, when succinic acid is added to the phenol-water mixture, its CST is lowered.

## DISTILLATION OF PARTIALLY MISCIBLE LIQUIDS

### Qualitative Features

A complete temperature-composition diagram of a binary mixture of partially miscible liquids also includes the vapour phase and gives a curve of the type shown in Fig. 4.18.7. The liquid-vapour curves will show a minimum as the two liquids show large positive deviations from Raoult's law. The upper and lower portions of the diagram shown in Fig. 4.18.7 can be discussed separately as we did earlier. If distillation of a binary mixture of partially miscible liquids is carried out at a low pressure, the temperature at which the mixture starts boiling is lowered. Consequently, the boiling point-composition curves in Fig. 4.18.7 are shifted downwards. If the pressure is low enough, a stage will be reached where the boiling point curves will intersect the liquid-liquid solubility curve. The nature of the resultant diagram depends on the corresponding vapour pressure-composition curves. Three types of vapour pressure-composition curves and hence also temperature-composition curves are possible. These are shown in Figs 4.18.8 to 4.18.10. The behaviour of the mixture during distillation will depend upon the characteristics of these curves.

### Distillation Involving no Azeotropic Mixture

Examples of such a system are aniline-water, ethyl acetate-water, and isobutyl alcohol-water mixtures. As usual, the curved lines AE and BE in Fig. 4.18.8b represent the vaporous curves and the lines AC and BD represent the liquidus curves. The horizontal line CD represents systems where two liquid phases C and D and the vapour phase E are in equilibrium with one another. The lines

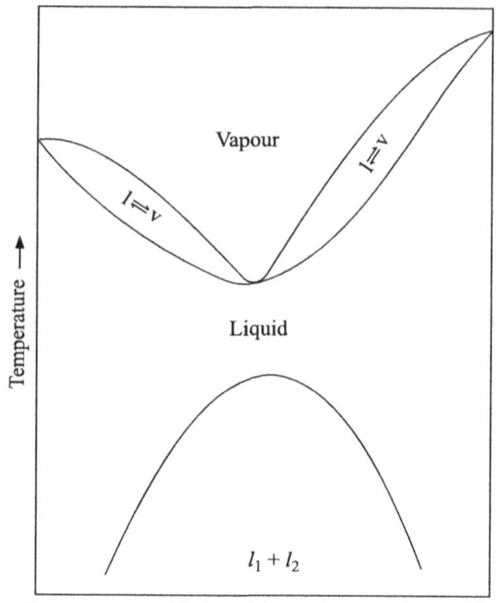


**Fig. 4.18.7 A complete temperature-composition diagram of a mixture of partially miscible liquids**

below the horizontal line CD represent the region of partial miscibility within which the two liquid phases are formed. The behaviour of the system during distillation can be described as usual. Consider, for example, a solution represented by the point a. This consists of a single unsaturated solution of B in A. If temperature of this solution is increased, it will start boiling at temperature $t_1$, the vapour which issues here has composition $v'$. The vapour phase is relatively richer in the component B (in accordance with Konowaloff's rule, the vapour is richer in the component whose addition to the mixture lowers the boiling point) and thus the remaining liquid will be richer in the component A. Consequently, composition of the liquid will move along bA and thus will boil at higher temperatures. If fractional distillation is carried out, it will eventually yield the pure component A. Now if the vapour which is formed at $t_1$ is cooled to the original temperature, it will form a mixture of two liquid phases c and d. If the temperature of this mixture is gradually increased, compositions of the two liquids will move along cC and dD, respectively. At temperature $t_2$, this mixture will start boiling with the vapour composition given by the point E. Since the vapour contains relatively more of B in comparison to the overall composition of liquids (point C′) and since the distillation at this stage takes place at a constant temperature, it follows that the overall composition of the remaining liquids will move along C′C. Thus during distillation, the relative amounts of the two liquid phases vary, the amount of liquid phase D gradually vanishes. The mixture continues to have two phases till the point C is reached. Further distillation beyond this stage completely vanishes the liquid phase of A in B and the composition of the liquid phase C moves along CA. The sequence of further distillation is the same as described earlier. Thus, we see that the end products of fractional distillation will consist of:

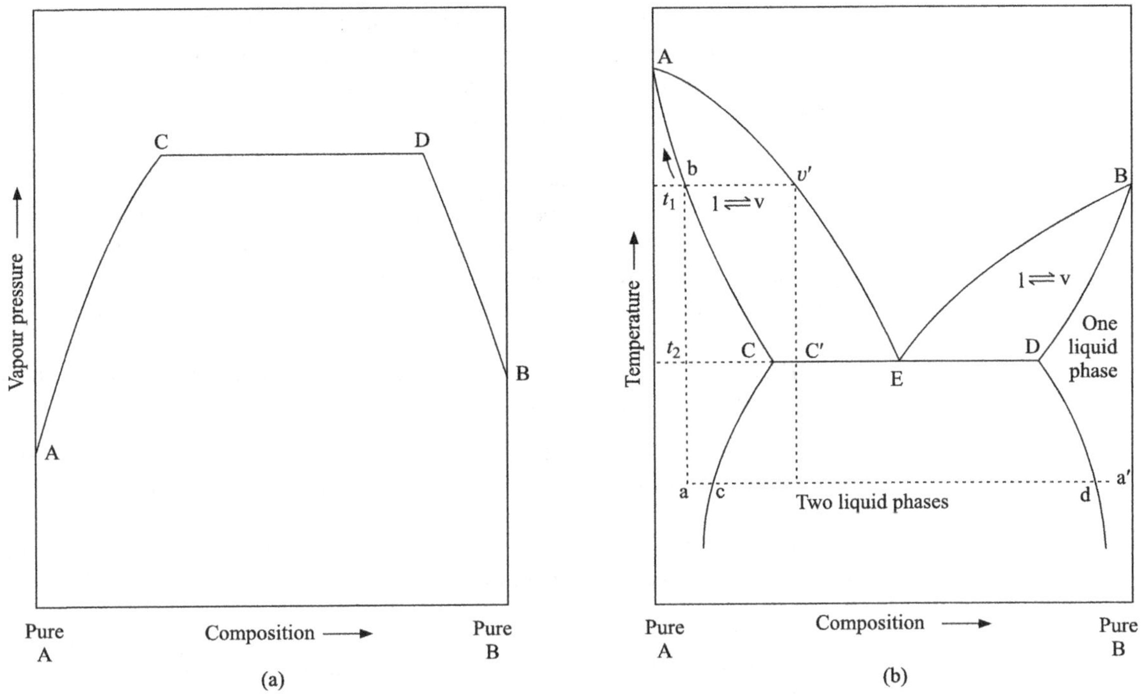


**Fig. 4.18.8 (a) Vapour pressure-composition and (b) temperature-composition diagrams for partial miscible liquids with no azeotropic mixture**

**Residue** The pure component A in the liquid form and can be collected from bottom of the column.

**Distillate** The vapour corresponding to the composition E, which can be collected from top of the column. This distillate cannot be fractionated further.

Similar results would be expected if the composition of the original solution lies at a′. Here the residue will consist of pure component B and the distillate of composition E. Thus, we see that the fractional distillation of this type of system will not give a complete separation of the two components, only one of them can be obtained in the pure form depending upon the original composition of the solution; pure A if the composition of the system lies left to the point E, and pure B if it lies right to the point E. It may, however, be noted that none of the two constituents can be obtained in pure form if the original composition of the mixture lies exactly on the point E, since the total composition of the liquid is the same as that of vapour with which it is in equilibrium.

## Distillation Involving Azeotropic Mixture

Examples of such a system are water-phenol, ethyl methyl ketone-water and triethylamine-water mixtures. The end products obtained in the fractional distillation of a mixture exhibiting Fig. 4.18.9b can be analyzed following the principles described above. Consider, for example, a solution as represented by the point a. This mixture will start boiling at temperature $t_1$. Since the vapour which issues here contains relatively more of B (point c in comparison to the point b), the composition of the remaining liquid will move along bD and hence its boiling point will rise. At point D, the single liquid splits into two liquids C and D, the former being present in a smaller amount. At this stage, distillation proceeds at a constant temperature and the vapour which appears has composition E. The vapour is relatively richer in B in comparison to either of the two liquid layers. Thus, in the remaining two liquid layers, the concentration of B decreases faster than that of A. In other words, the state of the mixture will move along DC. The net effect that is produced here is that the liquid layer D gradually decreases, whereas that of C gradually increases. At point C, practically entire liquid layer D has disappeared. Further distillation will change the liquid composition along CA and ultimately to pure A. Thus, the end products of the fractional distillation of the system are:

***Residue*** The residue consists of pure liquid A and can be collected from bottom of the fractionating column.

***Distillate*** The distillate consists of the azeotropic mixture of composition corresponding to the point M and can be collected from top of the fractionating column.

If in the beginning the composition of the solution lies to right of the point M, the end products would be

***Residue*** Pure liquid B.

***Distillate*** The azeotropic mixture of composition corresponding to the point M.

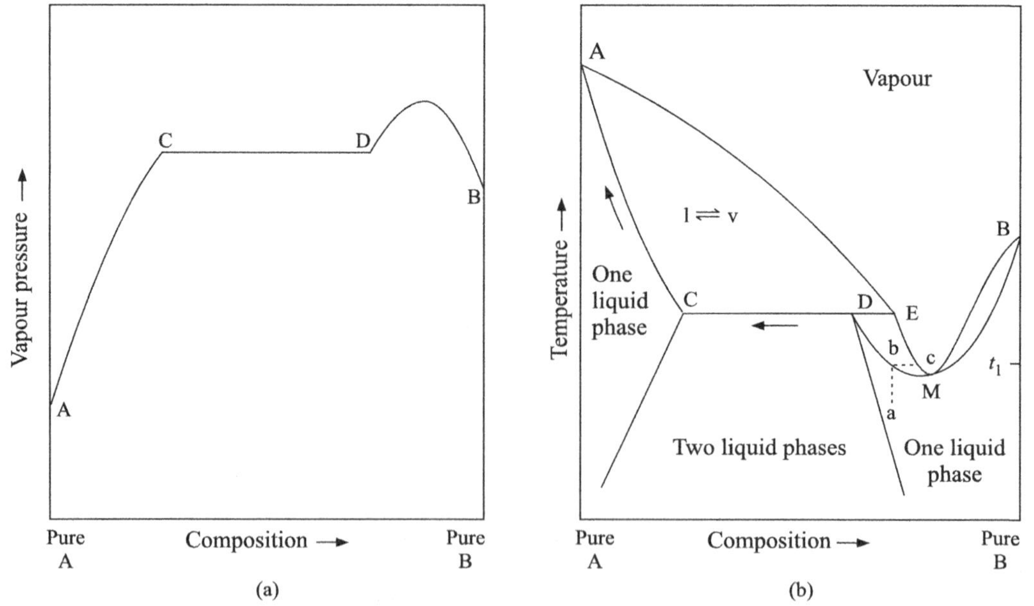


**Fig. 4.18.9 (a) Vapour pressure-composition and (b) temperature-composition diagrams for partial miscible liquids exhibiting azeotropic mixture**

**Distillation Involving Complete Separation**

Examples of such a system are isopentane-nitrobenzene, isopentane-dichloroacetic acid, and nicotine-water mixtures. In this case, complete separation of the two components is possible in spite of the formation of two layers during the course of fractional distillation (Fig. 4.8.10). The residue will consist of pure A in the liquid form and the distillate consisting of pure B in the vapour form.

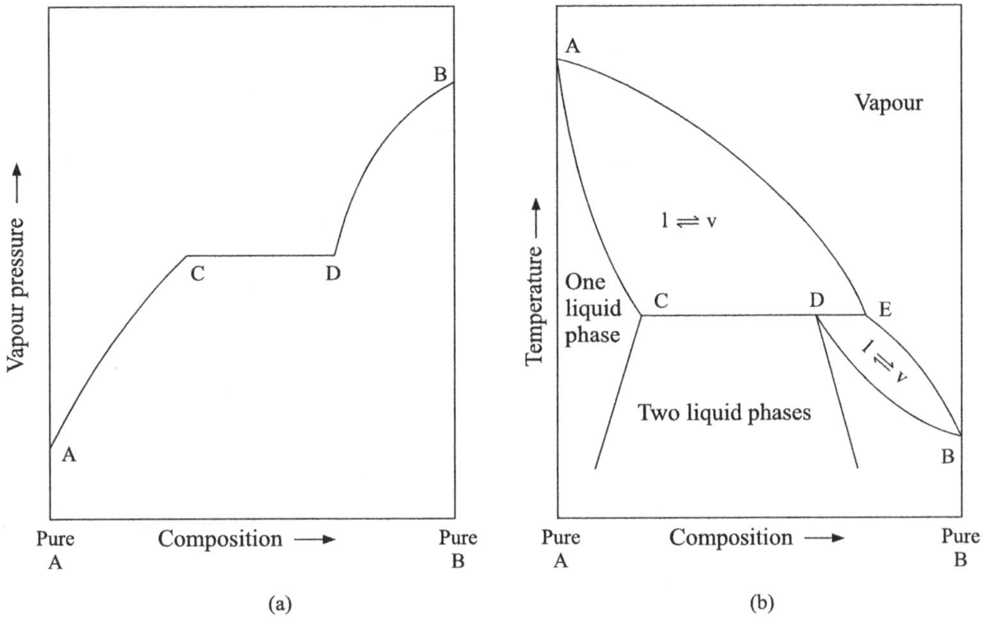


**Fig. 4.18.10 (a) Vapour pressure-composition and (b) temperature-composition diagrams for partially miscible liquids exhibiting complete separation of its constituents**

## 4.19 COMPLETELY IMMISCIBLE LIQUIDS (STEAM DISTILLATION)

If the two liquids are completely immiscible with each other, each one will behave independently of the other and will exert its own vapour pressure. The total vapour pressure of the system will be sum of vapour pressures of the two liquids. Thus

$$p = p_A^* + p_B^*$$

This total pressure will remain constant as long as both the liquids are present and will be independent of the relative amounts of the two liquids.

**Principle Underlying Boiling of Immiscible Liquids**

The boiling point of a liquid system is a temperature at which its vapour pressure becomes equal to the external pressure. Since the total vapour pressure of a mixture containing immiscible liquids is greater than that of either of the pure constituents, it follows that the mixture will boil at a temperature which is lower than boiling points of either of the pure constituents. Moreover, since the total vapour pressure is independent of the relative amounts of the two constituents, the boiling point will remain constant as long as both the layers are present.

The composition of vapour can be calculated by using Dalton's law of partial pressures. Thus, we have

$$\frac{p_A^*}{p_B^*} = \frac{y_A}{y_B} = \frac{n_A}{n_B} = \frac{m_A/M_A}{m_B/M_B} \quad \text{or} \quad \frac{m_A}{m_B} = \frac{M_A p_A^*}{M_B p_B^*} \tag{4.19.1}$$

The masses of the constituents in the distillate are thus proportional to their respective molar masses and vapour pressures.

**Process of Steam Distillation**

The process of distillation of immiscible liquids is generally utilized for the purification of those liquids which either boil at too high temperatures or decompose when heated to their normal boiling points. In laboratory and also in industry, water is used as one of the immiscible liquids and the process of distillation is known as *steam distillation.* The immiscible liquid is heated either directly along with water or by passing steam through it. Generally, the process of passing steam is preferred as it keeps the system agitating and thus equilibrium between the vapour and the two liquids is rapidly attained. The distillate is condensed and separated in the pure forms. It is evident from Eq. (4.19.1) that distillation of an immiscible mixture of an organic compound with water will yield a distillate containing a high proportion of compound provided the molar mass of the compound is larger and has appreciable vapour pressure at temperature near and just below the boiling water of water.

The process of steam distillation is often employed for the partially miscible mixtures. In this case, the efficiency of the process is slightly less than the theoretical value.

---

***Example 4.19.1***

The steam distillation of chlorobenzene is observed to occur at a temperature of 90.6 °C when the total pressure is 1.0 atm. Assuming complete immiscibility of these liquids, calculate the mass of chlorobenzene in 0.1 kg of distillate. The vapour pressure of water at 90.6 °C is 538.9 Torr.

***Solution***

Total vapour pressure = 760 Torr

Molar mass of chlorobenzene, $M_A = 112.5 \text{ g mol}^{-1}$

Vapour pressure of water at 90.6 °C = 538.9 Torr

Vapour pressure of pure chlorobenzene, $p_A^* = (760 - 538.9) \text{ Torr} = 221.1 \text{ Torr}$

Substituting the above data in the expression $m_A/m_B = p_A^* M_A / p_B^* M_B$,

we get $$\frac{m_A}{m_B} = \frac{(22.1 \text{ Torr})(112.5 \text{ g mol}^{-1})}{(538.9 \text{ Torr})(18 \text{ g mol}^{-1})} = 2.56 \quad \text{or} \quad \frac{m_B}{m_A} = \frac{1}{2.56}$$

Adding one on both sides, we have

$$\frac{m_A + m_B}{m_A} = \frac{1}{2.56} + 1 = \frac{3.56}{2.56} \quad \text{or} \quad \frac{m_A}{m_A + m_B} = \frac{2.56}{3.56} = 0.72$$

For 100 g of distillate, we have

$$m_A = (0.72)(m_A + m_B) = (0.72)(100 \text{ g}) = 72 \text{ g}$$

***Example 4.19.2***

When a liquid that is immiscible with water was steam distilled at 95.2 °C at a total pressure of 747.3 Torr, the distillate contained 1.27 g of the liquid per gram of water. Calculate the molar mass of the liquid. The vapour pressure of water is 638.6 Torr at 95.2 °C.

***Solution***

We are provided with

$$p_{\text{total}} = 747.3 \text{ Torr}$$

$$p_{\text{B}}^{*} = p_{\text{water}}^{*} = 638.6 \text{ Torr}$$

$$p_{\text{A}}^{*} = p_{\text{liquid}}^{*} = 747.3 \text{ Torr} - 638.6 \text{ Torr} = 108.7 \text{ Torr}$$

and $m_{\text{A}}/m_{\text{B}} = 1.27 \text{ g}/1 \text{ g}.$

We have

$$\frac{m_{\text{A}}}{m_{\text{B}}} = \frac{p_{\text{A}}^{*} M_{\text{A}}}{p_{\text{B}}^{*} M_{\text{B}}} \quad \text{or} \quad M_{\text{A}} = \left(\frac{m_{\text{A}}}{m_{\text{B}}}\right)\left(\frac{p_{\text{B}}^{*} M_{\text{B}}}{p_{\text{A}}^{*}}\right)$$

Substituting the given data, we get

$$M_{\text{A}} = (1.27)\left[\frac{(638.6 \text{ Torr})\,(18 \text{ g mol}^{-1})}{(108.7 \text{ Torr})}\right] = 134.3 \text{ g mol}^{-1}$$

## 4.20 DISTRIBUTION OF A SOLUTE BETWEEN TWO IMMISCIBLE LIQUIDS — THE NERNST DISTRIBUTION LAW

**Definition of Distribution Law**

If a solute A is added in a system containing two immiscible liquids α and β (or slightly miscible liquids) and if the solute is soluble in both the liquids, then it distributes itself between the two liquids in a definite manner, such that

$$\frac{[\text{A}]_{\alpha}}{[\text{A}]_{\beta}} = K_{\text{d}} \qquad (4.20.1)$$

where $[\text{A}]_{\alpha}$ and $[\text{A}]_{\beta}$ are the equilibrium molar concentrations of the solute A in the liquid phases α and β, respectively and $K_{\text{d}}$ is a constant known as the *distribution coefficient* or the *partition coefficient.* The value of the constant $K_{\text{d}}$ depends only on the temperature of the system and is independent of the relative amounts of the two layers and also that of the solute. Equation (4.20.1) is the defining equation for the distribution law or the partition law.

**Phase Rule Applied to Distribution Law**

The distribution law is in agreement with the phase rule requirements. We have

$$P = 2 \text{ and } C = 3$$

Thus $F = C - P + 2 = 3$

Of the three degrees of freedom, the two degrees of freedom, namely, temperature and pressure, are held constant. Thus stating the value of any third variable is sufficient to define the system completely. This implies that if the concentration of the solute in one of the layers is stated, then its value in the other layer will have a definite value and thus the ratio of the two concentrations will have a constant value.

## Thermodynamic Derivation of the Distributuon Law

**Derivation for Ideal Solutions.** The condition of equilibrium of the solute A between the two layers is

$$\mu_{A(\alpha)} = \mu_{A(\beta)}$$

where $\mu_{A(\alpha)}$ and $\mu_{A(\beta)}$ are the chemical potentials of the solute A in the two liquids $\alpha$ and $\beta$, respectively. If these liquids behave ideally, we have

$$\mu_{A(\alpha)} = \mu^*_{A(\alpha)} + RT \ln x_{A(\alpha)} \quad \text{and} \quad \mu_{A(\beta)} = \mu^*_{A(\beta)} + RT \ln x_{A(\beta)}$$

Substituting these in the previous expression, we get

$$\mu^*_{A(\alpha)} + RT \ln x_{A(\alpha)} = \mu^*_{A(\beta)} + RT \ln x_{A(\beta)}$$

or

$$\ln \frac{x_{A(\alpha)}}{x_{A(\beta)}} = \frac{1}{RT}(\mu^*_{A(\beta)} - \mu^*_{A(\alpha)})$$

At a given temperature and pressure, the right side of the above equation has a constant value. Thus, we have

$$\ln \frac{x_{A(\alpha)}}{x_{A(\beta)}} = \text{constant} \quad \text{or} \quad \frac{x_{A(\alpha)}}{x_{A(\beta)}} = \text{constant}$$

If the solutions are dilute, the amount fractions can be replaced by the corresponding molar concentrations. Thus, we have

$$\frac{[A]_\alpha}{[A]_\beta} = \text{constant} = K_d \qquad \text{(Eq. 4.20.1)}$$

### Derivation for Ideally Dilute Solutions

For ideally dilute solutions, we have

$$\mu_{A(\alpha)} = \mu^*_{A(\alpha,\,hs)} + RT \ln x_{A(\alpha)}$$

$$\mu_{A(\beta)} = \mu^*_{A(\beta,\,hs)} + RT \ln x_{A(\beta)}$$

The condition of equilibrium of A between the two phases gives

$$\mu_{A(\alpha)} = \mu_{A(\beta)}$$

i.e.

$$\mu^*_{A(\alpha,\,hs)} + RT \ln x_{A(\alpha)} = \mu^*_{A(\beta,\,hs)} + RT \ln x_{A(\beta)}$$

Hence

$$K_d = \frac{x_{A(\alpha)}}{x_{A(\beta)}} = \exp\left[\frac{\mu^*_{A(\beta,\,hs)} - \mu^*_{A(\alpha,\,hs)}}{RT}\right]$$

Since

$$\mu^*_{A(\alpha,\,hs)} = \mu^*_{A(v)} + RT \ln (k_{H,\,A(\alpha)}/p^\circ)$$

and

$$\mu^*_{A(\beta,\,hs)} = \mu^*_{A(v)} + RT \ln (k_{H,\,A(\beta)}/p^\circ)$$

we get

$$K_d = \frac{x_{A(\alpha)}}{x_{A(\beta)}} = \frac{k_{H,\,A(\beta)}}{k_{H,\,A(\alpha)}}$$

Thus, the amount of fraction of A in phase $\alpha$ is inversely proportional to its Henry's law constant in phase $\alpha$.

**Limitation of the Distribution Law**

The distribution expression as given by Eq. (4.20.1) will be strictly true provided (i) the solutions are ideal and dilute, (ii) the presence of solute does not change the mutual solubility of the two layers, and (iii) the solute has the same molar mass in both the layers.

A few examples which follow distribution law are the distribution of iodine between water and carbon tetrachloride (or chloroform or carbon disulphide), phenol between water and amyl alcohol, and bromine between water and carbon disulphide.

It was pointed out by Nernst that the distribution law as given by Eq. (4.20.1) is applicable only when the solute has the same molecular form (i.e. same molar mass) in both the layers. In some cases, solute does not have the same molar mass in the two solutions. This results because of association or dissociation or any other such changes which the solute can undergo in the solution. One of such examples is the distribution of benzoic acid in water and benzene. Benzoic acid in water is present in a partially ionized form whereas in benzene, it undergoes dimerization. Thus, we have

$$C_6H_5COOH \rightleftharpoons C_6H_5COO^- + H^+ \quad \text{in water}$$

and

$$2C_6H_5COOH \rightleftharpoons (C_6H_5COOH)_2 \quad \text{in benzene}$$

If the total concentrations of benzoic acid in the two layers are substituted in Eq. (4.20.1), it is found that the ratio does not remain constant as the amount of benzoic acid dissolved in both the layers is varied. According to Nernst, the distribution law will be applicable if the concentrations of the species common to both the layers (i.e. undissociated benzoic acid) are employed.

Thus

$$K_d = \frac{[A]_{aq} - \xi_d}{[A]_{org} - 2\xi_a} \tag{4.20.2}$$

where $\xi_d$ and $\xi_a$ are the extent of dissociation and association reaction, respectively.

## DISTRIBUTION LAW UNDER DIFFERENT CONDITIONS

A suitable modification in the original distribution law as given by Eq. (4.20.1) can be made in order to account for the association, dissociation, hydrolysis or any such changes which a solute undergoes in the solution. Some of these are described below.

**Normal Existence in α and Association in β**

In liquid α, solute A is present as it is, thus its concentration is $[A]_\alpha$. In liquid β, it associates according to the equation

$$nA \rightleftharpoons A_n \tag{4.20.3}$$

Let $[A]_{\beta,\ total}$ be the total concentration of A in liquid β and $\xi_a$ be its extent of association, then we have

$$[A]_{\beta,\ free} = [A]_{\beta,\ total} - n\xi_a$$

Substituting these in the equilibrium expression corresponding to the association reaction, we get

$$K_{eq} = \frac{[A_n]_\beta}{[A]^n_{\beta,\,free}} = \frac{\xi_a}{([A]_{\beta,\,total} - n\xi_a)^n} \tag{4.20.4}$$

or $$[A]_{\beta,\,free} = [A]_{\beta,\,total} - n\xi_a = \left(\frac{\xi_a}{K_{eq}}\right)^{1/n}$$

If the solute A in liquid $\beta$ is almost present in the associated form, then $\xi_a = [A]_{\beta,\,total}/n$. Thus, we have

$$[A]_{\beta,\,free} = [A]_{\beta,\,total} - n\xi_a = \left(\frac{[A]_{\beta,\,total}}{nK_{eq}}\right)^{1/n}$$

Substituting $[A]_\alpha$ and $[A]_{\beta,\,free}$ in the distribution law, we get

$$K_d = \frac{[A]_\alpha}{[A]_{\beta,\,free}} = \frac{[A]_\alpha}{([A]_{\beta,\,total}/n\,K_{eq})^{1/n}}$$

or $$\frac{[A]_\alpha}{([A]_{\beta,\,total})^{1/n}} = \frac{K_d}{(n\,K_{eq})^{1/n}} \tag{4.20.5}$$

At a given temperature, $K_{eq}$ has a constant value. Since $n$ is a constant for a given association reaction, it follows that the right side of Eq. (4.20.5) also has a constant value. Thus, we have

$$\frac{[A]_\alpha}{([A]_{\beta,\,total})^{1/n}} = \text{constant} = K'_d \tag{4.20.6}$$

where $K'_d$ is known as the apparent distribution coefficient.

The value of $n$ can be determined by plotting a graph between $\log\{[A]_\alpha/c^\circ\}$ and $\log\{[A]_{\beta,\,total}/c^\circ\}$, the slope of the resultant line is equal to $1/n$. Thus knowing $n$, the value of $K_d$ can be determined. For benzoic acid in benzene, $n$ is equal to 2. Thus, it dimerizes in benzene layer.

**Alternatively,** the value of $n$ can be determined from the given two sets of data as shown in the following.

Since $$\frac{[A]_\alpha}{([A]_{\beta,\,total})^{1/n}} = K'_d \quad \text{or} \quad \frac{[A]_\alpha/c^\circ}{([A]_{\beta,\,total}/c^\circ)^{1/n}} = \frac{K'_d}{(c^\circ)^{(n-1)/n}} \qquad (4.20.7)^\dagger$$

---

†Division by $c^\circ$ (= unit concentration) makes different quantities unitless.

Therefore

$$\log\left(\frac{K'_{\mathrm{d}}}{(c^{\circ})^{(n-1)/n}}\right) = \log\left(\frac{[\mathrm{A}]_{\alpha}}{c^{\circ}}\right) - \frac{1}{n}\log\left(\frac{[\mathrm{A}]_{\beta,\,\mathrm{total}}}{c^{\circ}}\right) \tag{4.20.8}$$

Representing $[\mathrm{A}]_{\alpha}/c^{\circ}$ and $[\mathrm{A}]_{\beta,\,\mathrm{total}}/c^{\circ}$ for the two sets of data as $x_1$, $y_1$ and $x_2$ and $y_2$, we get

$$\log K''_{\mathrm{d}} = \log x_1 - \frac{1}{n}\log y_1 \quad \text{and} \quad \log K''_{\mathrm{d}} = \log x_2 - \frac{1}{n}\log y_2$$

From these two expressions, we get

$$\log x_1 - \frac{1}{n}\log y_1 = \log x_2 - \frac{1}{n}\log y_2$$

or

$$n = \frac{\log y_1 - \log y_2}{\log x_1 - \log x_2} \tag{4.20.9}$$

---

***Example 4.20.1***

An organic substance X has a normal molar mass in water but gives abnormal molar mass in benzene. The following data were obtained during a distribution experiment.

| | I | II | III |
|---|---|---|---|
| $[\mathrm{X}]_{\mathrm{water}}/\mathrm{g\ dm^{-3}}$ | 0.01 | 0.12 | 0.24 |
| $[\mathrm{X}]_{\mathrm{benzene}}/\mathrm{g\ dm^{-3}}$ | $1.848 \times 10^{-5}$ | $2.661 \times 10^{-3}$ | $1.089 \times 10^{-2}$ |

Find without using the graphical method, the degree of complexity of the substance in benzene.

***Solution***

We know that the degree of complexity is given by

$$n = \frac{\log(y_1/\mathrm{g\ dm^{-3}}) - \log(y_2/\mathrm{g\ dm^{-3}})}{\log(x_1/\mathrm{g\ dm^{-3}}) - \log(x_2/\mathrm{g\ dm^{-3}})}$$

From the first two readings, we have

$$n = \frac{\log(1.848\times10^{-5}) - \log(2.661\times10^{-3})}{\log(0.01) - \log(0.12)} = \frac{\bar{5}.266\,7 - \bar{3}.425\,1}{\bar{2}.000\,0 - \bar{1}.079\,2} = 1.999 \simeq 2$$

Taking the second two readings, we have

$$n = \frac{\log(2.661\times10^{-3}) - \log(1.089\times10^{-2})}{\log(0.12) - \log(0.24)} = \frac{\bar{3}.425\,1 - \bar{2}.037\,0}{\bar{2}.000\,0 - \bar{1}.079\,2} = 2.033$$

Taking the first and the third readings, we have

$$n = \frac{\log(1.848\times10^{-5}) - \log(1.089\times10^{-2})}{\log(0.01) - \log(0.24)} = \frac{\bar{5}.266\,7 - \bar{2}.037\,0}{\bar{2}.000\,0 - \bar{1}.380\,2} = 2.007$$

The value of $n$ is equal to 2. Thus, the organic substance exists as dimer in benzene layer.

---

***Values of true distribution coefficient and equilibrium constant*** Equation (4.20.4) can be written as

$$K_{\mathrm{eq}} = \frac{[\mathrm{A}_n]_{\beta}}{[\mathrm{A}]^{n}_{\beta,\,\mathrm{free}}} \tag{4.20.10}$$

The total concentration of A in liquid β is given by

$$[A]_{\beta,\text{total}} = [A]_{\beta,\text{free}} + n[A_n]_\beta$$

Making use of Eq. (4.20.10), we get

$$[A]_{\beta,\text{total}} = [A]_{\beta,\text{free}} + n\,K_{eq}\,[A]^n_{\beta,\text{free}}$$

Now employing the distribution expression

$$K_d = \frac{[A]_\alpha}{[A]_{\beta,\text{free}}}$$

we get $$[A]_{\beta,\text{total}} = \frac{[A]_\alpha}{K_d} + nK_{eq}\left(\frac{[A]_\alpha}{K_d}\right)^n$$

or $$\frac{[A]_{\beta,\text{total}}}{[A]_\alpha} = \frac{1}{K_d} + \frac{nK_{eq}}{K_d^n}[A]_\alpha^{n-1} \tag{4.20.11}$$

Thus by plotting $[A]_{\beta,\text{total}}/[A]_\alpha$ versus $[A]_\alpha^{n-1}$, we can get $1/K_d$ and $nK_{eq}/K_d^n$ as intercept and slope, respectively. From these, the values of $K_d$ and $K_{eq}$ can be determined.

**Normal Existence in α and Dissoication in β**

Let the solute dissociate in Liquid β according to the following scheme.

$$A \rightleftharpoons B + C$$

Let $\xi$ be the extent of reaction. We will have

$$[A]_{\beta,\text{free}} = [A]_{\beta,\text{total}} - \xi$$

$$[B] = [C] = \xi$$

Substituting these in the equilibrium expression corresponding to the above dissociation, we get

$$K_{diss} = \frac{[B][C]}{[A]} = \frac{\xi^2}{[A]_{\beta,\text{total}} - \xi}$$

Thus $$[A]_{\beta,\text{free}} = [A]_{\beta,\text{total}} - \xi = \frac{\xi^2}{K_{diss}}$$

Hence, the distribution law becomes

$$K_d = \frac{[A]_\alpha}{[A]_{\beta,\text{free}}} = \frac{[A]_\alpha}{\xi^2/K_{diss}} \tag{4.20.12}$$

**Dissociation in α and Dimerization in β**

In this case, we have

$$K_d = \frac{[A]_{\alpha,\text{total}} - \xi_d}{[A]_{\beta,\text{total}} - 2\xi_a} \tag{4.20.13}$$

**Normal Existence in $\alpha$ and Undergoes a Reaction with $\beta$**

Let the reaction between the solute and the liquid $\beta$ be represented as

$$A + n\beta \rightleftharpoons A \cdot n\beta$$

Its equilibrium constant is given by

$$K_{eq} = \frac{[A \cdot n\beta]}{[A]_{\beta,\,free}\,[\beta]^n} \tag{4.20.14}$$

The total concentration of A in liquid $\beta$ is given by

$$[A]_{\beta,\,total} = [A]_{\beta,\,free} + [A \cdot n\beta]$$

Making use of Eq. (4.20.14), we get

$$[A]_{\beta,\,total} = [A]_{\beta,\,free} + K_{eq}[A]_{\beta,\,free}\,[\beta]^n$$

Hence,
$$[A]_{\beta,\,free} = \frac{[A]_{\beta,\,total}}{1 + K_{eq}\,[\beta]^n} \tag{4.20.15}$$

Substituting Eq. (4.20.15) in the distribution law, we get

$$K_d = \frac{[A]_\alpha}{[A]_{\beta,\,free}} = \frac{[A]_\alpha}{[A]_{\beta,\,total}/(1 + K_{eq}[\beta]^n)}$$

or
$$\frac{[A]_\alpha}{[A]_{\beta,\,total}} = \frac{K_d}{1 + K_{eq}[\beta]^n} \tag{4.20.16}$$

For a given temperature, $K_{eq}$ has a constant value. Since liquid $\beta$ is present in a very large amount, the concentration of $\beta$ practically remains constant. Thus, we have

$$\frac{[A]_\alpha}{[A]_{\beta,\,total}} = \text{constant} \tag{4.20.17}$$

**Study of Equilibria by Distribution Measurement**

Distribution measurements have been used to determine the equilibrium constants of various reactions such as complex formation, hydrolysis of ions, etc. In such studies, the second solvent which is generally chosen is the one that can dissolve one of the reactants or products of the reaction under study so that the distribution of this species occurs between the two layers. The underlying principle can be explained by taking a typical example of the complex formation between $I_2$ and KI. In aqueous solution, we have

$$KI + I_2 \rightleftharpoons KI_3$$

We choose a solvent in which $I_2$ is soluble, it may be carbon tetrachloride or carbon disulphide. First of all, the distribution coefficient $K_d$ of $I_2$ in water and $CCl_4$ is determined, such that

$$\frac{[I_2]_{CCl_4}}{[I_2]_{water}} = K_d$$

Now the equilibrium of $I_2$ between KI solution and $CCl_4$ is studied. If $[I_2]_{aq,\ total}$ and $[I_2]_{CCl_4}$ are the concentrations of iodine in aqueous and $CCl_4$ layers, respectively, then we have:

Concentration of free iodine in aqueous layer, $[I_2]_{aq,\ free} = \dfrac{[I_2]_{CCl_4}}{K_d}$

Concentration of combined $I_2$ in aqueous layer or the concentration of $KI_3$

= Total $I_2$ in aqueous layer – Free $I_2$ in aqueous layer

$$= [I_2]_{aq,\ total} - \frac{[I_2]_{CCl_4}}{K_d}$$

Concentration of free KI in aqueous solution, $[KI]_{free}$

= Total concentration of KI – Concentration of $KI_3$

$$= [KI]_{aq,\ total} - \left\{[I_2]_{aq,\ total} - \frac{[I_2]_{CCl_4}}{K_d}\right\}$$

The equilibrium constant of the reaction is thus given by

$$K_{eq} = \frac{[KI_3]}{[I_2]_{aq,\ free}[KI]_{free}} = \frac{[I_2]_{aq,\ total} - \dfrac{[I_2]_{CCl_4}}{K_d}}{\left(\dfrac{[I_2]_{CCl_4}}{K_d}\right)\left([KI]_{aq,\ total} - \left\{[I_2]_{aq,\ total} - \dfrac{[I_2]_{CCl_4}}{K_d}\right\}\right)}$$

Thus knowing $[I_2]_{aq,\ total}$, $[I_2]_{CCl_4}$, $[KI]_{aq,\ total}$ and $K_d$, the value of equilibrium constant $K_{eq}$ can be determined.

---

***Example 4.20.2*** The distribution coefficient of the base *p*-nitroaniline between benzene and water is 9.0 (in favour of benzene). 0.006 93 mol of *p*-nitroanilinium chloride was dissolved in one $dm^3$ of dil HCl solution of molarity 0.043 42 M. To this solution 60 $cm^3$ of benzene was added and the solution shaken up till the equilibrium was attained. Then 25 $cm^3$ of benzene solution was withdrawn and was analyzed for the base and thus found to contain 0.108 25 g *p*-nitroaniline. Calculate the hydrolysis constant of the salt.

***Solution*** Given that

$$\frac{\text{Concentration of } p\text{-nitroaniline in benzene layer}}{\text{Concentration of } p\text{-nitroaniline in aqueous layer}} = 9.0$$

$$\text{Amount of free base in 25 cm}^3 \text{ of benzene} = \frac{(0.108\ 25\ \text{g})}{(138\ \text{g mol}^{-1})}$$

$$\text{Thus in 60 cm}^3\text{, the amount} = \left(\frac{0.108\ 25\ \text{g}}{138\ \text{g mol}^{-1}}\right)\left(\frac{60\ \text{cm}^3}{25\ \text{cm}^3}\right) = 0.001\ 88\ \text{mol}$$

$$\text{Molar concentration of free base} = \left(\frac{0.001\ 88\ \text{mol}}{(60/1\ 000)\ \text{dm}^3}\right) = 0.031\ 33\ \text{mol dm}^{-3}$$

Using the distribution coefficient, we find that molar concentration of free base in aqueous

$$\text{solution} = \frac{(0.031\ 33\ \text{mol dm}^{-3})}{9} = 0.003\ 48\ \text{mol dm}^{-3}$$

The added amount of 0.006 93 mol of *p*-nitroanilinium chloride will be present in two forms, namely, (i) unhydrolyzed salt, and (ii) hydrolyzed salt. The hydrolysis reaction is

$$\underset{\text{Unhydrolyzed salt}}{p\text{-}O_2NC_6H_4NH_3^+Cl^-} + H_2O \rightleftharpoons \underset{\substack{\text{Hydrolyzed salt}\\ \text{(free base)}}}{p\text{-}O_2NC_6H_4NH_2} + H_3O^+ + Cl^-$$

Thus, the amount of unhydrolyzed cation

= Total amount of salt added – Total amount of free base

= Total amount of salt added – (Free base in benzene + Free base in water)

= 0.006 93 mol – (0.001 88 mol + 0.003 48 mol)

= 0.001 57 mol

Amount of acid set free because of hydrolysis

= Total amount of free base in benzene and water

= 0.001 88 mol + 0.003 48 mol = 0.005 36 mol

Total amount of acid = Acid already present + Acid set free

= 0.043 42 mol + 0.005 36 mol

= 0.048 78 mol

Thus, the concentrations of various species in water for the reaction

$$O_2NC_6H_4NH_3^+ + H_2O \rightleftharpoons O_2NC_6H_4NH_2 + H_3O^+$$

are

$$[O_2NC_6H_4NH_3^+] = 0.001\ 57\ \text{mol dm}^{-3}$$

$$[O_2NC_6H_4NH_2] = 0.003\ 48\ \text{mol dm}^{-3}$$

$$[H_3O^+] = 0.048\ 78\ \text{mol dm}^{-3}$$

and hence

$$K_h = \frac{[O_2NC_6H_4NH_2][H_3O^+]}{[O_2NC_6H_4NH_3^+]} = \frac{(0.003\ 48\ \text{mol dm}^{-3})(0.048\ 78\ \text{mol dm}^{-3})}{(0.001\ 57\ \text{mol dm}^{-3})}$$

$$= 0.108\ 1\ \text{mol dm}^{-3}$$

---

***Example 4.20.3*** The distribution ratio of ammonia between water and chloroform at 291 K is 25.8. When a 0.025 mol dm$^{-3}$ of copper sulphate solution was equilibrated with ammonia and chloroform, the aqueous and chloroform solutions contained 0.385 mol dm$^{-3}$ and 0.011 2 mol dm$^{-3}$ of ammonia, respectively. Determine how much of ammonia is complexed with the $Cu^{2+}$ ion and the value of $x$ in the formula $Cu(NH_3)_x^{2+}$.

***Solution*** We are provided that

$$\frac{\text{Concentration of } NH_3 \text{ in aqueous solution}}{\text{Concentration of } NH_3 \text{ in chloroform}} = 25.8$$

Concentration of $NH_3$ in aqueous $CuSO_4$ solution = 0.385 mol dm$^{-3}$

Concentration of $NH_3$ in chloroform = 0.011 2 mol dm$^{-3}$.

Using distribution coefficient, concentration of free $NH_3$ in aqueous layer

= (0.011 2 mol dm$^{-3}$)(25.8)

= 0.289 0 mol dm$^{-3}$

Concentration of combined $NH_3$ = 0.385 0 mol dm$^{-3}$ – 0.289 0 mol dm$^{-3}$
= 0.096 0 mol dm$^{-3}$

The value of $x$ in $Cu(NH_3)_x^{2+}$ $= \frac{[NH_3]_{combined}}{[CuSO_4]} = \frac{0.096\,0\text{ M}}{0.025\text{ M}} = 3.84 \approx 4$

**Solvent Extraction**

The Nernst distribution law is the basis for the solvent extraction process in which the dissolved solute in one solvent is extracted with the help of another solvent. When a solution containing the solute is shaken with another immiscible solvent, the solute distributes between the two solutions according to the Nernst distribution law. From the second solvent, the solute can be separated by simple processes such as distillation, evaporation, etc. For efficient use of the solvent, it is advantageous to carry out this process in a series of successive stages. This can be proved as follows.

Let $m$ be the mass of solute present in volume $V_1$ of the solvent 1 (or $V_1$ of the solution) and let the volume $V_2$ of the solvent 2 be added each time till the entire solvent 2 is exhausted. If the solvent 2 is added $n$ times, we have

Total volume of the solvent 2 used for extraction = $nV_2$

**Multi-stage Extraction**

If $m_1$ is the mass of the solute that remains unextracted in the solvent 1 after the addition of first lot of $V_2$ of the solvent 2, then we have

$$\frac{c_1}{c_2} = \frac{m_1/V_1}{(m - m_1)/V_2} = K_d$$

Solving for $m_1$, we get

$$m_1 = \left(\frac{K_dV_1}{K_dV_1 + V_2}\right) m \qquad (4.20.18)$$

If $m_2$ is the mass of the solute unextracted after the addition of the second lot, then

$$\frac{c_1'}{c_2'} = \frac{m_2/V}{(m_1 - m_2)/V_2} = K_d$$

or $$m_2 = \left(\frac{K_dV_1}{K_dV_1 + V_2}\right) m_1$$

Substituting $m_1$ from Eq. (4.10.18), we get

$$m_2 = \left(\frac{K_dV_1}{K_dV_1 + V_2}\right)^2 m \qquad (4.20.19)$$

Carrying out the above process $n$ times, we will get

$$m_n = \left(\frac{K_dV_1}{K_dV_1 + V_2}\right)^n m \qquad (4.20.20)$$

Thus, the fraction of the solute remaining unextracted in solvent 1 is

$$\frac{m_n}{m} = \left(\frac{K_dV_1}{K_dV_1 + V_2}\right)^n \qquad (4.20.21)$$

**Single-stage Extraction**

If the extraction were carried out in a single stage by adding whole of the solvent 2 in one lot, we would have

$$\frac{m'}{m} = \left(\frac{K_d V_1}{K_d V_1 + nV_2}\right) \qquad (4.20.22)$$

**Comparing Efficiency of the Two Processes**

Now we proceed to show that the multi-stage extraction is more effective than the single-stage extraction. Taking the inverse of Eq. (4.20.21), we get

$$\frac{m}{m_n} = \left(\frac{K_d V_1 + V_2}{K_d V_1}\right)^n = \left(1 + \frac{V_2}{K_d V_1}\right)^n \qquad (4.20.23)$$

The right side of Eq. (4.20.23) can be expanded by using binominal theorem. We have

$$\frac{m}{m_n} = \left(1 + \frac{V_2}{K_d V_1}\right)^n = 1 + \frac{nV_2}{K_d V_1} + \frac{n(n-1)}{2}\left(\frac{V_2}{K_d V_1}\right)^2 + \cdots \qquad (4.20.24)$$

The inverse of Eq. (4.20.22) is

$$\frac{m}{m'} = \frac{K_d V_1 + nV_2}{K_d V_1} = 1 + \frac{nV_2}{K_d V_1} \qquad (4.20.25)$$

Substituting Eq. (4.20.25) in Eq. (4.20.24), we get

$$\frac{m}{m_n} = \frac{m}{m'} + \frac{n(n-1)}{2}\left(\frac{V_2}{K_d V_1}\right)^2 + \cdots$$

Therefore it follows that

$$\frac{m}{m_n} > \frac{m}{m'} \quad \text{or} \quad m_n < m'$$

that is, the amount of the solute remaining unextracted is less in multi-stage extraction. In other words, the amount of extracted solute is larger in multi-stage extraction than in single-stage extraction. Hence the former is more effective for extraction of solute from a given solution.

It may, however, be mentioned that a complete extraction of the solute can never be achieved howsoever large number of extractions may be done, as something will always be left behind in the solution in accordance with the distribution law.

While calculating $m$ from Eqs. (4.20.21) and (4.20.22), one should be very careful about the expression for $K_d$. The latter represents the ratio of the amount of solute in the solvent in which it is already present to that of the extracting solvent. Sometimes, the value of $K_d$ is given as the ratio of the solute in the extracting solvent to that in which it is already present. In such cases, the reciprocal of the given value of $K_d$ should be employed while using Eqs (4.20.21) and (4.20.22).

***Example 4.20.4***

A solute A is distributed between two immiscible liquids B and C with the value of the distribution coefficient, $[A]_C/[A]_B$, equal to 10. The concentration units are expressed in terms of grams of solute per $dm^3$ of solvent. It is desired to remove 99% of the amount of A from a solution containing 1 g of A in 100 $cm^3$ of B by extraction with successive 10 $cm^3$ portions of solvent C. Calculate the approximate volume of solvent C required for this purpose.

***Solution***

We have

$$\frac{[A]_C}{[A]_B} = \frac{10}{1}$$

Therefore $$\frac{[A]_B}{[A]_C} = \frac{1}{10} = 0.1 = K_d$$

Since 99% of A is to be removed from a solution containing 1 g of A per 100 $cm^3$ of B, we have

$$m_n = 0.01 \text{ g} \quad m = 1.00 \text{ g} \quad V_B = 100 \text{ cm}^3 \quad V_C = 10 \text{ cm}^3$$

The number of times ($n$) solvent C to be added can be calculated from

$$\frac{m_n}{m} = \left(\frac{K_d V_B}{K_d V_B + V_C}\right)^n$$

Substituting the given data, we get

$$\frac{0.01 \text{ g}}{1 \text{ g}} = \left(\frac{0.1 \times 100 \text{ cm}^3}{0.1 \times 100 \text{ cm}^3 + 10 \text{ cm}^3}\right)^n$$

or $$0.01 = \left(\frac{1}{2}\right)^n \quad \text{or} \quad \log(0.01) = -n \log 2$$

or $$n = 7$$

Hence, volume of C to be used in 7 × 10 $cm^3$, i.e. 70 $cm^3$

***Example 4.20.5***

An aqueous solution contains 0.20 g aspirin in a 50 $cm^3$ solution. To this solution is added 20 $cm^3$ of ether, and the mixture is shaken and allowed to attain equilibrium at 298 K. At this temperature, the distribution coefficient of aspirin between ether and water is 4.7.

(i) How much aspirin remains in the aqueous phase?
(ii) If the extraction is carried out with two successive 10 $cm^3$ portions of ether, how much aspirin remains unextracted?

***Solution***

(i) We have

$$K_d = \frac{\text{Aspirin in g cm}^{-3} \text{ in ether}}{\text{Aspirin in g cm}^{-3} \text{ in water}} = 4.7$$

Let $m_1$ and $m_2$ be the respective masses of aspirin in ether and the aqueous phase after the addition of 20 $cm^3$ of ether in the given 50 $cm^3$ of aqueous solution. It follows that

$$\frac{m_1/20 \text{ cm}^3}{m_2/50 \text{ cm}^3} = 4.7$$

and $$m_1 + m_2 = 0.20 \text{ g}$$

Solving for $m_1$ and $m_2$, we get

$$m_1 = 0.130\,6 \text{ g} \quad \text{and} \quad m_2 = 0.069\,4 \text{ g}$$

Thus, the mass of aspirin remaining in aqueous phase = 0.069 4 g

(ii) For the multi-stage extraction, we have to use the distribution coefficient given below.

$$K_\text{d} = \frac{\text{Aspirin in water}}{\text{Aspirin in ether}} = \frac{1}{4.7} = 0.212\,8$$

We have

$$\frac{m_n}{m} = \left(\frac{K_\text{d}V_1}{K_\text{d}V_1 + V_2}\right)^n$$

where for the given problem, we have

$$m_n = ? \qquad m = 0.20 \text{ g} \qquad V_1 = 50 \text{ cm}^3 \qquad V_2 = 10 \text{ cm}^3 \qquad n = 2$$

Substituting the given data, we get

$$m_n = \left(\frac{0.212\,8 \times 50 \text{ cm}^3}{0.212\,8 \times 50 \text{ cm}^3 + 10 \text{ cm}^3}\right)^2 (0.20 \text{ g}) = \left(\frac{10.64}{20.64}\right)^2 (0.20 \text{ g})$$

$$= 0.053\,1 \text{ g}$$

Hence

Aspirin remains unextracted = 0.053 1 g

Aspirin extracted = 0.2 g – 0.053 1 g = 0.146 9 g

---

## REVISIONARY PROBLEMS

4.1 (a) Define the terms absorption coefficient $\alpha$ and coefficient of solubility $\beta$. Derive the expression relating $\alpha$ and $\beta$.
(b) What is the variance of a system involving a dissolved gas in a liquid?

4.2 Define Henry's law and show that the volume of an ideal gas dissolved in a given volume of solvent (measured at the pressure of the experiment) is independent of the pressure.

4.3 What is the effect of temperature on the solubility of a gas in liquids?

4.4 Show that Henry's law is a special case of the distribution law and Raoult's law.

4.5 Prove the statements that:
(i) Raoult's law may be considered as a special case of Henry's law; all systems which obey Raoult's law must satisfy Henry's law, but the reverse is not true unless Henry's law is applicable over the entire range of concentration.
(ii) Raoult's law is applicable to the solvent over the range where Henry's law in the form $x_2 = k\,p_2$ is applicable to solute.

4.6 What is an ideally dilute solution? Derive the expressions of chemical potential for solute and solvent in an ideally dilute solution.

4.7 Derive Henry's law from the viewpoint of kinetic-molecular theory.

4.8 Explain, what do you understand by an ideal solution. Show that for an ideal solution $\Delta_\text{mix}H = 0$ and $\Delta_\text{mix}V = 0$, where $\Delta_\text{mix}H$ and $\Delta_\text{mix}V$ are the changes in enthalpy and volume on mixing various constituents of the liquid solution. Suggest a possible reason for $\Delta_\text{mix}H$ and $\Delta_\text{mix}V$ being equal to zero.

4.9 (a) Following Raoult's law, derive an expression for the total vapour pressure of an ideal binary liquid solution in terms of the amount fraction of either of the two constituents in the liquids phase. Draw a schematic diagram for the variation of $p_A$ versus $x_A$, $p_B$ versus $x_B$ and $p$ versus $x_A$ (or $x_B$) for a system in which $p_A^* > p_B^*$.
(b) Derive an expression for the total vapour pressure of an ideal binary liquid solution in terms of the amount fraction of either of the two constituents in vapour phase. Draw a schematic plot of $p$ versus $y_A$ and $1/p$ versus $y_A$.

4.10 (a) What is the variance of a binary liquid solution? Show that the state of the solution at a constant temperature can be represented on a two dimensional plot of pressure versus composition. Draw such a plot and label different regions and lines of the plot.
(b) Explain, why a region instead a point is observed when a binary liquid mixture is in equilibrium with its vapour.
(c) Predict the sequence of steps that takes place when the pressure on the binary liquid mixture at a constant temperature is gradually lowered.
(d) What is the lever rule and deduce it from pressure versus composition diagram for the binary liquid mixture?
(e) Discuss briefly the principle underlying the isothermal fractional distillation of an ideal binary liquid mixture.

4.11 (a) Draw a plot of temperature versus composition diagram at constant pressure for an ideal binary liquid mixture. Explain qualitatively why the liquids and the vaporous curves are not straight lines.
(b) Explain, why you should have a region over which the liquid mixture is in equilibrium with its vapour.
(c) Predict the sequence of steps that takes place when the temperature of a binary liquid mixture is gradually increased; hence describe briefly the theory of the isobaric factional distillation of an ideal binary solution. What are the end products in such a distillation?
(d) Describe the theory of the bubble-cap fractionating column.

4.12 (a) What are nonideal solutions? How do they differ from ideal solutions? What are the main criteria for exhibiting positive sand negative deviations from ideality?
(b) Draw a schematic diagram for $p_A$ versus $x_A$, $p_B$ versus $x_B$ and $p$ versus $x_A$(or $x_B$) of a system showing (i) positive deviation, and (ii) negative deviation. Also, show in the diagram the limiting Henry's line and the limiting Raoult's law line.

4.13 (a) Starting from the Gibbs-Duhem equation, derive the Duhem-Margules equation as applicable to a binary liquid system. To which type of liquid mixtures (ideal or nonideal), the Duhem-Margules equation is applicable?
(b) Starting from the Duhem-Margules equation as applicable to a binary liquid system, show that: (i) if one component behaves ideally, then the other also does so, and (ii) if one component shows positive deviation (or negative deviation), so also must the other.

4.14 Konowaloff's rule states that:
The vapour phase is richer in the component whose addition to the liquid mixture results in an increase of total vapour pressure.

or

The liquid phase is richer in the component whose addition to the liquid mixture results in a decrease of total vapour pressure.

Derive the above statements from the Duhem-Margules equation.

4.15 How will you justify the following alternative statements of Konowaloff's rule.

The vapour phase is richer in the component whose addition to the liquid mixture causes a decrease in the boiling point or alternatively, the liquid is richer in the component whose addition to the liquid mixture causes an increase in its boiling point.

4.16 (a) Draw schematic plots of temperature versus composition for nonideal solutions showing very large positive and negative deviations from ideality.

(b) Show that at the maximum or minimum point of the above plots, the following relation holds good

$$\frac{y_A}{y_B} = \frac{x_A}{x_B}$$

that is, the composition of the vapour phase is identical with that of the liquid with which it is in equilibrium. On this basis justify that the liquids curve must meet with the vapour curve at the maximum or minimum point.

(c) Show with the help of diagrams that (i) both the components can be separated from a nonideal solution if the components show a little deviation from ideality, and (ii) only pure A or pure B and not both of them if the components show very large deviation from ideality.

(d) What is an azeotropic mixture? Show that it is not possible to separate the constituents from an azeotropic mixture following the ordinary fractional distillation. What other methods could be employed for the separation from the azeotropic mixture? How will you show that an azeotropic mixture is really a mixture and not a single compound resulted because of the definite union of the two constituents?

4.17 (a) Derive the relation

$$\Delta T_b = K_b m (1 - K)$$

as applicable to a dilute solution of a nonideal binary liquid mixture, i.e. one of the liquids is present in a very large amount (solvent) and the other one in a very small amount (solute). In the above expression, $\Delta T_b$ is the change in the boiling point of the solvent as a result of the addition of the volatile solute; $K_b$ is the boiling point elevation constant, and $K$ is the distribution coefficient of the solute between the vapour phase and the liquid phase and is defined as

$$K = \frac{\text{Amount fraction of the solute in the vapour phase}}{\text{Amount fraction of the solute in the liquid phase}}$$

(b) Show that $\Delta T_b$ can be positive or negative depending upon the value of $K$, and hence show that the change in $\Delta T_b$ is in agreement with Konowaloff's rule.

4.18 (a) What do you understand by partially miscible liquids? Show that the formation of partially miscible liquids have sound theoretical background and can be explained qualitatively from the Duhem-Margules equation.

(b) Using the phase rule expression, show that the compositions of the two conjugate solutions in the miscible gap are quite definite for a given temperature and are independent of the relative amounts of the two solutions. Draw a schematic plot connecting the compositions of the two conjugate solutions at different temperatures.

(c) What is the difference between the critical solution temperature and mutual solubility temperature?

(d) Show with the help of the phase rule that at a given pressure, the critical solution temperature is a nonvariant point.

(e) Describe the sequence of steps that takes place when (i) the temperature of a given overall composition of partially miscible liquids is changed.

(f) At a temperature lower than CST, if we sweep from one extreme composition to the other, we will have to pass through the miscibility gap. Suggest an alternative path where this miscibility gap could be avoided.
(g) Explain, why certain liquid pairs show lower CST.
(h) Explain, under what circumstances, a liquid pair shows both the upper and the lower CST. What effect the pressure has on the temperature-composition curve for such a type of liquid pairs?
(i) Explain why in some cases the CST (upper or lower) is not observed experimentally.
(j) What effects impurities have on the CST?
(k) Describe with the help of appropriate diagram, the end products that are obtained when the following categories of partially miscible binary liquid mixtures are subjected to fractional distillation: (i) Partially miscible binary liquid mixtures with no azeotropic mixture and also showing incomplete separation of constituents, (ii) Partially miscible binary liquid mixtures showing an azeotropic mixture, (iii) Partially miscible binary liquid mixtures exhibiting complete separation of constituents.

4.19 Describe the theory of steam distillation.

4.20 (a) State the Nernst distribution law and show that it is in agreement with the phase rule requirements.
(b) Derive thermodynamically the distribution law. What approximations are made in deriving this law?
(c) Show that for a solute distributed in two ideally dilute solutions, its solubility in a solution is inversely proportional to its Henry's law constant in that solution.
(d) One of the basic requirements of the distribution law as suggested by Nernst is that the solute must have the same molecular form in both the layers. However, in some cases the association, dissociation, etc., of the solute takes place. Under such conditions the original definition of the distribution law can be modified in order to account for such changes. Derive the following modified forms of the distribution law:
(i) Solute having normal existence in the liquid $\alpha$ and dimerizes in the liquid $\beta$.

$$[\mathrm{A}]_\alpha/([\mathrm{A}]_{\beta,\ \mathrm{total}})^{1/2} = \text{constant}$$

(ii) Solute A having normal existence in the liquid $\alpha$ and dissociates ($\mathrm{A} \rightarrow \mathrm{B} + \mathrm{C}$) in the liquid $\beta$

$$\frac{[\mathrm{A}]_\alpha}{\xi^2/K_{\mathrm{diss}}} = \text{constant}$$

(iii) Solute having normal existence in the liquid $\alpha$ and undergoes a chemical reaction with the liquid $\beta$ of the type $\mathrm{A} + n\beta \rightleftharpoons \mathrm{A}\cdot n\beta$

$$[\mathrm{A}]_\alpha/[\mathrm{A}]_{\beta,\ \mathrm{total}} = \text{constant}$$

(d) Describe how the distribution experiments can be used to determine the equilibrium constant of each of the following reactions.

(1) $\mathrm{KI} + \mathrm{I_2} \rightleftharpoons \mathrm{KI_3}$ (2) $\mathrm{C_6H_5NH_3^+} + \mathrm{H_2O} \rightleftharpoons \mathrm{C_6H_5NH_2} + \mathrm{H_3O^+}$

(3) $\mathrm{Cu^{2+}} + 4\ \mathrm{NH_3} \rightleftharpoons \mathrm{Cu(NH_3)_4^{2+}}$

(e) What is a solvent extraction? Show mathematically that multi-stage extraction is more profitable than single-stage extraction.

## TRY YOURSELF PROBLEMS

4.1 Hydrogen sulphide obeys Henry's law. What does this indicate about $H_2S$ dissolved in water?

4.2 (a) Henry's law can be written as $p = k_H x$ and Raoult's law as

$$p = k_R x$$

In general, Henry's law constant $k_H$ is different from Raoult's law constant $k_R$. Explain why it is so. Under what conditions the two constants have the same value?

(b) Show that Bunsen absorption coefficient is related to Henry's law constant by the expression $\alpha = \rho(22\ 414\ \text{cm}^3\ \text{mol}^{-1})/k_H M_1$.

[*Hint*: $p_2 = k_H x_2 = k_H n_2/(n_1 + n_2) \simeq k_H n_2/n_1$

$= k_H\ \{v_0/(22\ 414\ \text{cm}^3\ \text{mol}^{-1})\}/(V_1\ \rho/M_1]$

Hence, $\alpha = v_0/(p_2 V_1) = \rho(22\ 414\ \text{cm}^3\ \text{mol}^{-1})/k_H M_1$.]

4.3 (a) Figure 1 shows vapour pressure versus composition plots of a binary liquid system. Answer the following:

(i) Identify the two curves with the corresponding liquids,

(ii) By drawing limiting Henry's law and Raoult's law lines, show the regions of applicability of these laws.

(iii) How will you obtain the Henry's law constants for the liquids A and B?

(b) Explain why the types of vapour pressure versus composition plots of a binary-liquid system shown in Fig. 2 are theoretically not feasible.

(c) What is the difference between ideal and ideally dilute solutions?

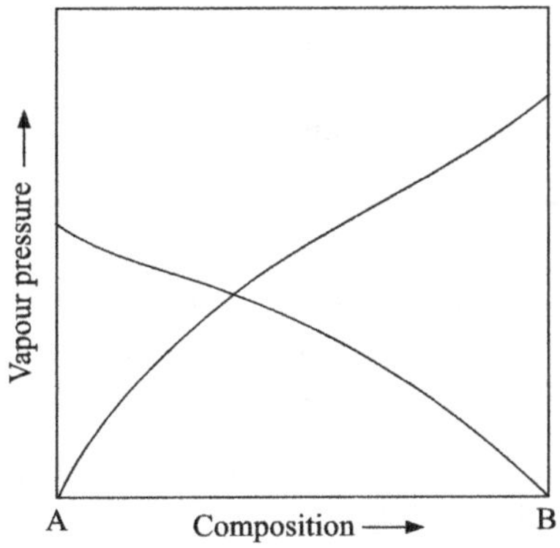


**Fig. 1**

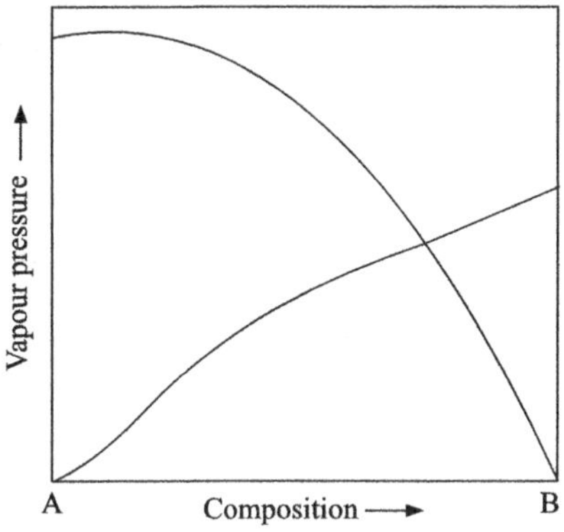


**Fig. 2**

4.4 Under what conditions are the following true?

(i) Henry's law constant = Raoult's law constant

(ii) Henry's law constant > Raoult's law constant

(iii) Henry's law constant < Raoult's law constant

4.5 For an ideal binary system, compute $\Delta_{mix}G$ and $\Delta_{mix}S$ at 298 K for $x_A = 0.1$ to 0.9 with a regular interval of 0.1 and then draw the plots of $\Delta_{mix}G$ versus $x_A$ and $\Delta_{mix}S$ versus $x_A$. Show mathematically or otherwise that $\Delta_{mix}G$ is minimum and $\Delta_{mix}S$ is maximum at $x_A = 0.5$.

4.6 Starting from the fact that for a binary system exhibiting positive deviation from Raoult's law, either solvent-solvent interactions or solute-solute interactions are strong compared to solvent-solute interactions, show that for such a system

$$\Delta_{mix}H > 0;\ \Delta_{mix}S < \Delta_{mix}S_{ideal};\ \Delta_{mix}G > \Delta_{mix}G_{ideal};\ p_A > p_{A,\text{ideal}}$$

(Hint: See Example 4.14.1.)

4.7 The chemical potential of a constituent in a nonideal system is given by $\mu_i = \mu_i^* + RT \ln a_i$, where $a_i$ is the activity of the constituent in the solution and is related to the amount fraction by the equation $a_i = x_i \gamma_i$ (where $\gamma_i$ is the activity coefficient of the constituent). For a very dilute solution, real solution behaves as an ideal solution, and, therefore, we have $a_i \to x_i$ and $\gamma_i \to 1$. Show that for a real solution

$$\Delta_{\text{mix}}G = RT \sum_i x_i \ln a_i \quad \text{and} \quad \Delta_{\text{mix}}S = -R \sum_i x_i \ln a_i$$

4.8 Suppose the distribution coefficient of $I_2$ is studied between an aqueous solution of $NH_4Cl$ and chloroform. Would you expect $K_d$ to be different or same as that between water and chloroform? (Ans. $K_d$ increases as $[I_2]_{NH_4Cl(aq)} > [I_2]_{water}$)

4.9 (a) Show that the nonideal binary liquid solutions may be classified into the following three categories based on the *K*s values as defined by Eq. (4.17.5). Designate $K$ as $K_{\text{A in B}}$, where A is solute and B is solvent, or $K_{\text{B in A}}$, where B is solute and A is solvent.

| *System* | *Ks values* |
|---|---|
| (i) Solution with a little deviation from ideality | $K_{\text{A in B}} > 1$ and $K_{\text{B in A}} < 1$ or $K_{\text{A in B}} < 1$ and $K_{\text{B in A}} > 1$ |
| (ii) Solution with large negative deviation from ideality | $K_{\text{A in B}} < 1$ and $K_{\text{B in A}} < 1$ |
| (iii) Solution with large positive deviation from ideality | $K_{\text{A in B}} > 1$ and $K_{\text{B in A}} > 1$ |

(b) Rationalize the following facts with the help of Eq. (4.17.5).
(i) The addition of ethyl bromide in ethyl iodide lowers the boiling point of the solvent whereas the addition of ethyl iodide in ethyl bromide raises the boiling point.
(ii) The addition of *n*-heptane in ethyl alcohol or ethyl alcohol in *n*-heptane lowers the boiling point of the solvent.
(iii) The addition of pyridine in acetic acid or acetic acid in pyridine raises the boiling point of the solvent.

4.10 Is it possible for one component of a binary solution to behave ideally but not the other component?

4.11 In aqueous solution of $NH_3$, the following equilibria exist:

$$NH_3 + H_2O \underset{}{\overset{K_{eq1}}{\rightleftharpoons}} NH_3 \cdot H_2O \underset{}{\overset{K_{eq2}}{\rightleftharpoons}} NH_4^+ + OH^-$$

Show that the expression for the distribution of $NH_3$ between water and chloroform is given by

$$\frac{[NH_3]_{\text{total, aq}}}{[NH_3]_{\text{org}}} = K_d(1 + K_{eq1} + K_{eq1}K_{eq2}/[OH^-])$$

For $NH_4OH$ solution, $K_{eq1} + K_{eq2}/[OH^-] << 1$. Hence show that

$$\frac{[NH_3]_{total,\,aq}}{[NH_3]_{org}} = \text{constant}$$

4.12 Show that for a given pressure, a binary mixture at the azeotropic point is an invariant system.

(*Hint:* Compositions of liquid and vapour phases are identical. Hence $F = (C - r) + 2 - P = (2 - 1) + 2 - 2 = 1$, which is the given pressure.)

## NUMERICAL PROBLEMS

**Solubility of Gases**

4.1 The Bunsen absorption coefficients of $O_2$ and $N_2$ in water at 298 K are 0.028 3 and 0.014 3 $atm^{-1}$, respectively. If air contains 20% $O_2$ and 80% $N_2$, by volume, how many cubic centimetre of gas measured at STP will be dissolved in 100 $cm^3$ of water in equilibrium with air at that temperature? How many, if pressure is 10 atm?
(Ans. 1.71 $cm^3$ and 17.1 $cm^3$)

4.2 Water saturated with air (20% $O_2$ and 80% $N_2$) at 298 K contains $8.9 \times 10^{-3}$ g $dm^{-3}$ of dissolved oxygen. Estimate the solubility of pure oxygen in water at a pressure of 25 atm and a temperature of 298 K. (Ans. 1.1125 g)

4.3 The solubility coefficient $S$ for $CO_2$ dissolved in water has the following values at different temperatures.

| $\theta_c/°C$ | 0 | 10 | 20 | 30 | 40 |
|---|---|---|---|---|---|
| $S$ | 1.713 | 1.238 | 0.943 | 0.766 | 0.608 |

Calculate the mean value of enthalpy of solution of $CO_2$ in water.
(Ans. $-18.12$ kJ $mol^{-1}$)

4.4 Given the following Bunsen absorption coefficients for CO at different temperatures, calculate the mean value of enthalpy of solution of CO in water.

| $\theta_c/°C$ | 0 | 10 | 20 | 30 | 40 |
|---|---|---|---|---|---|
| $S/atm^{-1}$ | 0.035 4 | 0.028 2 | 0.023 2 | 0.020 0 | 0.017 8 |

(Ans. $-11.87$ kJ $mol^{-1}$)

**Henry's Law**

4.5 The table below lists the vapour pressure of methyl chloride above the mixture with water at 298 K. Find Henry's law constant for methyl chloride.

| $m(CH_3Cl)/mol\ kg^{-1}$ | 0.029 | 0.051 | 0.106 | 0.131 |
|---|---|---|---|---|
| $p/mmHg$ | 205.2 | 363.2 | 756.1 | 945.9 |

(*Hint*: Convert molalities into amount fractions, plot pressure versus amount fraction and extrapolate the plot in the dilute solution region up to $x = 1$.)
(Ans. 40 000 mmHg)

4.6 At 0 °C the Henry's law constant for argon in water is $2.17 \times 10^4$ bar and at 30 °C the constant is $3.97 \times 10^4$ bar for $p = 1$ atm. Find the approximate enthalpy of solution over this temperature range. (Ans. 13.87 kJ $mol^{-1}$)

**Ideal Solution**

4.7 (a) The vapour pressures of *n*-hexane and *n*-heptane at 273 K are 45.5 and 11.4 mmHg. What is the composition of a solution of these two substances if its total vapour pressure at 273 K is 37.3 mmHg? (Ans. 0.759 5)
(b) The amount fraction of *n*-hexane in the vapour above a solution of *n*-hexane and *n*-heptane is 0.75 at 273 K. What is the composition of the liquid solution?
(Ans. 0.429)

4.8 Calculate the (i) enthalpy, (ii) entropy, (iii) volume, and (iv) Gibbs free energy of mixing of 1 mol of benzene and 2 mol of toluene at 298 K?
(Ans. (i) 0, (ii) 15.86 J $K^{-1}$, (iii) 0, and (iv) $-$ 4.732 kJ)

4.9 (a) Benzene and toluene form essentially ideal solutions, and a particular mixture which consists of two moles of benzene and three moles of toluene has a total vapour pressure of 280 mmHg at 333 K. If one additional mole of benzene is added to the solution, the new total vapour pressure is now 300 mmHg. Calculate the vapour pressures of pure benzene and toluene at 333 K?

(Ans. $p_b^* = 400$ mmHg and $p_t^* = 200$ mmHg)

4.10 At 413 K, the vapour pressure of $C_6H_5Cl$ is 939.4 Torr and that of $C_6H_5Br$ is 495.8 Torr. Assuming that these two liquids form an ideal solution, determine the composition of a mixture of the two which boils at 413 K under 1 atm pressure? What will be the composition of the vapour at this temperature?

(Ans. For $C_6H_5Cl$, $x = 0.596$, $y = 0.736$)

4.11 Ethanol and methanol form very nearly ideal solutions. The vapour pressures of ethanol and methanol are 44.5 Torr and 88.7 Torr, respectively. Calculate:
(a) the amount fraction of methanol in a solution obtained by mixing 100 g of each,
(b) the partial pressures and the total vapour pressure of the solution, and
(c) the amount fraction of methanol in the vapour.

(Ans. (a) 0.59, (b) $p_{\text{ethanol}} = 18.2$ Torr, $p_{\text{methanol}} = 52.3$ Torr, $p_{\text{total}} = 70.5$ Torr, (c) 0.741)

4.12 The normal boiling points of $C_6H_5Cl$ and $C_6H_5Br$ are 405.2 K and 429.2 K, respectively. Assuming that these two liquids form an ideal solution:
(a) Determine the composition of a mixture of the two which boils at 413 K under 1 atm pressure.
(b) What would be the corresponding composition of vapour which is liberated at 413 K?
Assume Trouton's law is applicable to both the above constituents.

(Ans. For $C_6H_5Cl$, $x = 0.606$, $y = 0.74$)

4.13 An ideal solution of two components with vapour pressures of 400 mmHg and 300 mmHg when pure contains two moles of the more volatile component and five moles of the less volatile one. Calculate:
(a) Total vapour pressure of the solution.
(b) Composition of the vapour in equilibrium with a solution of this composition.
(c) Composition of the last drop to vaporize from such a solution when none of the vapour is withdrawn from contact with the unvaporized mixture.
(d) Composition of the last drop to vaporize when the distillate from such a solution is continuously condensed elsewhere and is thus withdrawn from contact with the unvaporized mixture.

4.14 (a) Benzene and toluene form nearly an ideal solution. Their boiling points in pure state are 80.2 °C and 110 °C, respectively. The temperature-composition data for them are given below:

| | | | | | | |
|---|---|---|---|---|---|---|
| $x_{\text{benzene}}$ | 0.00 | 0.24 | 0.35 | 0.46 | 0.90 | 1.00 |
| $y_{\text{benzene}}$ | 0.00 | 0.44 | 0.62 | 0.77 | 0.95 | 1.00 |
| b.pt./°C | 110 | 101 | 93 | 88 | 83 | 82 |

(i) If a solution with 50 mol per cent in benzene is to be distilled, at what temperature will the distillation just start? What would be the composition of the distillate which just issues at the above temperature?
(ii) Suppose that the above distillate is condensed. Predict the temperature at which this liquid mixture will just start distilling and find out the corresponding composition of the vapour which issues at this temperature.

(iii) A liquid mixture containing 0.9 amount fraction of benzene is distilled without removing the distillate till it contains 0.24 amount fraction of benzene. What will be the composition of the resultant distillate?

(b) By employing Eqs. (4.12.8) and assuming Trouton's law to be applicable to both the species, calculate boiling point of the solution containing 50 mol per cent in benzene. After obtaining this value, calculate the composition of the vapour from Eq. (4.12.9).

4.15 At 363 K the vapour pressure of toluene is 400 mmHg, and that of *o*-xylene is 150 mmHg. What is the composition of a liquid mixture that will boil at 363 K when the pressure is 0.5 atm? What is the composition of the vapour produced?

(Ans. $x_{\text{toluene}} = 0.92$, $y_{\text{toluene}} = 0.97$)

4.16 Construct the temperature-composition diagram for the system ethanol-ethyl acetate from the following data available at 760 Torr external pressure.

| $x(C_2H_5OH)$ | $y(C_2H_5OH)$ | *b.pt.*/°C | $x(C_2H_5OH)$ | $y(C_2H_5OH)$ | *b.pt.*/°C |
|---|---|---|---|---|---|
| 0 | 0 | 77.15 | 0.563 | 0.507 | 72.0 |
| 0.025 | 0.070 | 76.7 | 0.710 | 0.600 | 72.8 |
| 0.100 | 0.164 | 75.0 | 0.833 | 0.735 | 74.2 |
| 0.260 | 0.295 | 72.6 | 0.942 | 0.880 | 76.4 |
| 0.360 | 0.398 | 71.8 | 0.982 | 0.965 | 77.7 |
| 0.462 | 0.462 | 71.6 | 1.000 | 1.000 | 78.3 |

From the obtained plot answer the following:

(a) What is the composition of the first vapour to come of from a solution containing 0.8 amount fraction of ethanol? What would be the boiling point of the liquid?

(b) What would be the composition of the last drop of liquid to evaporate?

**Nonideal Solution**

4.17 The following data were obtained for benzene(A)–ethanol(B) solutions at atmospheric pressure:

| | | | | | | |
|---|---|---|---|---|---|---|
| Boiling point/°C | 78 | 75 | 70 | 70 | 75 | 80 |
| $x_A$ | 0 | 0.04 | 0.21 | 0.86 | 0.96 | 1 |
| $y_A$ | 0 | 0.18 | 0.42 | 0.66 | 0.83 | 1 |

(a) Plot the temperature-composition diagram for this system and estimate the composition and boiling point of the azeotrope.

(b) For a solution containing 20 g of benzene and 50 g of ethanol, estimate the boiling point and the composition of the vapour first formed at this temperature.

(c) If the solution described in (b) is exhaustively fractionated, which component will be obtained as a pure substance? Is this component the distillate or residue? How much of this component will be obtained as a pure substance?

4.18 Given below are the partial vapour pressures of A and B in a binary liquid solutions at 30 °C.

| $x_A$ | 0 | 0.12 | 0.30 | 0.60 | 0.80 | 1.0 |
|---|---|---|---|---|---|---|
| $p_A$/Torr | 0 | 20.0 | 55.0 | 140.0 | 225.0 | 295.0 |
| $p_B$/Torr | 345.0 | 300.0 | 230.0 | 110.0 | 37.5 | 0 |

Calculate the Henry's law constants for A and B and also the activities and activity coefficients of A and B at the given compositions.

**Partial Miscible Liquids**

4.19 (a) Prepare a liquid-liquid phase diagram for the system of partially miscible liquids A and B, where the mass % of B at various temperatures are

| | | | | | |
|---|---|---|---|---|---|
| $\theta_C/°C$ | 0 | 10 | 20 | 30 | 40 |
| (mass % B)$_1$ | 30 | 37 | 45 | 53 | 64 |
| (mass % B)$_2$ | 94 | 90 | 87 | 84 | 80 |

What is the critical solution temperature?
(b) You are given a system containing 60 mass % of B at 20 °C. What is the nature of the system, whether it forms a single liquid or two liquids? If two, what are their compositions and the relative amounts? At what minimum temperature this system can be converted into the one liquid?

4.20 Methylethyl ketone and water are partially miscible. At 303 K the two layers contain 21.9 and 89.9% by mass of ketone. What will be the mass of each layer when 50 g each of methylethyl ketone and water are equilibrated at this temperature?
(Ans. 41.3 g methylethyl ketone—rich layer)

**Steam Distillation**

4.21 If the vapour pressure of water at 363 K is 526 mmHg, what mass of chlorobenzene will be distilled per gram of water collected in the distillate, if chlorobenzene is distilled at 363 K at a total pressure of 734 mmHg. (Ans. 8.876 g)

4.22 A mixture of an organic liquid A and water is distilled under one atmosphere pressure at 372.2 K. What mass of steam would be condensed to obtain 1.0 g of liquid A in the distillate? The vapour pressure of water at 372.2 K is 739 Torr; molar mass of A = 123 g mol$^{-1}$. (Ans. 5.15 g)

4.23 A mixture consisting of an organic liquid A and water (immiscible) distills at 368 K at 756 mmHg. The vapour pressure of $H_2O$ at 368 K is 634.6 Torr. Calculate the molar mass of A, if the distillate contains 65 per cent by mass of the liquid A.
(Ans. 174.8 g mol$^{-1}$)

**Distribution Law**

4.24 The following data were obtained when quinine $C_{20}H_{24}O_2N_2$ was distributed between ether and water at 273 K.

| | | | |
|---|---|---|---|
| (Water layer/g) per 100 cm$^3$ | 0.0547 | 0.059 0 | 0.062 2 |
| (Ether layer/g) per 100 cm$^3$ | 0.114 2 | 1.290 1 | 1.428 1 |

Quinine exists as a monomer in water. Determine its molecular complexity in ether.
(Ans. Dimeric)

4.25 When different amounts of phenol were equilibrated with a mixture of water (W) and chloroform (C) at 298 K the concentrations of the two solutions formed in each case had the following values.

| | | | | |
|---|---|---|---|---|
| $c_W$/mol dm$^{-3}$ | $7.85 \times 10^{-4}$ | $1.73 \times 10^{-3}$ | $2.64 \times 10^{-3}$ | $4.65 \times 10^{-3}$ |
| $c_C$/mol dm$^{-3}$ | $1.64 \times 10^{-3}$ | $2.10 \times 10^{-3}$ | $1.97 \times 10^{-2}$ | $5.77 \times 10^{-2}$ |

Phenol dissolved in water is almost completely monomeric.
(a) Is this also true of phenol dissolved in chloroform?
(b) Calculate the value of the distribution coefficient.
(Ans. (a) No, (b) $3.7 \times 10^{-4}$)

4.26 The distribution coefficient of succinic acid between water and ether is 5.7. How much ether is required to decrease the concentration of succinic acid in 50 cm$^3$ of water from 5 to 1 per cent? Suppose that the extraction had been done in 10 stages by employing the volume each time equal to the ($V$/10), where $V$ is the volume of ether obtained in the single-stage extraction. How much succinic acid remaining unextracted in water?
(Ans. 1 140 cm$^3$, 0.173%)

4.27 Solutions of ammonia in water were shaken with chloroform at 298 K. The following data were obtained when the equilibrium is reached.

| | | | |
|---|---|---|---|
| $[NH_3]_{water}$/mol dm$^{-3}$ | 0.011 0 | 0.08 5 | 0.64 2 |
| $[NH_3]_{CHCl_3}$/mol dm$^{-3}$ | 0.000 44 | 0.003 3 | 0.025 8 |

What is the distribution coefficient?

(Ans. 25.2 in favour of water)

4.28 Some iodine is dissolved in an aqueous solution of potassium iodide of concentration 0.102 mol dm$^{-3}$, and the solution is then shaken with carbon tetrachloride until equilibrium is reached (at 15 °C). The amount of iodine at equilibrium is found to be 0.048 mol dm$^{-3}$ in the aqueous layer and 0.089 mol dm$^{-3}$ in the carbon tetrachloride layer. The distribution coefficient of iodine between carbon tetrachloride and water is 85. Calculate the equilibrium constant at 15 °C for the reaction

$$I_3^-(aq) \rightleftharpoons I_2(aq) + I^-(aq)$$

(Ans. 1.23 × 10$^{-3}$ mol dm$^{-3}$)

4.29 0.062 4 mol of an amine hydrochloride is present per dm$^3$ of the aqueous solution. On shaking this solution with one dm$^3$ of benzene, 0.009 64 mol of the free base is extracted. The distribution coefficient of the free base for benzene and water is 24.1. Calculate the hydrolysis constant for the base hydrochloride.

(Ans. 7.67 × 10$^{-5}$ mol dm$^{-3}$)

4.30 An aqueous solution contains 0.20 g aspirin ($C_9H_8O_4$) in 50 cm$^3$ solution. To this solution 20 cm$^3$ of ether is added and the mixture is shaken and allowed to come to equilibrium at 298 K. At this temperature $K_d = c_{ether}/c_{water} = 4.7$. (a) How much aspirin remains in the aqueous phase? (b) If the extraction is carried out with two successive 10 cm$^3$ portions of ether, how much aspirin remains unextracted?

(Ans. (a) 0.000 38 mol, (b) 0.000 29 mol)

4.31 (a) A series of solutions of iodine in chloroform were shaken with water at 298 K until equilibrium was established. The following results were obtained.

| | | | | |
|---|---|---|---|---|
| $[I_2]_{water}$/mol dm$^{-3}$ | 0.000 25 | 0.001 20 | 0.001 84 | 0.002 42 |
| $[I_2]_{CHCl_3}$ mol dm$^{-3}$ | 0.033 8 | 0.154 6 | 0.231 8 | 0.320 7 |

What is the value of the distribution coefficient?

(b) How many times would it be necessary to equilibrate 1 dm$^3$ of a solution of iodine in water with 10 cm$^3$ samples of chloroform in order to reduce the iodine concentration to 10$^{-2}$ of an initial value?

(Ans. (a) 0.007 6, (b) six extractions)

4.32 Parke's process involves the extraction of silver from the molten argentiferrous lead with the help of molten zinc. Silver is more soluble in zinc than in lead ($K_d = [Ag]_{Zn}/[Ag]_{Pb} = 300$). Zn – Ag has a higher melting point and is thus solidified first. It is separated and subjected to distillation in a retort when Zn passes over leaving behind silver. Calculate the mass of Ag that can be separated from 1000 mL of argentiferrous lead containing 100 g of silver if 100 mL of Zn is added in (i) one lot, and (ii) two equal lots of 50 mL. (Ans. (i) 96.77 g, (ii) 99.61 g)

# 5 Phase Diagrams of One-Component Systems

## 5.1 APPLICATION OF THE PHASE RULE

According to the phase rule, the degrees of freedom of a one-component system is given by

$$F = C - P + 2 = 1 - P + 2 = 3 - P$$

Thus we have:

**System Consisting of One Phase**

The value of $F$ becomes

$$F = 3 - P = 3 - 1 = 2$$

that is, the system is bivariant. Therefore, we have to state the values of two variables in order to define the system completely. These variables are temperature and pressure. The given component may exist in any of the three phases, namely, solid, liquid, or vapour.

**System Consisting of Two Phases**

The value of $F$ in this case is

$$F = 3 - P = 3 - 2 = 1$$

and thus stating the value of either of the two variables (temperature or pressure) is sufficient to define the system completely. If temperature is stated then the other variable, namely, pressure, will have a definite value and vice versa. The two phases in equilibrium with each other may be solid-solid, solid-liquid, liquid vapour, and solid-vapour.

**System Consisting of Three Phases**

Here

$$F = 3 - P = 3 - 3 = 0$$

and thus the system is invariant, i.e. the three phases can exist in equilibrium only at definite values of temperature and pressure. The three phases in equilibrium with one another may be solid-liquid-vapour, solid-solid-liquid, and solid-solid-vapour.

It is obvious that for a one-component system, we cannot have more the three phases in equilibrium as otherwise the variance of the system would become negative.

## 5.2 QUALITATIVE DISCUSSION OF A PHASE DIAGRAM

Since the maximum degrees of freedom to define a one-component system are two, its phase diagram can be drawn in two dimensions by choosing rectangular axes representing temperature and pressure, respectively. A complete phase diagram of a one-component system will include the following aspects.

### Representation of a Single Phase

A single phase will be represented over a region (or area) since the given phase can exist under different conditions of temperature and pressure, i.e. temperature and pressure can both be varied independently over a range of values without affecting the existence of the single phase. In general, the solid phase is stable under conditions of low temperatures and high pressures, the gaseous phase at high temperatures and low pressures, and the liquid phase in between these two extreme conditions. Figure 5.2.1 shows the approximate regions of stability of the three phases.

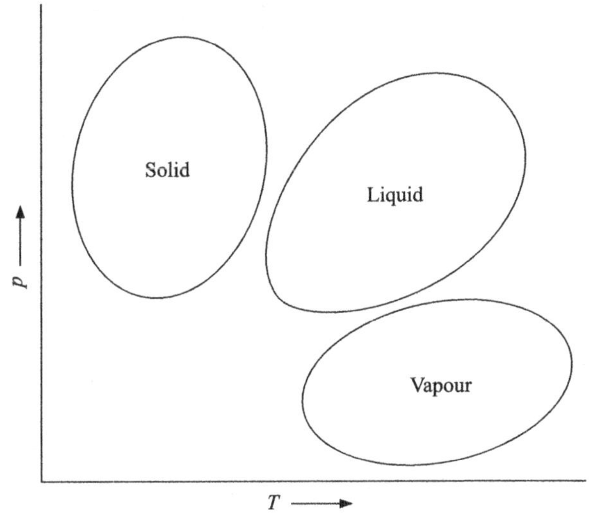


**Fig. 5.2.1 Regions of stability of the three phases**

### Representation of Two Phases in Equilibrium

Two phases in equilibrium will be represented on a line since for this type of equilibrium, one has to state only one variable (say, temperature), the other variable (say, pressure) has a definite value. Thus, if temperature is stated, the corresponding equilibrium value of pressure can be determined from the line and vice versa. The number of lines in a phase diagram will be equal to the number of two-phase equilibria which the system can possess. The slope d*p*/d*T* of these lines follows the Clapeyron equation, according to which, we have

$$\frac{dp}{dT} = \frac{\Delta_{trs}H_m}{\Delta_{trs}V_m} \qquad (5.2.1)$$

where $\Delta_{trs}H_m$ and $\Delta_{trs}V_m$ are the respective changes in the molar enthalpy and the molar volume during transformation of one phase into another at the equilibrium temperature *T*. In general, the slope d*p*/d*T* for solid to liquid transformation is much steeper than those of solid to vapour and liquid to vapour transformations, as $\Delta_{trs}V_m$ for the fusion process is usually very small. Since $\Delta_{trs}H_m$ for the fusion process is positive, the sign of the slope d*p*/d*T* of the melting point curve will depend on whether the term $\Delta_{trs}V_m$ is positive or negative. If the solid is denser than the liquid, as is normally the case, $\Delta_{trs}V_m$ is positive and the slope of the melting points curve is positive. If on the other hand, liquid is more dense than solid, then $\Delta_{trs}V_m$ is negative and so is the slope of the melting point curve. If the densities of liquid and solid are identical, the slope becomes infinite since $dp/dT$ becomes infinite in this case. These three cases along with the expected curves of solid to vapour and liquid to vapour equilibria are shown in Fig. 5.2.2.

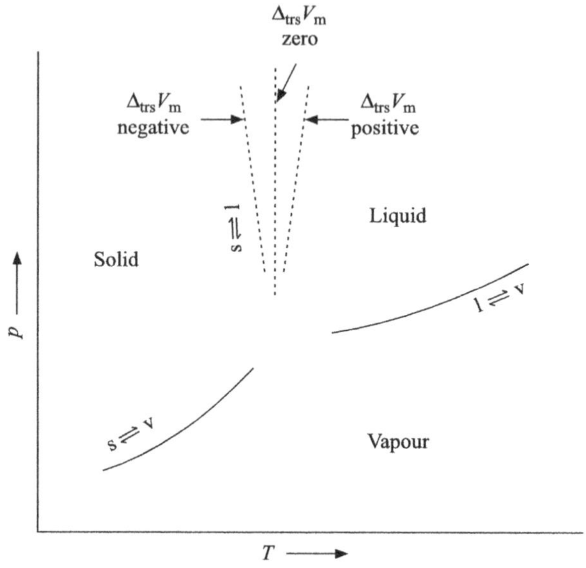


**Fig. 5.2.2 A schematic plot of three lines representing solid to liquid, solid to vapour and liquid to vapour equilibria**

## Representation of Three Phases in Equilibrium

Three phases in equilibrium will be represented by a point since both the variables (i.e. temperature and pressure) for such type of equilibrium have definite values. The three lines representing solid to liquid, solid to vapour and liquid to vapour equilibria meet one another at this point. This point is known as the *triple point of the system.* In general, we have

$$\Delta_{sub}H_m = \Delta_{fus}H_m + \Delta_{vap}H_m \tag{5.2.2}$$

where the subscripts sub, fus and vap stand respectively for sublimation, fusion and vaporization transformations. Since

$$\Delta_{sub}H_m > \Delta_{vap}H_m$$

it follows from the Clapeyron equation that at the triple point

$$\left(\frac{dp}{dT}\right)_{s \rightleftharpoons v} > \left(\frac{dp}{dT}\right)_{l \rightleftharpoons v} \tag{5.2.3}$$

Thus, at the triple point where the three lines representing solid to vapour, liquid to vapour and solid to liquid equilibria meet one another, the slope of the line representing solid to vapour transformation will be greater than that of the liquid to vapour transformation.

## 5.3 PHASE DIAGRAM OF WATER

**Experimentation**

Let an evacuated vessel fitted with a movable piston be placed in a thermostat whose temperature is well below 0 °C (say, – 20 °C). The position of the piston can be held anywhere against a set of stops. Let solid water (i.e. ice) be placed in the vessel. The behaviour of the above system at various stages of temperature and pressure as depicted in the phase diagram of water (Fig. 5.3.1) is described below.

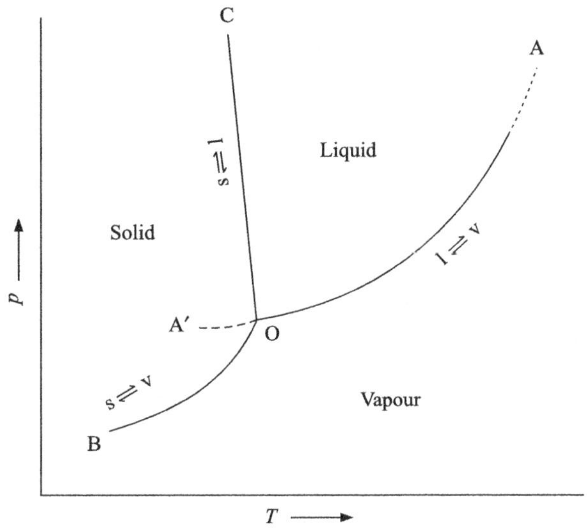


**Fig. 5.3.1 Phase diagram of water system**

**Solid-Vapour Equilibrium**

After some time when the system attains equilibrium, it is found that the pressure within the vessel attains a constant value. This pressure is the vapour pressure of solid water at – 20 °C. Let the point B in Fig. 5.3.1 represent the above system of ice in equilibrium with its vapour.

Suppose that the temperature of the system represented by the point B is raised slowly at constant volume. It is found that the vapour pressure of solid water also increases. If a plot of vapour pressure of solid water versus temperature is drawn, the curve BO as shown in Fig. 5.3.1 is obtained. The curve BO, known as the *sublimation curve,* describes the various conditions of temperature and pressure at which solid water and water vapour are in equilibrium with each other. The slope of the curve at any point as given by the Clapeyron equation is

$$\left(\frac{\mathrm{d}p}{\mathrm{d}T}\right)_{\mathrm{s}\rightleftharpoons\mathrm{v}} = \frac{\Delta_{\mathrm{sub}}H_{\mathrm{m}}}{T(V_{\mathrm{m,v}} - V_{\mathrm{m,s}})}$$

The variation of sublimation pressure with temperature (i.e. the curve BO may be represented by the Clausius-Clapeyron equation:

$$\ln\left(\frac{p}{p^{\circ}}\right) = -\frac{\Delta_{\mathrm{sub}}H_{\mathrm{m}}}{RT} + I$$

**Solid and Vapour Phases**

For a given temperature, the equilibrium between solid and vapour exists only at a specific vapour pressure which can be determined from the curve BO. Now suppose the volume of the vessel representing system at point B is increased isothermally by moving the piston in the outward direction. This, in turn will decrease the pressure of the vapour phase. Since at a given temperature vapour pressure of solid water has a fixed value, a little bit of ice will sublime to keep the vapour pressure to a constant value. If the above process of expansion at constant temperature is continued, more and more of ice sublimes till the entire solid phase disappears. At this stage, the system contains only the gaseous phase. The expansion of the latter merely causes a decrease in its pressure.

If the volume of system at point B is decreased isothermally, it will cause an increase in the pressure of the vapour phase. In order to keep the vapour pressure to a constant value, a little of vapour will condense to give solid water. If the above process of compressing the vapour phase is continued, eventually a stage is reached where the entire vapour phase is condensed and the system contains only the solid phase. The piston of the vessel will be directly in contact with the solid phase. The pressure on the solid phase can be increased by putting weights on the piston.

Thus, it may be concluded that the effect of isothermal increase and decrease of pressure on the solid-vapour equilibrium is to convert the two-phase equilibrium to that of one phase comprising of solid and vapour, respectively. Hence, regions above and below the curve BO represent solid and vapour phases, respectively.

**Melting of Solid Phase (Triple Point)**

If the temperature of the system at point B is slowly increased at constant volume, eventually a temperature is reached where the vapour pressure of solid water becomes equal to that of liquid water maintained at the same temperature. At this stage, solid water starts melting and the system contains three phases, viz., solid, liquid and vapour, in equilibrium with one another. According to the phase rule, as described earlier, the system now becomes invariant, i.e. so long as the three phases remain intact, the temperature and vapour pressure of system remain constant. The point O in Fig. 5.3.1 represents such a situation and is known as *triple point*. For water, this point lies at 0.0075 °C and 4.6 mmHg[†].

**Liquid-Vapour Equilibrium**

If the system at the triple point is slowly heated at constant volume, its temperature and vapour pressure do not change unless whole of solid is melted to give liquid. At this stage, the system contains two phases, viz., liquid and vapour, in equilibrium with each other. If now the system is gradually heated at constant volume, the temperature and the vapour pressure of the system change along the curve OA.

---

[†]Note that the pressure 4.6 mmHg is the vapour pressure of solid or liquid at temperature 0.0075 °C. Since the effect of external pressure on the vapour pressure of solid-liquid equilibrium is very small, the normal melting point of a substance is very near to its triple point and, for all practical purposes, the triple point of a substance may be referred to as its melting point.

The curve OA, known as the *vaporization curve,* represents the various conditions of temperature and pressure at which liquid and vapour are present in equilibrium with each other. The slope of the line OA at any point is given by the Clapeyron equation, such that

$$\left(\frac{dp}{dT}\right)_{l \rightleftharpoons v} = \frac{\Delta_{vap} H_m}{T(V_{m,v} - V_{m,l})}$$

The variation of vapour pressure of liquid phase with temperature (i.e. the curve OA) may be represented by the Clausius-Clapeyron equation:

$$\ln\left(\frac{p}{p^\circ}\right) = -\frac{\Delta_{vap} H_m}{RT} + I$$

where $I$ is the constant of integration. The curve OA has an upper limit at point A which represents the critical temperature of water. In accordance with Eq. (5.2.3), the slope of the line BO (representing solid $\rightleftharpoons$ vapour) at the triple point is larger than that of line OA (representing liquid $\rightleftharpoons$ vapour).

**Liquid and Vapour Phases**

For a given temperature, the equilibrium between liquid and vapour will exist only at a specific vapour pressure which can be determined from the curve OA. Now if the volume of a system anywhere on the line OA is increased at constant temperature by moving the piston in the outward direction, it will cause a decrease in the pressure of vapour phase. Since at a given temperature vapour pressure of liquid water has a fixed value, a little bit of water will evaporate to keep the vapour pressure to a constant value. If the above process of expansion is continued, more and more of liquid water evaporates till the entire liquid phase disappears. At this stage, the system contains only the gaseous phase. The expansion of the latter merely causes a decrease in its pressure. Similarly, if the volume of the system is gradually decreased, more and more of vapour is condensed in order to keep the vapour pressure of liquid water to a constant value. If the compression is continued, eventually a stage is reached where the entire vapour phase is condensed and the system contains only liquid phase. The piston of the vessel will be directly in contact with the liquid phase. The pressure on the liquid phase can be increased by applying more pressure on the piston.

Thus, it may be concluded that the regions above and below the curve OA represent liquid and vapour phases, respectively.

**Solid-Liquid Equilibrium**

If the system containing only solid phase at some high pressure is gradually heated, its temperature rises till a stage is reached where it starts melting. At this stage, temperature of the system remains constant till whole of the solid phase is converted into the liquid phase. The temperature at which solid melts depends on the pressure on the solid phase. The line OC in Fig. 5.3.1 represents the various conditions of temperature and pressure at which solid and liquid are in equilibrium with each other. The slope of this line follows the Clapeyron equation, such that

$$\left(\frac{\mathrm{d}p}{\mathrm{d}T}\right)_{\mathrm{s}\rightleftharpoons \mathrm{l}} = \frac{\Delta_{\mathrm{fus}}H_{\mathrm{m}}}{T(V_{\mathrm{m,l}} - V_{\mathrm{m,s}})}$$

Since $V_{\mathrm{m,l}} - V_{\mathrm{m,s}}$ is small, the slope of the line OC has a comparatively high value. Consequently, the line OC is nearly vertical. Since for water, $V_{\mathrm{m,s}} > V_{\mathrm{m,l}}$ the slope $\mathrm{d}p/\mathrm{d}T$ is negative and hence the line OC is slightly titled towards the left.

**Metastable Equilibrium**

If the system on the line AO is rapidly cooled, it may happen that ice fails to appear at the triple point and the vapour pressure of liquid continues to move along OA′ which now represents metastable equilibrium involving liquid and vapour phases. Similarly, if the system on the line BO is rapidly heated, liquid may not appear at the triple point and the vapour pressure of solid continues along the continuation of the curve BO where metastable equilibrium between solid and vapour phases exists.

**Comparison of Vapour Pressure of Stable and Metastable Phases**

Since at the triple point, the slope of the solid-vapour curve is greater than that of the liquid-vapour curve, the continuation of each of these curves beyond the triple point places them above the curve for the stable phase in that temperature interval. Consequently, the vapour pressure of the system in the metastable region is greater than that in the stable system at the same temperature. This generalization can also be derived thermodynamically as follows.

Since the transformation

Metastable phase → Stable phase

is a spontaneous process, we must have

$$\Delta G = \mu_{\mathrm{s}} - \mu_{\mathrm{ms}} = -\mathrm{ve}$$

where $\mu_{\mathrm{s}}$ and $\mu_{\mathrm{ms}}$ are the chemical potentials of stable and metastable phases, respectively. Since in both stable and metastable regions, the substance is in equilibrium with its vapour, we will have

$$\mu_{\mathrm{s}} = \mu_{\mathrm{vs}} = \mu_{\mathrm{v}}^{\circ} + RT \ln p_{\mathrm{s}}$$

$$\mu_{\mathrm{ms}} = \mu_{\mathrm{vm}} = \mu_{\mathrm{v}}^{\circ} + RT \ln p_{\mathrm{ms}}$$

where $\mu_{\mathrm{vs}}$ and $\mu_{\mathrm{vm}}$ are the chemical potentials of the vapour phase in stable and metastable regions, respectively. Hence

$$\Delta G = \mu_{\mathrm{s}} - \mu_{\mathrm{ms}} = RT \ln \frac{p_{\mathrm{s}}}{p_{\mathrm{ms}}} \tag{5.3.1}$$

Since $\Delta G$ is negative, it follows that

$$p_{\mathrm{ms}} > p_{\mathrm{s}}$$

that is, the vapour pressure of metastable phase is larger than that of the stable phase at the same temperature.

**Labelling of the Phase Diagram**

Table 5.3.1 describes, in brief, the phase diagram of water (Fig. 5.3.1) in the light of above analysis.

**Table 5.3.1** Description of the Phase Diagram of Water (Fig. 5.3.1)

| *Location of the system* | *Status of the system* | *Degrees of freedom* | |
|---|---|---|---|
| | | *Number* | *Kind* |
| BO | Sublimation curve | 1 | $T$ or $p$ |
| O | Triple point where solid ⇌ liquid ⇌ vapour | 0 | |
| OA | Vaporization curve | 1 | $T$ or $p$ |
| OC | Fusion curve | 1 | $T$ or $p$ |
| Area above AO and to the right of OC | Liquid phase | 2 | $T$ and $p$ |
| Area above OB and to the left of OC | Solid phase | 2 | $T$ and $p$ |
| Area, below the curve AOB | Vapour phase | 2 | $T$ and $p$ |
| OA′ | Metastable equilibrium liquid ⇌ vaour | 1 | $T$ or $p$ |

***Example 5.3.1***

Suppose solid water-liquid water equilibrium is maintained at triple point temperature (0.007 5 °C) and triple point pressure (4.6 mmHg). Calculate how much change in temperature be made so that the equilibrium is maintained at an external pressure of 1 atm. Given: $\rho$(ice) = 0.917 g cm$^{-3}$, $\rho$(liquid) = 0.999 87 g cm$^{-3}$ and $\Delta_{\text{fus}}H_{\text{m}}$ = 6 008.5 J mol$^{-1}$.

***Solution***

The temperature at which solid-liquid equilibrium is maintained, when the external pressure is raised from 4.6 mmHg to 1 atm, can be calculated by using the Clapeyron equation:

$$\frac{\mathrm{d}p}{\mathrm{d}T} = \frac{\Delta_{\text{fus}}H_{\text{m}}}{T(V_{\text{m,l}} - V_{\text{m,s}})}$$

Now
$$V_{\text{m,l}} - V_{\text{m,s}} = \frac{18\ \text{g mol}^{-1}}{0.999\ 87\ \text{g cm}^{-3}} - \frac{18\ \text{g mol}^{-1}}{0.917\ \text{g cm}^{-3}}$$
$$= 18.002\ \text{cm}^3\ \text{mol}^{-1} - 19.629\ \text{cm}^3\ \text{mol}^{-1}$$
$$= -1.627\ \text{cm}^3\ \text{mol}^{-1} = -1.627 \times 10^{-6}\ \text{m}^3\ \text{mol}^{-1}$$

Thus
$$\frac{\mathrm{d}p}{\mathrm{d}T} = \frac{6\ 008.5\ \text{J mol}^{-1}}{(273.15\ \text{K})(-1.627 \times 10^{-6}\ \text{m}^3\ \text{mol}^{-1})} = -1.352 \times 10^7\ \text{Pa K}^{-1}$$

or
$$\mathrm{d}T = -\frac{\mathrm{d}p}{1.352 \times 10^7\ \text{Pa K}^{-1}}$$

Now
$$\mathrm{d}p = 760\ \text{mmHg} - 4.6\ \text{mmHg} = 755.4\ \text{mmHg}$$
$$= \left(\frac{755.4}{760} \times 101.325 \times 10^3\ \text{Pa}\right) = 1.007 \times 10^5\ \text{Pa}$$

Hence,
$$\mathrm{d}T = -\frac{1.007 \times 10^5\ \text{Pa}}{1.352 \times 10^7\ \text{Pa K}^{-1}} = -0.007\ 5\ \text{K} \equiv -0.007\ 5\ °\text{C}$$

Thus, ice will melt at 0 °C (triple point temperature 0.007 5 °C) when the external pressure on it is 1 atm. This external pressure may be due to the atmospheric pressure. Thus, the normal melting point of ice is 0 °C.

***Example 5.3.2***

Suppose liquid water-water vapour equilibrium is maintained at triple point temperature and atmospheric pressure of 1 atm. Calculate the vapour pressure of water. Given: Triple point pressure = 4.6 mmHg.

***Solution***

In the absence of atmospheric gases, the vapour pressure of liquid water ($p^*$) would have been 4.6 mmHg. In the presence of air at atmospheric pressure ($P_2$), the vapour pressure of liquid water ($p_2$) can be calculated by using the expression

$$RT \ln \frac{p}{p^*} = V_{m,1}(P_2 - p^*) \tag{1}$$

The molar volume of liquid water is

$$V_{m,1} = \frac{M}{\rho} = \frac{18 \text{ g mol}^{-1}}{1 \text{ g cm}^{-3}} = 18 \text{ cm}^3 \text{ mol}^{-1} = 18 \times 10^{-6} \text{ m}^3 \text{ mol}^{-1}$$

The value of $p^*$ is

$$p^* = \left(\frac{4.6 \text{ mmHg}}{760 \text{ mmHg}}\right)(101\ 325 \text{ Pa}) = 613.28 \text{ Pa}$$

Substituting the values of $V_{m,1}$ and $p^*$ in Eq. (1), we get

$$(8.314 \text{ J K}^{-1} \text{ mol}^{-1})(273.15 \text{ K}) \ln \left(\frac{p_2}{4.6 \text{ mmHg}}\right)$$
$$= (18 \times 10^{-6} \text{ m}^3 \text{ mol}^{-1})(101\ 325 \text{ Pa} - 613.28 \text{ Pa})$$

or $$\ln \left(\frac{p_2}{4.6 \text{ mmHg}}\right) = \frac{(18 \times 10^{-6})(100\ 712)}{(8.314)(273.15)} = 7.98 \times 10^{-4}$$

or $$\frac{p_2}{4.6 \text{ mmHg}} = 1.0008 \quad \text{or} \quad p_2 = 4.604 \text{ mmHg}$$

## 5.4 POLYMORPHISM

Sometimes a substance can exist in more than one crystalline form and each form has its own characteristic vapour pressure curve. Such type of substances are said to exhibit the phenomenon of *polymorphism.* Two types of polymorphism are observed depending upon whether the change from one polymorphic form to the other can be carried out in either direction or only in one direction. The former is known as *enantiotropy* (Greek: opposite change) and the latter as *monotropy* (Greek: one change).

### ENANTIOTROPY

Two crystalline forms can be interconvertible and this conversion takes place at a definite temperature for a given external pressure. This temperature is the *transition point* and it is the only temperature at which the two solids can coexist in equilibrium at a given pressure. If the temperature of the system is greater than this temperature, there occurs a complete transformation of one form into the other, one being stable below the transition point and the other above it. Suppose, the two enantiotropic modifications are represented by α and β, then we have;

$$\alpha \underset{t < t_i}{\overset{t > t_i}{\rightleftharpoons}} \beta$$

where α-form is stable at lower temperatures while β-form is stable at higher temperatures.

Each polymorphic form has its own vapour pressure curve. The two vapour pressure curves meet each other at a common point as shown in Fig. 5.4.1. The curve AB is vapour pressure curve of the α-form and BC is that of the β-form. The curve AB comes before the curve BC as α-form is stable at lower temperatures. The point B is the intersection point and is the transition point where both α- and β-forms are at equilibrium with the vapour. Thus, if α-form is heated very slowly so that the equilibrium between solid and vapour is always established, the vapour pressure of the system will rise along the curve AB. When the point B is reached, the α-form is transformed into the β-form without any increase in temperature and vapour pressure. Heat supplied to the system is utilized in carrying out the transformation from the α-form to the β form. Once the transformation is complete, the vapour pressure of the system changes along the curve BC.

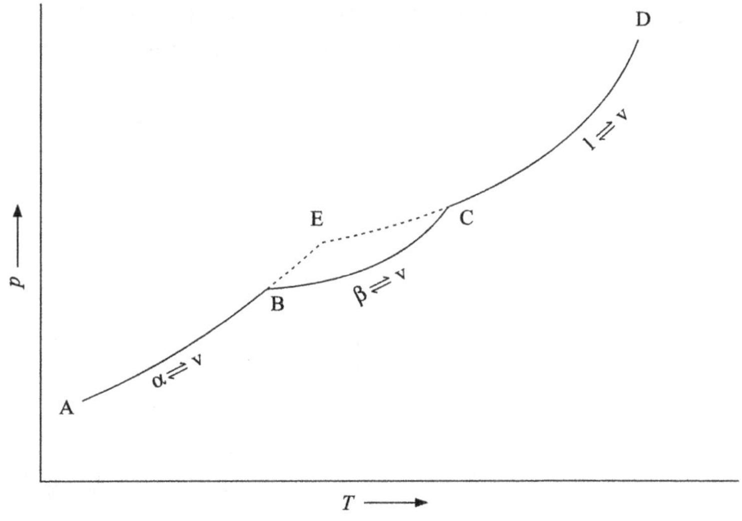


**Fig. 5.4.1 The phenomenon of enantiotropy**

Since at the triple point,

$$\Delta_{\text{sub}}H_{\text{m}}(\text{solid }\alpha) = \Delta_{\text{trs}}H_{\text{m}}(\text{solid }\alpha - \text{solid }\beta) + \Delta_{\text{sub}}H_{\text{m}}(\text{solid }\beta)$$

it follows that

$$\Delta_{\text{sub}}H_{\text{m}}(\text{solid }\alpha) > \Delta_{\text{sub}}H_{\text{m}}(\text{solid }\beta)$$

A direct consequence of the above relation is that the slope of solid α-vapour curve at the triple point is greater than that of solid β-vapour curve.

Consequently, the metastable extensions of the solid α-vapour and the solid β-vapour curves obey Eq. (5.3.1), according to which, the vapour pressure of component in the metastable form is larger than that in the stable form at same temperature.

Point C is the melting point of the β-form and thus the curve CD is the vaporization curve. If the heating of the α-form is not done slowly, it may be observed that at B no β-form is formed and the vapour pressure of the system continues along BE. Along this line, we have metastable equilibrium between α-form and its vapour. Similarly, if the liquid is cooled rapidly, then it may be observed that at C no β-form separates and the cooling of the liquid continues along the line CE, which now represents a metastable equilibrium between liquid and vapour. The two metastable curves meet each other at the point E, which represents the melting point of α-form.

Examples of substances exhibiting enantiotropy are sulphur, tin, ammonium nitrate, mercury iodide, carbon tetrachloride, and solid water(ice).

## Phase Diagram of Sulphur

Sulphur exists in two enantiotropic forms, namely, rhombic and monoclinic; the former being stable at lower temperatures. To describe the phase diagram shown in Fig. 5.4.2, we start with rhombic sulphur placed in an evacuated vessel maintained at some low temperature. The behaviour of the system at various stages of temperature and pressure is described below.

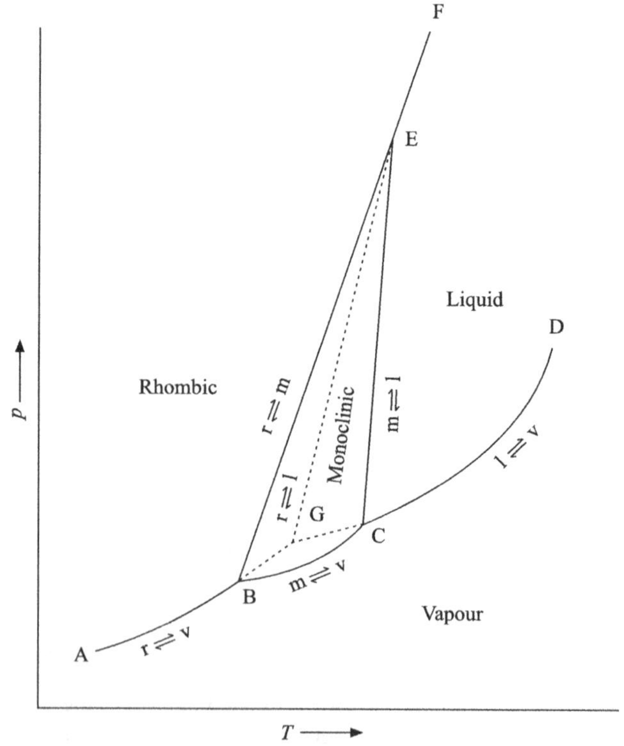


**Fig. 5.4.2 Phase diagram of sulphur**

**Rhombic Sulphur-Vapour Equilibrium**

After some time when the system attains equilibrium, the vapour pressure will attain a fixed value. Let the point A in Fig. 5.4.2 represent this state of the system.

Now suppose, the temperature of the system represented by the point A is raised at constant volume. The vapour pressure of rhombic sulphur is also increased. If a plot of vapour pressure versus temperature is drawn, we get the curve AB as shown in Fig. 5.4.2. The curve AB, known as *sublimation curve of rhombic sulphur,* describes the various conditions of temperature and pressure at which rhombic sulphur exists in equilibrium with its vapour.

If a system on the curve AB is subjected to isothermal expansion, more and more of solid sulphur is converted into sulphur vapour to keep the vapour pressure of rhombic sulphur to a constant value. Continued expansion will ultimately lead to the vapour phase only. Similarly, if the original system is subjected to isothermal compression, more and more of vapour phase is condensed to give rhombic sulphur. Continued compression will ultimately lead to only rhombic sulphur. Hence, the phases above and below the curve AB comprise of solid rhombic sulphur and vapour sulphur, respectively.

**Monoclinic Sulphur-Vapour Equilibrium**

If a system on AB is gradually heated at constant volume, eventually a stage is reached where the vapour pressure of rhombic sulphur becomes equal to that of monoclinic sulphur. Consequently, rhombic sulphur starts converting into monoclinic sulphur and the system contains three phases, viz., rhombic, monoclinic and vapour, in equilibrium with one another. According to the phase rule, the system becomes invariant ($F = C + 2 - P = 1 + 2 - 3 = 0$) and thus temperature and vapour pressure of the system remain constant so long as three phases remain intact. The point B in Fig. 5.4.2 represents such a situation which lies at 95.5 °C and 0.01 Torr.

If the system at point B is slowly heated at constant volume, its temperature and pressure do not change unless whole of rhombic sulphur is converted into monoclinic sulphur. At the latter stage, the system contains two phases, viz., monoclinic sulphur and vapour, in equilibrium with each other and thus the system becomes univariant ($F = C + 2 - P = 1 + 2 - 2 = 1$). If the heating is continued at constant volume, the temperature and vapour pressure of monoclinic sulphur change along the curve BC.

The slopes of curves AB and BC at point B are

$$\left(\frac{\mathrm{d}p}{\mathrm{d}T}\right)_{\mathrm{rhom} \rightleftharpoons \mathrm{v}} = \frac{\Delta_{\mathrm{sub}} H_{\mathrm{m, rhom}}}{T(V_{\mathrm{m, v}} - V_{\mathrm{m, rhom}})}$$

$$\left(\frac{\mathrm{d}p}{\mathrm{d}T}\right)_{\mathrm{mono} \rightleftharpoons \mathrm{v}} = \frac{\Delta_{\mathrm{sub}} H_{\mathrm{m, mono}}}{T(V_{\mathrm{m, v}} - V_{\mathrm{m, mono}})}$$

Now since $\Delta_{\mathrm{sub}}H_{\mathrm{m, rhom}} = \Delta_{\mathrm{trs}}H_{\mathrm{m, rhom} \to \mathrm{mono}} + \Delta_{\mathrm{sub}}H_{\mathrm{m, mono}}$, it follows that

$$\Delta_{\mathrm{sub}} H_{\mathrm{m, rhom}} > \Delta_{\mathrm{sub}} H_{\mathrm{m, mono}}$$

Hence $$\left(\frac{\mathrm{d}p}{\mathrm{d}T}\right)_{\mathrm{rhom} \rightleftharpoons \mathrm{v}} > \left(\frac{\mathrm{d}p}{\mathrm{d}T}\right)_{\mathrm{mono} \rightleftharpoons \mathrm{v}}$$

that is, the slope of the curve AB at point B is larger than that of the curve BC.

If a system on the curve BC is subjected to isothermal expansion and compression, it will eventually be converted into the system of one phase comprising of vapour and monoclinic sulphur, respectively. Thus, the system above and below the curve BC consists of solid monoclinic sulphur and vapour sulphur, respectively.

**Liquid-Vapour Equilibrium**

At point C, the vapour pressure of monoclinic sulphur becomes equal to that of liquid sulphur. Consequently, at this point three phases, viz., monoclinic sulphur, liquid sulphur and vapour sulphur, exist in equilibrium with one another. This point lies at 119.2 °C and 0.025 Torr.

If the system at the point C is heated gradually at constant volume, more and more of monoclinic sulphur converts into liquid sulphur. The temperature and vapour pressure of the system remain constant until whole of solid monoclinic sulphur is converted into liquid sulphur. If the heating is continued, the temperature and vapour pressure of liquid sulphur move along the curve CD.

At point C, the slope of the curve BC is larger than that of the curve CD. This is in accordance with Eq. (5.2.3).

If a system on the curve CD is subjected to isothermal expansion and compression, it would eventually lead to a system of one phase comprising vapour (below the curve CD) and liquid sulphur (above the curve CD), respectively.

**Rhombic-Monoclinic Sulphur in Equilibrium**

If a system of solid rhombic sulphur at some high pressure is heated slowly, its temperature increases till a stage is reached where it starts converting into monoclinic sulphur. At this stage, temperature of the system remains constant until whole of rhombic sulphur is converted into monoclinic sulphur. The temperature at which this transition takes place depends on the pressure of the system. The line BE in Fig. 5.4.2 represents the various conditions of temperature and pressure at which the above transition takes place. The slope of this line is positive since monoclinic sulphur is less dense than rhombic sulphur. On the left of the line BE, rhombic sulphur exists whereas to its right only monoclinic sulphur exists.

**Monoclinic Sulphur-Liquid in Equilibrium**

If the heating of monoclinic sulphur is continued, a stage is reached where it starts melting. This happens when the state point moves on the line CE. The temperature of the system remains constant until whole of monoclinic sulphur melts to give liquid sulphur. Thus, on the left and right of the line CE, the system consists of monoclinic sulphur and liquid sulphur, respectively.

The slope of the line CE is larger than that of BE. Consequently, these two lines meet each other at point E, which represents a triple point where rhombic, monoclinic and liquid sulphur are in equilibrium with one another. This point lies at 151°C and 1 290 atm.

**Rhombic Sulphur-Liquid Phases**

If the pressure of the system is higher than the triple point pressure at point E, then rhombic sulphur passes directly into liquid sulphur along the line EF.

**Metastable Equilibria**

If a system on the line AB is heated rapidly, it may happen that monoclinic sulphur does not appear at point B but the vapour pressure of the system continues along the line BG, which now represents metastable equilibrium between rhombic sulphur and vapour. Similarly, if a system on DC is cooled rapidly, monoclinic sulphur might not appear at point C and the system continues to move along CG, which represents metastable equilibrium between liquid sulphur and vapour. The point G where the curves BG and CG meet each other is another triple point (metastable) where rhombic, liquid and vapour sulphur are in equilibrium with one another. This point lies at 114.5 °C and 0.03 Torr.

Similarly, if a system of rhombic sulphur at some high pressure is heated rapidly, transition to monoclinic sulphur might not occur on the line BE but the system might pass over directly to liquid phase. The line GE in Fig. 5.4.2 represents such a type of metastable equilibrium.

It may be pointed out here that the stable and metastable transitions are to be discussed separately. For example, to the left and right of the metastable equilibrium line GE, rhombic and liquid sulphur exist whereas these regions actually are of monoclinic sulphur. The equilibrium involving rhombic and liquid sulphur will exist provided either rhombic sulphur does not pass over to monoclinic sulphur on the line BE or liquid sulphur does not pass over to the monoclinic sulphur on the line CE. If these stable transitions are observed, then the metastable transition involving rhombic to liquid sulphur will not have any existence. Metastable phases to the left and right of the line GE will have a tendency to pass spontaneously over to the stable phase, i.e. monoclinic sulphur.

**Labelling of the Phase Diagram**

Table 5.4.1 describes, in brief, the phase diagram of sulphur (Fig. 5.4.2) in the light of the above analysis.

## MONOTROPY

In the phenomenon of monotropy, one of the crystalline forms is stable while the other one is metastable over the whole range of its existence. In this case, transformation from one form into another can take place only in one direction; from metastable to stable form. The vapour pressure curve for the metastable form lies above that of the stable form as shown in Fig. 5.4.3. The curve AB is

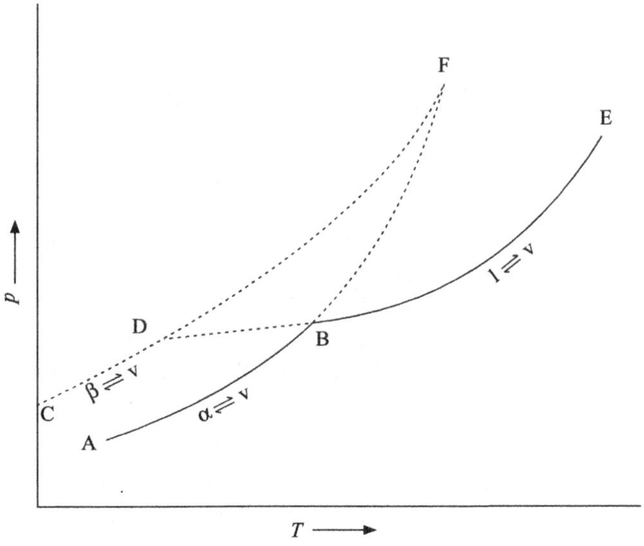


**Fig. 5.4.3 The phenomenon of monotropy**

**Table 5.4.1** Description of the Phase Diagram of Sulphur (Fig. 5.4.2)

| *Location of the system* | *Status of the system* | *Degrees of freedom* |
|---|---|---|
| AB | Sublimation curve of rhombic sulphur | 1 |
| B | Triple point where $\begin{pmatrix}\text{rhombic}\\ \text{sulphur}\end{pmatrix} \rightleftharpoons \begin{pmatrix}\text{monoclinic}\\ \text{sulphur}\end{pmatrix} \rightleftharpoons \begin{pmatrix}\text{vapour}\\ \text{sulphur}\end{pmatrix}$ | 0 |
| BC | Sublimation curve of monoclinic sulphur | 1 |
| C | Triple point where $\begin{pmatrix}\text{monoclinic}\\ \text{sulphur}\end{pmatrix} \rightleftharpoons \begin{pmatrix}\text{liquid}\\ \text{sulphur}\end{pmatrix} \rightleftharpoons \begin{pmatrix}\text{vapour}\\ \text{sulphur}\end{pmatrix}$ | 0 |
| CD | Vaporization curve of liquid sulphur | 1 |
| BE | Transition line of rhombic sulphur to monoclinic sulphur | 1 |
| CE | Fusion line of monoclinic sulphur to liquid sulphur | 1 |
| E | Triple point where $\begin{pmatrix}\text{rhombic}\\ \text{sulphur}\end{pmatrix} \rightleftharpoons \begin{pmatrix}\text{monoclinic}\\ \text{sulphur}\end{pmatrix} \rightleftharpoons \begin{pmatrix}\text{liquid}\\ \text{sulphur}\end{pmatrix}$ | 0 |
| EF | Fusion line of rhombic sulphur to liquid sulphur | 1 |
| BG | Metastable sublimative curve of rhombic sulphur | 1 |
| CG | Metastable varporization curve of liquid sulphur | 1 |
| G | Metastable triple point where $\begin{pmatrix}\text{rhombic}\\ \text{sulphur}\end{pmatrix} \rightleftharpoons \begin{pmatrix}\text{liquid}\\ \text{sulphur}\end{pmatrix} \rightleftharpoons \begin{pmatrix}\text{vapour}\\ \text{sulphur}\end{pmatrix}$ | 0 |
| GE | Metastable fusion line of rhombic sulphur to liquid sulphur | 1 |
| Area to the left of ABF | Rhombic sulphur | 2 |
| Area above CD and to the right of CEF | Liquid sulphur | 2 |
| Area BCEB | Monoclinic sulphur | 2 |
| Area below ABCD | Vapour sulphur | 2 |
| Area BGEB | Metastable rhombic sulphur | 2 |
| Area CGEC | Metastable liquid sulphur | 2 |

the vapour pressure curve of the stable α-form and the curve CD is that of the metastable β-form. BE is the vaporization curve. The point B is the stable triple point where we have

$$\alpha\text{-form} \rightleftharpoons \text{liquid} \rightleftharpoons \text{vapour}$$

The point D is the melting point of the β-form which can be obtained by extending the line EB up to the point D where it meets the curve CD. The system will have a hypothetical transition temperature which can be obtained by extending the lines AB and CD up to the point F where the two curves meet each other. The hypothetical transition temperature F has no real existence as it is above the melting points of the two modifications and the two solids do not exist.

In fact, points D and F are the metastable triple points which respectively represent

solid β ⇌ liquid ⇌ vapour

and solid β ⇌ solid α ⇌ vapour

Since the effect of pressure on the melting point and the transition point is usually very small, the two points D and F may be referred to as the melting point and the transition point, respectively. Similarly the point B may be referred to as the melting point of the α-form.

Since the β-form has higher vapour pressure, it is metastable with respect to α-form and thus β-form will pass over to the α-form and the reverse of this process (i.e. α-form to β-form) will not take place. The β-form, however, can be obtained from the α-form by an indirect method. First of all, it may be heated to get a liquid and then cooling the liquid rapidly will move the system along the line BD instead of BA and at D, the β-form will separate.

Examples of substances exhibiting monotropy are phosphorus, silica, iodine monochloride and benzophenone.

**Example Exhibiting both Monotropy and Enantiotropy**

Some substances show both enantiotropy and monotropy. A typical example which exhibits this behaviour is phosphorus, where we have

*Monotropy transition* White P(both α- and β-forms) → Violet P

*Enantiotropic transition* White P(β-form) → White P(α-form)

## Phase Diagram of Phosphorus

The phase diagram of phosphorus is shown in Fig. 5.4.4. Its description is given in Table 5.4.2. The vapour pressure curves of unstable phases are shown by dotted lines.

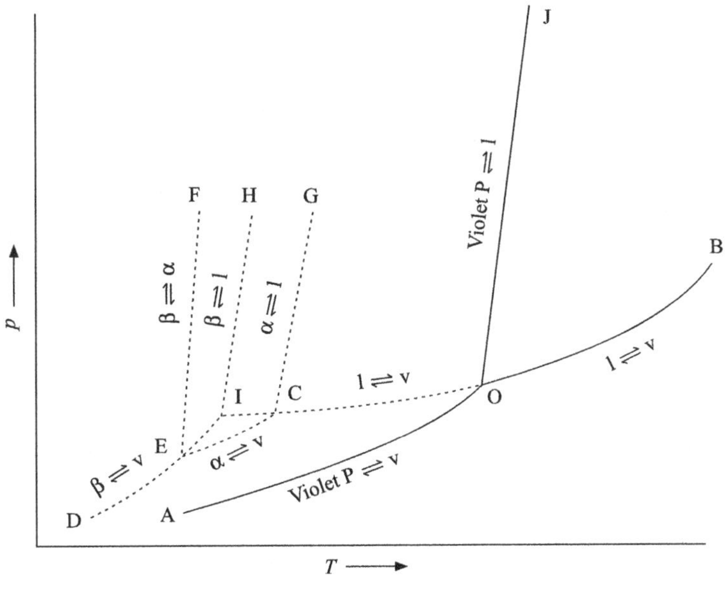


**Fig. 5.4.4 Phase diagram of phosphorus**

**Table 5.4.2** Description of the Phase Diagram Shown in Fig. 5.4.4.

| *Location of the system* | *Status of the System* |
|---|---|
| AO | Sublimation curve of stable solid violet phosphorus |
| OB | Vaporization curve of liquid phosphorus |
| OJ | Fusion point curve of violet phosphorus |
| O | Triple point where solid violet P $\rightleftharpoons$ liquid P $\rightleftharpoons$ vapour P |
| OCI | metastable vaporization curve of liquid phosphorus |
| DE | Sublimation curve of unstable white (β-form) phosphorus |
| EC | Sublimation curve of unstable white (α-form) phosphorus |
| EF | Enantiotropic transition point curve involving unstable white P(α-form) and unstable white P(β-form) |
| CG | Fusion point curve of unstable white P(α-form) |
| C | Metastable triple point where solid white P(α-from) $\rightleftharpoons$ liquid P $\rightleftharpoons$ vapour P |
| E | Metastable triple point, where solid white P(β-form) $\rightleftharpoons$ solid white P(α-form) $\rightleftharpoons$ vapour P |
| EI | Metastable sublimation curve of white P(β-form) |
| I | Metastable triple point, where solid white P(β-form) $\rightleftharpoons$ liquid P $\rightleftharpoons$ vapour P |
| IH | Metastable fusion point curve of white P(β-form) |

***Example 5.4.1***

Element X exists in three solid allotropic modifications, known respectively, as 1, 2 and 3. In the vicinity of the 1-2-3 triple point,

$$S_{\mathrm{m,\,2}} > S_{\mathrm{m,\,3}} > S_{\mathrm{m,\,1}} \quad \text{and} \quad V_{\mathrm{m,\,3}} > V_{\mathrm{m,\,1}} > V_{\mathrm{m,\,2}}$$

where $S$ and $V$ stand for entropy and volume, respectively. Sketch an arrangement of the three two-phase equilibrium lines around the 1-2-3 triple point on a pressure-temperature diagram consistent with the given information. Label all the lines and areas.

***Solution***

From the given information, we have

$$\left(\frac{\mathrm{d}p}{\mathrm{d}T}\right)_{1\rightleftharpoons 2} = \left(\frac{\Delta S_{\mathrm{m}}}{\Delta V_{\mathrm{m}}}\right)_{1\rightleftharpoons 2} = \frac{S_{\mathrm{m,\,2}} - S_{\mathrm{m,\,1}}}{V_{\mathrm{m,\,2}} - V_{\mathrm{m,\,1}}} = \frac{+\mathrm{ve}}{-\mathrm{ve}} = -\ \mathrm{ve}$$

$$\left(\frac{\mathrm{d}p}{\mathrm{d}T}\right)_{1\rightleftharpoons 3} = \left(\frac{\Delta S_{\mathrm{m}}}{\Delta V_{\mathrm{m}}}\right)_{1\rightleftharpoons 3} = \frac{S_{\mathrm{m,\,3}} - S_{\mathrm{m,\,1}}}{V_{\mathrm{m,\,3}} - V_{\mathrm{m,\,1}}} = \frac{+\mathrm{ve}}{+\mathrm{ve}} = +\ \mathrm{ve}$$

$$\left(\frac{\mathrm{d}p}{\mathrm{d}T}\right)_{2\rightleftharpoons 3} = \left(\frac{\Delta S_{\mathrm{m}}}{\Delta V_{\mathrm{m}}}\right)_{2\rightleftharpoons 3} = \frac{S_{\mathrm{m,\,3}} - S_{\mathrm{m,\,2}}}{V_{\mathrm{m,\,3}} - V_{\mathrm{m,\,2}}} = \frac{-\mathrm{ve}}{+\mathrm{ve}} = -\ \mathrm{ve}$$

The possible phase diagram (along with other information) consistent with the information is given in Fig. 5.4.5

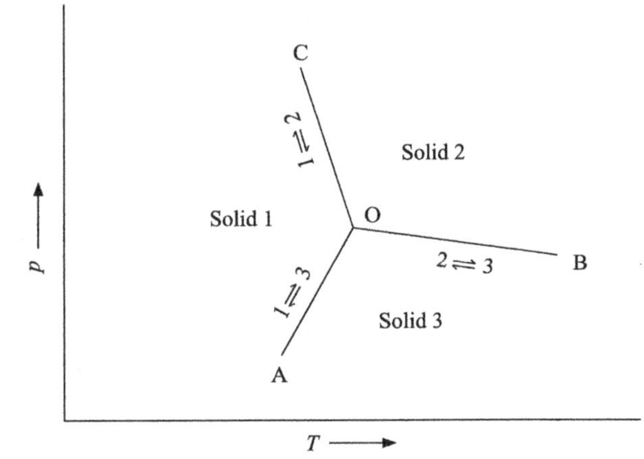


**Fig. 5.4.5**

---

***Example 5.4.2***

A certain substance exists in two solid modifications, α and β, as well as liquid and vapour. Under a pressure of one atmosphere, α is more stable at lower temperatures than β, which melts at a still higher temperature to form the liquid, α is denser than the liquid, but β is less dense than the liquid. No metastable equilibria are observed. Sketch the pressure-temperature phase diagram indicating the significance of each point, line and region. Include in your diagram every triple point that can be observed.

***Solution***

We are given that

(i) α is stable at lower temperatures
(ii) β is stable at higher temperatures and also that it melts at some higher temperature
(iii) $\rho_\alpha > \rho_l$ or $V_{m,\alpha} < V_{m,l}$
(iv) $\rho_\beta < \rho_l$ or $V_{m,\beta} > V_{m,l}$
(v) No metastable equilibria.

Taking points (iii) and (iv) together, we have

$$\rho_\alpha > \rho_l > \rho_\beta \qquad \text{or} \qquad V_{m,\alpha} < V_{m,l} < V_{m,\beta}$$

Since higher temperature favours the form with the greater enthalpy, we must have

$$H_{m,l} > H_{m,\beta} > H_{m,\alpha}$$

Now, we have

$$\left(\frac{dp}{dT}\right)_{\alpha\rightleftharpoons\beta} = \left(\frac{\Delta H_m}{T\Delta V_m}\right)_{\alpha\rightleftharpoons\beta} = \frac{H_{m,\beta} - H_{m,\alpha}}{T(V_{m,\beta} - V_{m,\alpha})} = \frac{+ve}{+ve} = +ve$$

$$\left(\frac{dp}{dT}\right)_{\alpha\rightleftharpoons l} = \left(\frac{\Delta H_m}{T\Delta V_m}\right)_{\alpha\rightleftharpoons l} = \frac{H_{m,l} - H_{m,\alpha}}{T(V_{m,l} - V_{m,\alpha})} = \frac{+ve}{+ve} = +ve$$

$$\left(\frac{dp}{dT}\right)_{\beta\rightleftharpoons l} = \left(\frac{\Delta H_m}{T\Delta V_m}\right)_{\beta\rightleftharpoons l} = \frac{H_{m,l} - H_{m,\beta}}{T(V_{m,l} - V_{m,\beta})} = \frac{+ve}{-ve} = -ve$$

The phase diagram consistent with the above information is given in Fig. 5.4.6.

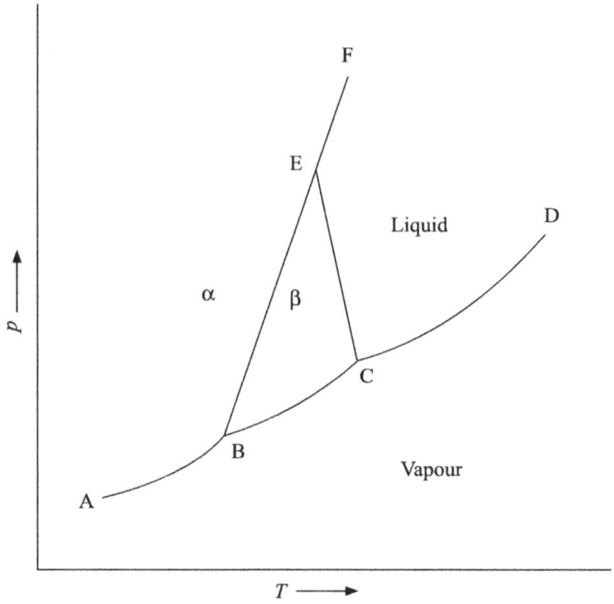


**Fig. 5.4.6**

## Description of the Diagram

The description of various lines, points and regions of Fig. 5.4.6 is given below.

(i) *Curve* AB: α-form ⇌ vapour.
(ii) *Curve* BC: β-form ⇌ vapour.
(iii) *Point* B: Triple point where α ⇌ β ⇌ v.
(iv) *Point* C: Triple point where β ⇌ 1 ⇌ v.
(v) *Curve* CD: Liquid ⇌ vapour.
(vi) *Curve* BE: α ⇌ β.
(vii) *Curve* CE: β ⇌ 1.
(viii) *Point* E: Triple point where α ⇌ β ⇌ 1.
(ix) *Curve* EF: α ⇌ 1.

---

***Example 5.4.3***

Show that on the basis of slope sign there are 8 and only 8 possible unique configurations of the two-phase lines around a triple point in a one-component pressure-temperature diagram. Draw the sketch also.

***Solution***

The slope of each line may be positive or negative. Labelling the three phases as 1, 2, and 3, the possible permutations of the signs of lines representing two phases in equilibrium are given below.

| | (I) | (II) | (III) | (IV) | (V) | (VI) | (VII) | (VIII) |
|---|---|---|---|---|---|---|---|---|
| 1-2 equilibrium line | + | + | + | + | − | − | − | − |
| 2-3 equilibrium line | + | + | − | − | + | + | − | − |
| 1-3 equilibrium line | + | − | + | − | + | − | + | − |

Thus there are, in all, eight possible configurations around the triple point. The corresponding diagrams are shown in Fig. 5.4.7.

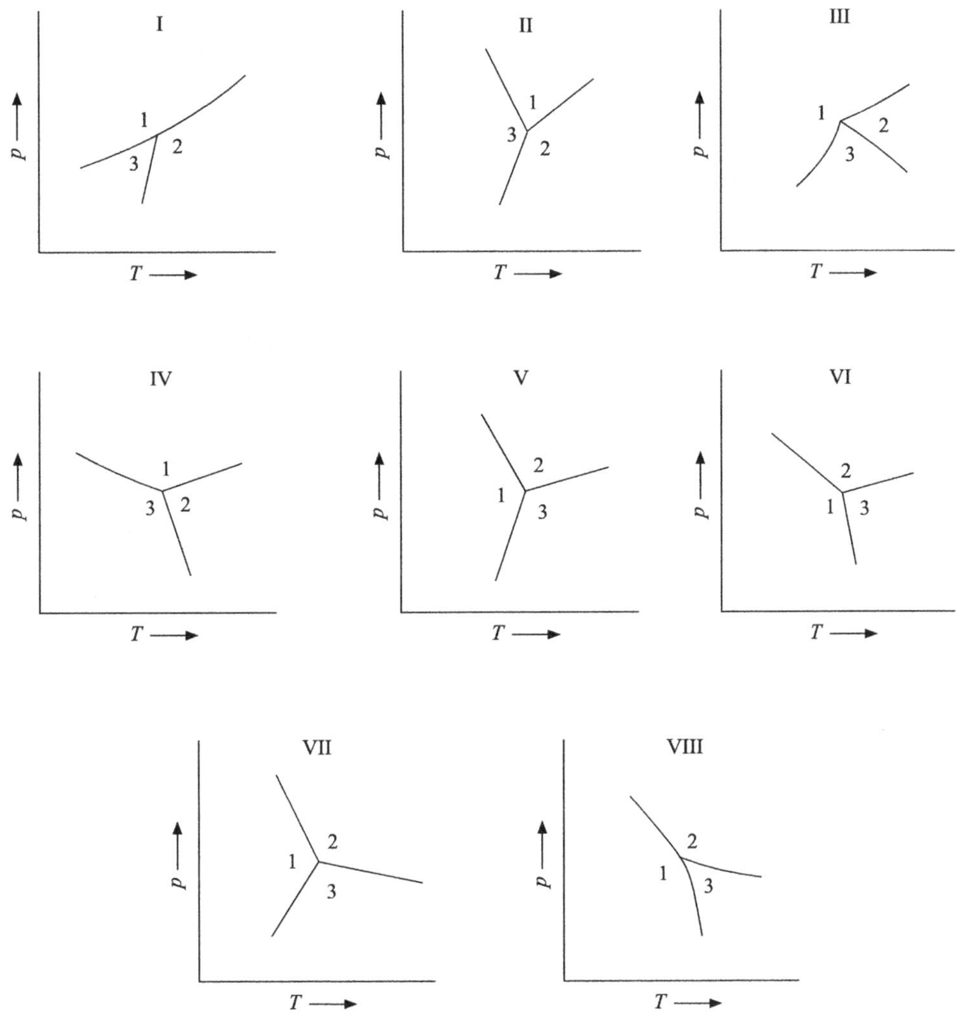


**Fig. 5.4.7**

***Example 5.4.4***

A substance S has a low-pressure form, $S_\alpha$, which melts at about 17 °C under its own pressure of about 8 mmHg. It has also a high-pressure form, $S_\beta$, and the phases $S_\alpha$, $S_\beta$ and liquid are in equilibrium at about 55 °C under a pressure of 2 000 atm. The β-form is denser than the α-form, but both are denser than the liquid. Draw the phase diagram.

***Solution***

We are given that

(i) Two allotropic form: $S_\alpha$ is stable at lower pressures, and $S_\beta$ is stable at higher pressures.

(ii) Melting pint of $S_\alpha$ is 17 °C at the pressure of 8 mmHg.

(iii) $S_\alpha \rightleftharpoons S_\beta \rightleftharpoons S_l$ at 55 °C and 2 000 atm.

(iv) $\rho_\beta > \rho_\alpha$.

(v) $\rho_\beta > \rho_l$ and $\rho_\alpha > \rho_l$.

Since higher temperature favours the form with greater enthalpy, we must have

$$H_{m,\alpha} < H_{m,\beta} < H_{m,l}$$

Taking point (iv) and (v) together, we have

$$\rho_\beta > \rho_\alpha > \rho_l \qquad \text{or} \qquad V_{m,\beta} < V_{m,\alpha} < V_{m,l}$$

Now, we have

$$\left(\frac{dp}{dT}\right)_{\alpha \rightleftharpoons \beta} = \left(\frac{\Delta H_m}{T\Delta V_m}\right)_{\alpha \rightleftharpoons \beta} = \frac{H_{m,\beta} - H_{m,\alpha}}{T(V_{m,\beta} - V_{m,\alpha})} = \frac{+ve}{-ve} = -ve$$

$$\left(\frac{dp}{dT}\right)_{\alpha \rightleftharpoons l} = \left(\frac{\Delta H_m}{T\Delta V_m}\right)_{\alpha \rightleftharpoons l} = \frac{H_{m,l} - H_{m,\alpha}}{T(V_{m,l} - V_{m,\alpha})} = \frac{+ve}{+ve} = +ve$$

$$\left(\frac{dp}{dT}\right)_{\beta \rightleftharpoons l} = \left(\frac{\Delta H_m}{T\Delta V_m}\right)_{\beta \rightleftharpoons l} = \frac{H_{m,l} - H_{m,\beta}}{T(V_{m,l} - V_{m,\beta})} = \frac{+ve}{+ve} = +ve$$

The phase diagram consistent with the given information is shown in Fig. 5.4.8.

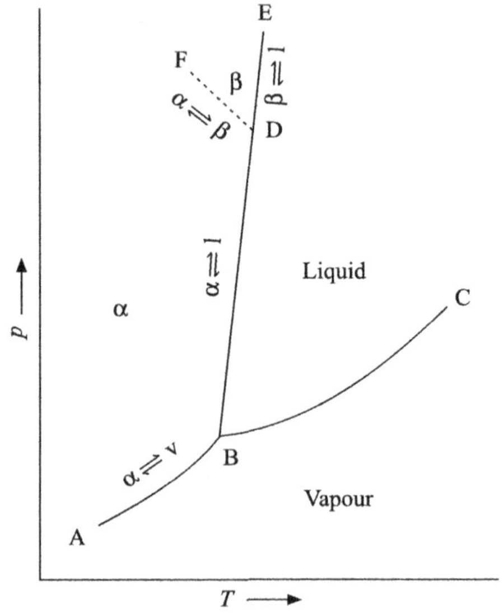


Fig. 5.4.8

The point B is at 17 °C and 8 mmHg where $S_\alpha \rightleftharpoons S_l \rightleftharpoons S_v$. The point D is at 55 °C and 2 000 atm where $S_\alpha = S_\beta = S_l$.

***Example 5.4.5***

From the following data, sketch the phase diagram of $N_2$ at low temperatures.

There are three crystal forms α, β, and γ which coexist at 4 650 atm and 44.5 K. At this triple point, the volume changes ($\Delta V$) are

| | |
|---|---|
| $\alpha \rightarrow \gamma$ | 0.165 $cm^3\ mol^{-1}$ |
| $\beta \rightarrow \gamma$ | 0.208 $cm^3\ mol^{-1}$ |
| $\beta \rightarrow \alpha$ | 0.043 $cm^3\ mol^{-1}$ |

At 1 atm and 36 K, $\beta \rightarrow \alpha$ has $\Delta V = 0.22\ cm^3\ mol^{-1}$. The $\Delta S$ values for the transition cited are 1.25, 5.88, 4.59 and 6.52 $J\ K^{-1}\ mol^{-1}$, respectively.

***Solution***

We have

| *Transition* | $\Delta V/\text{cm}^3\ \text{mol}^{-1}$ | $\Delta S/\text{J K}^{-1}\ \text{mol}^{-1}$ | $\dfrac{\text{d}p}{\text{d}T} = \dfrac{\Delta S}{\Delta V} \propto$ |
|---|---|---|---|
| $\alpha \rightarrow \gamma$ | 0.165 | 1.25 | $\dfrac{1.25}{0.165} = 7.58$ |
| $\beta \rightarrow \gamma$ | 0.208 | 5.88 | $\dfrac{5.88}{0.208} = 28.27$ |
| $\beta \rightarrow \alpha$ | 0.043 | 4.59 | $\dfrac{4.59}{0.043} = 106.74$ |
| $\beta \rightarrow \alpha$ at 1 atm and 36 K | 0.22 | 6.52 | $\dfrac{6.52}{0.22} = 29.64$ |

The schematic phase diagram consistent with the above information is shown in Fig. 5.4.9.

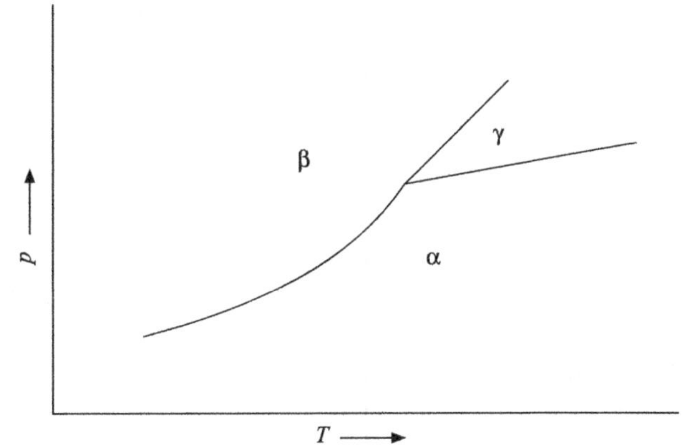


**Fig. 5.4.9**

***Example 5.4.6***

Substance A exists in two allotropic forms: white and red. White A melts at 44 °C and 0.2 mmHg. Red A melts at 490 °C and 43 atm pressure. White A is more dense than the liquid and the red form is less dense than the liquid. The vapour pressure of white form is everywhere greater than that of red. Sketch $p$ versus $T$ diagram, label the areas and explain the stable and unstable triple points.

***Solution***

From the given data, we conclude

(i) Melting point of white A: 44 °C at 0.2 mmHg.

(ii) Melting point of red A: 490 °C at 43 atm.

(iii) Vapour pressure of white A > Vapour pressure of red A. This indicates that white A is unstable with respect to red A.

(iv) $\rho_{\text{white A}} > \rho_1$ or $V_{\text{m, white A}} < V_{\text{m, 1}}$

Therefore

$$\left(\frac{\text{d}p}{\text{d}T}\right)_{\text{white A} \rightleftharpoons 1} = \frac{\Delta_{\text{fus}}H_{\text{m}}}{T(V_{\text{m,1}} - V_{\text{m, white A}})} = \frac{+\text{ve}}{+\text{ve}} = +\text{ve}$$

(v) $\rho_{\text{red A}} < \rho_{\text{l}}$ or $V_{\text{m, red A}} > V_{\text{m, l}}$

Therefore

$$\left(\frac{\mathrm{d}p}{\mathrm{d}T}\right)_{\text{red A} \rightleftharpoons \text{l}} = \frac{\Delta_{\text{fus}}H_{\text{m}}}{T(V_{\text{m, l}} - V_{\text{m, red A}})} = \frac{+\text{ve}}{-\text{ve}} = -\text{ve}$$

(vi) Since $\rho_{\text{white A}} > \rho_{\text{l}}$ and $\rho_{\text{l}} > \rho_{\text{red A}}$, it is obvious that

$$\rho_{\text{white A}} > \rho_{\text{red A}} \quad \text{or} \quad V_{\text{m, white A}} < V_{\text{m, red A}}$$

$$\left(\frac{\mathrm{d}p}{\mathrm{d}T}\right)_{\text{white A} \rightleftharpoons \text{red A}} = \frac{\Delta_{\text{trs}}H_{\text{m}}}{T(V_{\text{m, red A}} - V_{\text{m, white A}})} = \frac{+\text{ve}}{+\text{ve}} = +\text{ve}$$

The phase diagram consistent with the above information is given in Fig. 5.4.10.

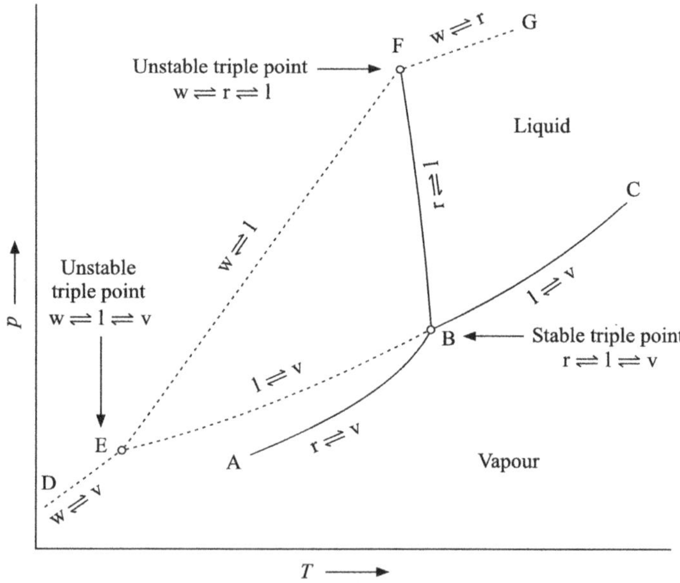


**Fig. 5.4.10 The required phase diagram. The symbols w, r, l and v stand for white, red, liquid and vapour, respectively**

***Example 5.4.7***

A substance X exists in two crystalline modifications, here called A and B. At temperature below 90 °C, the vapour pressure of the A-form is less than that of the B-form, and above 90 °C, the reverse is true. The A-form is 5% denser than liquid X and the B-form is 10% less dense than the A-form. There is an A-l-v triple point at 110 °C. Sketch the $p$-$T$ diagram for X.

***Solution***

From the given data, we conclude

(i) Since A-form is 5% denser than liquid X, we have

$$\rho_{\text{A}} = 1.05\rho_{\text{l}} \quad \Rightarrow \quad V_{\text{m, A}} < V_{\text{m, l}}$$

Therefore

$$\left(\frac{\mathrm{d}p}{\mathrm{d}T}\right)_{\text{A} \rightleftharpoons \text{l}} = \frac{\Delta_{\text{fus}}H_{\text{m}}}{T(V_{\text{m, l}} - V_{\text{m, A}})} = \frac{+\text{ve}}{+\text{ve}} = +\text{ve}$$

(ii) Since B-form is 10% less dense than the A-form, we have

$$\rho_B = 0.9\rho_A$$

or $$\rho_B = 0.9 \times (1.05\ \rho_l) = 0.945\ \rho_l$$

Hence $V_{m,\,B} > V_{m,\,l}$

and thus

$$\left(\frac{dp}{dT}\right)_{B \rightleftharpoons l} = \frac{\Delta_{fus}H_m}{T(V_{m,l} - V_{m,B})} = \frac{+ve}{-ve} = -\ ve$$

(iii) At temperature below 90 °C, the vapour pressure of A-form is less than that of B-form, It implies that the A-form is stable whereas B-form is unstable at temperatures below 90 °C. On the other hand, the vapour pressure of B-form is smaller than that of A-form at temperature above 90 °C implies that the B-form is stable. Thus, 90 °C is the transition point of A-form to B-form. It also follows that the molar enthalpy of B-form is greater than that of A-form.

(iv) Since $\rho_B = 0.9\ \rho_A$, we have

$$V_{m,\,B} > V_{m,\,A}$$

and hence

$$\left(\frac{dp}{dT}\right)_{A \rightleftharpoons B} = \frac{\Delta_{trs}H_m}{T(V_{m,B} - V_{m,A})} = \frac{+ve}{+ve} = +ve$$

(v) The triple point A $\rightleftharpoons$ l $\rightleftharpoons$ v exists at 110 °C.

The phase diagram consistent with the above information is shown in Fig. 5.4.11.

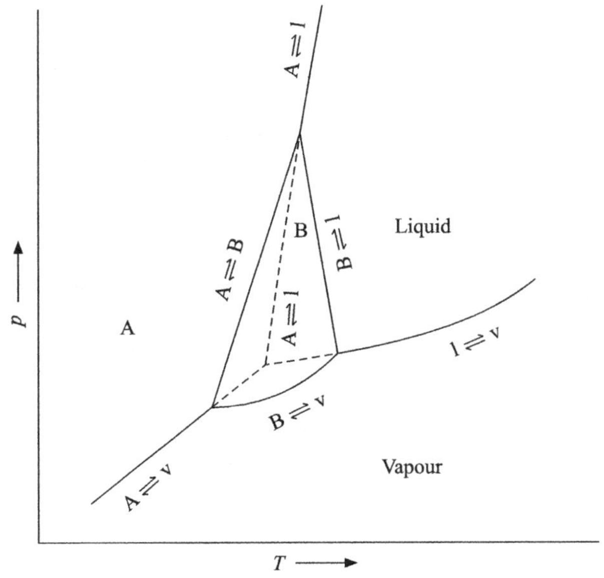


**Fig. 5.4.11**

## REVISIONARY PROBLEMS

5.1 (a) Determine the variance of a one-component system having number of phases equal to one, two and three, respectively.
(b) Show that it is not possible to have more than three phases in equilibrium with each other in a one-component system.
(c) Show how are the following systems represented in a one-component phase diagram:
  (i) System consisting of only one phase.
  (ii) System consisting of two phases in equilibrium.
  (iii) System consisting of three phases in equilibrium.

5.2 Describe the phase diagram of water with due emphasis on the following facts:
  (i) How many triple points does it have?
  (ii) Why is the solid-vapour line more steep than the liquid-vapour line at the triple point?
  (iii) Why is the solid-liquid line almost vertical and slightly tilted to the left?
  (iv) What is the upper limit of the liquid-vapour equilibrium line?

5.3 Illustrate with examples the meaning of the terms polymorphism, enantiotropy and monotropy. Describe the type of phase diagram that is expected for the phenomena of enantiotropy and monotropy.

5.4 (a) Describe the phase diagram of sulphur which exhibits the phenomenon of enantiotropy. How many triple points are possible in this phase diagram?
(b) The equilibrium line of rhombic sulphur $\rightleftharpoons$ monoclinic sulphur is slightly tilted to the right. Predict which enantiotropic form is more dense.
(c) The slope of lines representing rhombic sulphur $\rightleftharpoons$ monoclinic sulphur and monoclinic sulphur $\rightleftharpoons$ liquid sulphur are positive. Predict the slope of the line representing metastable equilibrium of rhombic sulphur $\rightleftharpoons$ liquid.

5.5 Describe briefly the phase diagram of phosphorus—an example which exhibit both monotropy as well as enantiotropy.

5.6 Starting from the fact that $\Delta G$ of the process

$$\text{metastable form} \rightarrow \text{stable form}$$

is negative (since the process occurs spontaneously), show that the vapour pressure of the metastable form is greater than that of the stable form at the same temperature.

## TRY YOURSELF PROBLEMS

5.1 Justify the phase diagram of $I_2$ as shown in Fig. 1 from the following information. Solid iodine ($A_1$) is in equilibrium with liquid and vapour at 114 °C and 90 mmHg pressure. Its second crystalline modification $A_2$ has a vapour pressure 110 mmHg at 114 °C and it has less enthalpy of sublimation than that of $A_1$ . $A_1$ is more dense than $A_2$ and is less dense than the liquid.
Identify each area, line and point shown in the diagram.

5.2 Figure 2 is the phase diagram of a one-component system. Identify the various points, lines and areas shown therein. Discuss the slopes of various lines meeting each other at the points $T_1$, $T_2$, $T_3$, $T_4$ and $T_5$ in terms of $\Delta H_m$ and $\Delta V_m$ of the involved transformations.

(*Hint:* Solid exists in three allotropic forms.)

5.3 Figure 3 is the phase diagram of a one-component system. Identify the various points, lines and areas shown therein. Discuss the slopes of various lines meeting

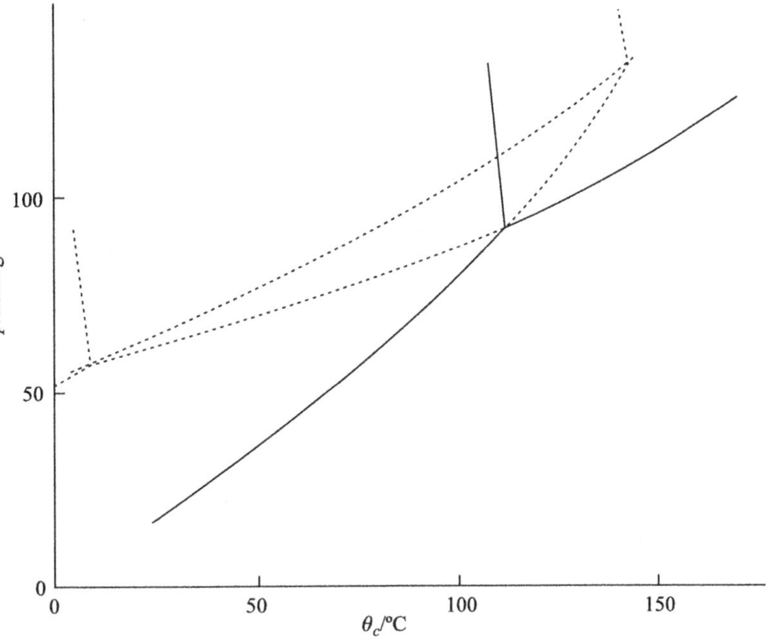


**Fig. 1**

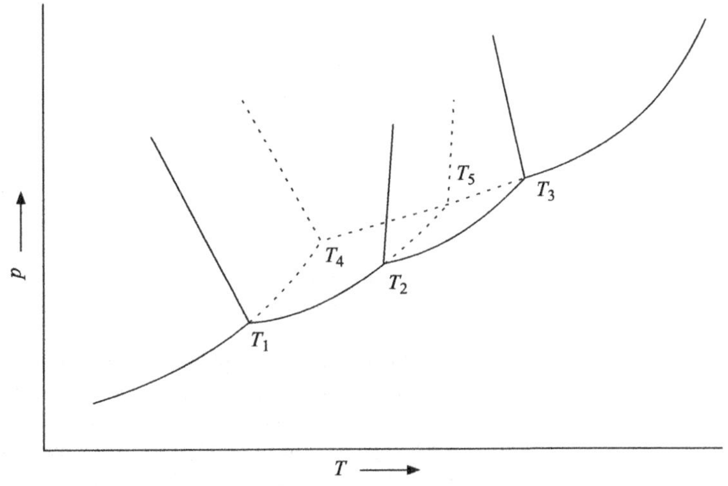


**Fig. 2**

each other at the points $T_1$, $T_2$, $T_3$, $T_4$ and $T_5$ in terms of $\Delta H_m$ and $\Delta V_m$ of the involved transformations.

(*Hint:* Solid has three unstable forms and one stable form.)

5.4 A substance S occurs in two solid modifications: $S_\beta$ and $S_\alpha$. $S_\beta$ has a stable melting point at about 119 °C (0.05 mmHg). $S_\beta$ and $S_\alpha$ are enantiotropic with respect to each other and the transition point involving $S_\alpha$, $S_\beta$ and vapour has the value 95.5 °C (0.007 mmHg). Phases $S_\alpha$, $S_\beta$ and liquid are in equilibrium at a triple point 151 °C and 1290 atm. The transition temperature of the equilibrium $S_\alpha \rightleftharpoons$ liquid increases with increase in pressure. Draw the phase diagram.

5.5 Explain why the phase diagram shown in Fig. 4 is not theoretically possible.
(*Hint:* Slope of s $\rightleftharpoons$ v line at triple point is smaller than that of l $\rightleftharpoons$ v line.)

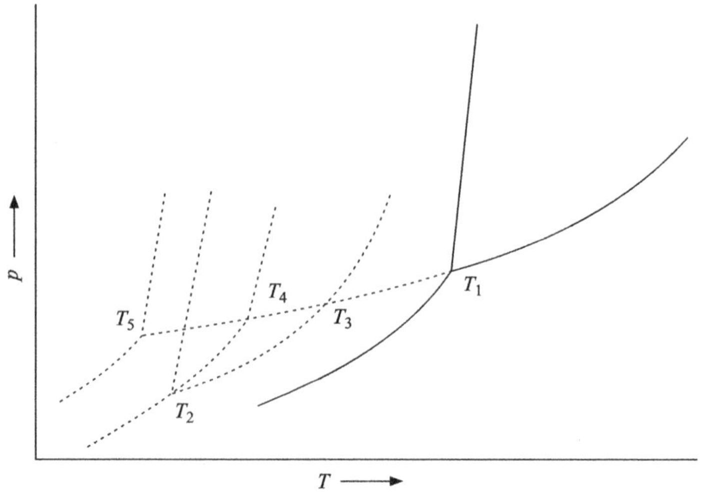


**Fig. 3**

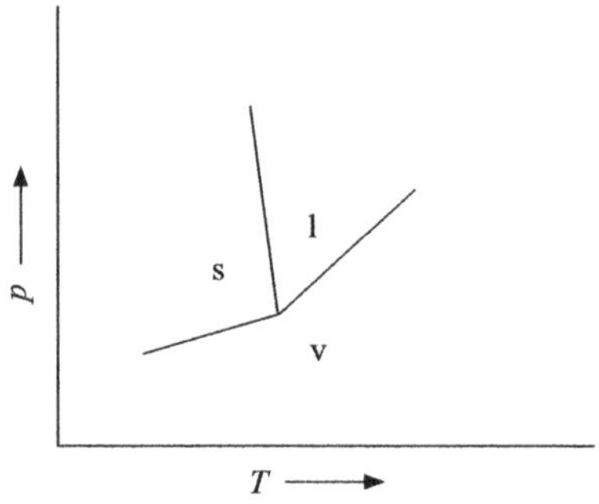


**Fig. 4**

5.6 Comment on the following statement.
The phase that is stable in equilibrium with the vapour is the phase that has the lower equilibrium vapour pressure.

5.7 Draw a schematic plot of chemical potential versus temperature for sulphur from the following information.
At 95.5 °C rhombic sulphur reversibly transforms to monoclinic sulphur.
At 120 °C monoclinic sulphur melts. From the plot, answer the following.
(i) Which has the higher melting point, rhombic or monoclinic sulphur?
(ii) Which form of solid sulphur has greater entropy?
(iii) Which of the two solids will exhibit more ideal solubility at 25 °C in a good solvent such as diethyl ether?
(iv) Does the melting point of rhombic sulphur lie above or below the rhombic-monoclinic transition temperature?

(Ans. (i) monoclinic, (ii) monoclinic, (iii) rhombic, (iv) above)
(*Hint:* Make use of Eq. (2.12.6) for part (iii)).

5.8 At higher pressures, ice is known to exist in six different forms and its phase diagram has an appearance as shown in Fig. 5. Ice 1 is the common form which exists under normal conditions. From the figure, answer the following.
(i) How many triple points are present?
(ii) Name the phases which are in equilibrium at each triple point.

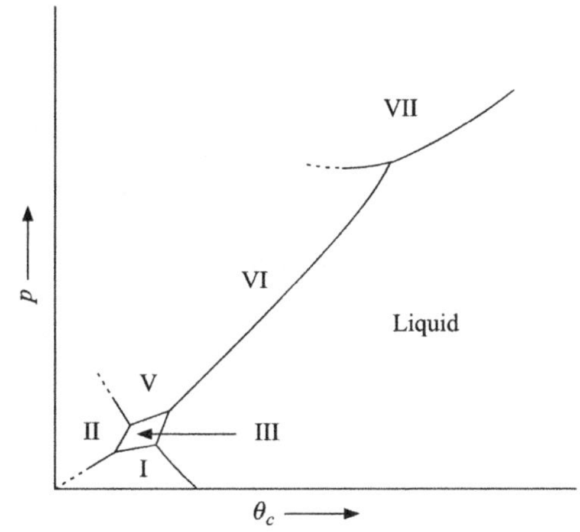


**Fig. 5**

(iii) From the slopes of various lines, predict the phase which is more dense than the other.

(iv) What is the effect of pressure on the melting point of each form of ice?

5.9 Solid ammonium nitrate exists in five crystalline forms with the following triple point temperatures:

| | | | | | |
|---|---|---|---|---|---|
| Tetragonal | ⇌ | β-Rhombic | ⇌ | Vapour | – 18 °C |
| β-Rhombic | ⇌ | α-Rhombic | ⇌ | Vapour | 32 °C |
| α-Rhombic | ⇌ | Rhombohedral | ⇌ | Vapour | 83 °C |
| Rhombohedral | ⇌ | Cubic | ⇌ | Vapour | 125 °C |
| Cubic | ⇌ | Liquid | ⇌ | Vapour | 169 °C |
| β-Rhombic | ⇌ | α-Rhombic | ⇌ | Rhombohedral | 50 °C |

Also, it is known that the order of increasing density is

Liquid < α-Rhombic < Rhombohedral < Cubic < β-Rhombic < Tetragonal

Sketch a reasonable $p$-$T$ diagram based on this information.

5.10 Describe the effects produced of (i) isothermal increase and decrease in pressure and (ii) isobaric increase and decrease in temperature on the following systems of water (Fig. 5.3.1).

(a) A system on sublimation curve

(b) A system on vaporization curve.

(c) A system on fusion line.

(d) A system at triple point.

(e) A system slightly at the left of fusion line.

(f) A system slightly at the right of fusion line.

(g) A system below the sublimation and vaporization curves.

# 6 Phase Diagrams of Two-Component Systems

## 6.1 APPLICATION OF THE PHASE RULE

According to the phase rule, the degrees of freedom of a two-component system is given by

$$F = C - P + 2 = 4 - P$$

Thus, we have:

**System Consisting of One Phase**

The phase can be gaseous or completely miscible liquid solution or solid solution. In this case, $F = 4 - 1 = 3$, thus three variables, namely, temperature, pressure and composition of the system, will have to be stated in order to define the system completely.

**System Consisting of Two Phases**

A few typical cases of two-phase equilibrium are given below.

***Liquid-vapour*** The two constituents will be completely miscible with each other in the liquid phase.

***Liquid-liquid*** The two liquids will be partially miscible in each other.

***Solid-liquid*** The solid will consist of either of the two constituents or the solid solution. In the liquid phase, both constituents are completely miscible.

Here since $P = 2$, we will have $F = 4 - 2 = 2$. Thus, stating the values of two variables is sufficient to define the system completely. These variables are temperature and composition of the liquid phase. All other variables will have fixed values.

**System Consisting of Three Phases**

A few typical cases of three-phase equilibrium are given below.

***Liquid-liquid-vapour*** The two constituents will be partially miscible in the liquid phase.

***Solid-liquid-liquid*** The solid will consist of either of the two constituents or the solid solution. The two constituents are partially miscible in the liquid phase.

***Solid-solid-liquid*** Both the constituents are present in the solid form and are completely miscible in the liquid phase.

Here since $P = 3$, we will have $F = 4 - 3 = 1$. Thus, stating the value of only one variable is sufficient to define the system completely.

**Comment on the Phase Diagram of Condensed Systems**

Generally various types of equilibria are studied at constant external pressure. Thus, the value of one of the degrees of freedom is already stated, and hence their number is reduced by one or alternatively, the degrees of freedom may be calculated by making use of the reduced phase rule $F = C - P + 1$. In such studies, therefore, we have

(i) $F = 2$ for $P = 1$,
(ii) $F = 1$ for $P = 2$, and
(iii) $F = 0$ for $P = 3$.

Since the maximum degrees of freedom to define a two-component system is two (for $p = 1$ atm), the phase diagram of such a system can be drawn on a sheet of paper, choosing the rectangular axes representing temperature and composition, respectively. A complete phase diagram of a system will have to include all the three phases, namely, gaseous, liquid and solid. However, it is more advantageous from the point of view of discussion to draw separately the phase diagram corresponding to the different types of phase equilibria. The equilibria involving liquid-vapour and liquid-liquid have already been considered in Chapter 4. In this chapter, we consider systems involving only solid-liquid equilibria. Such systems are known as *condensed systems*. Since in such systems, the gas phase is absent, the small variation in pressure does not affect the various solid-liquid equilibria of a system.

## 6.2 CLASSIFICATION OF DIAGRAMS

Solid-liquid equilibria of varying degree of complexity are known. These may be analyzed in terms of the following simpler cases.

1. The pure components only crystallize from the solution.
2. The pure components only crystallize from the solution and one of the solids exists in more than one polymorphic form.
3. The two components form a solid compound stable up to its melting point.
4. The two components form a solid compound which decomposes before its actual melting point.
5. The two components are completely miscible in the solid phase and form a complete series of solid solutions.
6. The two components are partially miscible in the solid phase and form stable solid solutions.
7. The two components form solid solutions which are stable only up to a transition temperature.
8. The two components are partially miscible in the liquid phase and only pure components crystallize from the solution.

   The above elementary systems are described in Sections 6.4 to 6.11.

## 6.3 THERMAL ANALYSIS

**General Procedure of Thermal Analysis**

The phase diagram representing different types of solid-liquid equilibria can be drawn with the method of thermal analysis (or cooling curves). In this method, a system of known composition is heated so as to obtain it in the molten state.

It is allowed to cool of its own and its temperature after regular interval of time (say, half a minute) is recorded. A cooling curve of temperature versus time is made and from the breaks and arrests in the cooling curve, the useful information regarding the initial and final solidification temperatures are determined. The above procedure is repeated with different compositions of the system.

For an illustration, we take an example where only pure components crystallize on cooling the system. The expected types of cooling curve are described below.

## Cooling Curve of a Pure Component

We start with the pure component A (or B) in the liquid state. The rate of cooling of liquid will be smooth till solidification starts. At this stage, the system contains two phases in equilibrium and, according to the phase rule, it becomes invariant ($F = C + 1 - P = 1 + 1 - 2 = 0$). Thus, solidification of A takes place at constant temperature in spite of the fact that heat is released during solidification. This is due to the fact that the rate of liberation of heat during solidification becomes equal to that of transfer of heat from the system to the surroundings. This gives rise to what is known as the complete arrest in the rate of cooling. When the entire liquid has solidified, the system becomes univariant and thus it shows a smooth cooling of solid phase.

***Schematic representation of cooling curve*** The above pattern of cooling is shown in Fig. 6.3.1.

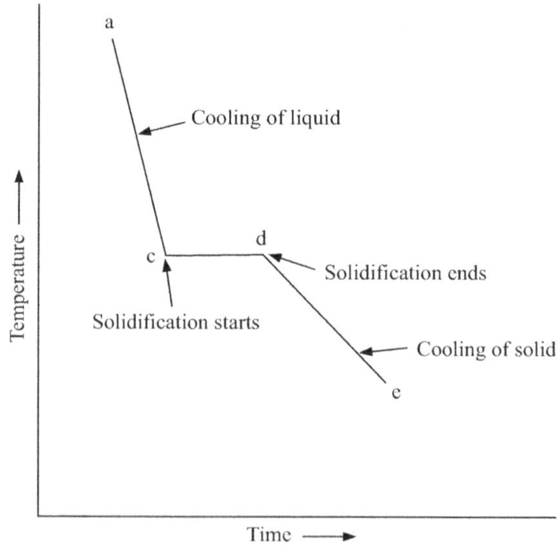


**Fig. 6.3.1 Cooling curve of a pure component**

## Cooling Curve of a Mixture

We start with a mixture of B in A of known composition. The system is heated to get it in the liquid phase. The degree of freedom of the system in liquid phase is

$$F = C + 1 - P = 2 + 1 - 1 = 2$$

Thus, two variables, viz., temperature and composition of liquid phase, have to be stated in order to define the system completely. In other words, it may be concluded that the liquid phase of known composition can exist at different temperatures. Thus, the liquid phase is expected to exhibit smooth cooling until solid A starts solidifying.

***Solidification of* A** When A starts solidifying, the system becomes univariant, ($F = C + 1 - P = 2 + 1 - 2 = 1$). Thus, the temperature at which solid A starts solidifying from the liquid mixture of known composition will have a fixed value. This temperature is expected to be lesser than the freezing point of A from pure liquid A since the addition of B in A lowers the freezing point of the latter. The separation of A in solid phase changes the composition of liquid phase and thus the separation of A from this liquid will take place at another fixed temperature. The latter is expected to be smaller than the previous temperature as the separation of A increases the molality of B in the solution. Thus, the system is expected to exhibit cooling during more and more solidification of A. However, the rate of cooling is retarded owing to the liberation of heat during solidification of A and thus a break in the cooling curve is observed. The break point indicates the temperature at which first crystal of A is obtained.

***Solidification of both* A *and* B** If cooling is continued, the solution may become saturated with respect to B and hence B also starts solidifying along with A. At this stage, the system becomes nonvariant ($F = C + 1 - P = 2 + 1 - 3 = 0$) and thus solidification from a solution of fixed composition (corresponding to the saturation solubility of B in A) takes place at constant temperature. This results in a complete arrest of cooling curve. Since at this temperature the saturation solubility of B in A has to be maintained, it follows that the composition of solid phase which separates will have the same composition as that of the liquid phase. The temperature of the system will remain constant until the entire liquid phase has solidified. After this, the system will exhibit smooth cooling of solid phases. Thus, the start and end of horizontal line in cooling curve indicate, respectively, the start of solidification of both components together and the complete solidification of the liquid phase.

***Schematic representation of cooling cure*** The above pattern of cooling is shown in Fig. 6.3.2. The description of Fig. 6.3.2 is also given in Table 6.3.1.

**Table 6.3.1** Description of Fig. 6.3.2

| *Location of system* | *Description* | *Degrees of freedom* |
|---|---|---|
| ab | Cooling of liquid mixture | 2 |
| b | Start of solidification of A | 1 |
| bc | Solidification of A continues | 1 |
| c | B also starts solidifying | 0 |
| cd | Solidification of A and B continues. The composition of solid mixture separated here is the same as that of the liquid phase | 0 |
| d | Almost complete solidification of liquid phase | 0 |
| de | Cooling of solid phases | 1 |

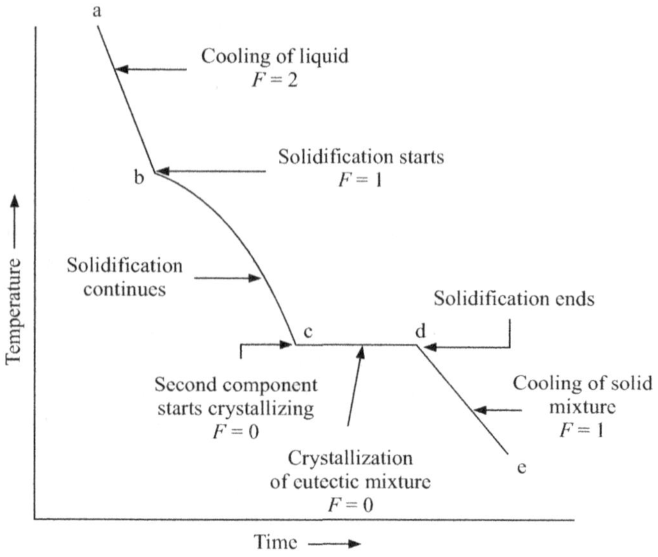


**Fig. 6.3.2 Cooling curve of a liquid mixture of a two-component system**

***Supercooling of liquid*** In some cases, the separation of solid phase does not occur readily and thus supercooling is observed. The cooling curve instead of showing a break continues along a continuation of ab. Since this represents an unstable state, soon one observes a sudden rise in temperature followed by a smooth cooling along b′c (Fig. 6.3.3). The correct freezing point may then be obtained by extrapolation back to b as shown in Fig. 6.3.3.

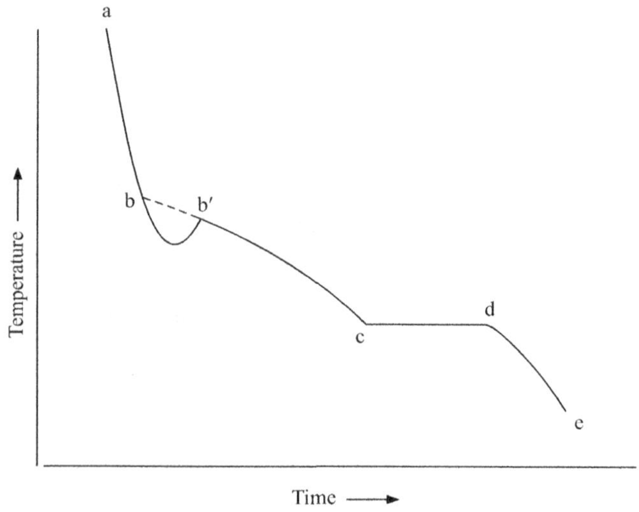


**Fig. 6.3.3 Supercooling phenomenon**

**Comment**

The pattern as shown in Fig. 6.3.2 is, however, not shown at a particular composition where on cooling both the components solidify simultaneously. In this case, the first break is not observed and one gets the cooling curve similar to that shown in Fig. 6.3.1. The cooling curve of a system in which solid solution and not the pure component separates on cooling is similar to that given in Fig. 6.3.2 with the exception that the horizontal portion is replaced by a break.

## 6.4 CRYSTALLIZATION OF PURE COMPONENTS — SIMPLE EUTECTIC PHASE DIAGRAM†

**Description of the Phase Diagram**

The general form of the phase diagram along with the expected cooling curves of such a system is shown in Fig. 6.4.1.

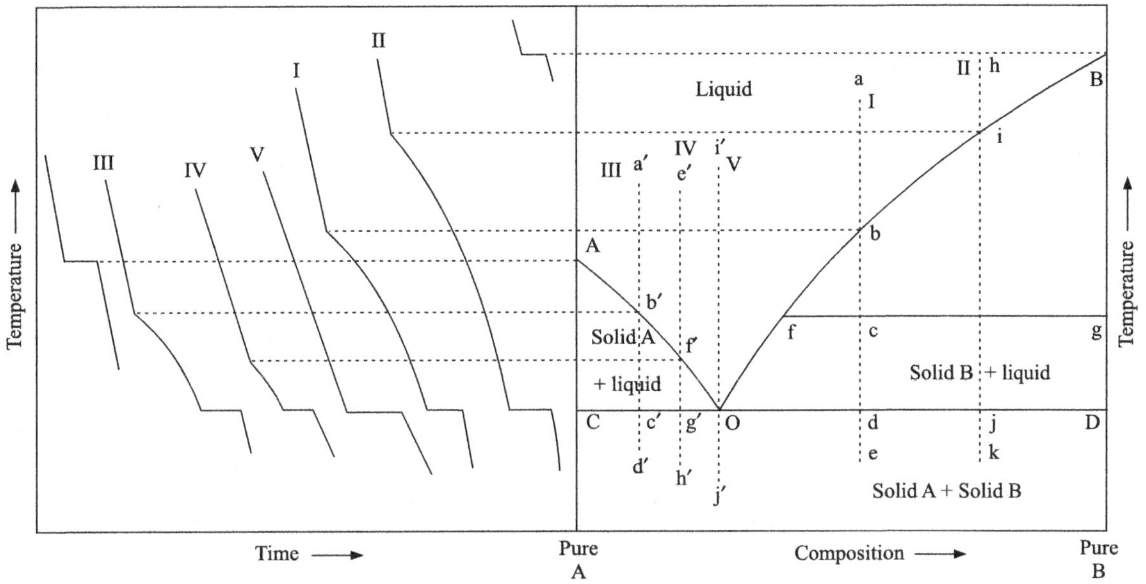


**Fig. 6.4.1 Phase diagram of a system exhibiting crystallization of pure components**

From the cooling curves shown in Fig. 6.4.1, the following information can be derived.

- From the cooling curves of pure components, the point A and B representing, respectively, the freezing points of the pure components A and B are obtained.
- From the cooling curves I and II of two systems of A in B, the points b and i representing, respectively, the start of solidification of B and the points d and j representing, respectively, the start of solidification of both A and B together are obtained. Since the addition of a solute in a solvent depresses the freezing point of the latter, the temperatures corresponding to the points b and i are smaller than the freezing point of the pure component B.
- Similarly, from the cooling curves III and IV of two systems of B in A, the points b′ and f′ representing respectively the start of solidification of A and the points c′ and g′ representing respectively the start of solidification of both A and B together are obtained.
- If a smooth curve passing through the points A, b′ and f′ is drawn, we get the curve AO which represents the start of freezing of pure component A from liquid solution of different compositions. Similarly, if a line passing through B, i and b is drawn, we get the freezing point curve of the component B.

†See Project 1 in vol 6 for the formation of eutectic phase diagram constructed from the consideration of free energies.

- If a line passing through the points c′, g′, d and j is drawn, we get a horizontal line CD which indicates the temperature at which both components start solidifying together from the liquid O irrespective of the original composition of the liquid phase. The point O where the freezing point curves AO and BO meet each other also lies on this horizontal line.

### Sequence of Effects Produced on Cooling a Liquid Mixture

The labelling of the phase diagram shown in Fig. 6.4.1 can be easily understood by first considering the expected behaviour of a liquid solution during its cooling. If, for example, a liquid solution represented by the point a is cooled, the state of the overall system will move vertically downward along abcde, known as **isopleth.**

- The system will remain in liquid phase till the point b is reached where the first crystal of B starts appearing.
- The removal of solid B from liquid phase makes the composition of liquid phase less rich in B and thus its composition moves towards the A-axis.
- If the crystallization of B is to be continued, the temperature of the liquid phase has to be decreased since its composition point moves on the line bfO. Thus on cooling, more and more of B solidifies and the composition of the liquid phase moves along the curve bfO. For example, if the system has been cooled up to the overall state point c, the system will consist of solid B (represented by the point g) in equilibrium with liquid phase whose composition corresponds to the point f. The point f is the intersection point of the tie line (horizontal line ) drawn from the point c with the curve BO. The tie line, by definition, is a line which connects different phases in equilibrium with one another. In the present case, the two phases in equilibrium are represented, respectively, by the points g (solid phase B) and f (liquid phase). The relative amounts of the two phases may be determined by using the lever rule, such that

$$\frac{\text{Amount of solid B}}{\text{Amount of liquid phase f}} = \frac{\text{fc}}{\text{cg}} \qquad (6.4.1)^{\dagger}$$

- During the cooling of the system from b to d, the relative length of the segment of tie line lying to the left of bcd (i.e. fc) increases indicating that the relative amounts of the two phases change in favour of solid B, i.e. more and more B solidifies during the cooling process.
- When the system has been cooled to the point d the liquid composition will move to the point O. Hence, here A also starts crystallizing along with B. From this, it follows that the yield of pure crystals of B will be maximum when the cooling of the system is carried out up to a point slightly above the point d.

---

† For derivation of the lever rule, see p. 259.

Throughout the Chapter, a system is represented by an alphabet printed in roman style while its composition is represented by the same alphabet printed in italics.

***Maximum amount of* B *recoverable*** We can compute the maximum amount of pure B that can be separated from the liquid a as shown in the following.

The maximum amount of pure B (which can be separated from the starting liquid a) relative to that of the liquid O is given by

$$\frac{\text{Amount of liquid O}}{\text{Maximum amount of solid B separated}} = \frac{\text{dD}}{\text{Od}} \tag{6.4.2}$$

Adding one on both sides and inverting the expression, we get

$$\frac{\text{Maximum amount of solid B separated}}{\text{Amount of (liquid O + solid B)}} = \frac{\text{Od}}{\text{Od + dD}} = \frac{\text{Od}}{\text{OD}}$$

Hence, the maximum amount of solid B separated

$$= \left(\frac{\text{Od}}{\text{OD}}\right) \text{(amount of starting liquid a)} \tag{6.4.3}$$

- At point d, the solid mixture which solidifies has the same composition as that of the liquid phase, vis-à-vis, the composition of the liquid phase remains unchanged. The temperature of the system also remains unchanged till whole of liquid phase is solidified. Since now solid phase contains both A and B, the point representing solid phases moves from D towards d along the line Dd as more and more of solid mixture is precipitated. When this point is reached at d, whole of liquid solution is solidified.
- Amount of solid eutectic mixture separated at any point before whole of liquid is solidified can be computed as follows.

At any point in between D to d (say at j), the amount of liquid O relative to the total amount of solids A and B, (that has separated from the liquid a) is given by

$$\frac{\text{Amount liquid O}}{\text{Amount of total solids (A + B) separated}} = \frac{\text{dj}}{\text{Od}} \tag{6.4.4}$$

Adding one on both sides, we get

$$\frac{\text{Amount of (liquid O + total solids separated)}}{\text{Amount of total solids separated}} = \frac{\text{Od + dj}}{\text{Od}} = \frac{\text{Oj}}{\text{Od}}$$

Taking inverse, we get

$$\text{Amount of total solids separated} = \left(\frac{\text{Od}}{\text{Oj}}\right)\left(\begin{matrix}\text{amount of starting} \\ \text{liquid a}\end{matrix}\right) \tag{6.4.5}$$

From this amount, if we subtract the pure solid B that has separated on reaching the point d (as given by Eq. 6.4.3), we get the total amount of solid mixture that has separated when the liquid is cooled beyond d till the composition of solid phase is given by the point j. If the composition of solid mixture is analysed, it will correspond to the point O.

Further cooling from d to e, merely decreases the temperature of the solid mixture.

**Conclusions of Phase Diagram**

From the above analysis, the following conclusions may be drawn.

- A point within the area BODB represents a system of solid B in equilibrium with a liquid phase.† The composition of the latter is given by the intersection point of the tie line drawn from the given point with the curve BO. The relative amounts of two phases can be determined by using the lever rule. Similarly, a point within the area ACOA represents a system of solid A in equilibrium with a liquid solution whose composition lies on the curve AO.
- Along the lines AO and BO and within the areas AOCA and BODB, the system is univariant ($F = C + 1 - P = 2 + 1 - 2 = 1$). Thus for a given temperature, the composition of liquid phase in equilibrium with solid A or solid B has a fixed value.
- A point on the line CD represents an invariant system ($F = C + 1 - P = 2 + 1 - 3 = 0$). For the point O the temperature is lowest at which any liquid of A and B will freeze out and, consequently, represents the lowest melting point of any solution of solids A and B. It is for this reason, the point O is known as the *eutectic point* ( Greek: easily melting).
- Line AOB is the *liquidus curve* as it gives composition of the liquid phase in equilibrium with the solid phase. The line ACODB is known as the *solidus curve*; lines AC, COD and DB represent, respectively, solid A, mixtures of solids A and B in equilibrium with liquid phase of composition O and solid B. The two curves AOB and ACODB meet each other at point O.
- Cooling of a system on the line CD will result into the solidification of A and B together from the liquid phase of eutectic composition *O*. The solid mixture which separates here has the same composition as that of the liquid phase.
- A system below the line CD represents a solid mixture of A and B.

**Labelling of Eutectic Phase Diagram**

Table 6.4.1 describes, in brief, the phase diagram shown in Fig. 6.4.1.

---

†In general, the two phases in equilibrium within an area are given by the two vertical end lines of the area. For example, for the area BDOB, the two vertical end lines are BD and BO. These lines represent solid phase B and liquid solution, respectively. Thus, within the area BDOB, solid B is in equilibrium with liquid solution whose composition lies on the line BO.

**Table 6.4.1** Description of Phase Diagram given in Fig. 6.4.1

| *Location of the system* | *Status of the system* | *Degrees of freedom* | |
|---|---|---|---|
| | | *number* | *kind* |
| A | Freezing point of A | 0 | fixed $T$ |
| B | Freezing point of B | 0 | fixed $T$ |
| AO | Start of crystallization of A | 1 | $T$ or composition |
| BO | Start of crystallization of B | 1 | $T$ or composition |
| O | Eutectic point | 0 | fixed $T$ and fixed composition |
| Area above AOB | Liquid phase | 2 | $T$ and composition |
| Area below COD | Solid mixture | 1 | $T$ |
| Area ACOA | Solid A in equilibrium with liquid phase whose composition lies on the curve AO. | 1 | $T$ or composition |
| Area BDOB | Solid B in equilibrium with liquid phase whose composition lies on the curve BO. | 1 | $T$ or composition |
| COD | Systems where both A and B crystallize together from liquid phase O. | 0 | fixed $T$ (eutectic temperature) and fixed composition (eutectic composition $O$) |

## Cooling Patterns

We describe below the expected pattern of cooling curves for a few typical cases.

*Pure components*

| | | | | |
|---|---|---|---|---|
| Rapid cooling of liquid | → | Halt; solidification of the compound | → | Cooling of solid compound |

*Composition lying anywhere between* AO *or* BO

| | | | | |
|---|---|---|---|---|
| Rapid cooling of liquid | → | Break; solidification of A or B | → | Slower cooling; solidification continues |
| | | | | ↓ |
| | | Cooling of solid mixture | ← | Horizontal portion; eutectic point |

*Eutectic composition*

| | | | | |
|---|---|---|---|---|
| Rapid cooling of liquid | → | Horizontal portion; eutectic point | → | Cooling of solid mixture |

These cooling patterns are also shown in Fig. 6.4.1.

***Example 6.4.1***

The solubility of KBr in water is

| | | | | | | |
|---|---|---|---|---|---|---|
| $\theta_c$/°C | 0 | 20 | 40 | 60 | 80 | 100 |
| g KBr/g $H_2O$ | 0.54 | 0.64 | 0.76 | 0.86 | 0.95 | 1.04 |

In a one molal solution, KBr depresses the freezing point of water by 3.29 °C. Estimate graphically the eutectic temperature for the KBr-$H_2O$ system.

***Solution***

In one molal solution of KBr, we have 110 g (molar mass of KBr) of KBr in 1 000 g of water or 0.110 g of KBr per g of water. We are given that ice separates from one molal solution of KBr at –3.29 °C, since ice separates from pure water at 0 °C. With these two temperatures, viz., 0 °C and –3.29 °C, we can draw the freezing point curve for ice as shown by the line AB in Fig. 6.4.2. On the other hand, we can draw the solubility curve CD from the given data on the solubility of KBr. If the two lines AB and CD are extended, we get the eutectic point at –13 °C.

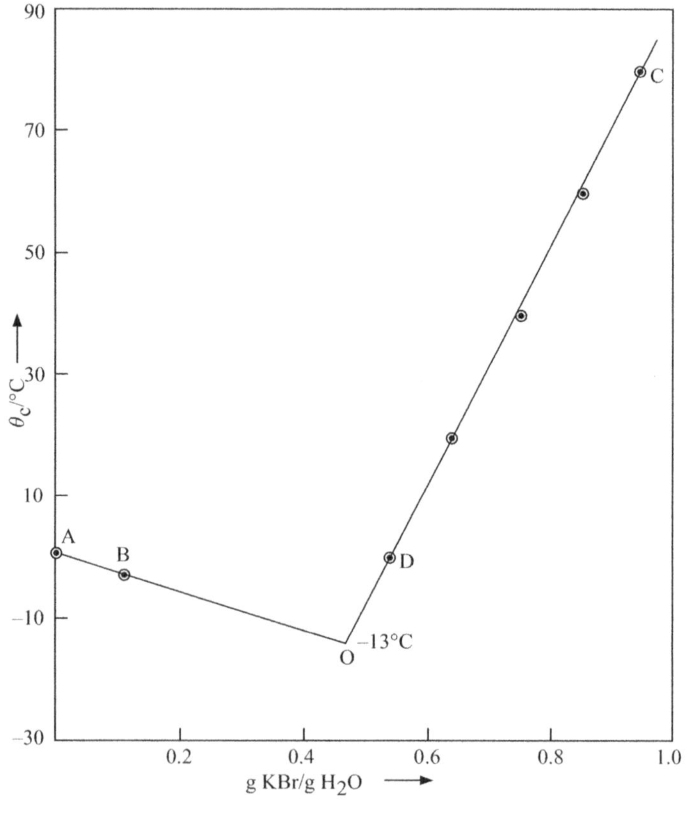


**Fig. 6.4.2 Required phase diagram**

***Example 6.4.2***

(a) KBr is recrystallized from water by saturating the solution at 100 °C, then cooling to 20 °C; the crystals obtained are redissolved in water and the solution evaporated until it is saturated at 100 °C. Cooling to 20 °C produces a second crop of crystals. What is the per cent yield of pure KBr after these two crystallizations? Use data given in Example 6.4.1.
(b) Two crops of KBr crystals are obtained as follows. A solution saturated at 100 °C is cooled to 20 °C; after filtering off the first crop, the mother liquor is evaporated until the

solution is saturated at 100 °C; cooling to 20 °C produces the second crop. What fraction of KBr is recovered in the two crops by this method? Use data of Example 6.4.1.

***Solution***

(**a**) At 100 °C, the solubility of KBr is 1.04 g/g $H_2O$. If this solution is cooled to 20 °C, solubility becomes 0.64 g KBr/g $H_2O$. Hence, the mass of KBr crystallized is (1.04 – 0.64) g, i.e. 0.4 g KBr. Now this 0.4 g KBr is again redissolved to give a saturated solution at 100 °C. The mass of water which need to be added is

$$\frac{1\ \text{g H}_2\text{O}}{1.04\ \text{g KBr}} \times 0.4\ \text{g KBr} = \frac{0.4}{1.04}\ \text{g H}_2\text{O}$$

If the resultant solution is cooled to 20 °C, the mass of KBr remaining in the solution will be equal to

$$\frac{0.64\ \text{g KBr}}{1\ \text{g H}_2\text{O}} \times \left(\frac{0.4}{1.04}\ \text{g H}_2\text{O}\right) = \frac{0.64 \times 0.4}{1.04}\ \text{g KBr}$$

Thus, the mass of KBr separated at this stage will be given by

$$0.4\ \text{g} - \frac{0.64 \times 0.4}{1.04}\ \text{g} = \frac{0.4 \times 0.4}{1.04}\ \text{g} = \frac{0.16}{1.04}\ \text{g}$$

Thus, out of 1.04 g of KBr which we had originally in the solution, (0.16/1.04) g of KBr has been separated after these two crystallizations. Hence per cent yield of pure KBr is

$$\frac{0.16}{1.04}\ \text{g} \times \frac{100}{1.04\ \text{g}} = 14.79$$

(**b**) At 100 °C, the solubility is 1.04 g KBr/g $H_2O$. If this solution is cooled to 20 °C, the solubility becomes 0.64 g KBr/g $H_2O$. Hence the mass of KBr crystallized is (1.04 – 0.64) g, i.e. 0.4 g KBr. Now the mother liquor is evaporated until the solution is saturated at 100 °C. The mass of water that remained in the above evaporation process is equal to

$$\frac{1\ \text{g H}_2\text{O}}{1.04\ \text{g KBr}} \times 0.64\ \text{g KBr} = \frac{0.64}{1.04}\ \text{g H}_2\text{O}$$

If the resultant solution is cooled to 20 °C, the mass of KBr that remained in the solution is equal to

$$\left(\frac{0.64\ \text{g KBr}}{1\ \text{g H}_2\text{O}}\right)\left(\frac{0.64}{1.04}\ \text{g H}_2\text{O}\right)$$

Thus, the mass of KBr separated in the second crop becomes

$$0.64\ \text{g} - \frac{0.64 \times 0.64}{1.04}\ \text{g} = \frac{0.64 \times 0.4}{1.04}\ \text{g}$$

The total mass of KBr separated in the two crops

$$0.4\ \text{g} + \frac{0.4 \times 0.64}{1.04}\ \text{g} = \frac{0.4 \times 1.68}{1.04}\ \text{g}$$

Hence

$$\text{Per cent of KBr recovered}\ \left(\frac{0.4 \times 1.68}{1.04}\text{g}\right)\left(\frac{100}{1.04\ \text{g}}\right) = 62.15$$

---

**Theoretical Aspects of Fig. 6.4.1**

The curve AO of Fig. 6.4.1 gives the freezing point curve of pure solid A and thus it tells us the first temperature at which pure solid A precipitates (alternatively, at which the last solid A melts) for a system whose starting composition lies between the points C and O. In other words, the curve AO is the saturation solubility curve for the component A. As mentioned earlier, the system anywhere on the line AO is univariant indicating that for a given composition, the temperature at which initial solidification starts has a fixed value.

For ideal systems, the variation of solubility of a component with temperature is given by the van't Hoff relation, which was derived thermodynamically in Section 2.12. The expression at $p = 1$ bar is:

$$\ln x = -\frac{\Delta_{\text{fus}}H_{\text{m}}^{\circ}}{R}\left(\frac{1}{T} - \frac{1}{T^*}\right)$$

where $x$ is the saturation solubility in amount fraction of the component in ideal solution at temperature $T$, and $T^*$ is the freezing point of the pure component. For the component A, we have

$$\ln x_{\text{A}} = -\frac{\Delta_{\text{fus}}H_{\text{m, A}}^{\circ}}{R}\left(\frac{1}{T_{\text{A}}} - \frac{1}{T_{\text{A}}^*}\right) \tag{6.4.1}$$

Now $x_{\text{A}}$ may have value anywhere between 0 and 1, thus $\ln x_{\text{A}}$ is negative. Since $\Delta_{\text{fus}}H_{\text{m, A}}$ is positive (fusion is an endothermic process), it follows that $T_{\text{A}}^*$ is greater than $T_{\text{A}}$, i.e. the temperature at which initial solidification of pure A starts from the liquid solution is smaller than the freezing point of pure A.

Similarly, the curve BO of Fig. 6.4.1 represents the solubility curve of the component B. For ideal systems, this solubility curve follows the relation

$$\ln x_{\text{B}} = -\frac{\Delta_{\text{fus}}H_{\text{m, B}}^{\circ}}{R}\left(\frac{1}{T_{\text{B}}} - \frac{1}{T_{\text{B}}^*}\right) \tag{6.4.2}$$

where $T_{\text{B}}$ is the temperature at which initial solidification of the pure component B starts from a liquid solution in which the amount fraction of the component B is $x_{\text{B}}$ and $T_{\text{B}}^*$ is the freezing point of the pure solid B.

At the eutectic point O, the two solubility curves AO and BO meet each other. At the eutectic temperature $T_{\text{e}}$, Eqs (6.4.1) and (6.4.2) become

$$\ln x_{\text{A}} = -\frac{\Delta_{\text{fus}}H_{\text{m, A}}^{\circ}}{R}\left(\frac{1}{T_{\text{e}}} - \frac{1}{T_{\text{A}}^*}\right) \tag{6.4.3}$$

$$\ln x_{\text{B}} = -\frac{\Delta_{\text{fus}}H_{\text{m, B}}^{\circ}}{R}\left(\frac{1}{T_{\text{e}}} - \frac{1}{T_{\text{B}}^*}\right) \tag{6.4.4}$$

Eliminating $T_{\text{e}}$, we get

$$-\frac{R}{\Delta_{\text{fus}}H_{\text{m, A}}^{\circ}}\ln x_{\text{A}} + \frac{1}{T_{\text{A}}^*} = -\frac{R}{\Delta_{\text{fus}}H_{\text{m, B}}^{\circ}}\ln x_{\text{B}} + \frac{1}{T_{\text{B}}^*}$$

Since $x_A + x_B = 1$, we have

$$\frac{1}{T_A^*} - \frac{1}{T_B^*} = \frac{R}{\Delta_{fus}H^\circ_{m,A}} \ln x_A - \frac{R}{\Delta_{fus}H^\circ_{m,B}} \ln(1 - x_A) \qquad (6.4.5)$$

Thus knowing $T_A^*$, $T_B^*$, $\Delta_{fus}H^\circ_{m,A}$ and $\Delta_{fus}H^\circ_{m,B}$, Eq, (6.4.5) can be solved for $x_A$ which represents the composition of the liquid mixture at the eutectic point. On substituting $x_A$ in Eq. (6.4.3) or Eq. (6.4.4), we can get the eutectic temperature of the system.

---

***Example 6.4.3***

Sodium chloride-sodium sulphate system exhibits the simple eutectic phase diagram with the following characteristics:

Melting point (m.pt.) of sodium chloride = 1 074 K
Enthalpy of fusion of sodium chloride = 28.87 kJ mol$^{-1}$
Enthalpy of fusion of sodium sulphate = 24.06 kJ mol$^{-1}$

Eutectic composition = 48.2 mol per cent of sodium sulphate. Assuming ideal behaviour, compute, (i) the m.pt. of sodium sulphate, and (ii) the eutectic temperature of the system,

***Solution***

(**i**) From Eq. (6.4.5), we get

$$\frac{1}{T_A^*} - \frac{1}{T_B^*} = \frac{R}{\Delta_{fus}H^\circ_{m,A}} \ln x_A - \frac{R}{\Delta_{fus}H^\circ_{m,B}} \ln x_B$$

where $T_A^*$ is the m.pt. of pure A, $T_B^*$ is the m.pt. of pure B, $x_A$ is amount fraction of A at the eutectic point, $\Delta_{fus}H^\circ_{m,A}$ is the molar enthalpy of fusion of A, and $\Delta_{fus}H^\circ_{m,B}$ is the molar enthalpy of fusion of B.

Identifying A with NaCl and B with $Na_2SO_4$ and substituting the given data in the above equation, we get

$$\frac{1}{1\,074\ \text{K}} - \frac{1}{T_B^*} = \frac{(8.314\ \text{J K}^{-1}\ \text{mol}^{-1})}{(28\,870\ \text{J mol}^{-1})} \times 2.303 \times \log(0.518)$$
$$- \frac{(8.314\ \text{J K}^{-1}\ \text{mol}^{-1})}{(24\,060\ \text{J mol}^{-1})} \times 2.303 \times \log(0.482)$$

$$0.000\,931\ \text{K}^{-1} - \frac{1}{T_B^*} = -0.000\,189\,4\ \text{K}^{-1} + 0.000\,252\,2\ \text{K}^{-1}$$

$$\frac{1}{T_B^*} = 0.000\,868\,2\ \text{K}^{-1}$$

Thus $\quad T_B^* = 1\,152\ \text{K}$

(**ii**) From Eq. (6.4.3), we get

$$\ln x_A = -\frac{\Delta_{fus}H^\circ_{m,A}}{R}\left(\frac{1}{T_e} - \frac{1}{T_A^*}\right) \quad \text{or} \quad \frac{1}{T_e} = -\frac{R}{\Delta_{fus}H^\circ_{m,A}} \ln x_A + \frac{1}{T_A^*}$$

Substituting the given data, we get

$$\frac{1}{T_e} = -\frac{(8.314\ \text{J K}^{-1}\ \text{mol}^{-1})}{(28\,870\ \text{J mol}^{-1})} \times 2.303 \log(0.518) + \frac{1}{1\,074\ \text{K}}$$

$$= 0.000\,189\,4\ \text{K}^{-1} + 0.000\,931\ \text{K}^{-1} = 0.001\,120\,4\ \text{K}^{-1}$$

Hence $\quad T_e = 892.5\ \text{K}$

---

**Derivation of the Lever Rule**

The area BODB (or AOCA) in Fig. 6.4.1 represents the region where solid B (or A) is at equilibrium with the liquid phase whose composition lies on the curve BO (or AO). The composition of the liquid phase corresponding to the given point can be obtained by drawing a horizontal line (known as tie line) though the given point. The point of intersection of the tie line with the curve BO (or AO) gives the composition of the liquid phase. For example, at the point c shown in Fig. 6.4.1, the composition of the liquid phase is given by the point f. In fact, a point anywhere on the line fg represents the liquid phase f in equilibrium with solid B. The only thing that varies along the line is the relative amount of solid and liquid phases. For example, at point c, we have

$$\frac{\text{Amount of solid B}}{\text{Amount in the liquid phase}} = \frac{\text{fc}}{\text{gc}} \tag{6.4.6}$$

Equation (6.4.6) is the expression of the lever rule and can be derived as follows.

Let $X_B$ be the amount fraction of B in the overall system at the point c. Since at c, only B is solidified, we will have

$$X_B = \frac{n_{B(s)} + n_{B(l)}}{n_{A(l)} + n_{B(s)} + n_{B(l)}} \tag{6.4.7}$$

If $x_B$ is the amount fraction of B in the liquid phase at the point f, we have

$$x_B = \frac{n_{B(l)}}{n_{A(l)} + n_{B(l)}} \tag{6.4.8}$$

Now from Fig. 6.4.1, we have

$$\text{fc} = X_B - x_B \tag{6.4.9}$$

$$\text{gc} = 1 - X_B \tag{6.4.10}$$

Substituting Eq. (6.4.8) in Eq. (6.4.9), we have

$$\text{fc} = X_B - \frac{n_{B(l)}}{n_{A(l)} + n_{B(l)}}$$

or

$$\{n_{A(l)} + n_{B(l)}\}\,(\text{fc}) = \{n_{A(l)} + n_{B(l)}\}\,X_B - n_{B(l)} \tag{6.4.11}$$

Multiplying Eq. (6.4.10) by $n_{B(s)}$, we get

$$n_{B(s)}\,(\text{gc}) = n_{B(s)} - n_{B(s)}\,X_B \tag{6.4.12}$$

Subtracting Eq. (6.4.12) from Eq. (6.4.11), we get

$$\{n_{A(l)} + n_{B(l)}\}\,(\text{fc}) - n_{B(s)}\,(\text{gc}) = X_B\{n_{A(l)} + n_{B(l)} + n_{B(s)}\} - \{n_{B(l)} + n_{B(s)}\}$$

According to Eq. (6.4.7), the right side of the above expression is zero. Hence, we have

$$\frac{(\text{fc})}{(\text{gc})} = \frac{n_{B(s)}}{n_{A(l)} + n_{B(l)}} = \frac{\text{Amount of solid B}}{\text{Amount of liquid phase}}$$

which is the required expression of the lever rule (Eq. 6.4.6).

***Example 6.4.4***

What mass of antimony crystallizes from 20 kg of Pb-Sb melt containing 35 mass % Pb after cooling to a temperature of 400 °C? What is the maximum mass of pure Sb that can be obtained from the melt? You are provided with the following data: m.pt. of Pb = 327 °C, m.pt. of Sb = 631 °C. Breaks on the cooling curve in thermal analysis were found for the following conditions:

| Mass % Pb | 30 | 44 | 66 | 80 | 86 (eutectic) |
|---|---|---|---|---|---|
| $\theta_c$/°C | 550 | 500 | 400 | 300 | 250 |

***Solution***

First of all, the phase diagram of the Pb-Sb system is plotted from the given data. It is shown in Fig. 6.4.3.

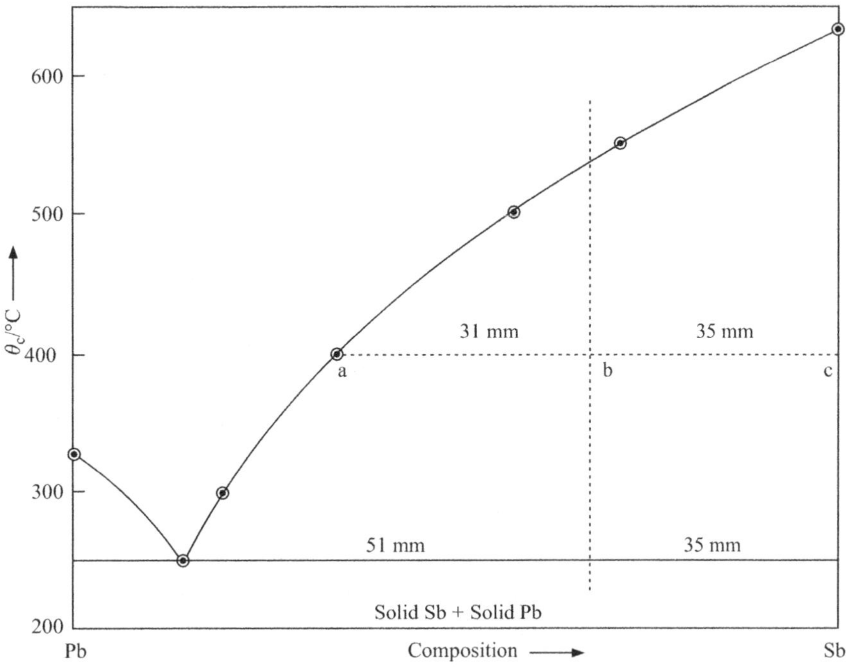


**Fig. 6.4.3**

At 400 °C the relative ratio of Sb and the melt in the given system of 35 mass % Pb is given by

$$\frac{\text{Mass of melt}}{\text{Mass of Sb}} = \frac{\text{bc}}{\text{ba}} = \frac{35\text{ mm}}{31\text{ mm}}$$

$$\frac{\text{Mass of (melt + Sb)}}{\text{Mass of Sb}} = \frac{35 + 31}{31}$$

Therefore

$$\frac{\text{Mass of Sb}}{\text{Total mass of melt}} = \frac{31}{66}$$

Hence $\quad \text{Mass of Sb} = \left(\frac{31}{66} \times 20\text{ kg}\right) = 9.39\text{ kg}$

Since maximum mass of Sb will be obtained just above the eutectic point, we have

$$\frac{\text{Mass of melt}}{\text{Mass of Sb}} = \frac{35\ \text{mm}}{51\ \text{mm}} \quad \text{or} \quad \frac{\text{Mass of Sb}}{\text{Mass of (Sb + melt)}} = \frac{51}{(51 + 35)}$$

Hence $\quad \text{Mass of Sb} = \frac{51}{86} \times 20\ \text{kg} = 11.86\ \text{kg}$

**Examples**

A few examples of systems exhibiting simple eutectic phase diagram are given in Table 6.4.2.

**Table 6.4.2** Systems Exhibiting Simple Eutectic Phase Diagrams

| *System* | | | | *Eutectic* | *Eutectic* |
|---|---|---|---|---|---|
| A | *m.pt.* °C | B | *m.pt.* °C | *temperature* °C | *composition* |
| Sodium sulphate | 884 | Sodium chloride | 800 | 628 | 48.2 mol % A |
| Naphthalene | 80.0 | Benzoic acid | 120 | 67–69 | 30 mol % B |
| Naphthalene | 80.0 | α-Naphthol | 95.0 | 54.1 | 48.7 mol % B |
| Succinimide | 123 | α-Naphthol | 95.0 | 57.0 | ≃ 70 mass % B |
| Benzamide | 124.8 | α-Naphthol | 92.0 | 38.1 | ≃ 60 mass % B |
| Benzamide | 124.8 | β-Naphthol | 122.0 | 54.1 | ≃ 60 mass % B |
| Urea | 131 | Benzoic acid | 121 | 76.5 | ≃ 70 mass % B |
| Acetanilide | 112 | Benzoic acid | 121.4 | 76.0 | 42.4 mol % B |
| Resorcinol | 115 | Succinic acid | 183 | 100 | 12.5 mass % B |
| α-Naphthol | 95 | Cinnamic acid | 133 | 68 | 37 mass % B |
| Diphenyl | 69.0 | Benzophenone | 47.7 | 25.2 | 60.7 mol % B |
| Stilbene | 120.4 | Benzil | 94.0 | 76.6 | 71.5 mol % B |
| β-Naphthol | 121 | Cinnamic acid | 133 | 87 | 47 mass % B |
| Benzoic acid | 121 | Cinnamic acid | 133 | 81 | 43.5 mol % B |
| Benzoic acid | 121.4 | Resorcinol | 110 | 86 | 40 mol % A |
| Resorcinol | 115 | Cinnamic acid | 133 | 87 | 41 mass % B |

†Temperatures mentioned here are the reported temperatures in literature.

## 6.5 CRYSTALLIZATION OF PURE COMPONENTS AND ONE OF THE SOLIDS EXISTS IN MORE THAN ONE CRYSTALLINE FORM

Let out of the two components A and B, the component B exist in two crystalline forms, namely, α-form and β-form. Let α-form be stable at lower temperatures and β-form at higher temperatures. If $T_i$ is the transition temperature, we will have

$$\alpha\text{-form} \underset{t < T_i}{\overset{t > T_i}{\rightleftharpoons}} \beta\text{-form}$$

The phase diagram of the present system along with the expected cooling curves is shown in Fig. 6.5.1.

**Description of the Phase Diagram**

From the cooling curves shown in Fig. 6.5.1, the following information can be derived.

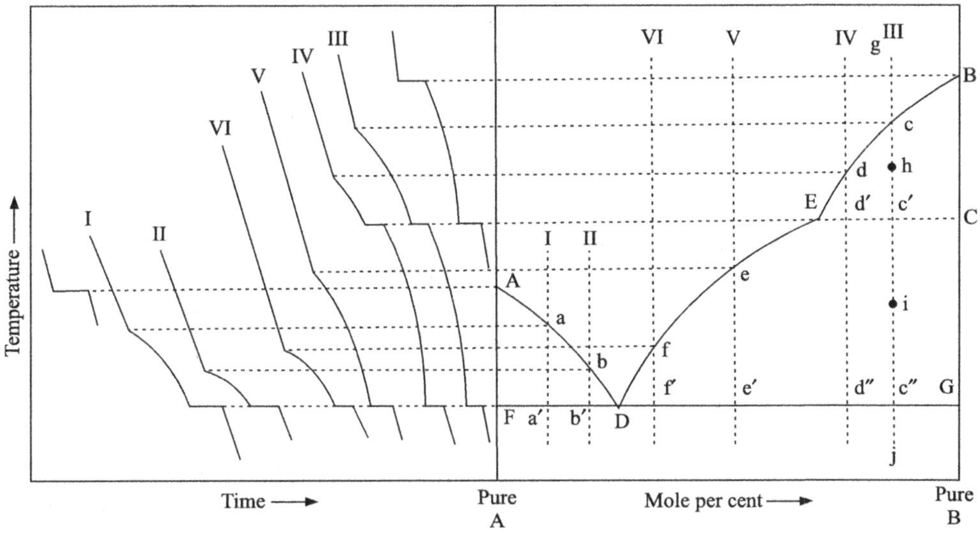


**Fig. 6.5.1 Simple eutectic phase diagram for a system exhibiting crystalline modification of one of the components**

- From the cooling curve of pure component A, the point A representing the freezing point of A is obtained.
- From the cooling curve of pure component B, the points B and C representing, respectively, the freezing point and the transition temperature of B are obtained.
- From the cooling curves I and II of two systems of B in A, the points a and b representing respectively the start of solidification of A and the points a′ and b′ representing respectively the start of solidification of both A and B ($\alpha$-form) together are obtained. Since the addition of B in A lowers the freezing point of the latter, the temperatures corresponding to the points a and b are smaller than the freezing point of pure A.
- From the cooling curves III and IV of two systems of A in B, the points c and d representing respectively the start of solidification of B($\beta$-form); the points c′ and d′ representing respectively the transition of B($\beta$-form) to B($\alpha$-form), and the points c″ and d″ representing respectively the start of solidification of both B($\alpha$-form) and A together are obtained.
- From the cooling curves V and VI of two liquid systems of A and B, the points e and f representing respectively the start of solidification of B($\alpha$-form) and the points e′ and f′ representing the start of solidification of both B($\alpha$-form) and A together are obtained.
- If a smooth curve passing through the points A, a and b is drawn, we get the curve Aab which represents the start of freezing of pure component A from liquid solutions of different compositions.
- Similarly, the curve passing through the points B, c and d, and that passing through the points e and f give the freezing point curves of B($\beta$-form) and B($\alpha$-form), respectively.

- A horizontal line FG passing through the points a′, b′, f′, e′, d″ and c″ gives the temperature at which solid A and solid B(α-form) start crystallizing together.
- The lines Aab and ef on extending meet at the point D. This point, which also lies on the line FG, gives the composition of liquid phase from which two solids A and B(α-form) crystallize together. The overall composition of the solids (i.e. amount of A plus amount of B) separated is the same as that of the liquid phase, vis-à-vis, the composition of the liquid phase remains unchanged during the solidification of A and B(α-form). The temperature of the system remains constant until the entire liquid phase is solidified. This is revealed by the complete arrest of cooling curves at this temperature.
- If a horizontal line passing through the points d′, c′ and C is drawn, we get the line Cc′d′ over which the transition of B(α-form) to B(β-form) occurs.
- If the lines Bcd, fe and Cc′d′ are extended, they meet at the point E which gives the composition of liquid phase at the transition temperature.

**Sequence of Effects Produced on Cooling Liquid Mixtures**

The phase diagram shown in Fig. 6.5.1 can be easily labelled by considering the behaviour of a system from which solid B crystallizes on cooling. If, for example, the liquid phase given by the point g is cooled, the state of the overall system is moved vertically downward along the isopleth gchc′ic″j. The system remains in liquid phase up to the point c. At c solid B(β-form) starts crystallizing. The separation of B from the liquid phase moves its composition towards A-axis. If the crystallization of B is to be continued, the liquid phase has to be cooled so that its composition moves on the line cdE. Thus, as cooling is continued, more and more of solid B(β-form) separates and the composition of liquid phase moves along cdE. For example, if the system has been cooled up to the overall state point h, the system now consists of solid B(β-form) in equilibrium with the liquid phase of composition corresponding to the point d. This point is obtained by drawing a tie line (horizontal line) from the point h which meets the curve BE at d. The relative amounts of two phases can be determined with the help of the lever rule.

If the cooling of the system is done up to the point c′, the system will contain three phases, viz., solid B(β-form), solid B(α-form) and the liquid phase E, in equilibrium with one another. Continued cooling of the system at c′ merely causes the polymorphic change of solid B(β-form) into B(α-form) without any change in temperature. This is revealed by an arrest of cooling curve at point c′. Once the transition is complete, the system exhibits cooling during which more and more of solid B(α-form) separates and the composition of liquid phase moves along ED. When the system is cooled up to the point c″, the composition of the liquid phase has moved up to the point D. Further cooling results into the solidification of B(α-form) and A together from the liquid phase D without any change in temperature. The composition of overall solids separated is the same as that of the liquid phase, vis-à-vis, the composition of the liquid phase remains constant throughout. Once the entire liquid phase has solidified, the system exhibits cooling of solid mixture of A and B(α-form).

For the system V (or VI), the system will remain liquid up to the point e (or f). At this point, solid B($\alpha$-form) starts solidifying and continues to solidify till the point e′ (or f′) is reached. The composition of liquid phase moves along efD. At e′, the liquid composition corresponds to the point D. Here solids B($\alpha$-form) and A start crystallizing together.

For the system I (or II), it is the solid A which starts crystallizing when the system reaches the point a (or b). Solid A continues to solidify till the point a′ (or b′) is reached. At a′ (or b′), solids A and B($\alpha$-form) crystallize together from the liquid phase represented by the point D.

### Conclusion of the Phase Diagram

From the above analysis, the following conclusions can be derived.

- A system within the area ADFA represents solid A in equilibrium with liquid phase whose composition lies on the line AD. The point at which the tie line drawn from the given point meets the curve AD gives the required composition. The relative amounts of the two phases can be determined by making use of the lever rule.
- A system within the area DECGD represents solid B($\alpha$-form) in equilibrium with a liquid phase whose composition lies on the line DE.
- A system within the area BECB represents solid B($\beta$-form) in equilibrium with a liquid phase whose composition lies on the line BE.
- A point within the area ADFA, DECGD and BECB represents an univariant system ($F = C + 1 - P = 2 + 1 - 2 = 1$). Thus for a given temperature, the composition of the liquid phase in equilibrium with solid A or solid B($\alpha$-form) or B($\beta$-form) has a fixed value.
- A point on the line FDG or EC represents a nonvariant system ($F = C + 1 - P = 2 + 1 - 3 = 0$).
- A point below the line FDG represents a mixture of solids A and B($\alpha$-form).
- The point D is the eutectic point at which solids A and B($\alpha$-form) are in equilibrium with liquid phase represented by the point D.

### Labelling of the Phase Diagram

Table 6.5.1 describes, in brief, the phase diagram shown in Fig. 6.5.1

### Cooling Patterns

The nature of cooling curve will depend upon the composition of the system. We now describe the expected pattern of cooling curves for a few typical systems shown in Fig. 6.5.1.

(i) *Composition lying between* AD *and* DE

Rapid cooling of liquid phase → Break at the appearance of solid → Slower cooling; soldification continues
↓
Horizontal portion; eutectic point → Cooling of solid mixture

(ii) *Eutectic composition:*

Rapid cooling of liquid phase → Horizontal portion; eutectic point → Cooling of solid mixture

**Table 6.5.1** Description of the Phase Diagram of Fig. 6.5.1

| *Location of the system* | *Status of the system* |
|---|---|
| A | Freezing temperature of A |
| B | Freezing temperature of B |
| C | Transition temperature of B |
| AD | Freezing point curve of A |
| DE | Freezing point curve of B(α-form) |
| BE | Freezing point curve of B(β-form) |
| A-axis | Solid A up to the point A, and liquid A beyond A |
| B-axis | Solid B(α-form) up to the point C, solid B(β-form) from C to B and liquid B beyond the point B |
| FDG | Solids A and B(α-form) in equilibrium with liquid phase D |
| EC | Solids B(α-form) and B(β-form) in equilibrium with liquid phase E |
| Area AFDA | Solid A ⇌ liquid (whose composition lies on AD) |
| Area DECGD | Solid B(α-form) ⇌ liquid (whose composition lies on DE) |
| Area BECB | Solid B(β-form) ⇌ liquid (whose composition lies on EB) |
| Area above ADEB | Liquid phase |
| Area below FDG | Solid mixtures |

(iii) *Composition lying between* EB

| | | | | |
|---|---|---|---|---|
| Rapid cooling of liquid phase | → | Break at the appearance of B(β-form) | → | Slower cooling, soldification of solid B(β-form) continues ↓ |
| Horizontal portion; eutectic point ↓ Cooling of solid mixture | → | Slower cooling; appearance of B (α-form) continues | ← | Horizontal portion; conversion of solid B(β-form) to B(α-form) |

(iv) *Composition corresponding to the Point* E

| | | | | |
|---|---|---|---|---|
| Repaid cooling of liquid phase | → | Break at the appearance of solid B(β-form) | → | Slower cooling; appearance of solid B(α-form) continues ↓ |
| | | Cooling of solid mixture | ← | Horizontal portion; eutectic point |

**Examples**

Examples in which one of the components exhibits more than one crystalline form are given below.

(i) Solid carbon tetrachloride exists in two forms, with a transition point at −48 °C. All systems with $CC1_4$ as one of the components show a break in the phase diagram at this transition temperature provided the eutectic temperature of the system lies below −48 °C.

(ii) Solid ammonium nitrate exists in five crystalline forms with the following transition temperatures:

$$\text{Cubic} \underset{125\ ^\circ\text{C}}{\rightleftharpoons} \text{Rhombohedral} \underset{48\ ^\circ\text{C}}{\rightleftharpoons} \alpha\text{-Rhombic} \underset{32\ ^\circ\text{C}}{\rightleftharpoons} \beta\text{-Rhombic} \underset{-18\ ^\circ\text{C}}{\rightleftharpoons} \text{Tetragonal}$$

Consequently, four breaks corresponding to these transitions are observed in the phase diagram of ammonium nitrate-water system.

**Theoretical Aspects of Fig. 6.5.1**

The entire phase diagram of Fig. 6.5.1 may be constructed by drawing the simple eutectic phase diagram AO′B of the system A – B(β) and then superimosping on it the simple eutectic phase diagram AOB′ of the system A – B(α) as shown in Fig. 6.5.2. In order to construct these two phase diagrams, we should have data on $T^*_\text{A}$, $\Delta_\text{fus}H^\circ_\text{m, A}$, $T^*_{\text{B}(\beta)}$, $\Delta_\text{fus}H^\circ_{\text{m, B}(\alpha)}$, $\Delta_\text{fus}H^\circ_{\text{m, B}(\beta)}$ and the melting point of the solid B(α).

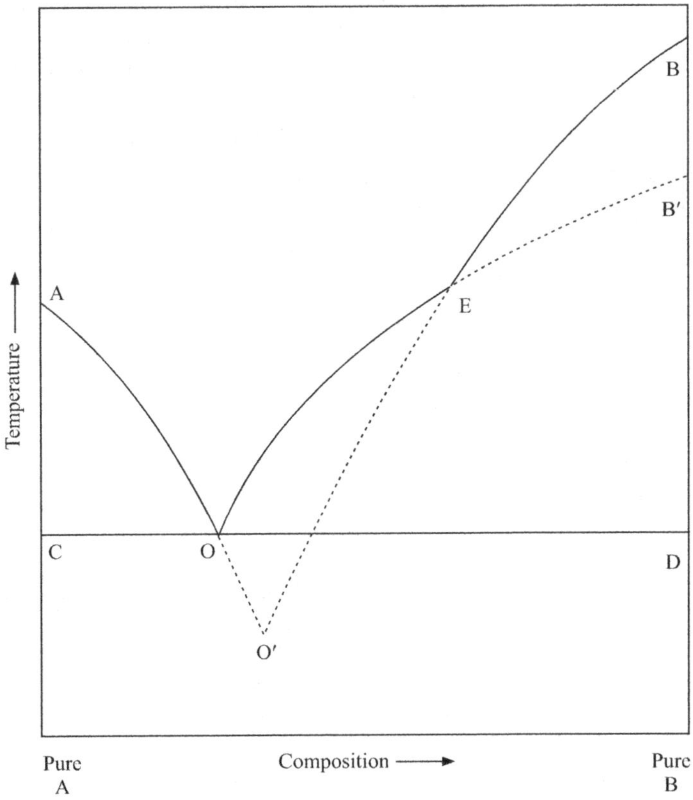


**Fig. 6.5.2 Construction of the phase diagram for a system exhibiting allotropic modification of one of the components**

## 6.6 FORMATION OF A COMPOUND STABLE UP TO ITS MELTING POINT

**Application of the Phase Rule**

Let the two components A and B form a stable compound AB. This compound will have its own characteristic melting point which may be greater or smaller than the melting points of pure components A and B. Since the compound is stable and melts only at its melting point, it is obvious that the compound on heating will remain in the solid phase up to its melting point where it melts to produce a liquid whose composition is the same as that of the compound. At this stage, the system becomes nonvariant. This may be proved with the help of the lever rule:

$$F = (C - r) + 1 - P$$

Here $r = 1$, since the compositions of solid and liquid phases are identical. Thus

$$F = (2 - 1) + 1 - 2 = 0$$

Hence, the temperature of the system remains constant till the entire solid compound melts. Such a melting point where both solid and liquid of the same composition can coexist is known as the *congruent melting point.*

Conversely, if a liquid of composition AB is cooled, it will exhibit uniform cooling up to the freezing point of solid AB. At this temperature, solid AB separates and continues to separate without any change in temperature until whole of the liquid phase is solidified. After this, the system will exhibit cooling of solid AB. Thus, the cooling pattern of the liquid phase of composition AB is similar to that of a pure component.

**Description of the Phase Diagram**

The phase diagram of the present system along with the expected cooling curves is shown in Fig. 6.6.1.

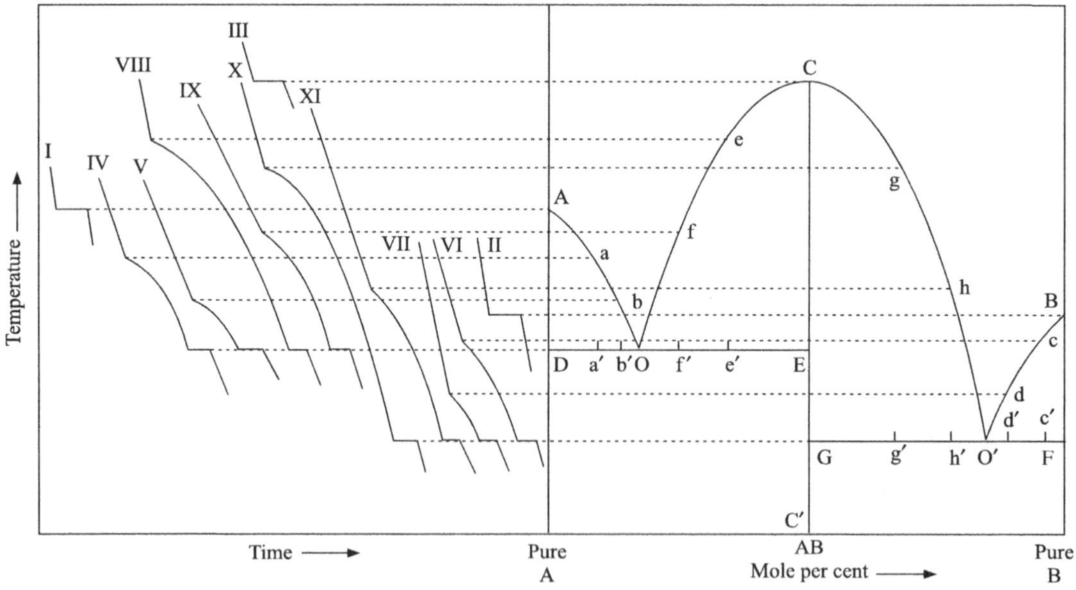


**Fig. 6.6.1 Phase diagram of a system exhibiting a congruent melting point**

From the cooling curves shown in Fig. 6.6.1, the following information can be derived.

- From the cooling curve I and II of pure components, the points A and B representing respectively the freezing points of A and B are obtained.

- The point C′ represents the composition of the compound AB. If this compound is heated, its state point will move vertically along the line C′C. At C, the compound melts to give a liquid of composition AB. Alternatively, the point C can be obtained from the cooling curve III of a liquid of composition AB.
- The addition of a solute in a solvent depresses the freezing point of the latter. Keeping this in mind, the following facts are revealed:

  (i) From the cooling curves IV and V of two liquid systems of B in A, the points a and b representing respectively the start of freezing of A are obtained.
  (ii) From the cooling curves VI and VII of two liquid systems of A in B, the points c and d representing respectively the start of freezing of B are obtained.
  (iii) If the component A or B is added to the liquid of composition AB, it will depress the freezing point of the compound AB. Thus, from the cooling curves VIII to XI, the points e, f, g and h representing respectively the start of freezing of AB are obtained.

- If a line passing through the points A, a and b is drawn, we get the freezing point curve of A.
- If a line passing through the points B, c and d is drawn, we get the freezing point curve of B.
- If a line passing through the points C, e and f, and C, g and h are drawn, we get the freezing point curves of AB.
- If the lines Aab and Cef are extended, they meet at the point O which gives the eutectic point where solid A and solid AB are present in equilibrium with liquid phase O.
- Through the point O, the tie-line DOE can be drawn. A point on this line represents solid A (point D) and solid AB (point E) in equilibrium with liquid phase of composition corresponding to the point O.
- The lines Cgh and Bcd on extending meet at O′ which gives the eutectic point where solid B and solid AB are present in equilibrium with liquid phase O′.
- Through the point O′, the tie line GO′F can be drawn. A point on this line represents solid B (point F), and solid AB (point G) in equilibrium with liquid phase O′
- The systems at the arrest of temperature on the cooling curves IV, V, VIII and IX are shown respectively by the points a′, b′, e′ and f′ on the line DE. Similarly, the points, c′, d′, g′, and h′ represent the systems at the arrest of temperature on the cooling curves VI, VII, X and XI, respectively.

**Labelling of the Phase Diagram**

The labelling of the phase diagram shown in Fig. 6.6.1 is described in Table 6.6.1

**Theoretical Aspects of Fig. 6.6.1**

The phase diagram as given in Fig. 6.6.1 may be considered to be made up of two simple eutectic diagrams placed side by side. To the left of the dividing line CC′, we have the simple eutectic phase diagram of A and AB, whereas to the right of this line we have that of B and AB. Theoretically, the curves OC and O′C

**Table 6.6.1** Labelling of Fig. 6.6.1

| *Location of the system* | *Status of the system* |
|---|---|
| A | Freezing point of A |
| B | Freezing point of B |
| C | Freezing point of AB |
| AO | Freezing point curve of A |
| BO′ | Freezing point curve of B |
| OC and O′C | Freezing point curves of AB |
| O | Eutectic point involving solid A, solid AB and liquid O |
| A-axis | Solid A up to the point A, liquid A beyond A |
| B-axis | Solid B up to the point B, liquid B beyond B |
| C′C | Solid AB, liquid AB beyond C |
| O′ | Eutectic point involving solid B, solid AB and liquid O′ |
| DOE | A system on this line represent solid A and solid AB in equilibrium with liquid O |
| GO′F | A system on this line represents solid B and solid AB in equilibrium with liquid O′ |
| Area AODA | Solid A $\rightleftharpoons$ liquid (whose composition lies on AO) |
| Area BO′FB | Solid B $\rightleftharpoons$ liquid (whose composition lies on BO′) |
| Area COEC | Solid AB $\rightleftharpoons$ liquid (whose composition lies on OC) |
| Area CGO′C | Solid AB $\rightleftharpoons$ liquid (whose composition lies on CO′) |
| Area below DOE | Solid mixtures of A and AB |
| Area below GO′F | Solid mixtures of B and AB |
| Area above AOCO′B | Liquid |

should meet at C to give a sharp point. Instead, we usually observe rounded maximum at this point. This is due to the fact that the compound AB is not very much stable at its melting point. It partly dissociates and the products of dissociation in the liquid phase depresses the actual melting point of the compound with the result that the curve has a flattened portion at the point C.

**Examples**

A few examples of the systems exhibiting one compound formation with their congruent melting points are given in Table 6.6.2.

## FORMATION OF MORE THAN ONE COMPOUND

In many cases, two components on combination form more than one compound. In such cases, the equilibrium phase diagram has a curve analogous to OCO′ for each and every compound. Figure 6.6.2 displays one such case where the two compounds $AB_2$ and $A_2B$ are formed. The melting points of the compounds are given by the points C and D, respectively. This phase diagram may by considered to be formed by the union of three simple eutectic diagrams. These are A – $A_2B$, $A_2B$ – $AB_2$ and $AB_2$ – B. The interpretation of the diagram can be carried out as usual and is shown in Fig. 6.6.2.

**Examples**

A few examples of systems exhibiting the formation of more than one congruent compound are given in Table 6.6.3.

**Table 6.6.2** Systems Exhibiting Compound Formation with Congruent Melting Point

| *System*† | | | | *Compound* | *Congruent m.pt.* | *Eutectic temperature* | *Eutectic composition* |
|---|---|---|---|---|---|---|---|
| A | *m.pt.* °C | B | *m.pt.* °C | | °C | °C | |
| Gold | 1 064 | Tin | 232 | AB | 425 | | |
| Calcium chloride | 777 | Potassium chloride | 776 | AB | 754 | | |
| Diphenylamine | 52.8 | Benzophenone | 47.7 | AB | 40.2 | 34.1 | 70 mol % A |
| | | | | | | 31.9 | 25.5 mol % A |
| Urea | 132 | Phenol | 42.5 | $AB_2$ | 61 | 60.0 | 65 mol % B |
| | | | | | | 34.0 | 90 mol % B |
| Acetamide | 79.4 | Phenol | 41 | $AB_2$ | 43 | 29 | 45 mol % B |
| | | | | | | 30.2 | 90 mol % B |
| Urea | 133 | Resorcinol | 110 | AB | ≃ 102.5 | 86.5 | 26 mol % B |
| | | | | | | ≃ 84 | ≃ 78 mol % B |
| Benzamide | 124.8 | Resorcinol | 108.5 | AB | ≃ 88.5 | 76.1 | ≃ 66.5 % B |
| | | | | | | 80.2 | ≃ 34.5 % B |
| Phenol | 42.5 | Picric acid | 122 | AB | 86 | 36.0 | |
| | | | | | | 80.5 | |

†Temperatures mentioned here are the reported temperatures in literature.

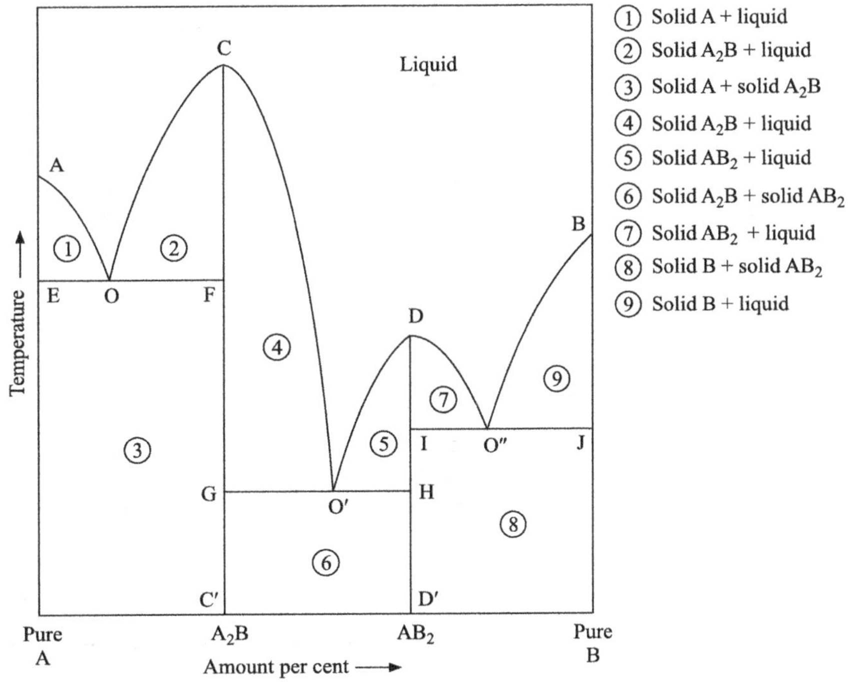


**Fig. 6.6.2 Formation of two congruent compounds**

**Table 6.6.3** System Exhibiting more than One Congruent Compound Formation

| *System* | | | | *Compound* | *Congruent* | *Eutectic temperatures* | |
|---|---|---|---|---|---|---|---|
| A | *m.pt./°C* | B | *m.pt./°C* | | *m.pt./°C* | °C | |
| K | 63.6 | Sb | 613 | | | 485; $(A + A_3B)$ | 34.9 % of Sb |
| | | | | $A_3B$ | 812 | | |
| | | | | AB | 605 | | |
| | | | | | | 400; $(B + AB)$ | 68 % Sb |
| Ag | 960.5 | Sr | 757 | | | | |
| | | | | | | 750 | |
| | | | | $A_4B$ | 781 | | |
| | | | | | | 693 | |
| | | | | $A_5B_3$ | 760 | | |
| | | | | | | 638 | |
| | | | | AB | 680 | | |
| | | | | | | 645 | |
| | | | | $A_2B_3$ | 665 | | |
| | | | | | | 436 | |

***Example 6.6.1***

For the water(A) and ethylene glycol(B) system, the following data were obtained from the cooling curves of different compositions. The solid phase or phases that first appear at the break or halt on cooling (that is, at the beginning of freezing processes) are given in the last column:

| *Solution composition mol % of ethylene glycol* | *Freezing point/°C* | *Solid phases that separate out at the beginning of freezing process* |
|---|---|---|
| 0 | 0 | A |
| 10.5 | –14.1 | A |
| 14.1 | –20.9 | A |
| 18.1 | –28.3 | A |
| 24.9 | –41.8 | A |
| 28.0 | –51.2 | A+B·2A |
| 31.8 | –49.6 | B·2A |
| 34.1 | –49.6 | B·2A |
| 44.0 | –58.6 | B·2A |
| 47.5 | –63.3 | B·2A + 3B·2A |
| 50.2 | –54.6 | 3B·2A |
| 60.9 | –40.7 | 3B·2A |
| 66.0 | –49.4 | 3B·2A + B |
| 67.8 | –45.3 | B |
| 75.3 | –36.4 | B |
| 90.1 | –22.4 | B |
| 100.0 | –12.8 | B |

Plot the phase diagram and label it.

***Solution***

First of all, the points corresponding to the given data were located in a temperature versus composition phase diagram. After this, the points corresponding to the given two compounds B · 2A and 3B · 2A were located on the composition axis and the vertical lines were drawn from these points as shown in Fig. 6.6.3.

From the given data, we find that the temperature at which solid A (ice) separates out from the solution decreases as the amount per cent of B (ethylene glycol) in the solution is increased. This decrease in temperature is continued up to the temperature –51.2 °C where the solution composition is 28 mol per cent of B. At this temperature, two solids, viz., A and B · 2A, are separated out. On increasing the amount per cent of B from 28 to 31.8, we find that there is a separation of only B · 2A and the freezing point of the solution increases from –51.2 to –49.6 °C. Obviously, the temperature –51.2 °C is the lowest temperature where the two solids are separated out and thus must represent eutectic temperature. So, we draw a horizontal line at this temperature starting from the pure solid A line to the vertical line of solid B · 2A. The solid B · 2A has a congruent melting point as the freezing curve of B · 2A has a maximum right above the vertical line of solid B · 2A. A similar analysis indicates that the solid 3B · 2A has a congruent melting point. The temperature –63.3 °C and –49.4 °C where two solid phases separate out are lowest ones and hence these are the eutectic temperatures. Thus, we draw horizontal lines through these points meeting vertical lines of solids B · 2A and 3B · 2A, and solids 3B · 2A and pure B, respectively.

Finally a smooth curve passing through the points of temperature versus composition is drawn. This gives us the phase diagram of the present system. Labelling of different areas is straightforward and is described in Fig. 6.6.3 itself.

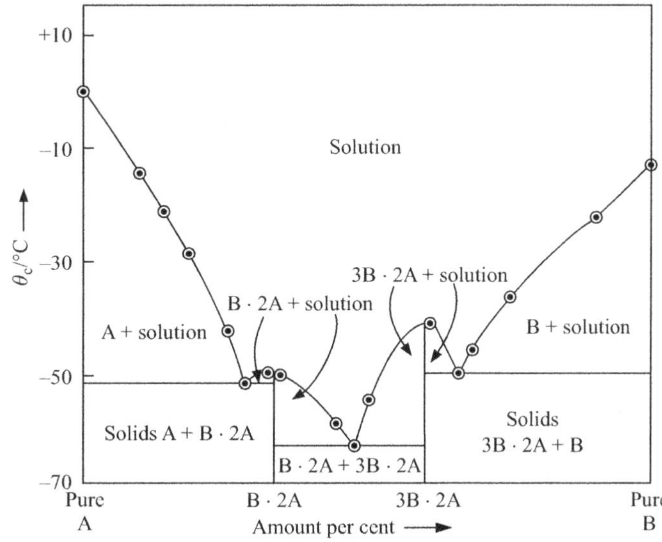


**Fig. 6.6.3 Required phase diagram**

## 6.7 FORMATION OF A COMPOUND WHICH DECOMPOSES BEFORE ATTAINING ITS MELTING POINT

**Definition of Incongruent Melting Point**

Sometimes a solid compound formed by the union of two components is not stable up to its true melting point. It decomposes before reaching the true melting point into another solid, whose composition is different from that of the compound, and a solution. In fact, the compound does not possess a true melting point. Whenever this happens, the compound is said to undergo a transition or *peritectic* (Greek: melting around) *reaction* or *incongruent fusion* which may be represented as

$$C_2 \rightleftharpoons C_1 + \text{soln (or melt)}$$

where $C_2$ is the compound and $C_1$ is new solid formed as a result of decomposition of $C_2$. This new solid may be a new congruent or incongruent compound or a pure component. Its melting point is always greater than that of $C_2$. The above reaction takes place at a definite temperature and is reversible in nature. The composition of the melt is also definite. These facts are in agreement with the phase rule requirement, as $C = 2$, $P = 3$ and $F = C - P + 1 = 2 - 3 + 1 = 0$. Thus, the system at this stage is invariant. The temperature at which this reaction takes place is known as the *incongruent melting point of* the compound or the *peritectic temperature* or the *transition temperature.*

## Expected Phase Diagram

Let the incongruent compound formed from A and B be $AB_2$ and let it undergo the following reaction

$$AB_2 \rightleftharpoons B + \text{melt}$$

The phase diagram of a system exhibiting above reaction has an appearance as shown in Fig. 6.7.1.

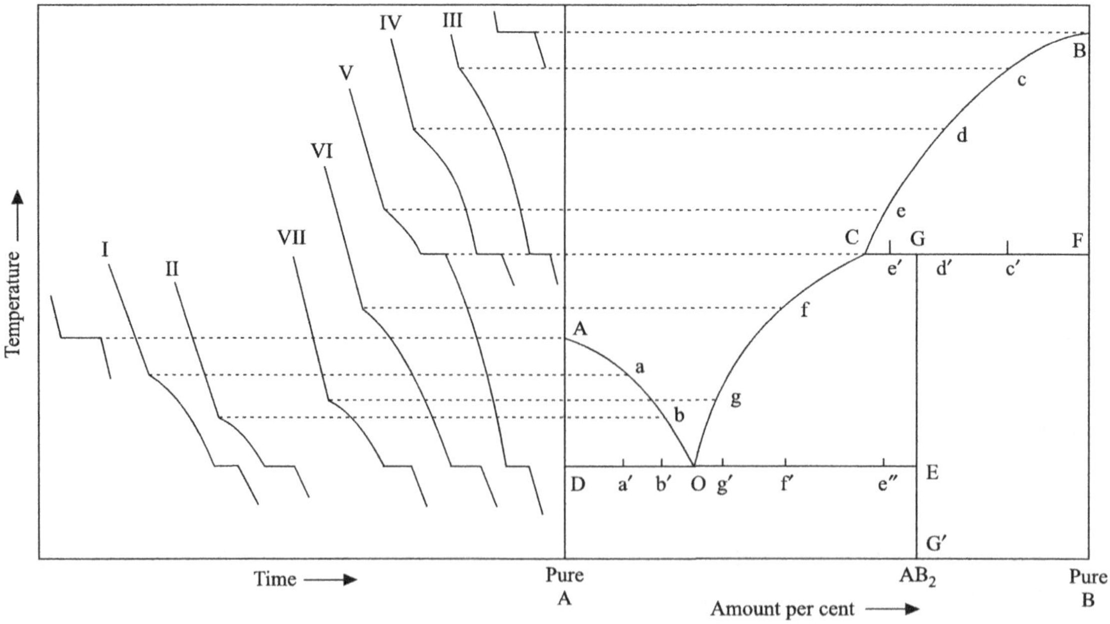


**Fig. 6.7.1 Phase diagram of a system exhibiting the formation of an incongruent compound $AB_2$**

**Description of the Phase Diagram**

From the cooling curves shown in Fig. 6.7.1, the following information may be drawn.

- From the cooling curves of pure components, the points A and B representing respectively the freezing points of A and B are obtained.
- From the cooling curves I and II of two liquid systems of B in A, the points a and b representing respectively the start of freezing of A are obtained.
- From the cooling curves III, IV and V of liquid systems of A in B, points c, d and e representing respectively the start of freezing of B are obtained.
- If a line passing through the points A, a and b is drawn, we get the freezing point curve of A.
- If a line passing through the points B, c and d is drawn, we get the freezing point curve of B.
- The point G′ on the composition axis corresponds to the composition of the compound $AB_2$. If this compound is heated, it remains in the solid phase until the point G at temperature $t_p$ (peritectic temperature) is reached. At the point G, the following peritectic reaction occurs:

$$AB_2 \underset{t < t_p}{\overset{t > t_p}{\rightleftharpoons}} B + \text{melt}$$

  Because of the separation of solid B from solid $AB_2$, the melt will be less rich in B as compared to the compound $AB_2$. Thus, the point representing the composition of the melt will lie left to the point G, say, at point C. Since a tie line connects different phases in equilibrium, such a line drawn at the point G (solid $AB_2$) will have it ends at point C (representing the melt) and F (representing solid B), respectively. Thus, a point anywhere on the tie line CGF represents solid $AB_2$ in equilibrium with solid B and the melt C. The system here will be invariant ($F = C + 1 - P = 2 + 1 - 3 = 0$).
- If heating of the compound $AB_2$ at G is continued, more and more of the compound decomposes into solid B and the melt C. The temperature of the system will remain constant till whole of the compound $AB_2$ has decomposed. At this stage, the system becomes univariant ($F = C + 1 - P = 2 + 1 - 2 = 1$) and thus heating will cause an increase in temperature, only solid B exists in equilibrium with liquid phase. Hence, the curve Bcd which represents the freezing point curve of B will end at the point C. The cooling curve III, IV and V will exhibit an arrest of temperature when the respective system reaches on the line CGF.
- The state of the system at a temperature below the peritectic temperature depends upon the overall composition of the system. We will have either solid B and solid $AB_2$ or solid $AB_2$ and the liquid phase. This can be explained on the basis of the peritectic reaction:

$$B + \text{melt} \underset{t > t_p}{\overset{t < t_p}{\rightleftharpoons}} AB_2$$

  The reaction between B and melt proceeds isothermally to form the compound $AB_2$ until either all the solid B or all the melt has been consumed. Thus, we have

(i) If solid B is present in excess amount than the requisite quantity that is required for the formation of the compound with whole of the melt, then solid B will remain in excess and hence the system will possess this excess quantity of solid B and the solid compound $AB_2$.

(ii) If solid B is present in deficient amount than the requisite quantity that is required for the formation of the compound with whole of the melt, then whole of solid B will disappear and the system will possess the excess quantity of the melt and the solid compound $AB_2$.

A point on the right of the line GG′, i.e. from G to F, represents excess of solid B and thus on cooling below the peritectic temperature, we would get a mixture of solid B and solid $AB_2$. The relative amounts of the two at a given point can be determined with the help of the lever rule. On the other hand, a point between CG represents the deficiency of solid B and thus on cooling just below the peritectic temperature, we would get the compound $AB_2$ in equilibrium with the melt.

- From the cooling curves VI and VII of the two liquid systems of compositions slightly left to the point C, we get the points f and g which represent respectively the start of freezing of the compound $AB_2$. If a smooth line passing through the points g and f is drawn, it will meet at the point C because just below the line CG, we have an equilibrium between solid $AB_2$ and the liquid phase.
- If the line Cfg is extended on the other side, it meets the freezing point curve Aab at the point O. This point represents an eutectic point involving solid A and solid $AB_2$ in equilibrium with liquid O. A tie line drawn from the point O will start from the point D (representing solid A) and will end at the point E (representing solid $AB_2$).
- The cooling curves I, II, VI and VII will exhibit an arrest of temperature (horizontal line) at the eutectic temperature corresponding to the point O.
- The cooling curve V besides exhibiting an arrest on the line CG will also exhibit one more arrest corresponding to the line OE.

**Sequence of Effects Produced on Cooling Liquid Mixtures**

The phase diagram shown in Fig. 6.7.1 can be easily labelled by considering the behaviour of the systems I, III and V during the cooling.

***System I*** The system will remain in the liquid phase till the point a is reached. Solid A starts crystallizing at a and continues to crystallize till the point a′ is reached. At a′, both A and $AB_2$ crystallize together from the liquid O.

***System III*** The system will remain in the liquid phase till the point c is reached. Solid B starts crystallizing at c and continues to crystallize till the point c′ is reached. At c′, the peritectic reaction starts and continues isothermally till whole of the liquid C is exhausted. The system now contains solid mixture of B and $AB_2$ which will show smooth cooling.

***System V*** The system will remain in the liquid phase till the point e is reached. Solid B starts crystallizing at e and continues to crystallize till the point e′ is reached. At e′, the peritectic reaction starts and continues isothermally till whole of solid B has disappeared. The system now contains solid $AB_2$ in equilibrium with the melt C. Further cooling will result in more and more separation of $AB_2$ with the composition of liquid moving along the line CO. This continues till the point e″ is reached where the liquid composition corresponds to the point O. Here, the solid mixture of A and $AB_2$ crystallize isothermally without any change in the composition of the liquid phase. This continues till whole of the liquid phase is exhausted. The system now consists of a solid mixture of A and $AB_2$ which will show smooth cooling.

**Labelling of the Phase Diagram**

With the help of above analysis, the phase diagram shown in Fig. 6.7.1 can be easily labelled and is described in Table 6.7.1.

**Table 6.7.1** Labelling of Fig. 6.7.1

| *Location of the system* | *Status of the system* |
|---|---|
| A | Freezing point of A |
| B | Freezing point of B |
| F | Peritectic temperature of $AB_2$ |
| AO | Freezing point curve of A |
| BC | Freezing point curve of B |
| CO | Freezing point curve of $AB_2$ |
| O | Eutectic point involving solid A, solid $AB_2$ and liquid O |
| C | Peritectic point involving solid B, solid $AB_2$ and liquid C |
| A-axis | Solid A up to the point A, liquid A beyond A |
| B-axis | Solid B up to the point B, liquid B beyond B |
| G′G | Solid $AB_2$ |
| CGF | A point on this line represents solid B and solid $AB_2$ in equilibrium with liquid C |
| DOE | A point on this line represents solid B and solid $AB_2$ in equilibrium with liquid O |
| Area ADOA | Solid A $\rightleftharpoons$ liquid (whose composition lies on AO) |
| Area BCFB | Solid B $\rightleftharpoons$ liquid (whose composition lies on BC) |
| Area COEGC | Solid $AB_2$ $\rightleftharpoons$ liquid (whose composition lies on CO) |
| Area below GF | Solid B and solid $AB_2$ |
| Area bellow DOE | Solid A and solid $AB_2$ |
| Area above AOCB | Liquid |

**Cooling Patterns**

The nature of the cooling curve will depend on the composition of the system. We describe below the expected pattern for a few typical systems.

(i) *Composition lying anywhere between* AO *or* OC

| | | | | |
|---|---|---|---|---|
| Rapid cooling of liquid | $\rightarrow$ | Break at the appearance of solid A or $AB_2$ | $\rightarrow$ | Slower cooling; solidification continues $\downarrow$ |
| | | Cooling of solid mixture of A and $AB_2$ | $\leftarrow$ | Horizontal portion; eutectic point |

(ii) *Eutectic composition*

| | | | | |
|---|---|---|---|---|
| Rapid cooling of liquid | $\rightarrow$ | Horizontal portion eutectic point | $\rightarrow$ | Cooling of the solid mixture of A and $AB_2$ |

(iii) *Composition corresponding to the peritectic point* C

| | | | | |
|---|---|---|---|---|
| Rapid cooling of liquid | → | Break at the appearance of solid | → | Slower cooling; appearance of more and more of $AB_2$ |
| | | | | ↓ |
| | | Cooling of solid mixture of A and $AB_2$ | ← | Horizontal portion; eutectic point |

(iv) *Composition anywhere between* C *and* G

| | | | | |
|---|---|---|---|---|
| Rapid cooling of liquid | → | Break; appearance of solid B | → | Slower cooling; appearance of more and more of solid B |
| | | | | ↓ |
| Horizontal portion; eutectic temperature | ← | Slower cooling appearance of more and more of $AB_2$ | ← | Horizontal portion at the peritectic temperature |
| ↓ | | | | |
| Cooling of solid mixture of A and $AB_2$ | | | | |

(v) *Composition anywhere between* G *and* F

| | | | | |
|---|---|---|---|---|
| Rapid cooling of liquid | → | Break; appearance of solid B | → | Slower cooling; appearance of more and more of solid B |
| | | | | ↓ |
| | | Cooling of solid mixture of B and $AB_2$ | ← | Horizontal portion at the peritectic temperature |

These cooling patterns are also shown in Fig. 6.7.1.

---

***Example 6.7.1*** Na and K melt at 98 °C and 65 °C, respectively. They form one compound NaK which decomposes at 10 °C to give a solid and a melt containing 60 mol % K. There is an eutectic at – 5 °C and the eutectic composition is 75 mol % of K. Sketch the phase diagram and level it.

***Solution*** Analysis of data

(i) At 10 °C, the compound NaK decomposes to give a new solid and melt. Thus, the temperature 10 °C must be a peritectic temperature. This also follows from the fact that the compositions of the compound (50 mol % K) and the melt (60 mol % K) are not identical. The new solid phase can be either Na or K. Since K content of the melt is larger than of the compound NaK, we may conclude that this new solid phase must be Na. This is also supported by the fact that K appears at the eutectic point. So we draw a horizontal line at 10 °C from the Na-axis right up to the melt composition. The vertical line representing NaK compound will have its end at this horizontal line (Fig. 6.7.2).

(ii) Since K separates out at the eutectic point, we joins the melting point of K with the point representing eutectic point.

(iii) Since Na appears on decomposition of the compound NaK, we join the melting point of Na with the peritectic point representing the melt composition at 10 °C.

(iv) Finally, we join the peritectic point representing the melt with the eutectic point and draw a horizontal line from the NaK-axis to the K-axis passing through the eutectic point.

Labelling of the obtained phase diagram is straightforward and is included in Fig. 6.7.2.

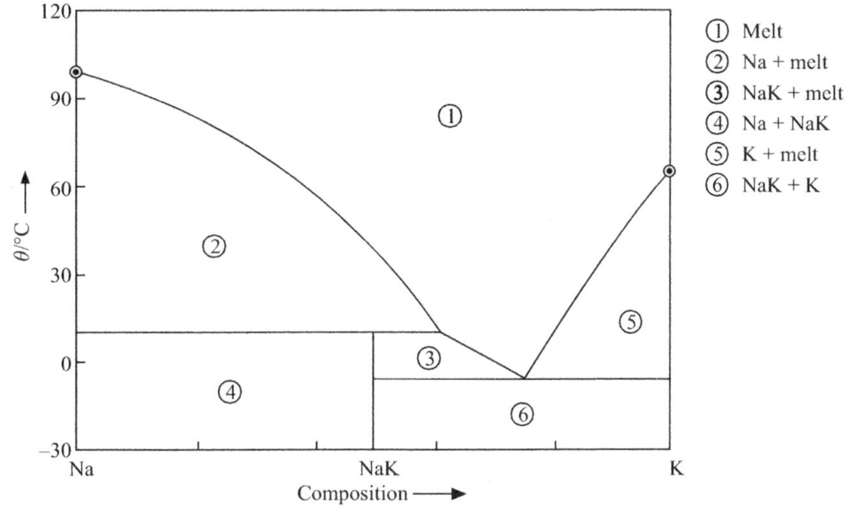


**Fig. 6.7.2 Required phase diagram**

## Theoretical aspect of Fig. 6.7.1

The curves AO, OC and CB of Fig. 6.7.1 are the solubility curves for substances A, $AB_2$ and B, respectively. The entire phase diagram shown in Fig. 6.7.1 may be considered to be made up of two systems, namely, A–B system and incomplete subsystem A–$AB_2$. Since B does not equilibrate with $AB_2$ at temperatures above the incongruent melting point $T_p$ of the latter, the solubility curve BC is determined by its composition in the system A–B. The solubility curves AO and OC are determined by their compositions in the subsystem A–$AB_2$. In other words, if the compound $AB_2$ is considered to exist as such in the form of a liquid at temperatures below $T_p$, then the composition of A in the solubility curve AO depends on the $AB_2$ composition and the composition of $AB_2$ in the solubility curve OC depends on the A composition. Thus from the given data on $T_A^*$, $T_B^*$, $\Delta_{fus}H^o_{m,A}$ and $\Delta_{fus}H^o_{m,B}$ we can construct the simple eutectic diagram of the system A-B and on this diagram we can superimpose the simple eutectic diagram of the subsystem A–$AB_2$ (after performing the appropriate amount fractions transformation) constructed from the given $T_A^*$, $\Delta_{fus}H^o_{m,A}$ $\Delta_{fus}H^o_{m,\,AB_2}$ and the hypothetical melting point $T^*_{AB_2}$ as shown in Fig. 6.7.3.

## Examples

A few examples of systems exhibiting incongruent melting points are given in Table 6.7.2

**Table 6.7.2** System Exhibiting Incongruent Melting Point

| *System* | | *Compound* | |
|---|---|---|---|
| A | B | | |
| Au | Sb | $AB_2$ | |
| KCl | $CuCl_2$ | 2A · B | |
| Picric acid | Benzene | AB | |
| Acetamide (79.9 °C) | Salicylic acid (156 °C) | AB | $t_p$ = 70 °C and 44 mol % salicylic acid |
| NaCl | $H_2O$ | A · 2B | |
| $Na_2SO_4$ | $H_2O$ | A · 10B | |

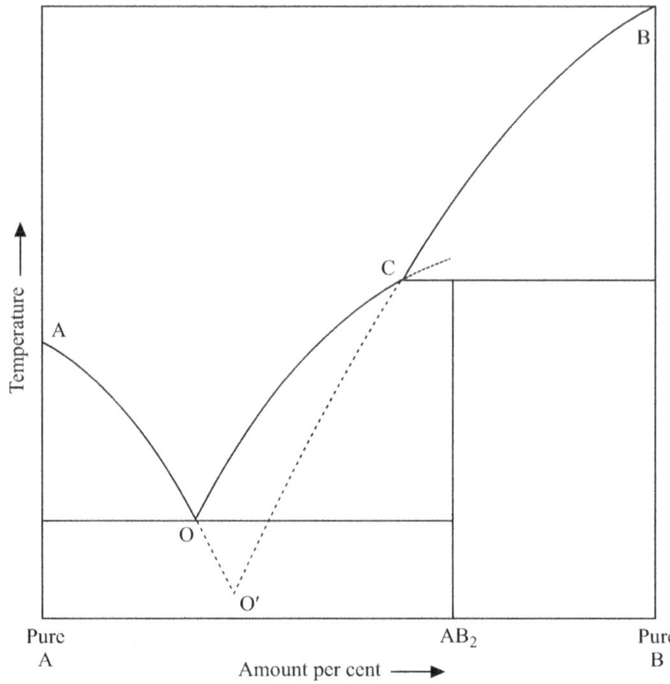


**Fig. 6.7.3 Construction of the phase diagram for a system exhibiting compound formation with an incongruent melting point**

***Example 6.7.2***

The Ca-Al system exhibits the following peritectic equilibrium:

$$CaAl_3 \rightleftharpoons CaAl_2 + \text{melt (14 mass \% Ca)}$$

Without using the lever rule, predict the nature and the respective masses of the system just below the incongruent point for the systems: (i) 40 g Ca + 70 g Al, and (ii) 20 g Ca + 90 g Al.

***Solution***

At temperature just below incongruent point, we will have either a mixture of solids $CaAl_2$ and $CaAl_3$ or solid $CaAl_3$ in equilibrium with the melt.

(**i**) Let the mass $x$ of Ca be present in the form of the compound $CaAl_2$ and let us assume that the remaining mass of Ca (i.e. 40 g – $x$) be present in the form of the compound $CaAl_3$. Thus, the total quantity of Al present in these two compounds is (54/40) $x$ + (81/40) (40 g – $x$) and this must be equal to the given quantity of Al, i.e. 70 g. Thus, we have

$$\left(\frac{54}{40}\right) x + \left(\frac{81}{40}\right) (40\text{ g} - x) = 70\text{ g}$$

Solving for $x$, we have $x = 16.3$ g

Thus the given system will consist of a mixture of $CaAl_2$ and $CaAl_3$ just below the incongruent point with the following compositions.

In $CaAl_2$ Ca = 16.3 g $\text{Al} = \left(\frac{54}{40}\right) \times 16.3\text{ g} = 22.0\text{ g}$

and in $CaAl_3$ $\quad$ Ca = (40 g – 16.3 g) = 23.7 g $\quad$ $\text{Al} = \left(\frac{81}{40}\right) \times 23.7\ \text{g} = 48.0\ \text{g}$

**(ii)** Let the mass $x$ of Ca be present in the form of the compound $CaAl_2$ and let us assume that the remaining mass of Ca (i.e. 20 g – $x$) be present in the form of the compound $CaAl_3$. Then the total quantity of Al present in these two compounds is (54/40) $x$ + (81/40) (20 g – $x$) and this must be equal to the given mass of Al, i.e. 90 g. Thus, we have

$$\left(\frac{54}{40}\right) x + \left(\frac{81}{40}\right) (20\ \text{g} - x) = 90\ \text{g}$$

Solving for, $x$ we get $\qquad x = -\ 73.33$ g

Since $x$ is coming out to be negative, our assumption that the system consists of $CaAl_2$ + $CaAl_3$ must be wrong. Let us now imagine that the system consists of melt + $CaAl_3$. Let us again suppose that the mass $x$ of Ca be present in the form $CaAl_3$ and the remaining in the melt containing 14 mass % Ca. The total quantity of Al present in these two forms is (81/40) $x$ + (86/14)(20 g – $x$), and this must be equal to the given quantity of 90g. Thus, we have

$$\left(\frac{81}{40}\right) x + \left(\frac{86}{14}\right) (20\ \text{g} - x) = 90\ \text{g}$$

Solving for $x$, we get $\qquad x = 7.98$ g

Thus, we have

Ca present in $CaAl_3$ = 7.98 g $\qquad$ Al present in $CaAl_3 = \left(\frac{81}{40}\right)(7.98\ \text{g}) = 16.16\ \text{g}$

Ca present in the melt = (20 g – 7.98 g) = 12.02 g

$$\text{Al present in the melt} = \left(\frac{86}{14}\right)(12.02\ \text{g}) = 73.84\ \text{g}$$

Hence the system just below the incongruent melting point consists of solid $CaAl_3$ and melt.

---

## More than one Incongruent Compounds

In some systems, more than one compound exhibiting peritectic temperature is formed. Many systems are known where all the compounds formed have incongruent melting points. A few examples are

| | |
|---|---|
| $K_2SO_4 - CdSO_4$ | $K_2SO_4 \cdot 2CdSO_4$, $K_2SO_4 \cdot 3CdSO_4$ |
| $MgSO_4 - H_2O$ | $MgSO_4 \cdot H_2O$, $MgSO_4 \cdot 6H_2O$ |
| | $MgSO_4 \cdot 7H_2O$, $MgSO_4 \cdot 12H_2O$ |

---

*Example 6.7.3*

Construct a phase diagram of Au-Pb system from the following data:

(a) Melting point of Au = 1 063 °C; m.pt. of Pb = 327 °C.
(b) Compound $Au_2Pb$ decomposes at 418 °C to give a liquid 44 mol % of Pb.
(c) Compound $AuPb_2$ has a peritectic at 254 °C and the melt is 71 mol % of Pb.
(d) Eutectic is at 215 °C and is 84 mol % of Pb.
(e) No solid solution is formed.

Label the areas of the diagram.

***Solution***

*Analysis of data*

(i) At 418 °C, the compound $Au_2Pb$ (33.3 mol % Pb) decomposes to give a liquid containing 44 mol % Pb. Since the compositions of solid and melt are different, the temperature 418 °C must be a peritectic temperature. The solid phase which appearas at this temperature must not be Pb as the Pb content in the melt is higher than that in the compound $Au_2Pb$. Since there is no other compound with Au content larger than in the compound $Au_2Pb$, it follows that this solid phase must be Au. So, we join the melting point of Au with the point representing the melt composition at 418 °C (Fig. 6.7.4). We also draw a horizontal line at 418 °C starting from the point representing melt composition to the Au-axis. The vertical line representing $Au_2Pb$ compound must end at this horizontal line.

(ii) At 254 °C, we have a peritectic reaction where the compound $AuPb_2$ decomposes to give the melt containing 71 mol % of Pb. The solid phase appearing here will not be pure Pb as the melt contains more of Pb than the compound $AuPb_2$. This also follows from the fact that Pb appears at the eutectic point. Since in the composition range $AuPb_2$-$Au_2Pb$, there is no other compound, we conclude that the new solid phase appearing at 254 °C must be the compound $Au_2Pb$. So, we join the points representing the two melts at 418 °C and 254 °C. We also draw a horizontal line starting from the point of the melt at 254 °C to the vertical line representing the compound $Au_2Pb$. The vertical line representing the compound $AuPb_2$ must end at this horizontal line.

(iii) It is given that at the eutectic point Pb separates out. The second solid which separates out here must be the compound $AuPb_2$ as there exists no other compound between the composition range of eutectic point and the compound $AuPb_2$. So, we join the melting point of Pb and the melt composition at 254°C with the eutectic point. We also draw a horizontal line through the eutectic point starting from Pb-axis to the vertical line representing the compound $AuPb_2$.

Now the labelling of the phase diagram is straightforward and is included in Fig. 6.7.4.

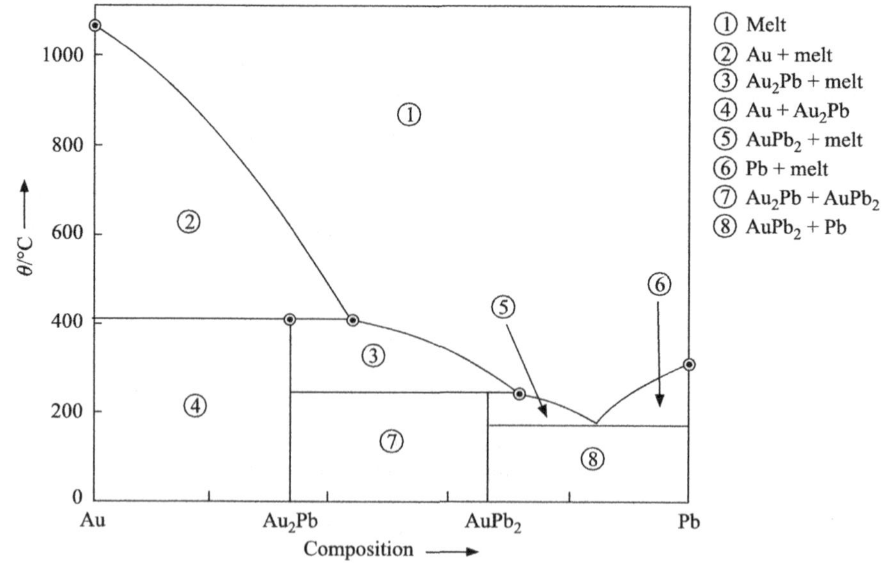


**Fig. 6.7.4 Required phase diagram**

## Systems Exhibiting both Congruent and Incongruent Compounds

In some systems, more than one compound is formed, some of them have congruent melting points while others have incongruent melting points. A few examples are given in Table 6.7.3. The phase diagram of such a system will exhibit characteristics of both congruent and incongruent melting points.

**Table 6.7.3** Systems Exhibiting both Congruent and Incongruent Compounds Formation

| A | B | *Congruent compound* | *Incongruent compound* |
|---|---|---|---|
| Ca | Al | $CaAl_2$ | $CaAl_3$ |
| Mg | Ni | $MgNi_2$ | $Mg_2Ni$ |
| Na | Bi | $Na_3Bi$ | NaBi |

Figure 6.7.5 displays such a diagram for the Ca-Al systems. The phase diagram may be considered to be made up of two simple systems of Ca-$CaAl_2$ and $CaAl_2$-Al. The former shows a simple eutectic whereas in the latter a compound $CaAl_3$ with the incongruent melting point is formed. This way the labelling of different areas becomes very simple and is shown in Fig. 6.7.5 itself.

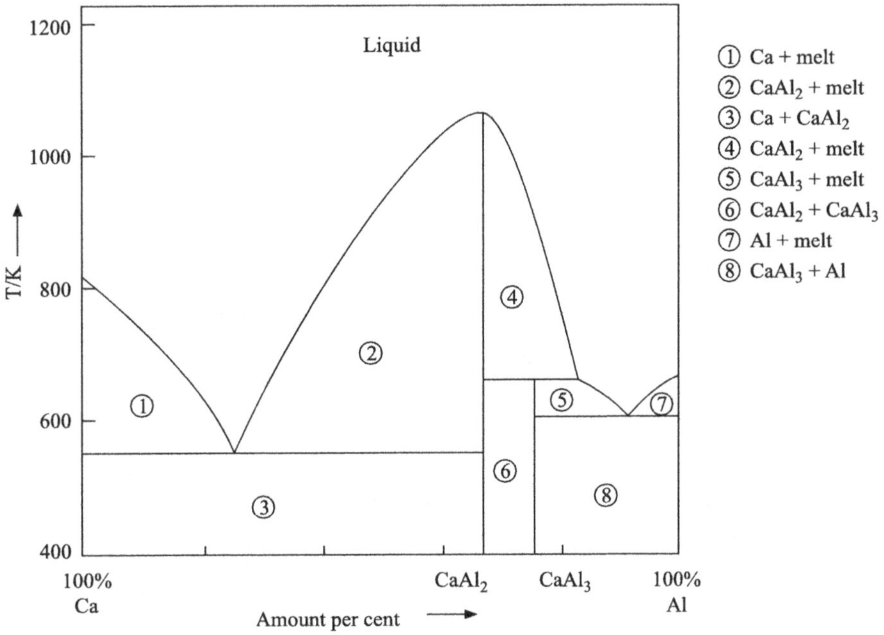


**Fig. 6.7.5 Phase diagram of Ca-Al system**

***Example 6.7.4***

A and B form two solid compounds $A_2B$ and $AB_2$. The compound $A_2B$ melts at 800 °C to give a liquid having the same composition as that of $A_2B$. The compound $AB_2$ melts with decomposition at 700 °C to give solid B and a liquid containing 60 mol % of B. The melting point of A is 500 °C and of B is 1 000 °C. No solid solutions are formed. Sketch temperature versus amount fraction of the system and label it. Draw cooling curves for a mixture containing 70 mol % of B.

***Solution***

*Analysis of data*

(i) The compound $A_2B$ is a congruent compound as on melting it produces the liquid of its own composition.

(ii) The compound $AB_2$ is an incongruent compound as on melting it produces a melt of composition different from its own composition.

(iii) Since solid B appears on melting $AB_2$ at 700 °C, we join the melting point of B with the melt composition at 700 °C. We also draw a horizontal line from this melt composition to the pure B axis. The vertical line representing the compound $AB_2$ must end at this horizontal line.

(iv) Since the compound $A_2B$ melts congruently, on either sides of the vertical line at $A_2B$ composition, we will have the solubility curves of $A_2B$ which will pass through the melting point of the compound $A_2B$.

(v) Since no other compound formation is there, we conclude that the systems A-$A_2B$ and $A_2B$-$AB_2$ form the eutectic pairs. So, we draw the curves characteristic of eutectic pairs as shown in Fig. 6.7.6.

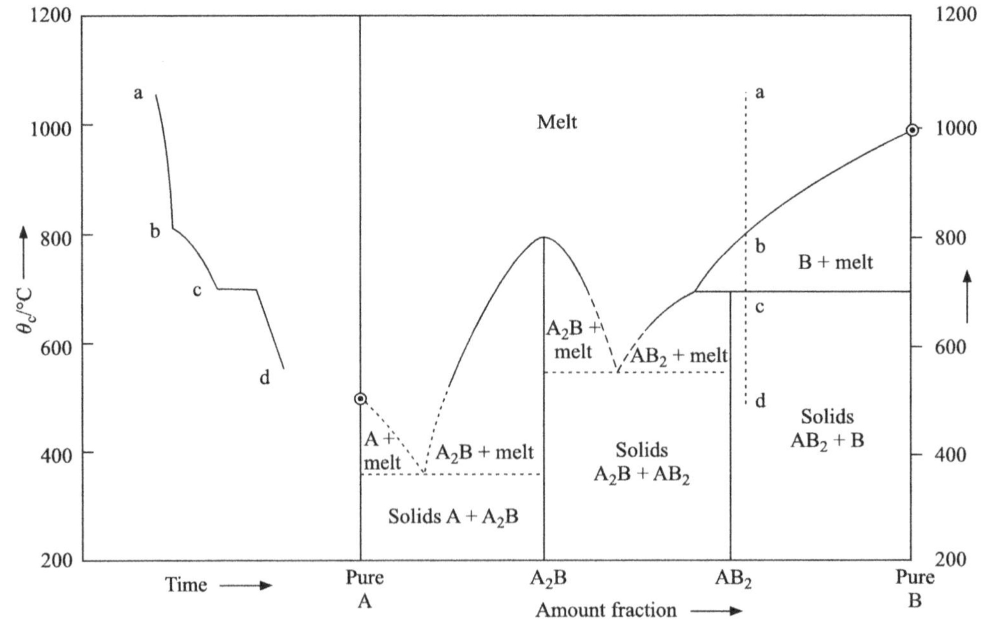


**Fig. 6.7.6**

The labelling of the phase diagram is straightforward and is shown in Fig. 6.7.6 itself. The expected cooling curve for the given composition is also shown in the phase diagram.

## 6.8 FORMATION OF A COMPLETE SERIES OF SOLID SOLUTIONS

A solid may dissolve in another solid to give a solid solution. This solution is perfectly homogeneous in character. X-ray examination of the lattice of solid solution reveals that one constituent enters the lattice of the other and is uniformly distributed throughout. There is no indication of two crystalline lattices as is in the case of solid mixture and thus only one crystal lattice characteristic of the solid solution is formed.

**Theoretical Aspects** Before drawing phase diagram of the present case, we consider the following two facts regarding the general nature of the diagram.

***(A) Shift in the freezing point of the solvent when a solid solution separates out on cooling the solution*** An expression for the shift in freezing point of the solvent when a solid solution separates out can be derived thermodynamically. If we assume that the solid solution also behaves like an ideal solution, then the chemical potentials of its constituents will be given by an expression analogous to that of an ideal gaseous mixture or ideal liquid solution. This expression is

$$\mu_i = \mu_i^* + RT \ln x_i$$

where $\mu_i^*$ is the chemical potential of the pure solid (at pressure $p$, if $p$ = 1 bar, $\mu_i^{*\circ}$ replaces $\mu_i^*$) and $x_i$ is the amount fraction of the constituent in the solid solution.

The condition of phase equilibrium between the solid solution and liquid solution for one of the components (say, solvent) is

$$\mu_{1(\mathrm{s})} = \mu_{1(\mathrm{l})} \tag{6.8.1}$$

where the subscripts s and l within the brackets stand for solid and liquid, respectively.

Now if it is assumed that both the solutions are ideal, we have

$$\mu_{1(\mathrm{s})}^* + RT \ln x_{1(\mathrm{s})} = \mu_{1(\mathrm{l})}^* + RT \ln x_{1(\mathrm{l})}$$

Rearranging the above relation, we have

$$R(\ln x_{1(\mathrm{s})} - \ln x_{1(\mathrm{l})}) = \frac{\mu_{1(\mathrm{l})}^* - \mu_{1(\mathrm{s})}^*}{T} = \frac{\Delta_{\mathrm{fus}}\mu_1^*}{T} \tag{6.8.2}$$

where $\Delta_{\mathrm{fus}}\mu_1^*$ is the molar free energy of fusion of the pure solvent at temperature $T$ and pressure $p$. Since

$$\Delta_{\mathrm{fus}}\mu_1^* = \Delta_{\mathrm{fus}}H_{1,\mathrm{m}}^* - T\,\Delta_{\mathrm{fus}}S_{1,\mathrm{m}}^*$$

and at the normal melting point $T_1^*$ of the pure solvent

$$\Delta_{\mathrm{fus}}S_{1,\mathrm{m}}^* = \Delta_{\mathrm{fus}}H_{1,\mathrm{m}}^*/T_1^*$$

we have

$$\Delta_{\mathrm{fus}}\mu_1^* = \Delta_{\mathrm{fus}}H_{1,\mathrm{m}}^* - T\frac{\Delta_{\mathrm{fus}}H_{1,\mathrm{m}}^*}{T_1^*} = \Delta_{\mathrm{fus}}H_{1,\mathrm{m}}^*\left(1 - \frac{T}{T_1^*}\right) \tag{6.8.3}$$

Substituting Eq. (6.8.3) in Eq. (6.8.2), we get

$$R(\ln x_{1(s)} - \ln x_{1(l)}) = \frac{\Delta_{fus}H^*_{1,m}}{T}\left(1 - \frac{T}{T_1^*}\right) = \Delta_{fus}H^*_{1,m}\left(\frac{-\Delta T_f}{TT_1^*}\right) \quad (6.8.4)$$

where $\Delta T_f = T - T_1^*$.

Equation (6.8.4) can be simplified under the following approximations.

(i) For a dilute solution, $T$ will be very near to $T_1^*$ so that the term $TT_1^*$ may be replaced by $T_1^{*2}$.

(ii) Again for dilute solution, we have

$$\ln x_{1(s)} = \ln(1 - x_{2(s)}) \simeq -x_{2(s)}$$

$$\ln x_{1(l)} = \ln(1 - x_{2(l)}) \simeq -x_{2(l)}$$

With these approximations, Eq. (6.8.4) reduces to

$$R(-x_{2(s)} + x_{2(l)}) = \Delta_{fus}H^*_{1,m}\left(\frac{-\Delta T_f}{T_1^{*2}}\right)$$

or
$$(-\Delta T_f) = \frac{RT_1^{*2}}{\Delta_{fus}H^*_{1,m}}x_{2(l)}\left(1 - \frac{x_{2(s)}}{x_{2(l)}}\right) \quad (6.8.5)$$

This expression can be written in terms of molality of the liquid solution by using the following approximation.

$$x_{2(l)} = \frac{n_{2(l)}}{n_{1(l)} + n_{2(l)}} \simeq \frac{n_{2(l)}}{n_{1(l)}} = \frac{n_{2(l)}}{m_{1(l)}/M_1} = \left(\frac{n_{2(l)}}{m_{1(l)}}\right)M_1 = mM_1$$

Thus, Eq. (6.8.5) modifies to

$$-\Delta T_f = \frac{RT_1^{*2}M_1}{\Delta_{fus}H^*_{1,m}}m\left(1 - \frac{x_{2(s)}}{x_{2(l)}}\right)$$

or
$$-\Delta T_f = K_f m(1 - K) \quad (6.8.6)$$

where $K$ is the distribution coefficient and is given by

$$K = \frac{\text{Amount fraction of the solute in the solid phase}}{\text{Amount fraction of the solute in the solution}} \quad (6.8.7)$$

In Eq. (6.8.6), $\Delta T_f = T - T_1^*$. The value of $\Delta T_f$ can be positive or negative depending upon the value of $K$. Thus, we have

**Condition for $\Delta T_f$ to be Negative**

If $K < 1$, then $T < T_1^*$ and thus a depression in freezing point of the solvent is observed. The presence of $1 - K$ factor in Eq. (6.8.6) makes the numerical value of $\Delta T_f$ smaller than the value where only the solvent separates on cooling.

**Condition for $\Delta T_f$ to be Positive**

If $K > 1$, then $T > T_1^*$ and an elevation in the freezing point of the solvent is observed.

These conclusions, in fact, are similar to those derived earlier for the liquid-vapour system and thus may be regarded as the expressions for the *Konowaloff's rule* as applicable to solid-liquid system. These conclusions may be stated as follows.

> *At any temperature, the liquid phase contains relatively more than solid phase of that component whose addition to the solvent (or solution) causes a decrease in the freezing point of the latter or the solid phase contains relatively more than liquid phase of the component whose addition to the liquid phase causes an increase in the freezing point of the latter.*

***(B) Nature of cooling curve when a solid solution is formed*** Since in a solid solution, the two components are completely miscible with each other and form a homogeneous solution, it is obvious that such a solution will always constitute a single solid phase. Thus, the system at the most can have two phases, namely, liquid and solid solutions, in equilibrium with each other. According to the phase rule, this system is univariant ($F = C - P + 1 = 2 - 2 + 1 = 1$). Thus, the value of one variable has to be stated in order to define the system completely. This variable may be temperature or composition of one of the solutions. In other words, the equilibrium between solid and liquid solutions can exist at different temperatures. For a given temperature, the compositions of the two solutions in equilibrium will have fixed values. The cooling curves will exhibit two breaks corresponding to the start of freezing of solid solution (where $P$ changes from 1 to 2 and $F$ changes from 2 to 1) and the end of freezing of solid solution (where $P$ changes from 2 to 1 and $F$ changes from 1 to 2), respectively.

If the composition of solid and liquid solutions in equilibrium are identical, the system becomes nonvariant ($F = (C - r) + 1 - P = (2 - 1) + 1 - 2 = 0$). Here, the freezing or melting of solid solution takes place at a constant temperature. Consequently, the cooling curve will exhibit an arrest at this temperature.

**Classification of Phase Diagrams**

For real systems, Eq. (6.8.6) is expected to be applicable only in very dilute solutions of A in B or B in A. Three types of phase diagrams may be distinguished depending upon the values of $K_{\text{A in B}}$ (i.e. $x_{\text{A(s)}}/x_{\text{A(l)}}$, A acting as a solute and B as a solvent) and $K_{\text{B in A}}$ (i.e. $x_{\text{B(s)}}/x_{\text{B(l)}}$, B acting as a solute and A as a solvent). These are:

***Ascending solid solution*** In this case $K_{\text{B in A}} > 1$ and $K_{\text{A in B}} < 1$ or vice versa. The freezing point of A (or B) gradually increases as more and more of B (or A) is added to it whereas that of B (or A) gradually decreases as more and more of A (or B) is added to it. In such a case, freezing point of a solution containing different amounts of A and B always lies between those of pure components.

***Minimum type solid solution*** In this case $K_{\text{B in A}} < 1$ and $K_{\text{A in B}} < 1$. The freezing points of both the components gradually decrease on addition of more and more of the second component. Thus, the freezing point curves show downward trends at both ends of the phase diagram and meet each other at a minimum point, i.e. the freezing point curve of such a system exhibits a minimum.

***Maximum type solid solution*** In this case $K_{\mathrm{B\,in\,A}} > 1$ and $K_{\mathrm{A\,in\,B}} > 1$. The freezing points of both the components gradually increase on addition of more and more of the second component. Thus, the freezing point curves show upward trends at both ends of the phase diagram and meet each other at a maximum point, i.e. the freezing point curve of such a system exhibits a maximum.

We now draw the expected phase diagrams for the above three cases and describe their main characteristics.

## ASCENDING TYPE SOLID SOLUTION

**Theoretical Aspects** Before drawing the expected phase diagram, it is worthwhile to consider a few general conclusions regarding the nature of liquidus and solidus curves. From Eq. (6.8.4), we have

$$R(\ln x_{1(\mathrm{s})} - \ln x_{1(\mathrm{l})}) = \Delta_{\mathrm{fus}} H^*_{1,\mathrm{m}} \left(\frac{1}{T} - \frac{1}{T_1^*}\right)$$

or

$$\ln\left(\frac{x_{1(\mathrm{l})}}{x_{1(\mathrm{s})}}\right) = -\frac{\Delta_{\mathrm{fus}} H^*_{1,\mathrm{m}}}{R}\left(\frac{1}{T} - \frac{1}{T_1^*}\right) \quad (6.8.8)$$

or

$$\frac{x_{1(\mathrm{l})}}{x_{1(\mathrm{s})}} = \mathrm{e}^{-a} \quad (6.8.9)$$

where $a = \dfrac{\Delta_{\mathrm{fus}} H^*_{1,\mathrm{m}}}{R}\left(\dfrac{1}{T} - \dfrac{1}{T_1^*}\right)$

Similarly, for the second component,

$$\frac{x_{2(\mathrm{l})}}{x_{2(\mathrm{s})}} = \mathrm{e}^{-b} \quad (6.8.10)$$

where $b = \dfrac{\Delta_{\mathrm{fus}} H^*_{2,\mathrm{m}}}{R}\left(\dfrac{1}{T} - \dfrac{1}{T_2^*}\right)$

Now in the liquid and solid phases, we will have

$$x_{1(\mathrm{l})} = 1 - x_{2(\mathrm{l})} \quad (6.8.11)$$

$$x_{1(\mathrm{s})} = 1 - x_{2(\mathrm{s})} \quad (6.8.12)$$

In Eq. (6.8.9), $x_{1(\mathrm{s})}$ may be rewritten in terms of $x_{1(\mathrm{l})}$ with the help of Eqs (6.8.10), (6.8.11) and (6.8.12). Substituting Eq. (6.8.12) in Eq. (6.8.9), we get

$$\frac{x_{1(\mathrm{l})}}{1 - x_{2(\mathrm{s})}} = \mathrm{e}^{-a}$$

In view of Eq. (6.8.10), the above equation becomes

$$\frac{x_{1(\mathrm{l})}}{1 - x_{2(\mathrm{l})}/\mathrm{e}^{-b}} = \mathrm{e}^{-a}$$

Now employing Eq. (6.8.11), we get

$$\frac{x_{1(\mathrm{l})}}{1 - (1 - x_{1(\mathrm{l})})/\mathrm{e}^{-b}} = \mathrm{e}^{-a}$$

Solving for $x_{1(\mathrm{l})}$, we have

$$x_{1(\mathrm{l})} = \frac{\mathrm{e}^{-a}(\mathrm{e}^{-b} - 1)}{\mathrm{e}^{-b} - \mathrm{e}^{-a}} \tag{6.8.13}$$

Substitution of Eq. (6.8.13) in Eq. (6.8.9) gives

$$x_{1(\mathrm{s})} = \frac{x_{1(\mathrm{l})}}{\mathrm{e}^{-a}} = \frac{\mathrm{e}^{-b} - 1}{\mathrm{e}^{-b} - \mathrm{e}^{-a}} \tag{6.8.14}$$

Finally, we have

$$x_{2(\mathrm{l})} = 1 - x_{1(\mathrm{l})} = \frac{\mathrm{e}^{-b}(\mathrm{e}^{-a} - 1)}{\mathrm{e}^{-a} - \mathrm{e}^{-b}} \tag{6.8.15}$$

$$x_{2(\mathrm{s})} = 1 - x_{1(\mathrm{s})} = \frac{\mathrm{e}^{-a} - 1}{\mathrm{e}^{-a} - \mathrm{e}^{-b}} \tag{6.8.16}$$

From Eq. (6.8.13)-(6.8.16), the following conclusions may be drawn:

- None of the relationship is linear.
- Since the relationship for $x_{1(\mathrm{s})}$ and $x_{1(\mathrm{l})}$ are different, it is evident that the liquidus and solidus curves do not coincide except at $T_1^*$ and $T_2^*$.
- For a case in which $T_1^* \le T \le T_2^*$, we will have:

  $a$ is negative (thus $\mathrm{e}^{-a}$ is greater than one)

and $b$ is positive (thus $\mathrm{e}^{-b}$ is less than one).

Hence from Eq. (6.8.13), we get

$$x_{1(\mathrm{l})} = \frac{\mathrm{e}^{-a}(\mathrm{e}^{-b} - 1)}{\mathrm{e}^{-b} - \mathrm{e}^{-a}} = \frac{\text{negative}}{\text{negative}} = \text{positive}$$

Similarly from Eqs (6.8.14)-(6.8.16), we get that $x_{1(\mathrm{s})}$, $x_{2(\mathrm{l})}$ and $x_{2(\mathrm{s})}$ all are positive quantities. The same conclusions are obtained for the case $T_2^* \le T \le T_1^*$. Thus, it may be concluded that if $T$ lies in the interval between the melting points of the two components, physically meaningful composition of the system is obtained. Conversely, it follows that if the amount fraction of the system lies between 0 and 1, then the temperatures of the liquidus and solidus curves lie between the melting points of the two components.

- For a case in which $T_1^* \leq T \leq T_2^*$. Eqs (6.8.9) and (6.8.10) yield

$$\frac{x_{1(l)}}{x_{1(s)}} > 1 \quad \text{and} \quad \frac{x_{2(l)}}{x_{2(s)}} < 1 \tag{6.8.17}$$

Similarly for a case $T_1^* \geq T \geq T_2^*$, we have

$$\frac{x_{1(l)}}{x_{1(s)}} < 1 \quad \text{and} \quad \frac{x_{2(l)}}{x_{2(s)}} > 1 \tag{6.8.18}$$

Equations (6.8.17) and (6.8.18) demonstrate that the liquid is richer than solid in the lower melting point component and the solid is richer than liquid in the higher melting point component.

## Depiction of the Phase Diagram

Keeping in mind the above conclusions the phase diagram of the ascending type solution is shown in Fig. 6.8.1.

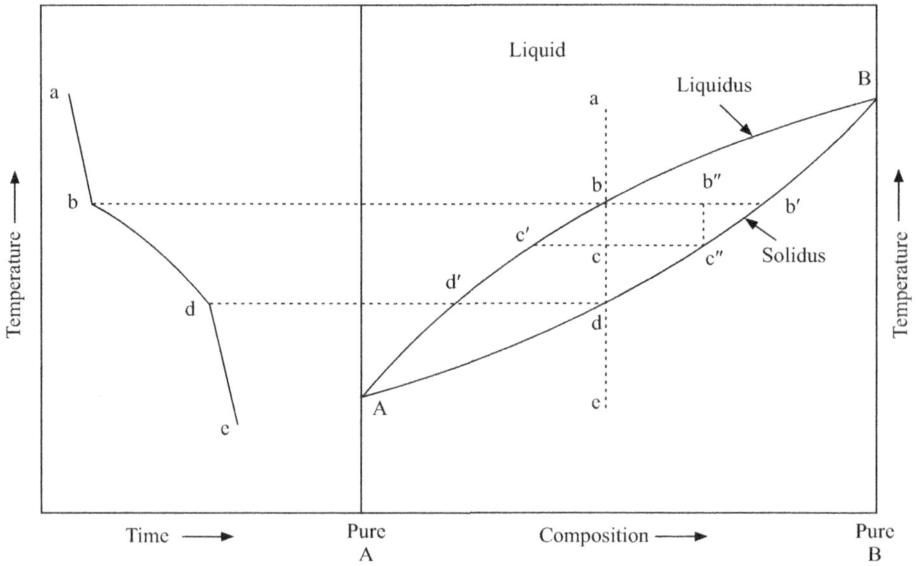


**Fig. 6.8.1 Phase diagram for a system exhibiting continuous series of solid solutions with no maximum or minimum**

## Cooling Behaviour of a Liquid System

The labelling of the phase diagram shown in Fig. 6.8.1 can be easily understood by first considering the expected cooling behaviour of the liquid solution represented by the point a. On cooling, the state of the overall system will move vertically downward along the isopleth abcde. The system remains in the liquid phase till the point b is reached where the first crystal of solid solution b′ starts appearing. The point b′ can be obtained by drawing a tie line from the point b. Since the obtained solid solution is more rich in B as compared to the liquid phase, the latter becomes less rich in B and thus its composition point moves towards the A-axis. If the crystallization of solid solution is to be continued, the temperature of the liquid phase has to be decreased so that its state point moves on the line bc′d′A. For example, if the system has been cooled up to the overall state point c, the system now consists of solid solution c″ in equilibrium with liquid solution c′. The points c′ and c″ are obtained by drawing a tie line from the point c. The relative amounts of solid solution c″ and liquid solution c′ at c can be obtained by making use of the lever rule.

During the cooling of the system from b to d, the relative length of the segment of tie line lying to the left of bcd increases indicating that the relative amounts of the two solutions change in favour of solid solution, i.e. more and more of solid solution solidifies during the cooling process. When the system has been cooled to the point d, the last trace of liquid solution d′ remains to be solidified and thus represents the point where practically whole of liquid phase is solidified to produce solid solution d whose composition is the same as that of starting liquid solution (i.e. point a). Further cooling from d to e, merely decreases the temperature of solid solution.

**Conclusion**

From the above analysis, the following conclusions can be drawn.

A point within the area AbBdA represents a solid solution in equilibrium with a liquid solution. Their compositions are given by the intersection points of the tie line (horizontal line) drawn from the given point with those of solidus (AdB) and liquidus (AbB) curves, respectively. The relative amounts of the two are given by the lever rule. On gradually cooling, the composition of liquid solution moves along BbA and that of solid solution along the curve BdA.

**Labelling of the Phase Diagram**

Table 6.8.1 describes, in brief, the phase diagram shown in Fig. 6.8.1.

**Table 6.8.1** Description of the Phase Diagram shown in Fig. 6.8.1

| *Location of the system* | *Status of the system* |
|---|---|
| A | Freezing point of A |
| B | Freezing point of B |
| Curve AbB | Freezing point curve of solid solution |
| Curve AdB | Fusion point curve of solid solution |
| Area above AbB | Liquid |
| Area below AdB | Solid solution |
| Area AbBdA | A point within this area represents a solid solution (whose composition lies on the curve AdB) in equilibrium with a liquid solution (whose composition lies on the curve AbB). |

**Fractional Crystallization**

Since the present system has phase diagram very similar to that of liquid-vapour system exhibiting ideality or little deviations from ideality, the two components can be separated from each other following the method of *fractional crystallization*—a method very similar to the fractional distillation employed in the liquid-vapour system. This method involves alternative cooling and heating repeatedly which results in the preferential separation of the two constituents—the solid carries more of the component of higher freezing point and the liquid carries lesser of this constituent (or larger of the other constituent of lesser freezing point). Finally, we can collect the pure component of larger freezing point in the solid form and the component of lesser freezing point in the liquid form.

## Zone Refining

*The technique of zone refining is based on the principle of fractional crystallization. It is employed to remove impurities from a given substance.* The end products obtained in fractional crystallization have been described above. The principle of fractional crystallization may be outlined again in order to facilitate the description of the technique of zone refining.

***Principle*** Consider a solid solution of composition e as shown in Fig. 6.8.1. Let it be heated to the point c. At this point, the system consists of two phases, liquid solution c′ and solid solution c″ in equilibrium with each other. Let the two phases be separated and the solid solution c″ be heated separately to the point b″. Again we get two phases, a liquid solution b and a solid solution b′. Again let the two phases be separated and the solid phase be heated separately and so on. Ultimately, we will get the solid phase consisting of pure B. The liquid solutions at different stages may be mixed together, solidified and subjected to the same sequence of steps as discussed above. At each stage, the preferential separation of the two constituents takes place, solid carries more of the constituent of higher freezing point and the liquid carries lesser of this constituent (or more of the other constituent of lesser freezing point). Carrying the above process repeatedly will result into the separation of the two constituents—the constituent of higher freezing point in the solid phase and that of lesser freezing point in the liquid phase.

***Process*** Now the process of zone refining may be described. The substance to be purified is taken in the form of a narrow cylinder and is heated in a narrow, disc-like zone which is swept from one end of the sample to the other (Fig. 6.8.2).

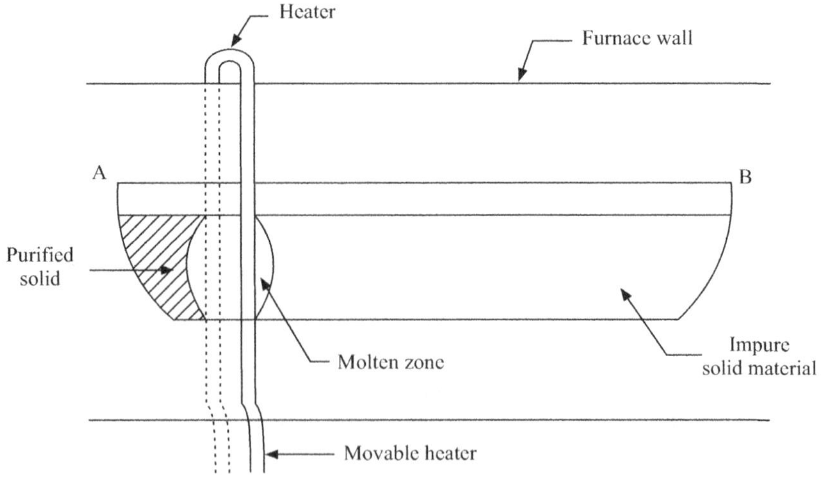


**Fig. 6.8.2 Schematic representation of a zone refining apparatus**

Immediately in front of the heater, the solid melts forming a liquid zone which carries more of impurity provided it has a lower melting point. As the heater moves forward this liquid zone also moves forward and immediately at the back of the heater, we get a solid which carries lesser of impurity. The liquid zone as it moves forward becomes more and more enriched in the impurity as it is preferentially partitioned into the liquid phase. One pass of the heater from one end of the substance to the other may have the effect of reducing the impurity content only slightly. In order to get rid of more and more of impurity, the heater is passed repeatedly from one end to the other.

## Examples

Substances of similar constitution or closely related elements usually form solid solutions of the the type discussed above. A few examples are: Co-Ni, Au-Ag, Au-Pt, AgCl-NaCl, $PbCl_2$-$PbBr_2$, and naphthalene-β-naphthol.

---

### *Example 6.8.1*

The following data were observed for Si-Ge system:

| % Si (by mass) | 0 | 10 | 25 | 40 | 45 | 62 | 75 | 80 | 90 | 100 |
|---|---|---|---|---|---|---|---|---|---|---|
| ($\theta_c$/°C) at which first solid appears | 940 | 1062 | 1160 | 1235 | 1252 | 1310 | 1350 | 1370 | 1395 | 1412 |
| ($\theta_c$/°C) at which complete solidification occurs | 940 | 697 | 1010 | 1070 | 1090 | 1170 | 1245 | 1275 | 1340 | 1412 |

(a) Plot the phase diagram and indicate the phases present in each region.

(b) What is composition of the phases present when the solid mixture containing 50 mass % Si is heated at 1200 °C.

(c) Explain, how you would attain a pure Si from a solid mixture containing 70 mass % Si.

### *Solution*

(**a**) The phase diagram plotted from the given data is shown in Fig. 6.8.3. The present system of Si-Ge will form a complete series of solid solutions.

(**b**) The phases present in equilibrium with each other when the solid mixture containing 50 mass % Si is heated to 1200 °C are:

Liquid solution containing 32.5 mass % Si

Solid solution containing 70 mass % Si.

(**c**) Pure Si can be obtained by the method of fractional crystallization. The residue will finally consist of pure Si.

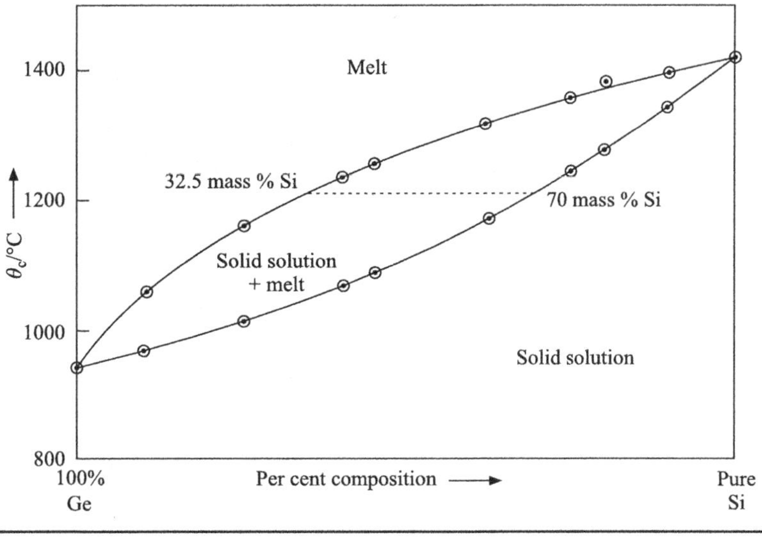


**Fig. 6.8.3 Required phase diagram**

---

## MINIMUM TYPE SOLID SOLUTION

**Depiction of the Phase Diagram**

Systems with these characteristics have phase diagram as shown in Fig. 6.8.4.

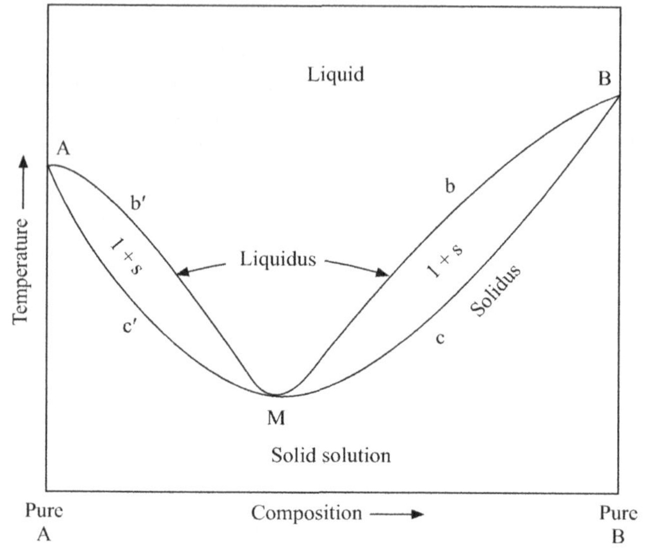


**Fig. 6.8.4 Phase diagram of a system exhibiting continuous series of solid solutions with a minimum freezing point**

**Labelling of the Phase Diagram**

The labelling of Fig. 6.8.4 is shown in Table 6.8.2.

**Table 6.8.2** Description of the Phase Diagram shown in Fig. 6.8.4

| *Location of the system* | *Status of the system* |
|---|---|
| A | Freezing point of A |
| B | Freezing point of B |
| M | Minimum freezing point of the liquid solution. Here, the composition of liquid and solid solutions are identical. The system here is nonvariant. Thus freezing or fusion takes place at constant temperature. This is exhibited by the arrest of temperature in the cooling curve |
| Area above Ab′MbB | Liquid |
| Area below Ac′McB | Solid solution |
| Area Ab′Mc′A | Solid solution (composition lies on Ac′M) in equilibrium with liquid solution (composition lies on Ab′M) |
| Area BbMcB | Solid solution (composition lies on BcM) in equilibrium with liquid solution (composition lies on BbM) |

**Comment on the Phase Diagram**

Phase diagram of this system is similar to that of the liquid-vapour system exhibiting positive deviation from ideality (Fig. 4.16.1) and thus has characteristics resembling those of the liquid-vapour system. Depending on the composition, the fractional crystallization of this system will yield either pure A or pure B as a solid phase (to the left to point M, pure A and to the right of M, pure B) and a liquid of composition corresponding to the minimum freezing point M.

**Examples**

Table 6.8.3 includes a few examples of a system exhibiting minimum type solid solution.

**Table 6.8.3** Systems Exhibiting Continuous Series of Solid Solutions with a Minimum Freezing Point

| System | | | | Minimum point M | |
|---|---|---|---|---|---|
| A | m.pt. °C | B | m.pt. °C | Temperature °C | Composition |
| Cu | | Mn | | | |
| Mn | | Ni | | | |
| Co | | Mn | | | |
| $HgBr_2$ | | $HgI_2$ | | | |
| KCl | 781 | AgCl | 748 | 716 | 60 mol % B |
| *p*-dichloro-benzene | 52.8 | *p*-chloro-iodobenzene | 53 | 41 | 50 mol % B |
| Chloro-benzene | −44 | iodobenzene | −29 | −51.3 | 39.02 mass % B |
| Stilbene | 123.5 | Hydrazo-benzene | 130.5 | 93 | 52.26 mass % B |
| Dibenzyl | 52.5 | Hydrazo-benzene | 130.5 | 45.9 | 13.8 mass % B |

In most cases, the two members of a system are found to have the same crystal structure (isostructural or isomorphic). Thus, in each of the solid solution-liquid fields shown in Fig. 6.8.4, it is the same solid phase that coexists with liquid. Besides, as indicated above, both the solidus and liquids curves are continuous, which meet each other only at a single point.

## MAXIMUM TYPE SOLID SOLUTION

**Depiction of the Phase Diagram**

In this case, the addition of one component into another elevates the freezing point of the latter and thus the phase diagram has an appearance as shown in Fig. 6.8.5. Description of the diagram is straightforward and is not given here.

**Comment on the Phase Diagram**

The phase diagram shown in Fig. 6.8.5 resembles that of a liquid-vapour system exhibiting negative deviation from ideality (Fig. 4.16.2) and thus has characteristics similar to those of a liquid-vapour system. Depending on the initial composition, the fractional crystallization will yield either pure A or pure B in the liquid phase (to the left of point M, pure A and to the right of point M pure B) and a solid phase of composition corresponding to the maximum freezing point M.

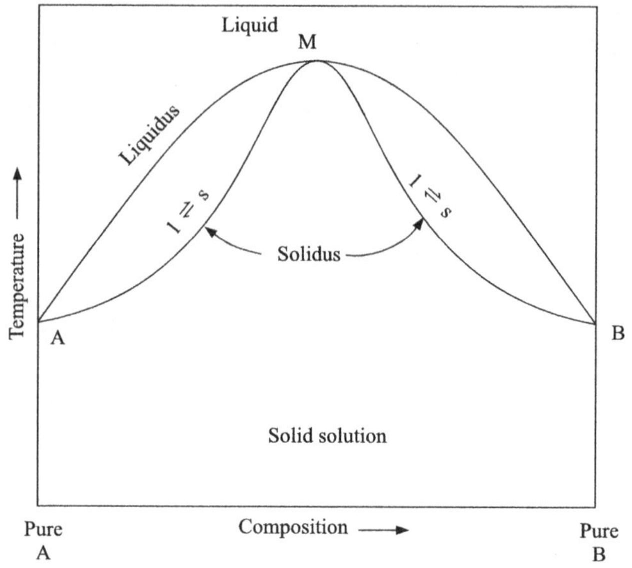


**Fig. 6.8.5 Phase diagram for a system exhibiting continuous series of solid solutions with a maximum**

## Examples

Very few examples of systems yielding a continuous solid solutions and having a freezing-point curve with a maximum are known. A few examples are *d*- and *l*-monobornyl esters of malonic acid.

The two constituents are the *d*- and *l*-forms of the same optically active compound. Since both of them have the same melting point, the phase diagrams are symmetrical with the maximum at a midway point. It has been suggested that the maximum may represent a racemic compound.

## Slope of *T* versus *x* Curve at Azeotropic Temperature

For a binary mixture, Gibbs-Duhem equation is

$$S\,\mathrm{d}T - V\,\mathrm{d}p + n_A \mathrm{d}\mu_A + n_B \mathrm{d}\mu_B = 0 \tag{6.8.19}$$

For a unit amount of mixture, we have

$$S_m \mathrm{d}T - V_m \mathrm{d}p + x_A\,\mathrm{d}\mu_A + x_B\,\mathrm{d}\mu_B = 0 \tag{6.8.20}$$

Using the expression $\mu = \mu^* + RT \ln a$ (where $a$ is activity), Eq. (6.8.20) may be written as

$$S_m \mathrm{d}T - V_m \mathrm{d}p + RT\,[x_A\,\mathrm{d}\ln a_A + x_B\,\mathrm{d}\ln a_B] = 0 \tag{6.8.21}$$

For a system of two binary solutions in equilibrium, Eq. (6.8.21) holds good separately for both the solutions. Hence, we write

$$S_m^{(\alpha)}\,\mathrm{d}T - V_m^{(\alpha)}\,\mathrm{d}p + RT\,[x_{A(\alpha)}\,\mathrm{d}\ln a_A + x_{B(\alpha)}\,\mathrm{d}\ln a_B] = 0 \tag{6.8.22}$$

$$S_m^{(\beta)}\,\mathrm{d}T - V_m^{(\beta)}\,\mathrm{d}p + RT\,[x_{A(\beta)}\,\mathrm{d}\ln a_A + x_{B(\beta)}\,\mathrm{d}\ln a_B] = 0 \tag{6.8.23}$$

Note that the chemical potential and thus activity of A (as well as those of B) in both the solutions will have the same value as the component A is in equilibrium between the two solutions.

Subtraction of Eq. (6.8.23) from Eq. (6.8.22) gives

$$\Delta_{trs}S_m\ dT - \Delta_{trs}V_m\ dT + RT\ [x_{A(\alpha)} - x_{A(\beta)}]\ d\ln a_A + RT[x_{B(\alpha)} - x_{B(\beta)}]\ d\ln a_B = 0 \qquad (6.8.24)$$

where $\Delta_{trs}S_m = S_m^{(\alpha)} - S_m^{(\beta)}$ and $\Delta_{trs}V_m = V_m^{(\alpha)} - V_m^{(\beta)}$

Since the solutions are at equilibrium at temperature $T$, we will have

$$\Delta_{trs}S_m = \frac{\Delta_{trs}H_m}{T}$$

With this, Eq. (6.8.24) may be written as

$$\frac{\Delta_{trs}H_m}{RT^2}dT - \frac{\Delta_{trs}V_m}{RT}dp + [x_{A(\alpha)} - x_{A(\beta)}]\ d\ln a_A + [x_{B(\alpha)} - x_{B(\beta)}]\ d\ln a_B = 0 \qquad (6.8.25)$$

At azeotropic temperature

$$x_{A(\alpha)} = x_{A(\beta)} \quad \text{and} \quad x_{B(\alpha)} = x_{B(\beta)}$$

With these, Eq. (6.8.25) become

$$\frac{\Delta_{trs}H_m}{RT^2}dT - \frac{\Delta_{trs}V_m}{RT}dp = 0 \qquad (6.8.26)$$

From Eq. (6.8.26) is follows that

$$\left(\frac{\partial T}{\partial x_B}\right)_p = 0 \quad \text{and} \quad \left(\frac{\partial p}{\partial x_B}\right)_T = 0$$

Thus, the slope of $T$ versus $x_B$ curves in Figs (6.8.4) and (6.8.5) will have zero slope at the point M.

## 6.9 FORMATION OF PARTIAL MISCIBLITY IN THE SOLID STATE LEADING TO STABLE SOLID SOLUTIONS

**Depiction of the Phase Diagram**

In some cases, solids are miscible with each other in limited range of concentrations. Outside this range, they form two solid solutions in equilibrium with each other. These are saturated solutions of A in B and B in A. This phenomenon is quite analogous to that of the partially miscible liquids and thus the phase diagram of the present system has an appearance similar to that of Fig. 4.18.8. This diagram can be relabelled by writing liquid and solid in places of vapour and liquid, respectively (Fig. 6.9.1).

This type of diagram is observed only when $K_{B\ in\ A} < 1$ and $K_{A\ in\ B} < 1$, i.e. the freezing point of A (or B) is lowered when B (or A) is added to it.

**Construction of the Phase Diagram**

- From the cooling curves of pure components, the points A and B representing respectively the freezing temperatures of A and B are obtained.
- From the cooling curves I and II of two systems of B in A, the points a and b indicating respectively the start of freezing of solid solution of B in A (hereafter referred to as $S_1$) and a′ and b′ indicating respectively the end of solidification of $S_1$ are obtained.
- From the cooling curve III, the point c representing the start of freezing of solid solution $S_1$ is obtained.

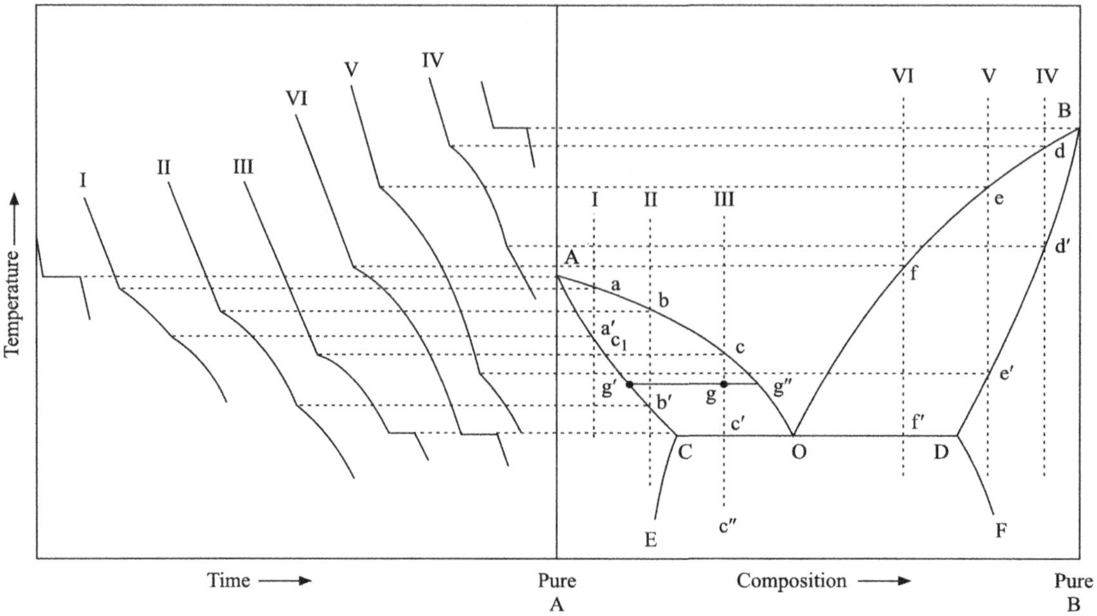


**Fig. 6.9.1 Phase diagram exhibiting partial miscibility in the solid state with the eutectic point**

- If a line passing through A, a, b and c is drawn, we get the freezing point curve Aabc of solid solution $S_1$. Similarly, if a line passing through A, a′ and b′ is drawn, we get the fusion point curve of solid solution $S_1$.
- From the cooling curves IV and V of two systems of A in B, the points d and e indicating respectively the start of freezing of solid solution of A in B (hereafter referred to as $S_2$) and d′ and e′ indicating respectively the end of solidification of $S_2$ are obtained.
- From the cooling curve VI, the point f representing the start of freezing of solid solution $S_2$ is obtained.
- If a line passing through B, d, e and f is drawn, we get the freezing point curve Bdef of solid solution $S_2$. Similarly, if a line passing through B, d′ and e′ is drawn, we get the fusion point curve Bd′e′ of solid solution $S_2$.
- Along the line Aabc, liquid solution is in equilibrium with solid solution $S_1$ whose composition lies on the line Aa′b′. For a given point on Aabc, the composition of solid solution in equilibrium is represented by the intersection point of the tie line (horizontal line) drawn from the given point with the line Aa′b′. Similarly, the lines Bdef and Bd′e′, respectively, describe the compositions of liquid solution and solid solution $S_2$ in equilibrium with each other.
- If the lines Aabc and Bdef are extended they meet at the point O. Here both solid solutions $S_1$ and $S_2$ will be in equilibrium with liquid phase O. The compositions of $S_1$ and $S_2$ can be obtained by extending the lines Aa′b′ and Bd′e′ up to the points where they meet the tie line (horizontal line) drawn from the point O. Thus, the points C and D shown in Fig. 6.9.1 represent the two solid solutions $S_1$ and $S_2$ in equilibrium with liquid O.

- A system on the line COD contains three phases in equilibrium and, according to the phase rule, is an invariant system. Hence, the temperature corresponding to the line COD will have a fixed value. Moreover, solidification of solid solutions $S_1$ and $S_2$ from the liquid O occurs isothermally. Consequently, the cooling curves III and VI exhibit arrest of temperature at points c′ and f′, respectively. The overall composition of solid phase (i.e. total amount of $S_1$ and $S_2$) separated on the line COD is the same as that of liquid phase, vis-à-vis, the composition of liquid phase remains unchanged.
- When the solidification of liquid O is complete, the system contains two solid solutions C and D in equilibrium with each other. According to the phase rule, the system becomes univariant ($F = C + 1 - P = 2 + 1 - 2 = 1$). Consequently, the system will exhibit cooling during which the compositions of solid solutions $S_1$ and $S_2$ change along the lines CE and DF, respectively.

**Effect produced on Cooling a System**

The labelling of the phase diagram can be done easily with the help of the above discussions. However, to make it more understandable, it is worth considering the sequence of effects that is produced during the cooling of the system III.

- Up to the point c, the system remains in the liquid phase.
- At c, the solid solution corresponding to the point $c_1$ starts precipitating. Since the solid solution $c_1$ is rich in A as compared to the liquid phase c, the composition of the latter moves towards B axis. If the crystallization of $S_1$ is to be continued, the temperature of the liquid phase has to be lowered so that its composition moves on the line cO. Thus, on cooling the system from c to c′, the composition of liquid phase moves along cO whereas that of solid phase along $c_1$C. The relative amounts of solid and liquid solutions at any stage of cooling can be determined with the help of the lever rule. For example, if the system is cooled up to the point g, then we have

$$\frac{\text{Amount of solid solution g}'}{\text{Amount of liquid solution g}''} = \frac{\text{gg}''}{\text{gg}'}$$

- When the system is cooled up to the point c′, the solid solution C is in equili -brium with liquid O. The relative amounts of the two solutions are given by

$$\frac{\text{Amount of solid solution C}}{\text{Amount of liquid solution O}} = \frac{\text{Oc}'}{\text{c}'\text{C}}$$

  If the cooling is continued, both $S_1$ and $S_2$ are separated isothermally from the liquid O. Since now solid contains both $S_1$ and $S_2$, the point representing solid phases moves from C towards c′ as more and more solid solutions are solidified. When this point is reached at c′, whole of liquid solution is solidified.
- From c′ to c″, only cooling of solid solutions $S_1$ and $S_2$ takes place. The composition of the former moves along CE whereas that of the latter moves along DF.

**Labelling of the Phase Diagram**

Table 6.9.1 describes the labelling of Fig. 6.9.1

**Table 6.9.1** Labelling of the Diagram shown in the Fig. 6.9.1

| *Location of the system* | *Status of the system* |
|---|---|
| A | Freezing point of A |
| B | Freezing point of B |
| AO | Freezing point curve of solid solution $S_1$ |
| AC | Fusion point curve of solid solution $S_1$ |
| BO | Freezing point curve of solid solution $S_2$ |
| BD | Fusion point curve of solid solution $S_2$ |
| O | Eutectic point where solid solutions $S_1$ and $S_2$ represented by C and D, respectively, are in equilibrium with the liquid O. |
| CE and DF | Compositions of the two conjugate solid solutions $S_1$ and $S_2$ |
| Area to the | |
| left of ACE | Solid solution $S_1$ |
| right of BDF | Solid solution $S_2$ |
| Area AOCA | A point in this area represents a solid solution $S_1$ in equilibrium with a liquid phase; composition of the former lies on the line AC, whereas that of the latter on the line AO. These compositions can be determined by drawing a tie line from the given point. |
| Area BODB | A point in this area represents a solid solution $S_2$ in equilibrium with a liquid phase; composition of the former lies on the line BD whereas that of the latter on the line BO. |
| Area within the lines CE and DF | A point within this area respresents two conjugate solutions $S_1$ and $S_2$ in equilibrium with each other. The compositions of these two solutions are given by the two intersection points of the tie line drawn from the given point with the lines CE and DF, respectively. |
| Area above AOB | Liquid |

## Cooling Patterns

The nature of the cooling curve will depend on the composition of the system. The following are a few typical patterns.

(i) *Composition anywhere between* AC (*or* BD):

Rapid cooling of liquid → Break at the appearance of solid solution → Slower cooling; solidification continues
↓
Break; solidification completes → Cooling of solid solution

(ii) *Composition anywhere between* CO (*or* DO):

Rapid cooling of liquid → Break at the appearance of solid solution → Slower cooling; solidification continues
↓
Horizontal portion at eutectic point → Cooling of the two conjugate solutions

(iii) *Eutectic composition:*

Rapid cooling of liquid → Horizontal portion at eutectic point → Cooling of the two conjugate solid solutions

These cooling patterns are shown in Fig. 6.9.1

### Fractional Crystallization

Results of fractional crystallization are also very similar to those of fractional distillation of the liquid-vapour system showing partial miscibility in the liquid phase. We give below the end products that are obtained in fractional crystallization of the present system.

(i) *When the original composition of the system lies to the left of the eutectic point* O:

Solid phase—pure A,
and liquid phase—a liquid of eutectic composition.

(ii) *When the original composition of the system lies to the right of the eutectic point* O:

Solid phase—pure B,
and liquid phase—a liquid of eutectic composition.

(iii) *When the original composition of the system is the eutectic composition:* In this case, no fractional crystallization is possible as the melt has the same composition as that of the solid phase.

### Examples

A few examples of system exhibiting the phase diagram of Fig. 6.9.1 are given below: Au–Ni, Bi–Pb, Cd–Sn, Fe–Cr, naphthalene-monochloroacetic acid, azobenzene-azoxybenzene, and dibenzyl and benzylphenate.

### Theoretical Aspects of Fig. 6.9.1

Phase diagram of Fig. 6.9.1, in fact, lies between those for components which are completely miscible in the solid state with the minimum freezing point and those which form completely immiscible mixture in the solid phase. If the solids in Fig. 6.9.1 are assumed to become more and more miscible, points C and D will eventually coincide and the phase diagram will resemble Fig. 6.8.4—a phase diagram in which solid solution with a minimum freezing point is formed. On the other hand, if the two solids become less miscible, points C and D will recede from each other and eventually lie on the axes of pure A and pure B, respectively, and the phase diagram will resemble Fig. 6.4.1, which respresents the separation of solid mixture.

Limited solid-phase solubility may result because of the contamination of each of the two members of a binary system by the other. If we assume that the two members A and B have different crystal structures, then if B contaminates A, the atoms of B in A conform to the crystal structure of A. Conversely, if A contaminates B, the atoms of A in B conform to the crystal structure of B. As mentioned above, if the contaminations are vanishingly small, phase diagram of the system will have the major characteristics of a simple eutectic system. If the two substances exhibit considerable solid solubility in each other, then the liquidus and solidus of A and B regions acquire the appearances of descending solid solutions of A in B, and B in A, respectively. Since the crystal structures of the two members are identical in the solid solution region, it is expected that the contaminated phase in this region is present in the metastable form whose crystal structure is identical with that of the primary phase. Thus, the entire phase diagram of Fig. 6.9.1 may be visualized to be formed as a result of two intersecting descending diagram as shown in Fig. 6.9.2.

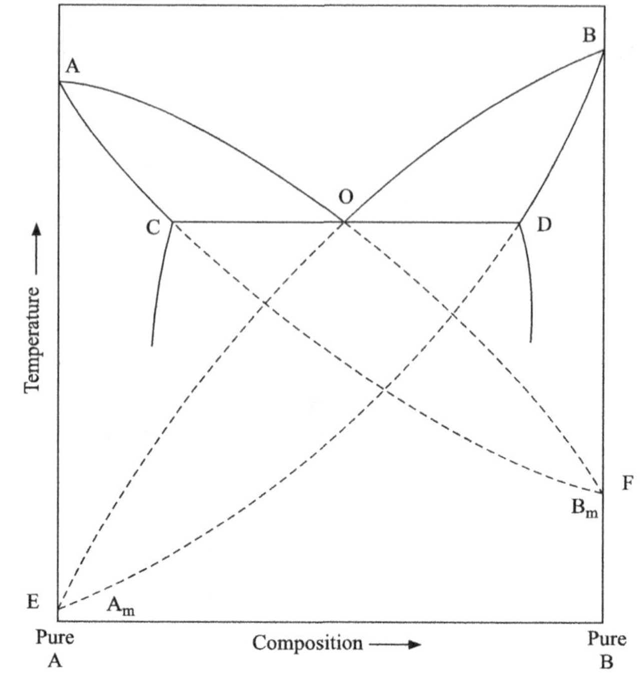


**Fig. 6.9.2 Metastable extensions of the liquids and solidus curves of a system exhibiting limited solubility with eutectic temperature**

To be specific, let the crystal structures of A and B be α and β, respectively. The solid solution in the regions ACE and ACO of Fig. 6.9.1 has α-structure of primary phase A. The component B in this solid solution has the metastable α-structure whose metastable melting point is represented by the point F in Fig. 6.9.2. Thus the region CAO may be considered to be a portion of complete descending solid solution diagram AOFCA. Similarly, the regions BOD and BDF of Fig. 6.9.1 have a solid solution of β-structure and is formed as a result of stable β-structure of B and metastable β-structure of A. The metastable melting point of the latter is shown by the point E in Fig. 6.9.2. Hence the region OBD may be considered to be a portion of complete descending solid solution diagram BOEDB. From Fig. 6.9.2, it is obvious that a system will exhibit the phase diagram of Fig. 6.9.1 only when the metastable melting points E and F lie below the stable melting points A and B, respectively.

## 6.10 FORMATION OF PARTIAL MISCIBLITY IN THE SOLID PHASE LEADING TO SOLID SOLUTIONS STABLE UP TO A TRANSITION TEMPERATURE

Instead of exhibiting an eutectic, two solid solutions may undergo a peritectic reaction of the type

$$S_1 \rightleftharpoons S_2 + \text{Liquid solution}$$

The above reaction takes place at a constant temperature since the system with three phases in equilibrium is an invariant system.

The phase diagram exhibiting the above peritectic reaction has an appearance as shown in Fig. 6.10.1. This type of diagram is observed only when $K_{\text{B in A}} > 1$

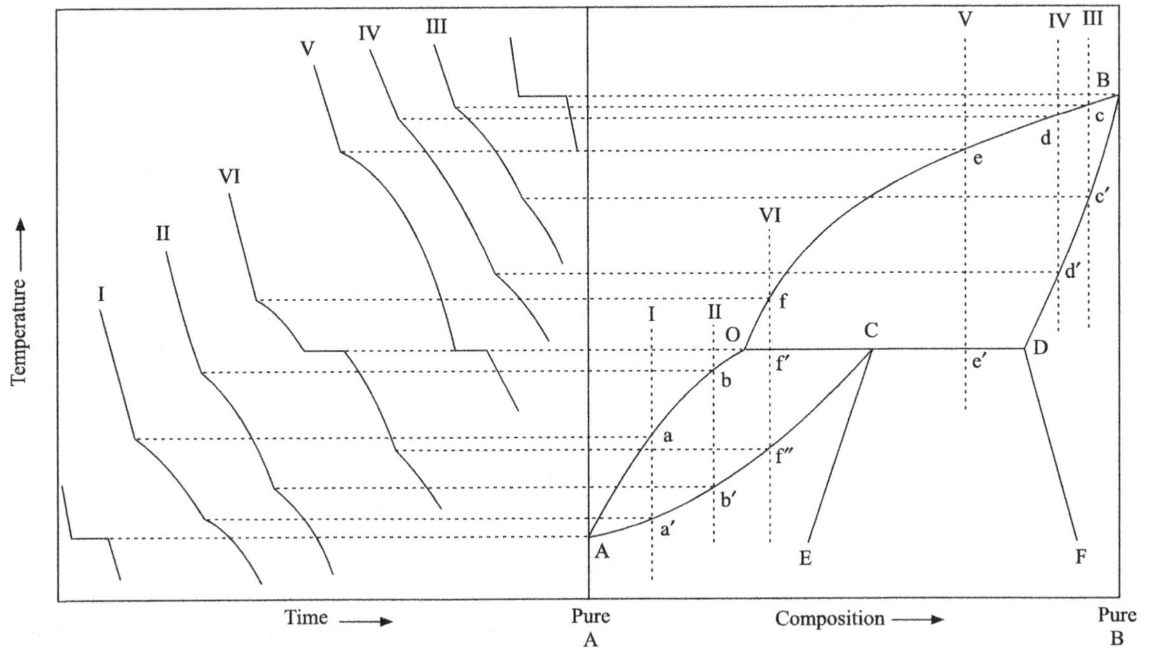


**Fig. 6.10.1 Phase diagram exhibiting partial miscibility in the solid state with a peritectic transition**

and $K_{\text{A in B}} < 1$, i.e., the freezing point A is elevated when B is added to it and that of B is lowered when A is added to it (or vice versa).

## Construction of the Phase Diagram

- From the cooling curves of pure components, the points A and B representing respectively the freezing temperatures of components A and B are obtained.
- From the cooling curves I and II of two systems of B in A, the points a and b representing respectively the start of freezing of solid solution $S_1$ and the points a′ and b′ representing respectively the end of freezing of solid solution $S_1$ are obtained.
- If a line passing through the points A, a and b is drawn, we get the freezing point curve Aab of solid solution $S_1$. Similarly, if a line passing through the points A, a′ and b′ is drawn, we get the fusion point curve Aa′b′ of solid solution $S_1$.
- From the cooling curves III and IV of two liquid systems of A and B, the points c and d representing respectively the start of freezing of solid solution $S_2$ and the points c′ and d′ representing respectively the end of freezing of solid solution $S_2$ are obtained.
- From the cooling curves V and VI of two systems of A and B, the points e and f representing respectively the start of freezing of solid solution $S_2$ are obtained.
- If a line passing through the point B, c, d, e and f is drawn, we get the freezing point curve Bcdef of solid solution $S_2$. Similarly, if a line passing through the points B, c′ and d′ is drawn, we get the fusion point curve Bc′d′ of solid solution $S_2$.

- Since the equilibrium

  Solid solution $S_1$ (on Aa′b′) $\rightleftharpoons$ liquid solution (on Aab)

  exists at lower temperatures whereas the equilibrium

  Solid solution $S_2$ (on Bc′d′) $\rightleftharpoons$ liquid solution (on Bcd)

  exists at higher temperatures, the peritectic reaction

  $S_1$ + solution $\rightleftharpoons$ $S_2$ + solution

  will take place at a temperature (peritectic) which will lie in between the freezing temperatures of pure components A and B. The composition of the associated liquid phase will have a fixed value since the peritectic reaction involves three phases in equilibrium ($F = C + 1 - P = 2 + 1 - 3 = 0$).
- Since the curves Aab and Bcdef describe respectively the start of freezing of solid solutions $S_1$ and $S_2$, these two curves on extending will meet at point O which describes the composition of the liquid phase in equilibrium with solid solutions $S_1$ and $S_2$. The compositions of two solid solutions will be given by the intersection points C and D of the tie line drawn from the point O with the extensions of the fusion point curves Aa′b′ and Bc′d′, respectively. Thus, a point on the line OCD represents a system where solid solution $S_1$ of composition *C*, solid solutions $S_2$ of composition *D* and liquid solution of composition *O* exist together in equilibrium with one another.
- Since a system on the line OCD has three phases in equilibrium, it will be invariant. Thus, so long three phases remain in equilibrium, the temperature of the system will remain constant. Thus, the cooling curves V and VI will exhibit a complete arrest of temperature at the points e′ and f′, respectively.
- The point C on the line OCD represents solid solution $S_1$ in equilibrium with liquid O. If this system is heated gradually, more and more of this solid solution is converted isothermally into solid solution $S_2$ and liquid solution O. Thus, at a temperature greater than the peritectic temperature the system will consist of a solid solution $S_2$ in equilibrium with a liquid solution. In fact, a system anywhere on the line OCD on heating will yield a system of solid solution $S_2$ (whose composition lies on the curve BD) in equilibrium with a liquid solution (whose composition lies on the curve BO).
- The state of the system at a temperature below the peritectic temperature depends upon the overall composition of the system. We will have either solid solutions $S_1$ in equilibrium with solid solution $S_2$ or solid solution $S_1$ in equilibrium with a liquid solution. This can be explained on the basis of the peritectic reaction:

$$S_2 + \text{liquid O} \underset{t > t_p}{\overset{t < t_p}{\rightleftharpoons}} S_1$$

The reaction between $S_2$ and liquid O proceeds isothermally to form the solid solution $S_1$ until either all the solid solution $S_2$ or all the liquid O has been consumed. We have

(i) If solid solution $S_2$ is present in excess amount than the requisite quantity that is required for the formation of the solid solution $S_1$ with whole of liquid O, then solid solution $S_2$ will remain in excess and hence the system will possess this excess quantity of $S_2$ and the solid solution $S_1$. This happens when the point representing $S_2 \rightleftharpoons$ solution lies in between the points C and D. Thus, a system below the line CD represents solid solution $S_1$ in equilibrium with solid solution $S_2$.

(ii) If solid solution $S_2$ is present in deficient amount than the requisite quantity that is required for the formation of $S_1$ with whole of the liquid O; then whole of solid solution $S_2$ will disappear and the system will possess the excess quantity of liquid phase and the solid solution $S_1$. This happens when the point representing $S_2 \rightleftharpoons$ solution lies in between the points O and C. Thus, the system below the line OC represents solid solution $S_1$ (whose composition lies on the line CA) in equilibrium with liquid phase (whose composition lies on the line OA).

## Labelling of the Phase Diagram

With the help of above analysis, the labelling of the phase diagram can be easily done and is described in Table 6.10.1.

**Table 6.10.1** Labelling of the Phase Diagram shown in Fig. 6.10.1

| *Location of the system* | *Status of system* |
|---|---|
| A | Freezing point of A |
| B | Freezing point of B |
| AO | Freezing point curve of solid solution $S_1$ |
| AC | Fusion point curve of solid solution $S_1$ |
| BO | Freezing point curve of solid solution $S_2$ |
| BD | Fusion point curve of solid solution $S_2$ |
| O | Composition of liquid phase at the peritectic temperature |
| OCD | A system on this line represents a solid solution $S_1$ of composition *C*, a solid solution $S_2$ of composition *D*, and a liquid solution O in equilibrium with one another |
| CE and DF | Compositions of two conjugate solutions $S_1$ and $S_2$ |
| Area AOCA | A point in this area represents a solid solution $S_1$ (whose composition lies on the line AC) in equilibrium with a liquid phase (whose composition lies on the line AO) |
| Area BODB | A point in this area represents a solid solution $S_2$ (whose composition lies on the line BD) in equilibrium with a liquid phase (whose composition lies on the line BO) |
| Area within the lines EC, CD and DF | A point in this area represents a solid solution $S_1$ (whose composition lies on the line EC) in equilibrium with a solid solution $S_2$ (whose composition lies on the line DF) |
| Area within the lines AC and CE | Solid solution $S_1$ |
| Area to the right of BDF | Solid solution $S_2$ |
| Area above the curve AOB | Liquid |

**Cooling Patterns**

The nature of cooling curve will depend upon the composition of the system. A few typical cases shown in Fig. 6.10.1 are described below.

(i) *Composition anywhere between* A *and* O

Rapid cooling of liquid → Break; appearance of solid solution $S_1$ → Slower cooling; solidification continues ↓ Break; solidification completes → Cooling of solid solution $S_1$

(ii) *Composition anywhere between* O *and* C

Rapid cooling of liquid → Break; appearance of solid solution $S_2$ → Slower cooling; solidification continues ↓ Horizontal portion; transformation of solid solution $S_2$ to solid solution $S_1$ → Slower cooling; solidification of $S_1$ continues → Break solidification completes ↓ Smooth cooling; cooling of solid solution $S_1$ → Break if second solid solution appears → Slower cooling of two solid solutions

(iii) *Composition anywhere between* C *and* D

Rapid cooling of liquid → Break appearance of solid solution $S_2$ → Slower cooling; solidification continues ↓ Horizontal portion; appearance of two solid solutions → Slower cooling of two solid solutions

(iv) *Composition anywhere between* D *and* B

Rapid cooling of liquid → Break; appearance of solid solution $S_2$ → Slower cooling; solidification continues ↓ Break; solidification completes → Slower cooling of solid solution $S_2$ → Break; if second solid solution appears ↓ Cooling of two solid solutions

**Examples**

A few examples of system exhibiting partial miscibility in the solid state with a peritectic transition are given below.

AgCl–LiCl, $AgNO_3$–$NaNO_3$, Co–Fe and *p*-iodochlorobenzene-*p*-diiodobenzene.

**Theoretical Aspects of Fig. 6.10.1**

Limited solid-phase solubility with a peritectic temperature is observed when the metastable melting point of one of the components (e.g., $B_m$) lies above the melting point of other stable component (e.g., A) so that the two ascending solid solution diagrams ascend in the same direction (Fig. 6.10.2)

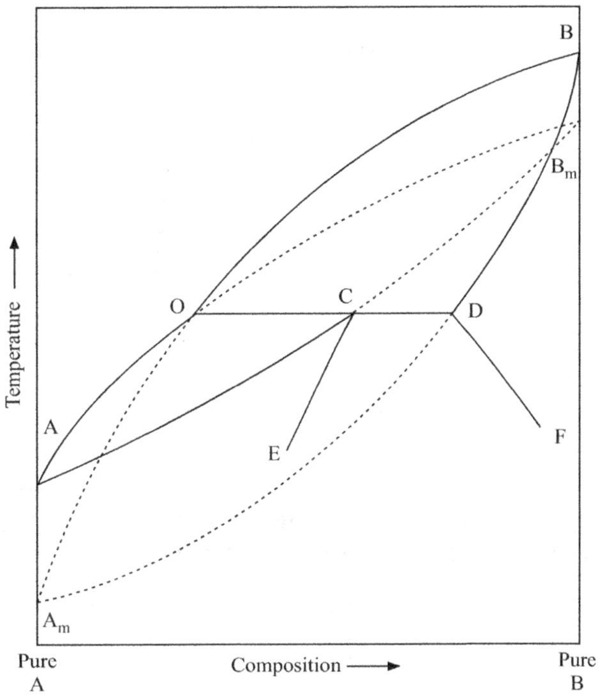


**Fig. 6.10.2 Metastable extensions of the liquidus and solidus curves of a system exhibiting limited solubility with peritectic temperature**

***Example 6.10.1***

Metals A and B melt at 1 200 °C and 1 600 °C, respectively. Thermal analysis shows presence of the following three-phase equilibria at 1 600 °C, melt containing 10 mol % B and two solid solutions containing 20 mol % B and 30 mol % B, respectively, at 1 250 °C, solid solution containing 65 mol % B, melt containing 75 mol % B and solid solution containing 95 mol % B. There is one compound $A_2B_3$ which melts at 1 700 °C (congruently). Construct the phase diagram and label the phase regions.

***Solution***

*Analysis of data*

(i) Since there is a formation of a congruently melting compound $A_2B_3$, we may draw the phase diagrams of systems A-$A_2B_3$ and $A_2B_3$-B separately (Fig. 6.10.1).

(ii) At 1 600 °C, the melt contains lesser of B as compared to those present in the two solid solutions. This implies the formation of two solid solutions with a peritectic temperature. Therefore, the nature of the phase diagram of A-$A_2B_3$ system will be similar to that shown in Fig. 6.10.1.

(iii) At 1 250 °C, composition of B in the melt is between those present in the two solid solutions. This implies the formation of stable solid solutions with an eutectic point and hence the nature of the phase diagram will be similar to that shown in Fig. 6.9.1.

Keeping in mind the above information, we obtain the phase diagram as shown in Fig. 6.10.3. The labelling of areas is also shown in the figure.

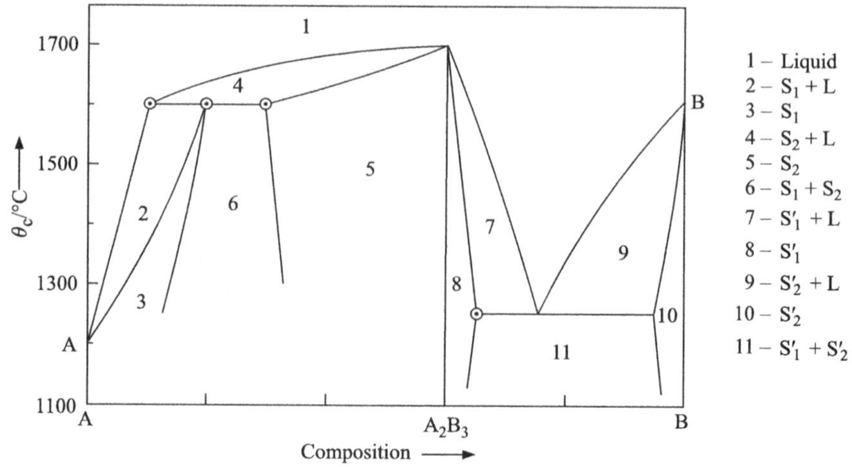


**Fig. 6.10.3 Required phase diagram**

## 6.11 FORMATION OF PARTIAL MISCIBILITY IN THE LIQUID PHASE AND CRYSTALLIZATION OF PURE COMPONENTS

The phase diagram of the present system has an appearance as shown in Fig. 6.11.1.

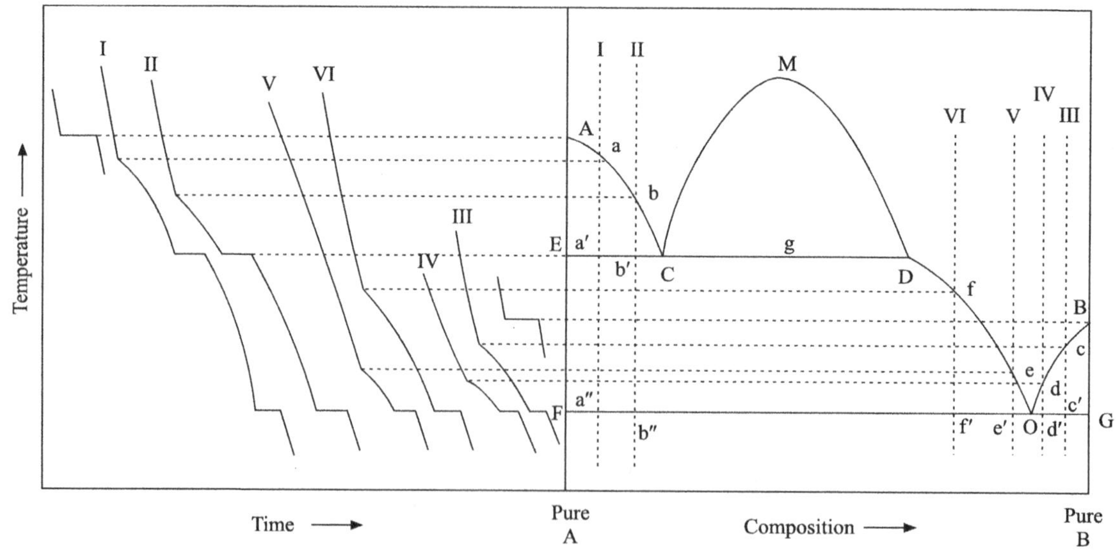


**Fig. 6.11.1 Phase diagram exhibiting partial miscibility in the liquid phase**

**Construction of the Phase Diagram**

- If the two components are partially miscible in the liquid phase, they will form two conjugate liquid solutions within certain range of concentrations. In the phase diagram, compositions of these two conjugate solutions will be represented by a dome-shaped curve such as CMD in Fig. 6.11.1. The curve CM represents compositions of saturated solutions of B in A whereas DM represents compositions of corresponding conjugate saturated solutions of A in B. The point M is the upper critical solution temperature.

- From the cooling curves of pure components, the points A and B representing respectively the freezing temperatures of A and B are obtained.
- From the cooling curves I and II of two systems of B in A, the points a and b representing respectively the start of freezing of A are obtained. If a line passing through A, a and b is drawn, we get the freezing point curve Aab of the component A.
- From the cooling curves III and IV of two systems of A in B, the points c and d representing respectively the start of solidification of B are obtained. If a line passing through B, c and d is drawn, we get the freezing point curve Bcd of the component B.
- If the line Aab is extended, it meets the saturation solubility curve CM of B in A at the point C. At this point, a single solution in equilibrium with solid A just splits into two conjugate solutions C and D, respectively. Thus, at the temperature corresponding to the point C, three phases, viz., solid A and two solutions C and D are in equilibrium with the one another. Since, the three phases in equilibrium constitutes an invariant system ($F = C + 1 - P = 2 + 1 - 3 = 0$), the temperature corresponding to the the point C will have a fixed value. Thus, the cooling curves I and II will exhibit an arrest of temperature at points a′ and b′, respectively.
- The tie line drawn at the point a′ or b′ starts from the point E (representing solid A) and after passing through the point C (representing one conjugate solution) ends at the point D (representing another conjugate solution). Thus, a point on the line ECD will represent solid A and two liquid solutions C and D in equilibrium with one another. The relative amounts of solid A and the two solutions taken together may be determined with the help of the lever rule. For example, at point a′, we have

$$\frac{\text{Amount of solid A}}{\text{Amount of solutions C and D}} = \frac{\text{a}'\text{C}}{\text{Ea}'}$$

- Further cooling of the system at a′ (or b′) causes more and more solidification of A without any change in temperature. The removal of A from the liquid phases moves the composition point C towards D along the line CD. As the point C moves towards D, the relative length of the segment of the tie line lying right to the point a′ increases indicating that more of A is precipitated. Besides this, the relative amounts of conjugate solutions change in favour of solution D. These facts are in agreement with the lever rule. For example, if the overall composition of solutions is moved to the point g, we will have

$$\frac{\text{Amount of solid A}}{\text{Amount of solutions C and D}} = \frac{\text{a}'\text{g}}{\text{Ea}'}$$

$$\frac{\text{Amount of solution C}}{\text{Amount of solutions D}} = \frac{\text{Dg}}{\text{Cg}}$$

  From the above expressions, the individual amounts of solid A, solution C and solution D can be determined for the given amount of the total system.
- When the overall composition of the solution is moved to the point D, virtually whole of solution C disappears and the system contains solid A and the solution D in equilibrium with each other. At this stage, the system a′

becomes univariant ($F = C + 1 - P = 2 + 1 - 2 = 1$) and thus it is expected to exhibit cooling along a′a″.

- If the cooling curves V and VI of two systems situated slightly right to the point D are studied, they will exhibit a break at the points e and f, respectively, where the solidification of A will just start. Thus, if a line passing through D, f and e is drawn, we get the freezing point curve Dfe of A from liquid whose compositions lie to the right of the point D. Hence, cooling of the system from a′ to a″, merely causes more and more solidification of A from a solution whose composition moves along the line Dfe. This fact is in agreement with the lever rule as the relative length of the segment of tie line right to the cooling line a′ a″ increases as temperature is lowered.
- When the system a′ has been cooled up to the point a″, the solution (on the extension of line Dfe) becomes saturated with respect to B and hence both solids A and B start solidifying together. This occurs at the eutectic point O where solid A and solid B are in equilibrium with the solution O. The system at this stage is invariant ($F = C + 1 - P = 2 + 1 - 3 = 0$). Thus, the temperature corresponding to the point O will have a fixed value. The cooling curves I, II, V and VI will exhibit an arrest of temperature at points a″, b″, e′ and f′, respectively.
- If the freezing point curve Bcd is extended, it will meet the freezing point curve Dfe at the eutectic point O. Thus, the cooling curves III and IV will exhibit an arrest of temperature at the points c′ and d′, respectively.

## Labelling of the Phase Diagram

With the help of the above analysis, the phase diagram shown in Fig. 6.11.1 can be easily labelled. Table 6.11.1 describes the labelling of Fig. 6.11.1.

**Table 6.11.1** Labelling of the Phase Diagram shown in Fig. 6.11.1

| *Location of the system* | *Status of the system* |
|---|---|
| A | Freezing point of A |
| B | Freezing point of B |
| Curve CMD | Composition of two conjugate solutions |
| M | Critical solution point |
| AabCDfe | Freezing point curve of A |
| BcD | Freezing point curve of B |
| O | Eutectic point involving solid A, solid B and liquid solution O |
| ECD | A system on this line represents solid A in equilibrium with the two conjugate solutions C and D |
| GOF | A system on this line represents solid A and solid B in equilibrium with solution O |
| Area ACDOFA | Solid A in equilibrium with liquid solution whose composition lies on the line ACDO |
| Area BGOB | Solid B in equilibrium with liquid solution whose composition lies on the line OB |
| Area below GOF | Solid mixtures of A and B |
| Area above ACMDOB | Liquid |
| Area CMDC | Two conjugate solutions (compositions lie on the curves CM and DM) in equilibrium with each other. |

**Cooling Patterns**

The nature of cooling curve will depend upon the composition of the system. A few typical cases shown in Fig. 6.11.1 are now described:

(i) *Composition anywhere between* AC

| | | | | |
|---|---|---|---|---|
| Rapid cooling of liquid | → | Break; appearance of solid A | → | Slower cooling; solidification of A continues and the composition of melt moves along AC |
| | | | | ↓ |
| Horizontal portion at the eutectic point O; solidification of solid mixture | ← | Slower cooling; solidification of A continues, composition of the melt moves along DO | ← | Horizontal portion; formation of two liquid solutions. Solidification of A continues till the melt has composition D |
| ↓ | | | | |
| Cooling of solid mixture | | | | |

(ii) *Composition corresponding to the point* C

| | | | | |
|---|---|---|---|---|
| Rapid cooling of liquid | → | Horizontal portion; formation of two liquid solutions; solidification of A continues till the melt has composition D | → | Slower cooling; solidification of A continues and the composition of the melt moves along DO |
| | | | | ↓ |
| | | Cooling of solid mixture | ← | Horizontal portion at the eutectic point O |

(iii) *Composition anywhere between* C *and* D

| | | | | |
|---|---|---|---|---|
| Rapid cooling of liquid | → | Break; formation of two liquid solutions | → | Cooling of two liquid solutions continues |
| | | | | ↓ |
| Horizontal portion at the eutectic point O | ← | Slower cooling; solidification of A continues, composition of the melt moves along DO | ← | Horizontal portion; solid A starts forming and continues till the composition of the melt is *D* |
| ↓ | | | | |
| Cooling of solid mixture | | | | |

(iv) *Composition anywhere between* B *and* O (*or* D *and* O)

| | | | | |
|---|---|---|---|---|
| Cooling of liquid | → | Break; solidification of B (or A) starts | → | Slower cooling; solidification of B (or A) continues and the composition of the melt moves along BO (or DO) |
| | | | | ↓ |
| | | Cooling of solid mixture | ← | Horizontal portion at the eutectic point |

**Examples**

A few examples which form two liquid layers are; benzoic acid-water, phenol-water, succinic nitrile-water, resorcinol-benzene and Zn-Bi.

## 6.12 PHASE DIAGRAMS OF AQUEOUS SOLUTIONS OF SALTS

An aqueous solution of a salt is an example of a two-component system where we may have

(1) Formation of a simple eutectic or cryohydrate.
(2) Formation of compounds (hydrates) with congruent melting points.
(3) Formation of compounds (hydrates) with incongruent or meritectic points.

At atmospheric pressure, the solubilities of most of salts in water are limited and, therefore, such systems are usually studied at high pressures. In this section, we shall discuss phase diagrams listed above by taking a typical example of each.

### FORMATION OF A SIMPLE EUTECTIC OR CRYOHYDRATE

An aqueous solution of KI constitutes a simple eutectic system the phase diagram of which is shown in Fig. 6.12.1. It has an appearance similar to that of Fig. 6.4.1. The description of the phase diagram is included in the figure itself. Along AO, ice separates and thus this curve represents the freezing point curve of water; along BO, solid KI separates and thus it represents the solubility curve of KI. Point O is the eutectic point where both ice and KI separate, the corresponding eutectic temperature is – 23 °C. Thus, this is the lowest temperature at which the system can exist in the liquid form.

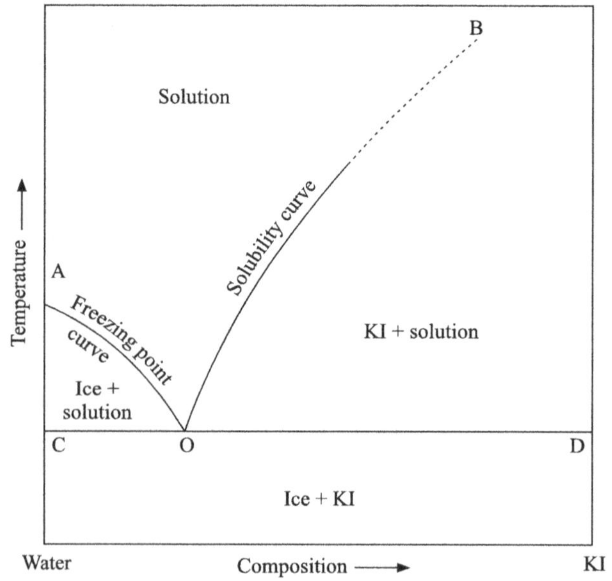


**Fig. 6.12.1 Required phase diagram**

Eutectic systems involving a salt and water were the first to be studied and were named as *cryohydrates* (Greek: frost water). Earlier, it was thought that at the cryohydric point (eutectic point) formation of a definite compound occurs. However, this view-point was rejected on the basis of following experimental facts:

- Composition of the eutectic varies with the external pressure and it seldom corresponds to a simple molecular proportion
- The heterogeneous structure of the solution can be seen under a microscope.

- Alcohol or acetone may dissolve the ice and leave behind a network of KI.
- X-ray analysis reveals the existence of two solid phases.
- The enthalpy of solution and density are the mean of the values for the two components.

Besides these facts, the phase rule also predicts that the formation of two solid phases at the cryohydric point is an invariant point.

**Production of Freezing Mixture with a Salt and Ice**

When a salt is added to ice and a little water at 0 °C, a freezing mixture is produced. The principle of the freezing mixture can be explained on the basis of the phase diagram shown in Fig. 6.12.1. Initially, we have three phases, namely, ice, solution and the salt; such a system can be stable only at the eutectic point which lies far below 0 °C. Thus at 0 °C, the system is not stable and it tries to attain the eutectic point by melting ice and dissolving more and more salt in the produced water. Since the melting of ice and in most cases the dissolution of the salt are endothermic processes, a fall in temperature of the system is observed. This is continued till either one of the phases is completely used up or the eutectic point is reached. The temperature of the system when this point is attained remains constant till sufficient heat from the surroundings is absorbed to cause all the ice to melt or all the salt to dissolve. The resultant system has two phases in equilibrium and, therefore, becomes univariant. Further absorption of heat from the surroundings will result in an increase in temperature of the system.

## FORMATION OF COMPOUNDS WITH CONGRUENT MELTING POINTS

**Ferric Chloride-Water System**

The best known example in which more than one congruent melting compound is formed is that of ferric chloride-water system. This salt forms four stable hydrates. These are

$$Fe_2Cl_6 \cdot 12H_2O; \quad Fe_2Cl_6 \cdot 7H_2O; \quad Fe_2Cl_6 \cdot 5H_2O \quad \text{and} \quad Fe_2Cl_6 \cdot 4H_2O$$

where double formula are used for ferric chloride in order to avoid a fractional number of molecules of water of crystallization. The phase diagram of this system has an appearance as shown in Fig. 6.12.2. The description of the diagram is given in Table 6.12.1.

**Isothermal Evaporation of Solution**

The cooling curve of a solution for a given composition can be drawn following the principles outlined earlier. For the present system it is of interest to consider isothermal evaporation of the solution. One such case is described below.

Let the solution be isothermally evaporated from the point a to m (Fig. 6.12.1). The sequence of events that takes place is as follows.

*Point* a to b: Unsaturated solution of ferric chloride.

*At* b: Solid $Fe_2Cl_6 \cdot 12H_2O$ commences.

*From* b to c: More and more of dodecahydrate separates, composition of the liquid remains unchanged and has a value corresponding to the point b. The volume of the solution decreases.

*Point* c: The solution completely disappears and the system consists of only solid $Fe_2Cl_6 \cdot 12H_2O$.

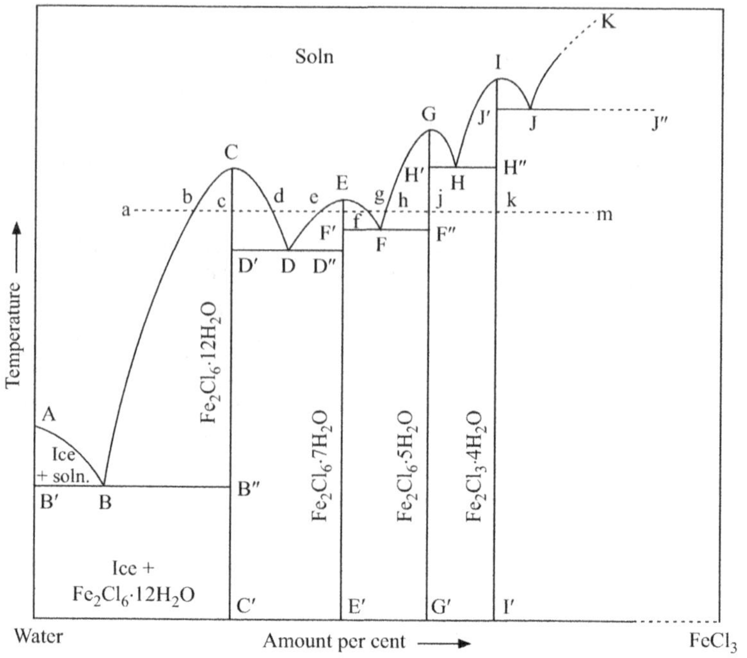


**Fig. 6.12.2 Phase diagram of $FeCl_3$-$H_2O$ system**

**Table 6.12.1** Description of the Phase Diagram shown in Fig. 6.12.2

| *Location of the system* | *Status of the system* |
|---|---|
| A | Freezing point of water |
| C′ | Composition point of dodecahydrate |
| C | Congruent melting point of dodecahydrate (37 °C) |
| C′C | Solid dodecahydrate |
| E′ | Composition point of heptahydrate |
| E | Congruent melting point of heptahydrate (32.5 °C) |
| E′E | Solid heptahydrate |
| G′ | Composition point of pentahydrate |
| G | Congruent melting point of pentahydrate (56 °C) |
| G′G | Solid pentahydrate |
| I′ | Composition point of tetrahydrate |
| I | Congruent melting point of tetrahydrate (78.5 °C) |
| I′I | Solid tetrahydrate |
| B | Eutectic point involving ice, dodecahydrate and solution B(–55 °C) |
| B′BB″ | Tie line at the point B. It starts from B′ (ice) and after passing through the point B (solution) ends at B″ (dodecahydrate) |
| D | Eutectic point involving dodecahydrate, heptahydrate and solution D (27.4 °C) |
| D′DD″ | Tie line at the point D. It starts from D′ (dodecahydrate) and after passing through the point D (solution) ends at D″ (heptahydrate) |
| F | Eutectic point involving heptahydrate, pentahydrate and solution F (30 °C) |

*(Contd.)*

**Table 6.12.1** *(Contd.)*

| *Location of the system* | *Status of the system* |
|---|---|
| F′FF″ | Tie line at the point F. It starts from F′ (heptahydrate) and after passing through the point F (solution) ends at F″ (pentahydrate) |
| H | Eutectic point involving pentahydrate, tetrahydrate and solution H (55 °C) |
| H′HH″ | Tie line at the point H. It starts from H′ (pentahydrate) and after passing through the point H (solution) ends at H″ (tetrahydrate) |
| J | Eutectic point involving tetrahydrate, anhydrous salt and solution J (66 °C) |
| J′JJ″ | Tie line at the point J. It starts from J′ (tetrahydrate) and after passing through the point J (solution) will end at J″ (anhydrous salt) |
| AB | Freezing point curve of ice |
| BCD | Solubility curve of dodecahydrate |
| DEF | Solubility curve of heptahydrate |
| FGH | Solubility curve of pentahydrate |
| HIJ | Solubility curve of tetrahydrate |
| JK | Solubility curve of anhydrous salt |
| Areas | |
| ABB′A | Ice $\rightleftharpoons$ solution (AB) |
| BCB″B | Dodecahydrate $\rightleftharpoons$ solution whose composition lies on BC |
| DCD′D | Dodecahydrate $\rightleftharpoons$ solution whose composition lies on DC |
| DED″D | Heptahydrate $\rightleftharpoons$ solution whose composition lies on DE |
| FEF′F | Heptahydrate $\rightleftharpoons$ solution whose composition lies on FE |
| FGF″F | Pentahydrate $\rightleftharpoons$ solution whose composition lies on FG |
| HGH′H | Pentahydrate $\rightleftharpoons$ solution whose composition lies on HG |
| HIH″H | Tetrahydrate $\rightleftharpoons$ solution whose composition lies on HI |
| JIJ′J | Tetrahydrate $\rightleftharpoons$ solution whose composition lies on IJ |
| J″JKJ″ | Anhydrous salt $\rightleftharpoons$ solution whose composition lies on JK |
| Area below B′BB″ | Solid mixtures of ice and dodecahydrate |
| D′DD″ | Solid mixtures of dodecahydrate and heptahydrate |
| F′FF″ | Solid mixtures of heptahydrate and pentahydrate |
| H′HH″ | Solid mixtures of pentahydrate and tetrahydrate |
| J′JJ″ | Solid mixtures of tetrahydrate and anhydrous salt |
| Area above the curve ABCDEFGHIJK | Aqueous solution |

*Between* c *and* d: There exists an equilibrium between dodecahydrate and the solution, but now composition of the latter corresponds to the point d. From c to d, more and more of the solution is produced and less and less of the solid is left.
*Point* d: Solid dodecahydrate disappeares and the system is completely a solution.
*Between* d *and* e: Once again the solution becomes unsaturated.
*At* e: Solid $Fe_2Cl_6 \cdot 7H_2O$ starts separating.
*Between* e *and* f: More and more of heptahydrate separates from the solution of e, volume of the solution decreases gradually.
*Point* f: Complete solidification of $Fe_2Cl_6 \cdot 7H_2O$ occurs at this point.
*Point* f *to* g: Again solution g appears, volume of the solution increases, and the amount of solid heptahydrate decreases.

*Point* g: Solid heptahydrate disappears and the system is completely a solution.
*Between* g *and* h: Again the solution becomes unsaturated.
*Point* h: $Fe_2Cl_6 \cdot 5H_2O$ starts appearing.
*Point* h *to* j: More and more of pentahydrate appears, volume of the solution h decreases.
*Point* j: Complete solidification to give $Fe_2Cl_6 \cdot 5H_2O$.
*Point* j *to* k: Transformation of more and more of the solid $Fe_2Cl_6 \cdot 5H_2O$ to $Fe_2Cl_6 \cdot 4H_2O$.
*At* k: Transformation of $Fe_2Cl_3 \cdot 5H_2O$ to $Fe_2Cl_3 \cdot 4H_2O$ is completed and we have only $Fe_2Cl_6 \cdot 4H_2O$.

*From* k *to* m: Transformation of solid $Fe_2Cl_6 \cdot 4H_2O$ to anhydrous salt.
If the evaporation is continued, then ultimately one would get the pure anhydrous salt.

**Other Examples of Hydrates**

Hydrates of congruent melting points have also been discovered in which one of the components is not a salt. A few examples are

| | | |
|---|---|---|
| | $N_2O_5 \cdot H_2O$ system: | Compounds are $N_2O_5 \cdot 7H_2O$ $N_2O_5 \cdot 3H_2O$ and $N_2O_5 \cdot H_2O$ |
| and | $SO_3 \cdot H_2O$ system: | Compounds are $SO_3 \cdot 5H_2O$, $SO_3 \cdot 3H_2O$ $SO_3 \cdot 2H_2O$, $SO_3 \cdot H_2O$ and $2SO_3 \cdot H_2O$ |

***Example 6.12.1***

Sulphuric acid forms three hydrates. All have congruent melting points. There are no solid solutions.

| *Compound* | *Melting point* / °C |
|---|---|
| $H_2SO_4$ (anhydrous) | 10.5 |
| $H_2SO_4 \cdot H_2O$ | 8.6 |
| $H_2SO_4 \cdot 2H_2O$ | –39.9 |
| $H_2SO_4 \cdot 4H_2O$ | –24.5 |
| Eutectic (ice + $H_2SO_4 \cdot 4H_2O$) | ≃ – 74 |
| Eutectic ($H_2SO_4 \cdot 4H_2O$ + $H_2SO_4 \cdot 2H_2O$) | ≃ – 51 |
| Eutectic ($H_2SO_4 \cdot 2H_2O$ + $H_2SO_4 \cdot H_2O$) | ≃ – 44 |
| Eutectic ($H_2SO_4 \cdot H_2O$ + $H_2SO_4$) | ≃ – 38 |

(i) Sketch the temperature-amount fraction phase diagram. Label each area to indicate the phases present.
(ii) Describe all the changes that occur when sulphuric acid is added in small increment to a sample of ice being maintained at –30 °C.

***Solution***

**(i)** The temperature-amount fraction phase diagram from the given data is shown in Fig. 6.12.3. Labelling of the phase diagram is shown in Table 6.12.2.
**(ii)** The changes that occur when sulphuric acid is added in small increment to a sample of ice maintained at – 30 °C can be described by drawing a horizontal line at –30 °C such as shown in Fig. 6.12.3. The effects produced will be as follows.

From a to b, the system consists of solid ice and solution of composition corresponding to the point b. The only things that variy from a to b are the relative amounts of the two; the amount of ice decreases whereas that of solution increases. At b, the last crystal of $H_2O$ disappears and from b to c, the system is in liquid state. At c, crystals of $H_2SO_4 \cdot 4H_2O$ start

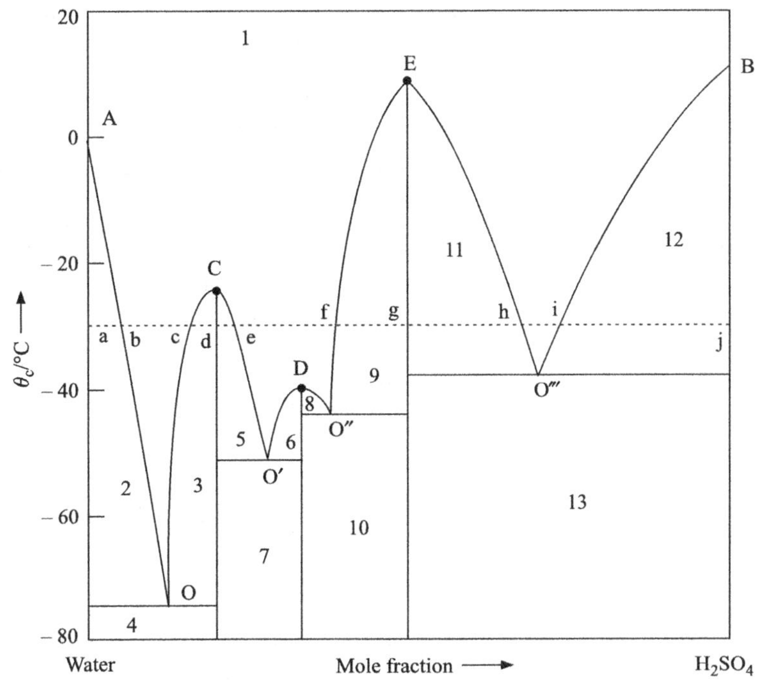


**Fig. 6.12.3 Required phase diagram of $H_2SO_4$ – $H_2O$ system**

**Table 6.12.2** Labelling of the Phase Diagram

| *Area* | *System* |
|---|---|
| 1 | solution |
| 2 | ice + solution |
| 3 | solid $H_2SO_4 \cdot 4H_2O$ + solution (OC) |
| 4 | ice + solid $H_2SO_4 \cdot 4H_2O$ |
| 5 | solid $H_2SO_4 \cdot 4H_2O$ + solution (CO′) |
| 6 | solid $H_2SO_4 \cdot 2H_2O$ + solution (O′D) |
| 7 | solid $H_2SO_4 \cdot 4H_2O$ + solid $H_2SO_4 \cdot 2H_2O$ |
| 8 | solid $H_2SO_4 \cdot 2H_2O$ + solution (O″D) |
| 9 | solid $H_2SO_4 \cdot H_2O$ + solution (O″E) |
| 10 | solid $H_2SO_4 \cdot 2H_2O$ + solid $H_2SO_4 \cdot H_2O$ |
| 11 | solid $H_2SO_4 \cdot H_2O$ + solution (EO‴) |
| 12 | solid $H_2SO_4 \cdot$ + solution (BO‴) |
| 13 | solid $H_2SO_4 \cdot H_2O$ + solid $H_2SO_4$ |

appearing and continue to appear till the point d is reached keeping all the while the solution composition at c. At d the system consists of only solid $H_2SO_4 \cdot 4H_2O$. From d to e, again we have solid $H_2SO_4 \cdot 4H_2O$ and solution, but now the solution composition is represented by the point e. The only thing that varies from d to e is that the amount of solid $H_2SO_4 \cdot 4H_2O$ decreases whereas the amount of solution increases. At e, the last crystal of solid $H_2SO_4 \cdot 4H_2O$ disappears and from e to f, the system is present in the solution form. From f to g, solid $H_2SO_4 \cdot H_2O$ is in equilibrium with the solution of composition corresponding to the point f. At g, the system consists of only solid $H_2SO_4 \cdot H_2O$. From g to h, the system contains solid

$H_2SO_4 \cdot H_2O$ in equilibrium with solution of composition corresponding to the point h. From h to i, it is present in the solution form and from i to j, the system consists of anhydrous $H_2SO_4$ and solution of composition corresponding to the point i. At j, the system consists of only solid anhydrous $H_2SO_4$.

## FORMATION OF COMPOUNDS WITH INCONGRUENT MELTING POINTS

### Sodium Sulphate-Water System

Formation of compounds (hydrates) with incongruent or meritectic melting points are more common, one of the well-known examples is the sodium sulphate and water system. The phase diagram of this system is shown in Fig. 6.12.4 and is described in Table 6.12.3.

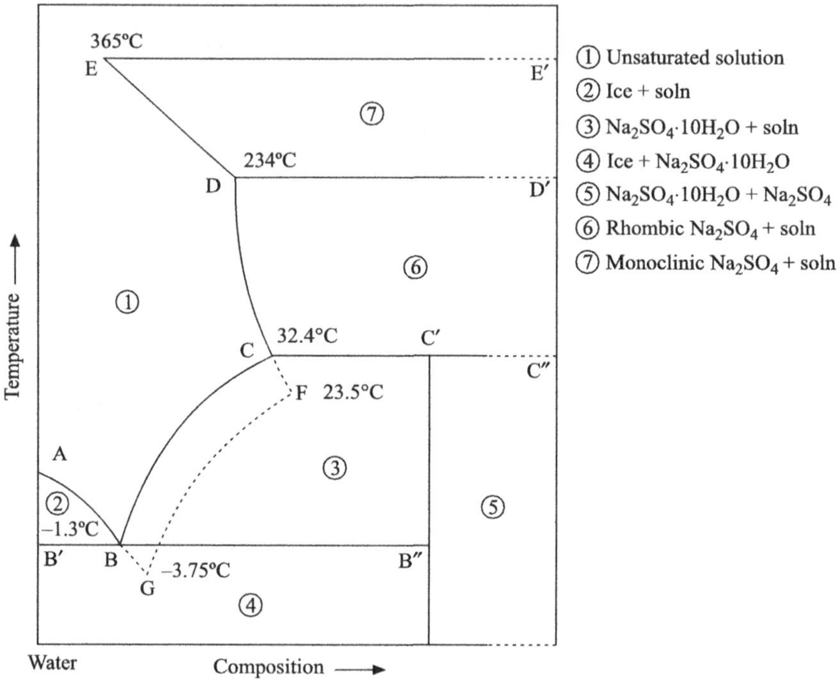


**Fig. 6.12.4 Phase diagram of $Na_2SO_4$ – $H_2O$ system**

**Table 6.12.3** Description of The Phase Diagram shown in Fig. 6.12.4

| *Location of the system* | *Status of system* |
|---|---|
| A | Freezing point of water |
| C′ | Incongruent point where solid $Na_2SO_4 \cdot 10H_2O$ undergoes the following peritectic reaction<br>$Na_2SO_4 \cdot 10H_2O \rightleftharpoons Na_2SO_4$(rhombic) + Solution C |
| C | Composition of solution which is produced at point C′ |
| CC′C″ | Tie line at the point C. It starts from C (solution) and after passing through C′ ($Na_2SO_4 \cdot 10H_2O$) ends at C″(rhombic $Na_2SO_4$) |
| B | Eutectic point where ice and $Na_2SO_4 \cdot 10H_2O$ are in equilibrium with solution B |

*(Contd.)*

**Table 6.12.3** *(Contd.)*

| | |
|---|---|
| B′BB″ | Tie-line at the point B. If starts from B′ (ice) and after passing through the point B (solution) ends at B″($Na_2SO_4 \cdot 10H_2O$) |
| AB | Freezing point curve of ice |
| BC | Solubility curve of $Na_2SO_4 \cdot 10H_2O$. |
| CD | Solubility curve of rhombic $Na_2SO_4$. This curve is slightly curved at left up to about 125 °C since solubility of rhombic $Na_2SO_4$ decreases with increase in temperature. Beyond this temperature, the solubility increases with increase in temperature |
| D | At temperature corresponding to the point D, polymorphic modification of rhombic $Na_2SO_4$ into monoclinic $Na_2SO_4$ takes place |
| DE | Solubility curve of monoclinic $Na_2SO_4$ (solubility of monoclinic $Na_2SO_4$ decreases with increase in temperature) |
| E | This point gives the maximum limit of the temperature up to which the system can be studied and represents critical temperature of water in solution. Above this temperature, liquid water does not exist. This temperature is very close to the critical temperature of pure water since the solubility of $Na_2SO_4$ at this temperature is very small |
| Areas | |
| ABB′A | ice $\rightleftharpoons$ solution (AB) |
| BCC′B″B | $Na_2SO_4 \cdot 10H_2O(s) \rightleftharpoons$ solution (BC) |
| Below the line B′BB″ | Solid mixtures of ice and $Na_2SO_4 \cdot 10H_2O$ |
| CDD′C″C | Rhombic $Na_2SO_4(s) \rightleftharpoons$ solution (CD) |
| Below the line C′C″ | Solid mixtures of $Na_2SO_4$ (rhombic) and $Na_2SO_4 \cdot 10H_2O$ |
| DEE′D′D | Monoclinic $Na_2SO_4(s) \rightleftharpoons$ solution (DE) |
| Line CF | Extension of the curve DC where metastable equilibrium between rhombic $Na_2SO_4$ and solution exists. This curve is continued up to the point F where metastable heptahydrate ($Na_2SO_4 \cdot 7H_2O$) is deposited |
| Line BG | Extension of the curve AB where metastable equilibrium between ice and solution exists. This curve is extended up to the point G where both ice and $Na_2SO_4 \cdot 7H_2O$ solidify together |
| GF | Metastable solubility curve of heptahydrate |
| G | Metastable eutectic point where ice and heptahydrate are in equilibrium with the solution G |
| F | The temperature corresponding to this point represents an incongruent melting point where the following equilibrium exists $Na_2SO_4 \cdot 7H_2O(s) \rightleftharpoons Na_2SO_4$ (rhombic) + Solution F |

**Solubilities of Stable and Metastable Forms**

The fact that the metastable curve GF lies to the right of the stable curve BC indicates that at a given temperature, the metastable form is more soluble than the stable form. This is a universal rule and can be derived thermodynamically. Since, the reaction

$$\underset{\text{(ms)}}{\text{Metastable solid form}} \rightarrow \underset{\text{(ss)}}{\text{Stable solid from}}$$

is a spontaneous process (and the reverse of this reaction can never occur), we must have

$$\mu_{ss} < \mu_{ms}$$

Since both these salts exist in equilibrium with the corresponding saturated solutions, we also have

$$\mu_{ss} = \mu_{ss(soln)}$$

$$\mu_{ms} = \mu_{ms(soln)}$$

With these, the previous expression modifies to

$$\mu_{ss(soln)} < \mu_{ms(soln)}$$

that is, the chemical potential of the stable solute $Na_2SO_4 \cdot 10H_2O$ in the saturated solution (represented by the curve BC) is less than that of the unstable solute $Na_2SO_4 \cdot 7H_2O$ in the corresponding saturated solution (represented by the curve GF). If it is assumed that both the solutions behave ideally (or depart from ideality to the same extent), then it is obvious that

$$x_{ss(soln)} < x_{ms(soln)}$$

**More Examples**

A few other examples in which compounds of incongruent melting points are formed are now described.

***Sodium Chloride-Water System*** In this system, we have

(i) Stable eutectic temperature at – 21.2 °C, where ice, $NaCl \cdot 2H_2O$ and the solution are in equilibrium.

(ii) Meritectic temperature at 0.15 °C where the following transformation takes place:

$$\text{Solid } NaCl \cdot 2H_2O \rightleftharpoons NaCl + \text{solution}$$

***Calcium Chloride-Water System*** In $CaCl_2$-$H_2O$ system, we have more than one compound formation. There are four stable hydrates and one metastable hydrate. These are

$$\left.\begin{array}{l} CaCl_2 \cdot 6H_2O \\ CaCl_2 \cdot 4H_2O\ (\alpha) \\ CaCl_2 \cdot 2H_2O \\ CaCl_2 \cdot H_2O \end{array}\right\} \text{Stable hydrate}$$

$$CaCl_2 \cdot 4H_2O(\beta) \quad \text{Unstable hydrate}$$

Four stable transition points will be present in the phase diagram corresponding to the following transformations.

$$CaCl_2 \cdot 6H_2O \xrightleftharpoons{29.8\,^\circ C} CaCl_2 \cdot 4H_2O(\alpha) + \text{solution}$$

$$CaCl_2 \cdot 4H_2O(\alpha) \xrightleftharpoons{45.3\,^\circ C} CaCl_2 \cdot 2H_2O + \text{solution}$$

$$CaCl_2 \cdot 2H_2O \xrightleftharpoons{175.5\,^\circ C} CaCl_2 \cdot H_2O + \text{solution}$$

$$CaCl_2 \cdot H_2O \rightleftharpoons CaCl_2 + \text{solution}$$

Besides these four points, two more points will exist corresponding to the following two unstable transformations.

$$CaCl_2 \cdot 6H_2O \xrightleftharpoons{29.2^\circ\ C} CaCl_2 \cdot 4H_2O(\beta) + \text{solution}$$

and $$CaCl_2 \cdot 4H_2O(\beta) \xrightleftharpoons{38.4\,^\circ C} CaCl_2 \cdot 2H_2O + \text{solution}$$

***Sodium Iodide-Water System*** We have

$$\text{Ice} + \text{NaI}\cdot 5H_2O \xrightleftharpoons{-31.5\ ^\circ C} \text{solution (39 mass \% NaI)}$$

$$\text{NaI}\cdot 5H_2O \xrightleftharpoons{-13.5\ ^\circ C} \text{NaI}\cdot 2H_2O + \text{solution (60.2 mass \% NaI)}$$

$$\text{NaI}\cdot 2H_2O \xrightleftharpoons{65\ ^\circ C} \text{NaI} + \text{solution (74.4 mass \% NaI)}$$

***Cupric Nitrate-Water System*** We have

$$\text{Ice} + Cu(NO_3)_2\cdot 9H_2O \xrightleftharpoons{-24\ ^\circ C} \text{solution (35.9 mass \% } Cu(NO_3)_2)$$

$$Cu(NO_3)_2\cdot 9H_2O \xrightleftharpoons{-20\ ^\circ C} Cu(NO_3)_2\cdot 6H_2O + \text{solution}$$
$$(39.9 \text{ mass \% } Cu(NO_3)_2)$$

$$Cu(NO_3)_2\cdot 6H_2O \xrightleftharpoons{24.5\ ^\circ C} Cu(NO_3)_2\cdot 3H_2O + \text{solution}$$
$$(61.4 \text{ mass \% } Cu(NO_3)_2)$$

***Example 6.12.2***

Given below are the data of the binary system ferrous sulphate-water. Construct the phase diagram and label it.

Composition of Saturated Solution at Various Temperatures

| *Temperature* / °C | $m(FeSO_4)$/g per 100 g *solution* | *Solid phase* |
|---|---|---|
| −0.7 | 5.18 | Ice |
| −1.3 | 9.47 | ” |
| −1.8 | 12.99 | Ice + $FeSO_4\cdot 7H_2O$ |
| 0 | 13.53 | $FeSO_4\cdot 7H_2O$ |
| + 10 | 17.02 | ” |
| 20 | 21.0 | ” |
| 30 | 24.78 | ” |
| 40 | 28.67 | ” |
| 50 | 32.7 | ” |
| 52 | 33.42 | ” |
| 54 | 34.25 | ” |
| 56.17 | 35.32 | $FeSO_4\cdot 7H_2O + FeSO_4\cdot 4H_2O$ |
| 60 | 35.46 | $FeSO_4\cdot 4H_2O$ |
| 64 | 35.65 | $FeSO_4\cdot 4H_2O + FeSO_4\cdot H_2O$ |
| 68 | 34.35 | $FeSO_4\cdot H_2O$ |
| 77 | 31.46 | ” |
| 80.4 | 30.35 | ” |
| 85 | 28.8 | ” |
| 90 | 27.15 | ” |

$FeSO_4\cdot 7H_2O$ = 54.6 mass % $FeSO_4$
$FeSO_4\cdot 4H_2O$ = 67.8 mass % $FeSO_4$
$FeSO_4\cdot H_2O$ = 89.4 mass % $FeSO_4$

***Solution***

First of all, the points corresponding to the given data on temperature versus composition were located in a two-dimension phase diagram. After this, the points corresponding to the given three compounds, viz., $FeSO_4 \cdot 7H_2O$, $FeSO_4 \cdot 4H_2O$ and $FeSO_4 \cdot H_2O$, were located on the composition axis and the vertical lines were drawn from these points as shown in Fig. 6.12.5.

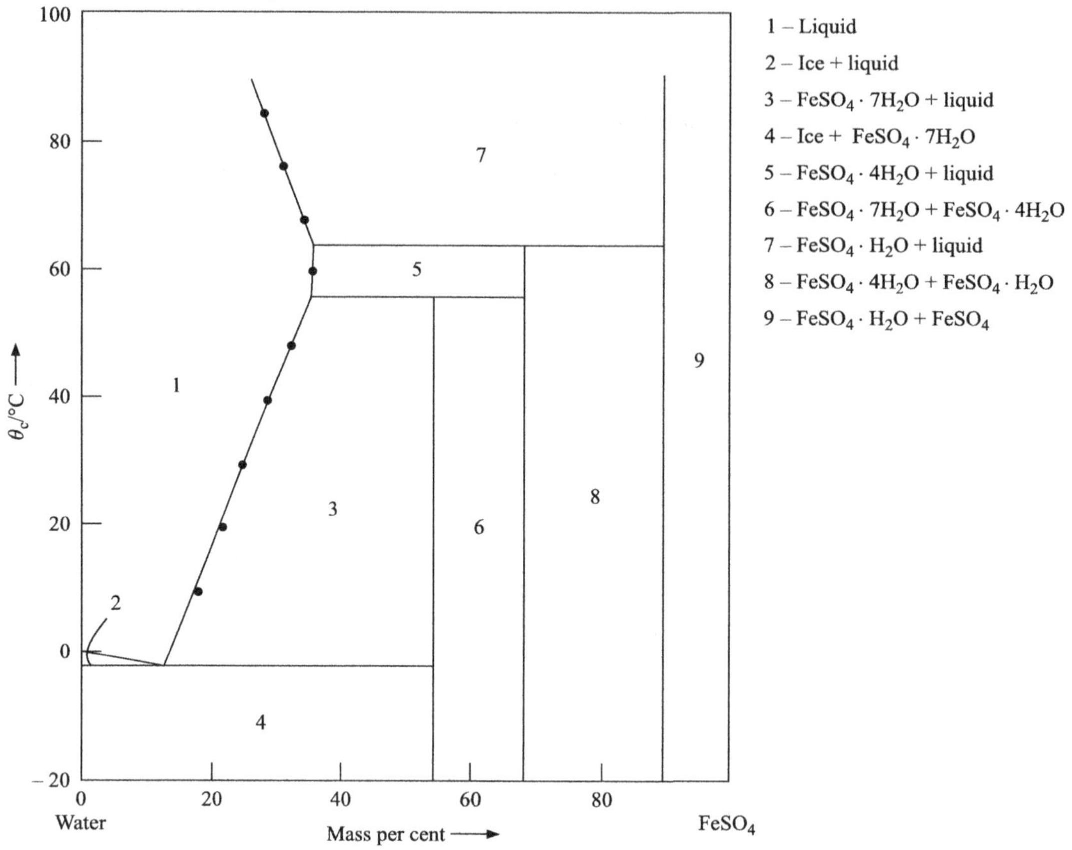


**Fig. 6.12.5 Required phase diagram of $FeSO_4$-$H_2O$ system**

From the given data, we find that the temperature at which ice separates out from the solution decreases as more and more of $FeSO_4$ is dissolved in it. This decrease of temperature is continued up to the temperature of –1.8 °C where the solution composition is 12.99 mass per cent of $FeSO_4$. At this temperature, the two solids, viz., ice and $FeSO_4 \cdot 7H_2O$ separate out. On increasing the mass per cent of $FeSO_4$ from 12.99 to 13.53, we find that there is a separation of only $FeSO_4 \cdot 7H_2O$ and the freezing point of the solution increases from –1.8 to 0 °C. Obviously, the temperature –1.8 °C is the lowest temperature at which the two solids separate out and thus must represent eutectic temperature. So, we draw a horizontal line at this temperature starting from the pure ice line to the vertical line of $FeSO_4 \cdot 7H_2O$. At temperature 56.17 °C, the two solids $FeSO_4 \cdot 7H_2O$ and $FeSO_4 \cdot 4H_2O$ are present in equilibrium with the solution which contains 35.32 mass per cent of $FeSO_4$. Since this composition is different from the mass per cent of $FeSO_4$ in the compound $FeSO_4 \cdot 7H_2O$, it is obvious that the compound $FeSO_4 \cdot 7H_2O$ has decomposed at a temperature lower than its actual melting point, i.e. the compound $FeSO_4 \cdot 7H_2O$ has an incongruent melting point of 56.17 °C. Thus, the vertical line representing this compound must end at this temperature. Since $FeSO_4 \cdot 4H_2O$ is separated out at temperature immediately higher than 56.17 °C, we draw a horizontal line from the peritectic point (56.17 °C and 35.32 mass per cent of $FeSO_4$) up to the vertical line representing the compound $FeSO_4 \cdot 4H_2O$ (Fig. 6.12.5). Similar arguements at temperature 64 °C, where again two phases $FeSO_4 \cdot 4H_2O$ and $FeSO_4 \cdot H_2O$ are separated out, indicates that the compound $FeSO_4 \cdot 4H_2O$ also has an incongruent melting point. So the vertical line representing this compound must end at the temperature 64 °C. Since $FeSO_4 \cdot H_2O$ separates out at temperature immediately above 64 °C, we draw a horizontal line from the peritectic point (64 °C and 35.65 mass percent of $FeSO_4$) up to the vertical line representing the compound $FeSO_4 \cdot H_2O$. At higher temperatures, we have only the solubility curve of $FeSO_4 \cdot H_2O$.

Finally a smooth curve passing through the points of temperature versus composition is drawn. This gives us the phase diagram of the present system. Labelling of different areas is straightforward and is described in Fig. 6.12.5 itself.

***Example 6.12.3***

Given below are the data of the binary system sodium hydroxide-water. Construct the phase diagram and label it.

Composition of Saturated Solutions at Various Temperatures

| *Temperature* / °C | $m$(NaOH)/g per 100 g *solution* | *Solid phase* |
|---|---|---|
| – 5.27 | 5.78 | Ice |
| – 10.29 | 10.03 | " |
| – 17.20 | 14.11 | " |
| – 25.2 | 18.17 | " |
| –28.0 | 19.0 | Ice + $NaOH \cdot 7H_2O$ |
| –26.0 | 19.98 | $NaOH \cdot 7H_2O$ |
| –25.2 | 21.1 | " |
| –24.0 | 22.1 | $NaOH \cdot 7H_2O$ + $NaOH \cdot 5H_2O$ |
| –21.7 | 23.31 | $NaOH \cdot 5H_2O$ |
| –19.95 | 23.97 | " |
| –18.0 | 24.7 | $NaOH \cdot 5H_2O$ + $NaOH \cdot 4H_2O$ |
| –12.60 | 25.47 | $NaOH \cdot 4H_2O$ |
| –8.45 | 26.91 | " |
| + 1.62 | 30.38 | " |
| 5.40 | 32.3 | $NaOH \cdot 4H_2O$ + $NaOH \cdot 3.5H_2O$ |
| 7.00 | 32.97 | $NaOH \cdot 3.5H_2O$ |
| 13.2 | 35.51 | " |
| 15.55 | 38.83 | " |
| 13.95 | 42.28 | " |
| 10.75 | 44.22 | " |
| 5.00 | 45.5 | $NaOH \cdot 3.5H_2O$ + $NaOH \cdot 2H_2O$ |
| 7.80 | 47.3 | $NaOH \cdot 2H_2O$ |
| 10.30 | 49.11 | " |
| 12.30 | 50.8 | $NaOH \cdot 2H_2O$ + $NaOH \cdot H_2O$ |
| 18.00 | 51.7 | $NaOH \cdot H_2O$ |
| 40.25 | 56.44 | $NaOH \cdot H_2O$ |
| 57.85 | 62.85 | " |
| 63.23 | 66.45 | " |

*(Contd.)*

*(Contd.)*

| *Temperature* °C | $m$(NaOH)/g per 100 g *solution* | *Solid phase* |
|---|---|---|
| 64.3 | 68.49 | ” |
| 63.0 | 71.17 | ” |
| 62.0 | 74.2 | $NaOH \cdot H_2O + NaOH$ |
| 80.0 | 75.83 | NaOH |
| 110.0 | 78.15 | NaOH |

$NaOH \cdot 7H_2O$ = 24.1 mass % NaOH
$NaOH \cdot 5H_2O$ = 30.7 mass % NaOH
$NaOH \cdot 4H_2O$ = 35.7 mass % NaOH
$NaOH \cdot 3.5\,H_2O$ = 38.9 mass % NaOH
$NaOH \cdot 2H_2O$ = 52.7 mass % NaOH
$NaOH \cdot H_2O$ = 69.0 mass % NaOH

***Solution***

*Analysis of data*

(i) The temperature –28.0 °C must be an eutectic temperature as the temperatures of data above and below this are higher than –28.0 °C.

(ii) The temperature –24.0 °C, where the two solids $NaOH \cdot 7H_2O$ and $NaOH \cdot 5H_2O$ appear must be a peritectic temperature as solution composition of 22.1 mass per cent of NaOH is different from the corresponding composition in the compound $NaOH \cdot 7H_2O$ (24.1 mass per cent). Thus, the compound $NaOH \cdot 7H_2O$ decomposes earlier than its actual melting point. The vertical line representing $NaOH \cdot 7H_2O$ must end at the temperature –24 °C, and the horizontal line representing the range of peritectic reaction at –24.0 °C will end at the vertical line of $NaOH \cdot 5H_2O$.

(iii) The temperature –18.0 °C, where the two solids $NaOH \cdot 5H_2O$ and $NaOH \cdot 4H_2O$ appear, is also a peritectic temperature as solution composition of 24.7 mass per cent of NaOH is different from the corresponding composition in the compound $NaOH \cdot 5H_2O$ (30.7 mass per cent). The vertical line representing $NaOH \cdot 5H_2O$ will end at the temperature –18.0 °C and the horizontal line representing peritectic reaction at – 18.0 °C will end at the vertical line of $NaOH \cdot 4H_2O$.

(iv) The temperature 5.4 °C, where the two solids $NaOH \cdot 4H_2O$ and $NaOH \cdot 3.5H_2O$ appear, is also a peritectic temperature as solution composition of 32.3 mass per cent of NaOH is less than that in the compound $NaOH \cdot 4H_2O$ (35.7 mass per cent). Thus, the compound $NaOH \cdot 4H_2O$ decomposes earlier than its actual melting point. The vertical line representing $NaOH \cdot 4H_2O$ will end at the temperature 5.4 °C, and the horizontal line representing peritectic reaction at 5.4 °C will end at the vertical line of $NaOH \cdot 3.5H_2O$.

(v) The compound $NaOH \cdot 3.5H_2O$ is a congruently melting compound as its solubility curve passes through a maximum around 15.55 °C.

(vi) The temperature 5.0 °C, where the two solids $NaOH \cdot 3.5H_2O$ and $NaOH \cdot 2H_2O$ appear, must be an eutectic temperature as it has a minimum value in the composition range under study (44.22–47.3 mass per cent of NaOH).

(vii) The temperature 12.30 °C, where the two solids $NaOH \cdot 2H_2O$ and $NaOH \cdot H_2O$ appear, is a peritectic temperature as the solution composition of 50.8 mass per cent of NaOH is less than that in the compound $NaOH \cdot 2H_2O$. Thus, the compound $NaOH \cdot 2H_2O$ decomposes earlier than its actual melting point. The vertical line representing $NaOH \cdot 2H_2O$ will end at the temperature 12.30 °C and the horizontal line representing peritectic reaction at 12.30°C will end at the vertical line of $NaOH \cdot H_2O$.

(viii) The compound $NaOH \cdot H_2O$ is a congruently melting compound as its solubility curve passes through a maximum around 64.3 °C.

(ix) The temperature 62.0 °C where the two solids $NaOH \cdot H_2O$ and NaOH appear must be an eutectic temperature as it has a minimum value in the composition range under study (71.17–75.83 mass per cent NaOH).

Keeping in mind the above analysis, the phase diagram of the present system can be drawn as shown in Fig. 6.12.6. The labelling of various areas is also shown in Fig. 6.12.6

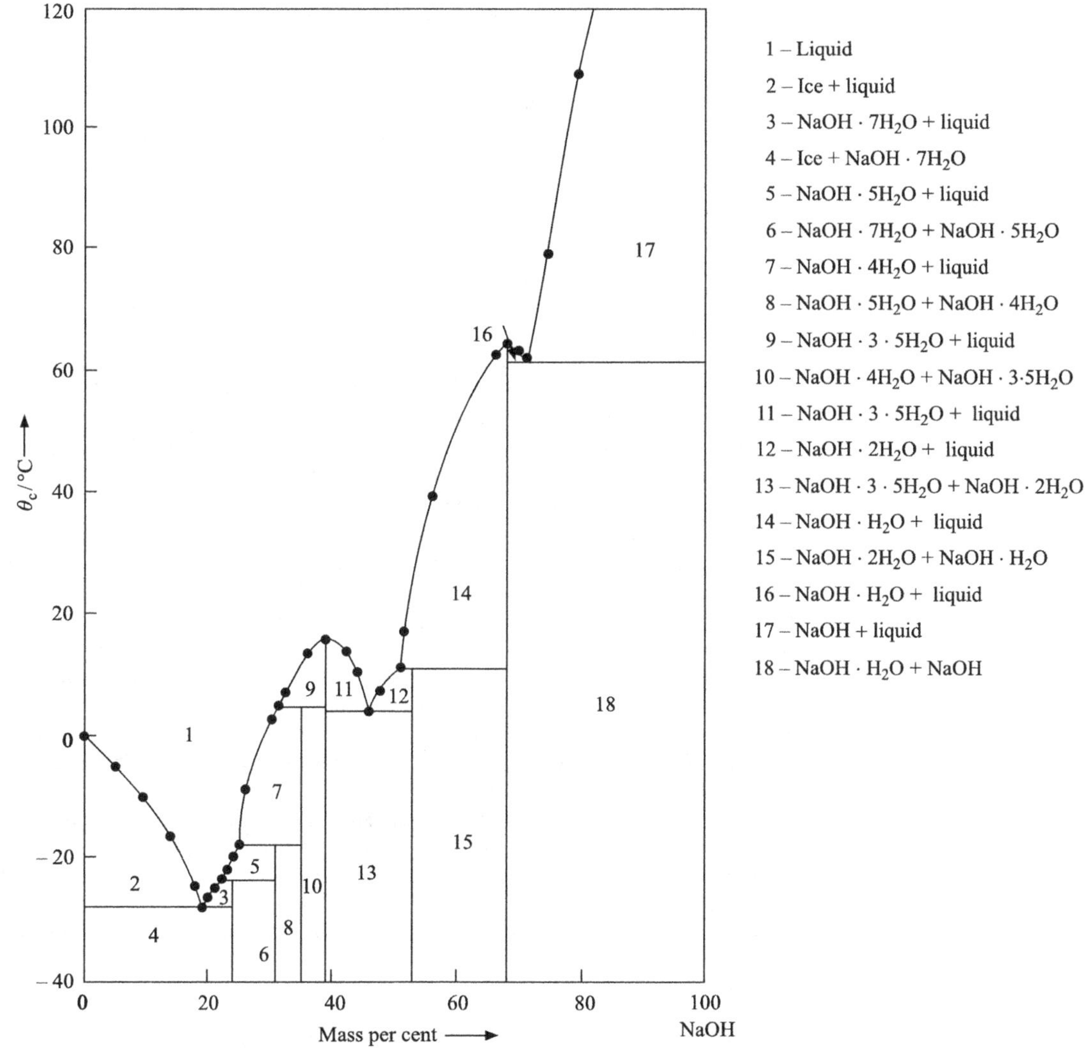


**Fig. 6.12.6 Required phase diagram of $NaOH - H_2O$ system**

## REVISIONARY PROBLEMS

6.1 (a) Determine the variance of a two-component system having number of phases equal to one, two and three, respectively.

(b) Show that in a system where the external pressure remains constant, the degrees of freedom can be calculated by employing the reduced phase rule $F = C - P + 1$.

6.2 (a) What do you understand by condensed systems? Classify various solid-liquid equilibria based on the nature of the liquid phase and the solid phase in two-component systems.

(b) Explain why we should exclude a case of forming a series of solid solutions when the two components are partially miscible in the liquid state?

6.3 Explain, how does thermal analysis help in drawing the phase-diagram of one and two-component systems? Support your answer by drawing a few typical cooling curves.

6.4 (a) Draw the phase diagram of a two-component system exhibiting simple eutectic behaviour.

Explain, why the melting point of a pure component and the eutectic point are invariant points. How will you establish that the solid which separates out at the eutectic point is a solid mixture and not a compound?

(b) Derive thermodynamically the relation governing the solubility curve of an ideal binary system.

(c) Explain, how the eutectic temperature and composition be determined by making use of the relation derived in part (b)?

(d) Derive the lever rule as applicable to the solid-liquid region of a simple eutectic phase diagram.

6.5 Explain, what modifications in the phase diagram of a simple eutectic system take place when one of the components exists in two allotropic forms. Show that the transition point is an invariant point.

6.6 (a) What do you understand by the congruent and incongruent melting points? What are the number of components and the variance at the congruent melting point of a system consisting of two constituents?

(b) Explain, why the curve passing through the congruent melting point shows a rounded maximum instead of a sharp intersecting point.

(c) Draw a typical phase diagram with complete labelling of a two-component system exhibiting a single congruent melting point. Draw also a few typical cooling curves.

6.7 (a) What do you understand by the peritectic reaction or incongruent fusion? Show with the help of phase rule that the above reaction takes place at a definite temperature, i.e. the system at the peritectic point is invariant.

(b) Draw a typical phase diagram of a system exhibiting incongruent compound formation. Explain in what respect it differs from that of a system exhibiting allotropic modification by one of its components.

6.8 What is meant by a binary solid solution? How does it differ from a solid mixture? Show that the minimum number of degrees of freedom when a solid solution is in equilibrium with a liquid solution is one. What do you conclude from this regarding the general nature of phase diagram of such a system?

6.9 (a) Show that the shift in the freezing point of solvent when a binary solid solution separates on cooling is given by

$$-\Delta T_{\mathrm{f}} = K_{\mathrm{f}} m_{2(\mathrm{l})}(1 - K)$$

where $m_{2(\mathrm{l})}$ is the molality of the solute in the liquid phase and $K$ is the distribution constant defined as

$$K = \frac{\text{Amount fraction of solute in the solid phase}}{\text{Amount fraction of solute in the solution}}$$

What approximations are involved in deriving the above expression?
(b) Under what conditions, (i) an increase in the freezing point of the solvent, and (ii) a decrease in the freezing point of the solvent is observed when a solute is added to the solvent?
From the derived conditions, deduce the statements of Konowaloff's rule as applicable to the solid-liquid equilibria.

6.10 Derive the following relations as applicable to the ascending type solid solution.

$$x_{1(1)} = \frac{\exp(-A)\,\{\exp(-B) - 1\}}{\exp(-B) - \exp(-A)}$$

$$x_{1(s)} = \frac{\exp(-B) - 1}{\exp(-B) - \exp(-A)}$$

where $$A = \frac{\Delta_{fus}H^{\circ}_{1,m}}{R}\left(\frac{1}{T} - \frac{1}{T_1^*}\right) \quad \text{and} \quad B = \frac{\Delta_{fus}H^{\circ}_{2,m}}{R}\left(\frac{1}{T} - \frac{1}{T_2^*}\right)$$

Derive the following conclusions from the above relations:
(1) None of the relationships is linear.
(2) The liquidus and solidus curves do not coincide except at $T_1^*$ and $T_2^*$.
(3) For any composition between 0 and 1, temperature of the liquid and solid curves lies between the melting points of the two components.

6.11 Classify phase diagrams involving solid solutions of nonideal binary systems on the basis of the distribution constants $K_{A\text{ in }B}$ (i.e. $x_{A(s)}/x_{A(1)}$, A is acting as a solute and B as a solvent) and $K_{B\text{ in }A}$ (i.e. $x_{B(s)}/x_{B(1)}$, B acting as a solute and A as a solvent). Draw their phase diagrams and label each area and line. Discuss the sequence of events that takes place on cooling a system in the above cases.

6.12 (a) What are the end products of fractional crystallization of a system exhibiting (i) ascending type solid solution, (ii) minimum type solid solution, and (iii) maximum type solid solution? (b) Describe the principle underlying the technique of zone refining.

6.13 Draw a typical phase diagram of a two-component system in which the two constituents are partially miscible in the solid state and form stable solid solutions. Is it possible to separate completely the two components in such a system by the method of fractional crystallization?

6.14 Draw a typical phase diagram of a two-component system in which the two constituents are partially miscible in the solid state and form solid solutions which are stable only up to a transition point. Is it possible to separate completely the two constituents in such a system through fractional crystallization?

6.15 Draw a typical phase diagram of a two-component system in which the two constituents are partially miscible in the liquid state and the solid phases consist of pure components.

6.16 (a) Draw a simple eutectic phase diagram as shown by KI-$H_2O$ system. Explain, why the name cryohydrate is given to this type of system. Explain, with the help of the phase diagram, the formation of a cooling mixture when a salt is added to ice containing a small amount of water.
(b) Ferric chloride forms four stable hydrates. These are $Fe_2Cl_6 \cdot 12H_2O$, $Fe_2Cl_6 \cdot 7H_2O$, $Fe_2Cl_6 \cdot 5H_2O$ and $Fe_2Cl_6 \cdot 4H_2O$. Draw the phase diagram of $FeCl_3$-$H_2O$ system with complete labelling. Explain, what do you understand by the isothermal evaporation of the solution. Selecting any point in the unsaturated solution region, describe the sequence of steps that takes place during the isothermal evaporation of water.

(c) Draw the phase diagram of $Na_2SO_4$-water system and describe its complete labelling. Show thermodynamically that at a given temperature the metastable form is more soluble in water than the stable form.

## TRY YOURSELF PROBLEMS

6.1 Derive the lever rule for the point c given in Fig. 6.4.1.

6.2 What types of cooling curves are expected for a one-component system?

6.3 Solid ammonium nitrate can exist in five crystalline forms with the following transition temperatures:

$$\text{Cubic} \underset{125\,^\circ\text{C}}{\rightleftharpoons} \text{Rhombohedral} \underset{84\,^\circ\text{C}}{\rightleftharpoons} \alpha\text{-Rhombic} \underset{32\,^\circ\text{C}}{\rightleftharpoons} \beta\text{-Rhombic} \underset{-18\,^\circ\text{C}}{\rightleftharpoons} \text{Tetragonal}$$

Explain the type of phase-diagram which you would expect for the $NH_4NO_3$–$H_2O$ system.

6.4 Explain, why the congruent and incongruent melting points are invariant points.

6.5 Explain, why we observe a region over which binary solid solution is in equilibrium with the melt. How are the relative amounts of the two solutions, namely, solid and liquid, in equilibrium with each other determined?

6.6 Figure 6.9.1 shows an eutectic phase diagram of a system exhibiting limited solubility in the solid state. Besides this, if a system also exhibits a compound formation, then the phase diagram has an appearance as shown in Fig. 1. Label the diagram with adequate explanation.

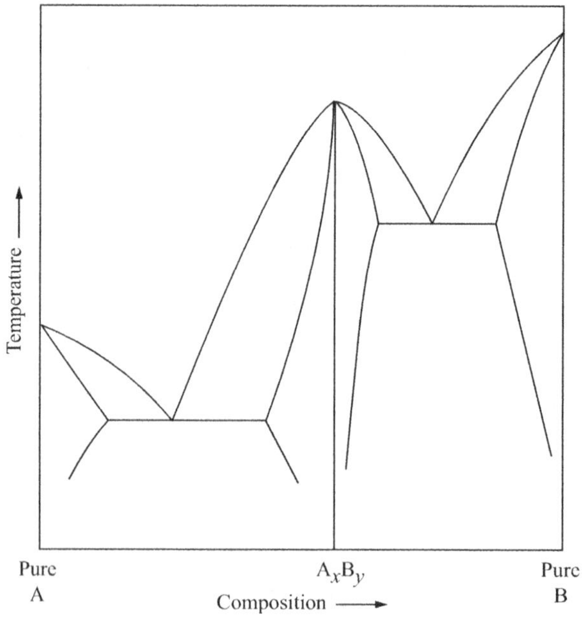


Fig. 1

6.7 Figure 6.10.1 is a peritectic phase diagram of a system exhibiting limited solubility in the solid state. Besides this, if a system also exhibits an incongruent compound formation, then the phase diagram has an appearance as shown in Fig. 2. Label the diagram with adequate explanation

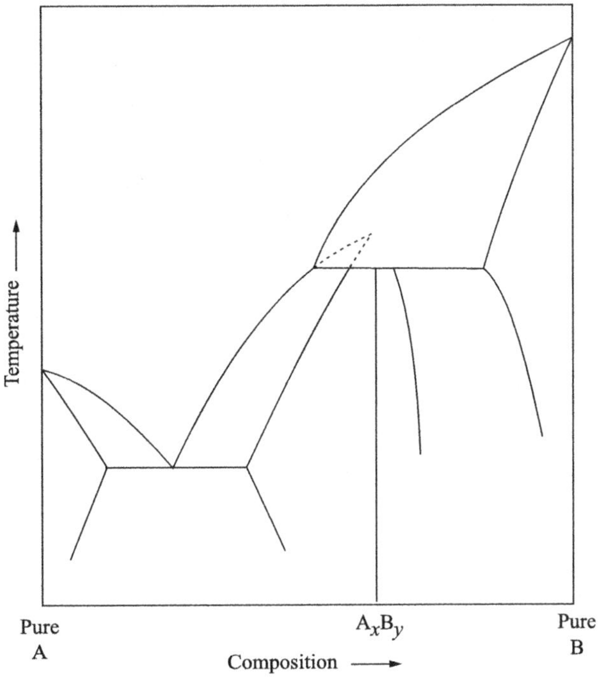


Fig. 2

6.8 Figure 3 depicts the phase diagram of a system where (i) the two components are partially miscible in the liquid phase over a certain range of compositions and (ii) the two components also form two solid solutions. Label the diagram and sketch the cooling curves for the system shown therein.

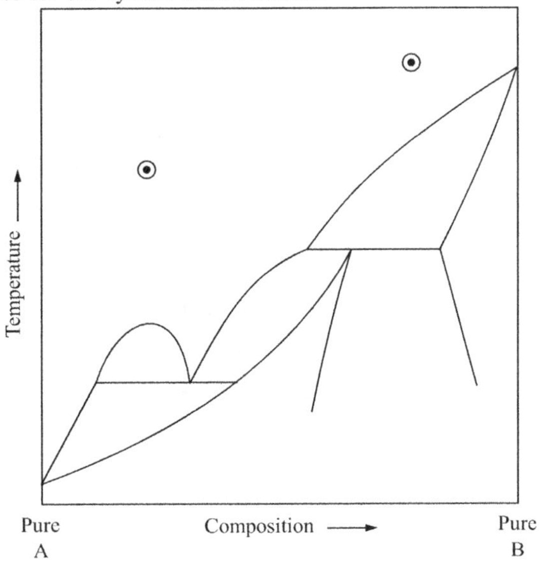


Fig. 3

6.9 (a) Given are the following equilibria:

(i) $CuSO_4 \cdot 5H_2O(s) \rightleftharpoons CuSO_4 \cdot 3H_2O(s) + 2H_2O(g)$

(ii) $CuSO_4 \cdot 3H_2O(s) \rightleftharpoons CuSO_4 \cdot H_2O(s) + 2H_2O(g)$

(iii) $CuSO_4 \cdot H_2O(s) \rightleftharpoons CuSO_4(s) + H_2O(g)$

With the help of the phase rule, show that each of the above equilibria is univariant. (b) Let the various equilibria given in part (a) be studied at a constant temperature. Since each equilibrium is univariant, it implies that the pressure of water vapour in equilibrium with two solid salts will have a definite value. Show that this conclusion is in agreement with that derived from a consideration of the equilibrium constant. (c) At 25 °C, the pressures of water vapour of various equilibria given in part (a) are found to be $p_{\mathrm{i}} = 7.8$ mmHg, $p_{\mathrm{ii}} = 5.6$ mmHg and $p_{\mathrm{iii}} = 0.8$ mmHg, respectively. The dehydration of $CuSO_4 \cdot 5H_2O$ is studied by enclosing it in a closed container (attached to a vacuum pump and manometer) followed by evacuation. The pressure of the system is determined from time to time and the results obtained are shown in Fig. 4. Explain the results shown therein.

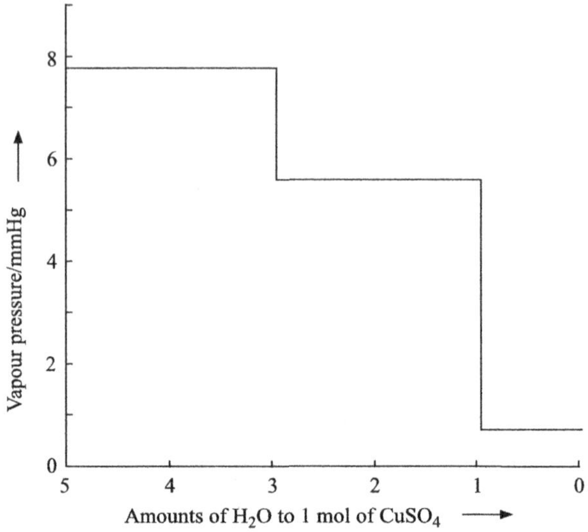


**Fig. 4**

6.10 What type of cooling curve do you expect for a pure substance?

6.11 The phase diagram of a binary system exhibiting partial miscibility both in liquid and solid phases has an appearance as shown in Fig. 5. Label the diagram and draw the cooling curves for the points shown therein.

(Ans. (1) $L_1$, (2) $L_2$, (3) $L_1 + L_2$, (4) solid solution α, (5) α and $L_1$, (6) α and $L_2$, (7) solid solution β, (8) β and $L_2$, and , (9) α and β)

6.12 The elements Ag and Sr form four stable congruent compounds. Construct the phase diagram of Ag-Sr system from the following data:

m.pt. of Ag 960.5 °C
m.pt. of $Ag_4Sr$ 781 °C
m.pt of $Ag_5Sr_3$ 760 °C
m.pt. of AgSr 680°C
m.pt. of $Ag_2Sr_3$ 665 °C
m.pt. of Sr 757 °C

$Eu_1$ 750 °C
$Eu_2$ 693 °C
$Eu_3$ 638 °C
$Eu_4$ 645 °C
$Eu_5$ 436 °C

Label the entire phase diagram.

6.13 (a) The phase diagram of the Ca-Al system has the following data:
m.pt. of Ca 840 °C
m.pt. of Al 660 °C
Congruent m.pt. of $CaAl_2$ 1 063 °C
Incongruent m.pt. of $CaAl_3$ 700 °C

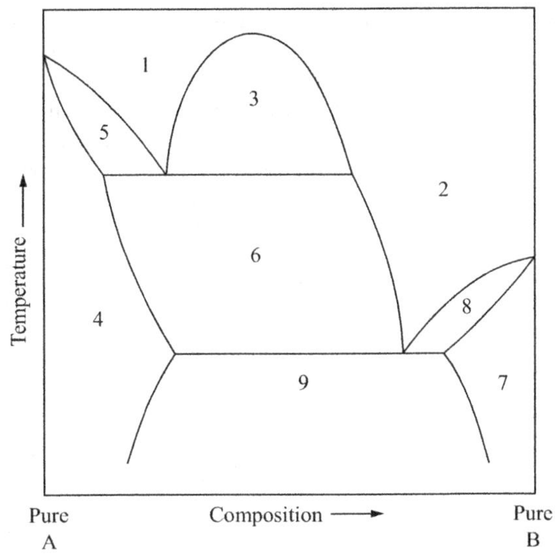


**Fig. 5**

Composition of the melt at 700 °C is 14 mass % Ca.
Eutectic point involving Ca and $CaAl_2$ is at 540 °C and has 75 mass % of Ca.
Eutectic point involving Al and $CaAl_3$ is at 600 °C and has 8 mass % Ca.
(b) Draw the phase diagram and from it predict the nature of the system for the following data:

| | |
|---|---|
| (i) 80 mass % Ca at 700 °C. | (vi) 25 mass % of Ca at 650 °C. |
| (ii) 60 mass % Ca at 800 °C. | (vii) 25 mass % of Ca at 500 °C. |
| (iii) 37 mass % of Ca at 800 °C. | (viii) 10 mass % of Ca at 650 °C. |
| (iv) 37 mass % of Ca at 600 °C. | (ix) 5 mass % of Ca at 640 °C. |
| (v) 25 mass % of Ca at 750 °C. | (x) 5 mass % of Ca at 500 °C. |

Also calculate wherever possible the relative amounts of the two phases.

6.14 Figure 6 depicts the phase diagram of $CuSO_4$-$H_2O$ system. Label the various

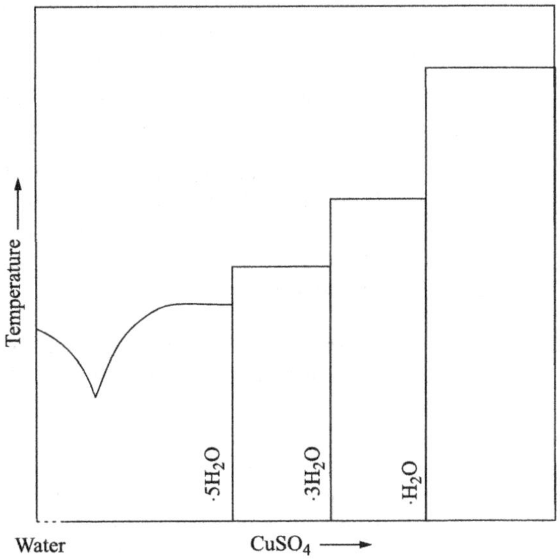


**Fig. 6**

points, lines and areas of the diagram. Describe the sequence of phase changes if a dilute solution of copper sulphate is dehydrated at 5 °C ending up with anhydrous copper sulphate.

6.15 Cooling curves for the binary alloy system, gold-antimony, give the following data:

| *Mass % Sb* | *Start of freezing* °C | *Completion of freezing* °C |
|---|---|---|
| 0 | 1 063 | 1 063 |
| 10 | 830 | 360 |
| 20 | 470 | 360 |
| 30 | 400 | 360 |
| 40 | 445 | 360 |
| 45 | 455 | 360 |
| 55 | 460 | 460 |
| 60 | 495 | 460 |
| 70 | 545 | 460 |
| 80 | 580 | 460 |
| 90 | 610 | 460 |
| 100 | 631 | 631 |

(a) Plot the phase diagram for the system and identify all areas and phases.
(b) Describe the sequence of events when liquid melts of 20, 50 and 80 amount per cent antimony are cooled from 800 °C to 300 °C.
(c) If a sample of overall composition 40% by mass Sb were examined at 200 °C. what components would be present and in roughly what proportions?

6.16 The following information is known about the two-component A-B system: m.pt. of pure A 660 °C, m.pt. of pure B 1 500 °C. Compound C having 24 mol % B decomposes at 950 °C to give D (50 mol % of B) and liquid containing 20 mol % of B. Compound D melts congruently at 1 650 °C. $(\text{Eutectic})_1$ at 600 °C. corresponding composition is 10 mol % B. $(\text{Eutectic})_2$ at 1 380 °C, corresponding composition is 80 mol % B. Draw the phase diagram. Give the form of compounds involved. A melt containing 40 mol % of B is cooled from 1 650 °C. Indicate the various phase changes which take place.

6.17 The following information is known about the two-component A-B system:

Molar mass of A = 24.3 g mol$^{-1}$, m.pt. of A = 651 °C

Molar mass of B = 59.7 g mol$^{-1}$, m.pt. of B = 450 °C

Two compounds C and D are formed.
C contains 54.7% by mass of B and melts incongruently at 770 °C yielding D and a liquid containing 38% by mass of B.
D melts sharply at 1 180 °C yielding a liquid containing 83% by mass of B.
Two eutectics at 510 °C with 28% by mass of B and at 1 080 °C with 88% by mass of B.
Solids are immiscible in the solid state and completely miscible in the liquid phase. Construct the phase diagram, and label it. Draw cooling curves for 45% and 60% by mass of B. Deduce the forms of C and D.

6.18 The system $CaF_2$(m.pt., 1 360 °C) and $CaCl_2$ (m. pt. 772 °C) shows an incongruent behaviour, forming a compound at 1 : 1 ratio. The compound melts at 737 °C giving a liquid containing 60 mol % of $CaCl_2$. The eutectic point is at 625 °C with the eutectic composition of 80 mol % of $CaCl_2$. Draw the phase diagram and label it. Draw cooling curves for 50% and 60% by mol of $CaCl_2$.

6.19 Given here are the data of the binary system copper nitrate-water. Construct the phase diagram and label it.

*Composition of Saturated Solution at Various Temperatures*

| *Temperature* °C | $Cu(NO_3)_2$ g *per* 100 g *solution* | *Solid phase* |
|---|---|---|
| – 2.35 | 8.0 | Ice |
| – 9.3 | 21.3 | " |
| – 20.0 | 34.3 | " |
| – 24.0 | 35.9 | Ice + $Cu(NO_3)_2 \cdot 9H_2O$ |
| – 21.0 | 37.4 | $Cu(NO_3)_2 \cdot 9H_2O$ |
| – 20.0 | 39.8 | $Cu(NO_3)_2 \cdot 9H_2O + Cu(NO_3)_2 \cdot 6H_2O$ |
| 0 | 45.0 | $Cu(NO_3)_2 \cdot 6H_2O$ |
| + 20.0 | 55.6 | " |
| 24.5 | 61.4 | $Cu(NO_3)_2 \cdot 6H_2O + Cu(NO_3)_2 \cdot 3H_2O$ |
| 40.0 | 61.5 | $Cu(NO_3)_2 \cdot 3H_2O$ |
| 60.0 | 64.2 | " |
| 80.0 | 67.5 | " |
| 114.5 | 77.6 | |

$Cu(NO_3)_2 \cdot 9H_2O$ = 53.7 mass % of $Cu(NO_3)_2$
$Cu(NO_3)_2 \cdot 6H_2O$ = 63.5 mass % of $Cu(NO_3)_2$
$Cu(NO_3)_2 \cdot 3H_2O$ = 77.7 mass % of $Cu(NO_3)_2$

6.20 Give below are the data of the binary system sodium iodide-water. Construct phase diagram and lable it.

*Composition of Saturated Solution at Various Temperatures*

| *Temperature* °C | NaI g *per* 100 g *solution* | *Solid phase* |
|---|---|---|
| – 4.25 | 14.6 | Ice |
| – 9.75 | 26.5 | " |
| – 31.4 | 39 | Ice + $NaI \cdot 5H_2O$ |
| – 15.2 | 57.2 | $NaI \cdot 5H_2O$ |
| – 13.5 | 60.2 | $NaI \cdot 5H_2O + NaI \cdot 2H_2O$ |
| 0 | 61.3 | $NaI \cdot 2H_2O$ |
| 10 | 62.8 | " |
| 20 | 64.2 | " |
| 30 | 65.5 | " |
| 40 | 67.2 | " |
| 50 | 69.5 | " |
| 60 | 72.0 | " |
| 65 | 74.4 | $NaI \cdot 2H_2O$ + NaI |
| 80 | 74.7 | NaI |
| 100 | 75.1 | " |
| 120 | 76.3 | " |
| 140 | 77.2 | " |

$NaI \cdot 5H_2O$ = 62.5 mass % NaI
$NaI \cdot 2H_2O$ = 80.6 mass % NaI

6.21 A system of A (1 100 °C) and B (600 °C) shows incongruent behaviour. It has a peritectic temperature at 700 °C where a solid containing 25 mol % of B melts giving a liquid of 48 mol % of B. The eutectic temperature is 400 °C and the corresponding composition is 85 mol % of B. Plot a phase diagram and label it. Draw cooling curves for solutions containing 25 mol % B, 48 mol % B and 56 mol % of B.

6.22 The following data were obtained for Mg-Ni system:

m.pt. of Mg = 651 °C
m.pt. of Ni = 1 450 °C

Compound $MgNi_2$ melts congruently at 1 180 °C.
Compound $Mg_2Ni$ melts incongruently at 770 °C giving a melt of 38 mass % of Ni.
First eutectic at 510 °C with eutectic composition of 28 mass % of Ni
Second eutectic at 1 080 °C with eutectic composition of 88 mass % of Ni
No solid solution formation
Construct phase diagram and label it.

6.23 Describe the sequence of effects that is produced when a 10% solution of ferrous sulphate of Fig. 6.12.5 is subjected to the isothermal dehydration at 10 °C.

6.24 The following data were obtained for the copper-lanthanum system:

Melting point of Cu = 1083 °C
Melting point of La = 812 °C
The compound $LaCu_6$ melts congruently at 913 °C
The compound $LaCu_4$ melts incongruently at 735 °C
The compound $LaCu_2$ melts congruently at 834 °C
The compound LaCu melts incongruently at 551 °C
Three eutectic temperatures were observed:

| | |
|---|---|
| 840 °C | ($Cu + LaCu_6$) |
| 725 °C | ($LaCu_2 + LaCu_4$) |
| 468 °C | (LaCu + La) |

Sketch the phase diagram and label each area.

6.25 The melting points and enthalpies of fusion of Pb and Sb are

| | | |
|---|---|---|
| Pb | 327.4 °C | 5.10 kJ $mol^{-1}$ |
| Sb | 630.5 °C | 20.08 kJ $mol^{-1}$ |

Calculate the solid-liquid equilibrium lines. Estimate the eutectic composition graphically and then calculate the eutectic temperature.

(Ans. $x_{Pb}$ = 0.855, 520 K)

6.26 Metals A and B form compounds $AB_3$ and $A_2B_3$. Solids A, B, $AB_3$ and $A_2B_3$ are immiscible in each other as solids but are completely miscible as liquids. Melting points of A and B are 600 °C and 1 100 °C, respectively. Compound $A_2B_3$ melts congruently at 900 °C and gives a simple eutectic with A at 450 °C. Compound $AB_3$ decomposes at 800 °C to give another compound and the melt. This gives another eutectic with B at 650 °C. Draw the simplest phase diagram with these facts and label it. Draw cooling curves for melts of 90 mol% A and 30 mol% A.

6.27 The following phase transitions are observed for iron:

$$\underset{(\alpha\text{-iron})}{\text{Body-centred cubic structure}} \underset{}{\overset{910\,°\text{C}}{\rightleftharpoons}} \underset{(\gamma\text{-iron})}{\text{Face-centred cubic structure}} \overset{1401\,°\text{C}}{\rightleftharpoons} \underset{(\delta\text{-iron})}{\text{Body-centred cubic structure}} \overset{1535\,°\text{C}}{\rightleftharpoons} \text{liquid}$$

The iron-rich portion of the phase diagram of iron-carbon system is of interest and is shown in Fig. 7

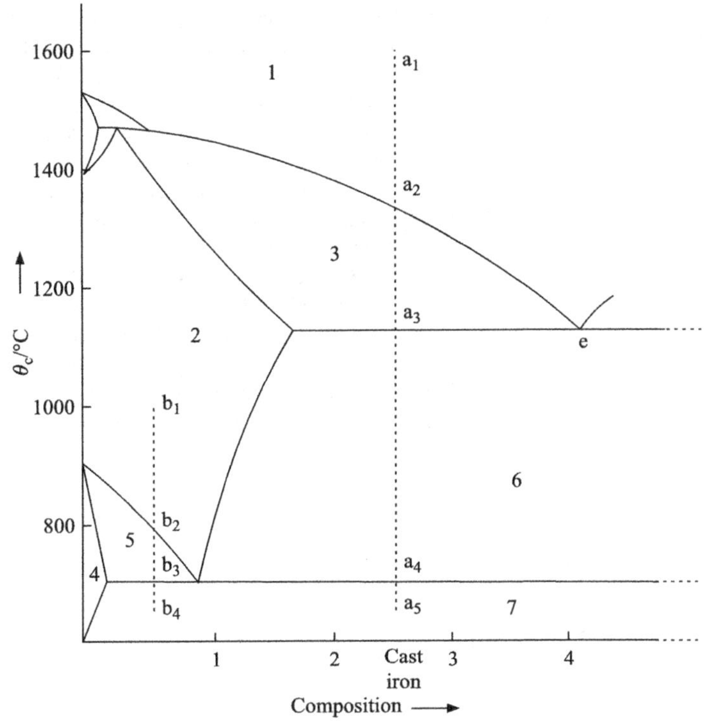


**Fig. 7**

The description of the phase diagram is as follows.

| *Location of the system* | *Status of the system* |
|---|---|
| Areas | |
| 1 | Liquid phase |
| 2 | Solid solution of Fe and C known as austenite |
| 3 | Austenite ⇌ Liquid |
| 4 | α-Fe containing some C |
| 5 | Austenite + α-Fe (with some C) |
| 6 | Austenite + $Fe_3C$ (called cementite) |
| 7 | α-Fe + cementite |
| Lines | |
| at 700 °C | A eutectic-type three phase line at 700 °C<br>Phases are α-Fe with some C, austenite and cementite |
| at 1 125 °C | A eutectic-type three phase line. Phases are austenite, liquid and cementite. |

Answer the following:

(a) Iron with about 2 per cent carbon forms a fairly crude steel known as cast iron. It has valuable corrosion-resistant properties. One such system in liquid phase is shown by the point $a_1$ in Fig. 7. Describe the sequence of events that takes place when this system is cooled along the isopleth $a_1a_2\,a_3a_4a_5$.

*(Note: The system of composition of cast iron does not become a single phase until its melting point is reached. If the composition is close to the eutectic, its melting point is low enough to allow a fairly easy casting procedure.)*

(b) When more of carbon is eliminated, a higher grade of steel is obtained. One such system in the austenite form is shown by the point $b_1$ in Fig. 7
Describe the sequence of events that takes place when this system is cooled along the isopleth $b_1b_2b_3b_4$.
*(Note: Steel when present in austenite form can be rolled very easily. At temperature below* 700 °C, *it contains* α-Fe *(with some* C) *and cementite. The rigidity of steel is due to extreme hardness of cementite.)*

6.28 In Mg-Zn system, there are four temperatures at which three phases exist with mass percentage of Zn as follows:

| | |
|---|---|
| 340 °C | solid solution α(8%) + liquid (53%) + MgZn |
| 354 °C | liquid (55%) + MgZn + $MgZn_2$ |
| 380 °C | liquid (96%) + $MgZn_2$ + $MgZn_5$ |
| 364 °C | liquid (97%) + $MgZn_5$ + Zn |

Melting points of Mg, Zn and $MgZn_2$ are 923 K, 693 K, and 863 K, respectively. Sketch the phase diagram and label it.

# 7 Phase Diagrams of Three-Component Systems

## 7.1 APPLICATION OF THE PHASE RULE

The expression of the degrees of freedom of a three-component system as given by the phase rule is

$$F = C - P + 2 = 3 - P + 2 = 5 - P$$

As in the case of a two-component system, various equilibria studied in this chapter do not involve any gaseous phase. Hence, the degrees of freedom of a system are reduced by one. Thus for a condensed three component system, we have

$$F = C - P + 1 = 3 - P + 1 = 4 - P$$

**System of One Phase only**

In this case, $F = 4 - 1 = 3$, and therefore, the values of three variables have to be stated in order to define the system completely. These are temperature and the amount fractions of any of the two components.

**System of Two Phases**

Here $F = 4 - 2 = 2$, and thus the values of two variables have to be specified in order to define the system completely. These are temperature and the amount fraction of any one component in either of the two phases. These two phases will be in equilibrium with each other and thus the amount fraction of the stated component in the second phase will have a definite value.

**System of Three Phases**

In this case, $F = 4 - 3 = 1$, and this degree of freedom is the temperature of the system. All the three phases will have definite compositions at a given temperature.

## 7.2 SCHEME OF TRIANGULAR PLOT

The most commonly used scheme for plotting the two-dimensional equilibrium diagram for three-component systems is the scheme of triangular plot as suggested by Stokes and Roozeboom. In this scheme, various equilibria are studied at a constant temperature and are plotted in an equilateral triangle as shown in Fig. 7.2.1. The sides of this triangle are divided into 10 (or sometimes more, say 100) equal parts and within the triangle, various lines parallel to the three sides are drawn. The concentrations of various species are represented according to the following scheme.

**System Involving One Component**

Each apex of the triangle represents a single component system and is designated by the symbol used for the component. For example, point A represents 100% of the component A, or at this point the composition of the ternary system is $x_A = 1$, $x_B = 0$ and $x_C = 0$. In terms of mass per cent, we have 100 mass % of A, 0 of mass % of B and 0 mass % of C.

**System Involving Two Components**

A point on the sides of the triangle represents a two-component system consisting of the constituents written on either ends of the line. For example, a point anywhere on the line BC in Fig. 7.2.1 represents a two-component system of B and C. The component A is completely absent here and thus has $x_A = 0$. Since the line BC is divided into 10 equal parts, the compositions of systems at the points shown on this line can be determined easily and are described in Table 7.2.1.

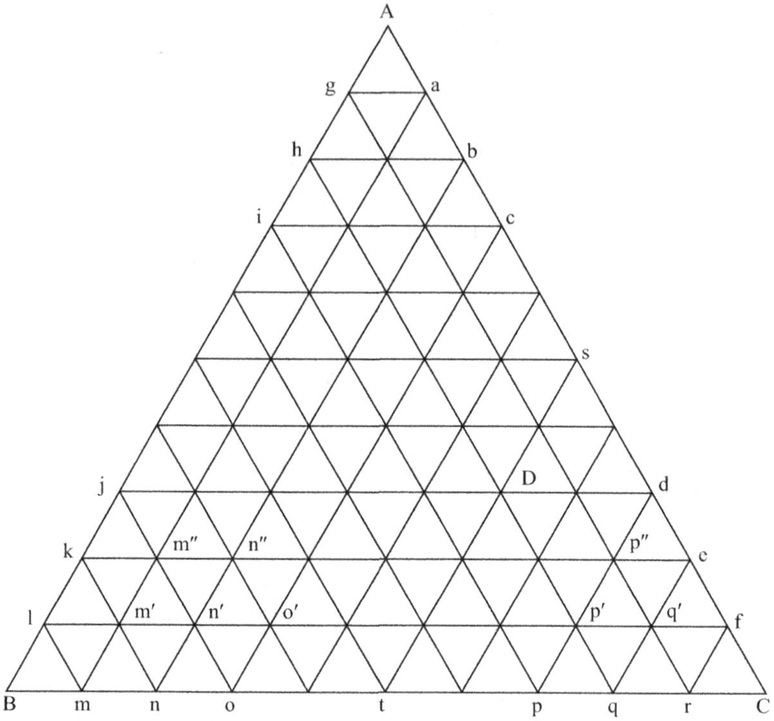


**Fig. 7.2.1 The triangle plot**

**Table 7.2.1** Compositions of Ternary System on the Line BC

| *Point* | *Amount fractions* | | |
|---|---|---|---|
| | $x_A$ | $x_B$ | $x_C$ |
| B | 0.0 | 1.0 | 0.0 |
| m | 0.0 | 0.9 | 0.1 |
| n | 0.0 | 0.8 | 0.2 |
| — | — | — | — |
| — | — | — | — |
| — | — | — | — |
| q | 0.0 | 0.2 | 0.8 |
| r | 0.0 | 0.1 | 0.9 |
| C | 0.0 | 0.0 | 1.0 |

Alternatively, the composition at any point on the line BC (or on the other two lines) can be determined with the help of the lever rule. For example, at the point n, we have

$$\frac{\text{Amount of B in the system}}{\text{Amount of C in the system}} = \frac{\text{nC}}{\text{Bn}} = \frac{8 \text{ units}}{2 \text{ units}}$$

If concentrations are expressed in amount fractions, it is obvious that at the point n, the amount fraction of B is equal to 0.8 whereas that of C is 0.2.

**System Involving Three Components**

The perpendicular distance from A to BC is also divided into ten equal parts with lines drawn parallel to the line BC, and thus these parallel lines represent systems with different proportions of the component A; $x_A$ being one at A and zero on the line BC. Points anywhere on one of the parallel lines represent systems with a constant amount fraction of the component A. For example, the line lf represents the systems with $x_A = 0.1$. It can be verified that the length of this line is 9/10 of the line BC and thus it represents a total amount fraction of 0.9 of both the components B and C. At the point 1, composition of the system is $x_A = 0.1$, $x_B = 0.9$ and $x_C = 0.0$ and that at the point f is $x_A = 0.1$, $x_B = 0.0$ and $x_C = 0.9$. The line lf is divided into nine equal parts. The compositions of systems at the points shown on this line are described in Table 7.2.2.

**Table 7.2.2** Compositions of Ternary Systems on the Line lf

| *Point* | *Amount fractions* | | |
|---|---|---|---|
| | $x_A$ | $x_B$ | $x_C$ |
| l | 0.1 | 0.9 | 0.0 |
| m′ | 0.1 | 0.8 | 0.1 |
| n′ | 0.1 | 0.7 | 0.2 |
| — | — | — | — |
| — | — | — | — |
| — | — | — | — |
| p′ | 0.1 | 0.2 | 0.7 |
| q′ | 0.1 | 0.1 | 0.8 |
| f | 0.1 | 0.0 | 0.9 |

On the line ke, compositions of systems represented by the points as shown in Fig. 7.2.1 are given in Table 7.2.3.

**Table 7.2.3** Compositions of Ternary Systems on the Line ke

| *Point* | *Amount fractions* | | |
|---|---|---|---|
| | $x_A$ | $x_B$ | $x_C$ |
| k | 0.2 | 0.8 | 0.0 |
| m″ | 0.2 | 0.7 | 0.1 |
| n″ | 0.2 | 0.6 | 0.2 |
| — | — | — | — |
| — | — | — | — |
| — | — | — | — |
| p″ | 0.2 | 0.1 | 0.7 |
| e | 0.2 | 0.0 | 0.8 |

The compositions of ternary systems on the other parallel lines can be similarly written down.

Thus the composition of a ternary system with non-zero amount fractions of components is represented by a point within the triangle. The composition corresponding to any given point (or the location of the point corresponding to any given composition) can be readily determined by the above method. Take, for example, the point D shown in Fig. 7.2.1, the composition corresponding to this point can be determined as follows:

(i) This point lies on the third parallel line from the bottom of BC and thus has $x_A = 0.3$.

(ii) This point also lies on the second parallel line from the line AC and thus has $x_B = 0.2$.

(iii) This point lies on the fifth parallel line from the line AB and thus has $x_C = 0.5$.

Hence, the composition of the system is

$$x_A = 0.3, \quad x_B = 0.2 \quad \text{and} \quad x_C = 0.5$$

## METHODS OF COMPUTING COMPOSITION OF A TERNARY SYSTEM

The following general methods may be employed for computing the composition of a ternary system.

### Method of Parallel Lines

From the given point (say, D in Fig. 7.2.1) draw three lines parallel to the three sides of the triangle. It can be proved that the sum of the lengths of these three such lines is always constant and is equal to the side of the triangle.† Moreover, distance of the line from the point D to any side is proportional to the amount fraction of the component occupied by the opposite corner. Thus, we have

Length of the line Dd (or Ds) $\propto$ amount fraction of B
Length of the line Dt (or Dq) $\propto$ amount fraction of A
Length of the line Dh (or Dj) $\propto$ amount fraction of C

The ratio of the lengths of these lines are

$$\text{Dd : Dh : Dt : : 2 : 5 : 3}$$

and thus the composition of the system is

$$x_B = 0.2, \quad x_C = 0.5 \quad \text{and} \quad x_A = 0.3$$

---

†The parallel lines drawn from the point D are Dh, Dt and Dd. We have the following equalities:

Dt = Dq, Dd = qC = eC, hq = Ae

Now the sum of the lengths of the three sides is

Dh + Dt + Dd = Dh + Dq + eC = hq + eC = Ae + eC = AC

**Method of Perpendicular Lines**

From the given point, draw three perpendiculars De, Df and Dg on the three sides of the triangle as shown in Fig. 7.2.2.

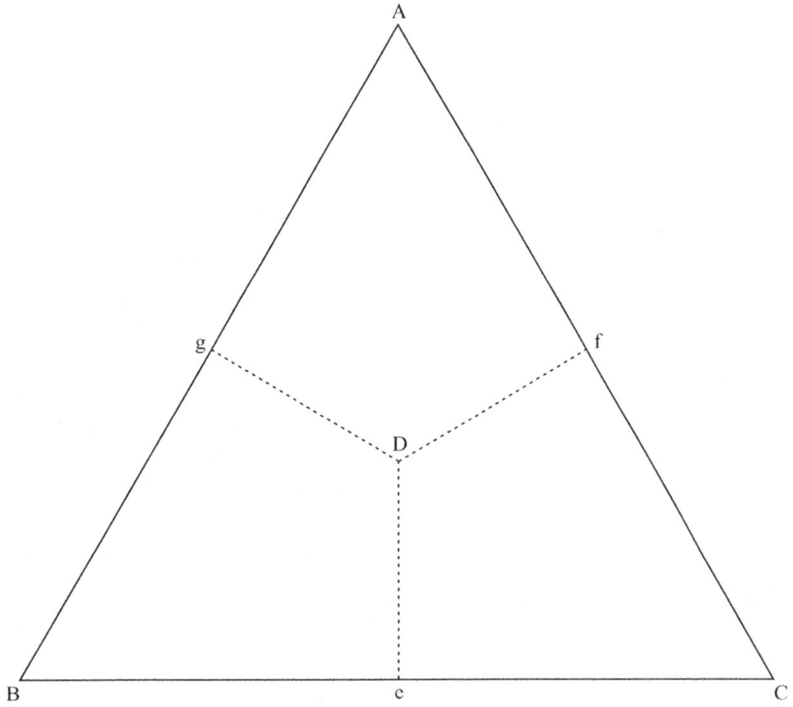


**Fig 7.2.2 Computation of composition of a ternary system**

The distances De, Df and Dg will be proportional to the amounts of A, B and C, respectively. Thus

$$\frac{\text{Length of line De}}{\text{Sum of the lengths of lines De, Df and Dg}} = x_A$$

$$\frac{\text{Length of line Df}}{\text{Sum of the lengths of lines De, Df and Dg}} = x_B$$

$$\frac{\text{Length of line Dg}}{\text{Sum of the lengths of lines De, Df and Dg}} = x_C$$

It can be shown that the sum of the three perpendiculars De, Df and Dg is equal to 0.867 × edge-length of the triangle. Thus, if edge-length is 100 mm, the above relations of amount fractions are modified as follows:

$$\frac{\text{Length of line De}}{86.7\ \text{mm}} = x_A$$

$$\frac{\text{Length of line Df}}{86.7\ \text{mm}} = x_B$$

$$\frac{\text{Length of line Dg}}{86.7\ \text{mm}} = x_C$$

**A Direct Method**

If the edge-length of the triangle is 100 mm, a direct method for computing composition for a given point D (or vice versa) may be employed. At point D, a scale parallel to the side BC is placed and the distances in mm between the point D and the sides AB and AC are measured. These distance yield directly the per cent of C and B, respectively.† The per cent of A at the point D will be 100 – (% B + % C).

For locating a point corresponding to the given composition, we adopt the following procedure. The distance of perpendicular from apex A to the side BC is

$$0.867 \text{ (edge-length)} = 0.867 \times 100 \text{ mm} = 86.7 \text{ mm}$$

This length represents 100 per cent of A. The per cent composition of A in the given ternary system is multiplied by 0.867 and let it be $x$. A scale at a distance of $x$ (in mm) parallel to the side BC is placed and a point at a distance (in mm) equal to the per cent of C from the side AB is located. This point represents the required composition of the given ternary system.

**Alternatively**, a point at a distance (in mm) equal to the given per cent of A along the side BA or CA (i.e. from B or C towards A) is located. A scale at this point parallel to the side BC is placed and a point at a distance (in mm) equal to the given per cent of C from the side AB as located. This point represents the required composition of the given ternary system.

**By Using the Lever Rule**

The composition corresponding to a given point can be determined by making use of the lever rule twice. For example, for the point a shown in Fig. 7.2.3, we can determine its composition as follows.

A line Aa joining the apex A and the given point a is drawn and is extended up to the side BC. Applying the lever rule at a along the line Ab, we get

$$\frac{\text{Total mass of B and C}}{\text{Mass of A}} = \frac{\text{Aa}}{\text{ab}} \tag{7.2.1}$$

Adding one on both sides and taking its inverse, we have

$$\frac{\text{Mass of A}}{\text{Total mass of A, B and C}} = \frac{\text{ab}}{\text{Ab}}$$

or

$$\text{Mass of A} = \left(\frac{\text{ab}}{\text{Ab}}\right) \text{(Total mass of A, B and C)} \tag{7.2.2}$$

Now applying the lever rule at b along the line BC, we get

$$\frac{\text{Mass of C}}{\text{Mass of B}} = \frac{\text{Bb}}{\text{bC}} \tag{7.2.3}$$

†If the edge-length is not equal to 100 mm (say, $x$) the lengths may be multiplied by a factor of 100 mm/$x$ to yield directly the per cent compositions C and B, respectively.

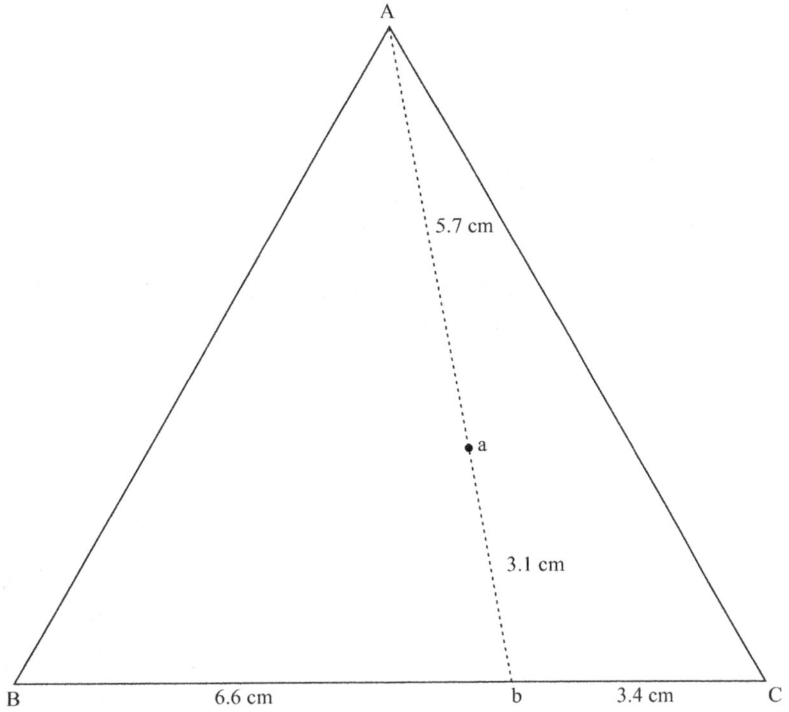


**Fig. 7.2.3 Use of the lever rule within a triangle**

Adding one on both sides and taking its inverse, we get

$$\frac{\text{Mass of B}}{\text{Mass of B and C}} = \frac{\text{bC}}{\text{BC}}$$

or $$\text{Mass of B} = \left(\frac{\text{bC}}{\text{BC}}\right)(\text{Mass of B and C}) \qquad (7.2.4)$$

Finally, we will have

$$\text{Mass of C} = (\text{Total mass of A, B and C}) - (\text{Mass of A and B}) \qquad (7.2.5)$$

For calculating per cent composition, total mass of A, B and C will be equal to 100 g. Thus, Eqs (7.2.2), (7.2.4) and (7.2.5) will be reduced to

$$\text{Mass of A} = \left(\frac{\text{ab}}{\text{Ab}}\right)(100\text{ g})$$

$$\text{Mass of B} = \left(\frac{\text{bC}}{\text{BC}}\right)(100\text{ g} - \text{mass of A})$$

$$\text{Mass of C} = 100\text{ g} - (\text{mass of A and B})$$

Hence, for the point a, shown in Fig. 7.2.3, we have

$$\text{Mass of A} = \left(\frac{3.1\text{ cm}}{8.8\text{ cm}}\right)(100\text{ g}) = 35.23\text{ g}$$

$$\text{Mass of B} = \left(\frac{3.4 \text{ cm}}{10 \text{ cm}}\right)(100 \text{ g} - 35.23 \text{ g}) = 22.02 \text{ g}$$

$$\text{Mass of C} = 100 \text{ g} - (35.23 \text{ g} + 22.02 \text{ g}) = 42.75 \text{ g}$$

**Method Based on the Two Features of an Equilateral Triangle**

The following two features of an equilateral triangle can also be used to locate the point corresponding to a given ternary system or vice versa.

**(1)** Various systems with a constant ratio of two components and the varying amount of the third component will lie on a line drawn from the apex representing the third component to the opposite side of the triangle. Take, for example, the perpendicular line AX from the corner A to the opposite side BC in Fig. 7.2.4.

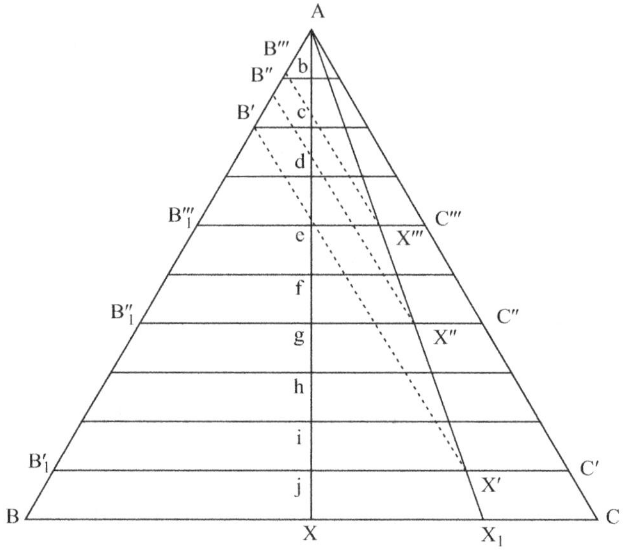


**Fig. 7.2.4 Triangular plot showing various systems of constant ratio of B and C and varying amount of A.**

This line intersects just in the middle of various lines drawn parallel to BC and hence a point anywhere on this line represents a system in which the ratio of the components B and C is always one (i.e. equal amount fractions of B and C). The compositions of systems represented by the points on this line as shown in Fig. 7.2.4 are given in Table 7.2.4.

**Table 7.2.4** Compositions of the Ternary System on the Line AX.

| | *Amount fractions* | | |
|---|---|---|---|
| *Point* | A | B | C |
| A | 1.0 | 0 | 0 |
| b | 0.9 | 0.05 | 0.05 |
| c | 0.8 | 0.1 | 0.1 |
| — | — | — | — |
| — | — | — | — |
| — | — | — | — |
| i | 0.2 | 0.4 | 0.4 |
| j | 0.1 | 0.45 | 0.45 |
| X | 0.0 | 0.5 | 0.5 |

The fact that a point on the line AX represents a constant ratio of B and C is also true for any line drawn from A to BC. For example, on the line $AX_1$ of Fig. 7.2.4, the constant ratio of B and C is

$$\frac{\text{Amount fraction of B}}{\text{Amount fraction of C}} = \frac{CX_1}{BX_1} = \frac{C'X'}{B_1'X'\ (\text{or } B'X')}$$
$$= \frac{C''X''}{B_1''X''\ (\text{or } B''X'')} = \frac{C'''X'''}{B_1'''X'''\ (\text{or } B'''X''')}$$

The above feature can also be used to locate the point corresponding to the composition of a given ternary system (say, for example, $x$ mass % A, $y$ mass % B and $z$ mass % C). First of all, a binary system (say B-C) having the given ratio of B and C in the ternary system is imagined. In the present case, the binary system B-C will contain $y/(y+z) \times 100$ mass per cent of B). A point on the line BC corresponding to this composition (say, b in Fig. 7.2.3) is chosen and it is joint with the apex A. The distance bA represents 100 mass % A. The length corresponding to the given mass per cent A is determined. In the present case, it is

$$\left(\frac{bA}{100 \text{ mass \% A}}\right)(\text{mass \% A})$$

A point at the above distance on the line bA from the side BC is located. This point represents the required composition of the ternary system.

(**2**) The fact that the binary systems A-B, B-C and A-C are represented by points on the lines AB, BC and AC, respectively, and the ternary system A-B-C by a point within the triangle ABC can be extended to any line or smaller triangle within bigger triangle ABC. For example, a point on the line DE in Fig. 7.2.5 represents a system that is composed of two systems D and E with their relative amounts as given by the lever rule. Take, for example, the point G on the line DE. The system corresponding to this point can be prepared from the two systems D and E by mixing them in the following ratio.

$$\frac{\text{Amount of D in the system}}{\text{Amount of E in the system}} = \frac{GE}{DG}$$

In other words, we can say that all systems that can be prepared from D and E will lie on the line DE. The same is also true on the lines EF and FD. Thus, all those systems which can be prepared by mixing E and F will lie on the line EF whereas those prepared from F and D will lie on FD. In a similar manner, all possible combinations that can be prepared from D, E and F will lie within the triangle DEF. Their relative amounts may be determined by employing the lever rule twice.

## Classification of Three-Component Systems

Many categories of three-component systems are possible depending on the characteristics of the components. For example, the given component may exist either in the liquid form or in the solid form and may be miscible or partially miscible or completely immiscible with other components. In the present chapter, we will consider only two important categories of these systems. These are:

(i) Systems composed of three liquid components which exhibit partial miscibility.
(ii) Systems composed of two solid components and a liquid.

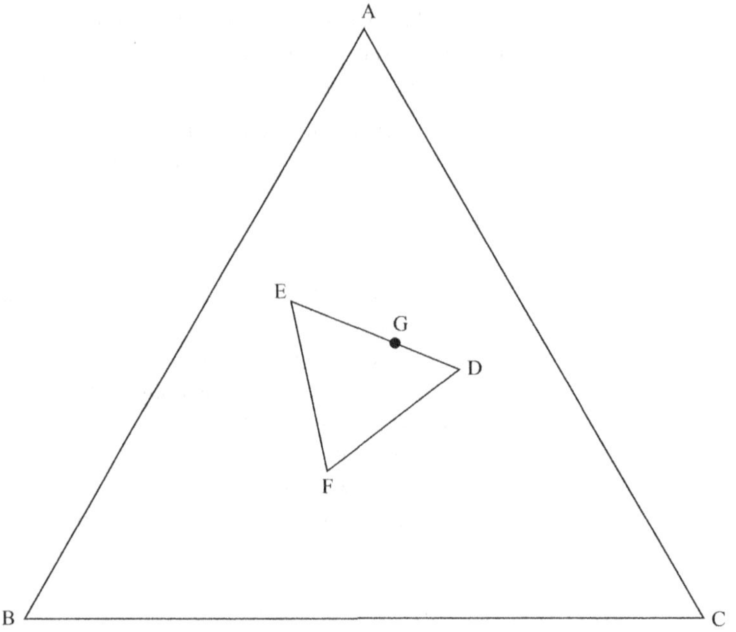


**Fig. 7.2.5 Triangular plot showing the application of the lever rule on any line drawn within the triangle**

In the next two sections, we shall consider the triangular plots of systems belonging to the above two categories.

## 7.3 SYSTEMS OF THREE LIQUID COMPONENTS EXHIBITING PARTIAL MISCIBILITY

In this section, we consider the following three possibilities based on the mutual solubilities of the three components.

- Formation of one pair of partially miscible liquids.
- Formation of two pairs of partially miscible liquids.
- Formation of three pairs of partially miscible liquids.

The general appearance of triangular plots of these three possibilities are discussed below one after the other.†

### FORMATION OF ONE PAIR OF PARTIALLY MISCIBLE LIQUIDS

#### System of Two Partially Miscible Liquids

We assume that out of the three components A, B and C, the components B and C show partial mutual solubility with each other and the third component A exhibits complete miscibility with both the components B and C. Let us start with only two components B and C. The state of such a system can be represented

---

†Throughout this Chapter, a system is represented by an alphabet, printed in roman type, and its composition is represented by the same alphabet printed in italics.

by a point on the line BC. The two components will be completely miscible in each other only in the limited range of concentrations and will exhibit partial miscibilities beyond these ranges with the formation of two conjugate saturated solutions, one composed of a solution of C in B and the other of B in C. These two solutions will have definite compositions at a given temperature and will be independent of individual amounts of the two liquids. Let these compositions be represented by the points b and c on the line BC as shown in Fig. 7.3.1. Thus, the two liquids B and C will be completely miscible in each other from the range B to b and from c to C and form two conjugate solutions b and c in the range b to c. The relative amounts of these two solutions can be determined with the help of the lever rule. For example, at the point X, we have

$$\frac{\text{Amount of the solution of C in B}}{\text{Amount of the solution of B in C}} = \frac{\text{Xc}}{\text{Xb}}$$

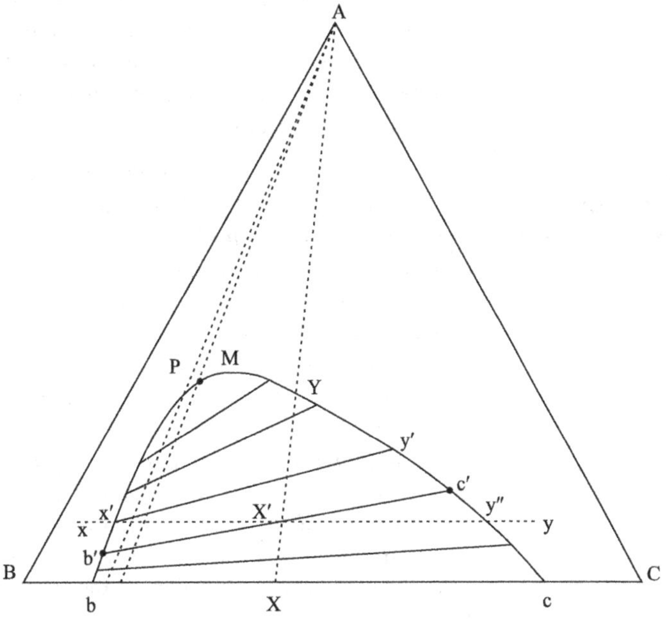


**Fig. 7.3.1 Triangular plot showing the formation of one pair of partially miscible liquids**

### Effect of Adding Third Component

If now the component A is added to a system represented by a point on the line BC, the state of the system will move along a line obtained by joining the given point with the apex A. Let us add this component to a system represented by the point X. Since A is miscible in both the conjugate solutions, the addition of A will cause an increase in the mutual solubilities of the two components B and C. Consequently, compositions of the two conjugate solutions will now become more close to each other.

### Nature of the Tie Line

As usual the compositions of the two conjugate solutions will be given by the two ends of a tie line. In general, the various tie lines are neither parallel to the line BC nor parallel to each other, if the added component A is not equally soluble in both the solutions. If the component A is relatively more soluble in the layer rich in C than it is in the layer rich in B, then obviously the state point of the layer rich in C will lie nearer to the apex A than that of the layer rich in B. The tie line becomes shorter and shorter as more of A is added to the system indicating that the compositions of the two solutions become closer and closer to each other. This is consistent with the fact that the mutual solubilities of the two components increase as more and more of A is added to the system.

**Plait Point**

At point P, compositions of the two solutions become identical and thus two solutions merge into each other and form a single solution. This is an invariant point (at constant $T$ and $p$, $F = C - r - P + 0 = 3 - 1 - 2 = 0$) and is known as *plait point.*

**Binodal Curve**

When a smooth curve passing through various points representing the compositions of the conjugate solutions is drawn, one gets a *binodal curve* bPc, with a maximum at M. In general, the plait point P lies either to the left or to the right of this maximum point M since various tie lines are not horizontal. The plait point will coincide with the maximum only in one particular case when the added constituent distributes equally in both the layers. In this particular case, various tie lines are horizontal and thus at the maximum point M, the two solutions have identical compositions and thus merge into each other to form a single solution.

**General Characteristics of the Phase Diagram**

The general characteristics of the phase diagram of a system exhibiting one pair of partially miscible liquids can now be summarized as follows:

***Point outside the binodal curve*** A point outside the binodal curve bPMc represents one liquid layer only. According to the phase rule, such a system can be completely defined if the values of two degrees of freedom, apart from temperature and pressure, are stated ($F = C - P + 2 = 3 - 1 + 2 = 4$, minus 2 of temperature and pressure, and thus $F = 4 - 2 = 2$). These two degrees of freedom are the two composition terms corresponding to any of the two components of the liquid phase.

***Point within the binodal curve*** A point inside the binodal curve represents a system in which the two partially miscible liquid solutions are formed. The compositions of these two solutions are the intersection points of the tie line drawn from the point under study with the binodal curve. The relative amounts of the two solutions are given by the lever rule. For example, at the point X′, we have

$$\text{Composition of the solution of B in C (i.e. richer in C)} = c'$$

$$\text{Composition of the solution of C in B (i.e. richer in B)} = b'$$

and

$$\frac{\text{Amount of the solution of B in C (i.e. richer in C)}}{\text{Amount of the solution of C in B (i.e. richer in B)}} = \frac{\text{b}'\text{X}'}{\text{c}'\text{X}'}$$

Applying the phase rule to a system within the binodal curve (where $P = 2$), we have

$$F = C - P + 2 = 3 - 2 + 2 = 3$$

Leaving two degrees of freedom corresponding to the constant temperature and pressure, we have

$$F = 1$$

and thus one concentration term, which allows the position of one layer to be fixed on the binodal curve, is sufficient to define the system completely.

## Effects Produced When the Component A is Added to a Binary System of B and C

As indicated earlier the state of the overall system will move along a line obtained by joining the state point, say X in Fig. 7.3.1, of the binary system with the apex A of the triangle. In general, the following two types of behaviour are observed depending on the original composition of the binary system.

***When the line* AX *does not pass through the plait point* P** Suppose the line AX lies to the right of the plait point P. Then as we move from X to A, we observe that the relative lengths of tie lines right to the line AX decrease steadily till the point Y is reached. The net effect that is produced is that the relative amounts of the two solutions change as more and more of A is added to the system; the amount of the solution of C in B decreases steadily whereas that of B in C steadily increases.† At the point Y, practically whole of the solution of C in B disappears and the system mainly consists of the solution of B in C. Above the point Y, the system remains that of one liquid only. Thus, the conversion of two liquid layers into one takes place through the vanishing of liquid layer of C in B. The same is also true if the line AX lies left to the point P. Here, it is the solution of B in C that gradually decreases and ultimately disappears completely.

***When the line* AX *passes through the plait point* P** The above behaviour of conversion of two liquid layers into one layer through the vanishing of one of the layers, however, is not observed when the line AX passes through the plait point P. In this case, both the solutions remain intact right up to the point P but their compositions approach each other and become identical at the point P. Consequently, the two liquid layers merge into each other at the point P to give a single liquid layer. Thus, the production of a single layer is not through the disappearance of one of the liquid layers but through the merger of the two solutions into each other at the point P. Similarly, the effects produced when component B (or C) is added to a mixture of A and C (or A and B) can be analysed.

## Effects Produced When the Proportion of A is Kept Constant and the Amounts of Other Two Components are Varied

The state of the system in which the proportion of A is kept constant and the

†Strictly speaking, after adding A, we get a ternary system of A. B and C. However, for the sake of discussion, we retain the original terminology of binary system. Thus for a ternary system, a solution of B in C means a solution of A and B in C or a solution that is richer in C.

amounts of the other two are varied will be represented by the points on a line drawn parallel to the side BC of the type xy shown in Fig. 7.3.1. The sequence of effects that is produced as one moves from x to y can be readily interpreted with the help of Fig. 7.3.1. The system continues to be a single layer till the binodal curve is reached where the single liquid splits into two layers x′ and y′; the former being the solution of C in B whereas the latter is of B in C. Now as one moves from x′ to y″, one finds that the line xy cuts various tie lines in different ratios as can be seen from the lengths of tie-lines below the line x′y″ and those lying above this line; the former increases gradually whereas the latter decreases. Consequently, the relative amounts of the two solutions vary as one moves from x′ to y″; the amount of the solution of B in C gradually increases whereas that of C in B gradually decreases. Besides this, composition of the solution of C in B moves along x′b′ whereas that of B in C moves along y′y″. At the point y″, the solution of C in B completely disappears and thus the system mainly consists of the solution of B in C. From y″ to y, the system continues to be a single liquid. The above sequence of conversion from one layer to two, and then back to one layer will be observed only for those liquids which contain lesser amount of component A as compared to that present in a solution corresponding to the maximum point M of the binodal curve.

If the original composition of the system happens to lie between P and M, it is obvious that the single solution that splits into two is of the one of B in C. The two solutions again merge on the other side of the binodal curve to the given back the solution of B in C. This behaviour is observed whenever the amount of A in the original system lies between *plait* point P and maximum point M. Such solutions are said to show *retrograde solubility.*

**Effect of Temperature**

Area within the binodal curve of a triangular plot represents the region of partial miscibility where two conjugate solutions are formed. If it be assumed that the mutual solubilities of these two solutions increase with increasing temperature, the compositions of the two solutions will come more close to each other when the temperature is increased. Consequently, the area enclosed by the binodal curve will become smaller and smaller. In order to express these effects adequately, we will have to construct triangular plots at different temperatures and then correlate them with each other or, alternatively, a 3-dimensional plot can be constructed with an equilateral triangle as a base and the vertical axis representing temperature of the system. This model, in fact, will be a triangular prism. Any horizontal section of such a three-dimensional diagram gives a triangular plot at a constant temperature. However, in the present case, the general behaviour can be understood by plotting a number of isothermal binodal curves in the same triangle as shown in Fig. 7.3.2. The binodal curve enclosing larger area corresponds to a lower temperature as the mutual solubilities of the two conjugate solutions are lesser at lower temperatures. At a sufficiently low temperature, the area covered may become so large that the binodal curve may cut the side AB as shown in Fig. 7.3.2 indicating that the two components A and B like B and C also become partially miscible with each other.

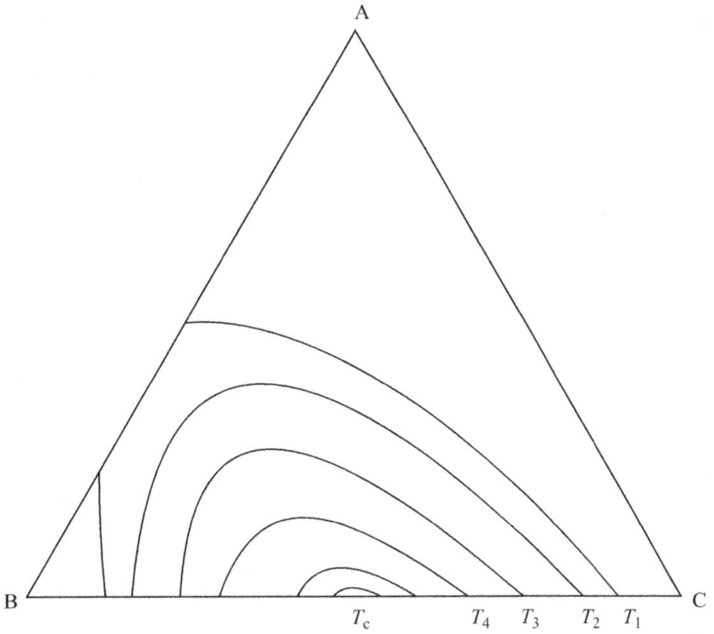


**Fig. 7.3.2 Effect of temperature on the binodal curve, $T_1 < T_2 < T_3 < T_4 < ... < T_c$**

## Critical Solution Temperature

When the temperature is raised, the area enclosed by the binodal curve becomes smaller and smaller and is ultimately reduced to a point at a sufficiently high temperature (say, at $T_c$). Thus, the three liquids will become completely miscible in all proportions at temperatures greater than $T_c$. The temperature $T_c$ for this reason is known as the *critical solution temperature* for the ternary system. This, however, is not true as on the line BC, where this point lies, we have a system of two components of B and C, and thus more precisely, temperature $T_c$ should be known as the *binary critical solution temperature.*

The true critical solution temperature for the ternary system is observed only when the various binodal curves have the general appearance as shown in Fig. 7.3.3. In this particular case, the binodal curves at high temperatures consists of closed curves and thus have two plaits points. The area of the closed curve decreases as the temperature is raised and ultimately reduces to a point K at some higher temperature. It is obvious that at temperatures higher than this, the three liquids will be completely miscible and thus the system consists of a single liquid layer only. The temperature corresponding to the point K is, in fact, the true critical solution temperature for the ternary system.

***Examples*** Examples of system yielding one pair of partial miscible liquids are

Acetic acid-chloroform-water (Fig. 7.3.1)
Aniline-phenol-water at 50 °C (Fig. 7.3.2)
Acetone-phenol-water (Fig. 7.3.3)

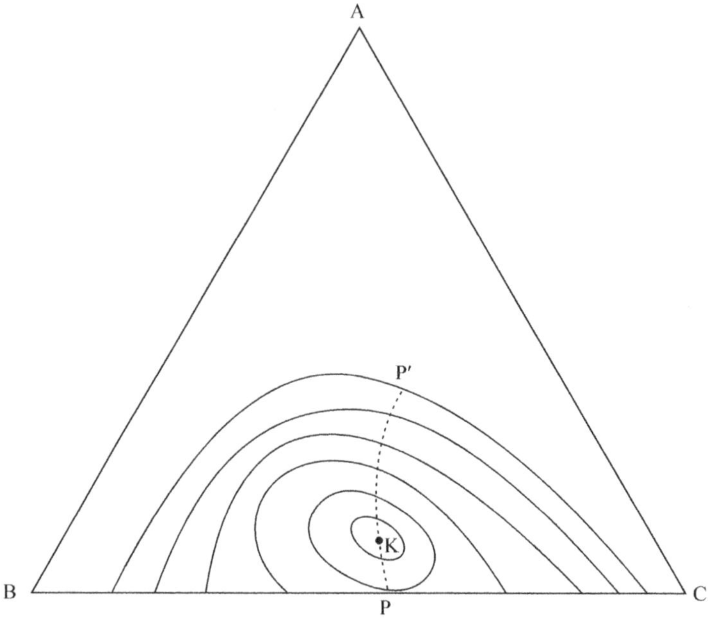


**Fig. 7.3.3 Triangular plot showing the critical solution temperature for the ternary system**

***Example 7.3.1***

For the ternary system benzene-isobutanol-water at 25 °C and 1 atm, the following compositions have been obtained for the two phases in equilibrium.

| *Water-rich phase* | | *Benzene-rich phase* | |
|---|---|---|---|
| *isobutanol* *mass %* | *water* *mass %* | *isobulanol* *mass %* | *benzene* *mass %* |
| 2.33 | 97.39 | 3.61 | 96.20 |
| 4.30 | 95.44 | 19.87 | 79.07 |
| 5.23 | 94.59 | 39.57 | 57.09 |
| 6.04 | 93.83 | 59.48 | 33.98 |
| 7.32 | 92.64 | 76.51 | 11.39 |

Plot these data on a triangular graph, indicating the tie lines.

(a) Estimate the compositions of phases that will be produced from a mixture of 20 mass % isobutanol, 55 mass % water and 25 mass % benzene and also calculate their relative amounts.

(b) What will be the composition of the principal phase if the first drop of the second phase separates when water is added to a solution of 80 mass % isobutanol in benzene?

***Solution***

The triangular plot obtained is shown in Fig. 7.3.4.

(**a**) The compositions of the two conjugate solutions produced at the given composition point a (20 mass % isobutanol + 55 mass % water and 25 mass % benzene) are

(i) Water-rich layer: 5.23 mass % isobutanol, 94.5 mass % water

(ii) Benzene-rich layer: 39.57 mass % isobutanol, 57.09 mass % benzene

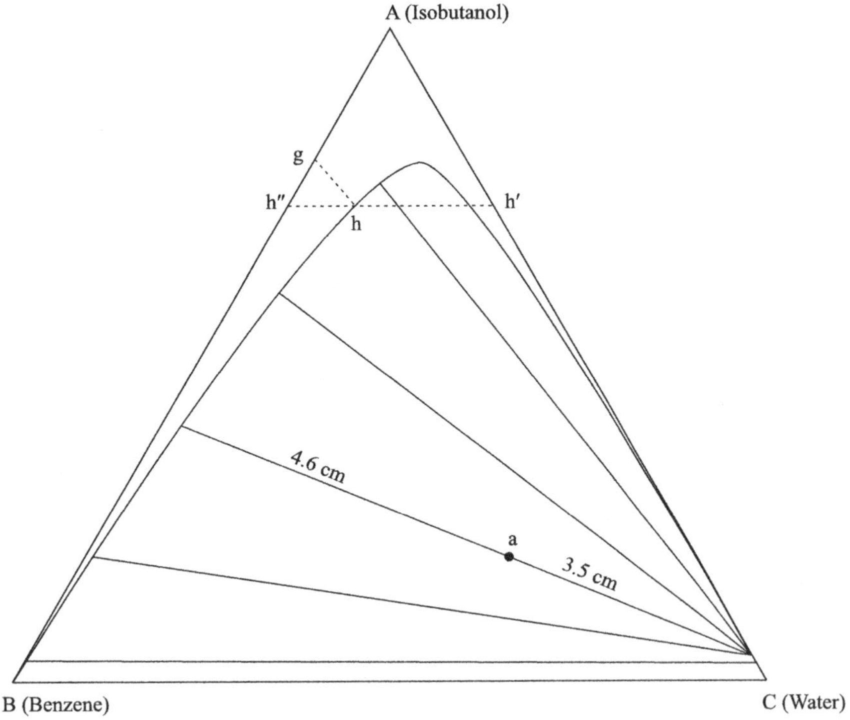


Fig. 7.3.4

The relative masses of the two solutions can be determined from the lever rule. Thus, we have

$$\frac{\text{Mass of water-rich layer}}{\text{Mass of benzene-rich layer}} = \frac{4.6 \text{ cm}}{3.5 \text{ cm}} = \frac{1.3}{1}$$

(b) Point g represents a solution of 80 mass % isobutyl alcohol in benzene. If water is added to this system, the state of the overall system will move along gh—a line obtained by joining g to the apex C. At the point h, the solution will just split into two. The composition of the system at h is

mass % benzene = hh′/mm = 18
mass % water = hh″/mm = 10
and mass % isobutanol = 100 − 18 − 10 = 72

## FORMATION OF TWO PAIRS OF PARTIALLY MISCIBLE LIQUIDS

### Expected Phase Diagram

Let us assume that out of the three components A, B and C, two pairs of liquids which show partial miscibilities be A and B, and B and C, and let the third pair of liquids A and C be completely miscible with each other. In such a case, the triangular plot will contain two binodal curves with their own plait points and various tie lines corresponding to each of these two pairs of partially miscible liquids as shown in Fig. 7.3.5. The binodal curve aQb′ is for the mixtures of two liquids A and B containing C and the curve bPc is for B and C containing A. Points Q and P are the respective plait points. The area enclosed within the curve aQb′ represents the region of compositions where two conjugate solutions of A

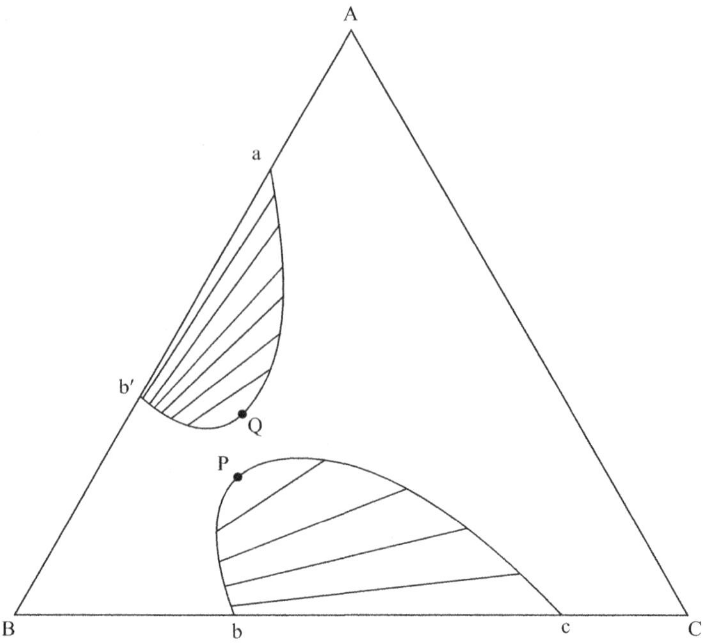


**Fig. 7.3.5 Triangular plot depicting the formation of two pairs of partially miscible liquids**

in B (containing C) and B in A (containing C) are formed; composition of the former lies on Qb′ and that of the latter on the line Qa. Compositions of the two conjugate solutions are joined with each other through various tie lines. At the point Q, the two solutions have identical compositions and thus merge into each other to give a single liquid layer. In a similar manner, the area enclosed within the curve bPc represents the region of compositions where two conjugate solutions of C in B (containing A) and B in C (containing A) are formed; composition of the former lies on bP and that of the latter on the line cP. Again compositions of the two conjugate solutions are joined with each other through various tie-lines drawn within this area. At point P, the two solutions have identical compositions and thus merge into each other to give a single liquid layer.

The area outside of these two binodal curves gives the region where the three components are completely miscible and thus form a single liquid layer.

**Example**

The system ethyl alcohol-water-succinic nitrile shows this type of behaviour between 18.5 °C to 31 °C and thus has a triangular plot similar to Fig. 7.3.5

**Effect of Temperature**

If it be assumed that the mutual solubilities decrease as temperature is lowered, the area covered by the two binodal curves will become larger and larger on decreasing the temperature. Ultimately a stage would be reached where the two curves may overlap with each other and thus form a band type diagram, such as shown in Fig. 7.3.6.

This diagram is very similar to the binodal curve of a system exhibiting one pair of partially miscible liquids at some low temperature (e.g. the binodal curve at $T_1$ of Fig. 7.3.2). It is thus obvious that if the binodal curve forms a band, it is not possible to predict the behaviour of the system when the temperature

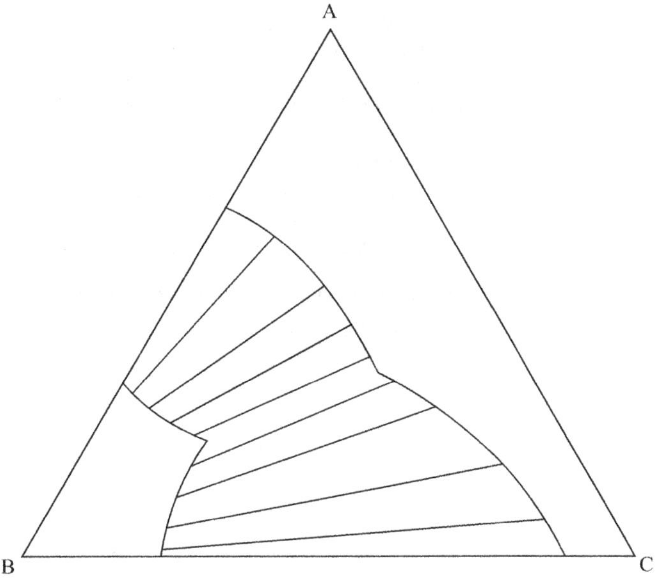


**Fig. 7.3.6 Overlapping of the two binodal curves**

is raised, that is, whether it will form a system containing only one pair or two pairs of partially miscible liquids.

Examples of systems exhibiting binodal bands are

Water-phenol-aniline
Water-ethyl acetate-*n*-butyl alcohol
Water-ethyl acetate-isobutyl alcohol
Water-ethyl acetate-*sec*-butyl alcohol

***Example 7.3.2***

Following are the compositions of the phases in equilibrium with each other in the system methylcyclohexane-aniline-*n*-heptane at 1 atm and 25 °C.

| *Hydrocarbon layer* | | *Aniline layer* | |
|---|---|---|---|
| *Methylcyclohexane* | *n-Heptane* | *Methylcyclohexane* | *n-Heptane* |
| *mass %* | *mass %* | *mass %* | *mass %* |
| 0.0 | 92.0 | 0.0 | 6.2 |
| 9.2 | 83.0 | 0.8 | 6.0 |
| 18.6 | 73.4 | 2.7 | 5.3 |
| 33.8 | 57.6 | 4.6 | 4.5 |
| 46.0 | 45.0 | 7.4 | 3.6 |
| 59.7 | 30.7 | 9.2 | 2.8 |
| 73.6 | 16.0 | 13.1 | 1.4 |
| 83.3 | 4.4 | 15.6 | 0.6 |
| 88.1 | 0.0 | 16.9 | 0.0 |

(a) Draw the triangular plot including tie lines.
(b) Characterize the system to which this triangle plot represents.
(c) Calculate the composition of the solution which just splits into two

obtained by adding sufficient amount of aniline to a solution of 40 mass % of methylcyclohexane in *n*-heptane.

(d) 100 g of a solution having composition 30 mass % methylcyclohexane + 28 mass % *n*-heptane was prepared. What is the nature of the system, whether it forms a single solution or two solutions? If two solutions are formed what will be their compositions and the respective amounts?

(e) If the system as given in part (d) represents two solutions, calculate the mass of methylcyclohexane which should be added so that the system becomes one of a single solution. What would be the composition of the resultant solution?

***Solution***

(**a**) The triangular plot obtained is shown in Fig. 7.3.7. The edge length of the triangle is 100 mm.

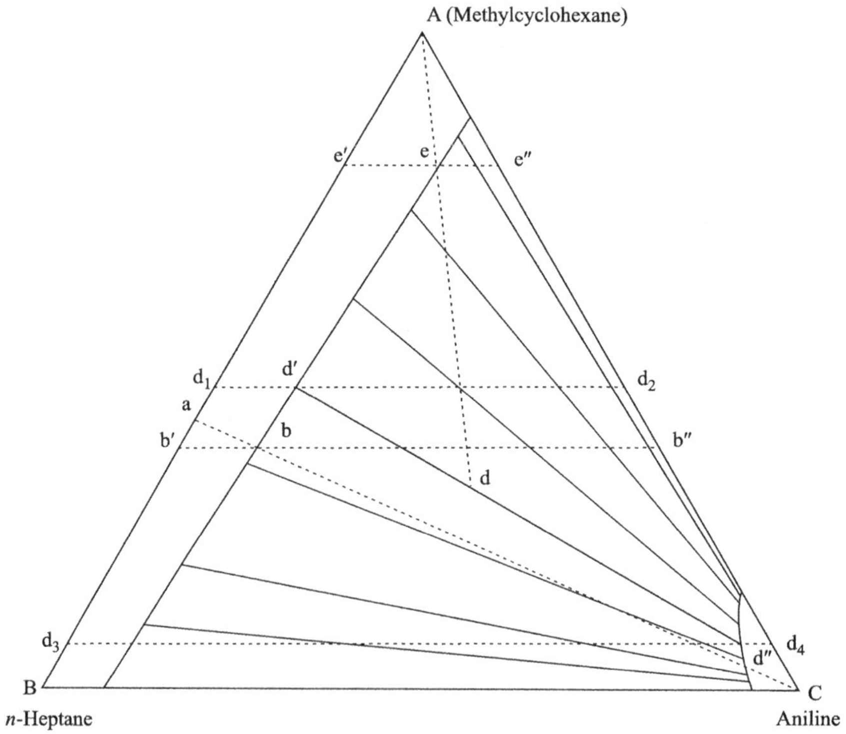


**Fig. 7.3.7 Required ternary plot**

(**b**) The present system represents a case where two pairs of partially miscible liquids are formed with an overlapping of the two binodal curves. The two pairs are

*n*-Heptane-Aniline
Methylcyclohexane-Aniline

The two liquids in the third pair, namely, methylcyclohexane and *n*-heptane, are completely miscible with each other.

(**c**) Point a on the line AB represents a binary system of 40 mass % methylcyclohexane in *n*-heptane. If aniline is added to this system, the state of the overall system will move along abC. At point b, the single solution will just split into two solutions. The composition of the system corresponding to the point b can be calculated as follows.

$$\text{Mass \% of aniline, } W_a = \frac{\text{length ab}}{\text{length aC}} \times 100 = \frac{7.5 \text{ mm}}{86.6 \text{ mm}} \times 100 = 8.662 = 9$$

Since the edge length of the triangle plot is 100 mm, the length bb′ in mm on the line b′bb″ will also represent the mass % of aniline. This length is 9 mm as may be verified from Fig. 7.3.7. Similarly, mass % of *n*-heptane will be equal to the length bb″, which is equal to 55 mm. Thus

$$\text{Mass \% } n\text{-heptane, } W_h = \text{length bb}''/\text{mm} = 55$$

and mass % methylcyclohexane, $W_m = 100 - W_a - W_h = 100 - 9 - 55 = 36$

(**d**) The point d represents the system of 30 mass % methylcyclohexane and 28 mass % *n*-heptane. The system will consist of two conjugate solution of compositions given below.

(i) *Hydrocarbon layer:* Its composition is given by the point d′ at which we have

$$\text{Mass \% aniline} = d_1d'/\text{mm} = 4$$
$$\text{Mass \% } n\text{-heptane} = d_2d'/\text{mm} = 45$$

(ii) *Aniline layer:* Its composition is given by the point d″ at which, we have

$$\text{Mass \% aniline} = d_3d''/\text{mm} = 89$$
$$\text{Mass \% } n\text{-heptane} = d_4d''/\text{mm} = 4$$

The relative amounts of the two solutions may be calculated by employing the lever rule as follows.

$$\frac{\text{Mass of hydrocarbon layer}}{\text{Mass of aniline layer}} = \frac{dd''}{dd'} = \frac{4.1 \text{ mm}}{2.8 \text{ mm}}$$

Adding one to both sides, we get

$$\frac{\text{Mass of hydrocarbon layer + Mass of aniline layer}}{\text{Mass of aniline layer}} = \frac{4.1 + 2.8}{2.8}$$

or
$$\frac{\text{Mass of aniline layer}}{\text{Total mass in the system}} = \frac{2.8}{6.9}$$

or
$$\text{Mass \% of aniline layer} = \frac{2.8}{6.9} \times 100 = 40.59$$

Thus mass % of hydrocarbon layer = 100 – 40.59 = 59.41

(**e**) When methylcyclohexane is added to a system represented by the point d, the state of the overall system will move along dA. At the point e, the two solutions will just merge into one. The composition of the solutions at e is

$$\text{Mass \% } n\text{-heptane} = ee''/\text{mm} = 8$$
$$\text{Mass \% } n\text{-aniline} = ee'/\text{mm} = 12$$
$$\text{Mass \% methylcyclohexane} = 100 - 12 - 8 = 80$$

The relative ratio of *n*-heptane and aniline at the point e will be the same as that of the point d. Thus 8 mass % of *n*-heptane at e represents 28 g of *n*-heptane (same as that at point d) and 12 mass % of aniline at e represents 42 g of aniline. Thus the mass of methylcyclohexane at the point d (80 mass %) is given as

$$\text{Mass of methylcyclohexane} = \frac{28\text{ g}}{8\%} \times 80\% = 280\text{ g}$$

Thus, the amount of methylcyclohexane that need to be added to get one solution

= 280 g – Mass of methylcyclohexane at the point d
= 280 g – 30 g = 250 g

## FORMATION OF THREE PAIRS OF PARTIALLY MISCIBLE LIQUIDS

### Expected Phase Diagram

Here all the three pairs A and B, B and C, and A and C show partial mutual solubilities with each other and thus the triangle plot will contain three binodal curves corresponding to each of these three pairs of partially miscible liquids (Fig. 7.3.8). Each binodal curve will have its own plait point and the various lies lines indicating the compositions of the two conjugate solutions that are formed.

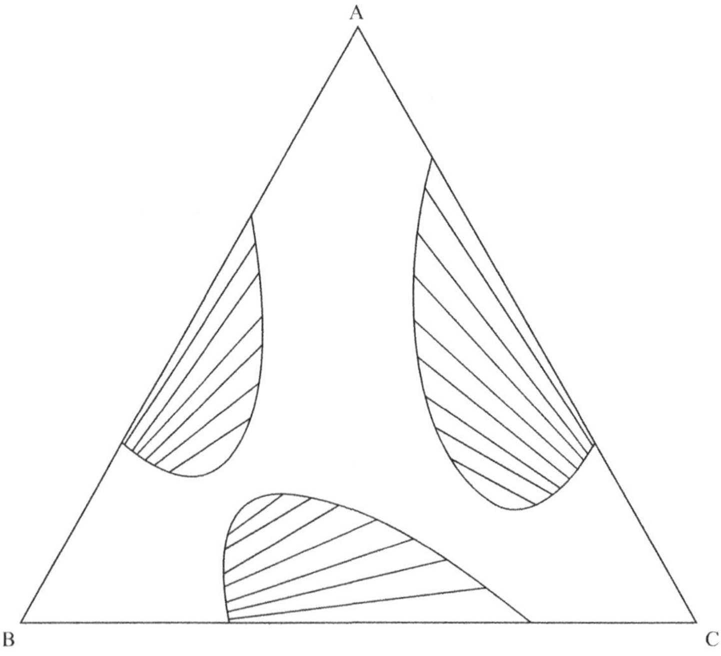


**Fig. 7.3.8 Triangular plot for the formation of three pairs of partially miscible liquids**

Interpretation of the phase diagram of Fig. 7.3.8 is straightforward. A point anywhere within the three binodal curves represents a system of the two conjugate solutions and that outside of these curves represents a single solution. At lower temperatures, the partial mutual solubilities of the two components become smaller, and thus the area enclosed by each of these binodal curved becomes larger as temperature of the system is lowered. Eventually a stage may be reached where the three binodal curves overlap each other and the triangular plot takes the form as given in Fig. 7.3.9. A point anywhere in the area (marked as 1) near each of the three corners of the triangle represents one liquid phase,

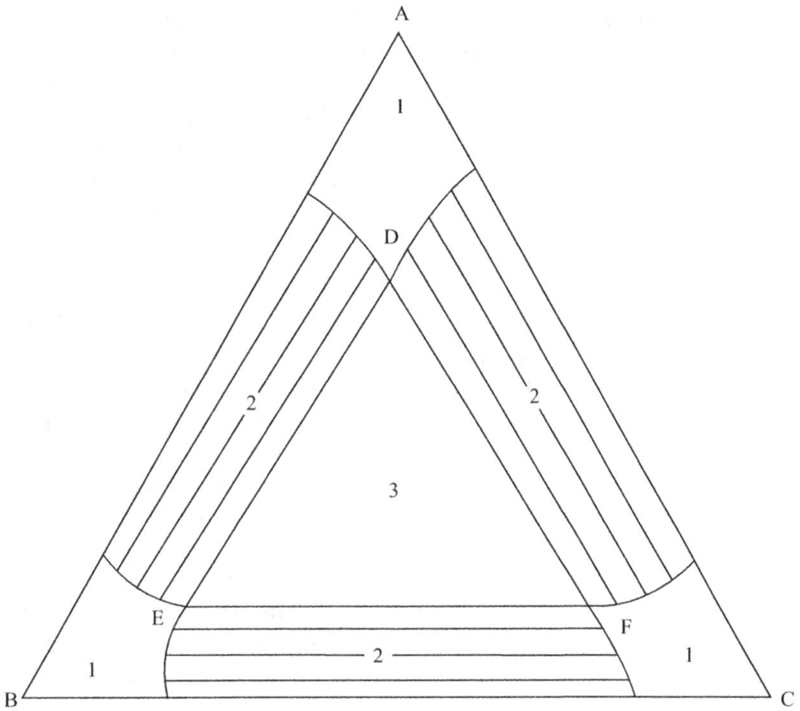


**Fig. 7.3.9 Overlapping of the three binodal curves**

whereas any point within the areas (marked as 2) traversed by tie lines implies two conjugate liquids whose compositions are given by the two ends of the tie line passing through the given point. Area within the smaller triangle DEF (marked as 3) represents the region where three liquid layers in equilibrium are formed, their compositions are given by the points D, E and F, respectively.

The compositions of these three liquids are definite at a given temperature and pressure; this fact is in agreement with the phase rule since for $P = 3$, $F = C - P + 0 = 3 - 3 + 0 = 0$.

An example of a system yielding three liquid phases in equilibrium is succinic nitrile-ether-water.

## 7.4 TRIANGULAR PLOTS OF A TERNARY SYSTEM DEPICTING CRYSTALLIZATION OF ITS COMPONENTS AT VARIOUS TEMPERATURES

It was stated in Section 7.3 that the triangular plots of a ternary system at various temperatures can be represented in a 3-dimensional model of a triangular prism. In this model, we have a base of an equilateral triangle and a vertical axis which represents temperature of the system. Thus, any horizontal section of this 3-dimensional model gives the triangular plot at that particular temperature. The understanding of the planar model involving the crystallization of the solid components becomes much simplified if the qualitative nature of the above 3-dimensional model is studied at a glance. In this model, we have the following characteristics.

**Depiction of Pure Components**

Each *vertical axis* represents the pure component at different temperatures. Thus, the axis AA′ in Fig. 7.4.1 represents the component A only, while BB′ and CC′ represent the components B and C, respectively. On each of these axes, the melting point of the corresponding component can be shown. The points a, b and c in Fig. 7.4.1 represent the melting points of the components A, B and C, respectively. Thus, the component above its melting point is present in the liquid form and that below this point is present in the solid form.

**Depiction of Two-Component System**

Each *vertical face* of the triangular prism represents a two-component system involving components occupied by the two corners of the triangle. Thus, the vertical faces AA′B′BA, BB′C′CB and AA′C′CA represent the two-component systems of A-B, B-C and A-C, respectively. On each of these faces, one can draw the binary equilibrium phase diagram of the respective two-component system. If, at present, it is assumed that the three two-component systems show simple phase diagrams involving only the formation of eutectics with no compound formation and solid solution, then the general appearance of the three two-component phase diagrams will be as shown in Fig. 7.4.1. The curve adb is the equilibrium phase diagram of A and B; along the line ad, solid A separates out and along bd solid B separates out. At point d, both A and B separate out simultaneously. It is thus the eutectic point of the system A-B. Similarly, the curves bec and afc are the equilibrium diagrams of the systems B-C and A-C, respectively, and the points e and f are the respective eutectic points.

**Depiction of Three-Component System**

If the third component, say A, is added to a system B-C, the resultant system will be represented by a point that lies within the triangular prism. Addition of the third component will also affect the equilibrium curves depicting the start of crystallization of the two components. In fact, the addition of A to B-C will shift the eutectic point e to g along the line eg while the addition of B to A-C will shift the eutectic point f to g along the line fg and that of C into A-B will shift the eutectic point d to g along the line dg.

**Crystallization of One component from Ternary System**

The three lines dg, eg and fg divide the total surface area of the prism into three distinct surfaces begdb, cegfc and adgfa. Previously, along the lines be and bd, we have the separation of solid B from the binary liquid mixtures of B-C and A-B, respectively, and now the separation of B in the ternary system is represented by the surface area begdb. Similarly, the separation of A and C is represented by the points on the surface areas adgfa and cegfc, respectively. Thus, these three surfaces describe the conditions of equilibrium of the melt and one of the solids; melt $\rightleftharpoons$ solid A on the surface adgfa, melt $\rightleftharpoons$ solid B on the surface begdb and melt $\rightleftharpoons$ solid C on the surface cegfc.

The fact that the separation of the solid in a ternary system is represented by the surface is consistent with the requirements of the phase rule since $P = 2$ and thus $F = C - P + 1 = 3 - 2 + 1 = 2$. Thus, two degrees of freedom must be stated in order to define the system completely. These are temperature and the amount fraction of one of the components.

At temperature above any one surface, the system consists of only melt and the moment the surface is reached on cooling the system, the pure component

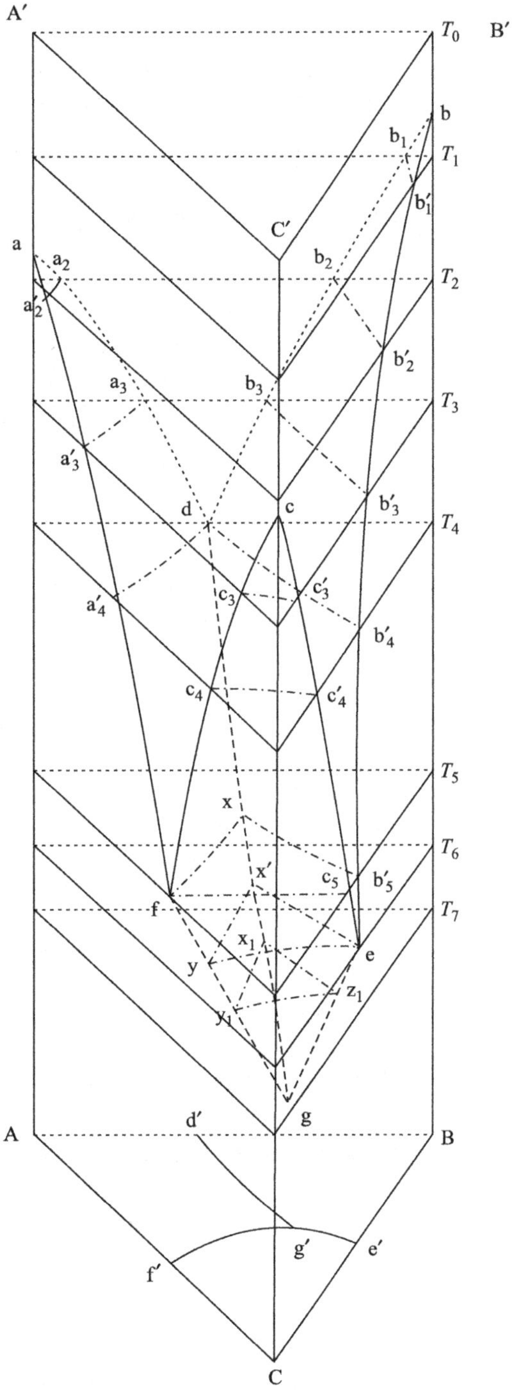


**Fig. 7.4.1 Triangular plots at various temperatures on a 3-dimensional triangular prism**

starts crystallizing. Thus, on cooling, the first crystal that separates out will consist of pure A if the original composition of the system lies above the surface of adgfa, it is pure B if the composition lies above the surface of begdb and is pure C if the composition lies above the surface cegfc.

**Crystallization of Two Components from a Ternary System**

The surfaces begdb and cegfc intersect each other along the line eg. Thus, the line eg represents various equilibria involving solid B, solid C and the melt. The line eg, in fact, is the ternary eutectic curve. In a similar manner the lines dg and gf are other eutectic curves where solids A and B, and solids A and C are in equilibrium with the melt, respectively. Thus, these three lines eg, dg and gf indicate the conditions for separation of the two solid phases from the liquid of different compositions. Systems along these lines are univariant as $P = 3$ and thus $F = C - 1 + 1 = 3 - 3 + 1 = 1$.

**Crystallization of all the Three Components**

The three ternary eutectic eg, dg and fg curves meet one another at the common point g which now represents a system when three solids A, B and C are in equilibrium with the melt of composition *g*. According to the phase rule, the point g is an invariant point since $P = 4$ and thus $F = C - P + 1 = 3 - 4 + 1 = 0$. Thus, temperature and composition of the melt corresponding to the point g are quite definite and are independent of the relative amounts of the three components. At all temperatures below the temperature corresponding to the point g, the system simply consists of a mixture of three solids A, B and C.

**Depiction of Crystallization on an Equilateral Triangle**

In order to show the process of initial crystallization from ternary melts of various compositions on a two-dimensional model of an equilateral triangle, we project out the boundaries eg, dg and fg separating the three surfaces on the base of the triangular prism as shown in Fig. 7.4.1. These projections are e′g′, d′g′ and f′g′ and they divide this phase triangle into three areas.

Crystals of pure component A freeze out first of all from systems whose compositions lie within the area Ad′g′f′A and components B and C freeze out respectively from systems whose compositions lie within the other two areas. If the composition of the system corresponds to any of the three lines, then on cooling two components will appear simultaneously; along d′g′, the components are A and B, along f′g′, the components are A and C and along e′g′, we have B and C. If composition of the system corresponds to the point g, then on cooling all the three components will crystallize simultaneously the moment point g is reached.

### Description of the System at Different Temperatures

In order to exhibit the behaviour of a ternary system at any particular temperature, we draw a horizontal plane passing through the three-dimensional triangular prism corresponding to the given temperature. This plane may or may not intersect the three surface areas adgfa, bdgeb and cfgec depending upon the temperature of the system. In Fig. 7.4.1, we have also drawn planes corresponding to the temperatures $T_0$, $T_1$, $T_2$, $T_3$, $T_4$, $T_5$, $T_6$ and $T_7$. We will now consider the behaviour of the system at these temperatures one by one.

**At Temperature $T_0$**

The temperature $T_0$ is greater than the respective melting points of the three components with the result that the horizontal plane at this temperature does not intersect any of the three surfaces. Therefore, no crystallization takes place and the ternary system consists of one liquid layer.

**At Temperature $T_1$**

The temperature $T_1$ is greater than the melting points of A and C and is smaller than that of B. Consequently, the plane drawn at this temperature intersects only the surface bdgeb along the curve $b_1b_1'$ as shown in Fig. 7.4.1. The equilateral triangle along with this intersection curve is also shown separately in Fig. 7.4.2. The curve $b_1b_1'$ gives the compositions of various systems at which initial solidification of B will start at the given temperature of $T_1$. Since the plane under study does not intersect the other two surfaces (i.e. adgfa and cegfc), crystallization of A and C will not take place. The portion $Ab_1b_1'CA$ of the triangle (Fig. 7.4.2) lies above the surface areas adgfa and cegfc and, therefore, represents the region where the system can exist only in the molten state consisting of only one liquid layer. The portion $b_1B_1b_1'b_1$ of the triangle lies below the surface area begdb but above the points e and d. Thus at a point anywhere in this region, we have solid B in equilibrium with the melt. The composition of the latter lies on the curve $b_1b_1'$. Take, for example, the point n in the region $b_1Bb_1'b_1$. This point represents a system in which solid B is in equilibrium with the melt of composition *m*. The latter can be obtained by joining the apex B with the point n and then extending the line Bn to m where it meets the curve $b_1b_1'$. The relative amounts of the two phases, viz., solid B and the melt, can be determined as usual with the help of the lever rule. For example, at n, we have

$$\frac{\text{Amount of solid phase B}}{\text{Amount of melt}} = \frac{\text{mn}}{\text{nB}}$$

It may be pointed out here that for an ideal ternary system, the line $b_1b_1'$

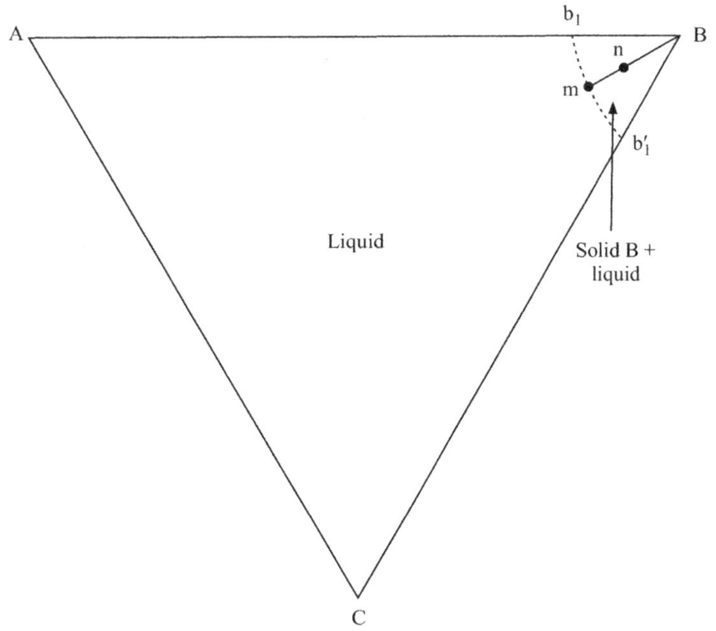


**Fig. 7.4.2 Triangular plot at the temperature $T_1$ just below the melting point of B but above the melting points of A and C**

of Fig. 7.4.2 would be a straight line parallel to the side AC of the equilateral triangle. The implication of this parallel line is that the precipitation of the solid B from the melt depends solely on the amount fraction of B in the solution and is independent of the nature of the components A and C. In other words, it is the total concentration of A and C that determines the melting point depression of B rather than the nature of A and C. Thus, if the amount fraction of B in the liquid phase of a binary solid-liquid equilibrium has the value of $x_{B(liquid)}$, its melting point will be lowered by an amount given by the simple van't Hoff equation. In a ternary system where the amount fraction of B is again $x_{B(liquid)}$, the melting point of B will be lowered by precisely the same amount as in the binary case, the sum of total of the amount fractions of A and C acting as if only a single additional component were present. In real cases, this situation is perturbed and thus the line $b_1b_1'$ may be curved as shown in Fig. 7.4.2.

## At Temperature $T_2$

The temperature $T_2$ is greater than the melting point of the component C and is smaller than those of A and B. Consequently, the plane drawn at this temperature intersects two surfaces begdb and adgfa along the curves $b_2b_2'$ and $a_2a_2'$, respectively, and does not intersect with the third surface cegfc. The curve $b_2b_2'$ gives compositions of various systems at which initial solidification of B will start at the given temperature of $T_2$ while the curve $a_2a_2'$ gives compositions at which the initial solidification of A will start. Since the plane at temperature $T_2$ lies above the melting point of the third component C. vis-à-vis, does not intersect the curve cegfc, crystallization of the component C will not take place at temperature $T_2$.

The equilibrium curve obtained at this temperature is also shown in Fig. 7.4.3. The interpretation of the diagram is described below

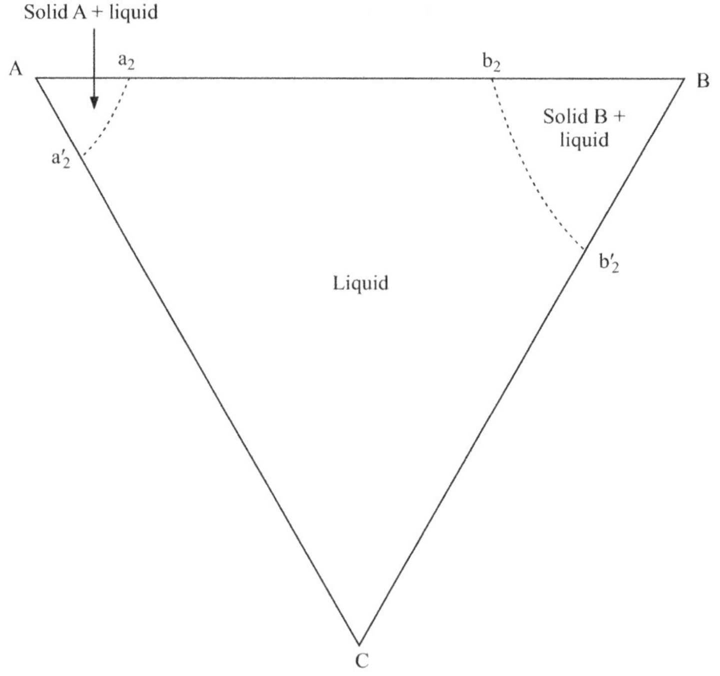


**Fig. 7.4.3 Triangular plot at temperature $T_2$ which lies below $T_A^*$ and $T_B^*$ but above $T_C^*$**

*Apex* A: Solid component A.
*Apex* B: Solid component B.
*Apex* C: Liquid component C.
*Side* $a_2b_2$: Binary liquid mixtures of A and B.
*Side* $b'_2C$: Binary liquid mixtures of B and C.
*Side* $a'_2C$: Binary liquid mixtures of A and C.

*Area* $Aa_2a'_2A$: A point in this region represents solid A in equilibrium with the melt whose composition lies on the line $a_2a'_2$.

*Area* $Bb_2b'_2B$: A point in this region represents solid B in equilibrium with the melt whose composition lies on the line $b_2b'_2$.

*Area* $Ca'_2a_2b_2b'_2C$: A point in this region represents a liquid system consisting of a single layer only.

**At Temperature $T_3$**

The temperature $T_3$ is smaller than all the three melting points of the three components, vis-à-vis, the plane drawn at this temperature intersects all the three surfaces adgfa, begdb and cegfc along the curves $a_3a'_3$, $b_3b'_3$ and $c_3c'_3$, respectively. The equilibrium curve obtained is also shown in Fig. 7.4.4. The interpretation of the diagram can be done as usual, only the main features are described below.

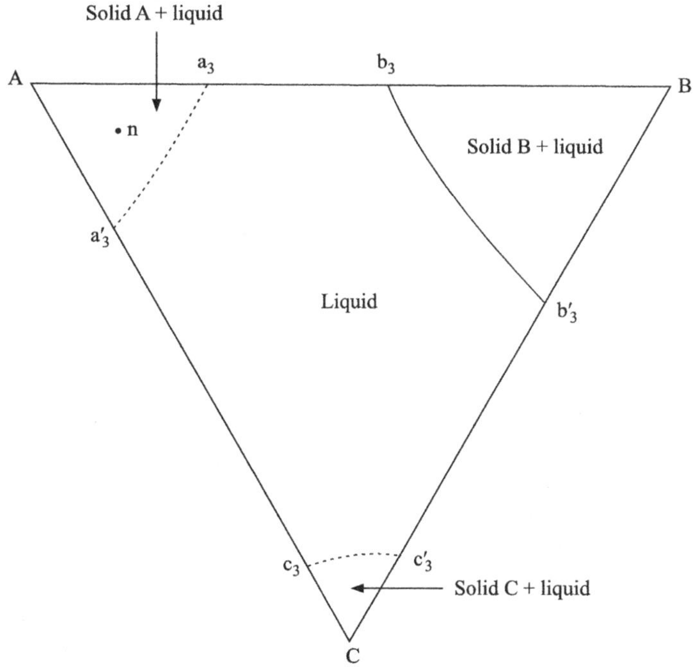


**Fig. 7.4.4 Triangular plot at the temperature $T_3$ which lies below $T_A^*$, $T_B^*$ and $T_C^*$**

*Curve* $a_3a'_3$: A point on this line represents a system where the crystallization of solid A has just started from the melt of composition corresponding to the given point.

*Area within the curve* $Aa_3a'_3A$: A point (say, n) in this region represents a system where again solid A is in equilibrium with the melt. The composition of the latter will lie on the line $a_3a'_3$ and can be obtained by extending the line An up to the curve $a_3a'_3$. The relative amounts of the two phases can be determined with the help of the lever rule.

In a similar manner, along the curves $b_3b'_3$ and $c_3c'_3$ the precipitation of solids B and C, respectively, has just started. A point within the curve $Bb_3b'_3B$ (or $Cc_3c'_3C$) represents a system where solid B (or solid C) is in equilibrium with the melt whose composition lies on the curve $b_3b'_3$ (or $c_3c'_3$). The latter can be obtained by joining the apex B (or C) with the given point and extending it to the curve $b_3b'_3$ (or $c_3c'_3$). The intersecting point gives the required composition of the melt.

*Area within the curve* $a_3a'_3c_3c'_3b'_3b_3a_3$: A point inside this region represents a system of a single liquid layer.

**At Temperature $T_4$**

This temperature corresponds to the eutectic temperature of the A-B system. The equilateral triangle at this temperature intersects the two surfaces adgfa and bdgeb along $a'_4d$ and $b'_4d$, respectively. The two intersection curves $a'_4d$ and $b'_4d$ meet each other at the eutectic point d which lies on the side AB of the triangle, the equilibrium curve at this temperature is also shown in Fig. 7.4.5. The interpretation of the diagram can be done as usual, though the main features are described below.

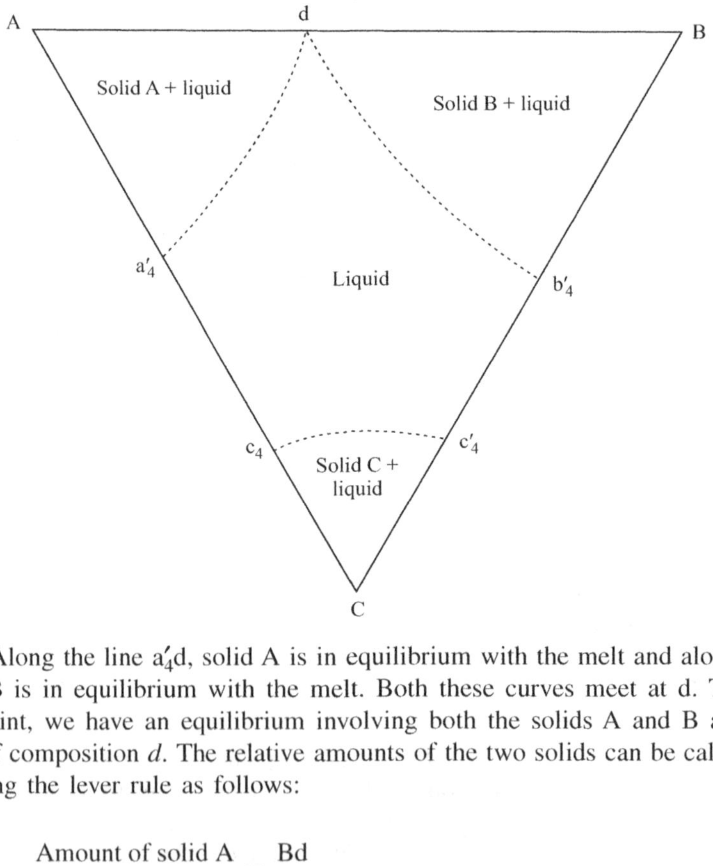


**Fig. 7.4.5 Triangular plot at the temperature $T_4$ which corresponds to the eutectic temperature of binary system A and B**

Along the line $a'_4d$, solid A is in equilibrium with the melt and along $b'_4d$, solid B is in equilibrium with the melt. Both these curves meet at d. Thus at this point, we have an equilibrium involving both the solids A and B and the melt of composition *d*. The relative amounts of the two solids can be calculated by using the lever rule as follows:

$$\frac{\text{Amount of solid A}}{\text{Amount of solid B}} = \frac{\text{Bd}}{\text{Ad}}$$

**At Temperature $T_5$**

This temperature lies below the eutectic temperature of the system A-B and also corresponds to the eutectic temperature of the system A-C. The equilateral triangle at this temperature intersects the two surfaces adgfa and begdb along fx and $b_5'x$, respectively. These two intersection curves meet each other at the point x which lies on the line dg. The triangle intersects the surface cegfc along the curve $fc_5$. The intersection curves fx and $fc_5$ meet each other at the eutectic point f. These three intersection curves are shown separately in Fig. 7.4.6. The interpretation of the diagram can be done as usual, however, the main features are now described.

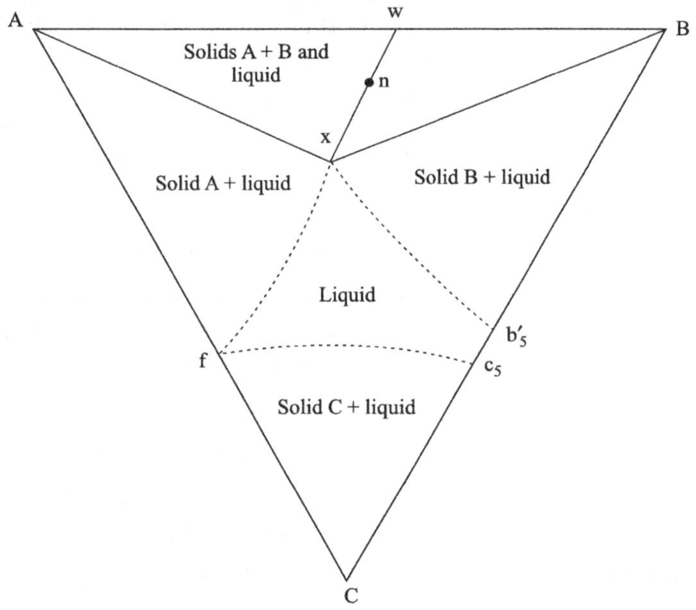


**Fig. 7.4.6 Triangular plot at the temperature $T_5$ which lies at the eutectic point of A and C system as well as below that of A and B**

(i) Along the curve fx, the initial solidification of A starts and within the area AxfA, we have solid A in equilibrium with the melt. The composition of the latter lies on the line fx and can be obtained by joining the apex A with the point under study and then extending it to the curve fx. The relative amounts of solid and melt, as usual, can be obtained with the help of the lever rule.

(ii) A point on the curve $b_5'x$. represents a system in which solidification of B has just started from the melt whose composition corresponds to the point under study. Within the area $Bb_5'xB$, we have solid B in equilibrium with the melt. The composition of the latter lies on the curve $b_5'x$ and can be obtained by joining apex B with the point under study and then extending it to the curve $b_5'x$.

(iii) The curve $c_5f$ represents systems where solidification of C has just started from the melt whose composition lies on the curve $c_5f$. A point anywhere within the area $fc_5Cf$ represents an equilibrium of solid C with the melt whose composition again lies on the curve $c_5f$.

(iv) The point f is the eutectic point of the system A and C. Thus at f both solids A and C are in equilibrium with the melt of composition *f*. The relative amounts of the two solids will be given by the lever rule.

(v) The point x lies on the line dg and thus represents a system where both solids A and B are in equilibrium with the melt of composition *x*. At this point, solidification of mixture of A and B have just started.

(vi) A point inside the region ABxA corresponds to a system where two solids A and B are in equilibrium with the melt of composition *x*. The amounts of the melt and the solid components in equilibrium can be determined with the help of the lever rule. Take, for example, the point n within this triangle (Fig. 7.4.6). This point will represent an equilibrium involving solid A, solid B and the melt of composition *x*. The relative amounts of the total solid present and the melt can be determined from the lever rule by drawing a line xn and then extending it to the side AB. If the line xn intersects AB at point w, then we have

$$\frac{\text{Total amount of solid separated}}{\text{Amount of melt}} = \frac{\text{nx}}{\text{nw}}$$

Knowing the total amount of crystals separated, we can determine the amount of each of the crystalline components by applying again the lever rule at w on the line AB. Thus

$$\frac{\text{Amount of A}}{\text{Amount of B}} = \frac{\text{wB}}{\text{wA}}$$

(vii) A point in the middle portion $b_5'xfc_5b_5'$ of the triangle (Fig. 7.4.6) represents a system of a single liquid layer.

**At Temperature $T_6$**

This temperature lies below the eutectic temperatures of systems A–B and A–C and also corresponds to the eutectic temperature of system B–C. The equilateral triangle at this temperature intersects the three surfaces adgfa, begdb and cegfc along x′y, ex′ and ey, respectively. The curves ex′ and ey meet each other at point e which lies on the side BC. Since the temperature $T_6$ lies below the eutectic temperatures of A–B and A–C systems, the intersection curve x′y does not cut the sides AB and AC but lies within the triangle. This curve meets the solubility curves of B and C (i.e. ex′ and ey) at points x′ and y, respectively. The three intersection curves are shown separately in Fig. 7.4.7. The interpretation of the diagram can be done as usual, however, the main features are described below.

| | |
|---|---|
| *Line* ex′ | Solubility curve of B. |
| *Line* ey | Solubility curve of C. |
| *Line* x′y | Solubility curve of A. |
| *Area* Bex′B | Solid B is in equilibrium with the melt whose composition lies on the line, ex′ |
| *Area* CeyC | Solid C is in equilibrium with the melt whose composition lies on the line ey. |
| *Area* Ax′yA | Solid A is in equilibrium with the melt whose composition lies on the line x′y. |

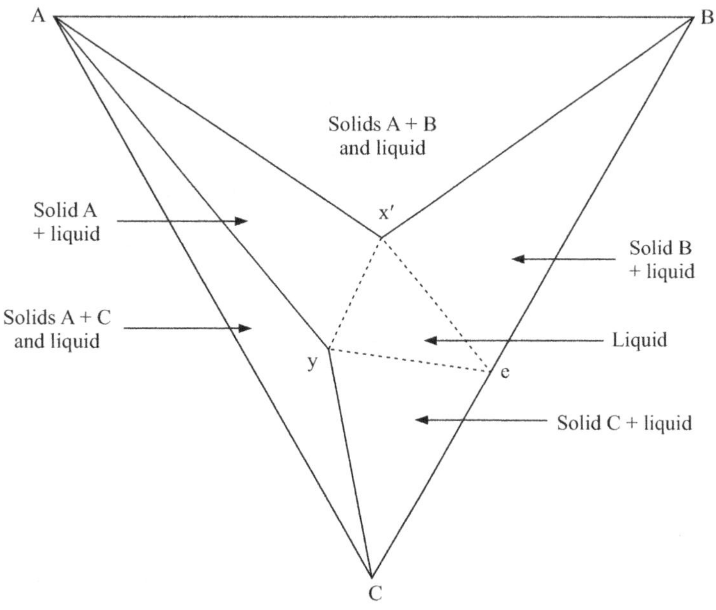


**Fig. 7.4.7 Triangular plot at temperature $T_6$ which lies at the eutectic point of B–C system as well as below those of A–B and A–C systems**

*Area* Ax′BA: Solids A and B are in equilibrium with the melt of composition corresponding to the point x′.

*Area* AyCA: Solids A and C are in equilibrium with the melt of composition corresponding to the point y.

*Area* x′yex′: Single liquid phase.

## At Temperature $T_7$

This temperature lies below all the three eutectic temperatures. Thus, the equilateral triangle at this temperature intersects the three surfaces adgfa, begdb and cegfc of Fig. 7.4.1 within the triangle along $x_1y_1$, $x_1z_1$ and $y_1z_1$, respectively. In Fig. 7.4.8 these three lines are shown by x′y′, x′z and y′z respectively. The main features of the diagram are described below.

*Area* Ax′y′A: Solid A is in equilibrium with melt whose composition lies on the curve x′y′.

*Area* Bx′zB: Solid B is in equilibrium with the melt whose composition lies on the curve x′z.

*Area* Cy′zC: Solid C is in equilibrium with the melt whose composition lies on the curve y′z.

*Area* Ax′BA: Solids A and B are in equilibrium with the melt of composition corresponding to the point x′.

*Area* Ay′CA: Solids A and C are in equilibrium with the melt of composition corresponding to the point y′.

*Area* BzCB: Solids B and C are in equilibrium with the melt of composition corresponding to the point z.

*Area* x′y′zx′: Single liquid phase.

## Crystallization Scheme At a Glance

Figures 7.4.2 to 7.4.8 have been shown together in Fig. 7.4.9 along with the projections of binary eutectic points d, e and f, ternary eutectic point g, and the three ternary eutectic curves dg, fg and eg. A few comments about the behaviour of the system at a particular temperature may be made here.

(i) If the temperature of the system is above the melting point of any one component, then the crystallization of that component will not take place at the given temperature, i.e. the triangular plot will not include the solubility curve for that component.

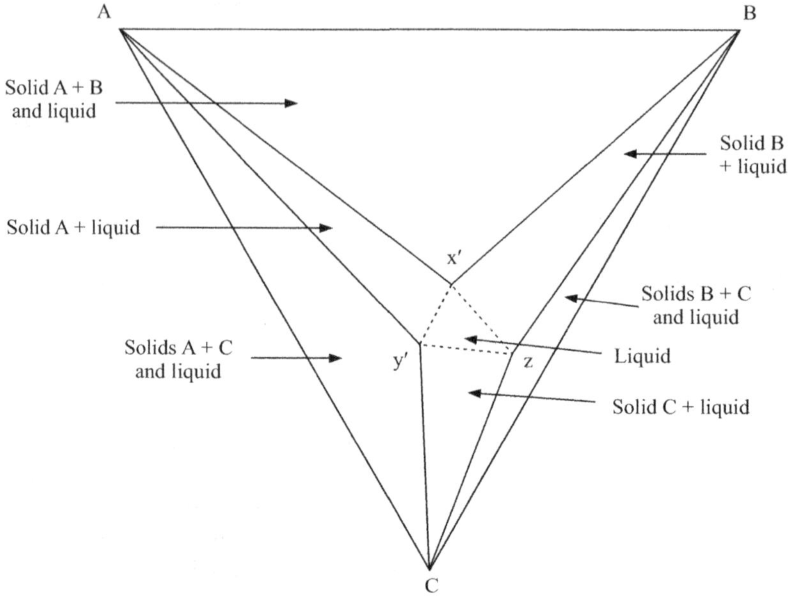


**Fig. 7.4.8 Triangular plot at temperature $T_7$ which lies below the eutectic temperatures of A–B, A–C and B–C systems**

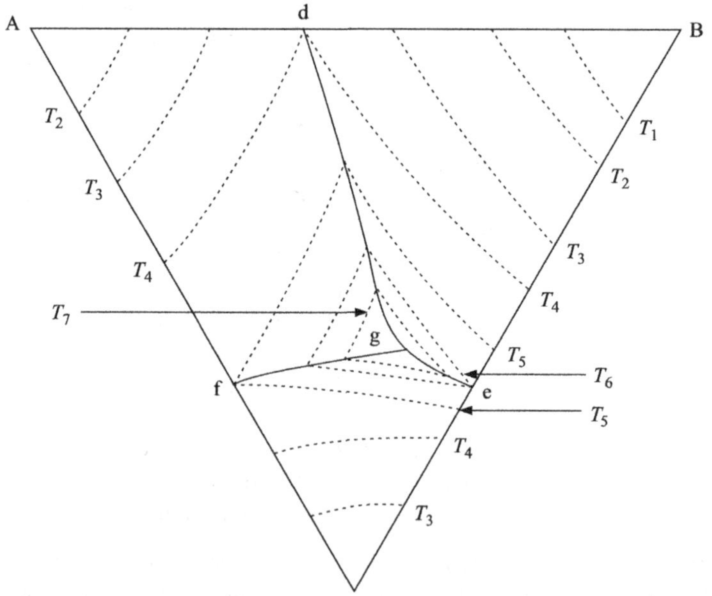


**Fig. 7.4.9 Triangular plots at different temperatures on a single triangle**

(ii) The triangular plot will contain as many solubility curves as the number of components having their freezing points above the temperature at which the system is being studied. These solubility curves will not meet each other if the temperature of the system is greater than the three binary eutectic temperatures and thus the triangular plot will have separate regions of crystallizations.

(iii) The solubility curves will meet each other on a side of the triangle if the temperature of the system is the same as that of the binary eutectic point for a system involving the two components which lie on either ends of the involved side. If the temperature of the system is lesser than the binary eutectic temperature but greater than the ternary eutectic temperature, the two solubility curves of components of the binary system will meet each other on the ternary eutectic curve which lies within the triangle. In such a case, the triangular plot will also include a region where the two solids can exist in equilibrium with a solution of constant composition.

(iv) If the temperature of the system lies below the eutectic temperatures of A–B, A–C and B–C systems but above the ternary eutectic temperature, the solubility curves of A, B and C will intersect each other on the respective ternary eutectic curves and thus lie within the triangle. In this case, the triangle plot will include the regions where (a) pure components are in equilibrium with the respective solubility curves, and (b) two components, viz., A and B, A and C, and B and C, are in equilibrium with the respective solution of fixed composition.

(v) If the temperature of the system is less than the ternary eutectic temperature, the system merely consists of a mixture of the three solids.

**Effects Produced on Cooling a Ternary System**

Suppose we have a ternary system represented by the point h of Fig. 7.4.10. Let the temperature of the system at the point h be $T_0$. Since the point h lies above the solid A-liquid surface afgda, the system is in the liquid state.

On lowering the temperature, the system moves along the line hi and it remains in the liquid state till the temperature $T_1$ is reached. At this temperature the system has reached to solid A-liquid surface afgda. Thus solid A starts precipitating.

On lowering the temperature further, the system moves below the surface afgda and thus represents solid A in equilibrium with the liquid. Since only solid A precipitates, the amounts of the components B and C in the liquid phase will remain unaffected. That is, so long as only solid A is in equilibrium with liquid, the ratio B : C in the liquid would remain constant. Thus during cooling of the system, the composition of the liquid phase is shifted on the surface afgda along the line ijkl which represents the constant ratio of B/C in the liquid phase. If the projections of the points i, j, k and l are made on the equilateral triangle, the projected points i, j, k and l would make a straight line which on extending will pass through the point A. This is shown at the bottom of Fig. 7.4.10. Each system on the line Aijkl represents the liquid composition with constant ratio of B/C and different amounts of A. Since the line ijkl passes through the apex A, it is the solid A which is in equilibrium with the liquid. At the point i, the temperature of the system is $T_1$, and at this point, solid A just starts precipitating. At the point j,

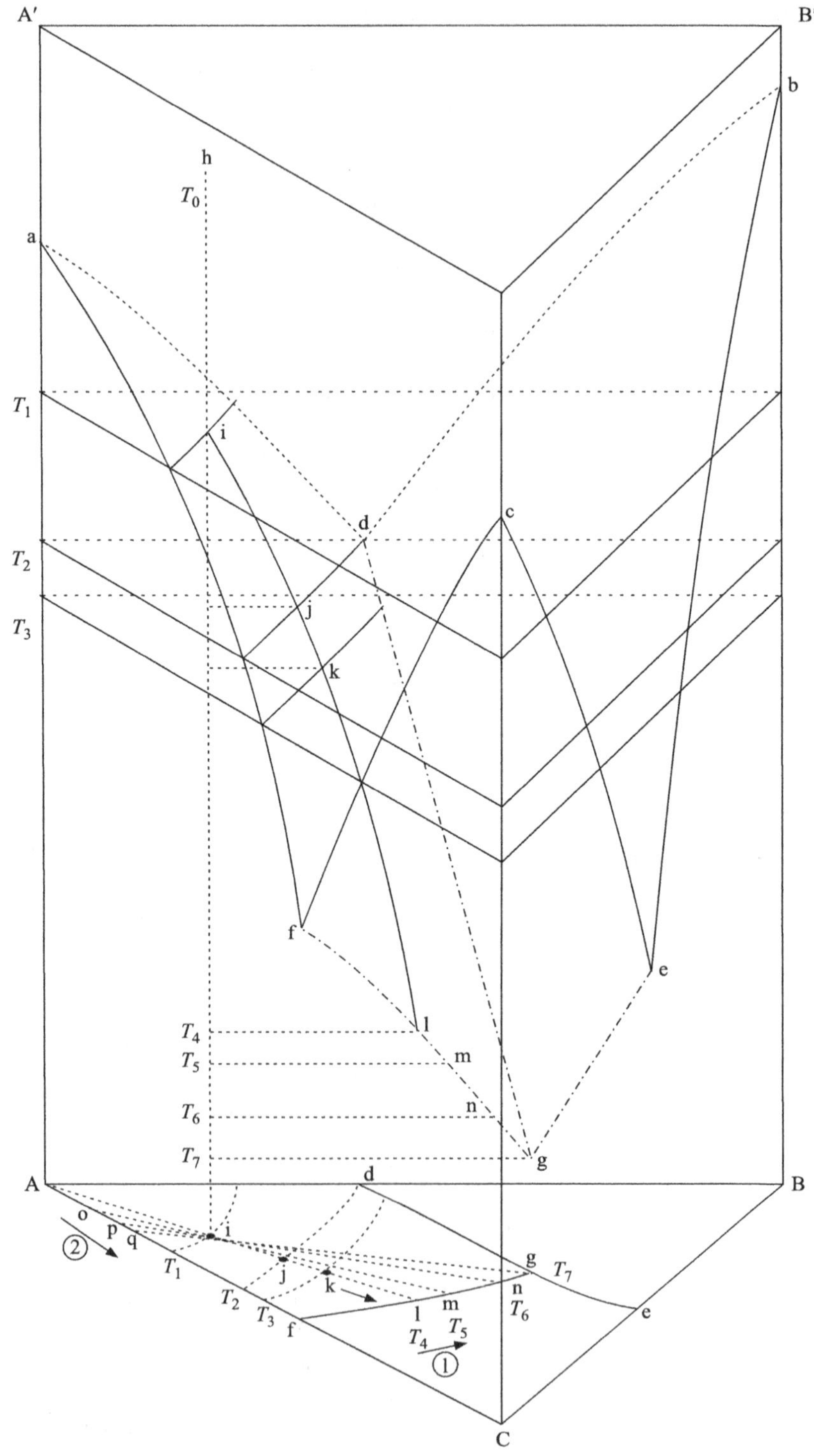


**Fig. 7.4.10 The sequence of effects produced on cooling a ternary system.**
**The tie-lines are as follows. Ai at $T_1$, Aij at $T_2$, Aik at $T_3$, Ail at $T_4$, oim at $T_5$, pin at $T_6$, and qig at $T_7$.**
**The arrows 1 and 2 indicate the directions of change in compositions of liquid and solid phases, respectively.**

temperature of the system is $T_2$ and the relative amounts of solid A and liquid as given by the lever rule are:

$$\frac{\text{Amount of solid A}}{\text{Amount of liquid}} = \frac{\text{ij}}{\text{Ai}}$$

At temperature $T_3$, the composition of liquid phase is given by the point k and the relative amounts of solid A and liquid are

$$\frac{\text{Amount of solid A}}{\text{Amount of liquid}} = \frac{\text{ik}}{\text{Ai}}$$

Obviously, the amount of solid A at $T_3$ is more than that at $T_2$.

Solid A continues to appear till the system has been cooled to temperature $T_4$, where the liquid composition lies on the intersection line fg of A and C liquidus surfaces. Thus, at temperature $T_4$, solid C also starts crystallizing along with solid A. As the solid C is also being removed from the liquid, the amount of C in the latter decreases, consequently, the C/B ratio in the liquid is decreased, and hence its composition will shift away from the apex C and towards the apex B, i.e. along the line lg shown at the bottom triangle of Fig. 7.4.10. Since the solid is no longer pure A but is a mixture of A and C, it is obvious that the line passing through the liquid composition and the point i does not pass through the apex A but will intersect the line AC. Thus on lowering the temperature further, the composition of liquid is shifted along lg and that of the solid along the line AC. For example, at temperature $T_5$, the liquid composition is represented by the point m and that of solid by the point o. At $T_6$, the liquid composition is represented by the point n and that of the solid by p. It may be seen that the length of the tie-line from the point i to the liquid composition increases while that from the point i to the solid composition decreases as a result of lowering the temperature of the system. This, according to the lever rule, implies that the amount of solid relative to that of liquid is increasing with the decrease in temperature. The composition of the solid mixture at any instant can be determined again with the help of the lever rule. For example, at temperature $T_6$, the relative amounts of A and C in the solid mixture is

$$\frac{\text{Amount of solid A}}{\text{Amount of solid C}} = \frac{\text{pC}}{\text{Ap}}$$

Finally, when the temperature of the system is lowered to $T_7$, the composition of liquid has reached the point g which represents the ternary eutectic point. Thus, at this temperature solid B also starts crystallizing along with A and C in such a manner that the composition of liquid remains unaltered. The process of crystallization at this point is isothermal in nature. The composition of solid must now reflect a composition increasing in B from its zero value on the line AC up to the original amount of B in the starting mixture at point i. Thus, the solid composition moves from point q to i keeping all the while temperature and composition of liquid to constant values. When the solid acquires the composition of point i, the entire liquid is solidified and further cooling results only in the cooling of the three-phase solid mixture.

## 7.5 TERNARY SYSTEMS OF TWO SOLID COMPONENTS AND A LIQUID

In this section, we shall describe phase diagrams of systems involving crystallization of solid phases from an aqueous solution containing two components. We consider these diagrams with reference to an important type of systems in which the two solid components are salts with an ion common between them. The condition of a common ion is essential to keep the system to a three-component one, since otherwise the number of components will become four. In the present treatment, it will be assumed that temperature at which the system is being studied lies above 0 °C, as otherwise one would also have to include an equilibrium curve corresponding to the separation of ice in the triangular plot.

The description of various equilibria diagrams can be conveniently discussed according to the following classification depending upon the nature of solids that crystallize at a given temperature.

(1) Crystallization of pure components only.
(2) Formation of binary compounds such as hydrates.
(3) Formation of a double salt.
(4) Formation of a ternary compound.
(5) Formation of a complete series of solid solutions.
(6) Formation of solid solutions with partial miscibility.

We discuss below phase diagrams of the above categories one after the other.

## 7.6 CRYSTALLIZATION OF PURE COMPONENTS ONLY

### Expected Triangular Plot

Figure 7.6.1 shows the triangular plot of a ternary system in which only pure components crystallize from the aqueous solution.

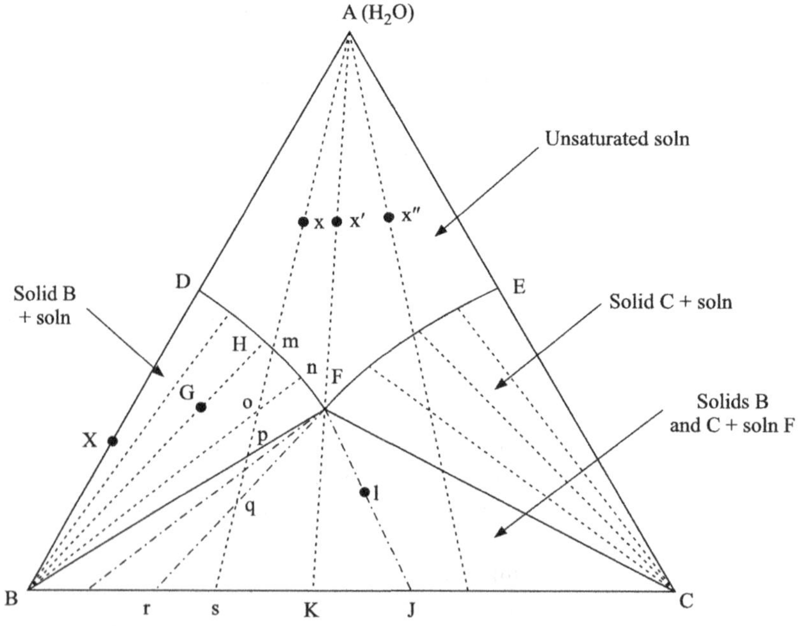


**Fig. 7.6.1 Triangular plot for a system where only pure components crystallize from the solution**

**Description of the Diagram**

*Point* D: This point represents the saturation solubility of component B in water. Thus from D to A($H_2O$), we have unsaturated solution of B in water and from D to B, we have saturated solution of B of composition *D* in equilibrium with the solid B; their relative amounts are given by the lever rule. For example, at the point X, this ratio is given as

$$\frac{\text{Amount of solid B}}{\text{Amount of saturated solution of B}} = \frac{\text{XD}}{\text{XB}}$$

*Line* DF: This line represents the saturation solubility of B in water in the presence of the component C.

*Point E:* This point represents the saturation solubility of the component C in water.

*Line EF:* This line represents the saturation solubility of C in water in the presence of the component B.

*Point* F: At the point F, the two lines DF and EF meet each other and thus represents a system where both solids B and C are in equilibrium with the solution F. According to the phase rule, the point F is an invariant point since $P = 3$ and $F = C - P + 0 = 3 - 3 + 0 = 0$.

*Area* ADFEA: A point within this area represents an unsaturated solution of B and C.

*Area* BDFB: A point within this area represents a system where solid B is in equilibrium with the solution. The composition of the solution lies on the line DF and can be obtained by joining the apex B with the point under study and then extending it to the line DF. For example, at the point G, we have solid B in equilibrium with the solution of composition corresponding to the point H. Their relative amounts as given by the lever rule are

$$\frac{\text{Amount of solid B}}{\text{Amount of solution H}} = \frac{\text{GH}}{\text{BG}}$$

The line BGH is, in fact, the tie line.

*Area* CEFC: A point in this region represents a system where solid C is in equilibrium with the solution whose composition lies on the line EF. The composition of the solution can be obtained by joining the apex C with the given point under study and then extending it to the line EF.

*Area* BFCB: A point in this area represents a system of two solids B and C in equilibrium with the melt of composition corresponding to the point F. In this area, each system is invariant as $P = 3$ and $F = C - P + 0 = 3 - 3 + 0 = 0$. The relative amounts of the two solids and those of total solid and the solution F can be obtained by making use of the lever rule twice as indicated below for the point I (Fig. 7.6.1).

The composition of the solid mixture at the point I can be obtained by joining the point F to the point I and then extending the line FI up to the side BC where it meets at the point J. The composition of the solid mixture is given by the point J. The ratio of the amounts of the two solids as given by the lever rule is

$$\frac{\text{Amount of solid B}}{\text{Amount of solid C}} = \frac{\text{CJ}}{\text{BJ}}$$

The ratio of amounts of the total solid and the solution as given by the lever rule is

$$\frac{\text{Amount of total solid}}{\text{Amount of solution F}} = \frac{\text{FI}}{\text{IJ}}$$

In fact, a point on the line JIF represents a system where the solid mixture of composition J is in equilibrium with the solution F. The only thing that varies from point to point is the relative amounts of the total solid and the solution.

**Process of Isothermal Evaporation**

The sequence of events that takes place when an unsaturated solution is subjected to isothermal evaporation (i.e. when more and more of water is gradually removed) can be easily predicted from Fig. 7.6.1. Let the unsaturated solution x be subjected to isothermal evaporation. The state of the overall system will move along the line xs during the isothermal evaporation. The line xs can be obtained by joining the apex representing water with the point x and then extending it to side BC where it meets at the point s. From the point x to m, the solution remains unsaturated. At m the solution becomes saturated with respect to the component B and thus the pure component B starts precipitating. The removal of B from the solution shifts the composition of the solution away from the apex B, i.e. along the line mF. As more of water is removed, the state of the system moves along mp with the removal of more and more of B. This may be verified from the increasing length of the portion of various tie lines lying right to the line mp relative to those lying left to this line. At the point m, the relative ratio of solid B and the solution is zero while at the point o, this ratio is

$$\frac{\text{Amount of solid B}}{\text{Amount of the solution n}} = \frac{\text{on}}{\text{Bo}}$$

and that at p is

$$\frac{\text{Amount of solid B}}{\text{Amount of the solution F}} = \frac{\text{pF}}{\text{Bp}}$$

At the point p, the removal of a very small amount of water further brings the state of the system within the area BFCB. Thus, now the system consists of two solids B and C in equilibrium with the solution F. The following effects are produced when the state of the system moves along the line ps as a result of the removal of more and more water from the system.

(i) The composition of the solid mixture steadily shifts from the point B to s and thus the proportion of C steadily increases.

(ii) The relative amounts of the total solid mixture and the solution are shifted more and more in favour of the solid mixture as may be seen from the increasing length of the portion of various tie lines lying right above the line ps relative to those lying below it.

At the point s, complete evaporation of water has taken place and the system merely consists of the mixture of solids B and C in the ratio given below.

$$\frac{\text{Amount of solid B}}{\text{Amount of solid C}} = \frac{\text{Cs}}{\text{Bs}}$$

This ratio will also be equal to the original ratio of B and C present in the unsaturated solution x.

If the composition of the original unsaturated solution lies at the point x″, then on gradual evaporation, it is the solid C that separates first as the line joining A to x″ passes through the saturation curve EF. If the original composition of the solution lies on the line obtained by joining the invariant point F to the apex representing $H_2O$, then both solids B and C will appear simultaneously. Moreover, the proportion of the two solids present in the mixture will be the same as that present in the solution F, since the composition of the latter remains unchanged during the process of evaporation.

**Process of Isothermal Addition of Water**

In a similar manner the effects produced during the isothermal addition of water to a mixture of salts can be predicted. In fact, the sequence of steps that is produced is exactly reverse of those seen during the isothermal evaporation. Take, for example, the system represented by the point s. The addition of water to this system will cause the system to move initially along the line sp and then along pm and finally along mx. From s to p, the addition of water will result in the formation of a saturated solution of constant composition corresponding to the point F keeping all the while both solids B and C in equilibrium with the solution; the relative amounts of the two solids increase in favour of solid B, i.e. the composition of the mixture will move along the line sB. At the point p, solid C has completely dissolved and we get a system of solid B in equilibrium with the saturated solution of composition corresponding to the point F. The addition of more water now causes the dissolution of more and more of solid B, and thus composition of the solution is shifted to the left along the line Fm. The relative amount of the remaining solid B decreases as may be seen from the decreasing portion of various tie lines lying above the line pm relative to those present below it. Finally from m to x, the solution becomes more and more unsaturated.

**Maximum Amount of B or C Recoverable**

The equilibrium diagram as given in Fig. 7.6.1 can also be utilized for determining the maximum amount of solid B (or C) that can be separated from a mixture of B and C of the given composition. Pure solid B will be obtained if the original composition of the solid mixture lies to the left of the point K, but pure solid C will be obtained if the composition lies to the right of this point. The maximum amount of solid B (or solid C) is obtained when enough water has been added so that the composition of the resultant system lies on the line BF (or CF). This amount can be calculated as follows.

Let the composition of the original solid mixture of B and C be represented by the point s. Let enough water be added so that the point p is reached. Applying the lever rule at the point p on the line BF, we have

$$\frac{\text{Amount of solution F}}{\text{Amount of solid B}} = \frac{\text{pB}}{\text{pF}}$$

Adding one to both sides, we have

$$\frac{\text{Amount of solution F} + \text{Amount of solid B}}{\text{Amount of solid B}} = \frac{\text{pB} + \text{pF}}{\text{pF}} = \frac{\text{BF}}{\text{pF}}$$

or $$\text{Amount of solid B} = \left(\frac{\text{pF}}{\text{BF}}\right)(\text{Total amount of the system}) \qquad (7.6.1)$$

Total amount of the system at the point p may be determined again through the application of the lever rule at the point p on the line sA($H_2O$). Thus, we have

$$\frac{\text{Amount of water to be added to reach the point p}}{\text{Amount of solid mixture is}} = \frac{\text{ps}}{\text{pA}}$$

Adding one on either sides, we have

$$\frac{\text{Amount of water + Amount of solid mixutre}}{\text{Amount of solid mixture}} = \frac{\text{ps + pA}}{\text{pA}}$$

i.e. $$\frac{\text{Total amount of the system}}{\text{Amount of solid mixture}} = \frac{\text{sA}}{\text{pA}}$$

Thus, Eq. (7.6.1)

$$\text{Amount of solid B} = \left(\frac{\text{pF}}{\text{BF}}\right)\left(\frac{\text{sA}}{\text{pA}}\right)(\text{amount of solid mixture}) \qquad (7.6.2)$$

It may also be seen from Fig. 7.6.1 that solid B can be recovered only when the original composition does not exceed the composition K. Once this composition is exceeded towards C, then the line joining the given point to the apex representing water will pass through the region of saturated solution of C and hence only C can be recovered from the mixture. If the original composition corresponds to the point K, the two substances cannot be recovered separately in pure forms from the system.

**Examples**

Examples of systems exhibiting triangular plot of Fig. 7.6.1 are

$NH_4Cl$–$NH_4NO_3$–Water
$NaCl$–$NaNO_3$–Water
$NH_4Cl$–$(NH_4)_2SO_4$–Water
$NaNO_3$–$KNO_3$–Water

***Example 7.6.1***

For the system $KNO_3$–$NaNO_3$–$H_2O$, a ternary point exists at 5 °C at which the two anhydrous salts are in equilibrium with a saturated solution containing 9.04% by mass of $KNO_3$ and 41.01% $NaNO_3$. Determine analytically the maximum mass of $KNO_3$ which could be recovered pure from a salt mixture containing 70 g of $KNO_3$ and 30 g of $NaNO_3$ by crystallization from an aqueous solution at 5 °C.

***Solution***

First of all, the point a corresponding to the given composition of the ternary point, viz., 9.04% by mass of $KNO_3$ and 41.01% $NaNO_3$, was located (Fig. 7.6.2). Then the point b corresponding to the given salt mixture, viz., 70 g of $KNO_3$ and 30 g of $NaNO_3$, was located. Points a and b were joined with the apexes B and A, respectively. The lengths bc, cA, Bc and ca were measured and found to be 2.3 cm, 6.6 cm, 4.1 cm and 3.7 cm, respectively. At the point c, we have

$$\frac{\text{Mass of water}}{\text{Total mass of salt}} = \frac{2.3}{6.6}$$

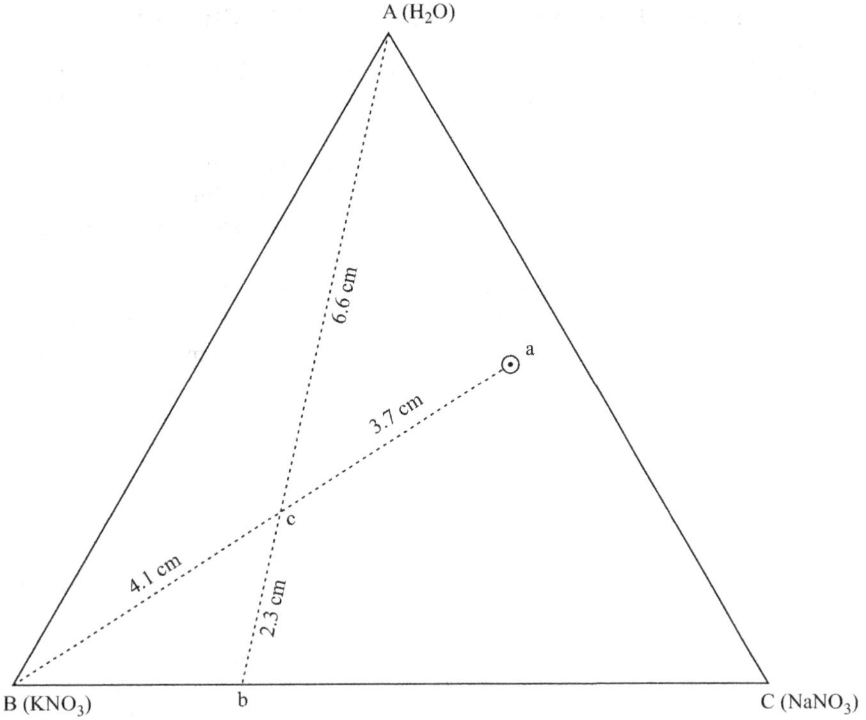


**Fig. 7.6.2**

$$\text{Mass of water} = \left(\frac{2.3}{6.6}\right)(\text{Total mass of salt}) = \frac{2.3}{6.6} \times 100 \text{ g}$$
$$= 34.83 \text{ g}$$

Total mass of the system at c = 100 g + 34.83 g = 134.83 g.

The point c represents solid $KNO_3$ in equilibrium with the ternary solution a. Applying the lever rule on the line Bca, we have

$$\frac{\text{Mass of the ternary solution}}{\text{Mass of KNO}_3} = \frac{4.1}{3.7}$$

or
$$\frac{\text{Mass of the ternary solution + mass of KNO}_3}{\text{Mass of KNO}_3} = \frac{7.8}{3.7}$$

or
$$\frac{\text{Mass of KNO}_3}{\text{Total mass of the system at c}} = \frac{3.7}{7.8}$$

or
$$\text{Mass of KNO}_3 = \left(\frac{3.7}{7.8}\right)(\text{Total mass of the system at c}) = \frac{3.7}{7.8} \times 134.83 \text{ g}$$
$$= 63.96 \text{ g}$$

$$\text{Per cent of KNO}_3 \text{ recovered} = \frac{63.96 \text{ g}}{70 \text{ g}} \times 100 = 91.37$$

## 7.7 FORMATION OF BINARY COMPOUNDS SUCH AS HYDRATES

### Description of the Diagram

Let B · $nH_2O$ be the stable hydrate which is formed by salt B in contact with water. The composition of the hydrate can be shown on the line $BA(H_2O)$ and is given by the point D in Fig. 7.7.1. Let the saturation solubility of this stable hydrate in the presence of a salt C varies along the line EF. The point G represents the saturation solubility of the salt C in water and the line GF is the corresponding solubility curve in the presence of the salt B. The two curves EF and GF meet each other at the point F and thus this point F gives a system where the stable hydrate D and the salt C are in equilibrium with the solution F. With these facts, the labelling of rest of the lines and various areas can be done without any difficulty as summarized below.

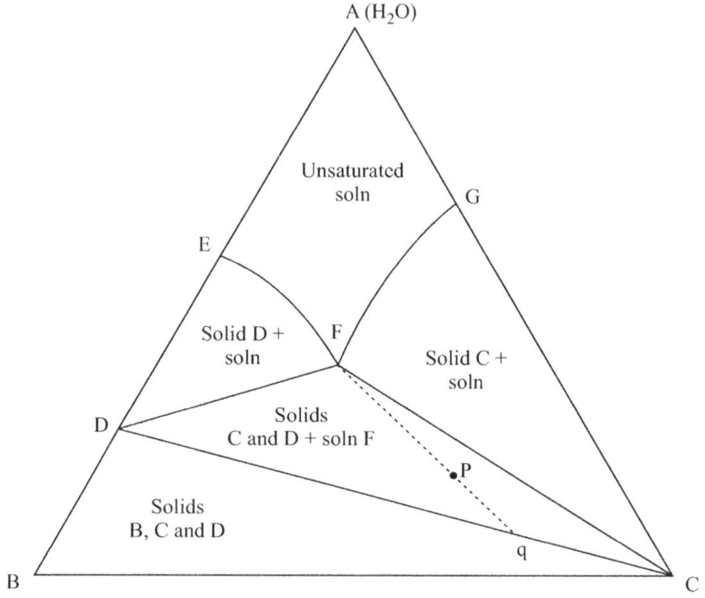


**Fig. 7.7.1 Triangular plot for system where the salt-hydrate is formed**

*Area* AEFGA: A point in this area represents an unsaturated aqueous solution of B and C.

*Area* DEFD: A point in this area represents a system in which stable hydrate D is in equilibrium with the solution. The composition of the solution lies on the line EF. Various tie lines start from the line EF and will merge at the point D.

*Area* CGFC: A point in this area represents a system in which solid C is in equilibrium with the solution. The composition of the solution lies on the line GF. Obviously, various tie lines start from the line FG and will merge at the apex C.

*Line* DC: All possible mixtures of D and C will lie on the line DC.

*Area* DFCD: Since the line DC represents the lowest limit up to which both solids D and C can exist, a point in this area represents a system in which solids D and C are in equilibrium with the solution of composition corresponding to the point F. Various tie lines in this region start from the line DC and will merge at the point F.

The relative amounts of solids D and C, and also of the total solid and the solution F can be determined by using the lever rule twice. For example, at the point p, we have

$$\frac{\text{Total amount of solid}}{\text{Amount of solution F}} = \frac{\text{Fp}}{\text{pq}}$$

$$\frac{\text{Amount of solid D}}{\text{Amount of solid C}} = \frac{\text{Cq}}{\text{Dq}}$$

From these, the respective amounts of solids C and D and the solution can be determined.

*Area* BDCB: Since the liquid phase disappears on the line DC, a point below the line DC will represent a solid mixture of B, D and C. The respective amounts of B, D and C can be determined by the method given above.

**Example**

The system sodium sulphate-sodium chloride-water at 15 °C exhibits phase diagram as given in Fig. 7.7.1. The stable salt formed is $Na_2SO_4 \cdot 10H_2O$.

## Formation of Two Hydrates

Sometimes both the salts form stable hydrates. For example, in the system magnesium chloride-calcium chloride-water at 0 °C, the two salts $MgCl_2 \cdot 6H_2O$ and $CaCl_2 \cdot 6H_2O$ are formed. The triangular plot of such a system will have an appearance as shown in Fig. 7.7.2. Labelling of various lines and areas can be done as usual, but attention must be paid to the lowest region of the plot where either the line CD (as shown) or the line BE (but not together) is observed depending upon the vapour pressure data of the system. If the vapour pressure of the system (E + C) has a value greater than that of the system (B + D), the hydrate E begins to dehydrate as soon as last drop of the solution disappears and

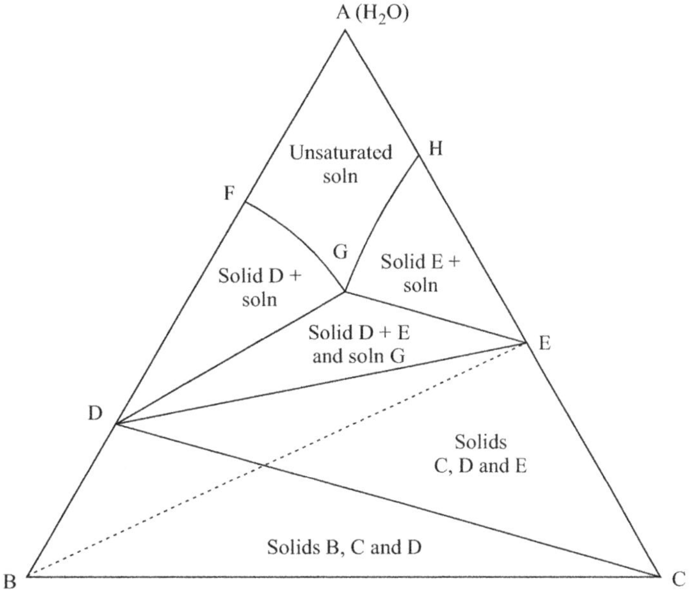


**Fig. 7.7.2 Triangular plot for system where two hydrates are formed**

we get a system consisting of (C + D + E). Consequently, the line CD is observed. When all of E has dehydrated, we reach the line CD and the pressure of the system falls to the vapour pressure of (B + D). After this, D starts dehydrating and we move into a region of (B + D + C). Conversely, if the vapour pressure of the system (B + D) has a value greater than that of (E + C), then D dehydrates first and E starts dehydrating on the dashed line BE.

### Formation of One Hydrate and Two Anhydrous Salts

Sometimes, under certain condition of temperature, the stable hydrate as well as the anhydrous salt can exist in equilibrium with the solution. In such cases, the triangular plot has an appearance as shown in Fig. 7.7.3. The labelling of various points, lines and areas is straightforward and is being summarized now.

*Point* E: Saturation solubility of the hydrate D.

*Line* EF: Saturation solubility curve of the hydrate D observed in the solution in the presence of C.

*Point* H: Saturation solubility of the component C.

*Line* HG: Saturation solubility curve of the component C observed in the solution in the presence of B.

*Line* FG: Saturation solubility curve for the pure component B.

*Point* F: Invariant point where solid hydrate D and solid B are in equilibrium with solution F.

*Point G:* Another invariant point where solid B and solid C are in equilibrium with the solution G.

*Area* AEFGHA: A point in this region represents an unsaturated solution.

*Area* DEFD: A point in this region represents solid hydrate D in equilibrium with the solution; composition of the solution lies on the line EF.

*Area* BFGB: A point in this region represents solid B in equilibrium with the solution; composition of the solution lies on the line FG.

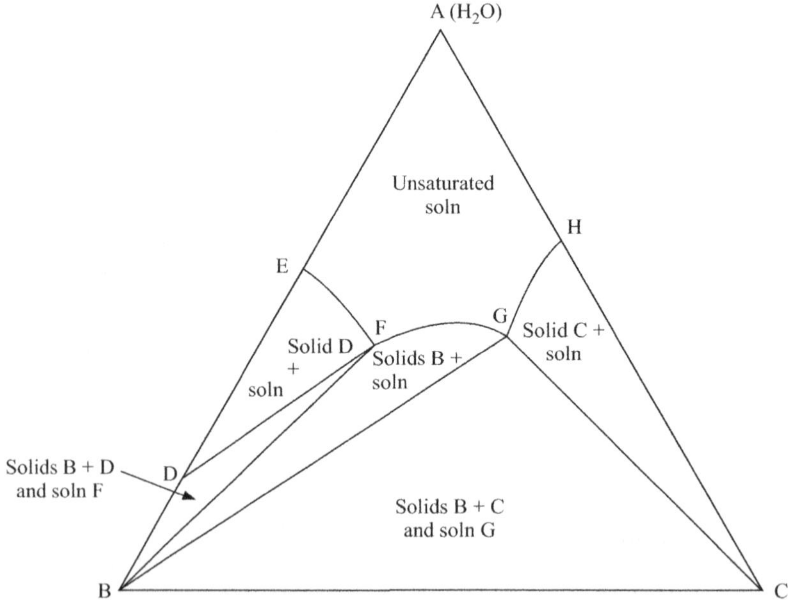


**Fig. 7.7.3 Triangular plot where both stable hydrate and the anhydrous salt crystallize together**

*Area* BDFB: Invariant region, where solid hydrate D and solid B are in equilibrium with solution of composition corresponding to the point F.
*Area* HGCH: A point in this region represents solid C in equilibrium with the solution; composition of the solution lies on the line HG.
*Area* BGCB; Another invariant region, where solid B and solid C are in equilibrium with solution of composition corresponding to the point G.

**Example** The system $Na_2SO_4$–NaCl–$H_2O$ between 17.9 °C and 32.4 °C exhibits the triangular plot as shown in Fig. 7.7.3.

## 7.8 FORMATION OF A DOUBLE SALT

**Description of Diagram**

In some cases the two salts B and C combine with each other to form a double salt of general formula $B_mC_n$. Its composition will be represented by a point such as D on the line BC as shown in Fig. 7.5.5. The triangular plot will now include three solubility curves corresponding to the salts B, D and C. These are shown in Fig. 7.8.1 by the curves EF, FG and GH, respectively.

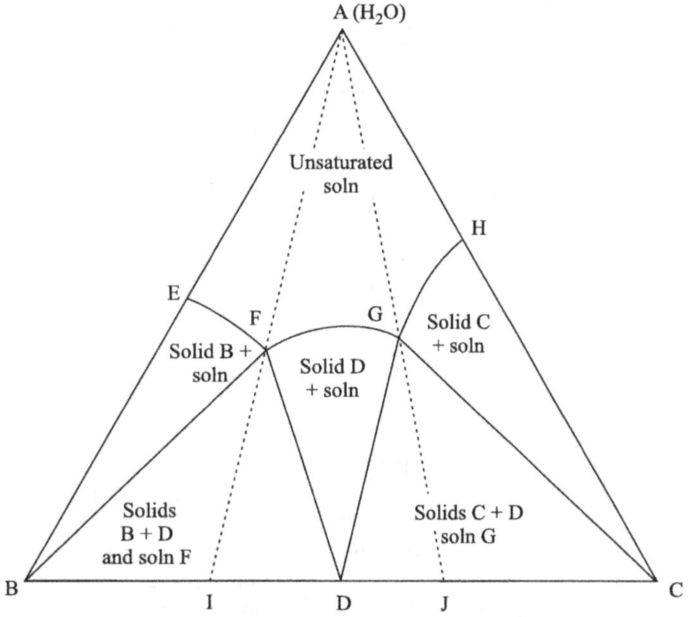


**Fig. 7.8.1 Triangular plot of a system where congruently saturating double salt is formed**

The points F and G are the two invariant points where we have
(i) *At point* F: Solid B and solid double salt D are in equilibrium with the solution of F.
(ii) *At point G*: Solid C and solid double salt D are in equilibrium with the solution of G.

Various tie lines start from the lines EF, FG and GH will meet at the points B, D and C, respectively. With this information, labelling of the triangular plot can be done without any difficulty and is summarized below.
*Area* AEFGHA; A point in this region represents an unsaturated solution of B and C.

*Area* BEFB: A point in this region represents a system in which solid B is in equilibrium with the solution whose composition lies on the line EF.

*Area* HGCH: A point in this region represents a system in which solid C is in equilibrium with the solution whose composition lies on the line HG.

*Area* DFGD: A point in this region represents a system in which solid double salt D is in equilibrium with the solution whose composition lies on the line FG.

*Area* BFDB: This region represents solids B and D in equilibrium with the solution F and, therefore, a point in this region represents an invariant system. Various tie lines in this region on starting from the line BD will merge at the point F. With the help of the lever rule, we can determine the relative amounts of the total solid (B + D) and the solution and those of solid B and solid D in the total solid mixture.

*Area* CDGC: A point in this area represents a system of solids C and D in equilibrium with solution G. Various tie lines start from the line CD and meet together at the point G.

***Example 7.8.1***

The following data are available for the system nickel sulphate-sulphuric acid-water at 25 °C. Sketch the triangular plot and label it.

| *Liquid phase* | | *Solid phase* |
|---|---|---|
| $NiSO_4$/*mass %* | $H_2SO_4$/*mass %* | |
| 28.13 | 0 | $NiSO_4 \cdot 7H_2O$ |
| 27.34 | 1.79 | $NiSO_4 \cdot 7H_2O$ |
| 27.16 | 3.86 | $NiSO_4 \cdot 7H_2O$ |
| 26.15 | 4.92 | $NiSO_4 \cdot 6H_2O$ |
| 15.64 | 19.34 | $NiSO_4 \cdot 6H_2O$ |
| 10.56 | 44.68 | $NiSO_4 \cdot 6H_2O$ |
| 9.65 | 48.46 | $NiSO_4 \cdot H_2O$ |
| 2.67 | 63.73 | $NiSO_4 \cdot H_2O$ |
| 0.12 | 91.38 | $NiSO_4 \cdot H_2O$ |
| 0.11 | 93.74 | $NiSO_4$ |
| 0.08 | 96.80 | $NiSO_4$ |

***Solution***

The state points corresponding to the given data are chosen and then a smooth curve passing through these points is drawn. On this curve, different regions of solubility curve are selected. Each solubility curve is then joined to the appropriate point representing composition of the corresponding solid phase (Fig. 7.8.2). Labelling of the triangular plot is shown in the figure itself.

## Stability of a Double Salt

The stability of a double salt in the presence of water depends on the placements of its composition point and its solubility curve. If the apex representing water in Fig. 7.8.1 is joined with the two invariant points F and G and the resultant lines are extended right up to the line BC, we get two lines AFI and AGJ.

***For congruently saturating salt*** If the composition of the double salt (represented by the point D) lies anywhere between I and J, then on gradually adding water to the double salt, the state of the overall system will move along DA which will

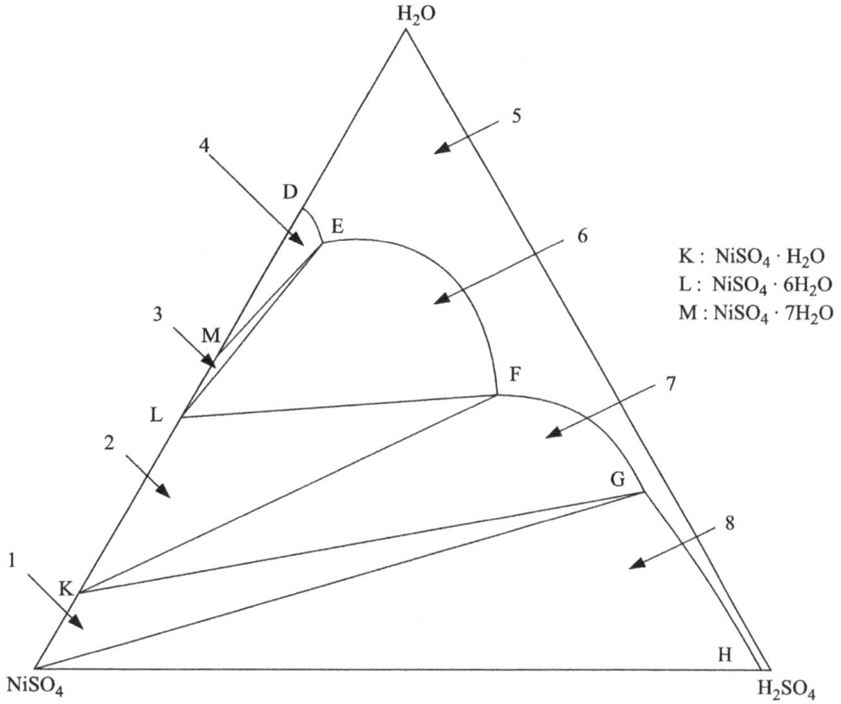


**Fig. 7.8.2 The required phase diagram:**
**(1) $NiSO_4$ and $NiSO_4 \cdot H_2O$ + soln. G**
**(2) $NiSO_4 \cdot 6H_2O$ and $NiSO_4 \cdot H_2O$ + soln. F;**
**(3) $NiSO_4 \cdot 7H_2O$ and $NiSO_4 \cdot 6H_2O$ + soln. E;**
**(4) $NiSO_4 \cdot 7H_2O$ + soln.(DE);**
**(5) unsaturated soln.;**
**(6) $NiSO_4 \cdot 6H_2O$ + soln.(EF);**
**(7) $NiSO_4 \cdot H_2O$ + soln.(FG), and**
**(8) $NiSO_4$ + soln.(GH)**

pass through the region DFGD. This implies that on adding water, a saturated solution of the double salt can be prepared and thus the double salt remains intact in the presence of water. In other words, the double salt is stable in the presence of water. Such a double salt is called *congruently saturating*.

***Incongruently Saturating Salt*** If composition of the salt lies outside the range IJ, then the line connecting the point under study with the apex A($H_2O$) will not pass through the region DFGD but will pass initially through the area BFDB or CGDC and then through BEFB or CGHC and finally through AEFGHA depending upon whether the point D lies on the left of the point I or to the right of the point J. In such a case, it is obvious that a saturated solution of the double salt alone cannot be prepared and the double salt on adding water will undergo partial or complete decomposition. Such a salt is known as *incongruently saturating*.

**Illustration**

Let us illustrate the stability of double salt with the help of a specific example as shown in Fig. 7.8.3.

The point D representing the composition of the double salt lies outside the region of stability. On adding water, we move into the region BDFB where both solid B and the double salt are in equilibrium with the solution F. Thus the double salt besides dissolving in water also dissociates to give solid salt B. In other words, the double salt is no longer stable in the presence of water. As more and more of water is added, composition of the solution will remain unaltered corresponding to the point F while the relative amounts of solid B and double salt are shifted more and more in favour of B. This implies that more and more of double salt decomposes as the state point moves from D to b. At the latter

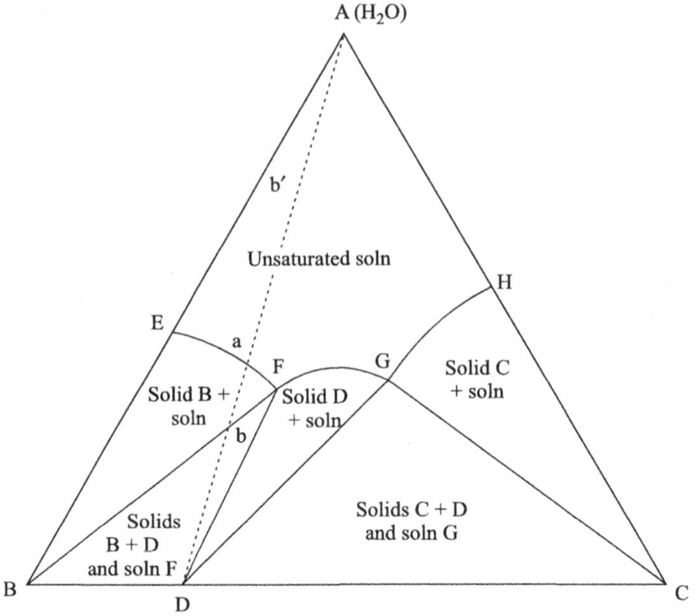


**Fig. 7.8.3 Triangular plot for a system where incongruently saturating double salt is formed**

point, the double salt completely decomposes to give a saturated solution of B of composition *F* in equilibrium with solid B only. On adding more water, we move along ba where more and more of solid B dissolves and the composition of the solution moves along Fa. At the point a, there is a complete dissolution of B and the addition of more of water gives the unsaturated solution of B and C.

## Process of Isothermal Evaporation

We now consider the process of isothermal evaporation in which water is gradually removed until the system becomes one of solid mixture. In fact, this process is just the reverse of the above process and the effects produced can be conveniently described with the help of a triangular plot. We consider some of the typical cases for both congruently saturating and incongruently saturating double salts. These are discussed below.

### For Congruently Saturating Double Salt

In general, the effects produced will depend on the original composition of the unsaturated solution and can be predicted by drawing lines from the apex representing water to the point under study and then extending it to a point where it meets the side BC. We shall consider four typical solutions $x_1$, $x_2$, $x_3$ and $x_4$ as shown in Fig. 7.8.4.

***For the solution*** $\mathbf{x_1}$ This solution contains the same proportion of B and C as that of the double salt. Thus the line $Ax_1$ will on extending meet at the point D corresponding to composition of the double salt. From $x_1$ to a, the solution remains unsaturated, at the point a, the double salt starts appearing and continues to appear till the point D is reached.

***For the solution*** $\mathbf{x_2}$ This system contains more than the requisite amount of B (corresponding to the point $x_1$) to form the double salt. Thus, the end-products

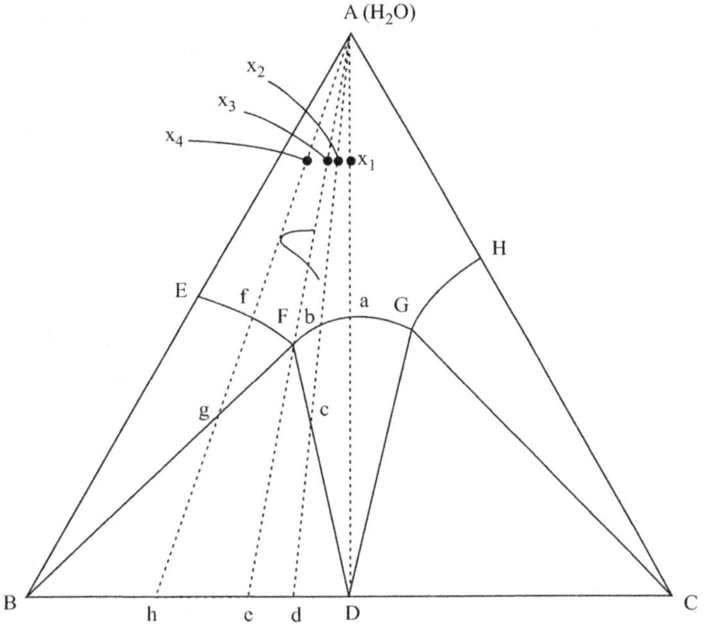


**Fig. 7.8.4 The process of isothermal evaporation as applied to four systems involving congruently saturating double salt**

consist of the double salt and the excessive amount of solid B. Form $x_2$ to b, the solution remains unsaturated, at b the double salt starts appearing and continues to appear till the point c is reached. Since the double salt contains more of C as compared to the amount present in the solution, removal of more and more of double salt along bc shifts the composition of the solution along the line bF. At c, the composition of the solution is *F.* From c to d both the solids, i.e. B and the double salt, start precipitating simultaneously from the solution, keeping all the while composition of the solution to a constant value of F.[†]

***For the solution*** $\mathbf{x_3}$ Here the line $Ax_3$ passes through the invariant point F. From $x_3$ to F, the solution remains unsaturated and from F onwards both the solids, B and the double salt, appear together till the point e is reached.

***For the solution*** $\mathbf{x_4}$ From $x_4$ to f, the solution remains unsaturated. At f, it is the solid B which appears first and continues to appear till the point g is reached, while composition of the solution has moved along fF. From g to h, again both the solids, B and the double salt, appear together keeping all the while constant composition of the solution at the point F.

The solution F is known as *congruently saturated solution.*

---

[†]Since the solution F contains more proportion of B than that present in the double salt, the appearance of solid B from c onwards is due to its precipitation from the solution. The removal of B from the solution will have to be associated with the removal of double salt since only then composition of the solution can be kept at a constant value of *F.* The appearance of B is not through dissociation of the double salt, since in that case the solution will become richer in the component C.

## For Incongruently Saturating Double Salt

Here again the effects produced will depend upon the original composition of the unsaturated solution. We shall consider six typical solutions $x_1$, $x_2$, $x_3$, $x_4$, $x_5$ and $x_6$ as shown in Fig. 7.8.5.

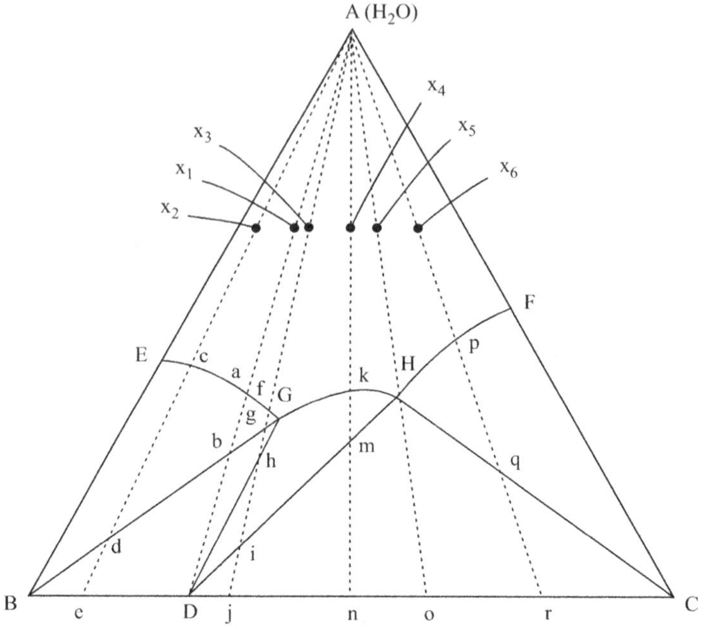


**Fig. 7.8.5 Process of isothermal evaporation as applied to six systems involving incongruently saturating double salt**

***For the solution*** $x_1$ The solution $x_1$ contains the same proportion of B and C as that of the double salt and thus the line $Ax_1$ on extending will meet at the point D corresponding to the composition of the double salt. From $x_1$ to a, the solution remains unsaturated. At a, the salt B starts appearing and continues to appear till the point b is reached. Removal of B from the solution makes the solution deficient in B and thus its composition moves along aG. At the point b, the solution has a composition of G, an invariant point where both the solids, B and the double salt D, are in equilibrium with the solution.

Since the double salt contains more proportion of B than is present in the solution, removal of the double salt from the solution will be associated with the dissolution of already precipitated B, since only then the composition of the solution can remain constant at the point G. In other words, the solution at G is *incongruently saturated.* Further removal of water causes the system to move on the line bD. The net result observed along the line bD is the dissolution of more and more solid B and precipitation of more and more double salt, keeping all the while the composition of the solution at the point G. The fact that the proportion of solid B decreases whereas that of solid D increases can be seen from the composition of the solid mixture which moves from B to D during the isothermal evaporation from point b to D. Finally, at the point D whole of water has been removed and the system consists of only double salt D; the composition of which is the same as that present in the original solution.

***For the solution*** $\mathbf{x_2}$ This system contains more of B than the requisite amount of B (corresponding to the point $x_1$) to form the double salt. Thus the end products of isothermal evaporation are the double salt and the excessive amount of solid B. From $x_2$ to c, the solution remains unsaturated, at c solid B starts appearing and continues to appear till the point d is reached. At d, composition of the solution corresponds to the point G and thus the system now contains solid B and double salt D in equilibrium with the solution. From d to e, more and more of double salt appears and simultaneously more and more of the salt B dissolves in order to keep the composition of the solution at the point G. Finally at e, the system contains solid mixture consisting of double salt and the excess amount of B. The relative amount of two solids is given as

$$\frac{\text{Amount of solid double salt}}{\text{Amount of solid B}} = \frac{\text{Be}}{\text{eD}}$$

***For the solution*** $\mathbf{x_3}$ The system contains more of C than the requisite amount of C to form the double salt and thus the end products of isothermal evaporation are the double salt and the excessive amount of C. From $x_3$ to f, the solution remains unsaturated. At f, the salt B starts appearing and continues to appear till the point g is reached. From g to h, more and more of the double salt is formed with the dissolution of more and more of the salt B. At h, practically whole of solid B has dissolved and the system consists of solid double salt of D and solution G. The system once again becomes univariant as only one solid phase is present in contact with the solution. The removal of more water causes the system to move along hi. Here more and more of the double salt is precipitated out. This may be verified from the increasing length of the portion of various tie lines (if drawn) lying right above the line hi relative to those lying left below it. The composition of the solution changes along the line GH. At the point i, the system again becomes invariant as now two solids, namely, double salt and C, are in contact with the solution H. From i to j, both the solids, C and the double salt, precipitate simultaneously from the solution, keeping all the while composition of the solution to a constant value *H*.[†] Thus the solution H is a *congruently saturated solution*.

***For the solution*** $\mathbf{x_4}$ From $x_4$ to k, the solution remains unsaturated. At k, solid double salt starts appearing and continues to appear till the point m is reached. The composition of the solution will move along kH. At the point m, the system becomes invariant and thus removal of more and more water causes the separation of more and more of the solids C and D from the solution, keeping all the while the composition of the solution H to a constant value. The composition of the solid mixture will move along Dn as one moves from m to n and finally at the point n, complete evaporation of water has taken place and the system consists of a solid mixture of C and D with the relative proportion as given below:

---

[†]The solution of composition *H* contains more proportion of C than is present in the double salt. Thus, the appearance of solid C from i onwards is due to its precipitation from the solution. The removal of solid C from the solution will have to be associated with the removal of double salt since only then the solution can be kept at constant composition *H*. The appearance of C cannot be through the dissociation of the double salt, since in that case the solution will become richer in the component B.

$$\frac{\text{Amount of solid C}}{\text{Amount of solid D}} = \frac{\text{Dn}}{\text{Cn}}$$

***For the solution*** $\mathbf{x_5}$ Here the line $Ax_5$ on extending will pass through the invariant point H. Thus from $x_5$ to H, the solution remains unsaturated. At the point H, both salts, C and D, appear together and this continues till the point o is reached where the system consists of a solid mixture of C and D.

***For the solution*** $\mathbf{x_6}$ From $x_6$ to p, the solution remains unsaturated, from p to q more and more of salt C appears and from q to r both the salts, that is, C and the double salt, appear together. Finally, we reach to the point r where only a solid mixture of C and D is present.

**Examples**

The system $NH_4NO_3$–$AgNO_3$–$H_2O$ at 30 °C forms congruently saturating double salt $NH_4NO_3 \cdot AgNO_3$ and the system $KNO_3$–$AgNO_3$–$H_2O$ at 30 °C forms incongruently saturating double salt.

## 7.9 FORMATION OF A TERNARY COMPOUND

**Expected Triangular Plot**

In some systems a ternary compound consisting of all three components (two salts and water) is formed which may be congruently saturating or incongruently saturating. The triangular plot for such type of systems will include a solubility curve corresponding to the ternary compound and the various tie lines start from this curve and meet together at a point D which represents the composition of the ternary compound. Obviously, such a point will lie within the triangle as it includes all the three components. The general appearance of the triangular plot is of the type shown in Fig. 7.9.1. The labelling of various areas is straightforward and is shown along with Fig. 7.9.1.

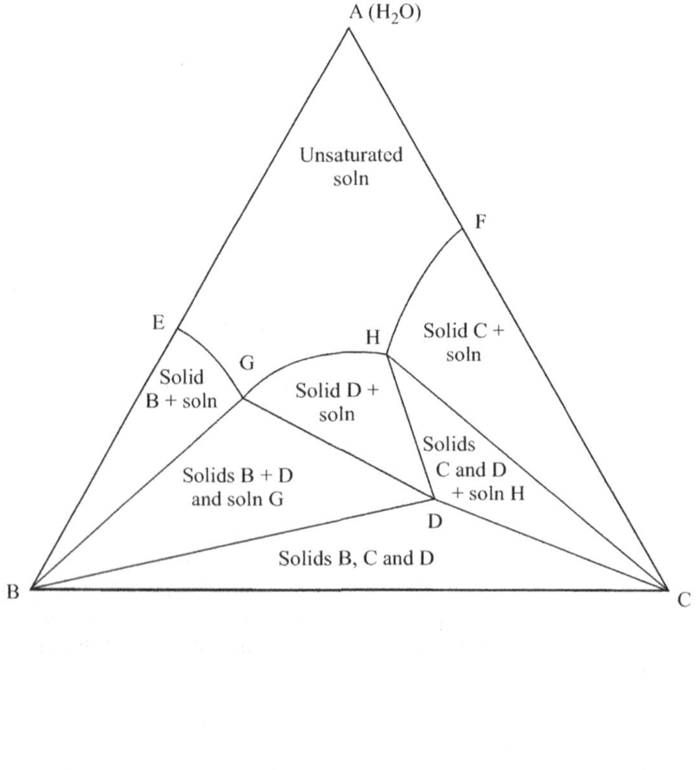


**Fig. 7.9.1 Triangular plot for a system in which a ternary compound is formed**

Many systems form binary as well as ternary compounds. In such cases the triangular plot takes the form as shown in Fig. 7.9.2.

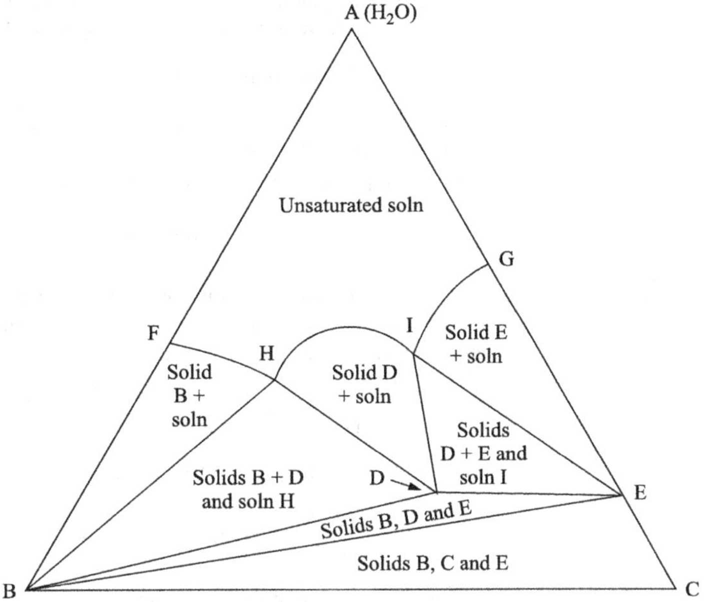


**Fig. 7.9.2 Description of different areas in the obtained triangular plot**

**Examples**

Examples of systems exhibiting congruently saturating ternary compound are the various alums with the general formula $B_2(SO_4) \cdot C_2(SO_4)_3 \cdot 24H_2O$ and examples of systems exhibiting incongruently saturating ternary compound are:

$CaCl_2$–$MgCl_2$–$H_2O$ at 25 °C (compound $CaCl_2 \cdot MgCl_2 \cdot 12H_2O$)
$MgSO_4$–$Na_2SO_4$–$H_2O$ at 25 °C (compound $MgSO_4 \cdot Na_2SO_4 \cdot H_2O$).

***Example 7.9.1***

The following data give compositions of the liquid and solid phase in equilibrium at 25 °C in the system ammonium sulphate-beryllium sulphate-water.

| *Liquid phase* | | *Wet solid phase* | |
|---|---|---|---|
| $(NH_4)_2SO_4$ / *mass %* | $BeSO_4$ / *mass %* | $(NH_4)_2SO_4$ / *mass %* | $BeSO_4$ / *mass %* |
| 43.45 | 0 | — | — |
| 38.60 | 8.62 | 82.29 | 2.82 |
| 35.88 | 16.04 | 71.09 | 7.59 |
| 37.56 | 19.72 | — | — |
| 34.26 | 22.08 | 42.93 | 32.83 |
| 32.20 | 24.04 | 42.84 | 33.64 |
| 28.09 | 26.79 | — | — |
| 24.99 | 26.68 | 9.12 | 46.90 |
| 15.66 | 27.26 | 3.07 | 52.18 |
| 5.77 | 28.56 | 1.37 | 51.16 |
| 0 | 29.94 | 0 | 59.35 |

In each case the difference from 100% represents water, (a) Plot the phase diagram identifying all phases, (b) What will be the composition of the first crystals to deposit when solutions containing 10 mass % $(NH_4)_2SO_4$ + 25 mass % $BeSO_4$ and 20 mass % $(NH_4)_2SO_4$ + 10 mass % $BeSO_4$ are evaporated?

***Solution***

**(a)** Plotting of the triangular diagram can be conveniently done as follows:

(i) Draw an equilateral triangle of an edge length of 100 mm.
(ii) Designate the corners as shown in Fig. 7.9.3.
(iii) The perpendicular distance A to BC is

$$100 \text{ mm} \times \sin 60° = 100 \text{ mm} \times \frac{\sqrt{3}}{2} = 86.6 \text{ mm}$$

Now this length 86.6 mm represents the various proportions of water. It is zero on the line BC and 100% at the corner A. First of all, the point on this perpendicular line is chosen corresponding to the given proportion of water. Taking, for example, the data of the liquid phase containing 38.60 mass % of $(NH_4)_2SO_4$ and 8.62 mass % $BeSO_4$, we have

$$\text{Water composition} = 100 - (38.60 + 8.62) = 52.78 \text{ mass } \%$$

The length representing this percentage on the perpendicular line

$$= \frac{52.78}{100} \times 86.6 \text{ mm} = 45.7 \text{ mm}$$

A point at a length 45.7 mm perpendicularly from the side BC is chosen and then a line parallel to the side BC passing through the chosen point is drawn (not shown in the figure). Then on this line, a point at a distance of 38.6 mm (mass % of $(NH_4)_2SO_4$ from

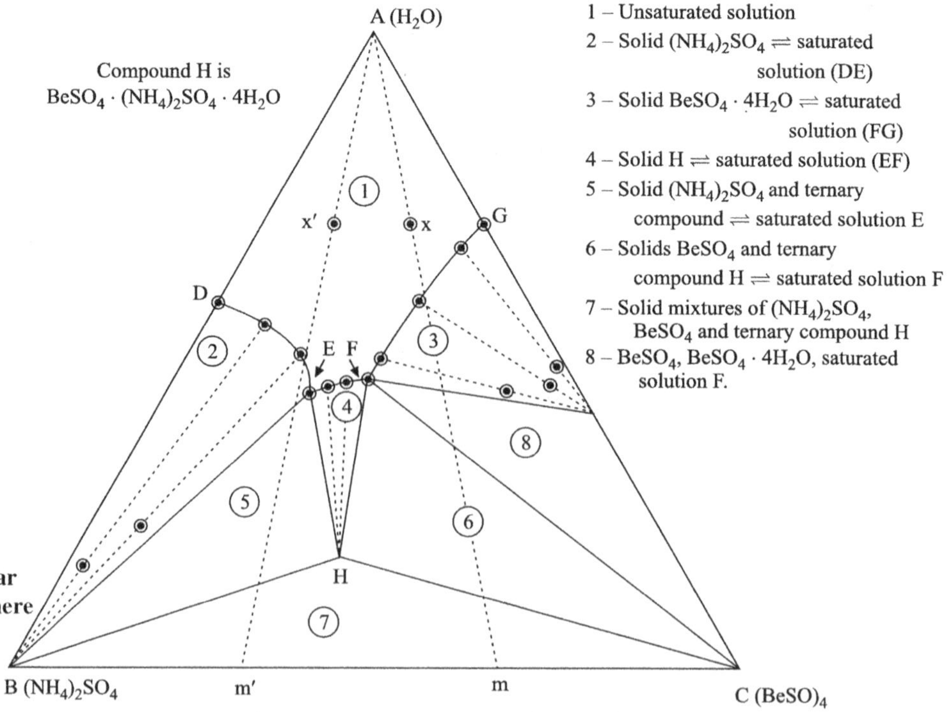


**Fig. 7.9.3 Triangular plot for a system where both binary as well ternary compounds are formed**

the side AC) or at a distance of 8.6 mm from the side AB is chosen. This point represents the required composition.

Alternatively, a point at a distance of 52.78 mm (equal to the water composition) on the side BA or CA (i.e. from B or C towards A) is chosen. A scale at this point parallel to the side BC is placed and a point at a distance of 38.6 mm (equal to the composition of ammonium sulphate) from the side AC is located. This point represents the required composition.

Following either of the above procedures, the various points corresponding to given compositions are chosen. The nature of the solid present in the wet solid phase can be determined by following the method of wet residues (*see*, page 398). Figure 7.9.3 depicts the triangular plot obtained from the given data along with the description of different areas.

**(b)** When the solution containing 10 mass % $(NH_4)_2SO_4$ + 20 mass% $BeSO_4$ (point x) is subjected to the isothermal evaporation, the system will move along xm, the first crystals which appear will be of salt hydrate $BeSO_4 \cdot 4H_2O$. For the system containing 20 mass % $(NH_4)_2SO_4$ + 10 mass % $BeSO_4$ (point x′), it is the solid $(NH_4)_2SO_4$ that separates first as shown in Fig. 7.9.3.

## 7.10 FORMATION OF SOLID SOLUTIONS

If the two solid components B and C are completely miscible in each other, they will form a continuous series of solid solutions ranging from pure B to pure C. If the components A and B and A and C form eutectic phase diagram and temperature of the system is below the melting points of B and C but above that of A, the ternary phase diagram will have an appearance as shown in Fig. 7.10.1. The system have only two phases in equilibrium and thus will never be invariant. Consequently, in the triangular plot, we have only one solubility curve corresponding to the solid solution of B and C as shown by the line DE in Fig. 7.10.1.

A point in the area below DE represents a system of unsaturated solution whereas that above DE represents a system in which the solid solution is in equilibrium with the corresponding saturated solution. The compositions of the two phases, as usual, are connected through various tie lines. The system $Ba(NO_3)_2$–$Pb(NO_3)_2$–$H_2O$ exhibits the phase diagram as shown in Fig. 7.10.1

Consider a case in which all the three pairs A-B, B-C and C-A form a complete series of solid solution both in the solid and liquid phases. Each of the three faces of a three-dimensional triangular prism will include the phase diagram of complete series of solid solution of the respective binary system as shown in Figs 7.10.2 to 7.10.4.

Figure 7.10.2 depicts the ternary triangular phase diagram in which $T_C^* > T_B^* > T > T_A^*$. The horizontal triangle at temperature $T$ intersects the binary phase diagrams of A-B and A-C as shown in Fig. 7.10.2. The solidus and liquidus lines obtained in this intersection when projected on the base of triangular prism gives the expected ternary phase diagram of the system. The labelling of the phase diagram is straight forward and is shown in the figure itself.

Figures 7.10.3 and 7.10.4 depict the ternary triangular phase diagrams in which (i) $T_C^* > T_B^* = T > T_A^*$ and (ii) $T_C^* > T > T_B^* > T_A^*$, respectively. The expected phase diagrams are shown in the bottom triangle of the respective triangular prisms along with the labelling of the phase diagrams.

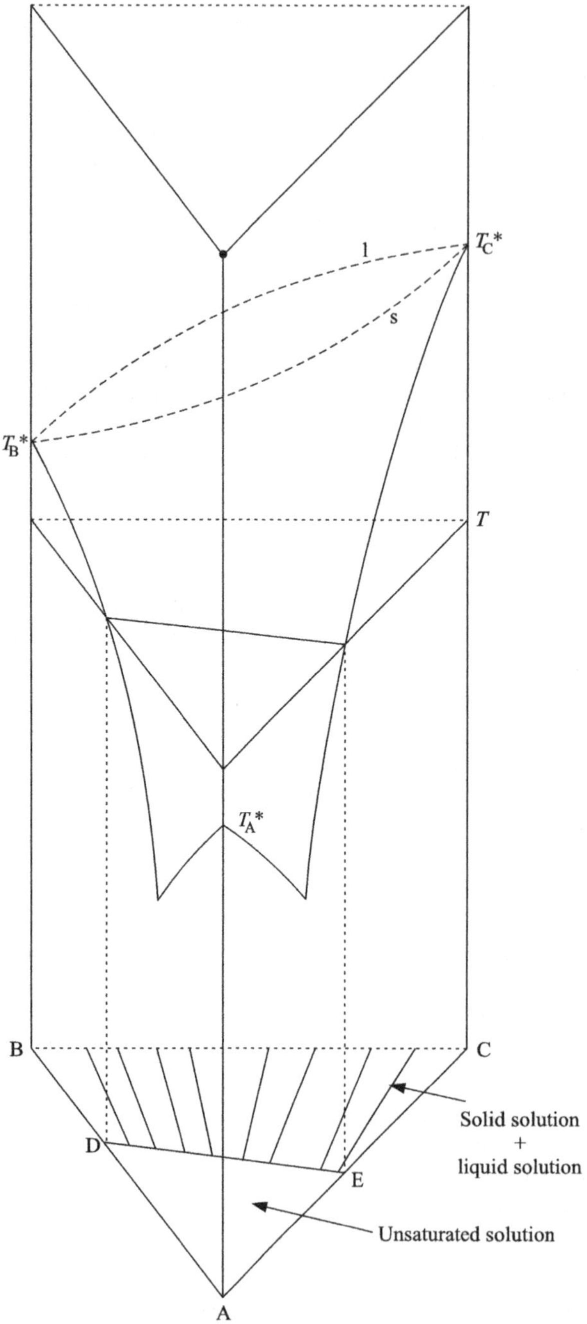


**Fig. 7.10.1 Triangular plot for a system exhibiting only one pair B-C of solid solution**

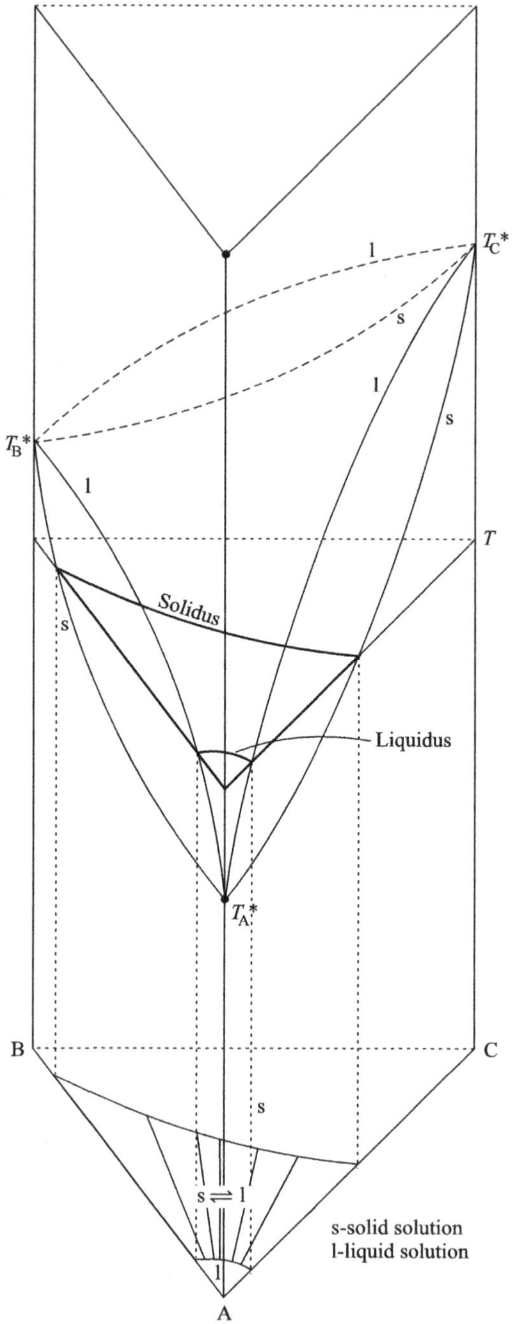


**Fig. 7.10.2 Triangular plot for a system exhibiting three pairs A-B, B-C and C-A of solid solutions with $T_C^* > T_B^* > T > T_A^*$**

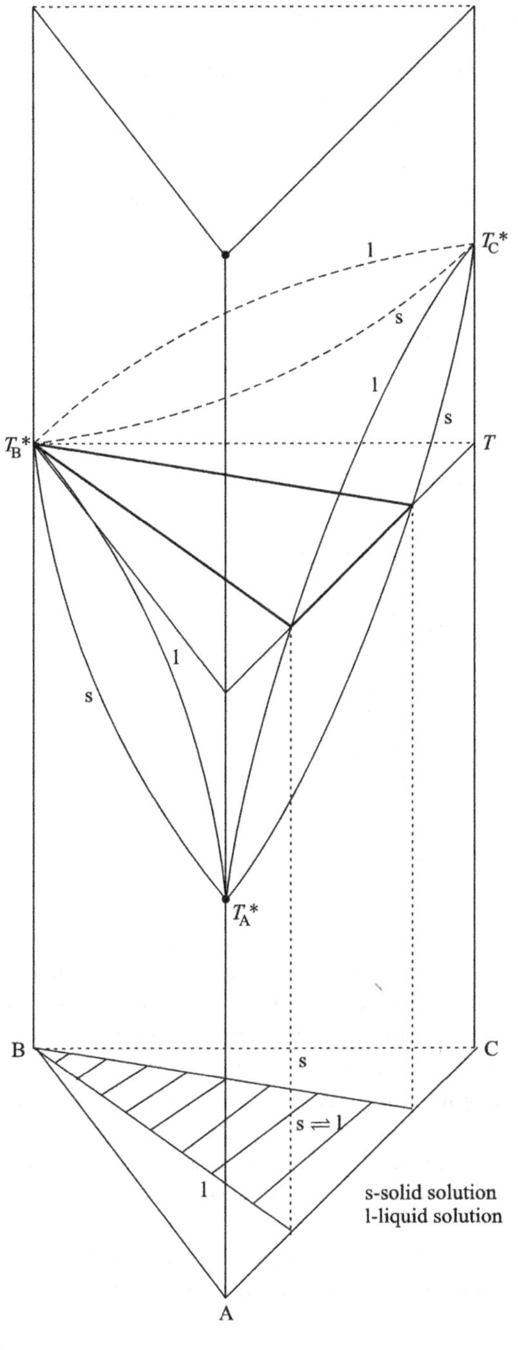


**Fig. 7.10.3 Triangular plot for a system exhibiting three pairs A-B, B-C and A-C of solid solutions with $T_C^* > T_B^* = T > T_A^*$**

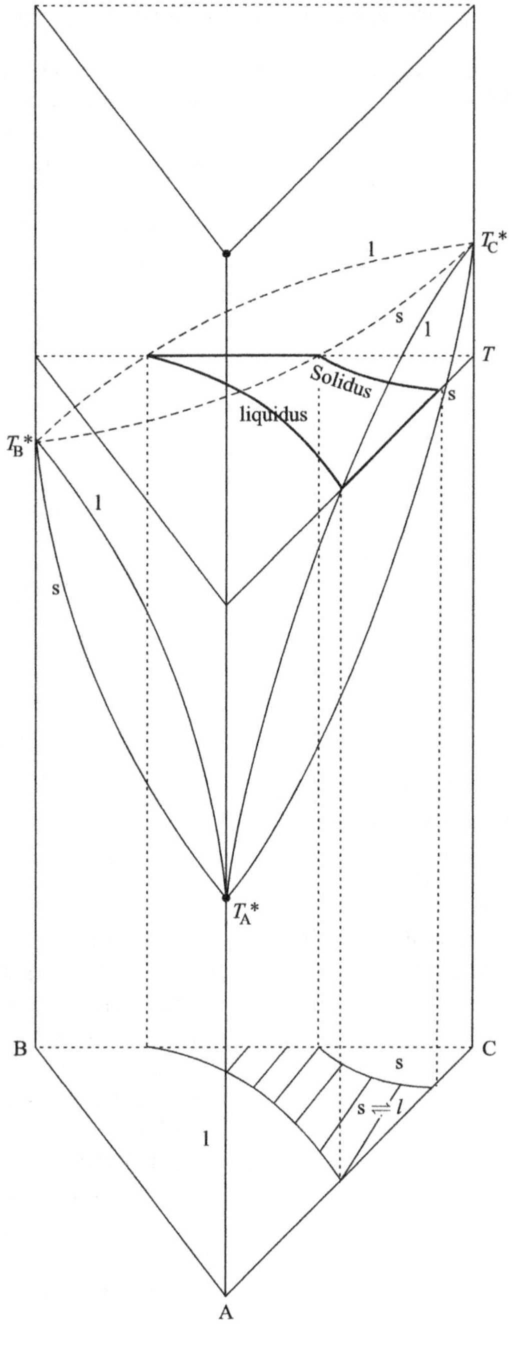


**Fig. 7.10.4 Triangular plot for a system exhibiting three pairs A-B, B-C and A-C of solid solutions with $T_C^* > T > T_B^* > T_A^*$**

## 7.11 FORMATION OF SOLID SOLUTIONS WITH PARTIAL MISCIBILITY

### Expected Triangular Plot

If the two components B and C form solid solutions with partial miscibility, then within certain range of concentrations they will form two solid solutions, one of C in B and the other of B in C. Both these solutions will have definite compositions. Beyond this range, both the components will be completely miscible and will form a single solid solution. Thus along the line BC of the triangular, we will have two points corresponding to the compositions of the two solid solutions such as shown by the points D and E in Fig. 7.11.1.

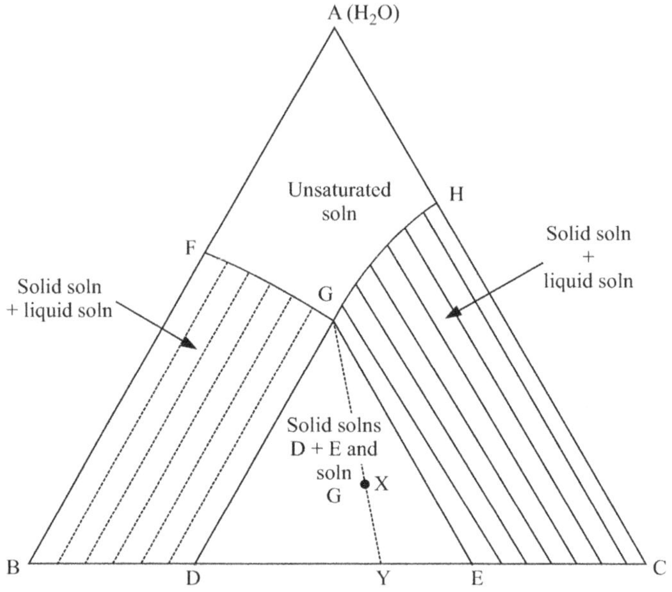


**Fig. 7.11.1 Triangular plot exhibiting solid solution with partial miscibility**

### Description of the Diagram

From B to D and E to C, the two components are completely miscible and from D to E they are partially miscible and thus form two solid solutions D and E, respectively. The lines FG and GH are the corresponding solubility curves of the two solid solutions. With this information the labelling of various areas of Fig. 7.11.1 can be done as usual and is summarized below.

*Area* AFGHA: A point in this region represents a system of an unsaturated solution.

*Area* FGDBF: A point in this region represents a system of solid solution of C in B in equilibrium with the saturated solution. The composition of the latter lies on the line FG.

*Area* GHCEG: A point in this region represents a system of solid solution of B in C in equilibrium with the saturated solution. The composition of the solution lies on the line GH.

*Area* DEGD: A point in this region represents an invariant system of solid solutions D and E in equilibrium with the saturated solution G. The relative amounts of the two solid solutions D and E and that of total solid and the liquid solution can be calculated as follows.

Join the point G with the given point (say, X) and then extend it to the point Y where it meets the side BC. Thus, we have

$$\frac{\text{Amount of solid solution rich in C}}{\text{Amount of solid solution rich in B}} = \frac{DY}{EY}$$

$$\frac{\text{Amount of total solid}}{\text{Amount of solution G}} = \frac{GX}{YX}$$

## 7.12 SALTING OUT PHENOMENON

Sometimes two liquids which show complete miscibility become partially miscible in the presence of a third component. This type of behaviour is common among organic solvents where a single solution of organic liquid in water splits into two layers in the presence of sufficient quantity of a salt. One such example is the ternary system of $H_2O$–$K_2CO_3$–$CH_3OH$ where $H_2O$ and $CH_3OH$ are completely miscible in each other but become partially miscible in the presence of sufficient quantity of $K_2CO_3$. The triangular plot of the above system is shown in Fig. 7.12.1. The points D and E represent the saturation solubility of $K_2CO_3$ in water and $CH_3OH$, respectively. Thus from D to A, we have an unsaturated solution of $K_2CO_3$ in water, whereas that from D to B, solid $K_2CO_3$ is in equilibrium with the saturated solution D. Similarly, from E to C, we have an unsaturated solution of $K_2CO_3$ in $CH_3OH$ and from E to B, a saturated solution E in equilibrium with solid $K_2CO_3$ is formed.

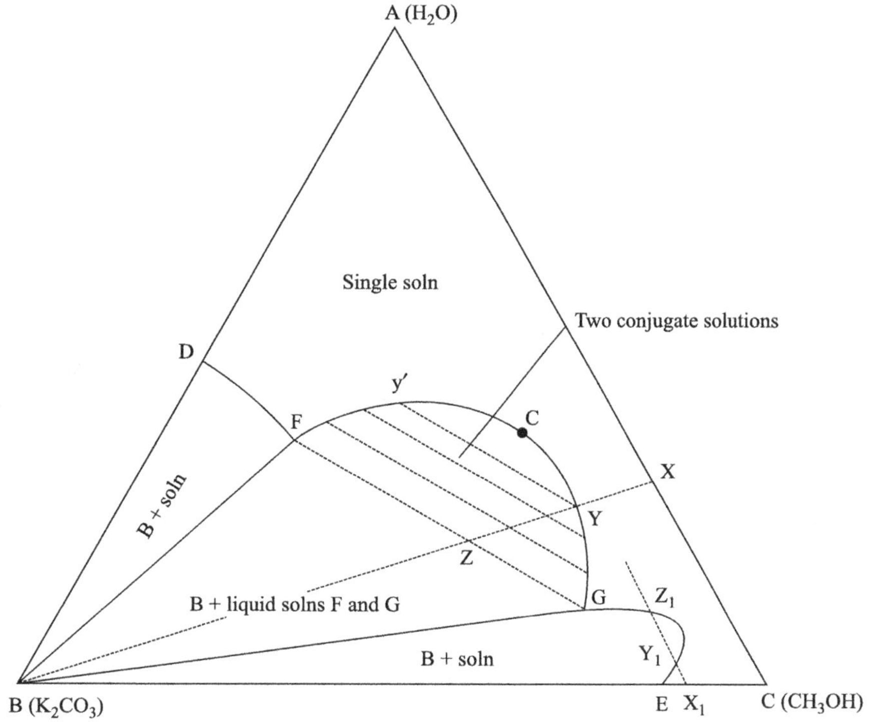


**Fig. 7.12.1 Triangle plot for the system $H_2O$–$K_2CO_3$–$CH_3OH$**

The line DF is the saturation solubility curve of $K_2CO_3$ in water in the presence of $CH_3OH$ and the line EG is the corresponding curve of $K_2CO_3$ in $CH_3OH$ in the presence of water. Between these two curves, we have the formation of two layers along the curve FCG. As usual, the compositions of the two conjugate layers are connected through various tie lines. With this information, the labelling of different areas can be done as usual and is described below.

*Area above the curve* DFCGE: A point in this region represents one liquid layer consisting of all the three components.

*Area* BDFB: A point in this region represents solid $K_2CO_3$ in equilibrium with a saturated solution whose composition lies on the line DF.

*Area* BGEB: A point in this region represents solid $K_2CO_3$ in equilibrium with a saturated solution whose composition lies on the line EG.

*Area* FCGF: A point in this area represents the two conjugate solutions.

*Area* BFGB: A point in this region represents an invariant system where solid $K_2CO_3$ is in equilibrium with the two conjugate liquid layers F and G, respectively.

**Effect of Adding Salt** With the help of Fig. 7.12.1 we can describe the behaviour of the binary system consisting of $H_2O$ and $CH_3OH$ when solid $K_2CO_3$ is added to it. Let us start with a system as represented by the point X. On adding $K_2CO_3$, the state of the system will move long X-B($K_2CO_3$). The system remains of one liquid layer up to the point Y, where it splits into two layers Y and y′. The former is present in major amount while the latter is just formed. On moving from Y to Z, the system continues to have two layers of varying compositions—the composition of the layer rich in water will move from y′ to F, whereas that rich in $CH_3OH$ will move along YG. At the point Z, $K_2CO_3$ no longer dissolves and thus the system from Z to B consists of three phases—solid $K_2CO_3$ and two liquid layers F and G, respectively.

**Effect of Adding Organic Solvent** In a similar manner, behaviour of the system when $CH_3OH$(or $H_2O$) is added to binary system of $K_2CO_3$ and $H_2O$ (or $K_2CO_3$ and $CH_3OH$) can be explained from Fig. 7.12.1. The addition of organic compound to an aqueous solution of a salt is often employed for salting out the dissolved salt in the solution. The principle underlying this method can be easily explained from the triangular plot of Fig. 7.12.1. We will have to simply enter into the region of BDFB or BFGB or BGEB where solid salt is present in equilibrium with the solution of different compositions. The behaviour of the system when $H_2O$ is added to a system represented by the point $X_1$ is interesting; from $X_1$ to $Y_1$, the system continues to be a single layer, at $Y_1$ sol id $K_2CO_3$ precipitates and remains in equilibrium with the solution up to the point $Z_1$ where it redissolves to give once again a single solution. From $Z_1$ onwards, the system continues to be of a single layer.

## 7.13 EXPERIMENTAL METHODS EMPLOYED FOR OBTAINING TRIANGULAR PLOTS

***For systems of three liquids showing partial mutual solubilities*** To a mixture of the two partially miscible liquids, the third component is added until the system previously consisting of two layers becomes homogeneous, i.e. forms a single layer. The amounts of three components are noted and the corresponding point in the triangle is determined. This point must lie on the binodal curve. The experiment is repeated with different proportions of the two partially miscible liquids and many more such points are determined. A smooth curve passing through these points gives the required binodal curve. To obtain the various tie lines, definite amounts of the three liquids are mixed up corresponding to any point within the binodal curve and then the amount of one of the components in either of the two solutions is determined. The point corresponding to this composition is determined on the binodal curve. The determined point is joined with the point representing the overall composition of the solution within the binodal curve. This gives one tie line. The experiment is repeated at different points and similar lines are then drawn.

***For systems consisting of two salts and water*** To plot a triangular plot for such a system, the method of *wet residues* of F.A.H. Schreinmakers is most commonly employed. In this method, first of all a system consisting of a saturated solution in equilibrium with solid is prepared. Then the saturated solution and the wet crystals are analysed and the corresponding points on the triangle are plotted and joined together by a line. The above experiment is repeated over many other systems of different compositions and various such lines are drawn. In the method of wet residues, the wet crystals are considered as the mixture of dry solid and the corresponding saturated solution. Hence the point representing the dry solid must lie on the extension of the line joining the state point of the saturated solution with that of the wet crystals. Moreover, as several solutions may have the same solid phase, all the tie lines for such solutions must intersect at a common point, which is the composition of the solid phase. Thus all lines are extended and the intersection points are determined. These intersection points give the compositions of various solid phases.

## REVISIONARY PROBLEMS

7.1 (a) Determine the variance of a three-component system having number of phases equal to one, two and three, respectively.
(b) Outline the salient features of the scheme of triangular plot employed for three-component systems.

7.2 Draw typical phase diagrams for the following cases:
(a) Systems exhibiting formation of one pair of partially miscible liquids,
(b) systems exhibiting formation of two pairs of partially miscible liquids, and
(c) systems exhibiting formation of three pairs of partially miscible liquids.

7.3 (a) What do you understand by the terms binodal curve and plait point?
(b) Explain why, in general, the various tie lines within the binodal curve are parallel neither to the sides of the triangle nor to each other? Under what conditions will the various tie lines be parallel to each other?
(c) Explain why, in general, the plait point lies either to the left or to the right of the maximum point in the binodal curve. Under what condition will the plait point coincide with the maximum point?
(d) What are the variances of the system within and outside the binodal curve?

7.4 (a) Describe the sequence of events that takes place when the component A (assumed to be in liquid form) is added to a binary system of two partially immiscible liquids B and C.
(b) What is the retrograde solubility?

7.5 (a) Describe the changes observed in the binodal curves when temperature of the system exhibiting one pair of partially miscible liquids is changed? Under what conditions will the above system exhibit true critical solution temperature?
(b) Justify the statement that it is not possible to predict the behaviour of a system exhibiting binodal band when its temperature is raised, that is, whether it will form a system containing only one pair or two pairs of partially miscible liquids.

7.6 Draw phase diagrams of a three-component system involving water and two solid salts with an ion in common for the following typical cases:
(a) Crystallization of pure components.
(b) Formation of binary compounds such as hydrates.
(c) Formation of a double salt.
(d) Formation of a ternary compound.
(e) Formation of complete series of solid solutions.
(f) Formation of solid solutions with partial miscibility.

7.7 Taking a few typical cases, discuss the sequence of events that takes place in the following cases when the system is subjected to isothermal evaporation and/or isothermal addition of water:
(a) Crystallization of pure components.
(b) Formation of a double salt with special emphasis on the stability of the double salt.

7.8 What do you understand by the congruently saturating double salt and incongruently saturating double salt? Support your answer by drawing typical phase diagrams.

7.9 What do you understand by the phenomenon of salting out? Describe this by drawing the triangular plot of $K_2CO_3$–$CH_3OH$–$H_2O$ system.

7.10 Describe the method of *wet residues* of the F.A.H. Schreinmakers employed in constructing the phase diagram of three-component systems involving two salts and water.

## TRY YOURSELF PROBLEMS

7.1 For the phase diagram given in Fig. 7.6.1 which of the following temperature conditions are true?
(i) $T$ > or < or = 0 °C
(ii) $T$ > or < or = eutectic temperature of A-B system.
(iii) $T$ > or < or = eutectic temperature of B-C system.
(iv) $T$ > or < or = eutectic temperature of A-C system.
(v) $T$ > or < or = ternary eutectic temperature of A-B-C system.

7.2 A ternary system consists of a solid and two partially miscible liquids. The solid is soluble in both the liquids. Construct the probable triangular plot when the solubility curves (i) overlap (ii) do not overlap the region of partial miscibility.

7.3 If a binary system A-B forms a congruently melting compound $A_xB_y$ and the systems A-C and B-C exhibit simple eutectic behaviour, the phase diagram of the ternary system A-B-C takes the form as shown in Fig. 1. Label the diagram with adequate explanation. (*Hint:* Label, as usual, the two triangles ACD and BCD separately.)

7.4 If in a ternary system A-B-C, a ternary congruently melting compound is formed, the phase diagram of the system takes the form as shown in Fig. 2. Label the diagram with adequate explanation.
(Hint: Label, as usual, the three triangles ABD, ACD and BCD separately.)

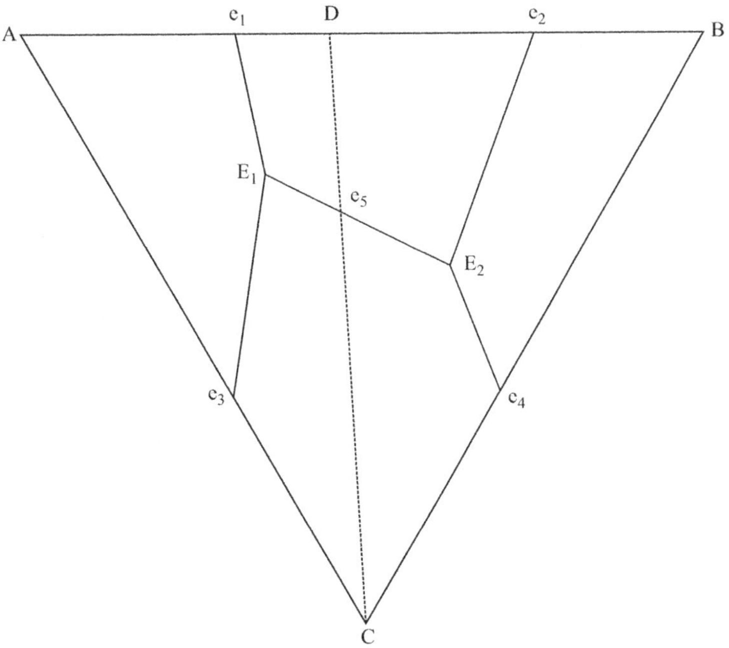


**Fig. 1**

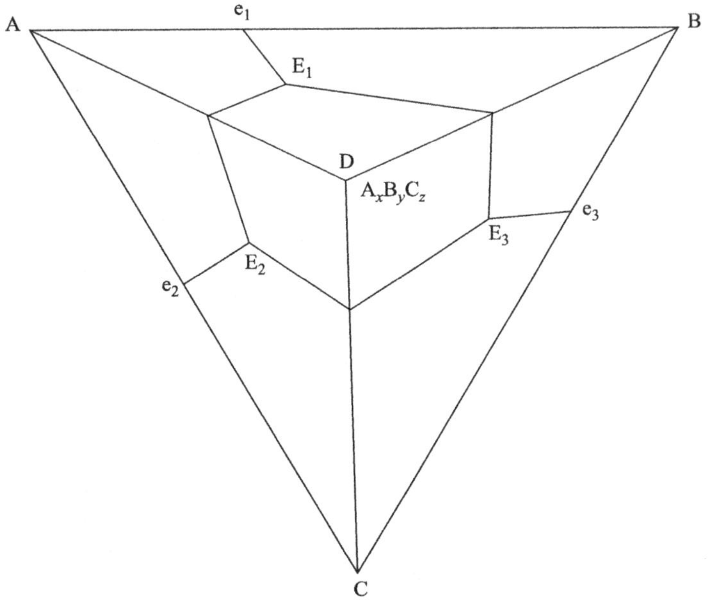


**Fig. 2**

7.5 If the temperature of the system exhibiting the phase diagram of Fig. 1 is greater than both the ternary eutectic temperatures, the phase diagram has an appearance of Fig. 3. Label the diagram with adequate explanation.

7.6 Construct a ternary plot of a three-component system exhibiting the crystallization of $A \cdot H_2O$, A, B and $A \cdot B$ substances.

7.7 Construct a ternary plot of a three-component system exhibiting the crystallization of $A \cdot H_2O$, $B \cdot 2H_2O$, A, B, $A \cdot 2B$ and $A \cdot B \cdot H_2O$ substances.

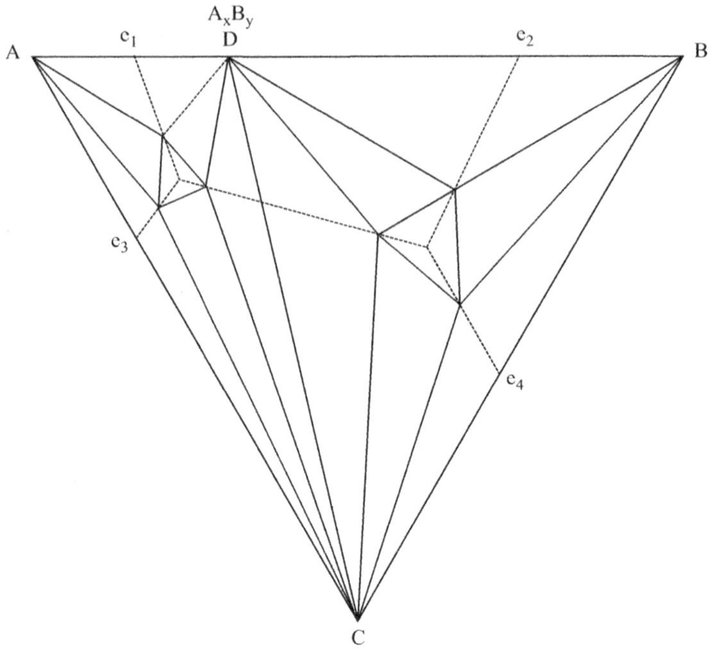


**Fig. 3**

7.8 Given a point within the triangle DEF of Fig. 7.3.9. How many phases are present at this point? What are their compositions? How will you determine the relative amounts of the different phases at this point?

## NUMERICAL PROBLEMS

7.1 The following data, in amount per cent, were obtained for the ternary system; aniline, toluene, *n*-heptane at 25 °C.

| *Aniline-rich phase* | | *Heptane-rich phase* | |
|---|---|---|---|
| *aniline* | *n-heptane* | *aniline* | *n-heptane* |
| 95.6 | 4.4 | 8.4 | 91.6 |
| 89.1 | 7.2 | 12.2 | 81.3 |
| 74.0 | 13.2 | 21.0 | 60.7 |
| 62.6 | 19.3 | 32.5 | 45.4 |

(a) Plot these data in a triangular graph and indicate 1- and 2-phase regions.
(b) Using this graph, indicate the changes that occur when successive increments of toluene are added to a system containing equal amounts of aniline and *n*-heptane.

7.2 The following hydrates may under proper conditions be crystallized from the system $Ba(OH)_2$–NaOH–$H_2O$ at 30 °C.

$Ba(OH)_2 \cdot 8H_2O$, $Ba(OH)_2 \cdot 3H_2O$, $Ba(OH)_2 \cdot H_2O$, $NaOH \cdot H_2O$

Sketch and interpret the phase diagram for this system.

7.3 The following data are available for the system $Na_2SO_4$–$Al_2(SO_4)_3$–$H_2O$ at 42 °C. Draw the phase diagram on triangular coordinate paper, and draw appropriate tie lines.

| *Liquid phase* | | *Solid phase* |
|---|---|---|
| *mass* % of $Na_2SO_4$ | *mass* % of $Al_2(SO_4)_3$ | |
| 33.20 | 0 | $Na_2SO_4$ |
| 32.00 | 1.52 | $Na_2SO_4$ |
| 31.79 | 1.87 | $Na_2SO_4$ |
| 28.75 | 1.71 | $Na_2SO_4 \cdot Al_2(SO_4)_3 \cdot 14H_2O$ |
| 24.47 | 2.84 | $Na_2SO_4 \cdot Al_2(SO_4)_3 \cdot 14H_2O$ |
| 16.81 | 5.63 | $Na_2SO_4 \cdot Al_2(SO_4)_3 \cdot 14H_2O$ |
| 10.93 | 10.49 | $Na_2SO_4 \cdot Al_2(SO_4)_3 \cdot 14H_2O$ |
| 4.72 | 17.11 | $Na_2SO_4 \cdot Al_2(SO_4)_3 \cdot 14H_2O$ |
| 1.75 | 18.59 | $Al_2(SO_4)_3$ |
| 0 | 16.45 | $Al_2(SO_4)_3$ |

7.4 Draw the triangular plot for the system $NH_4Cl$–$NH_4NO_3$–$H_2O$ at 25 °C from the following data.

| *Saturated solution* | | *Wet residue* | |
|---|---|---|---|
| *mass* % of $NH_4NO_3$ | *mass* % of $NH_4Cl$ | *mass* % of $NH_4NO_3$ | *mass* % of $NH_4Cl$ |
| 67.73 | 0 | — | — |
| 66.27 | 2.0 | 88.20 | 0.79 |
| 64.73 | 3.82 | 88.00 | 1.34 |
| 62.24 | 5.58 | 90.25 | 1.65 |
| 61.68 | 6.97 | 87.65 | 2.28 |
| 53.49 | 11.08 | 23.31 | 62.22 |
| 36.99 | 15.80 | 13.63 | 66.29 |
| 19.05 | 21.81 | 7.09 | 72.75 |
| 0 | 28.33 | — | — |

Answer the following with the help of the plot obtained.
(1) What is the ternary composition?
(2) Is there evidence of hydrate or double salt formation?
(3) Determine the maximum theoretical recovery of $NH_4Cl$ from a dry salt mixture of $NH_4Cl$–$NH_4NO_3$ containing 80% by mass of $NH_4Cl$.

7.5 The following data are given for the system $Na_2SO_4$–$MgSO_4$–$H_2O$ at 40 °C and 1 atm.

| *Composition of saturated solution* | *Nature of solid phase in equilibrium* |
|---|---|
| (1) 35 mass % $Na_2SO_4$<br>0 mass % $MgSO_4$ | $Na_2SO_4$ |
| (2) 0 mass % $Na_2SO_4$<br>30 mass % $MgSO_4$ | $MgSO_4$ |
| (3) 25 mass % $Na_2SO_4$<br>15 mass % $MgSO_4$ | $Na_2SO_4 + Na_2SO_4 \cdot MgSO_4 \cdot 4H_2O$ |
| (4) 5 mass % $Na_2SO_4$<br>28 mass % $MgSO_4$ | $MgSO_4 + Na_2SO_4 \cdot MgSO_4 \cdot 4H_2O$ |

Composition of $Na_2SO_4 \cdot MgSO_4 \cdot 4H_2O$ is 42.4 mass % $Na_2SO_4$ and 36 mass % $MgSO_4$.

(a) Sketch the phase diagram.

(b) If a solution containing 10 mass % $Na_2SO_4$ and 10 mass % $MgSO_4$ is evaporated, what solid will appear first, and what is the composition of the last solution to exist before the system becomes entirely solid?

(c) If equal masses of $Na_2SO_4$ and of solution (b) above are mixed, how many phases will be present at equilibrium and what will be their compositions?

7.6 (a) The following data were obtained for the ternary system sodium nitrate-potassium nitrate-water at 50 °C.

| *Composition of saturated solution* | | | *Solid phase* |
|---|---|---|---|
| $NaNO_3$ | $KNO_3$ | *Water* | |
| *mass %* | *mass %* | *mass %* | |
| 53.2 | 0 | 46.8 | $NaNO_3$ |
| 47.6 | 12.3 | 40.1 | $NaNO_3$ |
| 43.3 | 21.6 | 35.1 | $NaNO_3$ |
| 39.8 | 28.9 | 31.3 | $NaNO_3 + KNO_3$ |
| 31.7 | 31.3 | 37.0 | $KNO_3$ |
| 19.1 | 35.9 | 45.0 | $KNO_3$ |
| 0 | 46.2 | 53.8 | |

Construct the ternary phase diagram on an equilateral triangle. What is the nature of the system at each of the following compositions?

| *Constituents* | *Percentage compositions by mass* | | | | | |
|---|---|---|---|---|---|---|
| | (1) | (2) | (3) | (4) | (5) | (6) |
| Sodium nitrate | 20 | 43.7 | 15 | 70 | 16.0 | 47.0 |
| Potassium nitrate | 25 | 20.0 | 28 | 6 | 60.0 | 41.0 |
| Water | 55 | 36.3 | 57 | 24 | 24.0 | 12.0 |

If the system consists of more than one phase, determine also their compositions and relative amounts.

(b) Given is the system containing 10 mass % $NaNO_3$ and 30 mass % $KNO_3$. What will be the nature of first crystal obtained during the isothermal evaporation of the above system? Up to what limit the above evaporation should be carried out so that the yield of separated crystals is maximum?

(c) Given is the system containing 30 mass % $NaNO_3$ and 60 mass % $KNO_3$. Describe if it is possible to get the pure crystals of $KNO_3$ by adding more of water. If so, indicate up to what limit the dilution should be carried out to get the maximum amount of $KNO_3$ crystals.

(d) Describe the sequence of effects that are produced during the isothermal evaporation of the system 1 (i.e. 20 mass % $NaNO_3$, 25 mass % $KNO_3$).

(e) Given the system containing 30 mass % $NaNO_3$ and 60 mass % $KNO_3$. To this system a solution of 20 mass % $KNO_3$ is added. Indicate on which line composition of the resultant solution would move.

(Ans. (a)(l) unsaturated solution of $KNO_3$ and $NaNO_3$, (2) saturated solution of $NaNO_3$ in the presence of $KNO_3$, (3) saturated solution of $KNO_3$ in the presence of $NaNO_3$, (4) solid $NaNO_3$ in equilibrium with a solution of composition 48.5 mass % $NaNO_3$, 10.0 mass % $KNO_3$ and 41.5 mass % water; relative amounts are 41.7 parts by mass of solid $NaNO_3$ and 58.3 parts by mass of liquid, (5) solid $KNO_3$ in equilibrium with a solution of composition 27 mass % $NaNO_3$, 33 mass % $KNO_3$ and 40.0 mass % water; relative amounts are 40.3 parts by mass of solid $KNO_3$ and 59.7 parts by mass of water, and (6) solid mixture of 51.5 mass % $NaNO_3$ and 48.5 mass % $KNO_3$ in equilibrium with a solution of composition 39.8 mass % $NaNO_3$, 28.9 mass % $KNO_3$ and 31.3 mass % water; relative amounts are 38% by mass of liquid and 62% by mass of solid;

(b) $KNO_3$, 21 mass % $NaNO_3$, 63 mass % $KNO_3$ and 16 mass % water, and

(c) 25 mass % $NaNO_3$, 53 mass % $KNO_3$ and 22 mass % water.)

7.7 (a) In the ternary system potassium sulphate-sodium sulphate-water, a double salt of composition $3K_2SO_4 \cdot Na_2SO_4$ (known as Glaserite, 78.6 mass % $K_2SO_4$ and 21.4 mass % $Na_2SO_4$) is formed. The following compositions of the saturated solutions were obtained.

| *Composition of saturated solution* | | | *Solid phase* |
|---|---|---|---|
| $K_2SO_4$ | $Na_2SO_4$ | *Water* | |
| *mass %* | *mass %* | *mass %* | |
| 14.19 | 0 | 85.81 | $K_2SO_4$ |
| 13.8 | 5.68 | 80.52 | $K_2SO_4$-Glaserite |
| 5.85 | 29.4 | 64.75 | $Na_2SO_4$-Glaserite |
| 0 | 31.6 | 68.4 | $Na_2SO_4$ |

Construct the phase diagram and determine the nature of system at the following compositions.

| *Constituent* | *Percentage compositions by mass* | | | |
|---|---|---|---|---|
| | (1) | (2) | (3) | (4) |
| Potassium sulphate | 8 | 38 | 75 | 30 |
| Sodium sulphate | 17 | 22 | 10 | 45 |
| Water | 75 | 40 | 15 | 25 |

If the system consists of more than one phase, determine also their compositions and relative amounts.

(b) Is the double salt stable in the presence of water?

(c) Show that the solution at which $Na_2SO_4$ and glaserite are in equilibrium is a congruently saturated whereas the solution at which $K_2SO_4$ and glaserite are in equilibrium is an incongruently saturated.

(d) Suppose the apex representing water is joined with the point representing the composition of the double salt. Take a point on the line in the unsaturated solution region and describe the sequence of effects that is produced if the above system is subjected to the isothermal evaporation.

(e) Take a point corresponding to the binary system containing solid $Na_2SO_4$ and the saturated solution of $Na_2SO_4$. Describe the sequence of effects that is produced if $K_2SO_4$ is added to this binary system.

7.8 In the ternary system $NaCl–Na_2SO_4–H_2O$, a stable hydrate $Na_2SO_4 \cdot 10H_2O$ (44.1 mass % $Na_2SO_4$ and 55.9 mass % $H_2O$) is formed. The following compositions of saturated solutions were obtained at 25 °C.

| *Composition of saturated solution* | | | *Solid phase* |
|---|---|---|---|
| NaCl | $Na_2SO_4$ | *Water* | |
| *mass %* | *mass %* | *mass %* | |
| 0 | 21.6 | 78.4 | $Na_2SO_4 \cdot 10H_2O$ |
| 7.6 | 16.0 | 76.4 | $Na_2SO_4 \cdot 10H_2O$ |
| 14.5 | 14.5 | 71.0 | $Na_2SO_4 + Na_2SO_4 \cdot 10H_2O$ |
| 18.4 | 10.4 | 71.2 | $Na_2SO_4$ |
| 22.7 | 7.1 | 70.2 | $Na_2SO_4$ + NaCl |
| 24.65 | 3.35 | 72.0 | NaCl |
| 26.6 | 0 | 73.4 | NaCl |

Construct the phase diagram and determine the nature of system at the following compositions.

| *Constituent* | *Percentage composition by mass* | | | | | |
|---|---|---|---|---|---|---|
| | (1) | (2) | (3) | (4) | (5) | (6) |
| NaCl | 8 | 5 | 5 | 24.5 | 40 | 24 |
| $Na_2SO_4$ | 8 | 30 | 55 | 30.5 | 40 | 6 |
| Water | 84 | 65 | 40 | 45 | 20 | 70 |

If the system consists of more than one phase, determine also their compositions and relative amounts.

7.9 In the ternary system $NaCl–Na_2SO_4–H_2O$ at 17.5 °C, it is found that solid $Na_2SO_4$ is never in equilibrium with the solution. The following compositions of saturated solutions were obtained:

| *Composition of saturated solution* | | | *Solid phase* |
|---|---|---|---|
| $Na_2SO_4$ | NaCl | *Water* | |
| *mass %* | *mass %* | *mass %* | |
| 13.8 | 0 | 86.2 | $Na_2SO_4 \cdot 10H_2O$ |
| 8.5 | 8.1 | 83.4 | $Na_2SO_4 \cdot 10H_2O$ |
| 6.8 | 16.9 | 76.3 | $Na_2SO_4 \cdot 10H_2O$ |
| 7.4 | 22.3 | 70.3 | $Na_2SO_4 \cdot 10H_2O$ + NaCl |
| 0 | 26.5 | 73.5 | NaCl |

Construct the phase diagram and determine the nature of system at the following compositions.

| *Constituent* | *Percentage composition by mass* | | | | |
|---|---|---|---|---|---|
| | (1) | (2) | (3) | (4) | (5) |
| NaCl | 10 | 5 | 50 | 24 | 45 |
| $Na_2SO_4$ | 10 | 35 | 10 | 6 | 45 |
| Water | 80 | 60 | 40 | 70 | 10 |

If the system consists of more than one phase, determine also their compositions and relative amounts.

7.10 (a) In the ternary system $NaNO_3$–$Na_2SO_4$–$H_2O$, a simple hydrate $Na_2SO_4 \cdot 10H_2O$ and the hydrated double salt (ternary compound) $NaNO_3 \cdot Na_2SO_4 \cdot H_2O$ (darapskite 34.6 mass % $NaNO_3$, 58.0 mass % $Na_2SO_4$ and 7.4 mass % $H_2O$) are formed. The following compositions of saturated solutions were obtained:

| *Composition of saturated solution* | | *Solid phase* |
|---|---|---|
| $NaNO_3$ | $Na_2SO_4$ | |
| *mass %* | *mass %* | |
| 47.9 | 0 | $NaNO_3$ |
| 46.5 | 1.57 | $NaNO_3$ |
| 45.2 | 3.05 | $NaNO_3$ + Darapskite |
| 44.6 | 3.40 | Darapskite |
| 43.1 | 3.74 | Darapskite |
| 37.7 | 6.17 | Darapskite |
| 31.1 | 10.2 | Darapskite |
| 29.5 | 11.3 | Darapskite |
| 27.7 | 12.7 | Darapskite |
| 26.8 | 13.4 | Darapskite + $Na_2SO_4$ |
| 25.0 | 14.6 | $Na_2SO_4$ + $Na_2SO_4 \cdot 10H_2O$ |
| 21.2 | 14.8 | $Na_2SO_4 \cdot 10H_2O$ |
| 12.6 | 16.5 | $Na_2SO_4 \cdot 10H_2O$ |
| 0 | 21.6 | $Na_2SO_4 \cdot 10H_2O$ |

Construct the phase diagram and label it.

(b) Show that the ternary salt is unstable in the presence of water.

(c) Show that the solution at which $NaNO_3$ and the ternary compound are in equilibrium is a congruently saturated whereas the solution at which $Na_2SO_4$ and the ternary compound are in equilibrium is an incongruently saturated.

7.11 (a) Figure 4 shows the solubility curves of A · $H_2O$, A, A · B and B. Complete the phase diagram and label it.

(b) Given a point X in Fig. 4. What does it represent? Describe the sequence of events that is produced when the system X is subjected to the isothermal dehydration.

(c) Is it possible to obtain the crystals of pure A on evaporating the system X? If yes, at what point will the yield of crystal A be maximum?

(d) Is the double salt A · B congruently saturating?

(e) How does Fig. 4 help in obtaining the crystals of A · $H_2O$ from a solution containing both A and B?

(f) Is the temperature of the phase diagram of Fig. 4 greater than, equal to or less than 0 °C?

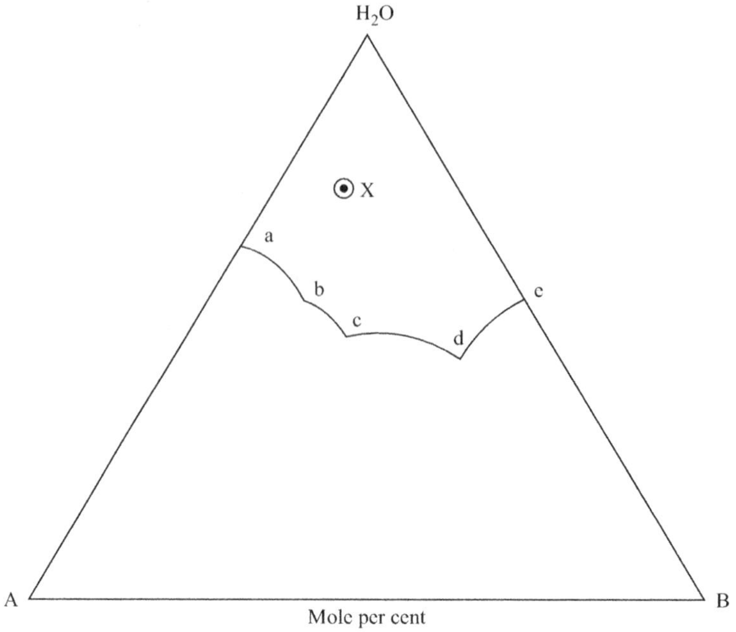


**Fig. 4**

7.12 The following data were obtained on the system $CaCl_2$–KCl–$H_2O$ at 75 °C.

| *Composition of saturated solution* | | *Solid phase* |
|---|---|---|
| $CaCl_2$ | KCl | |
| *mass %* | *mass %* | |
| 0 | 33.16 | KCl |
| 11.73 | 21.68 | KCl |
| 18.27 | 16.10 | KCl |
| 28.47 | 9.62 | KCl |
| 37.65 | 6.77 | KCl |
| 47.65 | 8.43 | KCl |
| 50.19 | 11.32 | KCl and $2KCl \cdot CaCl_2 \cdot 2H_2O$ |
| 50.92 | 9.36 | $2KCl \cdot CaCl_2 \cdot 2H_2O$ |
| 53.85 | 6.21 | $2KCl \cdot CaCl_2 \cdot 2H_2O$ |
| 56.33 | 4.51 | $2KCl \cdot CaCl_2 \cdot 2H_2O$ |
| 57.62 | 3.60 | $2KCl \cdot CaCl_2 \cdot 2H_2O$ and $CaCl_2 \cdot 2H_2O$ |
| 57.77 | 2.56 | $CaCl_2 \cdot 2H_2O$ |
| 58.58 | 0 | $CaCl_2 \cdot 2H_2O$ |

Construct the triangular plot and label each area. Describe what would happend on isothermal evaporation at 75 °C of a solution initially containing 85 mass % water, 10 mass % $CaCl_2$ and 5 mass % KCl. Assume 100 g of solution. What mass of $H_2O$ have been evaporated when half the mass of the system is in solids?

7.13 The following data refer to the system $K_2SO_4$–$(NH_4)_2SO_4$–$H_2O$ at 30 °C.

| *Composition of saturated solution* | | *Composition of moist solid solution* | |
|---|---|---|---|
| $K_2SO_4$ | $(NH_4)_2SO_4$ | $K_2SO_4$ | $(NH_4)_2SO_4$ |
| *mass %* | *mass %* | *mass %* | *mass %* |
| 0 | 44.2 | only $(NH_4)_2SO_4$ | |
| 1.2 | 42.7 | 1.6 | 90.3 |
| 2.4 | 40.9 | 22.1 | 69.3 |
| 4.1 | 37.8 | 33.9 | 57.8 |
| 4.9 | 33.5 | 50.8 | 42.3 |
| 6.4 | 31.0 | 62.2 | 30.0 |
| 9.1 | 18.5 | 84.0 | 12.7 |
| 10.7 | 8.4 | 93.2 | 3.1 |
| 11.2 | 0 | only $K_2SO_4$ | |

Draw the phase diagram and label each area. A system containing 90 g of water and 10 g each of $K_2SO_4$ and $(NH_4)_2SO_4$ is evaporated isothermally at 30 °C. (i) What is the composition of the solution when a precipitate first appears and what is the composition of the precipitate? (ii) What is the composition of the last solution to exist before the system solidifies? Give the compositions and amounts of the solid (s) present. (iii) When the system reaches the point containing 40% by mass of water what phases are present and in what amounts?

# 8 Electrochemical Cells

## 8.1 INTRODUCTION

Electrochemical cells can be broadly classified into two categories, viz., electrolytic cells and galvanic (or voltaic) cells. In the former, a chemical reaction (more precisely electrolysis) is carried out with the help of an electrical current, whereas in the latter an electrical current is produced as a result of some spontaneous chemical reaction. Thus, an electrolytic cell is a device to convert electrical energy into chemical energy whereas in the galvanic cell, the reverse of this process takes place. In this section, we will outline the working of these two types of cells along with their sign conventions.

### Electrolytic Cell

Consider an electrolytic cell involving the electrolysis of molten sodium chloride using platinum electrodes as shown in Fig. 8.1.1.

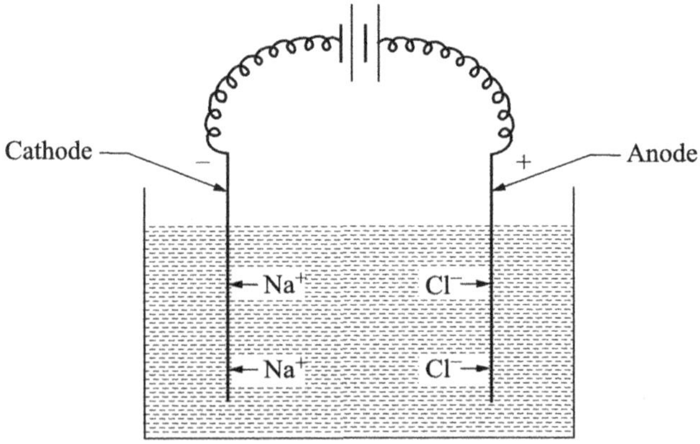


**Fig. 8.1.1 Electrolysis of molten sodium chloride**

The left-side electrode is connected to the negative terminal of an external battery which supplies electrons to this electrode. It is because of this, the left-side electrode becomes negatively charged. Electrons are being sucked but of the right-side electrode by the positive terminal of the battery and thus this electrode acquires a positive charge. Within the solution, electricity is carried out by the movements of ions. The positive ions (i.e. cations) are attracted towards the negative electrode and the negative ions (i.e. anions) towards the positive electrode. Since cations are attracted towards the negative electrode, the latter is known as the *cathode*. Similarly the positive electrode is known as the *anode*. Electrode reactions involve reduction at the cathode and oxidation at the anode. Thus in the present case, we have

*At cathode* $Na^+(l) + e^- \rightarrow Na(s)$

*At anode* $Cl^-(l) \rightarrow \frac{1}{2}Cl_2(g) + e^-$

The overall transformation is

$$Na^+(l) + Cl^-(l) \rightarrow Na(s) + \tfrac{1}{2}Cl_2(g)$$

## Galvanic Cell

As stated above, in the galvanic cell, electric current is generated as a result of some spontaneous chemical reaction that takes place in the cell.

## Working of Daniell Cell

To understand how electric current is generated, let us consider an example of Daniell cell where the following reaction is produced:

$$Zn(s) + CuSO_4(aq) \rightarrow Cu(s) + ZnSO_4(aq)$$

or, in ionic form

$$Zn(s) + Cu^{2+}(aq) \rightarrow Cu(s) + Zn^{2+}(aq)$$

This reaction occurs at a very fast speed, i.e. the moment zinc is added in copper sulphate solution, copper is immediately formed. In this reaction, we have basically the following two partial reactions:

$Zn(s) \rightarrow Zn^{2+}(aq) + 2e^-$ (oxidation)

and $Cu^{2+}(aq) + 2e^- \rightarrow Cu(s)$ (reduction)

If the two partial reactions are carried out in two physically separate places, the electrons will have to flow from a place where oxidation occurs to a place where reduction occurs through an external wire. This flow of electrons constitutes a current, which can be used to drive a motor, lift a mass, or do any other work.

## Formulation of Daniell Cell

The most obvious way to separate the partial reactions occurring in Daniell cell is to insert the Zn electrode into a solution containing $Zn^{2+}$ ions (but not $Cu^{2+}$ ions) and the Cu electrode (or some unreactive metal) into a solution containing $Cu^{2+}$ ions as shown in Fig. 8.1.2. From Zn electrode, a minute amount of Zn can go into the solution as $Zn^{2+}$ ions leaving a negative charge on the electrode (because of electrons left behind) and a positive charge in the solution. These charges stop the ionization process of zinc almost immediately; for every $Zn^{2+}$ ion entering the solution, it or another ion will be driven back to the Zn electrode by the attraction of the negative electrode and the repulsion of the positive solution. Similarly, on the Cu electrode a few $Cu^{2+}$ ions from the solution may be deposited as Cu atoms. This process will make the electrode positive and the solution negative and thus the deposition process is stopped.† However, these

†It may be that both Cu and Zn electrodes give up positive ions to solution and become negatively charged or they may both become positively charged by neutralization of positive ions from the solution. But from the knowledge that Zn is more electropositive than Cu, we can say that Zn has a greater tendency to dissolve as $Zn^{2+}$ than Cu has to dissolve as $Cu^{2+}$.

processes at the two electrodes, namely, oxidation at Zn electrode and reduction at Cu electrode, can be resumed provided we connect the two electrodes by a wire to allow the electrons to flow from Zn electrode to the electron-deficient Cu electrode and bring the two solutions into contact through a porous plug or a salt bridge† so as to neutralize the charges of the two solutions and prevent a direct flow of $Cu^{2+}$ ions towards Zn electrode.

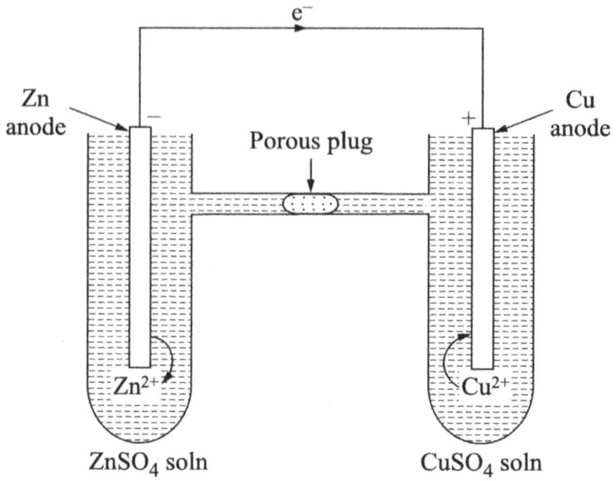


**Fig. 8.1.2 Daniell cell**

### Sign and Nomenclature of Electrodes

In the cell shown in Fig. 8.1.2, the Zn electrode becomes negatively charged as the electrons released in the oxidation reaction ($Zn \rightarrow Zn^{2+} + 2e^-$) reside at this electrode. Similarly, the Cu electrode becomes positively charged as the electrons are taken out of this electrode for the reduction reaction ($Cu^{2+} + 2e^- \rightarrow Cu$) which takes place at this electrode. As in the electrolytic cell, the electrode at which oxidation reaction takes place is known as the anode and that where reduction reaction occurs is known as the cathode. Thus, the negative electrode of the galvanic cell is known as anode and the positive one as the cathode.

### Galvanic Cell Without a Salt Bridge

In the above example of Daniell cell, we have two solutions which have to be connected so that no intermixing takes place. An example in which this difficulty does not arise is due to the following spontaneous reaction:

$$Zn(s) + 2AgCl(s) \rightarrow 2Ag(s) + Zn^{2+}(aq) + 2Cl^-(aq)$$

If the two solids Zn and AgCl are brought into contact, the reaction will take place very slowly owing to the difficulty of establishing sufficient intimate contact between them. But the reaction has a strong tendency to take place and can be carried out by using the cell shown in Fig. 8.1.3.

The reaction at the Zn electrode is

$$Zn(s) \rightarrow Zn^{2+}(aq) + 2e^-$$

---

†The salt bridge is an agar jelly saturated with either KCl or $NH_4NO_3$ and is used to connect the two electrode compartments. See also Section 8.21.

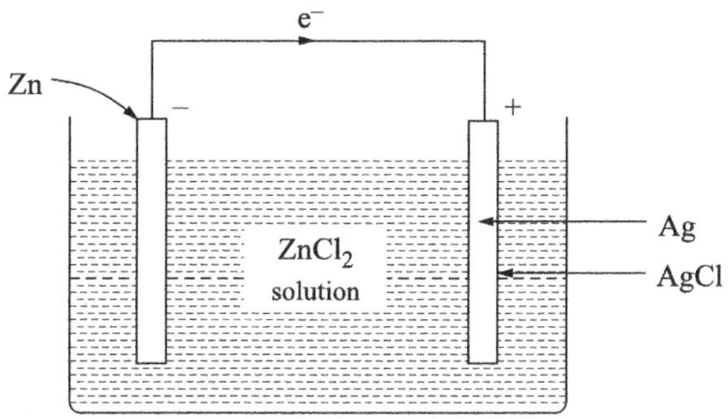


**Fig. 8.1.3 Another example of a galvanic cell**

The two electrons left behind on the Zn can travel through the external wire to the silver-silver chloride electrode where the following reaction occurs.

$$2AgCl(s) + 2e^- \rightarrow 2Ag(s) + 2Cl^-(aq)$$

Thus, the overall process is given by the reaction:

$$Zn(s) + 2AgCl(s) \rightarrow 2Ag(s) + Zn^{2+}(aq) + 2Cl^-(aq)$$

## Resume of Sign Conventions

In a galvanic cell, electrons are fed into the anode by the oxidation reaction that takes place at the electrode. The anode pushes these electrons into the external circuit and is, therefore, regarded as the negative electrode—the source of negative charge. The cathode is an electron sink and is, therefore, called the positive electrode.

In an electrolytic cell, the cathode is still the electron sink, but, this time, we are forcibly feeding electrons into the cathode by connecting it to the negative terminal of a generator or a galvanic cell. It is because of this that the cathode of an electrolytic cell is a negative electrode. Since electrons are being sucked out of the anode by the positive terminal of the external battery, the anode is considered a positive electrode.

Oxidation takes place at anode and reduction at cathode in a cell of either type. The only difference is about the sign of the electrodes. Table 8.1.1 gives the sign convention and electrode reactions in both cells.

**Table 8.1.1** Sign Convention and Electrode Reactions in a Cell

| | *Anode* | | *Cathode* | |
|---|---|---|---|---|
| *Cell* | *Sign* | *Type of reaction* | *Sign* | *Type of reaction* |
| Galvanic | Negative | Oxidation | Positive | Reduction |
| Electrolytic | Positive | Oxidation | Negative | Reduction |

## Diagrammatic Distinction between Galvanic and Electrolytic Cells

Figure 8.1.4 illustrates the distinction between the sign conventions of the two types of cells.

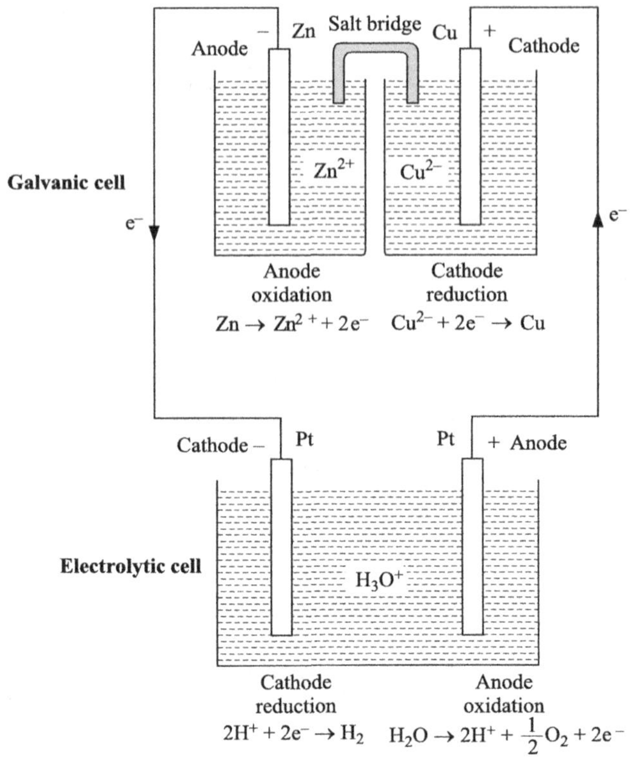


**Fig. 8.1.4 Sign conventions in the galvanic and the electrolytic cells**

## 8.2 REVERSIBLE AND IRREVERSIBLE CELLS

The use of thermodynamic principles when applied to a system is subject to one important restriction, namely, that the system must be reversible in the thermodynamic sense. This requires that:

(1) the driving and opposing forces be infinitesimally different from each other, and

(2) it should be possible to reverse any change taking place by applying a force infinitesimally greater than the acting one.

A cell satisfying the above two requirements constitutes a reversible cell. The potential difference of the cell can be substituted into the relevant thermodynamic relations and hence the values of thermodynamic properties such as free energy change, entropy change and enthalpy change of the cell reaction can be determined. When the above conditions are not satisfied, the cell is said to be irreversible, and thermodynamic relations do not apply.

The difference between reversible and irreversible cells may be illustrated with the following two examples.

### An Example of Reversible Cell

Consider a cell composed of Zn and Ag-AgCl electrodes dipping into an aqueous solution of zinc chloride. As seen earlier, the following reactions take place on connecting the electrodes externally

$$\textit{Anode} \qquad \tfrac{1}{2}Zn(s) \rightarrow \tfrac{1}{2}Zn^{2+}(aq) + e^-$$

$$\textit{Cathode} \qquad AgCl(s) + e^- \rightarrow Ag(s) + Cl^-(aq)$$

with the net reaction

$$\tfrac{1}{2}Zn(s) + AgCl(s) \rightarrow Ag(s) + \tfrac{1}{2}Zn^{2+}(aq) + Cl^-(aq)$$

The above process continues as long as the external opposing potential is infinitesimally smaller than that of the cell. However, if the opposing potential becomes slightly larger than that of the cell, the direction of current flow is reversed, and so is the cell reaction. Now zinc ions are converted to zinc at one electrode, silver chloride is formed from silver and chloride ions at the other, and the overall cell reaction becomes

$$Ag(s) + \tfrac{1}{2}Zn^{2+}(aq) + Cl^-(aq) \rightarrow \tfrac{1}{2}Zn(s) + AgCl(s)$$

Thus, it is obvious that the second condition of reversibility mentioned above is satisfied. The first condition can be satisfied by drawing from or passing through the cell a very minute current. Hence, the cell is reversible.

**An Example of Irreversible Cell**

Consider a cell composed of zinc and silver electrodes immersed in a solution of sulphuric acid. When the two electrodes are short circuited, (i.e. connected externally by a wire) zinc dissolves with the evolution of hydrogen to form zinc sulphate according to the reaction:

$$Zn(s) + H_2SO_4 \rightarrow ZnSO_4 + H_2(g)$$

However, when the cell is connected with an external source of potential slightly greater than its own, silver dissolves at one electrode, hydrogen is evolved at the other, and the cell reaction becomes

$$2Ag(s) + H_2SO_4 \rightarrow Ag_2SO_4 + H_2(g)$$

Thus, it is obvious that though the first condition of reversibility can be satisfied, the second cannot. Hence, the cell is an irreversible cell. The potential of such a cell does not have any definite thermodynamic significance.

## 8.3 ELECTROMOTIVE FORCE AND ITS MEASUREMENT

**Definition of Electromotive Force**

In the Daniell cell, electrons flow from Zn electrode to Cu electrode. This is due to the fact that Zn atom can be more easily oxidized to $Zn^{2+}$ ion than Cu atom to $Cu^{2+}$ ion. On the other side, $Cu^{2+}$ ion can be more easily reduced than the $Zn^{2+}$ ion. Consequently, Zn atom is oxidized to $Zn^{2+}$ ion and the electrons set free at the Zn electrode move towards Cu electrode where $Cu^{2+}$ ion is reduced to Cu atom. Basically, we can say that the flow of electrons is due to the difference of oxidation tendencies of the two atoms, or it is due to the difference of reduction tendencies of the two ions. The relative oxidation tendencies of atoms can be represented by the oxidation potentials. A more easily oxidizable atom will have a larger value of oxidation potential and a less easily oxidizable atom will have a low value of oxidation potential. Thus, Zn atom has a larger oxidation potential than the Cu atom. Electrons in the external circuit flow from the electrode of higher oxidation potential to the electrode of lower oxidation potential. The difference of potential which causes a current to flow from the electrode of higher potential to the lower one is known as the *electromotive force,* abbreviated as emf, of the cell and is expressed in volts. We will represent emf of a cell by the symbol $E_{\text{cell}}$. Thus

$$E_{\text{cell}} = \text{Higher oxidation potential} - \text{Lower oxidation potential} \qquad (8.3.1a)$$

## EMF in Terms of Reduction Potentials

By convention, $E_{\text{cell}}$ is expressed in terms of reduction potentials (known as *standard potentials)* of the two electrodes. The reduction potentials measure the relative reduction tendencies of ions and their values are simply negative of oxidation potentials of the corresponding atoms. This follows from the fact that an atom with the maximum tendency of oxidation (i.e. highest value of oxidation potential) will yield an ion with the least tendency of reduction (i.e. the minimum reduction potential) and vice versa. Thus, Eq. (8.3.1a) may be written as

$$E_{\text{cell}} = -(-\text{Higher oxidation potential}) + (-\text{ Lower oxidation potential})$$

$$= -\text{ Lower reduction potential} + \text{Higher reduction potential}$$

i.e. $$E_{\text{cell}} = \text{Higher reduction potential} - \text{Lower reduction potential} \qquad (8.3.1b)$$

## Measurement of EMF

To determine the accurate emf of a cell, direct use of voltmeter cannot be made, since it draws some current from the cell, which will change the emf due to the changes in the concentrations of species caused by the electrode reactions. Again, with appreciable current flow part of the emf will have to be utilized to overcome the internal resistance of the cell, and hence the potential measured on the voltmeter will not be the accurate cell emf.

The precise value of the emf of a cell can be determined by using *potentiometric method.* In this method, an unknown emf is opposed by another emf until the two are equal. Figure 8.3.1 illustrates the experimental set up for determining the emf of a cell with the potentiometric method, where we have:

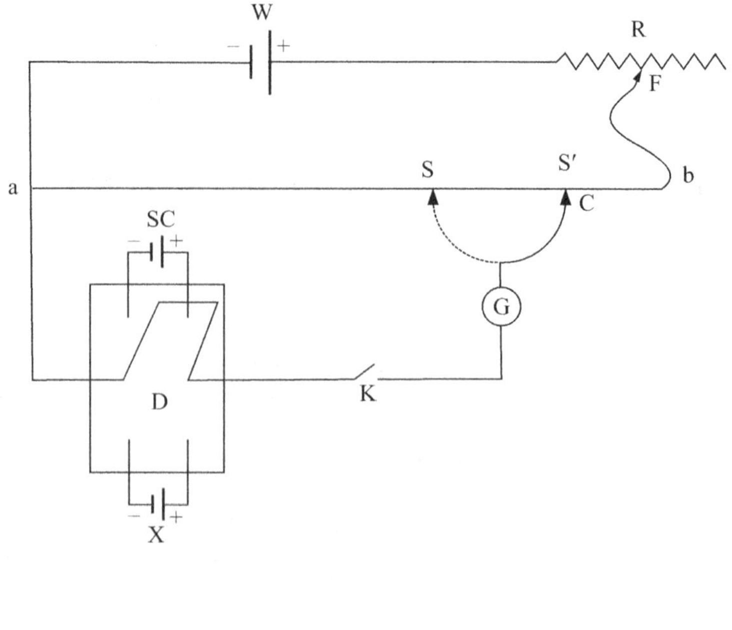


**Fig. 8.3.1 Experimental set up for determining the emf of a cell**

- ab is a uniform wire. The resistance offered by the wire is directly proportional to its length and is calibrated in volts.
- SC is the standard cell, whose emf is reproducible, constant with time. Usually a standard cadmium cell is used whose emf is 1.018 3 V at 20 °C.
- X is an unknown cell, whose emf is to be determined.
- W is a working cell, whose potential is larger than SC and X. All the three cells have their negative terminals attached to one end of the wire ab.
- R is a resistor of variable resistance.
- G is a galvanometer.

Because of the working cell W, there is a uniform potential gradient along the wire ab. First, the pointer C is set at a point S′ along ab corresponding to the value of the emf of the standard cell, say 1.018 3 V. Next SC is thrown into the circuit and F is moved along the resistor R until the galvanometer G shows no deflection. When this balance is established, the current flowing through ab is of such a magnitude as to make the potential drop between a and S′ exactly 1.0183 V and the voltage drop anywhere along ab is the same as the voltage marking on the slide wire. Now the given cell is thrown into circuit and the pointer C is moved along ab until a point S is found at which the galvanometer G again shows no deflection. The reading of the slide wire at S gives the voltage directly.

**Standard Cell**

The standard cell should be such that its potential should be reproducible, constant with time and also it should be reversible. It should not suffer permanent damage due to passage of current and should preferably have a low temperature coefficient of emf. The cells that most closely approximate these requirements are Weston unsaturated and saturated standard cells.

This type of cell consists of a H-shaped glass vessel containing in each arm one of the electrodes as shown in Fig. 8.3.2. The positive electrode consists of a platinum wire dipped into mercury that is covered with a paste of mercurous sulphate and mercury. The negative electrode consists of a platinum wire dipped into cadmium amalgam containing 12 to 14 mass % of cadmium. Over both electrodes are sprinkled some crystals of solid $CdSO_4\cdot$ (8/3) $H_2O$. The entire cell is filled with a saturated solution of cadmium sulphate, and the cell is closed with corks and sealing wax. The function of the solid crystals of $CdSO_4 \cdot (8/3)H_2O$ is to keep the solution saturated with $CdSO_4$ at all temperatures.

The working of the Weston saturated cell depends on the following reversible reactions:

$$Cd(s) + SO_4^{2-}(aq) + \tfrac{8}{3}H_2O(l) \rightarrow CdSO_4 \cdot \tfrac{8}{3}H_2O(s) + 2e^-$$

$$Hg_2SO_4(s) + 2e^- \rightarrow 2Hg(l) + SO_4^{2-}(aq)$$

with the overall reaction as

$$Cd(s) + Hg_2SO_4(s) + \tfrac{8}{3}H_2O(l) \rightarrow CdSO_4 \cdot \tfrac{8}{3}H_2O(s) + 2Hg(l)$$

The reaction occurs when the cell is acting as a source of current, while reverse reaction takes place when current is passed through the cell.

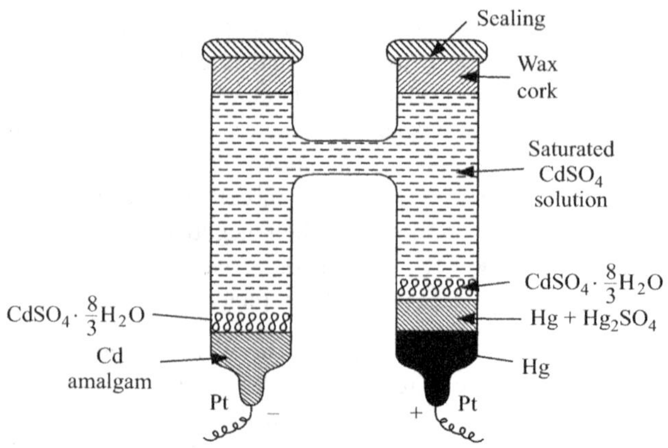


**Fig. 8.3.2 The standard Weston saturated cell**

The potential of the cell at any temperature is given by the equation

$$E/\text{V} = 1.018\ 30 - 4.06 \times 10^{-5}\ \{(\theta_c/°\text{C}) - 20\} - 9.5 \times 10^{-7}\ \{(\theta_c/°\text{C}) - 20\}^2$$

## 8.4 FORMULATION OF A GALVANIC CELL

### Half-Cell

An electrochemical cell requires two electrodes for a reaction to occur. The assembly comprising of the electrode and the reagents that are involved with it is called the *half-cell* and the component of the total chemical reaction that occurs in the half-cell is the *half-cell reaction* or *electrode reaction.* Thus, we require two half-cells to produce an electrochemical cell.

### Cell Diagram

A half-cell or an electrochemical cell can be briefly represented by a cell diagram following a few rules given below.

- The separation of two phases is shown by a verticle line (I).
- The various materials present in the same phase are shown together with the help of commas.
- The two half-cells are joined with the help of a dashed vertical line (⁞) if the two liquids at the junction are miscible and by a double dashed vertical lines (⁞⁞) if the liquid junction potential at the junction of the two liquids is assumed to be eliminated.
- The significant features of the substances such as fugacity of gas, activity of ion, etc., are indicated in brackets drawn immediately after writing the substance.

### Illustrations

Taking the examples of galvanic cells given in Figs 8.1.4 and 8.1.3, we represent respectively, the two cells as

$$\text{Zn} \mid \text{Zn}^{2+}(a_1) \ ⁞⁞\ \text{Cu}^{2+}(a_2) \mid \text{Cu}$$

$$\text{Zn} \mid \text{ZnCl}_2(m) \mid \text{AgCl} \mid \text{Ag}$$

The Weston cadmium cell shown in Fig. 8.3.2 will be written as

$$\text{Pt} \mid \text{Hg} \mid \text{Cd(Hg)} \mid \text{CdSO}_4 \cdot \tfrac{8}{3}\,\text{H}_2\text{O(s)} \mid \text{saturated solution of CdSO}_4$$

$$\mid \text{CdSO}_4 \cdot \tfrac{8}{3}\,\text{H}_2\text{O(s)} \mid \text{Hg}_2\text{SO}_4, \text{Hg} \mid \text{Pt}$$

## 8.5 ELECTRICAL AND ELECTROCHEMICAL POTENTIALS

**Definition of Electric Potential**

The electrical potential at a point in space is defined as the work involved in bringing a unit charge from infinity, where the electric potential is zero, to the point under study. If $w_{el}$ is the electrical work involved in bringing a charge $Q$ from infinity to a given point, then the potential $V$ at this point is given by

$$V = \frac{w_{el}}{Q} \qquad (8.5.1)$$

**Work Involved in Moving a Charge**

If a charge $dQ$ moves from a point at a potential $V_1$ to a point at a potential $V_2$ the work involved in this process is given by

$$dw_{el} = (V_2 - V_1)\, dQ \qquad (8.5.2)$$

If one coulomb of the charge is transferred under the potential difference of one volt, the work done is 1 volt-coulomb. The latter is numerically equal to 1 J of work.

**Electrochemical and Chemical Potentials of a Species**

The escaping tendency (i.e. chemical potential) of a charged particle (an ion or an electron) in a phase depends on the electrical potential of that phase. To find the relation between the electrical potential and the escaping tendency of a charged particle, we consider a system of two phases M and M′ of the same material at two different potentials $\phi$ and $\phi'$, respectively. If the amount $dn$ of the $i$th species carrying $z_i$ charge is transferred from M to M′, then

$$dQ = z_i F\, dn \qquad (8.5.3)$$

and

$$dw_{el} = (\phi' - \phi)\, dQ = (\phi' - \phi)\, z_i F\, dn \qquad (8.5.4)$$

If the transfer is carried out reversibly, the electrical work $dw_{el}$ will be equal to the change in the free energy of the system, i.e.

$$dG = dw_{el} \qquad (8.5.5)$$

Substituting the relation of $dw_{el}$ from Eq. (8.5.4), we get

$$dG = (\phi' - \phi)\, z_i F\, dn \qquad (8.5.6)$$

Let $\tilde{\mu}_{z_i}$ and $\tilde{\mu}'_{z_i}$ be the chemical potentials of the $i$th species in the phases M and M′, respectively. If the amount $dn$ of this species is transferred from M to M′, the free energy change of the system is given by

$$dG = \tilde{\mu}'_{z_i}\, dn - \tilde{\mu}_{z_i}\, dn \qquad (8.5.7)$$

Comparing Eqs (8.5.6) and (8.5.7), we get

$$\tilde{\mu}'_{z_i} - \tilde{\mu}_{z_i} = (\phi' - \phi)\, z_i F \qquad (8.5.8)$$

Let $\mu_{z_i}$ be the chemical potential of the $i$th species when the potential of the phase M′ is zero. With this, Eq. (8.5.8) becomes

$$\tilde{\mu}_{z_i} = \mu_{z_i} + z_i F \phi \qquad (8.5.9)$$

Equation (8.5.9) relates the chemical potential (or the escaping tendency) of the *i*th species of charge $z_i$ in a phase to its electrical potential. The escaping tendency is a linear function of $\phi$. Equation (8.5.9) divides the chemical potential $\tilde{\mu}_{z_i}$ of a charged species in two terms; the first term, $\mu_{z_i}$ is the *chemical contribution* and the second term $z_i F\phi$ is the *electrical contribution.* The former has the same value in two phases of the same chemical composition since its value is a function of $T$, $p$ and the composition. The electrical contribution of the phase depends upon the value of $\phi$. On account of these two contributions, the chemical potential $\tilde{\mu}_i$ is commonly referred to as the *electrochemical potential.*

**Electrochemical Potential of an Electron in a Metal**

For a metal, $z_i = 0$ and hence its electrochemical and chemical potentials have the same value. For an electron in a metal electrode, Eq. (8.5.9) becomes

$$\tilde{\mu}_{e(M)} = \mu_{e(M)} - F\phi \tag{8.5.10}$$

From Eq. (8.5.10), it follows that the electrochemical potential $\tilde{\mu}_{e(M)}$ and the electrical potential $\phi$ have opposite trends, if the value $\phi$ is increased, the value of $\tilde{\mu}_{e(M)}$ is decreased and vice versa.

**Electric Potential and Type of Reaction Occurring at the Electrode**

Since $\tilde{\mu}_{e(M)}$ in Eq. (8.5.10) represents the escaping tendency of electron from the electrode, it is obvious that if a negative electrical potential is applied to an electrode, the electrochemical potential becomes larger than the chemical potential indicating that the escaping tendency of an electron from the electrode is enhanced. This, in turn, implies that the tendency of an oxidation reaction to take place at the electrode is increased. On the other hand, the tendency of electrons to escape from the electrode is diminished if a positive potential is applied to it. Alternatively, we may state that the tendency of the electrode to accept electrons is increased and hence the process that can take place at the electrode is the reduction one. Now, if two electrodes of different electrical potentials (both positive or both negative or one positive and the other negative) are coupled to make a cell, it is obvious that the oxidation reaction will take place at the electrode of higher negative potential (or lesser positive potential) and the reduction at the electrode of lower negative potential (or higher positive potential).

**Reduction Potentials**

Since on increasing the value of electrical potential, the tendency of reduction reaction to occur at the electrode is increased, the electrical potential is commonly referred to as the *reduction* potential. It is for this reason, the equilibrium reaction at the electrode is also written as the reduction reaction.

## 8.6 DIFFERENT TYPES OF HALF-CELLS AND THEIR REDUCTION POTENTIALS

In this section, we consider various types of half-cells that are employed and the expressions which relate the half-cell potential with the activity and fugacity of the constituents of the half-cell.

### GAS-ION HALF-CELL

In the gas-ion half-cell, an inert collector of electrons, platinum or graphite is in contact with gas and a solution containing a specified ion.

## Hydrogen Gas-Hydrogen Ion Half-Cell

One of the most important gas-ion half-cell is the hydrogen gas-hydrogen ion half-cell (also commonly know as hydrogen electrode). In this half-cell, purified hydrogen gas at a constant pressure is passed over a platinum electrode which is in contact with an acid solution (Fig. 8.6.1).

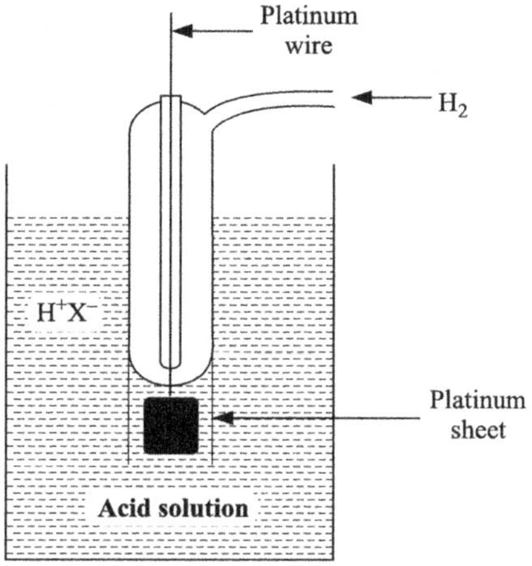


**Fig. 8.6.1 Hydrogen gas-hydrogen ion half-cell**

## Expression of Reduction Potential (Nernst Equation)

The expression of $E_{\text{half-cell}}$ can be derived by considering the reversible reduction reaction that occurs at the electrode. For the present case, we have

$$H^+(aq) + e^-(Pt) \rightarrow \tfrac{1}{2} H_2(g)$$

The equilibrium condition at the electrode is

$$\tilde{\mu}_{H^+(aq)} + \tilde{\mu}_{e^-(Pt)} = \tfrac{1}{2}\tilde{\mu}_{H_2(g)} \tag{8.6.1}$$

Making use of Eq. (8.5.9), the above expression becomes

$$\{\mu_{H^+(aq)} + F\phi_{aq}\} + \{\mu_{e^-(Pt)} - F\phi_{Pt}\} = \tfrac{1}{2}\mu_{H_2(g)} \tag{8.6.2}$$

Substituting the expressions

$$\mu_{H^+(aq)} = \mu^{\circ}_{H^+(aq)} + RT \ln a_{H^+(aq)} \qquad (8.6.3)^{\dagger}$$

and $$\mu_{H_2(g)} = \mu^{\circ}_{H_2(g)} + RT \ln (f_{H_2(g)}/f^{\circ}) \tag{8.6.4}$$

in Eq. (8.6.2) and rearranging it, we get

$$\phi_{Pt} - \phi_{aq} = \frac{\mu^{\circ}_{H^+(aq)} - \frac{1}{2}\mu^{\circ}_{H_2(g)} + \mu_{e^-(Pt)}}{F} - \frac{RT}{F} \ln \frac{(f_{H_2(g)}/f^{\circ})^{1/2}}{a_{H^+(aq)}}$$

---

† See Annexure I at the end of this chapter for the concept of activity.

By definition, $E_{\text{half-cell}} = \phi_{\text{Pt}} - \phi_{\text{aq}}$. Thus, we have

$$E_{\text{H}^+ \mid \text{H}_2 \mid \text{Pt}} = \frac{\mu^{\circ}_{\text{H}^+(\text{aq})} - \frac{1}{2}\mu^{\circ}_{\text{H}_2(\text{g})} + \mu_{\text{e}^-(\text{Pt})}}{F} - \frac{RT}{F} \ln \frac{(f_{\text{H}_2(\text{g})}/f^{\circ})^{1/2}}{a_{\text{H}^+}} \tag{8.6.5}$$

If the fugacity of the gas is unity and the activity of $H^+$ in solution is unity, the half-cell is said to be in the standard state and the corresponding potential, the standard reduction potential (or simply as the standard potential), is given by

$$E^{\circ}_{\text{H}^+ \mid \text{H}_2 \mid \text{Pt}} = \frac{\mu^{\circ}_{\text{H}^+(\text{aq})} - \frac{1}{2}\mu^{\circ}_{\text{H}_2(\text{g})} + \mu_{\text{e}^-(\text{Pt})}}{F} \tag{8.6.6}$$

With this, Eq. (8.6.5) modifies to

$$E_{\text{H}^+ \mid \text{H}_2 \mid \text{Pt}} = E^{\circ}_{\text{H}^+ \mid \text{H}_2 \mid \text{Pt}} - \frac{RT}{F} \ln \frac{(f_{\text{H}_2(\text{g})}/f^{\circ})}{a_{\text{H}^+}} \tag{8.6.7}$$

Equation of the above type is known as the *Nernst equation.*

## METAL-METAL ION HALF-CELL

Metal-metal ion half-cell consists of a bar of metal M in contact with a solution containing $M^{n+}$ ions. Examples include zinc-zinc ion, copper-cupric ion, silver-silver ion and gold-auric ion half-cells.

**Expression of Reduction Potential**

The equilibrium reaction at the electrode is

$$M^{n+}(\text{aq}) + ne^- \rightarrow M(\text{s})$$

Equality of electrochemical potentials yields

$$\tilde{\mu}_{\text{M}^{n+}(\text{aq})} + n\tilde{\mu}_{\text{e}^-(\text{M})} = \mu_{\text{M(s)}}$$

or $$\{\mu_{\text{M}^{n+}(\text{aq})} + nF\phi_{\text{aq}}\} + n\{\mu_{\text{e}^-(\text{M})} - F\phi_{\text{M}}\} = \mu_{\text{M(s)}} \tag{8.6.8}$$

Substituting the expression

$$\mu_{\text{M}^{n+}} = \mu^{\circ}_{\text{M}^{n+}} + RT \ln a_{\text{M}^{n+}(\text{aq})}$$

in Eq. (8.6.8) and rearranging the resultant expression, we get

$$\phi_{\text{M}} - \phi_{\text{aq}} = \frac{\mu^{\circ}_{\text{M}^{n+}(\text{aq})} + n\mu_{\text{e}^-(\text{M})} - \mu_{\text{M(s)}}}{nF} - \frac{RT}{nF} \ln \frac{1}{a_{\text{M}^{n+}(\text{aq})}} \tag{8.6.9}$$

If the activity of $M^{n+}$ is unity, we get the standard potential of M–$M^{n+}$ half-cell, such that

$$E^{\circ}_{\text{M}^{n+}\mid\text{M}} = \frac{\mu^{\circ}_{\text{M}^{n+}(\text{aq})} + n\mu_{\text{e}^-(\text{M})} - \mu_{\text{M(s)}}}{nF} \tag{8.6.10}$$

Substituting the above expression in Eq. (8.6.9), we get

$$E_{M^{n+}|M} = E^{\circ}_{M^{n+}|M} - \frac{RT}{nF} \ln \frac{1}{a_{M^{n+}}} \tag{8.6.11}$$

## METAL AMALGAM-METAL ION HALF-CELL

In this electrode, metal amalgam is placed in contact with a solution containing metal ion. Electrical contact is made by a platinum wire dipping into the amalgam pool.

**Expression of Reduction Potential**

The equilibrium reaction at the electrode is

$$M^{n+}(aq) + ne^-(Pt) \rightarrow M(Hg)$$

Equality of electrochemical potentials yields

$$\tilde{\mu}_{M^{n+}(aq)} + n\tilde{\mu}_{e^-(Pt)} = \mu_{M(Hg)} \tag{8.6.12}$$

or

$$\{\mu_{M^{n+}(aq)} + nF\phi_{aq}\} + n\{\mu_{e^-(Pt)} - F\phi_{Pt}\} = \mu_{M(Hg)} \tag{8.6.13}$$

Substituting the expressions

$$\mu_{M^{n+}(aq)} = \mu^{\circ}_{M^{n+}(aq)} + RT \ln a_{M^{n+}(aq)}$$

and

$$\mu_{M(Hg)} = \mu^{\circ}_{M} + RT \ln a_{M(Hg)}$$

in Eq. (8.6.13) and rearranging the resultant expression, we get

$$\phi_{Pt} - \phi_{aq} = \frac{\mu^{\circ}_{M^{n+}(aq)} + n\mu_{e^-(Pt)} - \mu^{\circ}_{M}}{nF} - \frac{RT}{nF} \ln \frac{a_{M(Hg)}}{a_{M^{n+}}} \tag{8.6.14}$$

The reduction potential of $M^{n+} | M(Hg) | Pt$ assembly will be given by

$$E_{M^{n+}|M(Hg)|Pt} = \{\phi_{Pt} - \phi_{M(Hg)}\} + \{\phi_{M(Hg)} - \phi_{aq}\}$$

$$= \phi_{Pt} - \phi_{aq}$$

Hence Eq. (8.6.14) becomes

$$E_{M^{n+}|M(Hg)|Pt} = \frac{\mu^{\circ}_{M^{n+}(aq)} + n\mu_{e^-(Pt)} - \mu^{\circ}_{M}}{nF} - \frac{RT}{nF} \ln \frac{a_{M(Hg)}}{a_{M^{n+}}} \tag{8.6.15}$$

If the activities of M(Hg) and $M^{n+}$(aq) are unity, we get the standard potential of M–$M^{n+}$ half-cell, such that

$$E^{\circ}_{M^{n+}|M|Pt} = \frac{\mu^{\circ}_{M^{n+}(aq)} + n\mu_{e^-(Pt)} - \mu^{\circ}_{M}}{nF}$$

With this, Eq. (8.6.15) modifies to

$$E_{M^{n+}|M(Hg)|Pt} = E^{\circ}_{M^{n+}|M|Pt} - \frac{RT}{nF} \ln \frac{a_{M(Hg)}}{a_{M^{n+}}} \tag{8.6.16}$$

Equation (8.6.16) may be written as

$$E_{M^{n+} | M(Hg) | Pt} = E^{\circ}_{M^{n+} | M | Pt} - \frac{RT}{nF} \ln a_{M(Hg)} - \frac{RT}{nF} \ln \frac{1}{a_{M^{n+}}}$$

$$= E^{\circ}_{M^{n+} | M(Hg) | Pt} - \frac{RT}{nF} \ln \frac{1}{a_{M^{n+}}} \tag{8.6.17}$$

where $E^{\circ}_{M^{n+} | M(Hg) | Pt}$ is the standard potential of the given metal amalgam. Its value may be determined by using a solution of known activity of $M^{n+}$.

**Alternatively,** the following procedure may be employed.

A cell is constructed with the given metal amalgam electrode and the pure metal electrode dipping into a solution of metal ion, i.e.

$$Pt | M(Hg) | M^{n+}(aq) | M$$

The emf of the cell is determined experimentally. It is given by the expression

$$E_{cell} = E_{M^{n+} | M} - E_{M^{n+} | M(Hg) | Pt} \tag{8.6.18}$$

where $$E_{M^{n+} | M} = E^{\circ}_{M^{n+} | M} - \frac{RT}{nF} \ln \frac{1}{a_{M^{n+}}} \tag{8.6.19}$$

and $$E_{M^{n+} | M(Hg) | Pt} = E^{\circ}_{M^{n+} | M(Hg) | Pt} - \frac{RT}{nF} \ln \frac{1}{a_{M^{n+}}} \tag{8.6.20}$$

Substituting Eqs (8.6.19) and (8.6.20) in Eq. (8.6.18), we get

$$E_{cell} = E^{\circ}_{M^{n+} | M} - E^{\circ}_{M^{n+} | M(Hg) | Pt} \tag{8.6.21}$$

Knowing $E^{\circ}_{M^{n+} | M}$ from the table of standard potentials and $E_{cell}$ experimentally, the value of $E^{\circ}_{M^{n+} | M(Hg) | Pt}$ may be determined from Eq. (8.6.21).

## METAL-INSOLUBLE SALT-ANION HALF-CELL

In this half-cell, a metal coated with its insoluble salt is in contact with a solution containing the anion of the insoluble salt. Three such half-cells are described below.

**Silver-Silver Chloride-Chloride Ion Half-Cell** This half-cell is represented as $Cl^- | AgCl | Ag$. The equilibrium reaction that occurs at the electrode is

$$AgCl(s) + e^- \rightarrow Ag(s) + Cl^-(aq)$$

**Expression of Reduction Potential**

The equality of electrochemical potentials gives

$$\mu_{AgCl(s)} + \tilde{\mu}_{e^-(AgCl\,|\,Ag)} = \mu_{Ag(s)} + \tilde{\mu}_{Cl^-(aq)}$$

Making use of Eq. (8.5.9), the above expression becomes

$$\mu_{AgCl(s)} + \{\mu_{e^-(AgCl\,|\,Ag)} - F\phi_{AgCl\,|\,Ag}\} = \mu_{Ag(s)} + \{\mu_{Cl^-(aq)} - F\phi_{aq}\} \quad (8.6.22)$$

Substituting

$$\mu_{Cl^-(aq)} = \mu^\circ_{Cl^-(aq)} + RT \ln a_{Cl^-}$$

in Eq. (8.6.22) and rearranging it, we get

$$\phi_{AgCl\,|\,Ag} - \phi_{aq} = -\frac{\mu_{Ag(s)} + \mu^\circ_{Cl^-(aq)} - \mu_{AgCl(s)} - \mu_{e^-(AgCl\,|\,Ag)}}{F} - \frac{RT}{F} \ln a_{Cl^-}$$

Defining the standard potential when $a_{Cl^-} = 1$, we get

$$E_{Cl^-\,|\,AgCl\,|\,Ag} = E^\circ_{Cl^-\,|\,AgCl\,|\,Ag} - \frac{RT}{F} \ln a_{Cl^-} \quad (8.6.23)$$

where $E_{Cl^-\,|\,AgCl\,|\,Ag} = \phi_{AgCl\,|\,Ag} - \phi_{eq}$

and $$E^\circ_{Cl^-\,|\,AgCl\,|\,Ag} = -\frac{\mu_{Ag(s)} + \mu^\circ_{Cl^-(aq)} - \mu_{AgCl(s)} - \mu_{e^-(AgCl\,|\,Ag)}}{F}$$

**Mercury-Mercuric Oxide-Hydroxide Ion Half-Cell**

In this half-cell, a pool of mercury is covered with a paste of solid HgO and a solution of a base. The equilibrium reaction that takes place at the platinum electrode is

$$HgO(s) + H_2O(1) + 2e^- \rightarrow Hg(1) + 2OH^-(aq)$$

**Expression of Reduction Potential**

The equality of chemical potential yields

$$\mu_{HgO(s)} + \mu_{H_2O(1)} + 2\tilde{\mu}_{e^-(Pt)} = \mu_{Hg(1)} + 2\tilde{\mu}_{OH^-(aq)}$$

Replacing $\tilde{\mu}$ in terms of $\mu$, we get

$$\mu_{HgO(s)} + \mu_{H_2O(1)} + \{2\mu_{e^-(Pt)} - 2F\phi_{Pt}\} = \mu_{Hg(1)} + \{2\mu_{OH^-(aq)} - 2F\phi_{aq}\}$$

Substituting

$$\mu_{OH^-(aq)} = \mu^\circ_{OH^-(aq)} + RT \ln a_{OH^-}$$

in the previous expression and rearranging it, we get

$$\phi_{Pt} - \phi_{aq} = -\frac{\mu_{Hg(l)} + 2\mu^{\circ}_{OH^-(aq)} - \mu_{HgO(s)} - \mu_{H_2O(l)} - 2\mu_{e^-(Pt)}}{2F} - \frac{RT}{F}\ln a_{OH^-}$$

Defining the standard potential when $a_{OH^-} = 1$, we get

$$\phi_{Pt} - \phi_{aq} = (\phi^{\circ}_{Pt} - \phi^{\circ}_{aq}) - \frac{RT}{F}\ln a_{OH^-} \tag{8.6.24}$$

where $$\phi^{\circ}_{Pt} - \phi^{\circ}_{aq} = -\frac{\mu_{Hg(l)} + 2\mu^{\circ}_{OH^-(aq)} - \mu_{HgO(s)} - \mu_{H_2O(l)} - 2\mu_{e^-(Pt)}}{2F}$$

The half-cell potential will be given by

$$E_{OH^- | HgO | Hg} = (\phi_{Pt} - \phi_{HgO | Hg}) + (\phi_{HgO | Hg} - \phi_{aq}) \equiv \phi_{Pt} - \phi_{aq}$$

Thus, Eq. (8.6.24) becomes

$$E_{OH^- | HgO | Hg} = E^{\circ}_{OH^- | HgO | Hg} - \frac{RT}{F}\ln a_{OH^-} \tag{8.6.25}$$

**Mercury-Mercurous Chloride-Chloride Ion Half-Cell**

This half-cell is known as calomel half-cell. The equilibrium reaction is

$$Hg_2Cl_2(s) + 2e^- \rightarrow 2Hg(l) + 2Cl^-(aq)$$

Its Nernst equation is

$$E_{Cl^- | Hg_2Cl_2 | Hg} = E^{\circ}_{Cl^- | Hg_2Cl_2 | Hg} - \frac{RT}{F}\ln a_{Cl^-} \tag{8.6.26}$$

where $$E^{\circ}_{Cl^- | Hg_2Cl_2 | Hg} = -\frac{2\mu_{Hg(l)} + 2\mu^{\circ}_{Cl^-} - \mu_{Hg_2Cl_2} - 2\mu_{e^-(Pt)}}{2F}$$

## OXIDATION-REDUCTION HALF-CELL

An oxidation-reduction half-cell has an inert metal collector, usually platinum, immersed in a solution which contains two ions of the same element in different states of oxidation.

**Ferric-Ferrous Half-Cell**

One of the examples is ferric-ferrous half-cell. The equilibrium reaction to be considered is

$$Fe^{3+}(aq) + e^-(Pt) \rightarrow Fe^{2+}(aq)$$

The Nernst equation is

$$E_{Fe^{3+}, Fe^{2+} | Pt} = E^{\circ}_{Fe^{3+}, Fe^{2+} | Pt} - \frac{RT}{F}\ln\frac{a_{Fe^{2+}}}{a_{Fe^{3+}}} \tag{8.6.27}$$

where $$E^{\circ}_{Fe^{3+}, Fe^{2+} | Pt} = -\frac{\mu^{\circ}_{Fe^{2+}} - \mu^{\circ}_{Fe^{3+}} - \mu_{e^-(Pt)}}{F}$$

**Stannic-Stannous Half-Cell**

Another example is stannic-stannous half-cell where the reduction reaction to be considered is

$$Sn^{4+}(aq) + 2e^- \rightarrow Sn^{2+}(aq)$$

and the Nernst equation is

$$E_{Sn^{4+}, Sn^{2+} | Pt} = E^{\circ}_{Sn^{4+}, Sn^{2+} | Pt} - \frac{RT}{2F} \ln \frac{a_{Sn^{2+}}}{a_{Sn^{4+}}} \tag{8.6.28}$$

**General Expression of Nernst Equation**

From the derived expressions of Nernst equation for the various half-cells (Eqs 8.6.5–8.6.28), it may be concluded that the form of Nernst equation for a given half-cell reaction can be written down without undergoing the explicit derivation by making use of the relation

$$E = E^{\circ} - \frac{RT}{nF} \ln \left\{ \prod_j (a_j)^{\nu_j} \right\}; \quad \nu_j \equiv \begin{cases} +\text{ve for product species in the reaction} \\ -\text{ve for reactant species in the reaction} \end{cases} \tag{8.6.29}$$

where $n$ is the number of electrons involved in the half-cell reaction and $\nu_j$ is the stoichiometry number of the $j$th species in the half-cell reaction. The latter has a positive/negative value for the species appeared on the right/left side of the cell reaction. The symbol $\Pi$ stands for the multiplication of activity or fugacity of all the $j$ species (except those appeared in the solid and liquid forms) involved in the cell reaction.

## 8.7 THE EMF OF A CELL AND ITS CELL REACTION

When the two half cells are combined and their electrodes are connected by an external electrical conductor, an electrical current will flow and chemical reactions will occur in the two half-cells. The nature of the reaction proceeding in a cell can be deduced from the manner in which the electrodes are connected to the standard cell in order to obtain a balance in the potentiometer.

- Electrode connected to the negative side of the standard cell is the negative electrode while the one connected to the positive side is the positive electrode.
- As stated earlier (Section 8.1), oxidation reaction occurs at the negative electrode and thus electrons are released at this electrode.
- The electrons are taken out from the negative electrode by the external conductor and are absorbed at the positive electrode for the reduction reaction.

**Cell Reaction**

The cell reaction can be determined by writing these oxidation and reduction reactions with equal number of electrons and then adding the two reactions. Thus

$$\text{Cell reaction} = \begin{pmatrix}\text{Reduction reaction}\\ \text{at cathode}\end{pmatrix} + \begin{pmatrix}\text{Oxidation reaction}\\ \text{at anode}\end{pmatrix} \quad (8.7.1)$$

Taking an example of Daniell cell, we have

*Anode: oxidation reaction* $Zn(s) \rightarrow Zn^{2+}(aq) + 2e^-$

*Cathode: reduction reaction* $Cu^{2+}(aq) + 2e^- \rightarrow Cu(s)$

Adding the two, we get

$$Zn(s) + Cu^{2+}(aq) \rightarrow Zn^{2+}(aq) + Cu(s)$$

**Recommendations for Cell Diagram**

The emf of the cell under consideration is due to the spontaneous reaction that takes place within the cell. Under balanced reversible conditions a cell reaction can proceed in either direction. Which process—whether forward or backward—corresponds to a given cell? This, of course, can be decided experimentally but not from the manner of writing cell on paper, unless the following recommended convention in this regard is accepted.

*If a cell is written on a piece of paper, then the cell reaction will be written so that electrons are accepted from the external circuit by the electrode written on the right and are given out by the electrode written on the left.*

In other words, once the cell has been written, it will be understood that the left electrode (represented by the symbol L) is the anode (where oxidation reaction takes place and thus electrons are set free) and the right electrode (represented by the symbol R) is the cathode (where reduction reaction occurs and thus electrons are absorbed). With this convention, Eq. (8.7.1) may be written as

$$\text{Cell reaction} = \begin{pmatrix}\text{Reduction reaction}\\ \text{at right electrode}\end{pmatrix} + \begin{pmatrix}\text{Oxidation reaction}\\ \text{at left electrode}\end{pmatrix} \quad (8.7.2)$$

**Illustrations**

To illustrate the procedure, we write the Daniell cell as

$$Zn \mid Zn^{2+}(aq) \;\vdots\vdots\; Cu^{2+}(aq) \mid Cu \quad (8.7.3)$$

Now, according to the above stated rule, the left electrode will be considered as the anode (or the negative terminal) and the right electrode as the cathode (or the positive terminal). Thus, the cell reaction would be

Right electrode *(reduction reaction)* $Cu^{2+}(aq) + 2e^- \rightarrow Cu(s)$

Left electrode *(oxidation reaction)* $Zn(s) \rightarrow Zn^{2+}(aq) + 2e^-$

*Sum* $Cu^{2+}(aq) + Zn(s) \rightarrow Cu(s) + Zn^{2+}(aq)$ (8.7.4)

Note that if the cell has been written as

$$Cu \mid Cu^{2+}(aq) \;\vdots\vdots\; Zn^{2+}(aq) \mid Zn \quad (8.7.5)$$

then while writing the cell reaction, the left electrode (i.e. Cu electrode) will be considered as the negative terminal and the right electrode (i.e. Zn electrode) as the positive terminal. The cell reaction would be

*Right electrode (reduction reaction)* $Zn^{2+}(aq) + 2e^- \rightarrow Zn(s)$

*Left electrode (oxidation reaction)* $Cu(s) \rightarrow Cu^{2+}(aq) + 2e^-$

*Sum* $Zn^{2+}(aq) + Cu(s) \rightarrow Zn(s) + Cu^{2+}(aq)$ (8.7.6)

**EMF of a Cell**

In Section 8.3, we have stated that the spontaneous cell reaction is due to the difference of relative reduction tendencies of the two ions at the respective electrodes. The ion with a higher reduction potential is reduced and the one with lower reduction potential is formed by the oxidation of the corresponding atom. Thus, electrons flow from the electrode of lower reduction tendency of ions to the one of higher reduction tendency of ions. The reduction tendency of an ion can be represented by reduction potential and hence emf of the cell can be written as the difference of the two reduction potentials, i.e.

emf = Higher reduction potential – Lower reduction potential

Keeping in view the convention of writing a cell, the above expression may be written as

$$\text{emf} = \begin{pmatrix}\text{Reduction potential}\\ \text{of right electrode}\end{pmatrix} - \begin{pmatrix}\text{Reduction potential}\\ \text{of left electrode}\end{pmatrix} \tag{8.7.7}$$

Now if a cell as written produces a spontaneous cell reaction, it is obvious that the reduction potential of the ion appeared at the right electrode (where reduction reaction occurs) will be greater than the ion appeared at the left electrode (where oxidation reaction occurs). Hence emf of such a cell will be positive. Conversely, if the cell as written has a positive emf, it implies that it would produce a spontaneous cell reaction with reduction at right electrode and oxidation at left electrode. On the other hand, if the cell reaction is not spontaneous, then the reduction potential of right electrode will be smaller than that of left electrode. Consequently, emf of the cell will have a negative value.

**Cell Reaction**

Equation (8.7.2) may be written in the form similar to that of Eq. (8.7.7). Since oxidation is just reverse of reduction, it is obvious that

Oxidation reaction = – Reduction reaction

Thus, Eq. (8.7.2) may be written as

$$\text{Cell reaction} = \begin{pmatrix}\text{Reduction reaction}\\ \text{at right electrode}\end{pmatrix} - \begin{pmatrix}\text{Reduction reaction}\\ \text{at left electrode}\end{pmatrix} \tag{8.7.8}$$

We shall adopt Eq. (8.7.8) in order to determine the cell reaction.

**Illustrations**

To illustrate the procedure of determining cell reaction and cell potential, we once again take the example of Daniell cell.

***Cell$_1$*** $Zn \mid Zn^{2+} \,\vdots\vdots\, Cu^{2+} \mid Cu$

We have

| *Electrode* | *Reduction reaction* | *Reduction potential* |
|---|---|---|
| Right | $Cu^{2+} + 2e^- \rightarrow Cu$ | $E_{Cu^{2+} \mid Cu} = E^\circ_{Cu^2 \mid Cu} - \frac{RT}{2F} \ln \frac{1}{a_{Cu^{2+}}}$ |
| Left | $Zn^{2+} + 2e^- \rightarrow Zn$ | $E_{Zn^{2+} \mid Zn} = E^\circ_{Zn^{2+} \mid Zn} - \frac{RT}{2F} \ln \frac{1}{a_{Zn^{2+}}}$ |

Subtracting reduction reaction of left electrode from that of right electrode, we get the cell reaction:

$$Cu^{2+} + Zn \rightarrow Cu + Zn^{2+}$$

Since the emf of the cell is

$$E_{cell_1} = E_R - E_L$$

we have

$$E_{cell_1} = E_{Cu^{2+} \mid Cu} - E_{Zn^{2+} \mid Zn}$$

$$= \left( E^\circ_{Cu^{2+} \mid Cu} - \frac{RT}{2F} \ln \frac{1}{a_{Cu^{2+}}} \right) - \left( E^\circ_{Zn^{2+} \mid Zn} - \frac{RT}{2F} \ln \frac{1}{a_{Zn^{2+}}} \right)$$

$$= (E^\circ_{Cu^{2+} \mid Cu} - E^\circ_{Zn^{2+} \mid Zn}) - \frac{RT}{2F} \ln \frac{a_{Zn^{2+}}}{a_{Cu^{2+}}}$$

or
$$E_{cell_1} = E^\circ_{cell_1} - \frac{RT}{2F} \ln \frac{a_{Zn^{2+}}}{a_{Cu^{2+}}} \tag{8.7.9}$$

***Cell***$_2$ If the cell were written as

$$Cu \mid Cu^{2+} \,\vdots\vdots\, Zn^{2+} \mid Zn$$

then we would have

| *Electrode* | *Reduction reaction* | *Reduction potential* |
|---|---|---|
| Right | $Zn^{2+} + 2e^- \rightarrow Zn$ | $E_{Zn^{2+} \mid Zn} = E^\circ_{Zn^{2+} \mid Zn} - \frac{RT}{2F} \ln \frac{1}{a_{Zn^{2+}}}$ |
| Left | $Cu^{2+} + 2e^- \rightarrow Cu$ | $E_{Cu^{2+} \mid Cu} = E^\circ_{Cu^{2+} \mid Cu} - \frac{RT}{2F} \ln \frac{1}{a_{Cu^{2+}}}$ |

Subtracting reduction reaction of left electrode from that of right electrode, we get the cell reaction as

$$Zn^{2+} + Cu \rightarrow Zn + Cu^{2+}$$

The emf of the cell is given by

$$E_{cell_2} = E_{Zn^{2+} \mid Zn} - E_{Cu^{2+} \mid Cu}$$

$$E_{\text{cell}_2} = E^{\circ}_{\text{cell}_2} - \frac{RT}{2F} \ln \frac{a_{\text{Cu}^{2+}}}{a_{\text{Zn}^{2+}}}$$

where $E^{\circ}_{\text{cell}_2} = E^{\circ}_{\text{Zn}^{2+} | \text{Zn}} - E^{\circ}_{\text{Cu}^{2+} | \text{Cu}}$

Obviously,

$$E_{\text{cell}_2} = - E_{\text{cell}_1}$$

and the reaction of $\text{cell}_2$ is just a reverse of that of $\text{cell}_1$.

**General Expression of Cell Potential**

The expression of cell potential, in general, can be written down directly from the cell reaction by making use of the relation

$$E_{\text{cell}} = E^{\circ}_{\text{cell}} - \frac{RT}{nF} \ln \left\{ \prod_j (a_j)^{\nu_j} \right\};\ \nu_j \equiv \begin{cases} +\text{ve for the right-side species} \\ -\text{ve for the left-side species} \end{cases} \tag{8.7.10}$$

where $n$ is the number of electrons involved in the cell reaction, $\nu_j$ is the stoichiometry of the $j$th species and $E^{\circ}_{\text{cell}}$ is given by

$$E^{\circ}_{\text{cell}} = E^{\circ}_{\text{R}} - E^{\circ}_{\text{L}} \tag{8.7.11}$$

The symbol $\prod_j$ stands for the multiplication of activity or fugacity raised to the corresponding stoichiometric number of all the $j$ species (except those appeared in the solid and liquid forms) involved in the cell reaction.

**Criterion for the Spontaneity of the Cell Reaction**

The connection between the sign of emf of a cell and the nature of the associated cell reaction may be derived thermodynamically. As stated above, the electrons flow spontaneously from left electrode to right electrode provided the reduction potential of right electrode is greater than that of left electrode. The amount of electrical work involved in moving the amount $n$ of electrons from left electrode to right electrode as given by Eq. (8.5.4) is

$$w = - nF(E_{\text{R}} - E_{\text{L}}) = - nFE_{\text{cell}}$$

where, $F$ is Faraday constant (= 96 500 C mol$^{-1}$), $E_{\text{R}}$ and $E_{\text{L}}$ are the reduction potentials of right electrode and left electrode, respectively, and $E_{\text{cell}}$ is the emf of the cell. Now, according to thermodynamics, the above work will be equal to the change in free energy of the cell (i.e. more precisely of the cell reaction). Hence, we have

$$\Delta G = - nFE_{\text{cell}} \tag{8.7.12}$$

From Eq. (8.7.12), we observe that $\Delta G$ and $E_{\text{cell}}$ have opposite signs, i.e. if $\Delta G$ is negative then $E_{\text{cell}}$ is positive and vice versa. Thus, we have the following criteria about the nature of cell reaction and its emf.

| $\Delta G$ | *Nature of cell reaction* | *Sign of* $E_{\text{cell}}$ |
|---|---|---|
| Negative | Spontaneous | Positive |
| Positive | Nonspontaneous | Negative |
| Zero | At equilibrium | Zero |

From Eq. (8.7.12), we observe that the cell emf is proportional to $\Delta G/n$. Since $\Delta G/n$ is an intensive property, it follows that $E_{\text{cell}}$ is also an intensive property and hence does not depend upon the size of the cell and the number of electrons that are associated while writing the half-cell reactions.

---

***Example 8.7.1***

Construct the galvanic cell for each of the following reactions and write down the corresponding expression for the cell potential.

(i) $2Cr(s) + 3Hg_2Cl_2(s) \rightarrow 2Cr^{3+}(aq) + 6Cl^-(aq) + 6Hg(l)$
(ii) $Ni^{2+}(aq) + H_2(g) + 2OH^-(aq) \rightarrow Ni(s) + 2H_2O(l)$
(iii) $AgCl(s) + I^-(aq) \rightarrow AgI(s) + Cl^-(aq)$
(iv) $Zn(s) + H_2SO_4(aq) \rightarrow ZnSO_4(aq) + H_2(g)$
(v) $Fe(s) + Cl_2(g) \rightarrow FeCl_2(aq)$
(vi) $Ag(s) + \frac{1}{2}Br_2(l) \rightarrow AgBr(s)$
(vii) $Ag^+(aq) + Cl^-(aq) \rightarrow AgCl(s)$
(viii) $HgO(s) + H_2(g) \rightarrow Hg(l) + H_2O(l)$

***Solution***

(i) $2Cr(s) + 3Hg_2Cl_2(s) \rightarrow 2Cr^{3+}(aq) + 6Cl^-(aq) + 6Hg(l)$

Cr is oxidized to $Cr^{3+}$ and $Hg_2^{2+}$ is reduced to Hg. Thus, we would have:

| *Electrode* | *Reduction reaction* | |
|---|---|---|
| Right | $Hg_2Cl_2(s) + 2e^- \rightarrow 2Hg(l) + 2Cl^-(aq)$ | (i) |
| Left | $Cr^{3+}(aq) + 3e^- \rightarrow Cr(s)$ | (ii) |

Multiplying Eq. (i) by 3 and Eq. (ii) by 2 and then subtracting Eq. (ii) from Eq. (i), we get

$$2Cr(s) + 3Hg_2Cl_2(s) \rightarrow 2Cr^{3+}(aq) + 6Hg(l) + 6Cl^-(aq)$$

The required cell would be

$$Cr \mid CrCl_3(aq) \,\vdots\vdots\, KCl(aq) \mid Hg_2Cl_2(s) \mid Hg(l) \mid Pt$$

and the cell potential will be given by

$$E_{\text{cell}} = E^\circ_{\text{cell}} - \frac{RT}{6F} \ln (a_{Cr^{3+}})^2 (a_{Cl^-})^6$$

i.e.

$$E_{\text{cell}} = E^\circ_{\text{cell}} - \frac{RT}{3F} \ln (a_{Cr^{3+}}) (a_{Cl^-})^3$$

where

$$E^\circ_{\text{cell}} = E^\circ_{Cl^- \mid Hg_2Cl_2 \mid Hg} - E^\circ_{Cr^{3+} \mid Cr}$$

**(ii)** $Ni^{2+}(aq) + H_2(g) + 2OH^-(aq) \rightarrow Ni(s) + 2H_2O(l)$

$Ni^{2+}$ is reduced to Ni and hydrogen in alkaline medium is oxidized to $H_2O$. Thus, we have:

| *Electrode* | *Reduction reaction* | |
|---|---|---|
| Right | $Ni^{2+}(aq) + 2e^- \rightarrow Ni(s)$ | (i) |
| Left | $2H_2O(l) + 2e^- \rightarrow H_2(g) + 2OH^-(aq)$ | (ii) |

Subtracting Eq. (ii) from Eq. (i), we get

$$Ni^{2+}(aq) + H_2(g) + 2OH^-(aq) \rightarrow Ni(s) + 2H_2O(l)$$

The required cell would be

$$Pt \mid H_2(g) \mid NaOH(aq) \,\vdots\vdots\, Ni^{2+}(aq) \mid Ni$$

and the cell potential will be given by

$$E_{cell} = E^\circ_{cell} - \frac{RT}{2F} \ln \frac{1}{(f_{H_2}/f^\circ)(a_{OH^-})^2 (a_{Ni^{2+}})}$$

where $E^\circ_{cell} = E^\circ_{Ni^{2+} \mid Ni} - E^\circ_{OH^- \mid H_2 \mid Pt}$

**(iii)** $AgCl(s) + I^-(aq) \rightarrow AgI(s) + Cl^-(aq)$

The given reaction may be considered to involve the following two half-cell reactions:

$$AgCl(s) + e^- \rightarrow Ag(s) + Cl^-(aq)$$

$$Ag(s) + I^-(aq) \rightarrow AgI(s) + e^-$$

Thus, we have:

| *Electrode* | *Reduction reaction* | |
|---|---|---|
| Right | $AgCl(s) + e^- \rightarrow Ag(s) + Cl^-(aq)$ | (i) |
| Left | $AgI(s) + e^- \rightarrow Ag(s) + I^-(aq)$ | (ii) |

Subtracting Eq. (ii) from Eq. (i), we have

$$AgCl(s) + I^-(aq) \rightarrow AgI(s) + Cl^-(aq)$$

The required cell would be

$$Ag \mid AgI \mid KI(aq) \,\vdots\vdots\, KCl(aq) \mid AgCl \mid Ag$$

and the cell potential will be given by

$$E_{cell} = E^\circ_{cell} - \frac{RT}{F} \ln \frac{a_{Cl^-}}{a_{I^-}}$$

where $E^\circ_{cell} = E^\circ_{Cl^- \mid AgCl \mid Ag} - E^\circ_{I^- \mid AgI \mid Ag}$

**(iv)** $Zn(s) + H_2SO_4(aq) \rightarrow ZnSO_4(aq) + H_2(g)$

Zn is oxidized to $Zn^{2+}$ and $H^+$ is reduced to $H_2$. Thus, we have:

| *Electrode* | *Reduction reaction* | |
|---|---|---|
| Right | $2H^+(aq) + 2e^- \rightarrow H_2(g)$ | (i) |
| Left | $Zn^{2+}(aq) + 2e^- \rightarrow Zn(s)$ | (ii) |

Subtracting Eq. (ii) from Eq. (i), we have

$$Zn(s) + 2H^+(aq) \rightarrow H_2(g) + Zn^{2+}(aq)$$

The cell would be

$$Zn(s) \mid Zn^{2+}(aq) \,\vdots\vdots\, H^+(aq) \mid H_2(g) \mid Pt$$

and the cell potential will be given by

$$E_{\text{cell}} = E^\circ_{\text{cell}} - \frac{RT}{2F} \ln \frac{(f_{H_2}/f^\circ)(a_{Zn^{2+}})}{(a_{H^+})^2}$$

where $E^\circ_{\text{cell}} = E^\circ_{H^+ \mid H_2 \mid Pt} - E^\circ_{Zn^{2+} \mid Zn}$

**(v)** $Fe(s) + Cl_2(g) \rightarrow FeCl_2(aq)$

Fe is oxidized to $Fe^{2+}$ and $Cl_2$ is reduced to $Cl^-$. Thus, we have:

| *Electrode* | *Reduction reaction* | |
|---|---|---|
| Right | $Cl_2(g) + 2e^- \rightarrow 2Cl^-(aq)$ | (i) |
| Left | $Fe^{2+}(aq) + 2e^- \rightarrow Fe(s)$ | (ii) |

Subtracting Eq. (ii) from Eq. (i), we have

$$Fe(s) + Cl_2(g) \rightarrow Fe^{2+}(aq) + 2Cl^-(aq)$$

The cell would be

$$Fe \mid FeCl_2(aq) \,\vdots\vdots\, KCl(aq) \mid Cl_2(g) \mid Pt$$

and the cell potential is given by

$$E_{\text{cell}} = E^\circ_{\text{cell}} - \frac{RT}{2F} \ln \frac{(a_{Fe^{2+}})(a_{Cl^-})^2}{(f_{Cl_2}/f^\circ)}$$

with $E^\circ_{\text{cell}} = E^\circ_{Cl^- \mid Cl_2 \mid Pt} - E^\circ_{Fe^{2+} \mid Fe}$

**(vi)** $Ag(s) + \frac{1}{2} Br_2(l) \rightarrow AgBr(s)$

Ag in the presence of $Br^-$ is changed to AgBr and $Br_2$ is changed to $Br^-$, thus, we have:

| *Electrode* | *Reduction reaction* | |
|---|---|---|
| Right | $\frac{1}{2} Br_2(l) + e^- \rightarrow Br^-(aq)$ | (i) |
| Left | $AgBr(s) + e^- \rightarrow Ag(s) + Br^-(aq)$ | (ii) |

Subtracting Eq. (ii) from Eq. (i), we have

$$Ag(s) + \frac{1}{2} Br_2(l) \rightarrow AgBr(s)$$

The required cell would be

$$Ag \mid AgBr(s) \mid KBr(aq) \mid Br_2(l) \mid Pt$$

and the cell potential is given by

$$E_{\text{cell}} = E^\circ_{\text{cell}} = E^\circ_{Br^- \mid Br_2 \mid Pt} - E^\circ_{Br^- \mid AgBr \mid Ag}$$

**(vii)** $Ag^+(aq) + Cl^-(aq) \rightarrow AgCl(s)$

This overall reaction may be produced through the following two half-cell reactions:

| *Electrode* | *Reduction reaction* | |
|---|---|---|
| Right | $Ag^+(aq) + e^- \rightarrow Ag(s)$ | (i) |
| Left | $AgCl(s) + e^- \rightarrow Ag(s) + Cl^-(aq)$ | (ii) |

Subtracting Eq. (ii) from Eq, (i), we have

$$Ag^+(aq) + Cl^-(aq) \rightarrow AgCl(s)$$

The required cell would be

$$Ag \mid AgCl(s) \mid KCl(aq) \mid \underset{\text{Bridge}}{NH_4NO_3} \mid AgNO_3 \mid Ag$$

and the cell potential is given by

$$E_{cell} = E^\circ_{cell} - \frac{RT}{F} \ln \frac{1}{(a_{Ag^+})(a_{Cl^-})}$$

where $E^\circ_{cell} = E^\circ_{Ag^+ \mid Ag} - E^\circ_{Cl^- \mid AgCl \mid Ag}$

**(viii)** $HgO(s) + H_2(g) \rightarrow Hg(l) + H_2O(l)$

This overall reaction may be produced through the following two half-cell reactions:

| *Electrode* | *Reduction reaction* | |
|---|---|---|
| Right | $HgO(s) + H_2O(1) + 2e^- \rightarrow Hg(l) + 2OH^-(aq)$ | (i) |
| Left | $2H_2O(1) + 2e^- \rightarrow H_2(g) + 2OH^-(aq)$ | (ii) |

Subtracting Eq. (ii) from Eq. (i), we have

$$HgO(s) + H_2(g) \rightarrow Hg(l) + H_2O(l)$$

The required cell would be

$$Pt \mid H_2(g) \mid \text{NaOH saturated with HgO} \mid HgO(s) \mid Hg(l) \mid Pt$$

and the cell potential is given by

$$E_{cell} = E^\circ_{cell} - \frac{RT}{2F} \ln \frac{1}{(f_{H_2}/f^\circ)}$$

with $E^\circ_{cell} = E^\circ_{OH^- \mid HgO \mid Hg} - E^\circ_{OH^- \mid H_2 \mid Pt}$

## 8.8 DETERMINATION OF STANDARD POTENTIALS

**Expression of Cell Potential**

At the very outset, it may be stated that the absolute value for the reduction potential of any single half-reaction cannot be determined experimentally or theoretically since the activity of single ionic species, such as hydrogen ion or chloride ion, can never be known exactly. However, difference between the two reduction potentials can be determined by constructing a suitable cell and then determining its emf experimentally. By definition, the emf of a cell is given by

$$E_{\text{cell}} = E_{\text{R}} - E_{\text{L}} \tag{8.8.1}$$

where $E_{\text{R}}$ and $E_{\text{L}}$ are the reduction potentials of right and left electrodes, respectively.

**Reference Half-Cell**

If $E_{\text{L}}$ is arbitrarily assigned some value, then the value of $E_{\text{R}}$ can be determined using the expression

$$E_{\text{R}} = E_{\text{cell}} + E_{\text{L}} \tag{8.8.2}$$

In the study of electrochemical cell, the *hydrogen-hydrogen ion half-cell has been adopted as the reference half-cell and its standard potential has been assigned the value zero at all temperatures.* By standard potential of hydrogen-hydrogen ion half-cell, we mean that the hydrogen ion and hydrogen gas involved in half-cell

$$H^+(aq) \mid H_2(g) \mid Pt$$

are present in their standard states of unit activity and unit fugacity (taken as 1 bar pressure), respectively.

Thus, if we have a cell in which the left half-cell is the standard hydrogen-hydrogen ion half-cell and the right half-cell constitutes the electrode system whose potential relative to that of standard hydrogen-hydrogen ion half-cell is required, then according to Eq. (8.8.2), we have

$$\begin{aligned} E_{\text{R}} &= E_{\text{cell}} + 0 \\ &= E_{\text{cell}} \end{aligned}$$

that is, the reduction potential of the given half-cell is numerically equal to the emf of the cell. Proceeding in this way, the standard potentials of other half-cell relative to the standard hydrogen-hydrogen ion half-cell can be determined.

**Illustrations**

To illustrate the procedure, we cite below two typical examples of silver-silver ion and zinc-zinc ion half-cells. If the silver-silver ion half-cell is coupled with the standard hydrogen-hydrogen ion half-cell (to be kept on left side), we get a cell

$$Pt \mid H_2(1\ \text{bar}) \mid H^+(a = 1) \ \vdots\vdots\ Ag^+(a = 1) \mid Ag$$

The standard emf of the cell is given by

$$E^\circ_{\text{cell}} = E^\circ_{Ag^+ \mid Ag} - E^\circ_{H^+ \mid H_2 \mid Pt}$$

and its value as determined experimentally is found to be 0.799 1 V. Hence

$$\begin{aligned} E^\circ_{Ag^+ \mid Ag} &= E^\circ_{\text{cell}} + E^\circ_{H^+ \mid H_2 \mid Pt} \\ &= E^\circ_{\text{cell}} = 0.799\ 1\ \text{V} \end{aligned}$$

Taking the example of zinc-zinc ion half-cell, we have

$$Pt \mid H_2(1 \text{ bar}) \mid H^+(a = 1) \,\vdots\vdots\, Zn^{2+}(a = 1) \mid Zn$$

Its emf as determined experimentally is found to be −0.763 V. Hence

$$E^{\circ}_{Zn^{2+} \mid Zn} = E^{\circ}_{cell} + E^{\circ}_{H^+ \mid H_2 \mid Pt}$$

$$= E^{\circ}_{cell} = -\,0.763 \text{ V}$$

Note that the cell emf is negative. It implies that the cell as written above will not produce spontaneous reaction. In fact, while determining emf of the cell, the electrode of zinc-zinc ion half-cell will serve as the negative terminal and Pt electrode of hydrogen-hydrogen ion half-cell as the positive terminal, in order to get a positive potential of 0.763 V. Since the cell has been written in the reverse direction (i.e. Pt electrode as the negative terminal and Zn as the positive terminal), it follows that the emf of the written cell will be −0.763 V.

**Values of Standard Potentials**

Table 8.8.1 records the standard potentials for some of the half-cell reactions at 298.15 K along with their respective half-cell assembly.[†]

**Important Comment**

From the two examples cited above, it may be concluded that if an electrode of a half-cell with a positive reduction potential is coupled with the standard hydrogen-hydrogen ion half-cell, it will constitute the positive terminal of the cell in order to have a spontaneous cell reaction. Similarly, an electrode of a half-cell with a negative reduction potential will constitute the negative terminal of the cell so as to get the spontaneous cell reaction. In other words, the nature of the electrode of half-cell (whether positive or negative) in a cell in which the other half-cell is the standard hydrogen-hydrogen ion half-cell is determined by the sign of the reduction potential of the given half-cell.

## 8.9 SIGNIFICANCE OF STANDARD HALF-CELL POTENTIALS

**Significance of Half-Cell Potentials**

The standard half-cell potential of the reaction

$$\frac{1}{n}M^{n+} + e^- \rightarrow \frac{1}{n}M \tag{8.9.1}$$

is a measure of reduction tendency of $M^{n+}$ to M relative to that of $H^+$ to $H_2$. The standard potential of the latter reaction, i.e.

$$H^+ + e^- \rightarrow \tfrac{1}{2}H_2 \tag{8.9.2}$$

is taken to be zero (reference level). A positive potential implies that the ion $M^{n+}$ can be more easily reduced to M relative to that of $H^+$ ion to $H_2$. On the other hand, a negative potential implies that the ion $M^{n+}$ is more difficult to reduce as compared to the $H^+$ ion.

[†]The representation of a half-cell corresponds to the right half-cell of the cell diagram in which the left half-cell is the hydrogen-hydrogen ion half-cell.

**Table 8.8.1** Standard Half-cell Potentials at 298.15 K

| *Half-cell reaction* | *E°/V* | *Half-cell representation* |
|---|---|---|
| $Li^+ + e^- \rightarrow Li$ | – 3.045 | $Li^+ \mid Li$ |
| $K^+ + e^- \rightarrow K$ | – 2.925 | $K^+ \mid K$ |
| $Na^+ + e^- \rightarrow Na$ | – 2.714 | $Na^+ \mid Na$ |
| $Mg^{2+} + 2e^- \rightarrow Mg$ | – 2.37 | $Mg^{2+} \mid Mg$ |
| $H_2 + 2e^- \rightarrow 2H^-$ | – 2.25 | $H^- \mid H_2 \mid Pt$ |
| $Al^{3+} + 3e^- \rightarrow Al$ | – 1.66 | $Al^{3+} \mid Al$ |
| $2H_2O + 2e^- \rightarrow H_2 + 2OH^-$ | – 0.828 | $OH^- \mid H_2 \mid Pt$ |
| $Zn^{2+} + 2e^- \rightarrow Zn$ | – 0.763 | $Zn^{2+} \mid Zn$ |
| $Cd(NH_3)_4^{2+} + 2e^- \rightarrow Cd + 4NH_3$ | – 0.61 | $Cd(NH_3)_4^{2+}, NH_3 \mid Cd$ |
| $2CO_2 + 2H^+ + 2e^- \rightarrow H_2C_2O_4$ | – 0.49 | $H_2C_2O_4, H^+ \mid CO_2 \mid Pt$ |
| $Fe^{2+} + 2e^- \rightarrow Fe$ | – 0.44 | $Fe^{2+} \mid Fe$ |
| $Cr^{3+} + e^- \rightarrow Cr^{2+}$ | – 0.41 | $Cr^{3+}, Cr^{2+} \mid Pt$ |
| $Cd^{2+} + 2e^- \rightarrow Cd$ | – 0.40 | $Cd^{2+} \mid Cd$ |
| $Ag(CN)_2^- + e^- \rightarrow Ag + 2CN^-$ | – 0.31 | $Ag(CN)_2^-, CN^- \mid Ag$ |
| $Cu(OH)_2 + 2e^- \rightarrow Cu + 2OH^-$ | – 0.224 | $OH^-, Cu(OH)_2 \mid Cu$ |
| $AgI + e^- \rightarrow Ag + I^-$ | – 0.151 | $I^- \mid AgI \mid Ag$ |
| $Sn^{2+} + 2e^- \rightarrow Sn$ | – 0.136 | $Sn^{2+} \mid Sn$ |
| $Pb^{2+} + 2e^- \rightarrow Pb$ | – 0.126 | $Pb^{2+} \mid Pb$ |
| $Cu(NH_3)_4^{2+} + 2e^- \rightarrow Cu + 4NH_3$ | – 0.12 | $Cu(NH_3)_4^{2+}, NH_3 \mid Cu$ |
| $Fe^{3+} + 3e^- \rightarrow Fe$ | – 0.036 | $Fe^{3+} \mid Fe$ |
| $2H^+ + 2e^- \rightarrow H_2$ | 0.000 | $H^+ \mid H_2 \mid Pt$ |
| $AgBr + e^- \rightarrow Ag + Br^-$ | 0.095 | $Br^- \mid AgBr \mid Ag$ |
| $Cu^{2+} + e^- \rightarrow Cu^+$ | 0.153 | $Cu^{2+}, Cu^+ \mid Pt$ |
| $Sn^{4+} + 2e^- \rightarrow Sn^{2+}$ | 0.15 | $Sn^{4+}, Sn^{2+} \mid Pt$ |
| $AgCl + e^- \rightarrow Ag + Cl^-$ | 0.222 | $Cl^- \mid AgCl \mid Ag$ |
| $Hg_2Cl_2 + 2e^- \rightarrow 2Hg + 2Cl^-$ | 0.267 6 | $Cl^- \mid Hg_2Cl_2 \mid Hg(Pt)$ |
| $Cu^{2+} + 2e^- \rightarrow Cu$ | 0.337 | $Cu^{2+} \mid Cu$ |
| $Ag(NH_3)_2^+ + e^- \rightarrow Ag + 2NH_3$ | 0.373 | $Ag(NH_3)_2^+, NH_3 \mid Ag$ |
| $O_2(g) + 2H_2O + 4e^- \rightarrow 4OH^-$ | 0.401 | $OH^- \mid O_2 \mid Pt$ |
| $Cu^+ + e^- \rightarrow Cu$ | 0.521 | $Cu^+ \mid Cu$ |
| $\frac{1}{2} I_2 + e^- \rightarrow I^-$ | 0.535 5 | $I_2, I^- \mid Pt$ |
| $Fe^{3+} + e^- \rightarrow Fe^{2+}$ | 0.771 | $Fe^{3+}, Fe^{2+} \mid Pt$ |
| $\frac{1}{2} Hg_2^{2+} + e^- \rightarrow Hg$ | 0.789 | $Hg_2^{2+} \mid Hg(Pt)$ |
| $Ag^+ + e^- \rightarrow Ag$ | 0.799 1 | $Ag^+ \mid Ag$ |
| $Br_2(l) + 2e^- \rightarrow 2Br^-$ | 1.086 2 | $Br^-, Br_2 \mid Pt$ |
| $O_2 + 4H^+ + 4e^- \rightarrow 2H_2O$ | 1.229 | $H^+ \mid O_2 \mid Pt$ |
| $Cr_2O_7^{2-} + 14H^+ + 6e^- \rightarrow 2Cr^{3+} + 7H_2O$ | 1.33 | $Cr_2O_7^{2-}, Cr^{3+}, H^+ \mid Pt$ |
| $\frac{1}{2} Cl_2(g) + e^- \rightarrow Cl^-$ | 1.359 5 | $Cl^- \mid Cl_2 \mid Pt$ |
| $Ce^{4+} + e^- \rightarrow Ce^{3+}$(1 mol dm$^{-3}$ $H_2SO_4$) | 1.44 | $Ce^{4+}, Ce^{3+} \mid Pt$ |
| $Au^{3+} + 3e^- \rightarrow Au$ | 1.455 | $Au^{3+} \mid Au$ |
| $MnO_4^- + 8H^+ + 5e^- \rightarrow Mn^{2+} + 4H_2O$ | 1.51 | $MnO_4^-, Mn^{2+}, H^+ \mid Pt$ |
| $Co^{3+} + e^- \rightarrow Co^{2+}$ | 1.82 | $Co^{3+}, Co^{2+} \mid Pt$ |

If $E^\circ_{M^+|M}$ is positive, then the cell

$$\text{Pt} \mid \text{H}_2(1\text{ bar}) \mid \text{H}^+(a = 1) \;\vdots\vdots\; \text{M}^{n+}(a = 1) \mid \text{M} \tag{8.9.3}$$

will have a positive emf and, therefore, the cell reaction

$$\frac{1}{n}\text{M}^{n+}(a = 1) + \frac{1}{2}\text{H}_2(1\text{ bar}) \rightarrow \frac{1}{n}\text{M} + \text{H}^+(a = 1)$$

will be spontaneous and hence $M^{n+}$ can be reduced to M by hydrogen gas. On the other hand, if $E^\circ_{M^+|M}$ is negative, the cell emf will also be negative and hence the cell reaction will not be spontaneous. Thus, $M^{n+}$ cannot be reduced to M by hydrogen gas. In the present case, if the cell as given by Eq. (8.9.3) is written in the reverse direction, i.e.

$$\text{M} \mid \text{M}^{n+}(a = 1) \;\vdots\vdots\; \text{H}^+(a = 1) \mid \text{H}_2(1\text{ bar}) \mid \text{Pt}$$

the cell will have a positive emf and hence the reaction

$$\frac{1}{n}\text{M} + \text{H}^+(a = 1) \rightarrow \frac{1}{n}\text{M}^{n+}(a = 1) + \frac{1}{2}\text{H}_2(1\text{ bar})$$

will be a spontaneous reaction. Thus, $H^+$ ions can be reduced to hydrogen gas by the metal M.

**Illustration**

Taking the typical examples of $Ag^+|Ag$ and $Zn^{2+}|Zn$ half-cells, we find that

$$E^\circ_{Ag^+|Ag} = 0.799\text{ V} \quad \text{and} \quad E^\circ_{Zn^{2+}|Zn} = -0.763\text{ V}$$

Hence we conclude that under standard conditions, $Ag^+$ ions can be reduced to Ag by hydrogen gas whereas $Zn^{2+}$ cannot be reduced. Alternatively, zinc metal can reduce $H^+$ ions to hydrogen gas whereas silver metal cannot reduce $H^+$ ions.

Active metals such as Zn, Na and Mg have highly negative standard potentials indicating that their compounds are not reduced by hydrogen but rather the metal itself can be oxidized by $H^+$ to yield $H_2$. Noble metals such as Cu, Ag and Au have positive $E^\circ$s and hence their compounds are readily reduced by $H_2$ gas; the metals themselves are not oxidized in the presence of $H^+$ ions.

**Making a Galvanic Cell**

Table 8.8.1 records the values of standard potentials in increasing order, i.e. at the top we have the most negative potential with the least tendency for reduction and at the bottom we have the most positive potential with maximum tendency for reduction. If a pair of half-cells is coupled to make a cell, the half-cell with more positive potential will constitute the positive terminal whereas the half-cell with less positive potential will constitute the negative terminal. In other words, the half-cell of the higher positive potential (which stands lower in the table) will constitute the right half-cell with the reduction half-cell reaction and the half-cell of less positive potential will constitute the left half-cell with the oxidation half-cell reaction. Consider, for example, the two half-cells $Ni^{2+} \mid Ni$ and $Ag^+ \mid Ag$. Their reduction potentials are

$$E^{\circ}_{Ni^{2+}Ni} = -2.250 \text{ V} \quad \text{and} \quad E^{\circ}_{Ag^+ \mid Ag} = 0.799 \text{ V}$$

Since $E^{\circ}_{Ag^+ \mid Ag}$ is more positive than $E^{\circ}_{Ni^{2+} \mid Ni}$, it is obvious that silver electrode will constitute the positive terminal (i.e. right half-cell) and nickel electrode as the negative terminal (i.e. left half-cell). Thus the required cell with the positive emf and hence spontaneous cell reaction would be

$$Ni \mid Ni^{2+}(a = 1) \,\vdots\vdots\, Ag^+ (a = 1) \mid Ag$$

---

***Example 8.9.1*** In acid solution the following are true under the standard conditions.

(a) $H_2S$ will react with oxygen to give $H_2O$ and sulphur.
(b) $H_2S$ will not react in the corresponding reaction with selenium and tellurium.
(c) $H_2Se$ will react with sulphur giving to $H_2S$ and Se but not react with tellurium.

Arrange the hydrides $H_2O$, $H_2S$, $H_2Se$ and $H_2Te$ in order of their tendency to lose electrons to form the corresponding elements.

***Solution*** (**a**) The reaction is

$$H_2S + \tfrac{1}{2} O_2 \rightarrow H_2O + S$$

This can be broken into two partial reactions:

$$\textit{Reduction:} \quad 2H^+ + \tfrac{1}{2} O_2 + 2e^- \rightarrow H_2O$$

$$\textit{Oxidation:} \quad H_2S \rightarrow 2H^+ + S + 2e^-$$

A cell corresponding to the given reaction can be constructed with the oxidation reaction at anode and reduction reaction at cathode. The cell potential would be

$$E^{\circ}_{cell} = E^{\circ}_{R} - E^{\circ}_{L} = E^{\circ}_{H^+ \mid O_2 \mid Pt} - E^{\circ}_{H^+ \mid H_2S \mid S}$$

Since the cell reaction is spontaneous, it follows that

$$E^{\circ}_{H^+ \mid O_2 \mid Pt} > E^{\circ}_{H^+ \mid H_2S \mid S}$$

(**b**) Proceeding similarly as in the part (a), we have

$$E^{\circ}_{H^+ \mid H_2S \mid S} > E^{\circ}_{H^+ \mid H_2Se \mid Se} \quad \text{and} \quad E^{\circ}_{H^+ \mid H_2S \mid S} > E^{\circ}_{H^+ \mid H_2Te \mid Te}$$

(**c**) Similarly in this case, we have

$$E^{\circ}_{H^+ \mid H_2S \mid S} > E^{\circ}_{H^+ \mid H_2Se \mid Se} \quad \text{and} \quad E^{\circ}_{H^+ \mid H_2Se \mid Se} > E^{\circ}_{H^+ \mid H_2Te \mid Te}$$

Arranging the above reduction potentials in the decreasing order, we get

$$E^{\circ}_{H^+ \mid O_2 \mid Pt} > E^{\circ}_{H^+ \mid H_2S \mid S} > E^{\circ}_{H^+ \mid H_2Se \mid Se} > E^{\circ}_{H^+ \mid H_2Te \mid Te}$$

The above order implies that the tendency for the reduction reaction

$$2H^+ + \frac{1}{n} X_n + 2e^- \rightarrow H_2X$$

to take place is the greatest for oxygen and least for Te. The tendency for the reverse reaction, i.e. the oxidation of hydride to the corresponding element will be greatest for $H_2Te$ and least for $H_2O$.

***Example 8.9.2***

Answer whether under standard conditions the following reactions are possible or not.

(i) Will Fe reduce $Fe^{3+}$ to $Fe^{2+}$?
(ii) Will the permanganate ion $MnO_4^-$ liberate $O_2$ from water in the presence of an acid?
(iii) Will $O_2$ oxidize gold to $Au(CN)_2^-$ in the presence of $CN^-$ and $OH^-$ ions?
(iv) Would you use silver spoon to stir a solution of $Cu(NO_3)_2$?
(v) Will ceric sulphate react with KI?
(vi) Will lead react with $AgNO_3$ solution?
(vii) Will permanganate ions in acidic medium oxidize oxalic acid?
(viii) Will dichromate ions in acidic medium oxidize $Fe^{2+}$ to $Fe^{3+}$?
(ix) Will $Fe^{3+}$ ions be reduced to $Fe^{2+}$ by $Sn^{2+}$ ions?
(x) Will $Fe^{3+}$ ions in acidic medium oxidize diphenylbenzidine and *N*-phenylanthranilic acid indicators?

***Solution***

**(i)** The reactions would be

*Reduction* $Fe^{3+} + e^- \rightarrow Fe^{2+}$

*Oxidation* $Fe \rightarrow Fe^{2+} + 2e^-$

The cell giving these half-cell reactions would be

$$Fe \mid Fe^{2+} \,\vdots\vdots\, Fe^{3+}, Fe^{2+} \mid Pt$$

with $E^\circ_{cell} = E^\circ_{Fe^{3+}, Fe^{2+} \mid Pt} - E^\circ_{Fe^{2+} \mid Fe} = 0.771\ V - (-0.440\ V) = 1.211\ V$

Since $E^\circ_{cell}$ is positive, the reduction of $Fe^{3+}$ to $Fe^{2+}$ by Fe is possible,

**(ii)** The reactions would be

*Reduction* $MnO_4^- + 8H^+ + 5e^- \rightarrow Mn^{2+} + 4H_2O$

*Oxidation* $H_2O \rightarrow 2H^+ + \frac{1}{2} O_2 + 2e^-$

The cell giving these half-cell reactions would be

$$Pt \mid O_2 \mid H^+, H_2O \,\vdots\vdots\, MnO_4^-, Mn^{2+}, H^+ \mid Pt$$

with $E^\circ_{cell} = E^\circ_{MnO_4^-, Mn^{2+}, H^+ \mid Pt} - E^\circ_{H^+ \mid O_2 \mid Pt} = 1.510\ V - 1.223\ V = 0.287\ V$

The reaction is possible.

**(iii)** The reactions would be

*Reduction* $O_2 + 2H_2O + 4e^- \rightarrow 4OH^-$

*Oxidation* $Au + 2CN^- \rightarrow Au(CN)_2^- + e^-$

The cell giving these half-cell reactions would be

$$Au \mid Au(CN)^-_2, CN^- \,\vdots\vdots\, OH^- \mid O_2 \mid Pt$$

with $$E^\circ_{cell} = E^\circ_{OH^- | O_2 | Pt} - E^\circ_{Au(CN)_2^-, CN^- | Au} = 0.401 \text{ V} - (-0.601 \text{ V}) = 1.002 \text{ V}$$

The reaction is possible.

**(iv)** The reactions would be

*Reduction* $Cu^{2+} + 2e^- \rightarrow Cu$

*Oxidation* $Ag \rightarrow Ag^+ + e^-$

The cell giving these half-cell reactions would be

$$Ag \mid Ag^+ \,\vdots\vdots\, Cu^{2+} \mid Cu$$

with $$E^\circ_{cell} = E^\circ_{Cu^{2+} | Cu} - E^\circ_{Ag^+ | Ag} = 0.337 \text{ V} - 0.799 = -0.462 \text{ V}$$

Since $E^\circ_{cell}$ is negative, the cell reaction will not be spontaneous. Hence, we can use the silver spoon to stir a solution of $Cu(NO_3)_2$.

**(v)** The reactions would be

*Reduction* $Ce^{4+} + e^- \rightarrow Ce^{3+}$

*Oxidation* $I^- \rightarrow \frac{1}{2} I_2 + e^-$

The cell giving these half-cell reactions would be

$$Pt \mid I_2 \mid I^- \,\vdots\vdots\, Ce^{4+}, Ce^{3+} \mid Pt$$

with $$E^\circ_{cell} = E^\circ_{Ce^{4+}, Ce^{3+} | Pt} - E^\circ_{I^- | I_2 | Pt} = 1.44 \text{ V} - 0.536 \text{ V} = 0.904 \text{ V}$$

The reaction is possible.

**(vi)** The reactions would be

*Reduction* $Ag^+ + e^- \rightarrow Ag$

*Oxidation* $Pb \rightarrow Pb^{2+} + 2e^-$

The cell giving these half-cell reactions would be

$$Pb \mid Pb^{2+} \,\vdots\vdots\, Ag^+ \mid Ag$$

with $$E^\circ_{cell} = E^\circ_{Ag^+ | Ag} - E^\circ_{Pb^{2+} | Pb} = 0.799 \text{ V} - (-0.126 \text{ V}) = 0.925 \text{ V}$$

The reaction is possible.

**(vii)** The reactions would be

*Reduction* $MnO_4^- + 8H^+ + 5e^- \rightarrow Mn^{2+} + 4H_2O$

*Oxidation* $\begin{array}{l} COOH \\ | \\ COOH \end{array} \rightarrow CO_2 + 2H^+ + 2e^-$

The cell giving these half-cell reactions would be

$$Pt \mid CO_2 \mid \text{oxalic acid}, H^+ \,\vdots\vdots\, MnO_4^-, Mn^{2+}, H^+ \mid Pt$$

with $E^\circ_{cell} = E^\circ_{MnO_4^-, Mn^{2+}, H^+ | Pt} - E^\circ_{oxalic\ acid, H^+ | CO_2 | Pt}$

$= 1.51\ V - (-0.49\ V) = 2.00\ V$

The reaction is possible.

**(viii)** The reactions would be

*Reduction* $Cr_2O_7^{2-} + 14H^+ + 6e^- \rightarrow 2Cr^{3+} + 7H_2O$

*Oxidation* $Fe^{2+} \rightarrow Fe^{3+} + e^-$

The cell giving these half-cell reactions would be

$Pt | Fe^{2+}, Fe^{3+} \vdots\vdots Cr_2O_7^{2-}, Cr^{3+}, H^+ | Pt$

with $E^\circ_{cell} = E^\circ_{Cr_2O_7^{2-}, Cr^{3+}, H^+ | Pt} - E^\circ_{Fe^{3+}, Fe^{2+} | Pt}$

$= 1.33\ V - 0.771\ V = 0.559\ V$

The reaction is possible.

**(ix)** The reactions would be

*Reduction* $Fe^{3+} + e^- \rightarrow Fe^{2+}$

*Oxidation* $Sn^{2+} \rightarrow Sn^{4+} + 2e^-$

The cell giving these half-cell reactions would be

$Pt | Sn^{4+}, Sn^{2+} \vdots\vdots Fe^{3+}, Fe^{2+} | Pt$

with $E^\circ_{cell} = E^\circ_{Fe^{3+}, Fe^{2+} | Pt} - E^\circ_{Sn^{4+}, Sn^{2+} | Pt}$

$= 0.771\ V - 0.150\ V = 0.621\ V$

The reaction is possible.

**(x)** The reactions for the diphenylbenzidine indicator would be

*Reduction* $Fe^{3+} + e^- \rightarrow Fe^{2+}$

*Oxidation* Diphenylbenzidine $\rightarrow$ Diphenylbenzidine violet $+ 2H^+ + 2e^-$
(DPB) (DPBV)

The cell producing these half-cell reactions would be

$Pt | DPB, DPBV, H^+ \vdots\vdots Fe^{3+}, Fe^{2+} | Pt$

with $E^\circ_{cell} = E^\circ_{Fe^{3+}, Fe^{2+} | Pt} - E^\circ_{DPB, DPBV, H^+ | Pt}$

$= 0.771\ V - 0.76\ V = 0.011\ V$

The reaction is possible.
The reactions for the *N*-phenylanthranilic acid would be

*Reduction* $Fe^{3+} + e^- \rightarrow Fe^{2+}$
*Oxidation* Reduced from of indicator $\rightarrow$ Oxidized form of indicator $+ 4H^+ + 4e^-$

The cell producing these half-cell reactions would be

Pt | Reduced and oxidized forms of indicator, $H^+ \vdots\vdots Fe^{2+}, Fe^{3+} | Pt$

with $E^\circ_{cell} = E^\circ_{Fe^{3+}, Fe^{2+} | Pt} - E^\circ_{indicator}$

$= 0.771 \text{ V} - 1.080 \text{ V} = -0.309 \text{ V}$

The reaction is not possible.

---

***Example 8.9.3***

Ascertain, with a little calculation, whether or not each of the following reactions will proceed to the forward or backward direction at 25 °C when all concentrations and pressures are unity.

(a) $2Cr + 3Cu^{2+} \rightarrow 2Cr^{3+} + 3Cu$ (b) $Zn + Mg^{2+} \rightarrow Zn^{2+} + Mg$

(c) $Fe + Cl_2(g) \rightarrow Fe^{2+} + 2Cl^-$ (d) $2Fe^{2+} + Cl_2(g) \rightarrow 2Fe^{3+} + 2Cl^-$

(e) $2Ag + Cu^{2+} \rightarrow 2Ag^+ + Cu$ (f) $2Ag + Cu^{2+} + 2Cl^- \rightarrow 2AgCl(s) + Cu$

***Solution***

**(a)** Cr is oxidized to $Cr^{3+}$ and $Cu^{2+}$ is reduced to Cu, the cell giving this reaction would be

$$Cr \mid Cr^{3+} \,\vdots\vdots\, Cu^{2+} \mid Cu$$

with $E^\circ_{cell} = E^\circ_{Cu^{2+} | Cu} - E^\circ_{Cr^{3+} | Cr} = 0.337 \text{ V} - (-0.774) = 1.111 \text{ V}$

Since $E^\circ$ is positive, the given reaction will proceed in the forward direction.

We can proceed similarly for other reactions. The given reaction, the corresponding cell, $E^\circ_{cell}$ and conclusion regarding whether the reaction proceeds in the forward or backward direction are tabulated below:

| *Reaction* | *Cell* | $E^\circ_{cell}$/V | *Direction of reaction* |
|---|---|---|---|
| (b) $Zn + Mg^{2+} \rightarrow Zn^{2+} + Mg$ | $Zn \mid Zn^{2+} \vdots\vdots Mg^{2+} \mid Mg$ | $-2.37 - (-0.763)$ | Backward |
| (c) $Fe + Cl_2(g) \rightarrow Fe^{2+} + 2Cl^-$ | $Fe \mid Fe^{2+} \vdots\vdots Cl^- \mid Cl_2 \mid Pt$ | $1.359\,5 - (-0.440)$ | Forward |
| (d) $2Fe^{2+} + Cl_2(g) \rightarrow 2Fe^{3+} + 2Cl^-$ | $Pt \mid Fe^{2+}, Fe^{3+} \vdots\vdots Cl^- \mid Cl_2 \mid Pt$ | $1.359\,5 - 0.771$ | Forward |
| (e) $2Ag + Cu^{2+} \rightarrow 2Ag^+ + Cu$ | $Ag \mid Ag^+ \vdots\vdots Cu^{2+} \mid Cu$ | $0.337 - 0.799\,1$ | Backward |
| (f) $2Ag + Cu^{2+} + 2Cl^- \rightarrow 2AgCl(s) + Cu$ | $Ag \mid AgCl \mid Cl^- \vdots\vdots Cu^{2+} \mid Cu$ | $0.337 - 0.222$ | Forward |

---

***Example 8.9.4***

(a) Which of the following half-reactions will oxidize Cu(s) to $Cu^{2+}$ under standard conditions?

(i) $Cl_2(g)$ to $Cl^-$; (ii) $MnO_4^-(aq)$ to $Mn^{2+}(aq)$; (iii) $Ag^+$ to Ag.

(b) Which of the following half-reactions will reduce $Fe^{3+}$ to $Fe^{2+}$ under standard conditions?

(i) $Sn^{2+}$ to $Sn^{4+}$; (ii) $Mn^{2+}(aq)$ to $MnO_4^-(aq)$; (iii) Pb(s) to $Pb^{2+}(aq)$

***Solution***

**(a)** Since Cu is to be oxidized to $Cu^{2+}$, it will constitute the left electrode. Each of the given reactions with which this oxidation reaction is to be carried out will constitute the right electrode. Thus, we have

(i) Right electrode $Cl_2 + 2e^- \rightarrow 2Cl^-$

Left electrode $Cu \rightarrow Cu^{2+} + 2e^-$

The cell reaction would be $Cl_2 + Cu \rightarrow Cu^{2+} + 2Cl^-$

and its emf would be $E^\circ_{cell} = E^\circ_{Cl^- | Cl_2 | Pt} - E^\circ_{Cu^{2+} | Cu}$

$= 1.359\,5 \text{ V} - 0.337 \text{ V} = 1.022\,5 \text{ V}$

Since $E^\circ_{cell}$ is positive, $Cl_2$ can oxidize Cu to $Cu^{2+}$.

(ii) Right electrode $2MnO_4^- + 16H^+ + 10e^- \rightarrow 2Mn^{2+} + 8H_2O$

Left electrode $5Cu \rightarrow 5Cu^{2+} + 10e^-$

The cell reaction would be $2MnO_4^- + 16H^+ + 5Cu \rightarrow 2Mn^{2+} + 8H_2O + 5Cu^{2+}$

and its emf would be $E^\circ_{cell} = E^\circ_{MnO_4^-, Mn^{2+}, H^+ | Pt} - E^\circ_{Cu^{2+} | Cu}$

$= 1.510 \text{ V} - 0.337 \text{ V} = 1.173 \text{ V}$

Thus oxidation of Cu to $Cu^{2+}$ is possible.

(iii) Right electrode $2Ag^+ + 2e^- \rightarrow 2Ag$

Left electrode $Cu \rightarrow Cu^{2+} + 2e^-$

The cell reaction would be $Cu + 2Ag^+ \rightarrow 2Ag + Cu^{2+}$

and its emf would be $E^\circ_{cell} = E^\circ_{Ag^+ | Ag} - E^\circ_{Cu^{2+} | Cu}$

$= 0.779\,1 \text{ V} - 0.337 \text{ V} = 0.442\,1 \text{ V}$

Thus oxidation of Cu to $Cu^{2+}$ is possible.

(**b**) In this case, we would have

(i) Right electrode $2Fe^{3+} + 2e^- \rightarrow 2Fe^{2+}$

Left electrode $Sn^{2+} \rightarrow Sn^{4+} + 2e^-$

The cell reaction would be $2Fe^{3+} + Sn^{2+} \rightarrow Sn^{4+} + 2Fe^{2+}$

and its emf would be $E^\circ_{cell} = E^\circ_{Fe^{3+}, Fe^{2+} | Pt} - E^\circ_{Sn^{4+}, Sn^{2+} | Pt}$

$= 0.771 \text{ V} - 0.150 \text{ V} = 0.621 \text{ V}$

Since $E^\circ_{cell}$ is positive, $Sn^{2+}$ can reduce $Fe^{3+}$ to $Fe^{2+}$.

(ii) Right electrode $5Fe^{3+} + 5e^- \rightarrow 5Fe^{2+}$

Left electrode $Mn^{2+} + 4H_2O \rightarrow MnO_4^- + 8H^+ + 5e^-$

The cell reaction would $Mn^{2+} + 4H_2O + 5Fe^{3+} \rightarrow MnO_4^- + 8H^+ + 5Fe^{2+}$

and its emf would be $E^\circ_{cell} = E^\circ_{Fe^{3+}, Fe^{2+} | Pt} - E^\circ_{MnO_4^-, Mn^{2+}, H^+ | Pt}$

$$= 0.771\ V - 1.51\ V = -0.739\ V$$

Thus, reduction of $Fe^{3+}$ to $Fe^{2+}$ by $Mn^{2+}$ ions is not possible.

(iii) Right electrode $2Fe^{3+} + 2e^- \rightarrow 2Fe^{2+}$

Left electrode $Pb \rightarrow Pb^{2+} + 2e^-$

The cell reaction would be $Pb + 2Fe^{3+} \rightarrow Pb^{2+} + 2Fe^{2+}$

and its emf would be $E^\circ_{cell} = E^\circ_{Fe^{3+}, Fe^{2+} | Pt} - E^\circ_{Pb^{2+} | Pb}$

$$= 0.771\ V - (-0.126\ V) = 0.897\ V$$

Thus reduction of $Fe^{3+}$ to $Fe^{2+}$ by Pb is possible.

***Example 8.9.5***

The four elements W, X, Y, Z form diatomic molecules and also form singly charged negative ions. The following observations are made in a series of experiments:

(a) $2X^- + Y_2 \rightarrow 2Y^- + X_2$ (b) $2W^- + Y_2 \rightarrow$ No reaction

(c) $2Z^- + X_2 \rightarrow 2X^- + Z_2$

Use these observations to write equation for each half-reaction. Arrange them in the decreasing reduction potential series. Which ion is the strongest reducing agent? Which molecule is the strongest oxidizing agent?

***Solution***

(**a**) The given reaction is

$$2X^- + Y_2 \rightarrow 2Y^- + X_2$$

Here $Y_2$ is reduced to $Y^-$ and $X^-$ is oxidized to $X_2$, therefore, the half-reactions would be

Right electrode $Y_2 + 2e^- \rightarrow 2Y^-$

Left electrode $2X^- \rightarrow X_2 + 2e^-$

The cell emf would be $E_{cell} = E_{Y^- | Y_2 | Pt} - E_{X^- | X_2 | Pt}$

Since the reaction is feasible, $E_{cell}$ will be positive. This yields

$$E_{Y^- | Y_2 | Pt} > E_{X^- | X_2 | Pt} \qquad (1)$$

(**b**) We have

$$2W^- + Y_2 \rightarrow \text{No reaction}$$

Had the reaction occurred, then $Y_2$ would have been reduced to $Y^-$ and $W^-$ would have been oxidized to $W_2$. The cell emf for these half-reaction would be

$$E_{cell} = E_{Y^- | Y_2 | Pt} - E_{W^- | W_2 | Pt}$$

Since the reaction is not possible, it means $E_{cell}$ is negative and thus

$$E_{W^- | W_2 | Pt} > E_{Y^- | Y_2 | Pt} \qquad (2)$$

(**c**) Here, we have

$$2Z^- + X_2 \rightarrow 2X^- + Z_2$$

$X_2$ is reduced to $X^-$ and $Z^-$ is oxidized to $Z_2$. The half-cell reactions are

Right electrode $X_2 + 2e^- \rightarrow 2X^-$

Left electrode $2Z^- \rightarrow Z_2 + 2e^-$

The cell emf would be $E_{cell} = E_{X^- | X_2 | Pt} - E_{Z^- | Z_2 | Pt}$

Since the reaction is possible, $E_{cell}$ will be positive and hence

$$E_{X^- | X_2 | Pt} > E_{Z^- | Z_2 | Pt} \quad (3)$$

Combining Eqs (1), (2) and (3), we get

$$E_{W^- | W_2 | Pt} > E_{Y^- | Y_2 | Pt} > E_{X^- | X_2 | Pt} > E_{Z^- | Z_2 | Pt}$$

Out of these, $W_2$ will have the greatest tendency for reduction and, therefore, will be the strongest oxidizing agent. $Z_2$ will have the least tendency for reduction or $Z^-$ will show the greatest tendency for oxidation. Thus, $Z^-$ is the strongest reducing agent.

***Example 8.9.6***

Sometimes a species in a given oxidation state undergoes self-oxidation reduction reaction (disproportionation) in aqueous medium to give products of lower and higher oxidation states. For example, when $Br_2$ is added in an alkaline solution ($[OH^-] = 1$ mol dm$^{-3}$), it disproportionates to give bromide and hypobromide ions. Rationalize the above results from the standard potentials given below:

(i) $\frac{1}{2} Br_2 + e^- \rightarrow Br^-$ $\qquad E° = 1.07$ V

(ii) $BrO^- + H_2O + e^- \rightarrow \frac{1}{2} Br_2 + 2OH^-$ $\qquad E° = 0.45$ V

(iii) $2BrO^- + 2H_2O + 4e^- \rightarrow 2Br^- + 4OH^-$ $\qquad E° = 0.71$ V

(iv) $BrO_3^- + 2H_2O + 4e^- \rightarrow BrO^- + 4OH^-$ $\qquad E° = 0.54$ V

The above data may be presented in a condensed form as

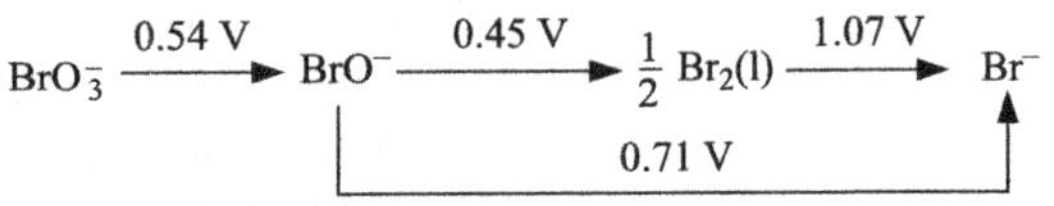


What would be the overall reaction of $Br_2$ in an alkaline solution?

***Solution***

The self-reduction oxidation reactions of $Br_2$ are

*Reduction* $\qquad \frac{1}{2} Br_2 + e^- \rightarrow Br^- \qquad (1)$

*Oxidation* $\qquad \frac{1}{2} Br_2 + 2OH^- \rightarrow BrO^- + H_2O + e^- \qquad (2)$

The cell producing the above two half-cell reactions would be

Pt | $Br_2$, $BrO^-$, $OH^-$ ⋮⋮ $Br^-$, $Br_2$ | Pt

with the cell potential given as

$$E°_{cell} = E°_R - E°_L = 1.07 \text{ V} - 0.45 \text{ V} = 0.62 \text{ V}$$

Since the cell potential is positive, it is obvious that $Br_2$ can be converted to $Br^-$ and $BrO^-$ simultaneously. Hence, $Br_2$ disproportionates in the alkaline solution to give $Br^-$ and $BrO^-$.

Similarly, the self-reduction oxidation reactions of $BrO^-$ are

*Reduction* $2BrO^- + 2H_2O + 4e^- \rightarrow 2Br^- + 4OH^-$ (3)

*Oxidation* $BrO^- + 4OH^- \rightarrow BrO_3^- + 2H_2O + 4e^-$ (4)

Thus, a cell producing the above two half-cell reactions would be

$$\text{Pt} \mid BrO^-, BrO_3^-, OH^- \;\vdots\vdots\; BrO^-, Br^-, OH^- \mid \text{Pt}$$

with the cell potential given as

$$E^\circ_{\text{cell}} = E^\circ_{\text{R}} - E^\circ_{\text{L}} \simeq 0.71\text{ V} - 0.54\text{ V} = 0.17\text{ V}$$

Since the cell potential is positive, it is obvious that $BrO^-$ can be converted to $Br^-$ and $BrO_3^-$ simultaneously. Hence $BrO^-$ disproportionates in alkaline solution to give $Br^-$ and $BrO_3^-$.

The overall reaction of $Br_2$ in an alkaline solution may be obtained by adding the two disproportionate reactions.

***Disproportionate reaction of*** $Br_2$ Adding Eqs (1) and (2), we get

$$2OH^- + Br_2 \rightarrow Br^- + BrO^- + H_2O \quad (5)$$

***Disproportionate reaction of*** $BrO^-$ Adding Eqs (3) and (4), we get

$$3BrO^- \rightarrow 2Br^- + BrO_3^- \quad (6)$$

Hence multiplying Eq. (5) by 3 and then adding it to Eq. (6), we have

$$6OH^- + 3Br_2 \rightarrow 5Br^- + BrO_3^- + 3H_2O$$

***Example 8.9.7***

Determine the cell reaction and $E_{\text{cell}}$ for each of the following cells at 25 °C. Assume activity and fugacity equal to molar concentration and pressure (bar), respectively.

(i) $\text{Mg} \mid MgI_2(0.1\text{ mol dm}^{-3}) \mid I_2(s) \mid \text{Pt}$
(ii) $\text{Ag} \mid AgBr(s) \mid HBr(0.02\text{ mol dm}^{-3}) \mid H_2(0.5\text{ bar}) \mid \text{Pt}$
(iii) $\text{Fe} \mid Fe(NO_3)_2\ (0.05\text{ mol dm}^{-3}) \;\vdots\vdots\; Cu(NO_3)_2(0.01\text{ mol dm}^{-3}) \mid \text{Cu}$
(iv) $\text{Pt} \mid H_2(g, 0.127\text{ bar}) \mid H^+Cl^- \mid H_2(g, 6.43\text{ bar}) \mid \text{Pt}$
(v) $\text{Pt} \mid Hg(l) \mid Hg_2Cl_2(s) \mid KCl(0.1\text{ mol dm}^{-3}) \;\vdots\vdots\; Zn^{2+}(0.733\text{ mol dm}^{-3}) \mid \text{Zn}$

***Solution***

Throughout this chapter $c^\circ$ represents unit concentration and $p^\circ$ represents unit pressure,

(i) $\text{Mg} \mid MgI_2(0.1\text{ mol dm}^{-3}) \mid I_2(s) \mid \text{Pt}$

We have

| *Electrode* | *Reduction reaction* | *Half-cell potential* |
|---|---|---|
| Right | $I_2(s) + 2e^- \rightarrow 2I^-(aq)$ | $E_{I^- \mid I_2} = E^\circ_{I^- \mid I_2} - \dfrac{RT}{2F} \ln ([I^-]/c^\circ)^2$ |
| Left | $Mg^{2+}(aq) + 2e^- \rightarrow Mg(s)$ | $E_{Mg^{2+} \mid Mg} = E^\circ_{Mg^{2+} \mid Mg} - \dfrac{RT}{2F} \ln \dfrac{1}{([Mg^{2+}]/c^\circ)}$ |

Since

cell reaction = reduction reaction at (right – left) electrode

and $E_{cell} = E_R - E_L$

we have

$$Mg(s) + I_2(s) \rightarrow Mg^{2+}(aq) + 2I^-(aq)$$

$$E_{cell} = (E^\circ_{I^- | I_2} - E^\circ_{Mg^{2+} | Mg}) - \frac{RT}{2F} \ln \left[ \left( \frac{[Mg^{2+}]}{c^\circ} \right) \left( \frac{[I^-]}{c^\circ} \right)^2 \right]$$

Substituting the data, we get

$$E_{cell} = \{0.535 \text{ V} - (-2.363\,0 \text{ V})\} - \left( \frac{0.059\,15 \text{ V}}{2} \right) \log (0.1 \times 0.2^2)$$

$$= 2.898\,5 \text{ V} + 0.070\,8 \text{ V} = 2.969 \text{ V}$$

**(ii)** Ag | AgBr(s) | HBr(0.02 mol $dm^{-3}$) | $H_2$(0.5 bar) | Pt

We have

| *Electrode* | *Reduction reaction* |
|---|---|
| Right | $H^+(aq) + e^- \rightarrow \frac{1}{2} H_2(g)$ |
| Left | $AgBr(s) + e^- \rightarrow Ag(s) + Br^-(aq)$ |

The cell reaction (reduction reaction at right electrode – reduction reaction at left electrode) is

$$Ag(s) + Br^-(aq) + H^+(aq) \rightarrow AgBr(s) + \frac{1}{2} H_2(g)$$

Hence $$E_{cell} = (E^\circ_{H^+ | H_2 | Pt} - E^\circ_{Br^- | AgBr | Ag}) - \frac{RT}{F} \ln \frac{(p_{H_2}/p^\circ)^{1/2}}{\{[H^+]/c^\circ\}\{[Br^-]/c^\circ\}}$$

Substituting the data, we get

$$E_{cell} = (0.0 \text{ V} - 0.071\,3 \text{ V}) - (0.059\,15 \text{ V}) \log \left( \frac{0.5^{1/2}}{0.02 \times 0.02} \right)$$

$$= -0.071\,3 \text{ V} - 0.192\,1 \text{ V} = -0.263\,4 \text{ V}$$

**(iii)** Fe | $Fe(NO_3)_2$(0.05 mol $dm^{-3}$) ⁞⁞ $Cu(NO_3)_2$(0.01 mol $dm^{-3}$) | Cu

We have

| *Electrode* | *Reduction reaction* |
|---|---|
| Right | $Cu^{2+}(aq) + 2e^- \rightarrow Cu(s)$ |
| Left | $Fe^{2+}(aq) + 2e^- \rightarrow Fe(s)$ |

The cell reaction is

$$Fe(s) + Cu^{2+}(aq) \rightarrow Fe^{2+}(aq) + Cu(s)$$

Hence $$E_{cell} = (E^{\circ}_{Cu^{2+}|Cu} - E^{\circ}_{Fe^{2+}|Fe}) - \left(\frac{0.059\ 15\ V}{2}\right) \log \left\{\frac{[Fe^{2+}]/c^{\circ}}{[Cu^{2+}]/c^{\circ}}\right\}$$

$$= \{0.337\ V - (-0.441\ V)\} - \left(\frac{0.059\ 15\ V}{2}\right) \log \left(\frac{0.05}{0.01}\right)$$

$$= 0.778\ V - 0.020\ 7\ V = 0.757\ 3\ V$$

**(iv)** Pt | $H_2(g)$ (0.127 bar) | $H^+Cl^-$ | $H_2(g)$ (6.43 bar) | Pt

We have

| *Electrode* | *Reduction reaction* |
|---|---|
| Right | $H^+(aq) + e^- \rightarrow \frac{1}{2} H_2(g)_R$ |
| Left | $H^+(aq) + e^- \rightarrow \frac{1}{2} H_2(g)_L$ |

The cell reaction is

$$\tfrac{1}{2}H_2(g)_L \rightarrow \tfrac{1}{2}H_2(g)_R$$

Hence $$E_{cell} = -\left(\frac{0.059\ 15\ V}{1}\right) \log \left\{\frac{p_{H_2(R)}/p^{\circ})}{p_{H_2(L)}/p^{\circ}}\right\}^{1/2}$$

$$= -\left(\frac{0.059\ 15\ V}{2}\right) \log \left\{\frac{p_{H_2(R)}/p^{\circ})}{p_{H_2(L)}/p^{\circ}}\right\}$$

Substituting the given data, we get

$$E_{cell} = -\left(\frac{0.059\ 15\ V}{2}\right) \log \left(\frac{6.43}{0.127}\right)$$

$$= -(0.029\ 57\ V)(1.704\ 4) = -0.503\ 7\ V$$

**(v)** Pt | Hg(l) | $Hg_2Cl_2(s)$ | KC1(0.1 mol $dm^{-3}$) ⁞⁞ $Zn^{2+}$(0.732 mol $dm^{-3}$) | Zn

We have

| *Electrode* | *Reduction reaction* |
|---|---|
| Right | $Zn^{2+}(aq) + 2e^- \rightarrow Zn(s)$ |
| Left | $Hg_2Cl_2(s) + 2e^- \rightarrow 2Hg(l) + 2Cl^-(aq)$ |

The cell reaction is

$$2Hg(l) + 2Cl^-(aq) + Zn^{2+}(aq) \rightarrow Hg_2Cl_2(s) + Zn(s)$$

Hence $$E_{cell} = (E^{\circ}_{Zn^{2+}|Zn} - E^{\circ}_{Cl^-|Hg_2Cl_2|Hg}) - \frac{RT}{2F} \ln \frac{1}{\{[Cl^-]/c^{\circ}\}^2\{[Zn^{2+}]/c^{\circ}\}}$$

$$= (E^{\circ}_{Zn^{2+}|Zn} - E^{\circ}_{Cl^-|Hg_2Cl_2|Hg}) + \frac{RT}{2F} \ln (\{[Cl^-]/c^{\circ}\}^2\{[Zn^{2+}]/c^{\circ}\})$$

Substituting the data, we get

$$E_{\text{cell}} = (-0.763\ \text{V} - 0.337\ \text{V}) + \left(\frac{0.059\ 15\ \text{V}}{2}\right) \log (0.1^2 \times 0.733)$$

$$= -1.1\ \text{V} - 0.063\ \text{V} = -1.163\ \text{V}$$

## 8.10 INFLUENCE OF IONIC ACTIVITY ON REDUCTION POTENTIAL

The reduction potential for the reaction

$$M^{n+} + ne^- \rightarrow M$$

as given by the Nernst equation is

$$E_1 = E^{\circ}_{M^{n+} | M} - \frac{RT}{nF} \ln \left(\frac{1}{a_{M^{n+}}}\right) \tag{8.10.1}$$

From Eq. (8.10.1), it follows that the reduction potential $E_{M^{n+} | M}$ depends on the activity of $M^{n+}$; it increases as the activity of $M^{n+}$ is increased. If the activity of $M^{n+}$ is decreased to one-tenth of its original activity, then the reduction potential becomes

$$E_2 = E^{\circ}_{M^{n+} | M} - \frac{RT}{nF} \ln \left(\frac{1}{0.1\ a_{M^{n+}}}\right) \tag{8.10.2}$$

The resulting change in potential is given by

$$E_2 - E_1 = -\frac{RT}{nF} \ln \left(\frac{1}{0.1\ a_{M^{n+}}}\right) + \frac{RT}{nF} \ln \left(\frac{1}{a_{M^{n+}}}\right) \tag{8.10.3}$$

$$= -\frac{RT}{nF} \ln 10 = -\frac{2.303RT}{nF}$$

At 298 K, we have

$$E_2 - E_1 = -\frac{(0.0592\ \text{V})}{n} \tag{8.10.4}$$

Thus every ten-fold decrease in the activity of cation produces a decrease of (0.059 $2/n$) volt of the reduction potential, where $n$ is the valence of the cation. For univalent cations $n$ is 1, and hence reduction potential decreases by a factor of 0.059 2 V. For bivalent cations $n$ is 2, and hence reduction potential decreases by a factor of (0.059 2/2) V, i.e. 0.029 6 V. The magnitude of reduction potential will change by the same factor (= (0.059 2 V)/$n$) for every ten-fold change in the activity of cation. For a hundred-fold change in the activity; the change of potential will be (2 × 0.059 2 V)/$n$) volt and for a thousand-fold change in activity, the potential will change by a factor of (3 × 0.059 2 V)/$n$) volt.

The change in reduction potential on dilution is determined by the ratio of activities of ions before and after dilution. Thus, if two solutions containing different amounts of a cation are diluted by the same factor (say, 10 times), the change in potential will be the same for both cases. For example, a change of concentration of a cation from 1.0 mol $dm^{-3}$ to 0.1 mol $dm^{-3}$ will produce the same change in potential as a decrease from $10^{-6}$ mol $dm^{-3}$ to $10^{-7}$ mol $dm^{-3}$ since the ratio of the two concentrations is the same, viz., 10 to 1.

**Molecular Formula of Mercurous Salts**

The mercurous chloride is written as $Hg_2Cl_2$ instead of HgCl and hence the mercurous cation is represented as $Hg_2^{2+}$. One argument for writing the double formula for mercurous ion comes from the effect of dilution on the reduction potential of the half-cell reaction

$$\text{Mercurous ion} + ne^- \rightarrow \text{Mercury}$$

It is found that the reduction potential changes by a factor of 0.029 6 V for a ten-fold change in concentration of mercurous ions. This result indicates that for mercurous ions $n$ is 2, so that they carry two unit charges per ion and hence its formula must be written as $Hg_2^{2+}$ and not as $Hg^+$.

The above principle can also be extended to other ions. For example, cuprous iodide is written as $Cu_2I_2$ and not as CuI. This can be derived from the change of reduction potential with dilution for the reaction

$$\text{Cuprous ion} + ne^- \rightarrow \text{Copper}$$

***Example 8.10.1***

The emf of the cell

$$\text{Hg} \mid \text{Mercurous nitrate (0.01 M)} \,\vdots\vdots\, \text{Mercurous nitrate (0.1 M)} \mid \text{Hg}$$

is found to be 0.029 5 V at 25 °C. Determine the molecular formula of mercurous nitrate.

***Solution***

For the given cell, we have

| *Electrode* | *Reduction reaction* | *Reduction potential* | |
|---|---|---|---|
| Right | $Hg_n^{n+} + ne^- \rightarrow nHg$ | $E_R = E_R^\circ - \dfrac{RT}{nF} \ln \dfrac{1}{([Hg_n^{n+}]_R/c^\circ)}$ | (i) |
| Left | $Hg_n^{n+} + ne^- \rightarrow nHg$ | $E_L = E_L^\circ - \dfrac{RT}{nF} \ln \dfrac{1}{([Hg_n^{n+}]_L/c^\circ)}$ | (ii) |

Since $E_{cell}^\circ = E_R^\circ - E_L^\circ$, on subtracting Eq. (ii) from Eq. (i), we get

$$E_{cell} = E_R - E_L = \frac{RT}{nF} \ln \frac{([Hg_n^{n+}]_R/c^\circ)}{([Hg_n^{n+}]_L/c^\circ)} = \frac{0.059\,15\ \text{V}}{n} \log\left(\frac{0.1}{0.01}\right)$$

$$= \frac{0.059\,15\ \text{V}}{n} = 0.029\,5\ \text{V}$$

Thus $\quad n = 2$

Hence, the molecular formula is $Hg_2(NO_3)_2$.

## 8.11 EFFECT OF COMPLEX FORMATION ON REDUCTION POTENTIAL

In general, a reduction reaction may be represented as

$$\text{oxidized form} + ne^- \rightarrow \text{reduced form} \tag{8.11.1}$$

The corresponding Nernst equation is

$$E = E^\circ - \frac{RT}{nF} \ln \left( \frac{[\text{reduced form}]/c^\circ}{[\text{oxidized form}]/c^\circ} \right) \tag{8.11.2}$$

Thus, if the concentration of the oxidized form is lowered by some means keeping the concentration of the reduced form unchanged, a decrease in the reduction potential will be observed. On the other hand, if the concentration of the reduced form is lowered, an increase in the reduction potential will be observed.

The number of electrons $n$ involved in Eq. (8.11.1) can be determined by plotting a graph between $E$ and ln ([reduced form]/[oxidized form]). According to Eq. (8.11.2), a straight line will be obtained with a slope equal to $-RT/nF$. Hence knowing the slope and temperature $T$, the value of $n$ can be determined.

Substance which forms the soluble complex (or precipitate) with the oxidized member of the couple (Eq. 8.11.1) will decrease the effective concentration of the oxidized form. Consequently, the reduction potential of the system is lowered. For example, if some $CN^-$ is added to the half-cell $Ag^+ \mid Ag$, its reduction potential is decreased owing to the formation of complex between $Ag^+$ and $CN^-$ ions. Thus, a metal which otherwise behaves as a noble metal may be converted into an active metal by lowering its reduction potential by some means. Hence whether a metal is a noble metal or an active metal, it all depends upon its environments. For example, ordinarily silver is a noble metal but in the presence of $I^-$, $S^{2-}$ and $CN^-$ ions, it becomes an active metal.

In an analogous way, if the concentration of the reduced form of the couple is lowered, the potential of the electrode is increased. If the reduced form is a metal, the most useful method of lowering its activity is to form an alloy of the metal with another metal. In amalgam electrodes, mercury is used for this purpose.

***Example 8.11.1***

A cell consists of a Pt electrode immersed in the redox system $A^{x+} \rightarrow A^{(x+n)+} + ne^-$ is combined with a saturated calomel electrode (right-side electrode). From the following data at 303 K, determine the value of $n$.

| | | | | |
|---|---|---|---|---|
| Per cent of reduced form | 24.4 | 48.8 | 73.2 | 84.5 |
| emf/V | 0.101 | 0.116 | 0.129 | 0.139 |

The standard potential of the calomel electrode at 303 K is 0.238 V.

***Solution***

For the given half-cells, we have

$$\begin{aligned} 2.303\, RT/F &= (2.303)\,(8.314\ \text{J K}^{-1}\ \text{mol}^{-1})\,(303\ \text{K})/(96\,500\ \text{C mol}^{-1}) \\ &= 0.060\,1\ \text{V} \end{aligned}$$

| *Electrode* | *Reduction reaction* | *Expression of reduction potential* | |
|---|---|---|---|
| Right | $Hg_2Cl_2 + 2e^- \rightarrow 2Hg + 2Cl^-$ | $E_R = E_R^\circ - \dfrac{0.060\,1\text{ V}}{2}\log([Cl^-]/c^\circ)^2$ | (i) |
| Left | $A^{(x+n)+} + ne^- \rightarrow A^{x+}$ | $E_L = E_L^\circ - \dfrac{0.060\,1\text{ V}}{n}\log\dfrac{([A^{x+}]/c^\circ)}{([A^{(x+n)+}]/c^\circ)}$ | (ii) |

Subtracting Eq. (ii) from Eq. (i), we get

$$E_{cell} = E_R - E_L$$

$$= \left(E_R^\circ - \frac{0.060\,1\text{ V}}{2}\log([Cl^-]/c^\circ)^2\right) - \left(E_L^\circ - \frac{0.060\,1\text{ V}}{n}\log\frac{([A^{x+}]/c^\circ)}{([A^{(x+n)+}]/c^\circ)}\right)$$

or $$E_{cell} = \left(E_R^\circ - E_L^\circ - \frac{0.060\,1\text{ V}}{2}\log([Cl^-]/c^\circ)^2\right) + \frac{0.060\,1\text{ V}}{n}\log\frac{([A^{x+}]/c^\circ)}{([A^{(x+n)+}]/c^\circ)}$$

or $$E_{cell} = \text{constant} + \frac{0.060\,1\text{ V}}{n}\log\frac{([\text{reduced form}]/c^\circ)}{([\text{oxidized form}]/c^\circ)}$$

If a plot is made between $E_{cell}$ and log ([reduced]/[oxidized]) a straight line will be obtained. The slope of the line will be equal to 0.060 1 V/$n$ and hence the value of $n$ can be calculated.

From the given data, we have

| | | | | |
|---|---|---|---|---|
| Per cent reduced form | 24.0 | 48.8 | 73.2 | 85.4 |
| $\log\left(\dfrac{[\text{reduced}]}{[\text{oxidized}]}\right)$ | – 0.491 1 | –0.020 9 | 0.436 4 | 0.767 1 |
| $E_{cell}$/V | 0.101 | 0.116 | 0.129 | 0.139 |

A plot between $E_{cell}$ and log ([reduced]/[oxidized]) is shown in Fig. 8.11.1

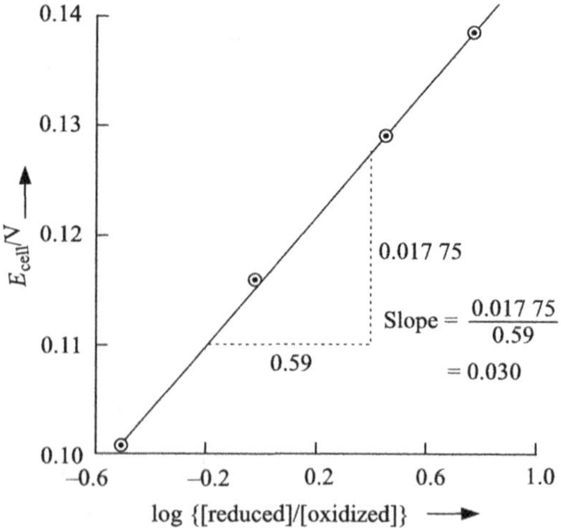


**Fig. 8.11.1 Graph between $E_{cell}$ and log ([reduced]/[oxidized])**

The slope of the line was found to be 0.030 V. Thus

$$\frac{0.060\,1\ \text{V}}{n} = 0.030\,0\ \text{V} \quad \text{which gives} \quad n = \frac{0.060\,1}{0.0300} = 2$$

## 8.12 RELATION BETWEEN METAL-METAL ION HALF-CELL AND THE CORRESPONDING METAL-INSOLUBLE SALT-ANION HALF-CELL

Consider a silver electrode dipped into a solution of silver nitrate. The potential of the electrode as given by the Nernst equation is

$$E_{\text{Ag}^+|\text{Ag}} = E^\circ_{\text{Ag}^+|\text{Ag}} - \frac{RT}{F}\ln\left(\frac{1}{a_{\text{Ag}^+}}\right) \tag{8.12.1}$$

with $E^\circ_{\text{Ag}^+|\text{Ag}} = 0.799$ V.

If now NaCl is added, the following equilibrium will be established:

$$\text{AgCl(s)} \rightleftharpoons \text{Ag}^+\text{(aq)} + \text{Cl}^-\text{(aq)}$$

with its standard solubility product $K^\circ_s$ as

$$K^\circ_s = (a_{\text{Ag}^+})(a_{\text{Cl}^-})$$

Multiplying and dividing the term within parantheses in Eq. (8.12.1) by $a_{\text{Cl}^-}$, we get

$$E_{\text{Ag}^+|\text{Ag}} = E^\circ_{\text{Ag}^+|\text{Ag}} - \frac{RT}{F}\ln\left(\frac{a_{\text{Cl}^-}}{a_{\text{Ag}^+}a_{\text{Cl}^-}}\right)$$

$$= E^\circ_{\text{Ag}^+|\text{Ag}} - \frac{RT}{F}\ln\left(\frac{a_{\text{Cl}^-}}{K^\circ_s(\text{AgCl})}\right)$$

or

$$E_{\text{Ag}^+|\text{Ag}} = \left(E^\circ_{\text{Ag}^+|\text{Ag}} + \frac{RT}{F}\ln K^\circ_s(\text{AgCl})\right) - \frac{RT}{F}\ln a_{\text{Cl}^-} \tag{8.12.2}$$

Now the electrode can also be treated as $\text{Cl}^- | \text{AgCl} | \text{Ag}$ since part of AgCl will deposit on solid Ag. Writing the reaction for this electrode, we have

$$\text{AgCl(s)} + \text{e}^- \rightarrow \text{Ag(s)} + \text{Cl}^-\text{(aq)}$$

The corresponding Nernst equation gives

$$E_{\text{Cl}^-|\text{AgCl}|\text{Ag}} = E^\circ_{\text{Cl}^-|\text{AgCl}|\text{Ag}} - \frac{RT}{F}\ln a_{\text{Cl}^-} \tag{8.12.3}$$

Equations (8.12.2) and (8.12.3) must be identical (i.e. $E_{\text{Ag}^+|\text{Ag}} = E_{\text{Cl}^-|\text{AgCl}|\text{Ag}}$) and hence on comparing these two equations, we get

$$E^\circ_{\text{Cl}^- | \text{AgCl} | \text{Ag}} = E^\circ_{\text{Ag}^+ | \text{Ag}} + \frac{RT}{F} \ln K^\circ_s(\text{AgCl})$$

or
$$E^\circ_{\text{Cl}^- | \text{AgCl} | \text{Ag}} = E^\circ_{\text{Ag}^+ | \text{Ag}} - \frac{2.303\,RT}{F} \text{p}K^\circ_s(\text{AgCl}) \quad (8.12.4)$$

where $\text{p}K^\circ_s(\text{AgCl}) = -\log K^\circ_s(\text{AgCl})$

**Alternative Procedures**

Equation (8.12.4) can also be derived by the following two methods.

(**i**) Consider the cell

Ag | saturated solution of AgCl | AgCl | Ag

Its cell reaction is

$$\text{AgCl(s)} \rightarrow \text{Ag}^+(\text{aq}) + \text{Cl}^-(\text{aq})$$

and the cell-potential is

$$E = (E^\circ_{\text{Cl}^- | \text{AgCl} | \text{Ag}} - E^\circ_{\text{Ag}^+ | \text{Ag}}) - \frac{RT}{F} \ln (a_{\text{Ag}^+})(a_{\text{Cl}^-})$$

For the reaction at equilibrium, we will have $E = 0$ (see Section 8.18) and thus

$$E^\circ_{\text{Cl}^- | \text{AgCl} | \text{Ag}} - E^\circ_{\text{Ag}^+ | \text{Ag}} = \frac{RT}{F} \ln (a_{\text{Ag}^+})(a_{\text{Cl}^-})$$

or
$$E^\circ_{\text{Cl}^- | \text{AgCl} | \text{Ag}} = E^\circ_{\text{Ag}^+ | \text{Ag}} + \frac{RT}{F} \{\ln K^\circ_s(\text{AgCl})\}$$

(**ii**) We have

$$\text{AgCl(s)} + \text{e}^- \rightarrow \text{Ag(s)} + \text{Cl}^-(\text{aq}); \quad \Delta G^\circ_1 = -FE^\circ_{\text{Cl}^- | \text{AgCl} | \text{Ag}} \quad (8.12.5)$$

$$\text{Ag}^+(\text{aq}) + \text{e}^- \rightarrow \text{Ag(s)}; \quad \Delta G^\circ_2 = -FE^\circ_{\text{Ag}^+ | \text{Ag}} \quad (8.12.6)$$

Subtracting Eq. (8.12.6) from Eq. (8.12.5), we get

$$\text{AgCl(s)} \rightarrow \text{Ag}^+(\text{aq}) + \text{Cl}^-(\text{aq});$$

with $\Delta G^\circ = -FE^\circ_{\text{Cl}^- | \text{AgCl} | \text{Ag}} + FE^\circ_{\text{Ag}^+ | \text{Ag}}$

Since $\Delta G^\circ = -RT \ln K^\circ_s$, we will have

$$E^\circ_{\text{Cl}^- | \text{AgCl} | \text{Ag}} = E^\circ_{\text{Ag}^+ | \text{Ag}} + \frac{RT}{F} \ln K^\circ_s(\text{AgCl})$$

**Computation of Potentials**

At 298 K, we have

$$E^\circ_{\text{Cl}^- | \text{AgCl} | \text{Ag}} = E^\circ_{\text{Ag}^+ | \text{Ag}} - (0.059\ 15\ \text{V})\,\text{p}K^\circ_s(\text{AgCl})$$

Now $E^{\circ}_{Ag^+ \mid Ag} = 0.799$ V and $pK^{\circ}_s(AgCl) = 9.75$. Thus, we have

$$E^{\circ}_{Cl^- \mid AgCl \mid Ag} = 0.799 \text{ V} - (0.059\ 15 \text{ V})(9.75)$$
$$= 0.222 \text{ V}$$

Equation (8.12.4) may also be used to calculate the solubility product of silver chloride by measuring the standard potentials of $Ag^+|Ag$ and $Cl^- \mid AgCl \mid Ag$ half-cells.

**Generalization of Eq. (8.12.4)**

Equation (8.12.4) may be generalized as

$$E^{\circ}_{X^- \mid AgX \mid Ag} = E^{\circ}_{Ag^+ \mid Ag} - \frac{2.303\ RT}{F}\{pK^{\circ}_s(AgX)\} \quad (8.12.7)$$

where $X^-$ stands for the halide ion ($Cl^-$, $Br^-$ and $I^-$). From Eq. (8.12.7) it follows that larger the $pK^{\circ}_s$, lesser the reduction potential of silver halide-silver electrode. Table 8.12.1 records the standard potentials for some of the metal-insoluble salt-anion electrodes.

**Table 8.12.1** Standard Potentials of Some of the Metal-Insoluble Salt-Anion Electrodes

| *Electrode reaction* | $E^{\circ}$/V |
|---|---|
| $AgCl(s) + e^- \rightarrow Ag + Cl^-$ | 0.222 |
| $AgBr(s) + e^- \rightarrow Ag + Br^-$ | 0.03 |
| $AgI(s) + e^- \rightarrow Ag + I^-$ | −0.151 |
| $Ag_2S(s) + 2e^- \rightarrow 2Ag + S^{2-}$ | −0.69 |

It follows from Table 8.12.1 that

$$pK^{\circ}_s(Ag_2S) > pK^{\circ}_s(AgI) > pK^{\circ}_s(AgBr) > pK^{\circ}_s(AgCl) \quad (8.12.8)$$

which implies that

$$K_s(AgCl) > K_s(AgBr) > K_s(AgI) > K_s(Ag_2S) \quad (8.12.9)$$

From Eq. (8.12.8), it follows that silver chloride is more soluble than silver bromide which, in turn, is more soluble than silver iodide. The fact that the silver-silver iodide couple has a negative potential means that silver should dissolve in HI with the release of hydrogen gas. This, in fact, occurs, but the action ceases promptly due to the layer of insoluble AgI which forms and protects the silver surface from further attack.

***Example 8.12.1***

Calculate $E^{\circ}$ for the following reactions at 298 K.

$$Ag(NH_3)_2^+ + e^- \rightarrow Ag + 2NH_3$$

$$Ag(CN)_2^- + e^- \rightarrow Ag + 2CN^-$$

Given: $E^\circ_{Ag^+ | Ag} = 0.799\,1$ V, $K^\circ_{Ins}(Ag(NH_3)_2^+) = 6.02 \times 10^{-8}$ and

$K^\circ_{Ins}(Ag(CN)_2^-) = 1.995 \times 10^{-19}$.

***Solution***

The equation which relates $E^\circ_{Ag^+ | Ag}$ with $E^\circ$ of the given expression is

$$E^\circ = E^\circ_{Ag^+ | Ag} - (0.059\,15 \text{ V})\, pK^\circ_{Ins}$$

Thus

$$\begin{aligned} E^\circ_{NH_3,\, Ag(NH_3)_2^+ | Ag} &= 0.799\,1 \text{ V} - (0.059\,15 \text{ V})(-\log 6.02 \times 10^{-8}) \\ &= 0.799\,1 \text{ V} - 0.427 \text{ V} \\ &= 0.372\,1 \text{ V} \end{aligned}$$

and

$$\begin{aligned} E^\circ_{CN^- | Ag(CN)_2^- | Ag} &= 0.799\,1 \text{ V} - (0.059\,15 \text{ V})(-\log 1.995 \times 10^{-19}) \\ &= 0.799\,1 \text{ V} - 1.106\,1 \text{ V} \\ &= -0.307 \text{ V} \end{aligned}$$

---

***Example 8.12.2***

Show that the standard potential of the reaction

$$H_2O + e^- \rightarrow \tfrac{1}{2}H_2 + OH^-$$

is given as

$$E^\circ = \frac{RT}{F} \ln K^\circ_w$$

Determine its value at 298 K.

***Solution***

The reduction potential of the reaction

$$H_2O + e^- \rightarrow \tfrac{1}{2}H_2 + OH^-$$

is given by $E = E^\circ - \dfrac{RT}{F} \ln \{(f_{H_2}/f^\circ)^{1/2} a_{OH^-}\}$ (1)

Since $K_w = (a_{H^+})(a_{OH^-})$, Eq. (1) may be written as

$$E = E^\circ - \frac{RT}{F} \ln \frac{(f_{H_2}/f^\circ)^{1/2}}{a_{H^+}} - \frac{RT}{F} \ln K^\circ_w \quad (2)$$

The cell in which the given half-cell reaction occurs may equally be considered as the hydrogen-hydrogen ion half cell. Hence, its potential will be given as

$$E = -\frac{RT}{F} \ln \frac{(f_{H_2}/f^\circ)^{1/2}}{a_{H^+}} \quad (3)$$

Comparing Eqs (2) and (3), we get

$$E^\circ = \frac{RT}{F} \ln K_w^\circ \tag{4}$$

At 298 K, the value of $E^\circ$ is

$$\frac{RT}{F} \ln K_w^\circ = \frac{(2.303)(8.314\ \text{J K}^{-1}\ \text{mol}^{-1})(298\ \text{K})}{(96\ 500\ \text{C mol}^{-1})} \log(10^{-14})$$
$$= -0.828\ \text{V}$$

## 8.13 CELL REACTION AND ITS RELATION WITH CELL POTENTIAL

The electrical work that is obtainable when the amount $n$ of electrons flow from left electrode to right electrode of a cell will be equal to the change in free energy of the cell reaction. The relation between the cell potential and the free energy change of the associated cell reaction was derived in Section 8.5 and may be derived from Eq. (8.5.6) by replacing d$n$ by $n$ and $z$ by –1. Thus, we have

$$\Delta G = -nF(E_R - E_L) = -nFE_{cell}$$

Superficially, it might appear that $\Delta G$ and $E$ differ only by a numerical factor, but actually they also differ in a qualitative sense. $\Delta G$ is uniquely determined by the overall cell reaction; $E_{cell}$ is determined by the overall reaction and the make up of the cell. To illustrate this difference, we consider the various ways in which the following three half-cell reactions

$$Fe^{2+} + 2e^- \rightarrow Fe; \qquad E^\circ_{Fe^{2+}|Fe} = -0.440\ \text{V} \tag{8.13.1}$$

$$Fe^{3+} + 3e^- \rightarrow Fe; \qquad E^\circ_{Fe^{3+}|Fe} = -0.036\ \text{V} \tag{8.13.2}$$

$$Fe^{3+} + e^- \rightarrow Fe^{2+}; \qquad E^\circ_{Fe^{3+}, Fe^{2+}|Pt} = -0.771\ \text{V} \tag{8.13.3}$$

may be combined to form a cell with the following overall cell reaction.

$$Fe + 2Fe^{3+} \rightarrow 3Fe^{2+} \tag{8.13.4}$$

**Combination of Eqs (8.13.1) and (8.13.2)**

The overall cell reaction (Eq. 8.13.4) may be obtained from Eqs (8.13.1) and (8.13.2) by multiplying the former by 3 and the latter by 2 (so that the electrons involved in the two half-cell reactions are same) and then subtracting the resultant Eq. (8.13.1) from the resultant Eq. (8.13.2). Thus, we have

Eq. (8.13.2): $2Fe^{3+} + 6e^- \rightarrow 2Fe$; $E^\circ_{Fe^{3+}|Fe} = -0.036\ \text{V}$

Eq. (8.13.1): $3Fe^{2+} + 6e^- \rightarrow 3Fe$; $E^\circ_{Fe^{2+}|Fe} = -0.440\ \text{V}$

with the overall reaction

$$Fe + 2Fe^{3+} \rightarrow 3Fe^{2+}; \qquad E^\circ_{cell} = 0.404\ \text{V}$$

The cell producing this reaction would be

$$\text{Fe} \mid \text{Fe}^{2+} \,\vdots\vdots\, \text{Fe}^{3+} \mid \text{Fe} \tag{8.13.5}$$

**Combination of Eqs (8.13.1) and (8.13.3)**

In this case, we multiply Eq. (8.13.3) by 2 and then subtract from it Eq. (8.13.1). Thus, we have

Eq. (8.13.3): $2Fe^{3+} + 2e^- \rightarrow 2Fe^{2+}$; $E^\circ_{Fe^{3+}, Fe^{2+} \mid Pt} = 0.771$ V

Eq. (8.13.1): $Fe^{2+} + 2e^- \rightarrow Fe$; $E^\circ_{Fe^{2+} \mid Fe} = -0.440$ V

with the overall reaction

$$Fe + 2Fe^{3+} \rightarrow 3Fe^{2+} \qquad E^\circ_{cell} = 1.211 \text{ V}$$

The cell producing this reaction would be

$$\text{Fe} \mid \text{Fe}^{2+} \,\vdots\vdots\, \text{Fe}^{3+}, \text{Fe}^{2+} \mid \text{Pt} \tag{8.13.6}$$

**Combination of Eqs (8.13.2) and (8.13.3)**

In this case, we multiply Eq. (8.13.3) by 3 and then subtract from it Eq. (8.13.2). Thus, we have

Eq. (8.13.3): $3Fe^{3+} + 3e^- \rightarrow 3Fe^{2+}$; $E^\circ_{Fe^{3+}, Fe^{2+} \mid Pt} = 0.771$ V

Eq. (8.13.2) $Fe^{3+} + 3e^- \rightarrow Fe$; $E^\circ_{Fe^{3+} \mid Fe} = -0.036$ V

with the cell reaction

$$Fe + 2Fe^{3+} \rightarrow 3Fe^{2+}; \qquad E^\circ_{cell} = 0.807 \text{ V}$$

The cell producing this reaction would be

$$\text{Fe} \mid \text{Fe}^{3+} \,\vdots\vdots\, \text{Fe}^{3+}, \text{Fe}^{2+} \mid \text{Pt} \tag{8.13.7}$$

Thus, we observe that the voltage developed by a cell depends on the half-cell reactions, i.e. what occurs at the electrodes (or in other words on the make up of the cell) as well as on the overall reaction. But $\Delta G$ depends only on the overall cell reaction as may be seen from the $\Delta G$ values recorded in Table 8.13.1.

**Conclusion**

From Table 8.13.1, it is obvious that for all the three cells, $\Delta G^\circ$ is the same but $n$ and $E^\circ_{cell}$ vary from cell to cell. The variations in $n$ and $E^\circ_{cell}$ are such that the

**Table 8.13.1** Cell, $E^\circ_{cell}$ and $\Delta G^\circ$ Values

| *Cell* | $E_{cell}$ / V | *Number of electrons associated in the half-cell reactions* $n$ | $\Delta G^\circ = -nFE^\circ$ |
|---|---|---|---|
| $Fe \mid Fe^{2+} \vdots\vdots Fe^{3+} \mid Fe$ | 0.404 | 6 | $-2.424\ F$ |
| $Fe \mid Fe^{2+} \vdots\vdots Fe^{3+}, Fe^{2+} \mid Pt$ | 1.211 | 2 | $-2.422\ F$ |
| $Fe \mid Fe^{3+} \vdots\vdots Fe^{3+}, Fe^{2+} \mid Pt$ | 0.807 | 3 | $-2.421\ F$ |

product of two always remains constant. Thus, we conclude that $\Delta G^\circ$ depends only on the cell reaction but $E^\circ_{cell}$ depends on the make up of the cell. It may be emphasized once again here that $E^\circ_{cell}$ is an intensive property (being equal to $\Delta G^\circ/n$) and thus its value does not depend upon the size of the cell and the extent to which the cell reaction is carried out. To make the latter point more clear, we take up the cell as given by Eq. (8.13.5). If the overall cell reaction

$$Fe + 2Fe^{3+} \rightarrow 3Fe^{2+}$$

is written as

$$2Fe + 4Fe^{3+} \rightarrow 6Fe^{2+}$$

then $E^\circ_{cell}$ remains the same but $\Delta G^\circ$ changes from $-$ 2.424 $F$ to $-$ 4.848 $F$. This change in $\Delta G^\circ$ will be brought about by the change in the number of electrons involved in the two-half reactions. Thus, $n$ changes from 6 to 12.

## 8.14 CALCULATION OF STANDARD POTENTIAL FOR AN UNKNOWN HALF-CELL REACTION

A given reaction may be obtained from other reactions by carrying out the simple mathematical manipulations such as addition, subtraction, multiplication by a number and so on. Since $G^\circ$ is a state function, the $\Delta G^\circ$ of the given reaction may be obtained by carrying out the corresponding manipulations on $\Delta G^\circ$s values of the reactions. This procedure, however, cannot be adopted while calculating $E^\circ$ of a half-cell reaction from the values of other half-cell reactions. Nevertheless, we can determine the potential of a half-cell reaction from the potentials of other half-cell reactions via the associated $\Delta G^\circ$ values. First of all, we write

$$\Delta G^\circ = -nFE^\circ$$

for each and every half-cell reaction and then carry out the appropriate manipulations of the reactions along with the corresponding $\Delta G^\circ$s values to give the given half-cell reaction and its $\Delta G^\circ$ value. Knowing the number $n$ of electrons involved in the given reaction and the obtained value of $\Delta G^\circ$, the value $E^\circ$ may be determined. To illustrate the procedure, we compute $E^\circ$ for the reaction

$$Fe^{3+} + e^- \rightarrow Fe^{2+} \qquad (8.14.1)$$

from $E^\circ$s of the reactions

$$Fe^{3+} + 3e^- \rightarrow Fe; \qquad E^\circ_{Fe^{3+}|Fe} = -0.036\ V \qquad (8.14.2)$$

$$Fe^{2+} + 2e^- \rightarrow Fe; \qquad E^\circ_{Fe^{2+}|Fe} = -0.440\ V \qquad (8.14.3)$$

The half-cell reaction of Eq. (8.14.1) can be obtained by subtracting the half-cell reaction of Eq. (8.14.3) from that of Eq. (8.14.2). Accordingly, $\Delta G^\circ$ for the reaction of Eq. (8.14.1) may be obtained by subtracting the $\Delta G^\circ$ of the reaction of Eq. (8.14.3) from that of Eq. (8.14.2). Thus, we have

Eq. (8.14.2): $Fe^{3+} + 3e^- \rightarrow Fe;$ $\Delta G_1^\circ = -3FE^\circ_{Fe^{3+} | Fe} = (0.108 \text{ V}) F$

Eq. (8.14.3): $Fe^{2+} + 2e^- \rightarrow Fe;$ $\Delta G_2^\circ = -2FE^\circ_{Fe^{2+} | Fe} = (0.880 \text{ V}) F$

Subtracting, we get

Eq. (8.14.1): $Fe^{3+} + e^- \rightarrow Fe^{2+};$ $\Delta G^\circ = \Delta G_1^\circ - \Delta G_2^\circ$

$$= (0.108 \text{ V}) F - (0.880 \text{ V}) F$$

$$= (-0.772 \text{ V}) F$$

Since the number of electrons involved in Eq. (8.14.1) is one, it follows that

$$\Delta G^\circ = (-0.772 \text{ V}) F = -1\, F\, E^\circ_{Fe^{3+}, Fe^{2+} | Pt}$$

Hence $E^\circ_{Fe^{3+}, Fe^{2+} | Pt} = 0.772 \text{ V}$

Note that if we had subtracted the two half-cell potentials, we would have got 0.404 V, a wrong result.

***Example 8.14.1***

From the standard potentials shown in the following diagram, calculate the potentials $E_1^\circ$ and $E_2^\circ$.

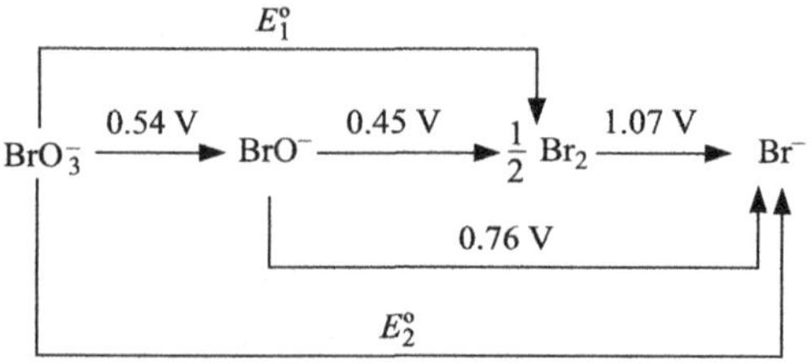


***Solution***

The reaction corresponding to the potential $E_1^\circ$ is

$$BrO_3^- + 3H_2O + 5e^- \rightarrow \tfrac{1}{2}Br_2 + 6OH^- \quad (1)$$

This reaction can be obtained by adding the following two reduction reactions:

$$BrO_3^- + 2H_2O + 4e^- \rightarrow BrO^- + 4OH^- \quad (2)$$

$$BrO^- + H_2O + e^- \rightarrow \tfrac{1}{2}Br_2 + 2OH^- \quad (3)$$

Hence the free energy change of reaction (1) will be

$$\Delta G^\circ_{\text{reaction (1)}} = \Delta G^\circ_{\text{reaction (2)}} + \Delta G^\circ_{\text{reaction (3)}}$$

Replacing $\Delta G^\circ$s in terms of potentials, we get

$$-5FE_1^\circ = -4F(0.54\ \text{V}) - 1F(0.45\ \text{V})$$
$$= (-2.61\ \text{V})\,F$$

Hence $$E_1^\circ = \frac{2.61\ \text{V}}{5} = 0.52\ \text{V}$$

Now the reaction corresponding to the potential $E_2^\circ$ is

$$BrO_3^- + 3H_2O + 6e^- \rightarrow Br^- + 6OH^- \qquad (4)$$

This reaction can be obtained by adding the following three reactions.

$$BrO_3^- + 2H_2O + 4e^- \rightarrow BrO^- + 4OH^- \qquad \text{(Eq. 2)}$$

$$BrO^- + H_2O + e^- \rightarrow \tfrac{1}{2}Br_2 + 2OH^- \qquad \text{(Eq. 3)}$$

$$\tfrac{1}{2}Br_2 + e^- \rightarrow Br^- \qquad (5)$$

Hence $$\Delta G^\circ_{\text{reaction (4)}} = \Delta G^\circ_{\text{reaction (2)}} + \Delta G^\circ_{\text{reaction (3)}} + \Delta G^\circ_{\text{reaction (5)}}$$

or $$-6F(E_2^\circ) = -4F(0.54\ \text{V}) - 1F(0.45\ \text{V}) - 1F(1.07\ \text{V})$$
$$= (-3.68\ \text{V})F$$

or $$E_2^\circ = \frac{3.68\ \text{V}}{6} = 0.61\ \text{V}$$

---

## 8.15 REFERENCE HALF-CELLS

**Desirability of a Reference Half-Cell**

In Section 8.8, we have seen that the standard potential of a given half-cell may be determined by constructing a cell in which the left half-cell is the standard hydrogen-hydrogen ion half-cell and the right side is the given half-cell and determining its emf experimentally. By definition, the standard potential of the given half-cell will be equal to the obtained value of emf of the cell. However, the hydrogen-hydrogen ion half-cell suffers from a large number of limitations. For example, it is very inconvenient to use it in practice as a constant supply of hydrogen gas at 1 bar pressure is required. Moreover, the potential of hydrogen-hydrogen ion half-cell is very sensitive to dissolved oxygen, hydrogen sulphide and the reducible species. The hydrogen-hydrogen ion half-cell is also very susceptible to electrode poisoning. It is for these reasons, the hydrogen-hydrogen ion half-cell may be replaced by a suitable reference half-cell whose potential has been accurately determined with respect to the hydrogen-hydrogen ion half-cell.

Reference half-cell systems are chosen to give reproducible potentials over a long period of time.

**Common Examples of Reference Half-Cell**

The most common type of reference half-cell has the following form:

saturated solution of sparingly soluble salt of metal + additional strongly ionized salt with a common anion | Metal

Examples include:

(1) Calomel half-cell: $KCl(aq) \mid Hg_2Cl_2(s) \mid Hg$
(2) Silver-silver chloride half-cell: $KCl(aq) \mid AgCl(s) \mid Ag$
(3) Mercury-mercurous sulphate half-cell: $K_2SO_4(aq) \mid Hg_2SO_4(s) \mid Hg$

The potential adopted by each of these reference half-cells is controlled by the activity of the anion in solution. We give below a brief discussion of these three half-cells.

## Calomel Half-Cell

The half-cell set up is shown in Fig. 8.15.1.

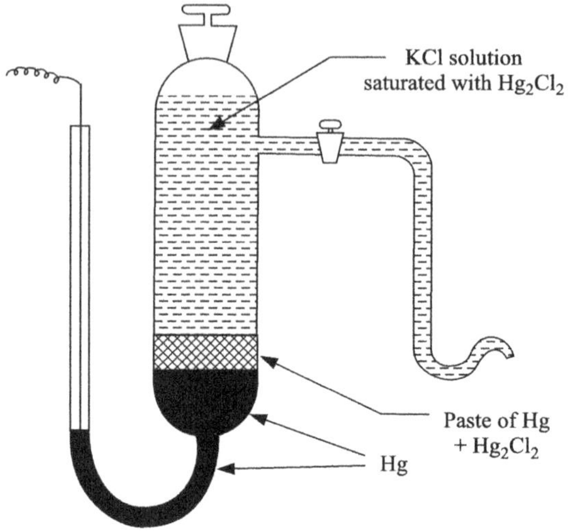


**Fig. 8.15.1 Calomel half-cell**

The half-cell reaction is

$$\tfrac{1}{2}\,Hg_2Cl_2(s) + e^- \rightarrow Hg(l) + Cl^-(aq)$$

and the Nernst equation has the form

$$E_{Cl^- \mid Hg_2Cl_2 \mid Hg} = E^\circ_{Cl^- \mid Hg_2Cl_2 \mid Hg} - \frac{RT}{F} \ln (a_{Cl^-})$$

with $E^\circ_{Cl^- \mid Hg_2Cl_2 \mid Hg} = 0.267\,6\ V$

Thus, the calomel half-cell functions as the reversible chloride electrode. Three types of calomel half-cell are commonly used. These are corresponding to 0.1 mol dm$^{-3}$, 1 mol dm$^{-3}$ and saturated KCl solution having potentials 0.336 V, 0.283 V and 0.244 V, respectively at 25 °C.

## Silver-Silver Chloride-Chloride Ion Half-Cell

The half-cell reaction is

$$AgCl(s) + e^- \rightarrow Ag(s) + Cl^-(aq)$$

Hence, the half-cell potential is given by

$$E_{Cl^- \mid AgCl \mid Ag} = E^\circ_{Cl^- \mid AgCl \mid Ag} - \frac{RT}{F} \ln (a_{Cl^-})$$

with $E^\circ_{Cl^- \mid AgCl \mid Ag} = 0.222\ V$

This half-cell cannot be used in a solution which contains $HNO_3$, $I^-$ and $CN^-$ ions.

**Mercury-Mercurous Sulphate-Sulphate Ion Half-Cell**

The half-cell set up is very similar to that of calomel half-cell. The cell-reaction is

$$Hg_2SO_4(s) + 2e^- \rightarrow 2Hg(l) + SO_4^{2-}(aq)$$

Hence the half-cell potential is given by

$$E_{SO_4^{2-} \mid Hg_2SO_4 \mid Hg} = E^{\circ}_{SO_4^{2-} \mid Hg_2SO_4 \mid Hg} - \frac{RT}{2F} \ln (a_{SO_4^{2-}})$$

with $\quad E^{\circ}_{SO_4^{2-} \mid Hg_2SO_4 \mid Hg} = 0.615\ 1\ \mathrm{V}$

This half-cell potential depends upon the activity of $SO_4^{2-}$ ions.

**Mercury-Mercuric Oxide-Hydroxide Ion Half-Cell**

Another half-cell which finds its utility in strongly alkaline solutions is the mercury-mercuric oxide half-cell. It consists of a pool of mercury in contact with a solution of NaOH or KOH which is saturated with HgO.
The half-cell reaction is

$$\tfrac{1}{2}HgO(s) + \tfrac{1}{2}H_2O + e^- \rightarrow \tfrac{1}{2}Hg(1) + OH^-(aq)$$

and thus the Nernst equation has the form

$$E_{OH^- \mid HgO \mid Hg} = E^{\circ}_{OH^- \mid HgO \mid Hg} - \frac{RT}{F} \ln (a_{OH-})$$

## 8.16 EXPRESSION OF $E_{cell}$ IN THE UNIT OF MOLALITY

If the emf of a cell has to be written in terms of mean activity of the electrolytes, we will have to consider the cell involving the electrolytes rather than the ions. For example, the Daniell cell

$$Zn \mid Zn^{2+}(aq) \,\vdots\vdots\, Cu^{2+}(aq) \mid Cu$$

will be written as

$$Zn \mid ZnSO_4(aq) \,\vdots\vdots\, CuSO_4(aq) \mid Cu$$

The emf of the cell as given by Eq. (8.7.9) is

$$E_{cell} = E^{\circ}_{cell} - \frac{RT}{2F} \ln \frac{a_{Zn^{2+}}}{a_{Cu^{2+}}}$$

Expressing $a$ in terms of molality and activity coefficient, we get

$$E_{cell} = E^{\circ}_{cell} - \frac{RT}{2F} \ln \frac{m'_{Zn^{2+}}}{m'_{Cu^{2+}}} - \frac{RT}{2F} \ln \frac{\gamma_{Zn^{2+}}}{\gamma_{Cu^{2+}}} \qquad (8.16.1)$$

†See Annexure I at the end of this Chapter for the discussion on activity and its relation with molality. Throughout this Chapter, $m'$ represents $m/m^{\circ}$, where $m$ is the molality and $m^{\circ}$ is the unit molality.

The expression of $m'$ and $\gamma$ in terms of mean molality and mean activity coefficient are as follows. In the present case

$$x = y = 1, \quad \nu = 2 \quad \text{and} \quad m'_+ = m'_-$$

Hence $\quad m'_{Zn^{2+}} = m'_{\pm}(ZnSO_4) \qquad m'_{Cu^{2+}} = m'_{\pm}(CuSO_4)$

$$\gamma_{Zn^{2+}} = \gamma_{\pm}(ZnSO_4) \qquad \gamma_{Cu^{2+}} = \gamma_{\pm}(CuSO_4) \tag{8.16.2}$$

Thus, Eq. (8.16.1) become

$$E_{cell} = E^{\circ}_{cell} - \frac{RT}{2F}\ln\frac{m'_{\pm}(ZnSO_4)}{m'_{\pm}(CuSO_4)} - \frac{RT}{2F}\ln\frac{\gamma_{\pm}(ZnSO_4)}{\gamma_{\pm}(CuSO_4)}$$

---

***Example 8.16.1***

Calculate the potential at 25 °C for the cell

$$Pt \mid H_2(p = 1 \text{ bar}) \mid HBr(a_{\pm} = 0.2) \mid Hg_2Br_2(s) \mid Hg$$

***Solution***

We have

| *Electrode* | *Reduction* |
|---|---|
| Right | $Hg_2Br_2(s) + 2e^- \rightarrow 2Hg(1) + 2Br^-(aq)$ |
| Left | $2H^+(aq) + 2e^- \rightarrow H_2(g)$ |

The cell reaction is

$$Hg_2Br_2(s) + H_2(g) \rightarrow 2Hg(1) + 2Br^-(aq) + 2H^+(aq)$$

Hence $\quad E_{cell} = (E^{\circ}_{Br^- \mid Hg_2Br_2 \mid Hg} - E^{\circ}_{H^+ \mid H_2 \mid Pt}) - \dfrac{RT}{2F}\ln\dfrac{(a_{H^+})^2(a_{Br^-})^2}{(p(H_2)/p^{\circ})}$

or $\quad E_{cell} = E^{\circ}_{Br^- \mid Hg_2Br_2 \mid H_2} - \dfrac{RT}{F}\ln\dfrac{(a_{H^+})(a_{Br^-})}{(p(H_2)/p^{\circ})^{1/2}}$

or $\quad E_{cell} = E^{\circ}_{Br^- \mid Hg_2Br_2 \mid Hg} - \dfrac{RT}{F}\ln\dfrac{\{(a_{\pm}(HBr)\}^2}{(p(H_2)/p^{\circ})^{1/2}}$

Substituting the data, we get

$$E_{cell} = 0.138\,5 \text{ V} - (0.059\,15 \text{ V})(2)\log(0.2)$$
$$= 0.138\,5 \text{ V} + 0.082\,7 \text{ V} = 0.221\,2 \text{ V}$$

---

***Example 8.16.2***

The potential of the cell

$$Cd \mid CdI_2(a_2) \mid AgI(s) \mid Ag$$

is 0.286 0 V at 25 °C. Calculate the mean ionic activity of the ions in the solution, and the activity of electrolyte.

***Solution***

We have

| *Electrode* | *Reduction reaction* |
|---|---|
| Right | $2AgI(s) + 2e^- \rightarrow 2Ag(s) + 2I^-(aq)$ |
| Left | $Cd^{2+}(aq) + 2e^- \rightarrow Cd(s)$ |

The cell reaction is

$$2AgI(s) + Cd(s) \rightarrow 2Ag(s) + Cd^{2+}(aq) + 2I^-(aq)$$

Hence $$E_{cell} = (E^\circ_{I^- | AgI | Ag} - E^\circ_{Cd^{2+} | Cd}) - \frac{RT}{2F} \ln (a_{Cd^{2+}})(a_{I^-})^2$$

or $$E_{cell} = (E^\circ_{I^- | AgI | Ag} - E^\circ_{Cd^{2+} | Cd}) - \frac{RT}{2F} \ln (a_\pm)^3$$

Substituting the data, we get

$$0.286\,0 \text{ V} = \{-0.152\,2 \text{ V} - (-0.403 \text{ V})\} - \left(\frac{3 \times 0.059\,15 \text{ V}}{2}\right) \log a_\pm$$

or $$0.286\,0 \text{ V} = 0.250\,8 \text{ V} - (0.088\,72 \text{ V}) \log a_\pm$$

or $$\log a_\pm = \frac{0.250\,8 - 0.286\,0}{0.088\,72} = -\frac{0.035\,2}{0.088\,72} = -0.396\,7$$

Thus $a_\pm = 0.401\,2$

Now since $a_2 = a_\pm^3$, we get

$$a_2 = (0.401\,2)^3 = 0.064\,55$$

***Example 8.16.3***

The potential of the cell

$$Zn(s) \mid ZnCl_2(0.010\,21 \text{ mol kg}^{-1}) \mid AgCl(s) \mid Ag(s)$$

is found to be 1.156 6 V. What is the mean ionic activity coefficient of $ZnCl_2$ in the solution?

***Solution***

We have

| *Electrode* | *Reduction reaction* |
|---|---|
| Right | $2AgCl(s) + 2e^- \rightarrow 2Ag(s) + 2Cl^-(aq)$ |
| Left | $Zn^{2+}(aq) + 2e^- \rightarrow Zn(s)$ |

The cell reaction is

$$2AgCl(s) + Zn(s) \rightarrow 2Ag(s) + Zn^{2+}(aq) + 2Cl^-(aq)$$

Hence $$E_{cell} = (E^\circ_{Cl^- | AgCl | Ag} - E^\circ_{Zn^{2+} | Zn}) - \frac{RT}{2F} \ln (a_{Zn^{2+}})(a_{Cl^-})^2$$

$$= (E^\circ_{Cl^- | AgCl | Ag} - E^\circ_{Zn^{2+} | Zn}) - \frac{RT}{2F} \ln a_\pm^3$$

$$= (E^\circ_{Cl^- | AgCl | Ag} - E^\circ_{Zn^{2+} | Zn}) - \frac{RT}{2F} \ln (m'_\pm)^3 \gamma_\pm^3$$

Now $(m'_\pm)^3 = (m'_+)(m'_-)^2$

$$= (0.010\,21)\,(2 \times 0.010\,21)^2$$

$$= 4 \times 0.010\,21^3 = 4.257 \times 10^{-6}$$

Substituting this along with other data in the expression of $E_{cell}$, we have

$$1.156\,6\text{ V} = \{0.222\text{ V} - (-0.762\text{ V})\} - \left(\frac{0.059\,15\text{ V}}{2}\right) \log m_\pm^3$$

$$- \left(\frac{0.059\,15\text{ V}}{2}\right) \log \gamma_\pm^3$$

$$= 0.984\text{ V} - (0.029\,57\text{ V}) \log (4.257 \times 10^{-6})$$

$$- 3(0.029\,57\text{ V}) \log \gamma_\pm$$

$$= 1.143\text{ V} - (0.088\,7\text{ V}) \log \gamma_\pm$$

or $$\log \gamma_\pm = -\frac{1.156\,6 - 1.143}{0.088\,7} = -\frac{0.013\,6}{0.088\,7} = -0.153\,3$$

Hence $\gamma_\pm = 0.702\,6$

## 8.17 DETERMINATION OF ACCURATE VALUE OF HALF-CELL POTENTIAL

The standard half-cell potentials are employed for calculating accurate values of various quantities such as equilibrium constant of a weak acid and ionic product of water. It is, therefore, essential to know their accurate values. For this purpose, the method of extrapolation is employed. To illustrate the method, we consider the following cell.

$$\text{Pt} \mid H_2(1\text{ bar}) \mid \text{HCl}(a) \mid \text{AgCl(s)} \mid \text{Ag} \qquad (8.17.1)$$

The cell reaction is

$$\text{AgCl} + \tfrac{1}{2}H_2 \rightarrow \text{Ag} + Cl^- + H^+ \qquad (8.17.2)$$

and the cell potential is given by

$$E = E^\circ - \frac{RT}{F} \ln \frac{(a_{H^+})(a_{Cl^-})}{(f_{H_2}/f^\circ)^{1/2}} \qquad (8.17.3)$$

Assuming ideal behaviour for the gas, we get

$$E = E^\circ - \frac{RT}{F} \ln (a_{H^+})(a_{Cl^-}) \tag{8.17.4}$$

Since individual ionic activities cannot be determined, we express $a_{H^+}$ and $a_{Cl^-}$ in terms of mean activity $a_\pm$ of hydrochloric acid by the relation

$$a_\pm^2 = (a_{H^+})(a_{Cl^-}) \tag{8.17.5}$$

Now the mean activity may be expressed in terms of mean molality $m_\pm$ of the ions and the mean activity coefficient $\gamma_\pm$ by the relation

$$a_\pm = m'_\pm \, \gamma_\pm \tag{8.17.6$^\dagger$}$$

For HCl solution

$$m'_\pm = m' = m/m^\circ \tag{8.17.7}$$

where $m$ is the molality of the solution. Using Eqs (8.17.5), (8.17.6) and (8.17.7) in Eq. (8.17.4), we get

$$E = E^\circ - \frac{2\,RT}{F} \ln m' - \frac{2\,RT}{F} \ln \gamma_\pm \tag{8.17.8}$$

If the measurements are done in dilute solutions, than $\gamma_\pm$ can be expressed in terms of the ionic strength of the solution using the Debye-Hückel limiting law[††]

$$\log \gamma_\pm = -\,A z_+ | z_- | \mu^{1/2} \tag{8.17.9}$$

where $A$ is equal to 0.514 (mol dm$^{-3}$)$^{-1/2}$, $z_+$ and $z_-$ are the charge numbers of ions, and $\mu$ is the ionic strength of the solution. The latter is defined as

$$\mu = \frac{1}{2} \sum_i c_i z_i^2 \tag{8.17.10}$$

where $c_i$ is the molar concentration of the $i$th ions. The summation is to be carried over all the ions present in the solution. For uni-univalent electrolyte such as the given one

$$\mu = c = m \tag{8.17.11}$$

Thus for HCl solution, Eq. (8.17.9) becomes

$$\log \gamma_\pm = -\,0.509\, m'^{1/2} \tag{8.17.12}$$

---

†Throughout this chapter, $m' = m/m^\circ$.

††See Annexure II for the derivation of Debye-Hückel limiting law.

Substituting Eq. (8.17.12) in Eq. (8.17.8), we get

$$E = E^\circ - \frac{2\,RT}{F}\ \ln m' - \frac{2.303 \times 2\,RT}{F}(-\,0.509\ m'^{1/2})$$

or
$$E + \frac{2\,RT}{F}\ \ln m' = E^\circ + \frac{2.303 \times 2\,RT}{F}\ (0.509\ m'^{1/2})$$

At 25 °C, the above equation becomes

$$E + (0.118\ 3\ \text{V})\ \log m' = E^\circ + (0.060\ 2\ \text{V})\sqrt{m'}$$

The value $E$ of the cell is determined at various values of $m'$ and then a plot between $E$ + (0.118 3 V) log $m'$ and $\sqrt{m'}$ is made. The plot will be a straight line at lower values of $\sqrt{m'}$ which can be extended to $\sqrt{m'} = 0$ as shown in Fig. 8.17.1. The intercept at $\sqrt{m'} = 0$ gives the value of $E^\circ$.

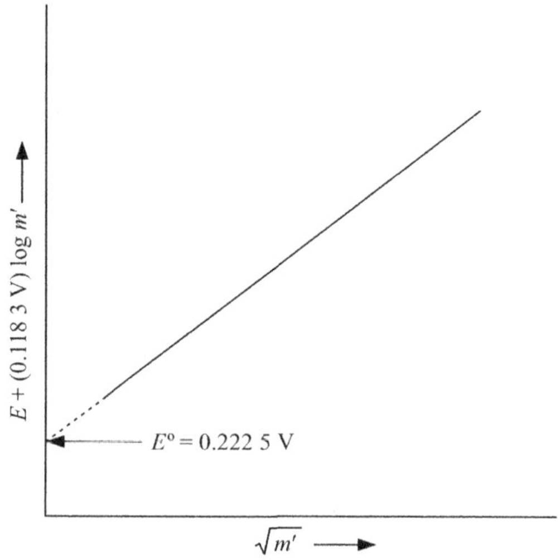


**Fig. 8.17.1 A graph of *E* + (0.118 3) log *m*′ versus $\sqrt{m'}$**

Since the standard potential of the cell (Eq. 8.17.1) is given by

$$E^\circ_{\text{cell}} = E^\circ_{\text{Cl}^- \mid \text{AgCl} \mid \text{Ag}} - E^\circ_{\text{H}^+ \mid \text{H}_2}$$

and the fact that

$$E^\circ_{\text{H}^+ \mid \text{H}_2} = 0$$

it follows that

$$E^\circ_{\text{Cl}^- \mid \text{AgCl} \mid \text{Ag}} = E^\circ_{\text{cell}} = 0.222\ 5\ \text{V}$$

Following the method outlined above, the accurate values of half-cell potentials of other electrodes can be determined.

## 8.18 APPLICATIONS OF ELECTROCHEMICAL CELLS

The emf measurement of electrochemical cells provides one of the convenient methods to determine the values of various physical quantities such as the equilibrium constant, ionization constant, ionic product of water, pH of a solution, and the values of thermodynamic properties of ions. Besides these, the quantitative estimation of species can be carried out with the help of potentiometric titration. In this section, we describe the above applications with specific examples.

### 1. DETERMINATION OF THERMODYNAMIC DATA

**Expression of Free Energy Change**

The free energy change accompanying a cell reaction can be obtained from the emf of the cell through the relation

$$\Delta G = -\,nFE_{\text{cell}} \tag{8.18.1}$$

**Expression of Enthalpy Change**

The changes in enthalpy and entropy of the cell reaction can also be determined if the temperature dependence of the emf of the cell is available. For this, we make use of the Gibbs–Helmholtz equation

$$\Delta G = \Delta H + T\left[\frac{\partial(\Delta G)}{\partial T}\right]_p \tag{8.18.2}$$

Now since

$$\Delta G = -\,nFE$$

we, therefore, get

$$\left[\frac{\partial(\Delta G)}{\partial T}\right]_p = -\,nF\left(\frac{\partial E}{\partial T}\right)_p \tag{8.18.3}$$

Substituting Eqs (8.18.1) and (8.18.3) in Eq. (8.18.2), we get

$$-nFE = \Delta H - nFT\left(\frac{\partial E}{\partial T}\right)_p$$

Rearranging, we have

$$\Delta H = -\,nF\left[E - T\left(\frac{\partial E}{\partial T}\right)_p\right] \tag{8.18.4}$$

Thus, $\Delta H$ of the cell reaction can be determined from the knowledge of $E$ and $(\partial E/\partial T)_p$. Since $E$ and $F$ are expressed respectively in volt and coulomb, the units of $\Delta H$ is in volt-coulomb and is equal to the joule unit.

**Expression of Entropy Change**

Now since $\Delta G = \Delta H - T\Delta S$, therefore

$$\Delta S = \frac{\Delta H - \Delta G}{T} \tag{8.18.5}$$

Substituting the expressions of $\Delta H$ and $\Delta G$ from Eqs (8.18.4) and (8.18.1), we get

$$\Delta S = nF\left(\frac{\partial E}{\partial T}\right)_p \tag{8.18.6}$$

Thus knowing $(\partial E/\partial T)_p$, $\Delta S$ of the cell reaction can be computed from Eq. (8.18.6).

## 1. STANDARD FREE ENERGIES AND ENTROPIES OF AQUEOUS IONS

**Free Energy of $H^+(aq)$**

The assignment of zero potential to the standard hydrogen-hydrogen ion half-cell provides the conventional zero for free energy of formation of hydrogen ions in aqueous solution. This follows immediately from the potential expression of the standard hydrogen-hydrogen ion half-cell (Eq. 8.6.6). We have

$$E^\circ_{H^+ | H_2} = \frac{\mu^\circ_{H^+} - \frac{1}{2}\mu^\circ_{H_2} + \mu_{e^-(Pt)}}{F} \qquad \text{(Eq. 8.6.6)}$$

Since the electrons are not emitted as such from the neutral metal, we set $\mu_{e^-(Pt)} = 0$. Also $\mu^\circ_{H_2}$ at 298 K and 1 bar is zero. Hence

$$E^\circ_{H^+ | H_2} = \frac{\mu^\circ_{H^+}}{F}$$

Now since $E^\circ_{H^+ | H_2}$ is assigned a zero value, it follows that

$$\mu^\circ_{H^+} = 0$$

*(Alternatively, we may assume $\mu^\circ_{H^+} = 0$ and proceed to prove that $E^\circ_{H^+|H_2|Pt} = 0$.)*

Thus the reference standard for free energies of formation of ions in solution is the hydrogen ion at $a = 1$, which is assigned a zero value.

**Free Energy of Other Ions**

The standard free energies of formation of ions relative to the above reference may be determined as follows.

Let the free energy of formation of $Cd^{2+}$ ion be required. We construct a cell involving $Cd^{2+}|Cd$ half-cell and the standard hydrogen-hydrogen ion half-cell:

$$Cd \mid Cd^{2+}(a = 1) \;\vdots\vdots\; H^+\ (a = 1) \mid H_2(1\ \text{bar}) \mid Pt$$

The cell reaction is

$$Cd + 2H^+ \rightarrow Cd^{2+} + H_2$$

The cell potential is given by

$$\begin{aligned} E^\circ_{cell} &= E^\circ_{H^+ | H_2 | Pt} - E^\circ_{Cd^{2+} | Cd} \\ &= 0 - (-0.403\ \text{V}) = 0.403\ \text{V} \end{aligned}$$

The change in the standard free energy of the cell reaction

$$Cd + 2H^+ \rightarrow Cd^{2+} + H_2$$

is given by

$$\Delta G^\circ = -nFE^\circ_{cell}$$

$$= -2\,(96\,487\text{ C mol}^{-1})(0.403\text{ V}) = -77\,769\text{ J mol}^{-1}$$

$\Delta G^\circ$ will also be given by

$$\Delta G^\circ = \mu^\circ_{Cd^{2+}} + \mu^\circ_{H_2} - \mu^\circ_{Cd} - 2\mu^\circ_{H^+}$$

By convention,

$$\mu^\circ_{Cd} = 0; \quad \mu^\circ_{H_2} = 0 \text{ and } \mu^\circ_{H^+} = 0$$

Thus $\quad \Delta G^\circ = \mu^\circ_{Cd^{2+}} = -77\,769\text{ J mol}^{-1}$

Proceeding this way, we can determine the free energy of formation or other ions in the solution.

## Standard Entropy of Formation of Ions

The standard ionic entropies in solution is also determined relative to that of hydrogen ion at unit activity. By convention, the latter is assigned a zero value†, i.e.

$$S^\circ_{H^+} = 0$$

To illustrate the procedure of calculating standard ionic entropy, we take an example of $Cd^{2+}$. Consider again the reaction

$$Cd + 2H^+ \rightarrow Cd^{2+} + H_2$$

The standard entropy change is

$$\Delta S^\circ = S^\circ_{Cd^{2+}} + S^\circ_{H_2} - 2S^\circ_{H^+} - S^\circ_{Cd}$$

Since $S^\circ_{H^+} = 0$, we have

$$\Delta S^\circ = S^\circ_{Cd^{2+}} + S^\circ_{H_2} - S^\circ_{Cd}$$

The standard entropies of Cd and $H_2$ at 298 K are 51.46 and 130.58 J $K^{-1}$ $mol^{-1}$, respectively.

Hence $\quad S^\circ_{Cd^{2+}} = \Delta S^\circ + S^\circ_{Cd} - S^\circ_{H_2} = \Delta S^\circ - 79.12\text{ J K}^{-1}\text{ mol}^{-1}$

The value of $\Delta S^\circ$ can be obtained from the relation

---

†The absolute value of third law entropy of proton in water is – 20.99 J $K^{-1}$ $mol^{-1}$.

$$\Delta G^\circ = \Delta H^\circ - T\Delta S^\circ$$

i.e. $$\Delta S^\circ = \frac{\Delta H^\circ - \Delta G^\circ}{T}$$

The values of $\Delta G^\circ$ of the reaction as determined above is – 77.769 kJ $mol^{-1}$. The standard enthalpy change $\Delta H^\circ$ can be determined experimentally by dissolving 1 mol of $Cd^{2+}$ in a large excess of dilute acid. The value obtained is – 69.83 kJ $mol^{-1}$. Hence

$$\Delta S^\circ = \frac{-69\ 830\ \text{J mol}^{-1} + 77\ 769\ \text{J mol}^{-1}}{298\ \text{K}} = 26.64\ \text{J K}^{-1}\ \text{mol}^{-1}$$

Thus $$S^\circ_{Cd^{2+}} = \Delta S^\circ - 79.12\ \text{J K}^{-1}\ \text{mol}^{-1} = (26.64 - 79.12)\ \text{J K}^{-1}\ \text{mol}^{-1}$$

$$= -52.48\ \text{J K}^{-1}\ \text{mol}^{-1}$$

It may be pointed here that the above standard entropy of formation of $Cd^{2+}$ is not the absolute entropy of the $Cd^{2+}$ ion in aqueous solution, but refers to a neutral combination of the given ion with $H_3O^+$ ion, i.e. $S^\circ_{Cd^{2+}} - S^\circ_{H^+}$.

***Example 8.18.1***

The emf of the cell

Ag | AgCl(s) | KCl(aq) | $Hg_2Cl_2$(s) | Hg

is 0.045 5 V at 298 K and the temperature coefficient is $3.38 \times 10^{-4}$ V $K^{-1}$. What is the reaction taking place in the cell and what are free energy, enthalpy and entropy changes at 298 K?

***Solution***

For the given cell, we have

| *Electrode* | *Reduction reaction* | |
|---|---|---|
| Right | $Hg_2Cl_2(s) + 2e^- \rightarrow 2Hg(l) + 2Cl^-(aq)$ | (i) |
| Left | $2AgCl(s) + 2e^- \rightarrow 2Ag(s) + 2Cl^-(aq)$ | (ii) |

Subtracting Eq. (ii) from Eq. (i), we get

$$Hg_2Cl_2(s) + 2Ag(s) \rightarrow 2Hg(l) + 2AgCl(s)$$

The number of electrons involved at the electrode reactions is 2.

Thus $$\Delta G = -nFE$$

$$= -2(96\ 500\ \text{C mol}^{-1})\ (0.045\ 5\ \text{V}) = -8\ 782\ \text{J mol}^{-1}$$

$$\Delta H = -nF\left[E - T\left(\frac{\partial E}{\partial T}\right)_p\right]$$

$$= -2(96\ 500\ \text{C mol}^{-1})\ [0.045\ 5\ \text{V} - (298\ \text{K})\ (3.38 \times 10^{-4}\ \text{V K}^{-1})]$$

$$= 2(96\ 500\ \text{C mol}^{-1})\ (0.055\ 2\ \text{V}) = 10\ 650\ \text{J mol}^{-1}$$

and $$\Delta S = nF\left(\frac{\partial E}{\partial T}\right)_p = 2(96\ 500\ \text{C mol}^{-1})\ (3.38 \times 10^{-4}\ \text{V K}^{-1})$$

$$= 65.23\ \text{J K}^{-1}\ \text{mol}^{-1}$$

***Example 8.18.2***

At 298 K, the emf of the cell

$$\mathrm{Hg}\left|\begin{array}{c}0.01\ \mathrm{mol\ dm^{-3}\ KCl}\\ \mathrm{sat.\ with\ Hg_2Cl_2}\end{array}\right.\vdots\begin{array}{c}1\ \mathrm{mol\ dm^{-3}\ KNO_3}\\ \mathrm{salt\ bridge}\end{array}\vdots\left.\begin{array}{c}0.01\ \mathrm{mol\ dm^{-3}\ KOH}\\ \mathrm{sat.\ with\ HgO}\end{array}\right|\mathrm{Hg}$$

is found to be 0.163 4 V and the temperature coefficient of the emf to be 0.000 837 V K$^{-1}$. Calculate the enthalpy and entropy changes of the reaction. What is the reaction that occurs in the cell?

***Solution***

For the given cell, we have

| *Electrode* | *Reduction reaction* | |
|---|---|---|
| Right | $HgO(s) + H_2O(l) + 2e^- \rightarrow Hg(l) + 2OH^-(aq)$ | (i) |
| Left | $Hg_2Cl_2(s) + 2e^- \rightarrow 2Hg(l) + 2Cl^-(aq)$ | (ii) |

Subtracting Eq. (ii) from Eq. (i), we get

$$HgO(s) + H_2O(l) + Hg(l) + 2Cl^-(aq) \rightarrow Hg_2Cl_2(s) + 2OH^-(aq)$$

The number of electrons involved in the electrode reactions is 2.

Thus $\Delta G = -nFE = -2(96\ 500\ \mathrm{C\ mol^{-1}})\ (0.163\ 4\ \mathrm{V}) = -31\ 536.2\ \mathrm{J\ mol^{-1}}$

$$\Delta H = -nF\left[E - T\left(\frac{\partial E}{\partial T}\right)_p\right]$$

$$= -2(96\ 500\ \mathrm{C\ mol^{-1}})\ [(0.163\ 4\ \mathrm{V}) - (298\ \mathrm{K} \times 0.000\ 837\ \mathrm{V\ K^{-1}})]$$

$$= 16\ 603\ \mathrm{J\ mol^{-1}}$$

$$\Delta S = nF\left(\frac{\partial E}{\partial T}\right)_p$$

$$= 2(96\ 500\ \mathrm{C\ mol^{-1}})\ (0.000\ 837\ \mathrm{V\ K^{-1}}) = 161.54\ \mathrm{J\ K^{-1}\ mol^{-1}}$$

***Example 8.18.3***

For the cell

$$Ag(s)\ |\ AgBr(s)\ |\ KBr(aq)\ |\ Hg_2Br_2(s)\ |\ Hg(l)$$

the emfs at several temperatures are

| $T$/K | 293 | 298 | 303 |
|---|---|---|---|
| emf/V | 0.066 3 | 0.068 39 | 0.070 48 |

(a) Write the cell reaction for the passage of two-faraday of charge.
(b) Calculate $\Delta G$, $\Delta H$ and $\Delta S$ at 298 K.

***Solution***

First of all a graph between $E$ and $T$ was drawn and its slope was determined to yield the value of $(\partial E/\partial T)_p$ as shown in Fig. 8.18.1. The value was found to be 0.000 42 V K$^{-1}$

(**a**) The cell reaction may be determined as follows:

| *Electrode* | *Reduction reaction* | |
|---|---|---|
| Right | $Hg_2Br_2(s) + 2e^- \rightarrow 2Hg(l) + 2Br(aq)$ | (i) |
| Left | $2AgBr(s) + 2e^- \rightarrow 2Ag(s) + 2Br^-(aq)$ | (ii) |

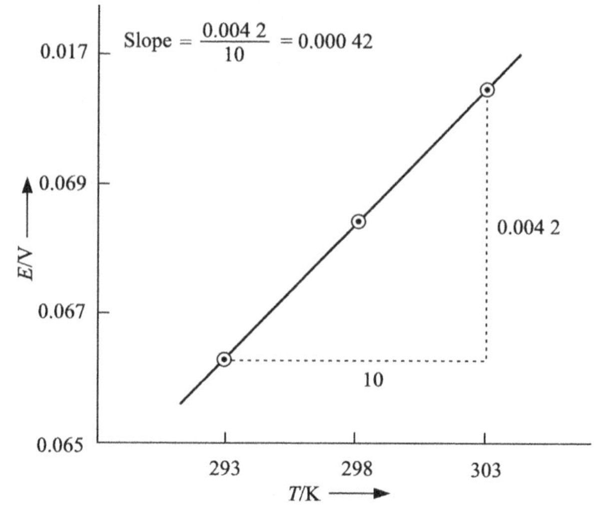


**Fig. 8.18.1 The graph of *E* versus *T***

Subtracting Eq. (ii) from Eq. (i), we get

$$Hg_2Br_2(s) + 2Ag(s) \rightarrow 2Hg(1) + 2AgBr(s)$$

(**b**) Now

$$\Delta G = -nFE = -2(96\ 500\ \text{C mol}^{-1})\ (0.068\ 39\ \text{V}) = -13\ 199.3\ \text{J mol}^{-1}$$

$$\Delta H = -nF\left[E - T\left(\frac{\partial E}{\partial T}\right)_p\right]$$

$$= -2(96\ 500\ \text{C mol}^{-1})\ [(0.068\ 39\ \text{V}) - (298\ \text{K})\ (0.000\ 42\ \text{V K}^{-1})]$$

$$= 10\ 956.6\ \text{J mol}^{-1}$$

$$\Delta S = nF\left(\frac{\partial E}{\partial T}\right)_p$$

$$= 2(96\ 500\ \text{C mol}^{-1})\ (0.000\ 42\ \text{V K}^{-1}) = 81.06\ \text{J K}^{-1}\ \text{mol}^{-1}$$

***Example 8.18.4*** For the following cell

$$Pb \mid PbCl_2(s) \mid PbCl_2(soln.) \mid AgCl(s) \mid Ag$$

the potential at 298 K is 0.490 V and the variation of emf with temperature is given by

$$E = a - (1.86 \times 10^{-4}\ \text{V K}^{-1})\ (T - 25\ \text{K})$$

Write the equation for the cell reaction and calculate $\Delta G$, $\Delta H$ and $\Delta S$ for the reaction at 298 K.

***Solution*** For the given cell, we have

| *Electrode* | *Reduction reaction* | |
|---|---|---|
| Right | $2AgCl(s) + 2e^- \rightarrow 2Ag(s) + 2Cl^-(aq)$ | (i) |
| Left | $PbCl_2(s) + 2e^- \rightarrow Pb(s) + 2Cl^-(aq)$ | (ii) |

Subtracting Eq. (ii) from Eq. (i), we get

$$2AgCl(s) + Pb(s) \rightarrow 2Ag(s) + PbCl_2(s)$$

Now since $E = a - (1.86 \times 10^{-4}\ V\ K^{-1})\ (T - 25\ K)$, therefore

$$\left(\frac{\partial E}{\partial T}\right)_p = -1.86 \times 10^{-4}\ V\ K^{-1}$$

Hence
$$\Delta G = -nFE = -2(96\ 500\ C\ mol^{-1})\ (0.490\ V)$$
$$= -94\ 570\ J\ mol^{-1}$$
$$\Delta H = -nF\left[E - T\left(\frac{\partial E}{\partial T}\right)_p\right]$$
$$= -2(96\ 500\ C\ mol^{-1})\ [(0.490\ V) - (298\ K)\ (-1.86 \times 10^{-4}\ V\ K^{-1})]$$
$$= -105\ 267.6\ J\ mol^{-1}$$
$$\Delta S = nF\left(\frac{\partial E}{\partial T}\right)_p = 2(96\ 500\ C\ mol^{-1})\ (-1.86 \times 10^{-4}\ V\ K^{-1})$$
$$= -35.9\ J\ K^{-1}\ mol^{-1}$$

---

***Example 8.18.5***

At 273 K, the calorimetric determination of $\Delta H$ for the reaction

$$Zn(s) + 2AgCl(s) \rightarrow ZnCl_2(aq) + Ag(s)$$

yielded $-$ 217.78 kJ $mol^{-1}$, while the emf of the corresponding cell was 1.015 V. What was $(\partial E/\partial T)_p$ of the cell?

***Solution***

From the expression

$$\Delta H = -nF\left[E - T\left(\frac{\partial E}{\partial T}\right)_p\right]$$

we get
$$\left(\frac{\partial E}{\partial T}\right)_p = \frac{1}{T}\left[E + \frac{\Delta H}{nF}\right]$$

Substituting the given data, we get

$$\left(\frac{\partial E}{\partial T}\right)_p = \frac{1}{(273\ K)}\left[(1.015\ V) + \frac{(-217\ 780\ J\ mol^{-1})}{2(96\ 500\ C\ mol^{-1})}\right]$$
$$= -4.154 \times 10^{-4}\ V\ K^{-1}$$

---

***Example 8.18.6***

(a) At 298 K the standard free energy of formation of $H_2O(l)$ is $-$ 237.23 kJ $mol^{-1}$, while that of its ionization to hydrogen and hydroxyl ions is 79.71 kJ $mol^{-1}$. What will be the reversible emf at 298 K of the cell

$$\mathrm{Pt|H_2(g,\ 1\ bar)\ |\ H^+(1\ M)\ ⁞⁞\ OH^-(1\ M)\ |\ O_2(g,\ 1\ bar)\ |\ Pt}$$

(b) The enthalpy of formation at 298 K of $H_2O(1)$ is – 285.85 kJ mol$^{-1}$, while its enthalpy of ionization is 56.90 kJ mol$^{-1}$. Calculate the temperature coefficient of the standard emf at 298 K for the cell given in part (a).

***Solution*** (**a**) The given data are

$$\mathrm{H_2(g) + \tfrac{1}{2}\,O_2(g) \rightarrow H_2O(l);} \qquad \Delta_f G^\circ_{298K} = -\,237.23\ \mathrm{kJ\ mol^{-1}} \qquad \text{(i)}$$

$$\mathrm{H_2O(l) \rightarrow H^+(aq) + OH^-(aq);} \qquad \Delta G^\circ_{298K} = 79.71\ \mathrm{kJ\ mol^{-1}} \qquad \text{(ii)}$$

$$\mathrm{Pt\ |\ H_2(g,\ 1\ bar)\ |\ H^+(1\ M)\ ⁞⁞\ OH^-(1\ M)\ |\ O_2(g,\ 1\ bar)|Pt;} \quad E^\circ_{298\,K} = ?$$

The cell reaction may be obtained as follows:

| *Electrode* | *Reduction reaction* | |
|---|---|---|
| Right | $H_2O(1) + \frac{1}{2}O_2(g) + 2e^- \rightarrow 2OH^-(aq)$ | (iii) |
| Left | $2H^+(aq) + 2e^- \rightarrow H_2(g)$ | (iv) |

Subtracting Eq. (iv) from Eq. (iii), we get

$$\mathrm{H_2O(l) + \tfrac{1}{2}\,O_2(g) + H_2(g) \rightarrow 2H^+(aq) + 2OH^-(aq)}$$

The above cell reaction can be obtained from the given reactions (i) and (ii) as follows.

$$\mathrm{H_2(g) + \tfrac{1}{2}\,O_2(g) \rightarrow H_2O(l)}$$

$$\mathrm{2 \times [H_2O(1) \rightarrow H^+(aq) + OH^-(aq)]}$$

Adding, we get $\quad \mathrm{H_2O(1) + \tfrac{1}{2}\,O_2(g) + H_2(g) \rightarrow 2H^+(aq) + 2OH^-(aq)}$

Hence, $\Delta G^\circ$ of the cell reaction is given by

$$\Delta G^\circ = -\,237.23\ \mathrm{kJ\ mol^{-1}} + 2(79.71\ \mathrm{kJ\ mol^{-1}}) = -\,77.81\ \mathrm{kJ\ mol^{-1}}$$

Since $\quad \Delta G^\circ = -\,nFE^\circ$, it follows that

$$E^\circ = -\frac{\Delta G^\circ}{nF} = -\frac{(-\,77\,810\ \mathrm{J\ mol^{-1}})}{2(96\,500\ \mathrm{C\ mol^{-1}})} = 0.403\ \mathrm{J\ C^{-1}} = 0.403\ \mathrm{V}$$

(**b**) The given data are

$$\text{(v)}\quad \mathrm{H_2(g) + \tfrac{1}{2}\,O_2(g) \rightarrow H_2O(1);} \qquad \Delta_f H^\circ_{298K} = -\,285.85\ \mathrm{kJ\ mol^{-1}}$$

$$\text{(vi)}\quad \mathrm{H_2O(1) \rightarrow H^+(aq) + OH^-(aq);} \qquad \Delta H^\circ_{298K} = 56.9\ \mathrm{kJ\ mol^{-1}}$$

The $\Delta H^\circ$ of the cell reaction is

$$\Delta H^\circ_{298K} = -\,285.85\ \mathrm{kJ\ mol^{-1}} + 2(56.9\ \mathrm{kJ\ mol^{-1}}) = -\,172.05\ \mathrm{kJ\ mol^{-1}}$$

Now since

$$\Delta H^\circ = -\,nF\left[E^\circ - T\left(\frac{\partial E^\circ}{\partial T}\right)_p\right] = \Delta G^\circ + nFT\left(\frac{\partial E^\circ}{\partial T}\right)_p$$

therefore

$$\left(\frac{\partial E^\circ}{\partial T}\right)_p = \frac{\Delta H^\circ - \Delta G^\circ}{nFT}$$

Substituting the data, we get

$$\left(\frac{\partial E^\circ}{\partial T}\right)_p = \frac{-172\ 050\ \text{J mol}^{-1} - (-77\ 810\ \text{J mol}^{-1})}{2(96\ 500\ \text{C mol}^{-1})(298\ \text{K})}$$

$$= -0.001\ 64\ \text{V K}^{-1}$$

## 2. DETERMINATION OF EQUILIBRIUM CONSTANT FROM STANDARD CELL POTENTIAL

### Derivation of the Expression

The standard equilibrium constant of a reaction is related to the standard free energy of the reaction by the relation

$$\Delta G^\circ = -RT \ln K^\circ_{\text{eq}}$$

Now since $\Delta G^\circ = -nFE^\circ_{\text{cell}}$, it follows that

$$-nFE^\circ_{\text{cell}} = -RT \ln K^\circ_{\text{eq}}$$

or $$E^\circ_{\text{cell}} = \frac{RT}{nF} \ln K^\circ_{\text{eq}} = \frac{2.303\ RT}{nF} \log K^\circ_{\text{eq}} \qquad (8.18.7)$$

At 298 K, the above expression becomes

$$E^\circ_{\text{cell}} = \left(\frac{0.059\ 13\ \text{V}}{n}\right) \log K^\circ_{\text{eq}} \qquad (8.18.8)$$

For a given chemical reaction, we can construct the appropriate cell and then determine its $E^\circ$ value from the tabulated values of standard half-cell potentials. The obtained value of $E^\circ$ can be substituted in Eq. (8.18.8) to give the value of the standard equilibrium constant $K^\circ_{\text{eq}}$. The following problems illustrate the procedure.

***Example 8.18.7***

Determine the standard equilibrium constant of the following reaction at 298 K.

$$2Fe^{3+} + Sn^{2+} \rightarrow 2Fe^{2+} + Sn^{4+}$$

***Solution***

In the given reaction, $Fe^{3+}$ is reduced to $Fe^{2+}$ (and hence constitutes right half-cell) and $Sn^{2+}$ is oxidized to $Sn^{4+}$ (constitutes left half-cell). Hence the cell producing the given reaction is

$$\text{Pt} \mid Sn^{2+}, Sn^{4+} \ \vdots\vdots\ Fe^{3+}, Fe^{2+} \mid \text{Pt}$$

Its standard potential is

$$E^\circ_{\text{cell}} = E^\circ_{Fe^{3+}, Fe^{2+} \mid \text{Pt}} - E^\circ_{Sn^{4+}, Sn^{2+} \mid \text{Pt}} = 0.771\ \text{V} - 0.150\ \text{V} = 0.621\ \text{V}$$

Therefore $\log K^\circ_{eq} = \dfrac{nFE^\circ_{cell}}{2.303\,RT} = \dfrac{2(96\,500\text{ C mol}^{-1})(0.621\text{ V})}{2.303\,(8.314\text{ J K}^{-1}\text{ mol}^{-1})(298\text{ K})}$

$= 21.005$

Hence $K^\circ_{eq} = 1.0 \times 10^{21}$

The above calculations show that the equilibrium constant of the given reaction has a very large value. It implies that $Sn^{2+}$ can reduce $Fe^{3+}$ quantitatively.

---

***Example 8.18.8*** Determine the standard equilibrium constant of the following reaction at 298 K.

$$2MnO_4^- + 6H^+ + 5H_2C_2O_4 \rightarrow 2Mn^{2+} + 8H_2O + 10CO_2$$

***Solution*** In the given reaction, $MnO_4^-$ is reduced to $Mn^{2+}$ (and hence constitutes the right half-cell) and $H_2C_2O_4$ is oxidized to $CO_2$ (left half-cell). Thus, the cell producing the given reaction is

$$Pt \mid CO_2(g) \mid H_2C_2O_4(aq) \,\vdots\vdots\, H^+(aq),\ MnO_4^-(aq),\ Mn^{2+}(aq) \mid Pt$$

Its standard potential is

$$E^\circ_{cell} = E^\circ_{MnO_4^-,\,Mn^{2+},\,H^+\mid Pt} - E^\circ_{H_2C_2O_4\mid CO_2\mid Pt} = 1.51\text{ V} - (-\,0.49\text{ V})$$

$$= 2.0\text{ V}$$

Now since $n = 10$, we have

$$\log K^\circ_{eq} = \frac{nFE^\circ_{cell}}{2.303\,RT} = \frac{10\,(96\,500\text{ C mol}^{-1})(2.0\text{ V})}{2.303\,(8.314\text{ J K}^{-1}\text{ mol}^{-1})\,298\text{ K})}$$

$$= 331$$

Hence $K^\circ_{eq} = 10^{331}$

Thus, the oxidation of oxalic acid by permanganate occurs quantitatively.

---

***Example 8.18.9*** Determine the standard equilibrium constant of the following reaction at 298 K.

$$Cd^{2+} + 4NH_3 \rightarrow Cd(NH_3)_4^{2+}$$

***Solution*** The two half-cell reactions of a cell producing the above reaction are

Right half-cell $\quad Cd^{2+} + 2e^- \rightarrow Cd$

Left half-cell $\quad Cd(NH_3)_4^{2+} + 2e^- \rightarrow Cd + 4NH_3$

Hence the cell is

$$Cd \mid Cd(NH_3)_4^{2+},\ NH_3 \,\vdots\vdots\, Cd^{2+}(aq) \mid Cd$$

and its standard potential is

$$E^\circ_{cell} = E^\circ_R - E^\circ_L = -\,0.40\text{ V} - (-\,0.61\text{ V})$$

$$= 0.21\text{ V}$$

Thus $\log K^\circ_{eq} = \dfrac{nFE^\circ_{cell}}{2.303\,RT} = \dfrac{2\,(96\,500\ \text{C mol}^{-1})(0.21\ \text{V})}{2.303\,(8.314\ \text{J K}^{-1}\ \text{mol}^{-1})(298\ \text{K})} = 7.10$

Hence $K^\circ_{eq} = 1.26 \times 10^7$

---

***Example 8.18.10*** Determine the standard equilibrium constant of the following reaction at 298 K.

$$Cu(OH)_2 \rightarrow Cu^{2+}(aq) + 2OH^-(aq)$$

***Solution*** The two half-cell reactions of a cell producing the above reaction are:

Right half-cell $Cu(OH)_2 + 2e^- \rightarrow Cu + 2OH^-(aq)$

Left half-cell $Cu^{2+}(aq) + 2e^- \rightarrow Cu$

Hence, the cell is

$$Cu \mid Cu^{2+}(aq) \,\vdots\vdots\, OH^-(aq) \mid Cu(OH)_2(s) \mid Cu$$

with $E^\circ_{cell} = E^\circ_R - E^\circ_L = -0.224\ \text{V} - (0.337\ \text{V})$

$= -0.561\ \text{V}$

Thus $\log K^\circ_{eq} = \dfrac{nFE^\circ_{cell}}{2.303\,RT} = \dfrac{2\,(96\,500\ \text{C mol}^{-1})(-0.561\ \text{V})}{2.303\,(8.314\ \text{J K}^{-1}\ \text{mol})(298\ \text{K})} = -18.97$

Hence $K^\circ_{eq} = 1.1 \times 10^{-19}$

---

***Example 8.18.11*** How would you use the arrangement

$$Ag \mid AgI(\text{sat. soln.}) \mid AgI(s) \mid Ag$$

to measure the solubility product of silver iodide at 298 K?

***Solution*** The half-cell reactions are

Right half-cell (*Reduction*) $AgI + e^- \rightarrow Ag + I^-(aq)$

Left half-cell (*Oxidation*) $Ag \rightarrow Ag^+(aq) + e^-$

The cell reaction is

$$AgI \rightarrow Ag^+(aq) + I^-(aq)$$

The cell potential is given as

$$E_{cell} = E^\circ_{cell} - \frac{RT}{F}\ln\,(a_{Ag^+})(a_{I^-}) = E^\circ_{cell} - \frac{RT}{F}\ln K^\circ_s(AgI)$$

At equilibrium, we will have

$$\frac{RT}{F}\ln K^\circ_s(AgI) = E^\circ_{cell}$$

At 298 K, we have

$$\log K^\circ_s(AgI) = \frac{E^\circ_{cell}}{(2.303\,RT/F)} = \frac{(-0.151\ \text{V}) - (0.799\ \text{V})}{0.059\,13\ \text{V}} = \frac{-0.950\ \text{V}}{0.059\,13\ \text{V}}$$

$$= -16.07$$

Thus $K^\circ_s(AgI) = 8.58 \times 10^{-17}$

---

***Example 8.18.12***

An excess of solid AgCl is added to a 0.1 mol $dm^{-3}$ solution of $Br^-$ ions. Assuming activity equal to ionic concentration, calculate the concentration of $Cl^-$ and $Br^-$ ions at equilibrium.

***Solution***

When AgCl is added to a solution containing $Br^-$ ions, the possible reaction that can take place is

$$AgCl(s) + Br^-(aq) \rightarrow AgBr(s) + Cl^-(aq)$$

The two partial half-cell reactions of the above reaction are

Reduction $\quad AgCl + e^- \rightarrow Ag + Cl^-$

Oxidation $\quad Ag + Br^- \rightarrow AgBr + e^-$

Hence the above reaction will be produced by the cell

$$Ag \mid AgBr \mid Br^-(aq) \,\vdots\vdots\, Cl^-(aq) \mid AgCl \mid Ag$$

Its standard emf would be

$$E^\circ_{cell} = E^\circ_{Cl^- \mid AgCl \mid Ag} - E^\circ_{Br^- \mid AgBr \mid Ag}$$

Consulting the emfs table, we get

$$E^\circ_{cell} = 0.222 \text{ V} - 0.095 \text{ V} = 0.127 \text{ V}$$

Now since $\Delta G^\circ = -nFE^\circ_{cell} = -RT \ln K^\circ_{eq}$, we get

$$\log K^\circ_{eq} = \frac{nFE^\circ_{cell}}{2.303\,RT}$$

Substituting the data, we get

$$\log K^\circ_{eq} = \frac{(1)(96\,500 \text{ C mol}^{-1})(0.127 \text{ V})}{(2.303)\,(8.134 \text{ J K}^{-1} \text{ mol}^{-1})\,(298 \text{ K})} = 2.148$$

Hence $\quad K^\circ_{eq} = 140.6$

Let $x$ be the concentration of $Br^-$ that has converted to $Cl^-$ at equilibrium. Thus, we will have

$$K_{eq} = \frac{a_{Cl^-}}{a_{Br^-}} = \frac{[Cl^-]}{[Br^-]} = \frac{x}{(0.1 \text{ mol dm}^{-3}) - x} = 140.6$$

Solving for $x$, we get

$$x = 0.009\,928 \text{ mol dm}^{-3}$$

Hence at equilibrium, we have

$$[Cl^-] = 0.009\,928 \text{ mol dm}^{-3}$$

$$[Br^-] = 0.1 \text{ mol dm}^{-3} - 0.009\,928 \text{ mol dm}^{-3} = 0.090\,072 \text{ mol dm}^{-3}$$

**Alternative Method of Computing Equilibrium Constant**

If no electrons flow when a cell is short-circuited, then its emf is zero. At this stage, the cell reaction is said to be at equilibrium. The emf of a cell can be zero only when the electrode potentials of the two electrodes are equal. Thus for a cell reaction at equilibrium, we have

$$E_R = E_L \tag{8.18.9}$$

From the above equality, the value of $K^\circ_{eq}$ may be computed. The following example illustrates the procedure.

---

***Example 8.18.13***

Compute the equilibrium constant of the following reaction at 298 K:

$$Zn + Cu^{2+} \rightarrow Zn^{2+} + Cu$$

***Solution***

The two half-cell reduction reactions are

Right half-cell $\quad Cu^{2+} + 2e^- \rightarrow Cu$

Left half-cell $\quad Zn^{2+} + 2e^- \rightarrow Zn$

and the half-cell potentials are given as

$$E_{Cu^{2+}\,|\,Cu} = E^\circ_{Cu^{2+}\,|\,Cu} - \frac{RT}{2F}\ln\frac{1}{a_{Cu^{2+}}}$$

$$E_{Zn^{2+}\,|\,Zn} = E^\circ_{Zn^{2+}\,|\,Zn} - \frac{RT}{2F}\ln\frac{1}{a_{Zn^{2+}}}$$

To start with, let $a_{Cu^{2+}} > (a_{Cu^{2+}})_{eq}$ and $a_{Zn^{2+}} < (a_{Zn^{2+}})_{eq}$ where the subscript eq represents the activity prevailing at equilibrium. At this stage, the emf of the cell is positive indicating that $E_{Cu^{2+}\,|\,Cu} > E_{Zn^{2+}\,|\,Zn}$. If the cell is allowed to operate reversibly, $a_{Cu^{2+}}$ decreases whereas $a_{Zn^{2+}}$ is increased. Consequently, $E_{Cu^{2+}\,|\,Cu}$ decreases whereas $E_{Zn^{2+}\,|\,Zn}$ is increased. Eventually a stage will be reached where $E_{Cu^{2+}\,|\,Cu}$ becomes equal to $E_{Zn^{2+}\,|\,Zn}$ and the cell stops working. The cell reaction at this stage is at equilibrium. Hence equating the two reduction potentials, we get

$$E^\circ_{Cu^{2+}\,|\,Cu} - \frac{RT}{2F}\ln\frac{1}{(a_{Cu^{2+}})_{eq}} = E^\circ_{Zn^{2+}\,|\,Zn} - \frac{RT}{2F}\ln\frac{1}{(a_{Zn^{2+}})_{eq}}$$

On rearranging the above equation, we get

$$\frac{RT}{2F}\ln\frac{(a_{Zn^{2+}})_{eq}}{(a_{Cu^{2+}})_{eq}} = E^\circ_{Cu^{2+}\,|\,Cu} - E^\circ_{Zn^{2+}\,|\,Zn}$$

or
$$\log K^\circ = \frac{2F}{2.303\,RT}E^\circ_{cell}$$

where $K^\circ$ is the required standard equilibrium constant and is given as

$$K^\circ = \left(\frac{a_{Zn^{2+}}}{a_{Cu^{2+}}}\right)_{eq}$$

Substituting $E^\circ_{Cu^{2+}|Cu}$ and $E^\circ_{Zn^{2+}|Zn}$, we get

$$\log K^\circ = \frac{2 \times (96\,500\ \text{C mol}^{-1})(0.337\ \text{V} - (-0.763\ \text{V})]}{2.303 \times (8.314\ \text{J K}^{-1}\ \text{mol}^{-1})(298\ \text{K})} = 37.207$$

Hence $K^\circ = 1.6 \times 10^{37}$

## 3. DETERMINATION OF ACCURATE VALUE OF DISSOCIATION CONSTANT OF A WEAK ACID

The accurate value of the dissociation constant of a weak acid HA can be determined by using the cell

$$\text{Pt} \mid \text{H}_2(1\ \text{bar}) \mid \text{HA}(m_1),\ \text{NaA}(m_2),\ \text{NaCl}(m_3) \mid \text{AgCl(s)} \mid \text{Ag} \qquad (8.18.10)$$

where NaA is a salt formed from the given acid HA and the strong base NaOH. In addition to NaA, the electrolytic solution also contains a salt which contains same cation as the salt NaA and anion common with that of the electrode anion. In this particular case, the chosen salt is NaCl. The emf of the above cell is given by

$$E_{cell} = E_{Cl^- | AgCl | Ag} - E_{H^+ | H_2 | Pt}$$

or

$$E_{cell} = E^\circ_{Cl^- | AgCl | Ag} - \frac{RT}{F} \ln a_{Cl^-} + \frac{RT}{F} \ln \frac{1}{a_{H^+}}$$

$$= E^\circ_{Cl^- | AgCl | Ag} - \frac{RT}{F} \ln (a_{H^+})(a_{Cl^-})$$

Replacing the activities in terms of molalities and activity coefficients, we get

$$E_{cell} = E^\circ_{Cl^- | AgCl | Ag} - \frac{RT}{F} \ln (m'_{H^+} m'_{Cl^-})(\gamma_{H^+} \gamma_{Cl^-})$$

which on rearranging gives

$$\frac{F(E_{cell} - E^\circ_{Cl^- | AgCl | Ag})}{RT} = -\ln m'_{H^+} m'_{Cl^-} - \ln \gamma_{H^+} \gamma_{Cl^-} \qquad (8.18.11)$$

Now

$$K^\circ_a = \frac{(a_{H^+})(a_{A^-})}{a_{HA}} = \frac{(m'_{H^+}\gamma_{H^+})(m'_{A^-}\gamma_{A^-})}{(m'_{HA}\gamma_{HA})}$$

Taking logarithm, we get

$$\ln K^\circ_a = \ln \frac{m'_{H^+} m'_{A^-}}{m'_{HA}} + \ln \frac{\gamma_{H^+}\gamma_{A^-}}{\gamma_{HA}} \qquad (8.18.12)$$

Substituting Eq. (8.18.12) in Eq. (8.18.11), we get

$$\frac{F(E_{cell} - E^\circ_{Cl^- | AgCl | Ag})}{RT} = -\ln \frac{m'_{HA} m'_{Cl^-}}{m'_{A^-}} - \ln \frac{\gamma_{HA}\gamma_{Cl^-}}{\gamma_{A^-}} - \ln K^\circ_a$$

or
$$\frac{F(E_{\text{cell}} - E^\circ_{\text{Cl}^- | \text{AgCl} | \text{Ag}})}{RT} + \ln \frac{m'_{\text{HA}} m'_{\text{Cl}^-}}{m'_{\text{A}^-}} = -\ln \frac{\gamma_{\text{HA}} \gamma_{\text{Cl}^-}}{\gamma_{\text{A}^-}} - \ln K^\circ_{\text{a}} \tag{8.18.13}$$

The emf of the cell is measured at various values of $m_{\text{HA}}$, $m_{\text{Cl}^-}$ and $m_{\text{A}^-}$. The left side of Eq. (8.18.13) is evaluated and plotted as a function of the ionic-strength $\mu$ of the solution. The latter can be calculated from the expression

$$\mu = \frac{1}{2} \sum_i c_i z_i^2$$

where the summation is to be carried over all ions in the solution. The graph obtained is extrapolated to zero ionic strength. As $\mu$ approaches zero, all activity coefficients approach unity. Now according to Eq. (8.18.13), the extrapolated value (or intercept) is equal to $-\ln K^\circ_{\text{a}}$ and hence $K^\circ_{\text{a}}$ can be determined.

## 4. DETERMINATION OF ACCURATE VALUE OF IONIC PRODUCT OF WATER

The accurate value of ionic product of water can be determined by using the cell

$$\text{Pt} \,|\, \text{H}_2(1 \text{ bar}) \,|\, \text{MOH}(m_1), \text{MCl}(m_2) \,|\, \text{AgCl(s)} \,|\, \text{Ag} \tag{8.18.14}$$

The two half-cell reduction reactions are

Right electrode $\quad \text{AgCl} + e^- \rightarrow \text{Ag} + \text{Cl}^-$

Left electrode $\quad \text{H}^+ + e^- \rightarrow \frac{1}{2}\text{H}_2$

and thus the cell reaction is

$$\text{AgCl} + \tfrac{1}{2}\text{H}_2 \rightarrow \text{Ag} + \text{H}^+ + \text{Cl}^-$$

The emf of the cell is given by

$$E = E^\circ - \frac{RT}{F} \ln (a_{\text{H}^+})(a_{\text{Cl}^-})$$

Replacing $a_{\text{H}^+}$ by $K^\circ_{\text{w}}/a_{\text{OH}^-}$, we get

$$E = E^\circ - \frac{RT}{F} \ln K^\circ_{\text{w}} - \frac{RT}{F} \ln \frac{a_{\text{Cl}^-}}{a_{\text{OH}^-}}$$

$$= E^\circ - \frac{RT}{F} \ln K^\circ_{\text{w}} - \frac{RT}{F} \ln \frac{m'_{\text{Cl}^-}}{m'_{\text{OH}^-}} - \frac{RT}{F} \ln \frac{\gamma_{\text{Cl}^-}}{\gamma_{\text{OH}^-}}$$

or
$$\frac{E - E^\circ}{RT/F} + \ln \frac{m'_{\text{Cl}^-}}{m'_{\text{OH}^-}} = -\ln K^\circ_{\text{w}} - \ln \frac{\gamma_{\text{Cl}^-}}{\gamma_{\text{OH}^-}} \tag{8.18.15}$$

If the left side of Eq. (8.18.15) is plotted as a function of ionic strength (or $\sqrt{\text{ionic strength}}$, which gives a rather better plot) and $-\ln K_w^\circ$ is determined as the intercept at $\mu = 0$ where $\gamma_{Cl^-}/\gamma_{OH^-} = 1$. From the value obtained for $-\ln K_w^\circ$, $K_w^\circ$ can be determined.

***Example 8.18.14***

The following emfs refer to the cell at 298 K:

$$\text{Pt} \mid H_2(1 \text{ bar}) \mid \text{LiOH}(0.01 \text{ mol dm}^{-3}), \text{LiCl}(m) \mid \text{AgCl(s)} \mid \text{Ag}$$

| $m$/mol dm$^{-3}$ | 0.01 | 0.02 | 0.05 | 0.10 | 0.20 |
|---|---|---|---|---|---|
| $E$/V | 1.049 5 | 1.031 5 | 1.007 3 | 0.988 5 | 0.969 4 |

The standard potential of silver-silver chloride half-cell is 0.222 5 V. Determine the ionic product of water.

***Solution***

The emf of the given cell is

$$E_{\text{cell}} = E_{Cl^- \mid AgCl \mid Ag} - E_{H^+ \mid H_2 \mid Pt}$$

$$= E^\circ_{Cl^- \mid AgCl \mid Ag} - \frac{RT}{F} \ln (a_{H^+})(a_{Cl^-})$$

Since $K_w^\circ = (a_{H^+})(a_{OH^-})$, we have

$$E_{\text{cell}} = E^\circ_{Cl^- \mid AgCl \mid Ag} - \frac{RT}{F} \ln K_w^\circ - \frac{RT}{F} \ln \frac{a_{Cl^-}}{a_{OH^-}}$$

or

$$E_{\text{cell}} = E^\circ_{Cl^- \mid AgCl \mid Ag} - \frac{RT}{F} \ln K_w^\circ - \frac{RT}{F} \ln \frac{m'_{Cl}}{m_{OH^-}} - \frac{RT}{F} \ln \frac{\gamma_{Cl^-}}{\gamma_{OH^-}}$$

On rearranging, we get

$$\frac{E_{\text{cell}} - E^\circ_{Cl^- \mid AgCl \mid Ag}}{2.303\, RT/F} + \log \frac{m'_{Cl^-}}{m'_{OH^-}} = -\log K_w^\circ - \log \frac{\gamma_{Cl^-}}{\gamma_{OH^-}} \qquad (8.18.16)$$

Thus if the left side of Eq. (8.18.16) is plotted against ionic strength and extrapolated to zero ionic strength (where $\gamma_{Cl^-}$ and $\gamma_{OH^-}$ approach unity), the extrapolated value will be equal to $-\log K_w^\circ$. Table 8.18.1 includes the values of left side of Eq. (8.18.16) and the corresponding ionic strengths for the given values of $m$ of LiCl.

**Table 8.18.1** Values of $m$ and the left side of Eq. (8.18.16) For the Various Values of $m$

| $m$/mol dm$^{-3}$ | $\mu$/mol dm$^{-3}$ | $\dfrac{E_{\text{cell}} - E^\circ_{Cl^- \mid AgCl \mid Ag}}{0.059\ 13 \text{ V}} + \log \left(\dfrac{m}{0.01 \text{ mol dm}^{-3}}\right)$ |
|---|---|---|
| 0.01 | 0.02 | 13.986 + 0.000 = 13.986 |
| 0.02 | 0.03 | 13.682 + 0.301 = 13.983 |
| 0.05 | 0.06 | 13.273 + 0.699 = 13.972 |
| 0.10 | 0.11 | 12.955 + 1.000 = 13.955 |
| 0.20 | 0.21 | 12.631 + 1.301 = 13.932 |

The graph between $(E_{cell} - E^{o}_{Cl^- | AgCl | Ag})/0.059\ 13\ V + \log(m/0.01\ mol\ dm^{-3})$ and $\mu$ is shown in Fig. 8.18.2.

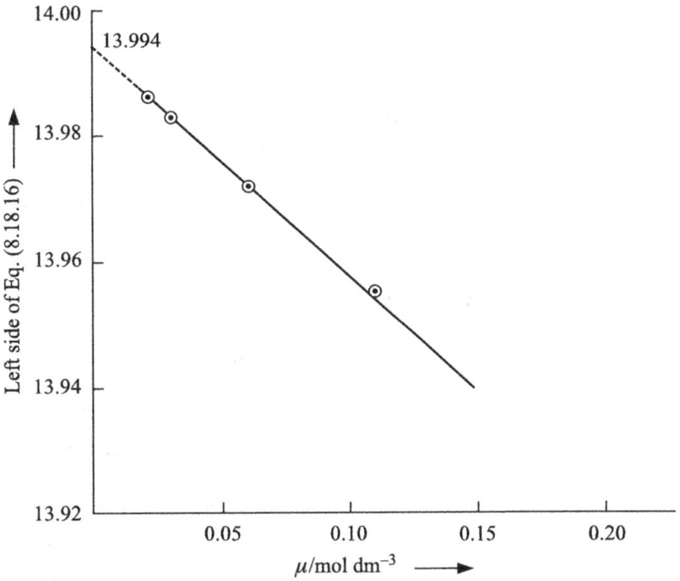


**Fig. 8.18.2 Graph between $\{(E_{cell} - E^{o}_{Cl^- | AgCl | Ag})/0.059\ 13\ V\} + \log(m/0.01\ mol\ dm^{-3})$ and $\mu$**

From the graph, we observe that

$$\text{Intercept} = -\log K^{\circ}_{w} = pK^{\circ}_{w} = 13.994$$

Hence $\quad K^{\circ}_{w} = 1.01 \times 10^{-14}$

## 5. DETERMINATION OF pH OF A SOLUTION

If the half-cell reaction involves either $H^+$ ion or $OH^-$ ion, the electrode potential depends on the activity of $H^+$ ions in the solution. The latter is usually represented on the pH scale† which is defined as

$$pH = -\log a_{H^+} \simeq -\log\{[H^+]/mol\ dm^{-3}\}$$

Alternatively, if the electrode potential and the activities (or fugacities if gases are involved) of species other than $H^+$ ions involved in the half-cell reaction are known, we can calculate the activity of $H^+$ ions or pH of the solution. Among the electrodes used to measure pH are the hydrogen electrode, the quinhydrone electrode and the glass electrode. We now describe these electrodes one by one.

**Hydrogen Electrode** The equilibrium reaction at the electrode is

$$H^+ + e^- = \tfrac{1}{2}H_2$$

†In potentiometric method, we involve activities (or effective concentrations) rather than the concentrations. We can extend the definition to $pH = -\log a_{H^+}$ in spite of the fact that the activity of individual ions has no practical significance. For the operational definition of pH, See Annexure III.

The potential adopted is given by

$$E = -\frac{RT}{F} \ln \frac{(f_{H_2}/f^\circ)^{1/2}}{(a_{H^+})}$$

Assuming $H_2$ gas to be an ideal gas with its pressure equal to 1 bar, the potential of the electrode becomes

$$E = -\frac{RT}{F} \ln \frac{1}{a_{H^+}} = -\frac{2.303\ RT}{F} \text{pH}$$

At 298 K, we have

$$E = -(0.059\ 13\ \text{V})\ \text{pH} \tag{8.18.17}$$

The hydrogen electrode is coupled with a reference electrode to form the cell.

Reference electrode ⁞⁞ $H_3O^+$ (?) | $H_2$ | Pt

and its emf, $E_{cell}$, is determined experimentally. At 298 K, we have

$$E_{cell} = E_R - E_L = -(0.059\ 13\ \text{V})\ \text{pH} - E_{reference}$$

Hence $$\text{pH} = -\frac{E_{cell} + E_{reference}}{(0.059\ 13\ \text{V})} \tag{8.18.18}$$

In practice, the use of hydrogen electrode for determination is severely limited in that it may not be used in solutions containing reducible materials and is easily poisoned by catalytic poisons. The latter hinders the establishment of the equilibrium at the surface and the electrode no longer behaves reversibly.

**Quinhydrone Electrode**

Quinhydrone is the name given to the molecular crystal formed between quinone (Q) and hydroquinone ($QH_2$). When dissolved in water, the crystal decomposes into its constituent compounds. The reduction reaction at the electrode is

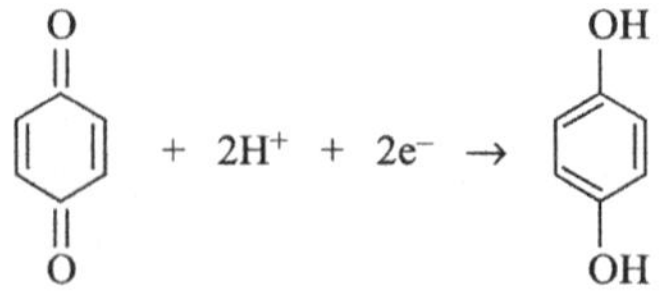


The quinhydrone electrode consists of a spiral of gold wire immersed in the solution whose pH is to determined. The solution is saturated with quinhydrone. The reduction potential of the electrode is given by

$$E_{Q,\,QH_2,\,H^+\,|\,Au} = E^\circ_{Q,\,QH_2,\,H^+\,|\,Au} - \frac{RT}{2F} \ln \frac{a_{QH_2}}{(a_Q)(a_{H^+})^2} \tag{8.18.19}$$

where $E^\circ_{Q,\,QH_2,\,H^+\,|\,Au}$ has a value of 0.699 6 V.

In solution, the activities of Q and $QH_2$ are equal. Thus Eq. (8.18.19) modifies to

$$E_{Q,\, QH_2,\, H^+ \mid Au} = E^{\circ}_{Q,\, QH_2,\, H^+ \mid Au} - \frac{RT}{2F} \ln \frac{1}{(a_{H^+})^2}$$

or
$$E_{Q,\, QH_2,\, H^+ \mid Au} = E^{\circ}_{Q,\, QH_2,\, H^+ \mid Au} - \frac{2.303\, RT}{F} \mathrm{pH} \qquad (8.18.20)$$

The quinhydrone electrode can be coupled with a reference electrode to form a cell:

Reference electrode ⁞⁞ $H_3O^+$ (?), Q, $QH_2$ | Au

and its emf is determined experimentally. We have

$$E_{\text{cell}} = E_R - E_L$$

or
$$E_{\text{cell}} = E^{\circ}_{Q,\, QH_2,\, H^+ \mid Au} - \frac{2.303\, RT}{F} \mathrm{pH} - E_{\text{reference}}$$

Hence
$$\mathrm{pH} = \frac{E^{\circ}_{Q,\, QH_2,\, H^+ \mid Au} - E_{\text{reference}} - E_{\text{cell}}}{2.303\, RT/F}$$

At 298 K, have

$$\mathrm{pH} = \frac{E^{\circ}_{Q,\, QH_2,\, H^+ \mid Au} - E_{\text{reference}} - E_{\text{cell}}}{0.059\ 13\ \mathrm{V}} \qquad (8.18.21)$$

**Limitation of using Eq. (8.18.21)**

The linear relation between electrode potential and pH of the solution (Eq. 8.18.20) holds good only up to certain pH value (≃ up to 8.5). Beyond this pH, the relation is no longer linear as may be seen from the following considerations.

The hydroquinone is a weak acid and it dissociates in a stepwise manner as shown below:

$$H_2Q \rightleftharpoons H^+ + HQ^- \qquad K_1 = \frac{[H^+][HQ^-]}{[H_2Q]} \qquad (8.18.22)$$

$$HQ^- \rightleftharpoons H^+ + Q^{2-} \qquad K_2 = \frac{[H^+][Q^{2-}]}{[HQ^-]} \qquad (8.18.23)$$

Mass-balance expression gives

$$[H_2Q]_0 = [H_2Q] + [HQ^-] + [Q^{2-}] \qquad (8.18.24)$$

The $[HQ^-]$ and $[Q^{2-}]$ are related to $[H_2Q]$ through the equilibrium constants $K_1$ and $K_2$. Thus

$$[H_2Q]_0 = [H_2Q] + \frac{K_1[H_2Q]}{[H^+]} + \frac{K_1K_2[H_2Q]}{[H^+]^2} \quad (8.18.25)$$

or

$$[H_2Q] = \frac{[H_2Q]_0}{\left(1 + \frac{K_1}{[H^+]} + \frac{K_1K_2}{[H^+]^2}\right)} = \frac{[H_2Q]_0[H^+]^2}{[H^+]^2 + K_1[H^+] + K_1K_2} \quad (8.18.26)$$

Since the solubility of quinhydrone is low, the activities of Q and $QH_2$ may be replaced by their concentrations, Substituting the concentration of $QH_2$ from Eq. (8.18.26) into Eq. (8.18.19), we get

$$E_{Q,\,QH_2,\,H^+\,|\,Au} = E^\circ_{Q,\,QH_2,\,H^+\,|\,Au} - \frac{RT}{2F}\ln\left[\frac{[H_2Q]_0\,c^{\circ 2}}{[Q]_0([H^+]^2 + K_1[H^+] + K_1K_2)}\right] \quad (8.18.27)$$

where $c^\circ$ is the unit concentration. Since $[H_2Q]_0 = [Q]_0$, therefore

$$E_{Q,\,QH_2,\,H^+\,|\,Au} = E^\circ_{Q,\,QH_2,\,H^+\,|\,Au} + \frac{RT}{2F}\ln\left[\{[H^+]^2 + K_1[H^+] + K_1K_2\}/c^{\circ 2}\right] \quad (8.18.28)$$

**Applicability of Eq. (8.18.28)**

- In acid solution, we have

$$[H^+]^2 >> K_1K_2 + K_1[H]^+$$

and thus Eq. (8.18.28) reduces to

$$E_{Q,\,QH_2,\,H^+\,|\,Au} = E^\circ_{Q,\,QH_2,\,H^+\,|\,Au} + \frac{RT}{F}\ln\{[H^+]/c^\circ\}$$

or

$$E_{Q,\,QH_2,\,H^+\,|\,Au} = E^\circ_{Q,\,QH_2,\,H^+\,|\,Au} - \frac{2.303RT}{F}\text{pH} \quad (8.18.29)$$

If $K_1[H^+] >> [H^+]^2 + K_1K_2$, then Eq. (8.18.28) becomes

$$E_{Q,\,QH_2,\,H^+\,|\,Au} = E^\circ_{Q,\,QH_2,\,H^+\,|\,Au} + \frac{RT}{2F}\ln K_1^\circ - \frac{2.303RT}{2F}\text{pH} \quad (8.18.30)$$

- In alkaline solution where

$$K_1K_2 >> K_1[H^+] + [H^+]^2$$

we have

$$E_{Q,\,QH_2,\,H^+\,|\,Au} = E^\circ_{Q,\,QH_2,\,H^+\,|\,Au} + \frac{RT}{2F}\ln K_1^\circ K_2^\circ \quad (8.18.31)$$

that is, the potential is independent of pH.

Equation (8.18.29) holds for pH < 7; above this pH hydroquinone undergoes partial ionization and atmospheric oxidation. The quinhydrone electrode is not useful if the solution contains oxidizing or reducing substances.

***Example 8.18.15***

Determine at 298 K for the cell

$$\text{Pt} \mid \text{Q, QH}_2\text{, H}^+ \,\vdots\vdots\, 1\ \text{M KCl} \mid \text{Hg}_2\text{Cl}_2\text{(s)} \mid \text{Hg(1)} \mid \text{Pt}$$

(a) its emf when pH = 5.0,
(b) the pH when $E_{\text{cell}} = 0$,
(c) the positive electrode when pH = 7.5.

***Solution***

For the given cell, we have

| *Electrode* | *Reaction* | *Potential* |
|---|---|---|
| Right (1 M calomel) | | $E_{\text{calomel}} = E^\circ_{\text{calomel}}$ |
| Left (quinhydrone) | $\frac{1}{2}\text{Q} + \text{H}^+ + \text{e}^- = \frac{1}{2}\text{QH}_2$ | $E_{\text{Q, QH}_2\text{, H}^+ \mid \text{Pt}} = E^\circ_{\text{Q, QH}_2\text{, H}^+ \mid \text{Pt}} - \dfrac{RT}{F}\ln\dfrac{1}{a_{\text{H}^+}}$ |

The emf of the cell is

$$\begin{aligned} E_{\text{cell}} &= E^\circ_{\text{calomel}} - E_{\text{Q, QH}_2\text{, H}^+ \mid \text{Pt}} \\ &= (E^\circ_{\text{calomel}} - E^\circ_{\text{Q, QH}_2\text{, H}^+ \mid \text{Pt}}) + \frac{RT}{F}\ln\frac{1}{a_{\text{H}^+}} \end{aligned}$$

At 298 K, we have

$$\begin{aligned} E_{\text{cell}} &= (E^\circ_{\text{calomel}} - E^\circ_{\text{Q, QH}_2\text{, H}^+ \mid \text{Pt}}) + (0.059\ 13\ \text{V})\ \text{pH} \\ &= (0.280\ \text{V} - 0.699\ 6\ \text{V}) + (0.059\ 13\ \text{V})\ \text{pH} \\ &= -\,0.419\ 6\ \text{V} + (0.059\ 13\ \text{V})\ \text{pH} \end{aligned}$$

Thus, we have

**(a)** $E_{\text{cell}} = -\,0.419\ 6\ \text{V} + (0.059\ 13\ \text{V}) \times 5 = -\,0.123\ 9\ \text{V}$

**(b)** $\text{pH} = \dfrac{1}{(0.059\ 13\ \text{V})}(E_{\text{cell}} + 0.419\ 6\ \text{V}) = \dfrac{0.419\ 6\ \text{V}}{0.059\ 13\ \text{V}} = 7.1$

**(c)** $E_{\text{cell}} = -\,0.419\ 6\ \text{V} + (0.059\ 13\ \text{V}) \times 7.5 = 0.023\ 9\ \text{V}$

Since $E_{\text{cell}}$ is positive, the cell as written would give a spontaneous reaction. Hence the calomel electrode acts as the positive terminal of the cell.

## Glass Electrode

The glass electrode consists of a glass tube terminating in a thin-walled bulb A as shown in Fig. 8.18.3. A special glass of relatively low melting point and high electrical conductance is used for this purpose. The bulb contains a solution of constant hydrogen-ion concentration and an electrode of definite potential. Usually either silver-silver chloride electrode dipped in 0.1 mol $\text{dm}^{-3}$ solution of hydrochloric acid or a platinum wire inserted in a pH 4.0 buffer solution containing a small quantity of quinhydrone is employed. The bulb is inserted in a solution whose pH is to be determined. The resulting half-cell is combined with a reference electrode (say, for example, the calomel electrode) through a salt bridge. This entire assembly constitutes a cell which may be represented as

Ag | AgCl(s) | 0.1 mol $dm^{-3}$ KCl | Glass | Solution of unkown pH ¦¦ Calomel electrode (8.18.32)

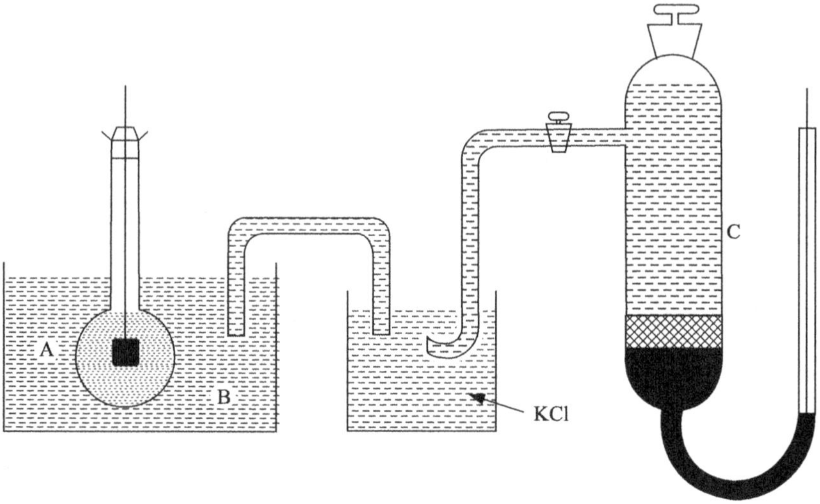


**Fig. 8.18.3 Determination of pH of the solution with the cell containing galss electrode**

The emf of the cell is given by

$$E_{\text{cell}} = E_{\text{R}} - E_{\text{L}} \qquad (8.18.33)$$

where $E_{\text{R}}$ is the reduction potential of the reference electrode and $E_{\text{L}}$ is that of the rest of unit, i.e.

Ag | AgCl(s) | 0.1 mol $dm^{-3}$ HCl | Glass | Experimental solution of unknown pH (8.18.34)

The value of $E_{\text{L}}$ is experimentally found to be dependent on the difference of pH of solutions on either side of glass. Since pH of the solution within the bulb remains constant, the potential $E_{\text{L}}$ depends on pH of the experimental solution. This dependence is given by

$$\begin{aligned} E_{\text{glass}} &= E^{\circ}_{\text{glass}} - \frac{RT}{F} \ln a_{\text{H}^+} \\ &= E^{\circ}_{\text{glass}} + \frac{2.303\, RT}{F} \text{pH} \end{aligned} \qquad (8.18.35)$$

where $a_{\text{H}^+}$ is the activity of hydrogen-ions in the experimental solution. The standard reduction potential of the glass electrode is usually determined experimentally as all glass electrodes do not have the same value. For this, a buffer solution of known pH is taken and the glass electrode is dipped into this solution and the emf of the cell (Eq. 8.18.32) is determined. From this emf, $E_{\text{glass}}$ can be determined as follows:

$$E_{\text{cell}} = E_{\text{calomel}} - E_{\text{glass}}$$

$$= E_{\text{calomel}} - E^{\circ}_{\text{glass}} - \frac{2.303\,RT}{F}\text{pH}$$

Thus $$E^{\circ}_{\text{glass}} = E_{\text{calomel}} - E_{\text{cell}} - \frac{2.303\,RT}{F}\text{pH}$$

After this, the glass electrode is dipped into the solution whose pH is to be determined and then pH is calculated with $E^{\circ}_{\text{glass}}$ found above.

**Theory Underlying Glass Electrode**

The theory of glass electrode has been explained on the basis of the transfer of hydrogen ions through the glass. The potential set up at the glass-solution interface is analogous to junction potential between two solutions containing hydrogen ions (Section 8.20), the value of such a potential is given by

$$E = (t_{-} - t_{\text{H}^+})\frac{RT}{F}\ln\frac{a_2}{a_1} \tag{8.18.36}$$

where $a_2$ and $a_1$ are the hydrogen-ion activities of outer and inner solutions, respectively. Since in the glass electrode only hydrogen ions can move across the boundary, we have

$$t_{-} = 0 \text{ and } t_{\text{H}^+} = 1 \tag{8.18.37}$$

Thus, Eq. (8.18.36) reduces to

$$E = -\frac{RT}{F}\ln\frac{a_2}{a_1}$$

Since $a_1$ is constant (inner solution), it follows that

$$E = \text{constant} - \frac{RT}{F}\ln a_2$$

or $$E = \text{constant} + \frac{2.303\,RT}{F}\text{pH} \tag{8.18.38}$$

Equation (8.18.38) is identical in form with the experimentally derived Eq. (8.18.35).

**Limitations of Glass Electrode**

Because of the high internal resistance of the glass electrode, which may amount to as much as 100 million ohms, ordinary potentiometers cannot be used to measure emf of cell. Instead, special vacuum tube potentiometers are employed. With the glass electrode, it is possible to measure with an accuracy of about 0.000 5 volt or 0.01 pH unit.

The glass electrode can be used up to pH 9. It is unaffected by poisoning and oxidizing and reducing agents. If the solution pH is greater than 9, particularly when sodium ions are present, the experimentally obtained pH value is usually lower than the true value. This is probably due to infiltration of sodium ions into the glass lattice. However, new glasses have been developed with which pH range can be extended up to 13 or 14.

***Example 8.18.16***

In a cell consisting of glass electrode in solution with pH of 7.0 and a calomel electrode, the emf of glass electrode at 298 K was 0.062 V. When a solution of unknown pH replaced the standard buffer, the glass electrode potential was 0.145 V with the glass electrode negative. Calculate the unknown pH if during caliberation, the glass electrode was (a) negative, and (b) positive.

***Solution***

(**a**) *Glass electrode as the negative terminal* The emf of the glass electrode at 298 K is given by

$$E_{\text{glass}} = E^{\circ}_{\text{glass}} + (0.059\ 13\ \text{V})\ \text{pH}$$

where $E^{\circ}_{\text{glass}}$ is a constant. Its value can be determined from the given data that $E_{\text{glass}}$ = 0.062 V at pH = 7. Thus

$$E^{\circ}_{\text{glass}} = E_{\text{glass}} - (0.059\ 13\ \text{V})\ \text{pH} = 0.062\ \text{V} - (0.059\ 13\ \text{V}) \times 7 = -\,0.352\ \text{V}$$

Thus when $E_{\text{glass}}$ = 0.145 V

$$\text{pH} = \frac{(E_{\text{glass}} - E^{\circ}_{\text{glass}})}{0.059\ 13\ \text{V}} = \frac{0.145\ \text{V} + 0.352\ \text{V}}{0.059\ 13\ \text{V}} = 8.4$$

(**b**) *Glass electrode as the positive terminal* Here

$$E_{\text{glass}} = -\,0.062\ \text{V}$$

Therefore

$$E^{\circ}_{\text{glass}} = E_{\text{glass}} - 0.0591\ 13\ \text{V pH} = -\,0.062\ \text{V} - (0.059\ 13\ \text{V}) \times 7$$
$$= -\,0.476\ \text{V}$$

Hence $$\text{pH} = \frac{(E_{\text{glass}} - E^{\circ}_{\text{glass}})}{0.059\ 13\ \text{V}} = \frac{0.145\ \text{V} + 0.476\ \text{V}}{0.059\ 13\ \text{V}} = 10.5$$

## 6. POTENTIOMETRIC TITRATIONS

### Principle Underlying Potentiometric Titrations

The potential measurements provide one of the most important analytical techniques for carrying out acid-base, redox and precipitation titrations.[†] The equivalence point of titrations can be determined accurately without making use of any visual indicators. The principle underlying the potentiometric titrations is the Nernst equation according to which the potential of an electrode depends upon the activities of reduced and oxidized forms of the involved redox couple. During titration, the activities of these species vary, which in turn, varies the potential of the electrode. The variation in potential is gradual in the beginning but becomes faster as the equivalence point is reached. The variation is maximum at the equivalence point and becomes lesser and lesser beyond the equivalence point. Thus at the equivalence point, one observes a point of inflexion.

[†]See Section 8.19 for the construction of potentiometric titration curves.

The potential of the electrode under study (known as indicator electrode) is determined as usual by combining it with a reference electrode and then determining the emf of the cell. Calomel electrode is usually employed as the reference electrode. The equivalence point of titration is determined graphically. A plot between the potential of the electrode (or the emf of the constructed cell) and the volume of added titrant is drawn. Figure 8.18.4 displays a typical variation in the potential of electrode against the volume of added titrant. The equivalence point of the titration which corresponds to the point of inflexion can be determined from the graph and the corresponding abscissa gives the volume of added titrant at the equivalence point.

## Location of Equivalence Point

A more exact method to locate the equivalence point is to construct a graph between $\Delta E/\Delta V$ and $V$, where $V$ stands for the volume of titrant added. Figure 8.18.5 displays the typical graph of $\Delta E/\Delta V$ versus $V$. The maximum on the curve gives the equivalence point of the titration. In addition to the above procedure, a graph of $\Delta^2 E/\Delta V^2$ versus $V$ or $(\Delta V/\Delta E)/\bar{V}$ versus $\bar{V}$, where $\bar{V}$ is the mean volume at each point, may be drawn (Figs 8.18.6 and 8.18.7). In the graph of second derivative, the second derivative has a zero value at the equivalence point and has different signs before and after the equivalence point. In the graph between $(\Delta V/\Delta E)/\bar{V}$ and $\bar{V}$, two straight lines are observed whose intersection point gives the equivalence point.

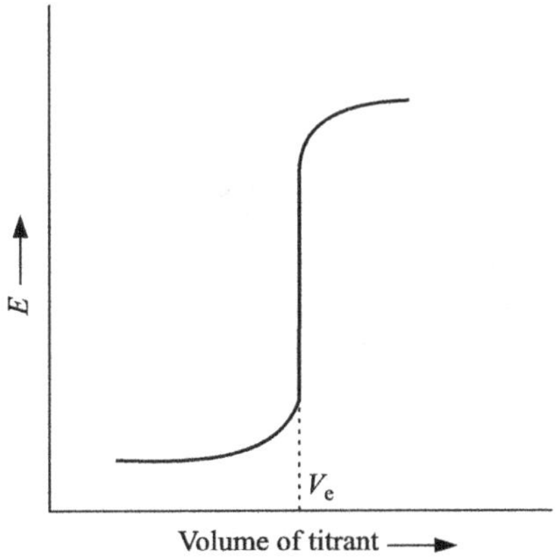


**Fig. 8.18.4 A typical potentiometric titration curve**

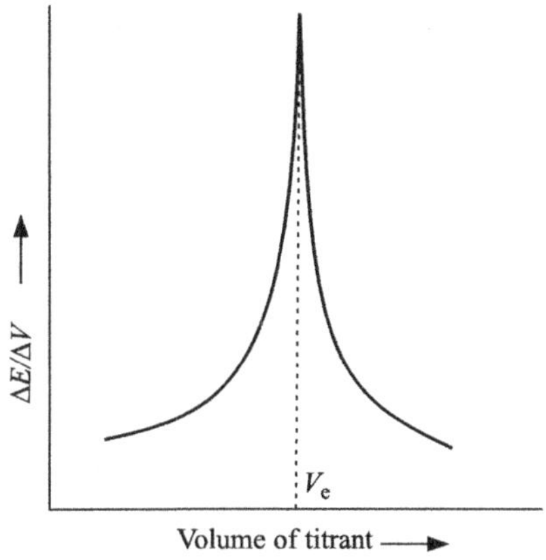


**Fig. 8.18.5 A typical first derivative graph**

## Applicability of Potentiometric Titrations

Potentiometric titrations constitute the most versatile method of electrochemical analysis and practically every analytical determination of both inorganic and organic systems carried out with classical method could be performed potentiometrically. The technique offers notable advantages in extending the range of usefulness of conventional methods. For example, the titration between weak acid and weak base, titrations in coloured solutions, titrations in non-aqueous media (where indicators cannot be used), and differential determinations of two or more than two substances with different solubility products or with different redox potentials can be carried out by using the potentiometric method.

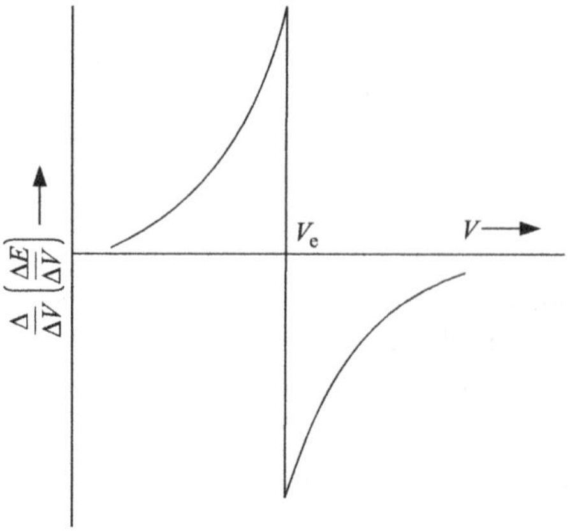


**Fig. 8.18.6 A typical second derivative graph**

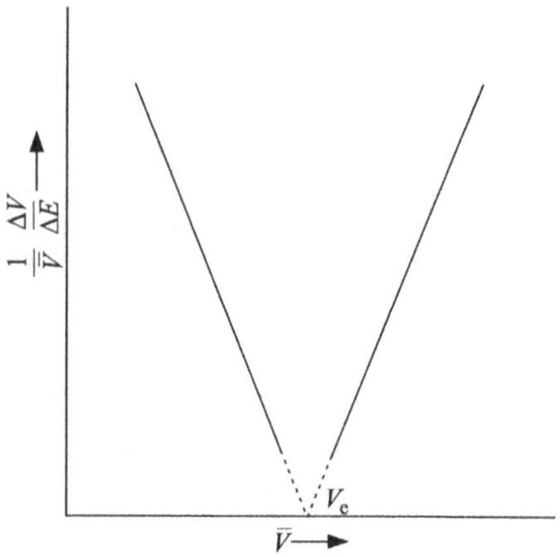


**Fig. 8.18.7 A typical graph between $(\Delta V/\Delta E)/\bar{V}$ and $\bar{V}$ (Gran's method)**

## 8.19 CONSTRUCTION OF POTENTIOMETRIC TITRATION CURVES

In this section, we describe the procedure for constructing potentiometric titration curves for (1) acid-base titrations, (2) redox titrations, and (3) precipitation titrations.

Throughout this section, activity of an ion is replaced by the numerical value of its concentration.

### ACID-BASE TITRATIONS

In acid-base titrations, the hydrogen-ion concentration varies during the course of titration. The indicator electrode employed in titration is either the quinhydrone electrode or the glass electrode. The potential of the electrode is given as follows.

***Quinhydrone Electrode***

$$E_{\mathrm{Q,\,QH_2,\,H^+\,|\,Pt}} = E^{\circ}_{\mathrm{Q,\,QH_2,\,H^+\,|\,Pt}} - \frac{2.303\,RT}{F}\,\mathrm{pH} \qquad (8.19.1)$$

***Glass Electrode***

$$E_{\mathrm{glass}} = E^{\circ}_{\mathrm{glass}} + \frac{2.303\,RT}{F}\,\mathrm{pH} \qquad (8.19.2)$$

Saturated calomel electrode is almost universally employed as the reference electrode. Thus, the assembly used in potentiometric acid-base titration is

Reference electrode ⁞⁞ Indicator electrode

i.e. $\mathrm{Pt \mid Hg \mid Hg_2Cl_2 \mid Cl^-}$ ⁞⁞ $\mathrm{H^+, Q, QH_2 \mid Pt}$

or $\mathrm{Pt \mid Hg \mid Hg_2Cl_2 \mid Cl^-}$ ⁞⁞ $\mathrm{H^+ \mid}$ Glass electrode

The emf of the cell is measured at different volume $V$ of added titrant and then graphs of $E$ of indicator electrode versus $V$ and $\Delta E/\Delta V$ versus $V$ are drawn. From these graphs, the equivalence point of the titration is determined. From Eqs (8.19.1) and (8.19.2), it may be observed that the potential of either of electrodes depends linearly on pH of the solution. Thus, the general appearance of graphs of $E_{\text{cell}}$ versus $V$ and pH versus $V$ will be identical and will differ only in the scaling factor on the $y$-axis. Since the titration curves of pH versus $V$ have already been constructed in Chapter 4 of Vol. 1, construction of the corresponding potentiometric curves can be carried out following Eq. (8.19.1) or (8.19.2). For illustration, we construct the potentiometric titration curve of the titration of a strong acid with a strong base.

## TITRATION OF HCl VERSUS NAOH

Let 50 cm$^3$ of 0.1 M HCl solution be titrated against 0.1 M NaOH. The pH of the solution at different stages of titration as calculated in Chapter 4 of Volume 1 are given in Table 8.19.1. It also includes the calculated values of $E_{\text{Q, QH}_2,\text{H}^+|\text{Pt}}$ at 25 °C, $\Delta E/\Delta V$, $\Delta^2 E/\Delta V^2$ and $(\Delta V/\Delta E)/\bar{V}$.

**Table 8.19.1** Calculated Data of Potentiometric Titration of 50 cm$^3$ of 0.1 M HCl versus 0.1 M NaOH

| *Total volume of base added* / cm$^3$ | pH | $E_{\text{Q, QH}_2,\text{H}^+|\text{Pt}}$ / V | $(\Delta E/\Delta V)\times 10^2$ / V cm$^{-3}$ | $\bar{V}$ / cm$^3$ | $(\Delta^2 E/\Delta V^2)\times 10^2$ / V$^2$ cm$^{-6}$ | $(\Delta V/\Delta E)\times 10^3/\bar{V}$ / V$^{-1}$ |
|---|---|---|---|---|---|---|
| 0 | 1.0 | 0.699 6 | | | — | |
| | | | 0.60 | 5 | | 33 400 |
| 10.0 | 1.18 | 0.639 8 | | | — | |
| | | | 0.21 | 15 | | 31 400 |
| 20.0 | 1.37 | 0.618 6 | | | — | |
| | | | 0.14 | 25 | | 29 400 |
| 30.0 | 1.60 | 0.605 0 | | | 0.007 | |
| | | | 0.21 | 35 | | 13 500 |
| 40.0 | 1.95 | 0.583 9 | | | 0.050 | |
| | | | 0.68 | 44.5 | | 3 306 |
| 49.0 | 2.99 | 0.522 7 | | | 0.202 | |
| | | | 6.64 | 49.45 | | 304.6 |
| 49.9 | 4.00 | 0.462 9 | | | 120 | |
| | | | 66.00 | 49.945 | | 30.3 |
| 49.99 | 5.0 | 0.403 8 | | | 11 920 | |
| | | | 656 | 49.994 5 | | 3.05 |
| 49.999 | 6.00 | 0.344 6 | | | 918 000 | |
| | | | 5 200 | 49.999 45 | | 0.38 |
| 49.999 9 | 6.79 | 0.297 9 | | | 14 400 000 | |
| | | | 12 400 | 49.999 95 | | 0.16 |
| 50.0 | 7.0 | 0.285 5 | | | | |
| | | | 12 400 | 50.000 05 | | 0.16 |
| 50.000 1 | 7.21 | 0.273 1 | | | −14 400 000 | |
| | | | 5 200 | 50.000 55 | | 0.38 |
| 50.001 | 8.00 | 0.226 3 | | | −918 000 | |
| | | | 656 | 50.005 5 | | 3.05 |
| 50.01 | 9.00 | 0.167 2 | | | −11 920 | |
| | | | 66 | 50.055 | | 30.3 |
| 50.1 | 10.0 | 0.108 0 | | | −120 | |
| | | | 6.5 | 50.55 | | 304.6 |
| 51.0 | 10.99 | 0.049 5 | | | −4.9 | |
| | | | 1.84 | 51.5 | | 1 055 |
| 52.0 | 11.30 | 0.031 1 | | | −0.84 | |
| | | | 1.00 | 52.5 | | 1 904 |
| 53.0 | 11.47 | 0.021 1 | | | −0.22 | |
| | | | 0.78 | 53.4 | | 2 396 |
| 54.0 | 11.60 | 0.013 3 | | | | |

Figure 8.19.1 displays the graphs of $E$ versus $V$, $\Delta E/\Delta V$ versus $V$ and $\Delta^2E/\Delta V^2$ versus $V$. All the three graphs give 50 cm$^3$ of base at the equivalence point of the titration.

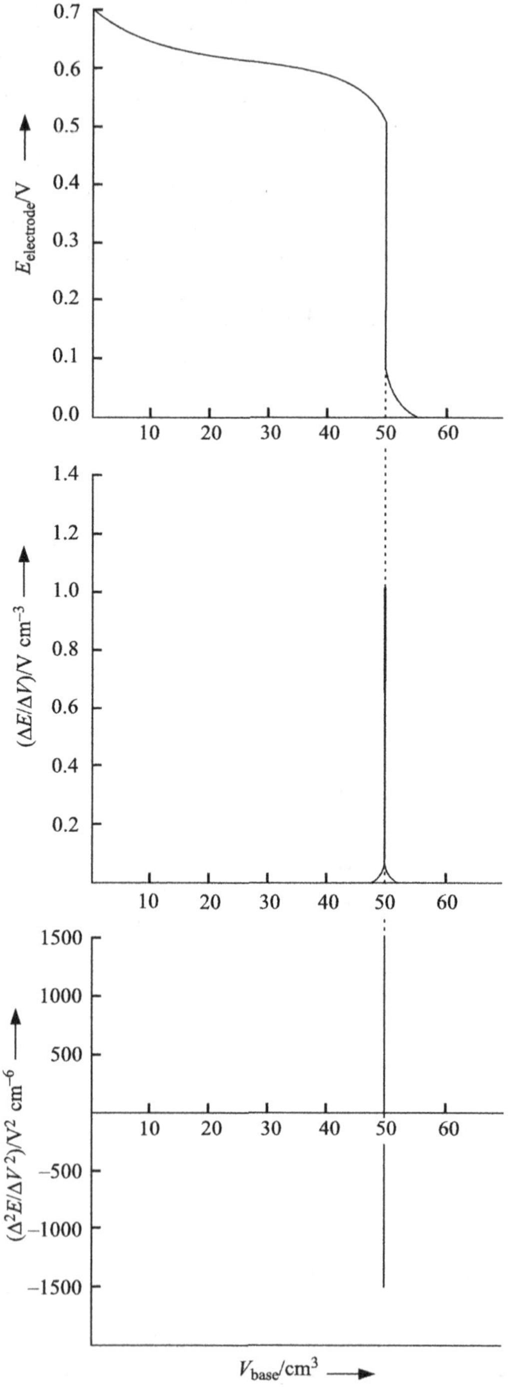


**Fig. 8.19.1 Graphs of $E$ versus $V$, $\Delta E/\Delta V$ versus $V$ and $\Delta^2E/\Delta V^2$ versus $V$**

## REDOX TITRATIONS

In redox titrations, the addition of titrant changes the concentrations of reduced and oxidized forms of reactant; the concentration of one of these increases while that of the other decreases. This results into the change in the potential of the electrode which can be determined experimentally or theoretically. The platinum electrode, which is usually employed as the indicator electrode, is dipped into the solution of reactant. The potential of the electrode relative to a reference electrode is measured at different stages of titration and then graphs between $E$ versus $V$, $\Delta E/\Delta V$ versus $V$, etc., are plotted to determine the equivalence point. The above curves can also be drawn theoretically following the procedure given below.

At any stage of titration, the two half reactions for the given redox titration are separately at equilibrium and their reduction potentials are identical. The potential of the electrode at any point after the start and up to, but not including, the equivalence point can be conveniently determined by considering the reduction reaction involving reduced and oxidized forms of reactant. Beyond the equivalence point, potential can be determined by considering the reduction reaction involving the two redox species of titrant.

In order to illustrate the procedure for constructing the titration curve, we describe in detail two typical titrations, namely, (i) ferrous ions versus ceric ions, and (ii) acidified ferrous ions versus dichromate ions. These two titrations have been selected with specific objectives in mind. The titration of $Fe^{2+}$ versus $Ce^{4+}$ is perhaps the simplest titration for which the potential data can be calculated easily. The titration of $Fe^{2+}$ with $CrO_7^{2-}$ has been chosen to justify the use of diphenylamine in the presence of phosphoric acid and *N*-phenylanthranilic acid as the common redox indicators in volumetric analysis of ferrous ions. We now describe these titrations.

### Titration of Ferrous Ions versus Ceric Ions

Let 50 $cm^3$ of 0.1 M $Fe^{2+}$ be titrated against 0.1 M $Ce^{4+}$ in a medium 0.5 M with respect to $H_2SO_4$. The values of $E^\circ_{Ce^{4+}, Ce^{3+} | Pt}$ and $E^\circ_{Fe^{3+}, Fe^{2+} | Pt}$, at 25 °C in this medium can be taken as 1.459 V and 0.700 V, respectively. The redox reaction is

$$Fe^{2+} + Ce^{4+} \rightarrow Fe^{3+} + Ce^{3+} \tag{8.19.3}$$

**Degree of conversion of $Fe^{2+}$ to $Fe^{3+}$ at the Equivalence Point**

Before computing the potential of indicator electrode at various stages of titration, it is worthwhile to explore the degree of conversion of $Fe^{2+}$ to $Fe^{3+}$ at the equivalence point. This may be done as follows.

The equilibrium constant of the above redox reaction is given by

$$K_{eq} = \frac{[Fe^{3+}][Ce^{3+}]}{[Fe^{2+}][Ce^{4+}]} = 10^{(E^\circ_{Ce^{4+}, Ce^{3+} | Pt} - E^\circ_{Fe^{3+}, Fe^{2+} H^+ | Pt})/0.059\,13\text{ V}}$$

$$= 10^{(1.459 - 0.700)/0.059\,13} = 10^{12.84} = 6.9 \times 10^{12} \tag{8.19.4}$$

At the equivalence point, we will have

$$[Fe^{3+}] = [Ce^{3+}] \tag{8.19.5}$$

and $$[Fe^{2+}] = [Ce^{4+}] \tag{8.19.6}$$

With these, Eq. (8.19.4) gives

$$\frac{[Fe^{3+}]^2}{[Fe^{2+}]^2} = 6.9 \times 10^{12} \quad \text{or} \quad \frac{[Fe^{3+}]}{[Fe^{2+}]} = 2.6 \times 10^6 \tag{8.19.7}$$

Since the volume of the solution at the equivalence point will be twice as the original volume of reactant, we will have

$$[Fe^{3+}] + [Fe^{2+}] = \frac{0.1\ M}{2} = 0.05\ M \tag{8.19.8}$$

From Eqs (8.19.7) and (8.19.8), we get

$$[Fe^{2+}] = 1.9 \times 10^{-8}\ M \tag{8.19.9}$$

Thus, the concentration of $Fe^{2+}$ is reduced from 0.1 M to $1.9 \times 10^{-8}$ M indicating that the conversion is practically complete.

The potential of indicator electrode at various stages of titration may be computed as follows.

**Potential at the Start of Titration**

To start with, the concentration of $Fe^{2+}$ in solution is given by

$$[Fe^{2+}] = 0.1\ M$$

Since the solution does not contain $Fe^{3+}$ (or the concentration of $Fe^{3+}$ which may have formed due to the aerial oxidation is not known), it is customary to leave undefined the value of $E_{electrode}$ at the start.

**Potential Before the Equivalence Point**

The concentration of $Fe^{2+}$ and $Fe^{3+}$ ions at any stage of titration before the equivalence point may be determined as follows.

The extent to which the reaction has proceeded may be represented in terms of fraction $f$ given as

$$f = \frac{\upsilon}{V} \tag{8.19.10}$$

where $\upsilon$ is the volume of 0.1 M solution of $Ce^{4+}$ that has been added in the volume $V$ (= 50 $cm^3$) of 0.1 M solution of $Fe^{2+}$. The total volume of the solution will become $V + \upsilon$. The concentration of $Fe^{3+}$ and $Fe^{2+}$ will be given by

$$[Fe^{3+}] = (0.1\ M)\frac{V}{V+\upsilon}f \tag{8.19.11}$$

and

$$[Fe^{2+}] = (0.1\ M)\frac{V}{V+\upsilon}(1-f) \tag{8.19.12}$$

The potential of indicator electrode will be given by

$$E_{Fe^{3+}, Fe^{2+} | Pt} = E^{\circ}_{Fe^{3+}, Fe^{2+} | Pt} - \frac{RT}{F}\ln\frac{[Fe^{2+}]}{[Fe^{3+}]} \tag{8.19.13}$$

Substituting $[Fe^{3+}]$ and $[Fe^{2+}]$ from Eqs (8.19.11) and (8.19.12), we have

$$E_{Fe^{3+}, Fe^{2+} | Pt} = E^{\circ}_{Fe^{3+}, Fe^{2+} | Pt} - \frac{RT}{F} \ln \frac{(0.1 \text{ M}) \dfrac{V}{V+v}(1-f)}{(0.1 \text{ M}) \dfrac{V}{V+v} f}$$

or

$$E_{Fe^{3+}, Fe^{2+} | Pt} = E^{\circ}_{Fe^{3+}, Fe^{2+} | Pt} - \frac{RT}{F} \ln \frac{1-f}{f} \tag{8.19.14}$$

From Eq. (8.19.14), it follows that the potential of electrode depends on the fraction of ferrous ion titrated and not on its initial concentration.

**Potential at the Equivalence Point**

The two half-cell potentials are given by

$$E_{Fe^{3+}, Fe^{2+} | Pt} = E^{\circ}_{Fe^{3+}, Fe^{2+} | Pt} - \frac{RT}{F} \ln \frac{[Fe^{2+}]}{[Fe^{3+}]} \tag{8.19.15}$$

$$E_{Ce^{4+}, Ce^{3+} | Pt} = E^{\circ}_{Ce^{4+}, Ce^{3+} | Pt} - \frac{RT}{F} \ln \frac{[Ce^{3+}]}{[Ce^{4+}]} \tag{8.19.16}$$

Adding Eqs (8.19.15) and (8.19.16), we get

$$E_{Fe^{3+}, Fe^{2+} | Pt} + E_{Ce^{4+}, Ce^{3+} | Pt} = E^{\circ}_{Fe^{3+}, Fe^{2+} | Pt} + E^{\circ}_{Ce^{4+}, Ce^{3+} | Pt} - \frac{RT}{F} \ln \frac{[Fe^{2+}][Ce^{3+}]}{[Fe^{3+}][Ce^{4+}]} \tag{8.19.17}$$

As indicated earlier, the two half-cell potentials will be identical at any stage of titration. Thus Eq. (8.21.17) becomes

$$E_{\text{electrode}} = \frac{1}{2}(E^{\circ}_{Fe^{3+}, Fe^{2+} | Pt} + E^{\circ}_{Ce^{4+}, Ce^{3+} | Pt}) - \frac{1}{2}\frac{RT}{F} \ln \frac{[Fe^{2+}][Ce^{3+}]}{[Fe^{3+}][Ce^{4+}]} \tag{8.19.18}$$

Now at the equivalence point, Eqs (8.19.5) and (8.19.6) will also hold good. With these, Eq. (8.19.18) becomes

$$E_{\text{equiv. point}} = \frac{1}{2}[E^{\circ}_{Fe^{3+}, Fe^{2+} | Pt} + E^{\circ}_{Ce^{4+}, Ce^{3+} | Pt}] \tag{8.19.19}$$

**Potential After the Equivalence Point**

The potential of the electrode may be determined from the expression

$$E_{Ce^{4+}, Ce^{3+} | Pt} = E^{\circ}_{Ce^{4+}, Ce^{3+} | Pt} - \frac{RT}{F} \ln \frac{[Ce^{3+}]}{[Ce^{4+}]} \tag{8.19.20}$$

Now the concentration of $Ce^{3+}$ will vary only due to the dilution effect whereas that of $Ce^{4+}$ will increase as more and more $Ce^{4+}$ solution is added. Thus, we have

$$[Ce^{3+}] = [Fe^{2+}]_{\text{original}} \left(\frac{V}{2V+v}\right) = (0.1 \text{ M})\left(\frac{V}{2V+v}\right) \tag{8.19.21}$$

and $$[\mathrm{Ce}^{4+}] = (0.1\ \mathrm{M})\left(\frac{\upsilon}{2V + \upsilon}\right) \tag{8.19.22}$$

where $\upsilon$ is the volume of $Ce^{4+}$ solution added after the equivalence point and $V$ (= 50 $cm^3$) is the initial volume of the reactant. Substituting Eqs (8.19.21) and (8.19.22) in Eq. (8.19.20), we get

$$E_{\mathrm{Ce}^{4+},\,\mathrm{Ce}^{3+}\,|\,\mathrm{Pt}} = E^{\circ}_{\mathrm{Ce}^{4+},\,\mathrm{Ce}^{3+}\,|\,\mathrm{Pt}} - \frac{RT}{F}\ln\frac{V}{\upsilon} \tag{8.19.23}$$

**Display of Computed Potentials**

Table 8.19.2 shows the data obtained from Eqs (8.19.14), (8.19.19) and (8.19.23) at various stages of titration of $Fe^{2+}$ versus $Ce^{4+}$.

**Table 8.19.2** Calculated Data of Potentiometric Titration of 50 $cm^3$ of 0.1 M $Fe^{2+}$ versus 0.1 M $Ce^{4+}$

| *Volume of* $Ce^{4+}/cm^3$ | $f$ | $\dfrac{E_{\mathrm{Electrode}}}{\mathrm{V}}$ | $\dfrac{(\Delta E/\Delta V)\times 10^2}{\mathrm{V\ cm^{-3}}}$ | $\bar{V}/cm^3$ | $\dfrac{(\Delta^2 E/\Delta V^2)\times 10^2}{\mathrm{V^2\ cm^{-6}}}$ |
|---|---|---|---|---|---|
| 0.0 | | | | | |
| 10.0 | 0.2 | 0.66 | | | |
| 20.0 | 0.4 | 0.69 | | | |
| | | | 0.2 | 25 | |
| 30.0 | 0.6 | 0.71 | | | 0.01 |
| | | | 0.3 | 35 | |
| 40.0 | 0.8 | 0.74 | | | 0.013 |
| | | | 0.4 | 42.5 | |
| 45.0 | 0.9 | 0.76 | | | 0.13 |
| | | | 1.0 | 47.0 | |
| 49.0 | 0.98 | 0.80 | | | 1.3 |
| | | | 4.0 | 49.25 | |
| 49.5 | 0.99 | 0.82 | | | 13 |
| | | | 10 | 49.7 | |
| 49.90 | 0.998 | 0.86 | | | 140 |
| | | | 40 | 49.925 | |
| 49.95 | 0.999 | 0.88 | | | 7 200 |
| | | | 400 | 49.975 | |
| 50.00 | 1.0 | 1.08 | | | 0 |
| | | | 400 | 50.025 | |
| 50.05 | — | 1.28 | | | −7 200 |
| | | | 40 | 50.075 | |
| 50.10 | — | 1.30 | | | −140 |
| | | | 10 | 50.3 | |
| 50.50 | — | 1.34 | | | −13 |
| | | | 4.0 | 50.75 | |
| 51.00 | — | 1.36 | | | −1.3 |
| | | | 1.0 | 53.0 | |
| 55.00 | — | 1.40 | | | −0.013 |
| | | | 0.4 | 57.5 | |
| 60.00 | — | 1.42 | | | |

Figure 8.19.2 displays the graphs obtained for $E$ versus $V$, $\Delta E/\Delta V$ versus $V$ and $\Delta^2 E/\Delta V^2$ versus $V$. Note that the graph of $E$ versus $V$ is symmetrical around the equivalence point.

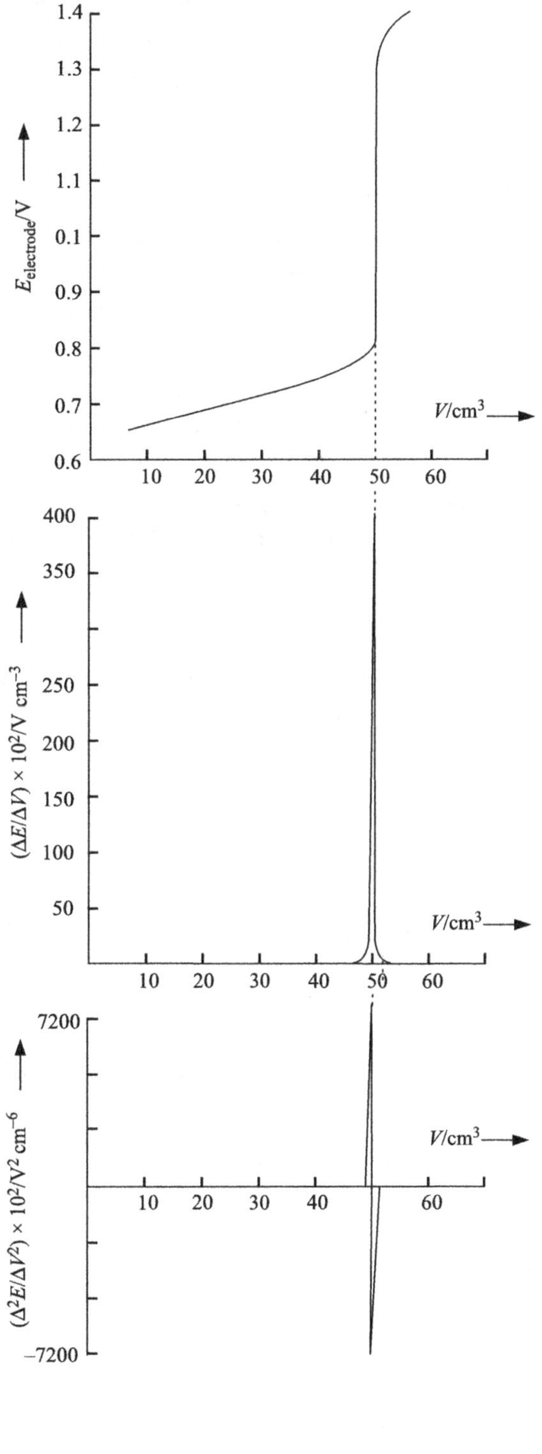


**Fig. 8.19.2 Graphs of $E$ versus, $V$, $\Delta E/\Delta V$ versus $V$ and $\Delta^2 E/\Delta V^2$ versus $V$**

### Titration of Ferrous Ions versus Dichromate Ions

Let 50 $cm^3$ of an acid solution (pH = 1.0) of 0.1 M $Fe^{2+}$ be titrated against (0.1 M/6), i.e. 0.1 N dichromate solution. The two half-cell reduction reactions are:

$$Cr_2O_7^{2-} + 14H^+ + 6e^- \rightarrow 2Cr^{3+} + 7H_2O$$

$$Fe^{3+} + e^- \rightarrow Fe^{2+}$$

The values of $E^\circ_{Fe^{3+}, Fe^{2+} | Pt}$ and $E^\circ_{C_2O_7^{2-}, Cr^{3+}, H^+ | Pt}$ are 0.770 V and 1.330 V, respectively.

The potential of the indicator electrode at various stages of titrations may be computed as follows.

**Potential at the Start of Titration**

To start with, the concentration of $Fe^{2+}$ in solution is given by

$$[Fe^{2+}] = 0.1 \text{ M}$$

Since the solution does not contain $Fe^{3+}$ (or the concentration of $Fe^{3+}$ which may have formed due to the aerial oxidation is not known), it is customary to leave undefined the value of $E_{electrode}$ at the start.

**Potential Before the Equivalence Point**

The concentration of $Fe^{2+}$ and $Fe^{3+}$ may be determined as follows.

The extent to which the reaction has proceeded may be represented in terms of fraction $f$ given as

$$f = \frac{v}{V}$$

where $v$ is the volume of 0.1 N solution of $Cr_2O_7^{2-}$ that has been added in the volume $V$ (= 50 $cm^3$) of 0.1 M solution of $Fe^{2+}$. The total volume of the solution will become $V + v$. The concentrations of $Fe^{3+}$ and $Fe^{2+}$ will be given by

$$[Fe^{3+}] = (0.1 \text{ M})\frac{V}{V + v}f \tag{8.19.24}$$

$$[Fe^{2+}] = (0.1 \text{ M})\frac{V}{V + v}(1 - f) \tag{8.19.25}$$

The potential of indicator electrode will be given by

$$E_{Fe^{3+}, Fe^{2+} | Pt} = E^\circ_{Fe^{3+}, Fe^{2+} | Pt} - \frac{RT}{F} \ln \frac{[Fe^{2+}]}{[Fe^{3+}]} \tag{8.19.26}$$

Substituting $[Fe^{3+}]$ and $[Fe^{2+}]$ from Eqs (8.19.24) and (8.19.25), we have

$$E_{Fe^{3+}, Fe^{2+} | Pt} = E^\circ_{Fe^{3+}, Fe^{2+} | Pt} - \frac{RT}{F} \ln \frac{1 - f}{f} \tag{8.19.27}$$

From Eq. (8.19.27), it follows that the potential of electrode depends on the fraction of ferrous ions titrated and not on its initial concentration.

**Potential at the Equivalence Point**

The two half-cell potentials are given by

$$E_{Fe^{3+}, Fe^{2+} | Pt} = E^\circ_{Fe^{3+}, Fe^{2+} | Pt} - \frac{RT}{F} \ln \frac{[Fe^{2+}]}{[Fe^{3+}]} \tag{8.19.28}$$

$$E_{Cr_2O_7^{2-}, Cr^{3+}, H^+ | Pt} = E^\circ_{Cr_2O_7^{2-}, Cr^{3+}, H^+ | Pt} - \frac{RT}{6F} \ln \frac{[Cr^{3+}]^2 c^{\circ 13}}{[Cr_2O_7^{2-}][H^+]^{14}} \tag{8.19.29}$$

where $c^\circ = 1 \text{ mol cm}^{-3}$.

Multiplying Eq. (8.19.29) by 6 and then adding to it Eq. (8.19.28), we get

$$E_{Fe^{3+}, Fe^{2+} | Pt} + 6E_{Cr_2O_7^{2-}, Cr^{3+}, H^+ | Pt} = E^\circ_{Fe^{3+}, Fe^{2+} | Pt} + 6E^\circ_{Cr_2O_7^{2-}, Cr^{3+}, H^+ | Pt} - \frac{RT}{F} \ln\left(\frac{[Fe^{2+}][Cr^{3+}]^2 c^{\circ 13}}{[Fe^{3+}][Cr_2O_7^{2-}][H^+]^{14}}\right) \tag{8.19.30}$$

The two half-cell potentials will be identical at any stage of titration. Thus, Eq. (8.19.30) becomes

$$E_{electrode} = \frac{1}{7}[E^\circ_{Fe^{3+}, Fe^{2+} | Pt} + 6E^\circ_{Cr_2O_7^{2-}, Cr^{3+}, H^+ | Pt}] - \frac{1}{7}\frac{RT}{F} \ln\left(\frac{[Fe^{2+}][Cr^{3+}]^2 c^{\circ 13}}{[Fe^{3+}][Cr_2O_7^{2-}][H^+]^{14}}\right) \tag{8.19.31}$$

From the reaction

$$Cr_2O_7^{2-} + 6Fe^{2+} + 14H^+ \rightarrow 2Cr^{3+} + 6Fe^{3+} + 7H_2O$$

it follows that at the equivalence point

$$[Fe^{2+}] = 6[Cr_2O_7^{2-}] \text{ and } [Fe^{3+}] = 3[Cr^{3+}]$$

Substituting these in Eq. (8.19.31), we get

$$E_{equiv.\ point} = \frac{1}{7}[E^\circ_{Fe^{3+}, Fe^{2+} | Pt} + 6E^\circ_{Cr_2O_7^{2-}, Cr^{3+}, H^+ | Pt}] + \frac{1}{7}\frac{RT}{F} \ln([H^+]/c^\circ)^{14} - \frac{1}{7}\frac{RT}{F} \ln(2[Cr^{3+}]/c^\circ) \tag{8.19.32}$$

**Potential After the Equivalence Point**

The potential of the electrode may be determined from the expression

$$E_{Cr_2O_7^{2-}, Cr^{3+}, H^+ | Pt} = E^\circ_{Cr_2O_7^{2-}, Cr^{3+}, H^+ | Pt} - \frac{RT}{6F} \ln \frac{[Cr^{3+}]^2 c^{\circ 13}}{[Cr_2O_7^{2-}][H^+]^{14}} \tag{8.19.33}$$

Now the concentration of $Cr^{3+}$ will vary only due to the dilution effect whereas that of $Cr_2O_7^{2-}$ will increase as more and more $Cr_2O_7^{2-}$ solution is added. Thus, we have

$$[Cr^{3+}] = \frac{1}{3}[Fe^{2+}]_{original}\left(\frac{V}{2V + \upsilon}\right) = \left(\frac{0.1\text{ M}}{3}\right)\left(\frac{V}{2V + \upsilon}\right) \tag{8.19.34}$$

$$[Cr_2O_7^{2-}] = \left(\frac{0.1\text{ M}}{6}\right)\left(\frac{\upsilon}{2V + \upsilon}\right) \tag{8.19.35}$$

where $\upsilon$ is the volume of $Cr_2O_7^{2-}$ solution added after the equivalence point and $V(= 50 \text{ cm}^3)$ is the initial volume of reactant.

Substituting Eqs (8.19.34) and (8.19.35) in Eq. (8.19.33), we get

$$E_{\mathrm{Cr_2O_7^{2-},\,Cr^{3+},\,H^+\,|\,Pt}} = E^{\circ}_{\mathrm{Cr_2O_7^{2-},\,Cr^{3+},\,H^+\,|\,Pt}} - \frac{RT}{6F}\ln\left\{\frac{\left(\dfrac{0.1}{3}\,\dfrac{V}{2V+v}\right)^2}{\left(\dfrac{0.1}{6}\,\dfrac{v}{2V+v}\right)([\mathrm{H^+}]/c^{\circ})^{14}}\right\} \tag{8.19.36}$$

**Display of Computed Potentials**

Table 8.19.3 shows the data obtained from Eqs (8.19.27), (8.19.32) and (8.19.36) at various stages of titration of an acid solution of $Fe^{2+}$ versus $Cr_2O_7^{2-}$.

**Table 8.19.3** Calculated Data of Potentiometric Titration of 50 $cm^3$ of an Acid Solution (pH = 1.0) of 0.1 M $Fe^{2+}$ versus (0.1 M/6) $Cr_2O_7^{2-}$

| *Volume of* $Cr_2O_7^{2-}$ / $cm^3$ | $f$ | $E_{electrode}$ / V | $(\Delta E/\Delta V)\times 10^2$ / V $cm^{-3}$ | $\bar{V}$ / $cm^3$ | $(\Delta E^2/\Delta V^2)\times 10^2$ / $V^2$ $cm^{-6}$ |
|---|---|---|---|---|---|
| 0.0 | | | | | |
| 10.0 | 0.2 | 0.74 | | | |
| | | | | 15 | |
| 20.0 | 0.4 | 0.76 | | | |
| | | | | 25 | |
| 30.0 | 0.6 | 0.78 | | | |
| | | | 0.20 | 35 | |
| 40.0 | 0.8 | 0.80 | | | 0.053 |
| | | | 0.60 | 42.5 | |
| 45.0 | 0.9 | 0.83 | | | 0.088 |
| | | | 1.0 | 47.0 | |
| 49.0 | 0.98 | 0.87 | | | 1.3 |
| | | | 4.0 | 49.25 | |
| 49.5 | 0.99 | 0.89 | | | 13 |
| | | | 10 | 49.7 | |
| 49.9 | 0.998 | 0.93 | | | 130 |
| | | | 40 | 49.925 | |
| 49.95 | 0.999 | 0.95 | | | 6 800 |
| | | | 380 | 49.975 | |
| 50.0 | 1.0 | 1.14 | | | –6 040 |
| | | | 76 | 50.025 | |
| 50.05 | — | 1.178 | | | –1 400 |
| | | | 4.0 | 50.075 | |
| 50.10 | — | 1.18 | | | –20.0 |
| | | | 1.75 | 50.3 | |
| 50.5 | — | 1.187 | | | –2.0 |
| | | | 0.60 | 50.75 | |
| 51.0 | — | 1.190 | | | –0.19 |
| | | | 0.17 | 53.0 | |
| 55.0 | — | 1.197 | | | –0.024 |
| | | | 0.06 | 57.5 | |
| 60.0 | — | 1.200 | | | |

Figure 8.19.3 displays the graphs of $E$ versus $V$, $\Delta E/\Delta V$ versus $V$ and $\Delta^2 E/\Delta V^2$ versus $V$. Note that the graph of $E$ versus $V$ is not symmetrical around the equivalence point.

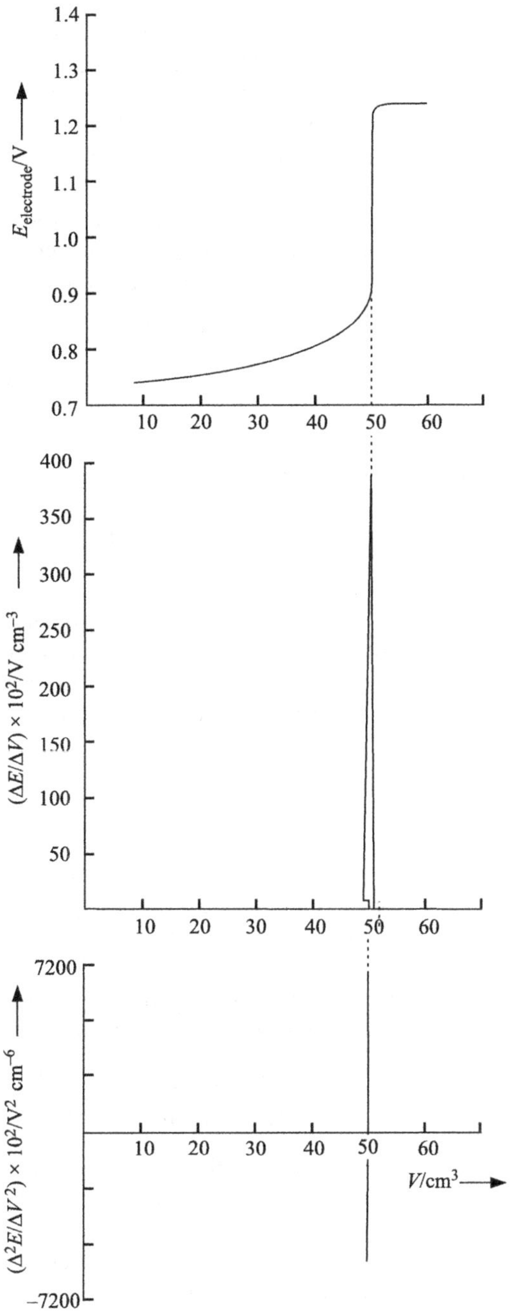


**Fig. 8.19.3 Graphs of $E$ versus $V$, $\Delta E/\Delta V$ versus $V$ and $\Delta^2E/V^2$ versus $V$**

**Use of Redox Indicators**

In the routine volumetric titration of $Fe^{2+}$ versus $Cr_2O_7^{2-}$ in laboratories, we make use of diphenylamine in the presence of phosphoric acid and *N*-phenylanthranilic acid as the internal indicators. It is worthwhile to consider here the theory underlying the use of internal indicators in the redox titrations.

In general, a redox indicator is an organic compound which can undergo oxidation-reduction reaction. The colours of oxidized and reduced forms of the indicator in solution differ significantly. For example, diphenylamine or diphenylbenzidine (reduced form) is colourless whereas diphenylbenzidine violet (oxidized form) is violet in colour. In general, we have two types of redox indicators depending upon whether $H_3O^+$ participitates or does not participitate in the indicator redox reaction. Thus, we have

(i) $\mathrm{In_{ox}} + ne^- \rightleftharpoons \mathrm{In_{red}}$

$$E_{\text{ox, red}\,|\,\text{Pt}} = E^\circ_{\text{ox, red}\,|\,\text{Pt}} - \frac{RT}{nF}\ \ln\left\{\left(\frac{\mathrm{In_{red}}}{\mathrm{In_{ox}}}\right)\right\} \quad (8.19.37a)$$

(ii) $\mathrm{In_{ox}} + nH_3O^+ + ne^- \rightleftharpoons \mathrm{In_{red}}$

$$E_{\text{ox, red, H}^+\,|\,\text{Pt}} = E^\circ_{\text{ox, red, H}^+\,|\,\text{Pt}} - \frac{RT}{nF}\ \ln\left\{\frac{[\mathrm{In_{red}}]\,c^{\circ n}}{[\mathrm{In_{ox}}][H_3O^+]^n}\right\} \quad (8.19.37b)$$

The colour of the solution depends on the ratio $[\mathrm{In_{red}}]/[\mathrm{In_{ox}}]$. On an average, the solution acquires a distinct colour characteristic of $\mathrm{In_{red}}$ if concentration of the latter is approximately 10 times (or more) greater than that of $\mathrm{In_{ox}}$ and vice versa. Thus at 298 K, we have:

(i) At

$$E_{\text{ox, red}\,|\,\text{Pt}} = E^\circ_{\text{ox, red}\,|\,\text{Pt}} - \frac{(0.059\ \text{V})}{n} \quad (8.19.38a)$$

or $$E_{\text{ox, red, H}^+\,|\,\text{Pt}} = E^\circ_{\text{ox, red, H}^+\,|\,\text{Pt}} - \frac{(0.059\ \text{V})}{n} + (0.059\ \text{V})\log([H_3O^+]/c^\circ)$$

(8.19.38b)

the ratio $[\mathrm{In_{red}}]/[\mathrm{In_{ox}}]$ in the solution is (10/1). For this potential, the percentage conversion of reduced form to oxidized form is

$$\frac{[\mathrm{In_{ox}}]}{[\mathrm{In_{ox}}] + [\mathrm{In_{red}}]} = \frac{1}{(1 + 10)} \times 100 = 0.91$$

Thus, most of the indicator is present in the reduced form and the solution gets the colour characteristic of this form. In fact, the value of $E$ as given by Eq. (8.19.38a) or (8 19.38b) is the maximum potential up to which the solution has a distinct colour characteristic of the reduced form of the indicator. At potentials smaller than this potential still more of the indicator is present in the reduced form. Thus at

$$E_1 \leq E_{\text{ox, red}\,|\,\text{Pt}} - \frac{(0.59\ \text{V})}{n}$$

or $$E_1 \le E_{\text{ox, red, H}^+ \mid \text{Pt}} - \frac{(0.059\ \text{V})}{n} + (0.059\ \text{V}) \log ([\text{H}_3\text{O}^+]/c^\circ)$$

the solution has a colour characterstic of the reduced form of the indicator.
(ii) At

$$E_{\text{ox, red} \mid \text{Pt}} = E^\circ_{\text{ox, red} \mid \text{Pt}} + \frac{(0.059\ \text{V})}{n} \tag{8.19.39a}$$

or $$E_{\text{ox, red, H}^+ \mid \text{Pt}} = E^\circ_{\text{ox, red, H}^+ \mid \text{Pt}} + \frac{(0.059\ \text{V})}{n} + (0.059\ \text{V}) \log ([\text{H}_3\text{O}^+]/c^\circ) \tag{8.19.39b}$$

the ratio of $[\text{In}_{\text{red}}]/[\text{In}_{\text{ox}}]$ in the solution is (1/10). For this potential, the percentage conversion of reduced form to oxidized form is

$$\frac{[\text{In}_{\text{ox}}]}{[\text{In}_{\text{ox}}] + [\text{In}_{\text{red}}]} = \frac{10}{(10+1)} \times 100 \simeq 91$$

Thus, most of the indicator is present in the oxidized form and the solution gets the colour characteristic of this form. In fact, the value of $E$ as given by Eq. (8.19.39a) or (8.19.39b) is the minimum potential up to which the solution has a distinct colour characteristic of the oxidized form of the indicator. At potentials greater than this potential, still more of the indicator is present in the oxidized form. Thus at

$$E_2 \ge E_{\text{ox, red} \mid \text{Pt}} + \frac{(0.059\ \text{V})}{n}$$

or $$E_2 \ge E_{\text{ox, red, H}^+ \mid \text{Pt}} + \frac{(0.059\ \text{V})}{n} + (0.059\ \text{V}) \log ([\text{H}_3\text{O}^+]/c^\circ)$$

the solution has a colour characteristic of the oxidized form of the indicator.

**Indicator Transition Range**

The potential range $E_1$ to $E_2$ is known as *indicator transition range* and has a value of $(0.118/n)$ V. Within this range, transition of colour of the solution takes place. Since the transition interval is usually very small (it is maximum when $n = 1$), it is usually sufficient to know the value of $E_{\text{ox, red} \mid \text{Pt}}$, called the *transition potential*, around which the indicator changes colour. A list of indicators is given in Table 8.19.4, which includes the colours of reduced and oxidized forms of each indicator.

**Selection of Indicators**

At any stage of titration, the potential $E_{\text{red, ox} \mid \text{Pt}}$ will have the same value as that of the indicator electrode. This, in turn, will vary the ratio [red]/[ox] in the solution.

We have seen earlier that the solution potential changes very rapidly near the equivalence point of the titration. The curve is almost steep near the equivalence point. Consequently, the ratio [red]/[ox] also changes rapidly near the equivalence point of the titration. It is obvious that if the steep change in the potential of the solution near the equivalence point covers the transition potential of the indicator a change of colour of the solution is expected. Thus, the selection of indicator must satisfy the following requirement.

*The transition potential of redox indicator must coincide with the steep portion of the titration curve.*

**Table 8.19.4** Oxidation-Reduction Indicators

| *Indicator* | *Colour of* | | *Transition potential at approximately* pH = 0 |
|---|---|---|---|
| | *reduced form* | *oxidized form* | |
| 5-Nitro-l, 10-phenanthroline iron(II) (nitroferroin) | Red | Pale blue | 1.25 |
| 1, 10-Phenanthroline iron(II) ferroin | Red | Pale blue | 1.11 |
| *N*-phenylanthranilic acid | Colourless | Pink | 1.08 |
| Erioglaucine A | Green | Red | 1.02 |
| 2, 2′-Dipyridineiron(II) | Red | Pale blue | 0.97 |
| Sodium diphenylamine sulphonic acid and sodium diphenylbenzidine sulpnonic acid | Colourless | Purple | 0.85 |
| Diphenylamine and diphenylbenzidine | Colourless | Violet | 0.76 |
| Methylene blue | Colourless | Blue | 0.53 |
| Nile blue | Colourless | Blue | 0.41 |
| Indigo tetrasulphonic acid | Colourless | Blue | 0.36 |
| Indigo monosulphonic acid | Colourless | Blue | 0.26 |

## Use of Diphenylamine or Diphenylbenzidine in the Volumetric Estimation of Ferrous Ions

Now the use of diphenylamine or diphenylbenzidine in the presence of phosphoric acid as the common indicator in the redox titration of $Fe^{2+}$ versus $Cr_2O_7^{2-}$ may be explained as follows.

Diphenylamine undergoes the following oxidation reactions in strong oxidizing media:

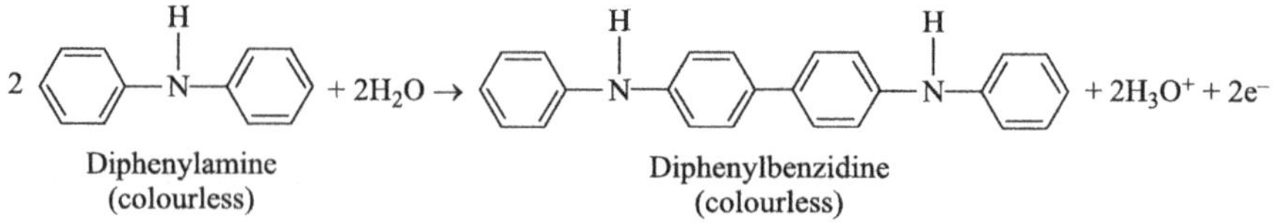


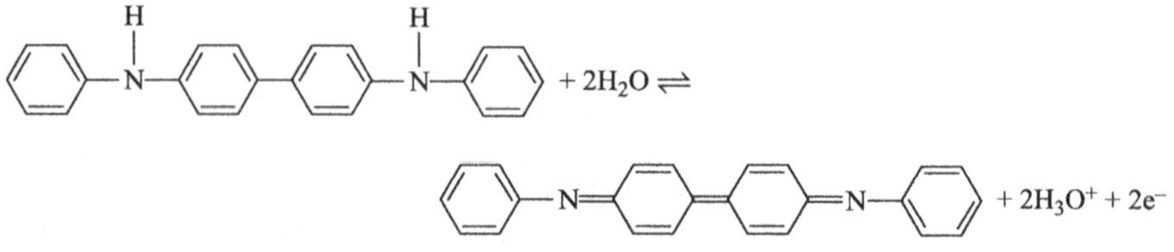


Diphenylbenzidine violet (purple)

The reaction transforming diphenylamine (DPA) to diphenylbenzidine (DPB) is irreversible, so that the use of diphenylbenzidine instead of diphenylamine eliminates this step and avoids the consumption of additional titrant. The indicator reaction is the transformation of diphenylbenzidine to diphenylbenzidine violet (DPVB). The transition potential in a solution of pH = 0 is 0.76 V at 298 K.

The potential of ferric-ferrous system is 0.77 V. Hence, ferric ions can oxidize diphenylbenzidine to diphenylbenzidine violet and thus result into a premature end point What we want is a complete transformation of $Fe^{2+}$ to $Fe^{3+}$ by $Cr_2O_7^{2-}$ and then a slight excess of the latter should oxidize the indicator. This is achieved by adding phosphoric acid in the reactant solution. The ferric ion forms complex $Fe(H_2PO_4)^{2+}$ and thus its concentration in solution is very much diminished. This, in turn, lowers the potential of ferric-ferrous system as may be seen from the potential expression

$$E_{Fe^{3+}, Fe^{2+} | Pt} = E^\circ_{Fe^{3+}, Fe^{2+} | Pt} - \frac{RT}{F} \ln \frac{[Fe^{2+}]}{[Fe^{3+}]}$$

Thus, the ability of ferric ions to oxidize diphenylamine benzidine indicator is completely lost. When $Fe^{2+}$ is completely oxidized to $Fe^{3+}$ and a slight excess of $Cr_2O_7^{2-}$ is present, the potential of the solution suffers a steep change which encompasses the transition potential of the indicator, and hence a change of colour of the solution takes place.

## PRECIPITATION TITRATIONS

In precipitation titrations, the concentration of reactant decreases as a result of formation of a precipitate with the titrant. If an electrode reversible with the reactant is dipped into the solution, its potential will vary during the course of titration. Hence precipitation titration can be carried out potentiometrically. In order to illustrate the procedure for constructing the titration curve, we describe below the precipitation titration involving chloride and silver ions.

### Titration of Chloride Ions Versus Silver Ions

Let 50 $cm^3$ of 0.1 M $Cl^-$ be titrated against 0.1 M $Ag^+$. The silver-silver chloride electrode can be employed as the indicator electrode. This electrode is reversible with chloride ions according to the following reaction.

$$AgCl(s) + e^- \rightarrow Ag(s) + Cl^-$$

Its potential is given by the expression

$$E_{Cl^- | AgCl | Ag} = E^\circ_{Cl^- | AgCl | Ag} - \frac{RT}{F} \ln ([Cl^-]/c^\circ) \qquad (8.19.40)$$

with $E^\circ_{Cl^- | AgCl | Ag} = 0.222 \text{ V}$

The potential of the electrode at various stages of titrations may be computed as follows.

**Potential at the Start of Titration** To Start with, the concentration of $Cl^-$ in solution is given by

$$[Cl^-] = 0.1 \text{ M}$$

Thus, the potential of electrode at 298 K as given by Eq. (8.19.40) is

$$E_{\mathrm{Cl^- \mid AgCl \mid Ag}} = 0.222\ \mathrm{V} - (0.059\ \mathrm{V}) \times \log (0.1)$$
$$= 0.281\ \mathrm{V}$$

**Potential Before the Equivalence Point**

During titration, the concentration of $Cl^-$ decreases. The progress of titration can be described by the fraction $f$ which is given as

$$f = \frac{v}{V}$$

where $v$ is the volume of 0.1 M solution of $Ag^+$ that has been added in $V(= 0.50\ \mathrm{cm}^3)$ of 0.1 M solution of $Cl^-$. The total volume of the solution will become $V + v$. The concentration of $Cl^-$ during the course of titration will be given as

$$[\mathrm{Cl^-}] = (0.1\ \mathrm{M}) \left(\frac{V}{V + v}\right) (1 - f)$$

The potential of the electrode will be given by

$$E_{\mathrm{Cl^- \mid AgCl \mid Ag}} = E^{\circ}_{\mathrm{Cl^- \mid AgCl \mid Ag}} - \frac{RT}{F} \ln \left\{\frac{(0.1)\,V}{V + v}\ (1 - f)\right\}$$

**Potential at the Equivalence Point**

The concentration of $Cl^-$ ions at the equivalence point may be calculated from the solubility product of silver chloride. At 25 °C, the latter has a value of $1.56 \times 10^{-10}\ (\mathrm{mol\ dm^{-3}})^2$. Since at equivalence point, the $[Cl^-]$ is identical to $[Ag^+]$ it is obvious that

$$[\mathrm{Cl^-}]_{\text{equiv. point}} = \sqrt{K_s(\mathrm{AgCl})} = \sqrt{1.56 \times 10^{-10}\ \mathrm{M}^2}$$
$$= 1.24 \times 10^{-5}\ \mathrm{M}$$

The above concentration can be substituted in Eq. (8.19.40) to get the potential value at the equivalence point.

**Potential Beyond the Equivalence Point**

Beyond the equivalence point, the concentration of $Cl^-$ ions can again be determined through the solubility product of silver chloride and is given by

$$[\mathrm{Cl^-}] = \frac{K_s}{[\mathrm{Ag^+}]} = \frac{1.56 \times 10^{-10}\ \mathrm{M}^2}{[\mathrm{Ag^+}]} \tag{8.19.41}$$

The concentration of $Ag^+$ will be given by

$$[\mathrm{Ag^+}] = (0.1\ \mathrm{M}) \left(\frac{v}{2V + v}\right) \tag{8.19.42}$$

where $v$ is the volume of $Ag^+$ solution added after the equivalence point and $V$ $(= 0.50\ \mathrm{cm}^3)$ is the initial volume of reactant.

Alternatively, the potential of the electrode may be computed from the expression:

$$E_{\mathrm{Ag^+ \mid Ag}} = E^{\circ}_{\mathrm{Ag^+ \mid Ag}} - \frac{RT}{F} \ln \left(\frac{1}{([\mathrm{Ag^+}]/c^{\circ})}\right) \tag{8.19.43}$$

where $E^{\circ}_{\mathrm{Ag^+ \mid Ag}} = 0.799\ \mathrm{V}$

Table 8.19.5 shows the data obtained from Eq. (8.19.40) at various stages of titration of 0.1 M $Cl^-$ versus 0.1 M $Ag^+$.

**Table 8.19.5** The Calculated Data of Potentiometric Titration of 50 $cm^3$ of 0.1 M $Cl^-$ versus 0.1 M $Ag^+$

| *Volume of* $Ag^+/cm^3$ | $f$ | $\frac{E_{electrode}}{V}$ | $\frac{(\Delta E/\Delta V)\times 10^2}{V\ cm^{-3}}$ | $\frac{\bar{V}}{cm^3}$ | $\frac{(\Delta^2 E/\Delta V^2)\times 10^4}{V^2\ cm^{-6}}$ |
|---|---|---|---|---|---|
| 0.0 | 0 | 0.282 | | | |
| | | | 0.10 | 5.0 | |
| 10.0 | 0.2 | 0.292 | | | 0.10 |
| | | | 0.11 | 15.0 | |
| 20.0 | 0.4 | 0.303 | | | 0.30 |
| | | | 0.14 | 25.0 | |
| 30.0 | 0.6 | 0.317 | | | 0.7 |
| | | | 0.21 | 35.0 | |
| 40.0 | 0.8 | 0.338 | | | 2.3 |
| | | | 0.38 | 42.5 | |
| 45.0 | 0.9 | 0.357 | | | 10.5 |
| | | | 0.80 | 46.5 | |
| 48.0 | 0.96 | 0.381 | | | 50 |
| | | | 1.8 | 48.5 | |
| 49.0 | 0.98 | 0.399 | | | 240 |
| | | | 3.6 | 49.25 | |
| 49.5 | 0.99 | 0.417 | | | 1 100 |
| | | | 8.0 | 49.65 | |
| 49.8 | 0.996 | 0.441 | | | 4 500 |
| | | | 17.0 | 49.85 | |
| 49.9 | 0.998 | 0.458 | | | 37 000 |
| | | | 54.0 | 49.95 | |
| 50.0 | 1.0 | 0.512 | | | 0.0 |
| | | | 54.0 | 50.05 | |
| 50.10 | — | 0.566 | | | −37 000 |
| | | | 17.0 | 50.15 | |
| 50.20 | — | 0.583 | | | −4 500 |
| | | | 8.0 | 50.35 | |
| 50.50 | — | 0.607 | | | −1 100 |
| | | | 3.6 | 50.75 | |
| 51.0 | — | 0.625 | | | −240 |
| | | | 1.8 | 51.5 | |
| 52.0 | — | 0.643 | | | |

Figure 8.19.4 displays the graphs of $E$ versus $V$, $(\Delta E/\Delta V)$ versus $V$ and $(\Delta^2 E/\Delta V^2)$ versus $V$.

### Titration Involving Multiple Redox-System

It is possible to titrate a solution containing two (or more) reductants with a standard solution of a suitable oxidant. The curve of $E_{electrode}$ versus volume of titrant will show two (or more) distinct breaks, one at each equivalence point, provided that the separation of the $E^{\circ}_{red,\ ox\ |\ Pt}$ values for the two reductants is large enough and that the $E^{\circ}_{red,\ ox\ |\ Pt}$ values for the oxidant is sufficiently higher than the larger $E^{\circ}_{red,\ ox\ |\ Pt}$ value of the reductant system. For example, an acid solution of $Ti^{3+}$ and $Fe^{2+}$ can be titrated against a standard solution of $Ce^{4+}$. The values of $E^{\circ}_{TiO_2^{2+},\ Ti^{3+}\ |\ Pt}$, $E^{\circ}_{Fe^{3+},\ Fe^{2+}\ |\ Pt}$ and $E^{\circ}_{Ce^{4+},\ Ce^{3+}\ |\ Pt}$ at 25 °C in 2 M $H_2SO_4$ are 0.10, 0.70 and 1.459 V, respectively.

Thus, first of all, $Ti^{3+}$ is oxidized to $TiO_2^{2+}$ and when it is completely oxidized, $Fe^{2+}$ starts oxidizing to $Fe^{3+}$. The entire potentiometric curve can be calculated following the procedure given above.

Similarly, multiple acid-base and precipitation titrations can be carried out potentiometrically. For example, a solution containing $Cl^-$ and $I^-$ ions can be titrated against a standard solution of silver nitrate. Two breaks in the potentiometric curve will be observed; the first break corresponds to the complete precipitation of $I^-$ and the second one to that of $Cl^-$.

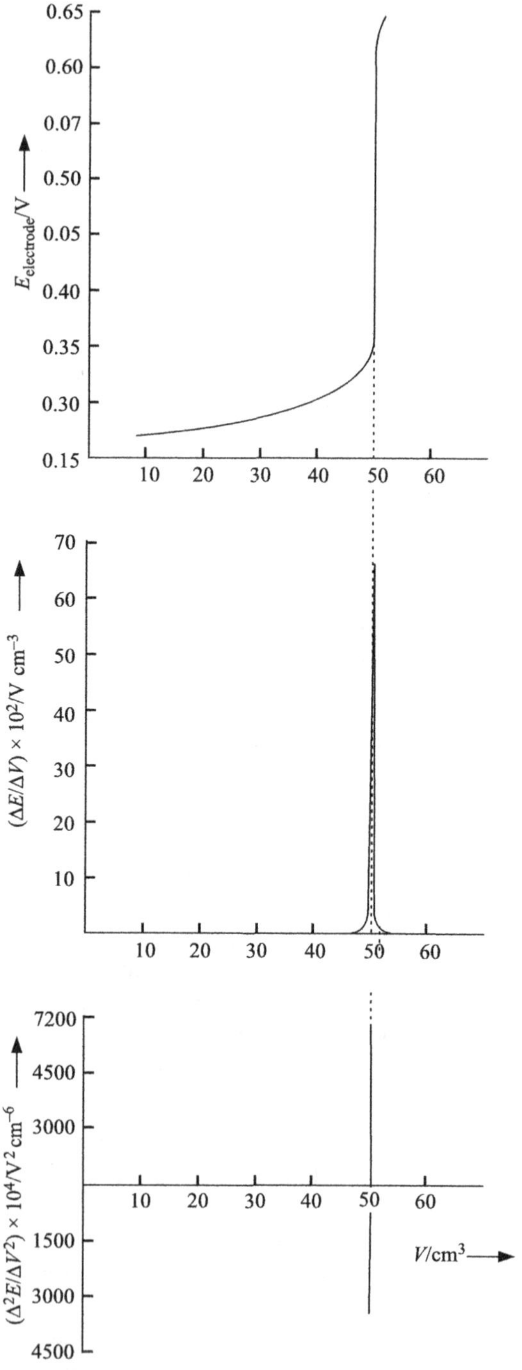


**Fig. 8.19.4 Graphs of $E$ versus $V$, $\Delta E/\Delta V$ versus $V$ and $\Delta^2E/\Delta V^2$ versus $V$**

## 8.20 CONCENTRATION CELLS WITHOUT LIQUID JUNCTION POTENTIAL

Concentration cells are made up of two half-cells which are similar chemically but differ in the activity of some common component. The common component may be the electrode or the electrolytic solution. The emf of the cell is due to the difference in activity of the common component. We describe below three categories of concentration cells without liquid junction.

### Cells with Amalgam Electrodes

Such a cell is formed by two metal amalgam electrodes of different metal activity dipping into a common solution of a soluble salt of the metal. For example, the cell

$$\mathrm{Pb(Hg)}(a_{\mathrm{Pb}} = a_1) \mid \mathrm{Pb^{2+}}(a_{\mathrm{Pb^{2+}}}) \mid \mathrm{Pb(Hg)}(a_{\mathrm{Pb}} = a_2) \quad (8.20.1)$$

includes two lead amalgams with activities of lead equal to $a_1$ and $a_2$ immersed in a solution of lead ions of activity $a_{\mathrm{Pb^{2+}}}$. For this cell, we have

| *Electrode* | *Reduction reaction* | *Reduction potential* | |
|---|---|---|---|
| Right | $\mathrm{Pb^{2+}} + 2e^- \rightarrow \mathrm{Pb(Hg)}(a_2)$ | $E_R = E° - \dfrac{RT}{2F} \ln \dfrac{a_2}{a_{\mathrm{Pb^{2+}}}}$ | (i) |
| Left | $\mathrm{Pb^{2+}} + 2e^- \rightarrow \mathrm{Pb(Hg)}(a_1)$ | $E_L = E° - \dfrac{RT}{2F} \ln \dfrac{a_1}{a_{\mathrm{Pb^{2+}}}}$ | (ii) |

Subtracting Eq. (ii) from Eq. (i), we get

$$\mathrm{Pb(Hg)}(a_1) \rightarrow \mathrm{Pb(Hg)}(a_2) \quad (8.20.2)$$

and $$E_{\mathrm{cell}} = \frac{RT}{2F} \ln \frac{a_1}{a_2} \quad (8.20.3)$$

### Cells with Gas Electrodes Operating at Different Pressures

Consider a cell consisting of two hydrogen electrodes operating at different pressures dipping into a common solution of hydrochloric acid:

$$\mathrm{Pt} \mid \mathrm{H_2}(p_1) \mid \mathrm{H^+}(a_{\mathrm{H^+}}) \mid \mathrm{H_2}(p_2) \mid \mathrm{Pt} \quad (8.20.4)$$

We have

| *Electrode* | *Reduction reaction* | *Reduction potential* | |
|---|---|---|---|
| Right | $2\mathrm{H^+} + 2e^- \rightarrow \mathrm{H_2}(p_2)$ | $E_R = -\dfrac{RT}{2F} \ln \dfrac{(p_2/p°)}{a^2_{\mathrm{H^+}}}$ | (i) |
| Left | $2\mathrm{H^+} + 2e^- \rightarrow \mathrm{H_2}(p_1)$ | $E_L = -\dfrac{RT}{2F} \ln \dfrac{(p_1/p°)}{a^2_{\mathrm{H^+}}}$ | (ii) |

Subtracting Eq. (ii) from Eq. (i), we get

$$\mathrm{H_2}(p_1) \rightarrow \mathrm{H_2}(p_2) \quad (8.20.5)$$

and $$E_{\mathrm{cell}} = \frac{RT}{2F} \ln \frac{p_1}{p_2} \quad (8.20.6)$$

**Cells with Different Electrolytic Activities**

This type of cells can be formed by making a composite cell out of two cells differing only in the activity of the electrolytic solution. For example, the cell

$$\mathrm{Pt} \mid \mathrm{H_2}(p) \mid \mathrm{H^+Cl^-}(a_\pm)_1 \mid \mathrm{AgCl} \mid \mathrm{Ag} \tag{8.20.7}$$

may be combined with a cell

$$\mathrm{Pt} \mid \mathrm{H_2}(p) \mid \mathrm{H^+Cl^-}(a_\pm)_2 \mid \mathrm{AgCl} \mid \mathrm{Ag} \tag{8.20.8}$$

to give the following composite cell.

$$\mathrm{Pt} \mid \mathrm{H_2}(p) \mid \mathrm{H^+Cl^-}(a_\pm)_1 \mid \mathrm{AgCl(s)} \mid \mathrm{Ag} - \mathrm{Ag} \mid \mathrm{AgCl} \mid \mathrm{H^+Cl^-}(a_\pm)_2 \mid \mathrm{H_2}(p) \mid \mathrm{Pt} \tag{8.20.9}$$

The emf of the above cell is the sum of the emfs of two cells separately, i.e. the emf of the left cell (L) plus the emf of the right cell (R). Thus

$$E_{\text{cell}} = E_{\mathrm{L}} + E_{\mathrm{R}} \tag{8.20.10}$$

$$= (E_{\mathrm{Cl^-} \mid \mathrm{AgCl} \mid \mathrm{Ag}} - E_{\mathrm{H^+} \mid \mathrm{H_2} \mid \mathrm{Pt}})_{\mathrm{L}} + (E_{\mathrm{H^+, H_2} \mid \mathrm{Pt}} - E_{\mathrm{Cl^-} \mid \mathrm{AgCl} \mid \mathrm{Ag}})_{\mathrm{R}} \tag{8.20.11}$$

Writing the Nernst equation for each potential, we obtain

$$E_{\text{cell}} = \left[ E^\circ_{\mathrm{Cl^-} \mid \mathrm{AgCl} \mid \mathrm{Ag}} - \frac{RT}{F} \ln (a_{\mathrm{Cl^-}})_1 + \frac{RT}{F} \ln \frac{(p/p^\circ)^{1/2}}{(a_{\mathrm{H^+}})_1} \right] + \left[ -\frac{RT}{F} \ln \frac{(p/p^\circ)^{1/2}}{(a_{\mathrm{H^+}})_2} - E^\circ_{\mathrm{Cl^-} \mid \mathrm{AgCl} \mid \mathrm{Ag}} + \frac{RT}{F} \ln (a_{\mathrm{Cl^-}})_2 \right]$$

or

$$E_{\text{cell}} = \frac{RT}{F} \ln \frac{(a_{\mathrm{H^+}})_2 (a_{\mathrm{Cl^-}})_2}{(a_{\mathrm{H^+}})_1 (a_{\mathrm{Cl^-}})_1} = \frac{2RT}{F} \ln \frac{(a_\pm)_2}{(a_\pm)_1} \tag{8.20.12}$$

The net cell reaction is obtained by adding the individual cell reactions. Thus, we have

| *Cell* | *Cell reaction* | |
|---|---|---|
| Left | $\mathrm{AgCl} + \frac{1}{2}\mathrm{H_2} \rightarrow \mathrm{Ag} + \mathrm{H^+}(a_1) + \mathrm{Cl^-}(a_1)$ | (i) |
| Right | $\mathrm{Ag} + \mathrm{H^+}(a_2) + \mathrm{Cl^-}(a_2) \rightarrow \mathrm{AgCl} + \frac{1}{2}\mathrm{H_2}$ | (ii) |

Adding Eqs (i) and (ii), we get

$$\mathrm{H^+}(a_2) + \mathrm{Cl^-}(a_2) \rightarrow \mathrm{H^+}(a_1) + \mathrm{Cl^-}(a_1) \tag{8.20.13}$$

Note that the emf of the cell (Eq. 8.20.12) may be derived directly from the cell reaction (Eq. 8.20.13).

If one faraday of electricity is withdrawn from the cell, the net result that is produced is the transfer of 1 mol of each of hydrogen and chloride ions from the right-side cell to the left-side cell (Eq. 8.20.13). A cell of this type is called a *concentration cell without transference.* The operation of cell is not accompanied by the direct transfer of electrolyte from one solution to the other. The transference occurs indirectly as the result of chemical reactions at the electrodes. A comparison of the concentration cell with transference (Section 8.21) with that for the same solutions without transference shows that the change of type is achieved by introducing an intermediate electrode AgCl(s) | Ag | AgCl(s) between the two solutions. In general, a concentration cell with transference, in which the electrodes are reversible with respect to the cation, can be converted into one without transference by separating the two solutions by means of an electrode that is reversible with the anion of the electrolyte. If the end electrodes were the anion electrodes, then a cell without transference would be obtained by using a suitable cation electrode to separate the two solutions. One such example is

$$\mathrm{Ag} \mid \mathrm{AgCl(s)} \mid \mathrm{ZnCl_2}(a_1) \mid \mathrm{Zn} - \mathrm{Zn} \mid \mathrm{ZnCl_2}(a_2) \mid \mathrm{AgCl(s)} \mid \mathrm{Ag}$$

***Example 8.20.1***

Find the cell reaction and calculate the potential of the following cell at 298 K.

$$\mathrm{Cd(s)} \mid \mathrm{CdSO_4}(0.01\ \mathrm{mol\ kg^{-1}},\ \gamma = 0.383),\ \mathrm{PbSO_4(s)} \mid \mathrm{Pb} - \mathrm{Pb} \mid \mathrm{PbSO_4(s)},\ \mathrm{CdSO_4}(1.00\ \mathrm{mol\ kg^{-1}},\ \gamma = 0.042) \mid \mathrm{Cd(s)}$$

***Solution***

The cell reaction can be obtained by adding individual cell reactions.

*Cell reaction of left cell* We have

| *Electrode* | *Reduction reaction* | |
|---|---|---|
| Right | $\mathrm{PbSO_4(s)} + 2e^- \rightarrow \mathrm{Pb(s)} + (\mathrm{SO_4^{2-}})_L$ | (i) |
| Left | $(\mathrm{Cd^{2+}})_L + 2e^- \rightarrow \mathrm{Cd(s)}$ | (ii) |

Subtracting Eq. (ii) from Eq. (i), we get

$$\mathrm{PbSO_4(s)} + \mathrm{Cd(s)} \rightarrow \mathrm{Pb(s)} + (\mathrm{Cd^{2+}})_L + (\mathrm{SO_4^{2-}})_L \tag{iii}$$

*Cell reaction of right cell* We have

| *Electrode* | *Reduction reaction* | |
|---|---|---|
| Right | $(\mathrm{Cd^{2+}})_R + 2e^- \rightarrow \mathrm{Cd(s)}$ | (iv) |
| Left | $\mathrm{PbSO_4(s)} + 2e^- \rightarrow \mathrm{Pb(s)} + (\mathrm{SO_4^{2-}})_R$ | (v) |

Subtracting Eq. (v) from Eq. (iv), we get

$$\mathrm{Pb(s)} + (\mathrm{Cd^{2+}})_R + (\mathrm{SO_4^{2-}})_R \rightarrow \mathrm{PbSO_4(s)} + \mathrm{Cd(s)} \tag{vi}$$

Adding Eqs (iii) and (vi), we get the overall reaction as given below.

$$(\mathrm{Cd^{2+}})_R + (\mathrm{SO_4^{2-}})_R \rightarrow (\mathrm{Cd^{2+}})_L + (\mathrm{SO_4^{2-}})_L$$

The cell potential is given by

$$E_{\mathrm{cell}} = -\frac{RT}{2F} \ln \left\{ \frac{(a_{\mathrm{Cd^{2+}}})_L (a_{\mathrm{SO_4^{2-}}})_L}{(a_{\mathrm{Cd^{2+}}})_R (a_{\mathrm{SO_4^{2-}}})_R} \right\}$$

$$= -\frac{RT}{F} \ln \frac{(a_\pm)_L}{(a_\pm)_R} = -\frac{RT}{F} \ln \frac{(m_\pm)_L (\gamma_\pm)_L}{(m_\pm)_R (\gamma_\pm)_R}$$

Substituting the given data, we get

$$E_{\text{cell}} = -(0.059\,13\text{ V})\log\left\{\frac{(0.01)(0.383)}{(1.0)(0.042)}\right\} = 0.061\,5\text{ V}$$

***Example 8.20.2*** Show that the emf of the cell

$$\text{M(s)} \mid \text{M}_x\text{X}_y(a_{\pm 1}) \mid \text{A}_x\text{X}_n\text{(s)} \mid \text{A} - \text{A} \mid \text{A}_x\text{X}_n\text{(s)} \mid \text{M}_x\text{X}_y(a_{\pm 2}) \mid \text{M(s)}$$

is given by

$$E = -\left(\frac{x+y}{x}\right)\frac{RT}{yF}\ln\left(\frac{a_{\pm 1}}{a_{\pm 2}}\right)$$

***Solution*** The cell reaction can be obtained by adding the individual cell reactions.

***Cell reaction of left cell*** We have

| *Cell* | *Cell reaction* | |
|---|---|---|
| Right | $\frac{1}{xn}\text{A}_x\text{X}_n + \text{e}^- \rightarrow \frac{1}{n}\text{A} + \frac{1}{x}(\text{X}^{x-})_\text{L}$ | (i) |
| Left | $\frac{1}{y}(\text{M}^{y+})_\text{L} + \text{e}^- \rightarrow \frac{1}{y}\text{M}$ | (ii) |

Subtracting Eq. (ii) from Eq. (i), we get the cell reaction

$$\frac{1}{xn}\text{A}_x\text{X}_n + \frac{1}{y}\text{M} \rightarrow \frac{1}{n}\text{A} + \frac{1}{y}(\text{M}^{y+})_\text{L} + \frac{1}{x}(\text{X}^{x-})_\text{L}$$

***Cell reaction of right cell*** Here, the cell reaction will be just reverse of the above cell reaction, i.e.

$$\frac{1}{n}\text{A} + \frac{1}{y}(\text{M}^{y+})_\text{R} + \frac{1}{x}(\text{X}^{x-})_\text{R} \rightarrow \frac{1}{xn}\text{A}_x\text{X}_n + \frac{1}{y}\text{M}$$

Adding the two cell reactions, we get

$$\frac{1}{y}(\text{M}^{y+})_\text{R} + \frac{1}{x}(\text{X}^{x-})_\text{R} \rightarrow \frac{1}{y}(\text{M}^{y+})_\text{L} + \frac{1}{x}(\text{X}^{x-})_\text{L}$$

The cell potential will be given by

$$E = -\frac{RT}{F}\ln\frac{(a_{\text{M}^{y+}})_\text{L}^{1/y}(a_{\text{X}^{x-}})_\text{L}^{1/x}}{(a_{\text{M}^{y+}})_\text{R}^{1/y}(a_{\text{X}^{x-}})_\text{R}^{1/x}}$$

Multiplying and dividing the right side of the above expression by $xy$, we get

$$E = -\frac{1}{x}\frac{RT}{yF}\ln\frac{(a_{\text{M}^{y+}})_\text{L}^{x}(a_{\text{X}^{x-}})_\text{L}^{y}}{(a_{\text{M}^{y+}})_\text{R}^{x}(a_{\text{X}^{x-}})_\text{R}^{y}}$$

Now making use of the expression

$$a_{\pm}^{x+y}(M_xX_y) = (a_{M^{y+}})^x \, (a_{X^{x-}})^y$$

we get $$E = -\left(\frac{x+y}{x}\right)\frac{RT}{yF}\ln\left(\frac{a_{\pm 1}}{a_{\pm 2}}\right)$$

***Example 8.20.3***

Calculate the potential at 298 K of the following cell.

$$\text{Zn} \mid \text{ZnCl}_2(m = 0.01\ \text{mol kg}^{-1},\ \gamma_{\pm} = 0.708) \mid \text{AgCl} \mid \text{Ag} \mid \text{AgCl} \mid \text{ZnCl}_2(m = 0.10\ \text{mol kg}^{-1},\ \gamma_{\pm} = 0.502) \mid \text{Zn}$$

***Solution***

The cell reaction can be obtained by adding the individual cell reactions.

***Cell reaction of left cell*** We have

| *Electrode* | *Reduction reaction* | |
|---|---|---|
| Right | $2AgCl(s) + 2e^- \rightarrow 2Ag(s) + 2(Cl^-)_L$ | (i) |
| Left | $(Zn^{2+})_L + 2e^- \rightarrow Zn(s)$ | (ii) |

Subtracting Eq. (ii) from Eq. (i), we get

$$2AgCl(s) + Zn(s) \rightarrow 2Ag(s) + (Zn^{2+})_L + 2(Cl^-)_L \qquad \text{(iii)}$$

***Cell reaction of right cell*** We have

| *Electrode* | *Reduction reaction* | |
|---|---|---|
| Right | $(Zn^{2+})_R + 2e^- \rightarrow Zn(s)$ | (iv) |
| Left | $2AgCl(s) + 2e^- \rightarrow 2Ag(s) + 2(Cl^-)_R$ | (v) |

Subtracting Eq. (v) from Eq. (iv), we get

$$(Zn^{2+})_R + 2(Cl^-)_R + 2Ag(s) \rightarrow Zn(s) + 2AgCl(s) \qquad \text{(vi)}$$

Adding Eqs (iii) and (vi), we get the overall cell reaction:

$$(Zn^{2+})_R + 2(Cl^-)_R \rightarrow (Zn^{2+})_L + 2(Cl^-)_L$$

The cell potential is given by

$$E_{\text{cell}} = -\frac{RT}{2F}\ln\left\{\frac{(a_{Zn^{2+}})_L(a^2_{Cl^-})_L}{(a_{Zn^{2+}})_R(a^2_{Cl^-})_R}\right\} = -\frac{RT}{2F}\ln\left\{\frac{(a_{\pm})^3_L}{(a_{\pm})^3_R}\right\}$$

$$= -\frac{RT}{2F}\ln\left\{\frac{(4m'^3)_L(\gamma_{\pm})^3_L}{(4m'^3)_R(\gamma_{\pm})^3_R}\right\} = -\frac{3RT}{2F}\ln\left\{\frac{(m')_L(\gamma_{\pm})_L}{(m')_R(\gamma_{\pm})_R}\right\}$$

where $m' = m/m^\circ$. Substituting the given data, we get

$$E_{\text{cell}} = -\frac{3}{2}\times(0.059\,13\ \text{V})\times\log\left\{\frac{(0.01)(0.708)}{(0.10)(0.502)}\right\}$$

$$= 0.075\,5\ \text{V}$$

## 8.21 CONCENTRATION CELLS WITH LIQUID JUNCTION POTENTIAL

**Development of Liquid Junction Potential**

In a cell if two electrolytic solutions of different concentrations are in contact with each other, a potential difference develops across the boundary of the two solutions. This potential difference is called the *liquid junction potential* or the *diffusion potential.* It arises because of the difference in the rates of diffusion of positive and negative ions from more concentrated solution to less concentrated solution. The rate of diffusion of an ion is determined by its transference number. To illustrate how the liquid junction potential arises, we consider two hydrochloric acid solutions of different concentrations separated from each other by a boundary. If the boundary between the two solutions is an open one, then $H^+$ and $Cl^-$ ions will diffuse from more concentrated solution to less concentrated solution. Since $H^+$ ions can diffuse more rapidly than $Cl^-$ ions, an electrical double layer is formed at the boundary. This gives rise to the liquid junction potential. Once the electrical double layer is formed, it has an effect of decreasing the speed of faster moving ion and increasing the speed of slower moving ion. A steady state is established where both the ions migrate with the same speed; the ion which moved faster initially leads the march.

**Computation of Cell Potential**

The emf $E_{\text{cell(wlj)}}$ of a cell with liquid junction potential may be computed from the changes that take place in the cell when one faraday of electricity is withdrawn from the cell. We can determine the sum of the free-energy changes of various processes that take place in the cell and then equate it to $-FE_{\text{cell(wlj)}}$, such that

$$-FE_{\text{cell(wlj)}} = \sum_i \Delta G_i \qquad (8.21.1)$$

**Computation of Liquid Junction Potential**

The liquid junction potential may be determined as follows.

The given cell with liquid junction potential is converted into a cell without liquid junction by connecting the two electrolytic solutions through a salt bridge. The emf $E_{\text{cell(wolj)}}$ of the resultant cell without liquid junction potential is determined. The difference of $E_{\text{cell(wlj)}}$ and $E_{\text{cell(wolj)}}$ gives the value of liquid junction potential $E_{\text{lj}}$, i.e.

$$E_{\text{lj}} = E_{\text{cell(wlj)}} - E_{\text{cell(wolj)}} \qquad (8.21.2)$$

We now consider the following two types of cells with liquid junction and derive the expressions of $E_{\text{cell(wlj)}}$, $E_{\text{cell(wolj)}}$ and $E_{\text{lj}}$.

### 1. CELL IN WHICH ELECTRODES ARE REVERSIBLE WITH RESPECT TO CATION

**A Typical Example**

Consider the cell

$$\text{Pt} \mid H_2(1\ \text{bar}) \mid \text{HCl}(a_1) \,\vdots\, \text{HCl}\ (a_2) \mid H_2(1\ \text{bar}) \mid \text{Pt}$$

The above cell is also shown in Fig. 8.21.1.

**Working of the Cell**

We know that electrons are given out from the left electrode and are fed back at the right electrode, i.e. electrons flow externally from left to right. Inside the electrochemical cell, the current is carried by ions. In order to have a complete

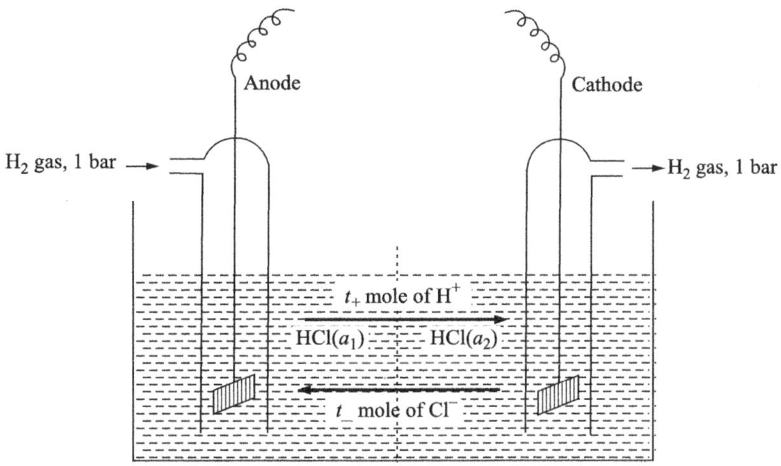


**Fig. 8.21.1 Cell in which electrodes are reversible with respect to cations**

circuit of flow of electrons, it is obvious that negative ions (which carry the surplus electrons) will move from right to left and positive ions will move from left to right. Let $t_+$ and $t_-$ be the transference numbers of positive and negative ions, respectively. If one faraday of electricity is withdrawn, it is obvious that $t_+$ mole of positive ions will migrate across the liquid junction from left to right and $t_-$ mole of negative ions from right to left. The changes that take place when one faraday of electricity is withdrawn from the cell given in Fig. 8.21.1 are described below.

(i) *Electrode reaction at anode* $\quad \frac{1}{2}H_2(1\ \text{bar}) \rightarrow H^+(a_1) + e^-$

(ii) *Electrode reaction at cathode* $\quad H^+(a_2) + e^- \rightarrow \frac{1}{2}H_2(1\ \text{bar})$

(iii) *Transfer of* $t_+$ mole *of* $H^+$ *from left to right* $\quad t_+H^+(a_1) \rightarrow t_+H^+(a_2)$

(iv) *Transfer of* $t_-$ mole *of* $Cl^-$ *from right to left* $\quad t_-Cl^-(a_2) \rightarrow t_-Cl^-(a_1)$

The net change in the cell is obtained by adding the above four changes. Thus, we have

$$\frac{1}{2}H_2(1\ \text{bar}) + H^+(a_2) + t_+H^+(a_1) + t_-Cl^-(a_2) \rightarrow H^+(a_1) + \frac{1}{2}H_2(1\ \text{bar}) + t_+H^+(a_2) + t_-Cl^-(a_1)$$

Cancelling the common term, we get

$$H^+(a_2) + t_+H^+(a_1) + t_-Cl^-(a_2) \rightarrow H^+(a_1) + t_+H^+(a_2) + t_-Cl^-(a_1)$$

which on rearranging gives

$$H^+(a_2) - t_+H^+(a_2) + t_-Cl^-(a_2) \rightarrow H^+(a_1) - t_+H^+(a_1) + t_-Cl^-(a_1)$$

or $$(1 - t_+)H^+(a_2) + t_-Cl^-(a_2) \rightarrow (1 - t_+)H^+(a_1) + t_-Cl^-(a_1)$$

or $$t_-H^+(a_2) + t_-Cl^-(a_2) \rightarrow t_-H^+(a_1) + t_-Cl^-(a_1)$$

Thus, the net cell reaction is to transfer $t_-$ mole of HCl from the solution of activity $a_2$ to that of activity $a_1$.

**Free-Energy of Cell Reaction**

The total free energy change of the net cell reaction is

$$\Delta G = t_-[\mu_{H^+(a_1)} + \mu_{Cl^-(a_1)} - \mu_{H^+(a_2)} - \mu_{Cl^-(a_2)}]$$

$$= t_-[\{\mu^\circ_{H^+} + RT \ln (a_1)_{H^+}\} + \{\mu^\circ_{Cl^-} + RT \ln (a_1)_{Cl^-}\}$$

$$- \{\mu^\circ_{H^+} + RT \ln (a_2)_{H^+}\} - \{\mu^\circ_{Cl^-} + RT \ln (a_2)_{Cl^-}\}]$$

$$= t_- RT \ln \left\{ \frac{(a_1)_{H^+}(a_1)_{Cl^-}}{(a_2)_{H^+}(a_2)_{Cl^-}} \right\}$$

$$= 2t_-\ RT \ln \frac{(a_{\pm 1})_{HCl}}{(a_{\pm 2})_{HCl}} \tag{8.21.3}$$

Substituting Eq. (8.21.3) in Eq. (8.21.1), we get

$$E_{cell(wlj)} = -2t_- \frac{RT}{F} \ln \frac{(a_{\pm 1})_{HCl}}{(a_{\pm 2})_{HCl}} \tag{8.21.4}$$

**Cell Without Liquid Junction Potential**

Now if the two solutions of the cell given in Fig. 8.21.1 are connected through a salt bridge, the liquid junction potential is eliminated. If now one faraday of electricity is withdrawn, the changes in the two half-cells will only be due to the electrode reactions. Thus, we have

*Electrode reaction at anode* $\quad \frac{1}{2}H_2(1\ \text{bar}) \rightarrow H^+(a_1) + e^-$

*Electrode reaction at cathode* $\quad H^+(a_2) + e^- \rightarrow \frac{1}{2}H_2(1\ \text{bar})$

The net cell reaction is

$$H^+(a_2) \rightarrow H^+(a_1)$$

and the cell potential is

$$E_{cell(wolj)} = -\frac{RT}{F} \ln \frac{(a_1)_{H^+}}{(a_2)_{H^+}} = -\frac{RT}{F} \ln \frac{(a_{\pm 1})_{HCl}}{(a_{\pm 2})_{HCl}} \tag{8.21.5}$$

where $a_{H^+}$ has been replaced by the mean ionic activity $a_\pm$.

**Expression of Liquid Junction Potential**

Making use of Eq. (8.21.2), we get

$$E_{lj} = E_{cell(wlj)} - E_{cell(wolj)}$$

$$= -2t_- \frac{RT}{F} \ln \frac{(a_{\pm 1})_{HCl}}{(a_{\pm 2})_{HCl}} + \frac{RT}{F} \ln \frac{(a_{\pm 1})_{HCl}}{(a_{\pm 2})_{HCl}}$$

$$= (1 - 2t_-) \frac{RT}{F} \ln \frac{(a_{\pm 1})_{HCl}}{(a_{\pm 2})_{HCl}} \tag{8.21.6}$$

Since $t_+ + t_- = 1$, Eq. (8.21.6) may be written as

$$E_{lj} = (t_+ - t_-) \frac{RT}{F} \ln \frac{(a_{\pm 1})_{HCl}}{(a_{\pm 2})_{HCl}} \tag{8.21.7}$$

Comparing Eqs (8.21.4) and (8.21.5), we get

$$E_{cell(wlj)} = 2t_- \, E_{cell(wolj)} \tag{8.21.8}$$

If the cell without liquid junction is to function spontaneously, we must have

$$(a_{\pm 2})_{HCl} > (a_{\pm 1})_{HCl}$$

since only then $E_{cell(wolj)}$ as given by Eq. (8.21.5) will have a positive value. Since $t_{H^+}$ is greater than $t_{Cl^-}$, it follows from Eq. (8.21.7) that the liquid junction potential $E_{lj}$ will have a negative value.

**Comment on Liquid Junction Potential**

In general, the sign and magnitude of the liquid junction potential depends on the transport numbers of the involved cations and anions. In a cell with $(a_{\pm 2})_{HCl} > (a_{\pm 1})_{HCl}$, we will have

$$E_{lj} \text{ positive} \quad \text{if } t_- > t_+$$

$$E_{lj} \text{ negative} \quad \text{if } t_- < t_+$$

and

$$E_{lj} \text{ zero} \quad \text{if } t_- = t_+ \tag{8.21.9}$$

From Eq. (8.21.8), we get

$$E_{cell(wlj)} > E_{cell(wolj)} \text{ if } t_- > t_+$$

$$E_{cell(wlj)} < E_{cell(wolj)} \text{ if } t_- < t_+$$

and

$$E_{cell(wlj)} = E_{cell(wolj)} \text{ if } t_- = t_+ \tag{8.21.10}$$

The results of Eq. (8.21.10) also follow from Fig. 8.21.2 where the separation of charges across the boundary of the two solutions is shown. The separation of charges results from the diffusion of ions from a more concentrated solution (right electrode) to a less concentrated solution (left electrode). When $t_- > t_+$, the separation of charges across the boundary matches with that of the electrodes and thus the two potentials, viz., liquid junction $E_{lj}$ and cell potential $E_{cell(wolj)}$ have the same sign. Thus on adding the two potentials, we get $E_{cell(wlj)}$ which is greater than $E_{cell(wolj)}$. On the other hand, when $t_- < t_+$, the separation charges across the boundary is opposite to that of electrodes and thus the two potentials have different signs which on adding gives $E_{cell(wlj)} < E_{cell(wolj)}$.

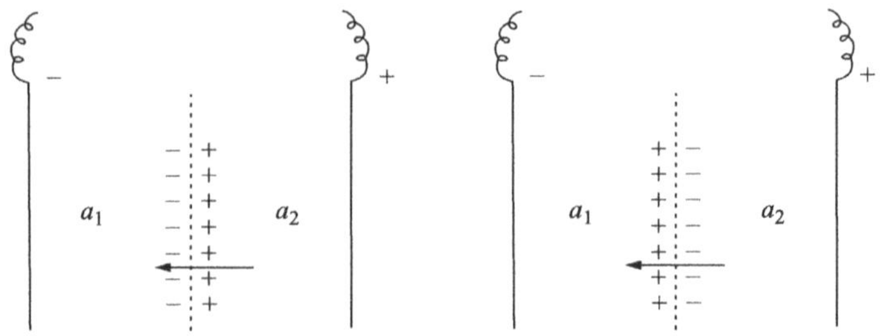


**Fig. 8.21.2 Correlation of the sign of liquid junction potential with the separation of charges at the boundary;**
**(i) $t_- > t_+$, $E_{lj}$ is positive and $E_{cell(wlj)} > E_{cell(wolj)}$ and**
**(ii) $t_- < t_+$, $E_{lj}$ is negative and $E_{cell(wlj)} < E_{cell(wolj)}$**

## 2. CELL IN WHICH ELECTRODES ARE REVERSIBLE WITH RESPECT TO ANIONS

### A Typical Example

Consider the cell

$$\text{Ag} \mid \text{AgCl} \mid \text{HCl}(a_1) \,\vdots\, \text{HCl}(a_2) \mid \text{AgCl} \mid \text{Ag}$$

The above call is also shown in Fig. 8.21.3.

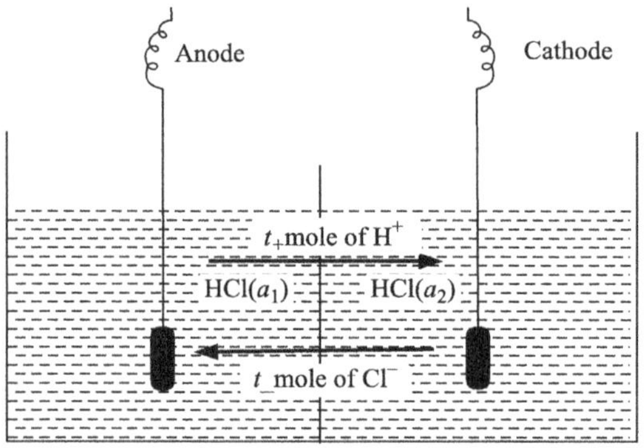


**Fig. 8.21.3 A cell in which electrodes are reversible with respect to anions**

### Working of the Cell

The changes that take place when one faraday of electricity is withdrawn from the cell are shown below:

| | | |
|---|---|---|
| (i) | Electrode reaction at anode | $Ag(s) + Cl^-(a_1) \rightarrow AgCl(s) + e^-$ |
| (ii) | Electrode reaction at cathode | $AgCl(s) + e^- \rightarrow Ag(s) + Cl^-(a_2)$ |
| (iii) | Migration of $H^+$ ions | $t_+H^+(a_1) \rightarrow t_+H^+(a_2)$ |
| (iv) | Migration of $Cl^-$ ions | $t_-Cl^-(a_2) \rightarrow t_-Cl^-(a_1)$ |

The net effect is obtained by adding the above four changes, which gives

$$Cl^-(a_1) + t_+H^+(a_1) + t_-Cl^-(a_2) \rightarrow Cl^-(a_2) + t_+H^+(a_2) + t_-Cl^-(a_1)$$

or

$$t_+H^+(a_1) + Cl^-(a_1) - t_-Cl^-(a_1) \rightarrow t_+H^+(a_2) + Cl^-(a_2) - t_-Cl^-(a_2)$$

$$t_+H^+(a_1) + t_+Cl^-(a_1) \rightarrow t_+H^+(a_2) + t_+Cl^-(a_2)$$

Thus, the net cell reaction is to transfer $t_+$ mole of HCl from the solution of activity $a_1$ to that of activity $a_2$.

The free energy change of the net cell reaction is

$$\Delta G = t_+ \, RT \ln \frac{(a_2)_{H^+}(a_2)_{Cl^-}}{(a_1)_{H^+}(a_1)_{Cl^-}} = 2t_+ \, RT \ln \frac{(a_{\pm 2})_{HCl}}{(a_{\pm 1})_{HCl}}$$

Hence $$E_{cell(wlj)} = -\frac{\Delta G}{F} = -2t_+ \frac{RT}{F} \ln \frac{(a_{\pm 2})_{HCl}}{(a_{\pm 1})_{HCl}} \tag{8.21.11}$$

**Cell Without Liquid Junction Potential**

If the cell given by Fig. 8.21.3 functions without liquid junction potential, its cell reaction would be

$$Cl^-(a_1) \rightarrow Cl^-(a_2)$$

The cell potential would be

$$E_{cell(wolj)} = -\frac{RT}{F} \ln \frac{(a_2)_{Cl^-}}{(a_1)_{Cl^-}} = -\frac{RT}{F} \ln \frac{(a_{\pm 2})_{HCl}}{(a_{\pm 1})_{HCl}} \tag{8.21.12}$$

where $a_{Cl^-}$ has been replaced by the mean ionic activity $a_\pm$.

**Expression of Liquid Junction Potential**

Now since

$$E_{lj} = E_{cell(wlj)} - E_{cell(wolj)}$$

we get

$$E_{lj} = -2t_+ \frac{RT}{F} \ln \frac{(a_{\pm 2})_{HCl}}{(a_{\pm 1})_{HCl}} + \frac{RT}{F} \ln \frac{(a_{\pm 2})_{HCl}}{(a_{\pm 1})_{HCl}}$$

or $$E_{lj} = (1 - 2t_+) \frac{RT}{F} \ln \frac{(a_{\pm 2})_{HCl}}{(a_{\pm 1})_{HCl}} \tag{8.21.13}$$

or $$E_{lj} = (t_- - t_+) \frac{RT}{F} \ln \frac{(a_{\pm 2})_{HCl}}{(a_{\pm 1})_{HCl}} \tag{8.21.14}$$

Combining Eqs (8.20.11) and (8.20.12), we get

$$E_{cell(wlj)} = 2t_+ \, E_{cell(wolj)} \tag{8.21.15}$$

If the cell without liquid junction is to function spontaneously, we must have

$$(a_{\pm 1})_{HCl} > (a_{\pm 2})_{HCl}$$

since only then $E_{cell(wolj)}$ as given by Eq. (8.21.12) will have a positive value.

Since $t_{H^+}$ is greater than $t_{Cl^-}$, it follows from Eq. (8.21.14) that the liquid junction potential $E_{lj}$ will have a positive value.

**Comment on Liquid Junction Potential**

In general, the sign and magnitude of the liquid function potential depends on the transference numbers of involved cations and anions. In a cell with $(a_{\pm 1})_{HCl} > (a_{\pm 2})_{HCl}$, we will have

$$E_{lj} \text{ positive if } t_+ > t_-$$
$$E_{lj} \text{ negative if } t_+ < t_-$$
and $$E_{lj} \text{ zero} \quad \text{if } t_+ = t_- \tag{8.21.16}$$

From Eq. (8.21.15) we get

$$E_{cell(wlj)} > E_{cell(wolj)} \text{ if } t_+ > t_-$$
$$E_{cell(wlj)} < E_{cell(wolj)} \text{ if } t_+ < t_-$$
and $$E_{cell(wlj)} = E_{cell(wolj)} \text{ if } t_+ = t_- \tag{8.21.17}$$

The results of Eq. (8.21.17) also follows from Fig. 8.21.4 where the separation of charges across the boundary of the two solutions is shown as a result of migration ions.

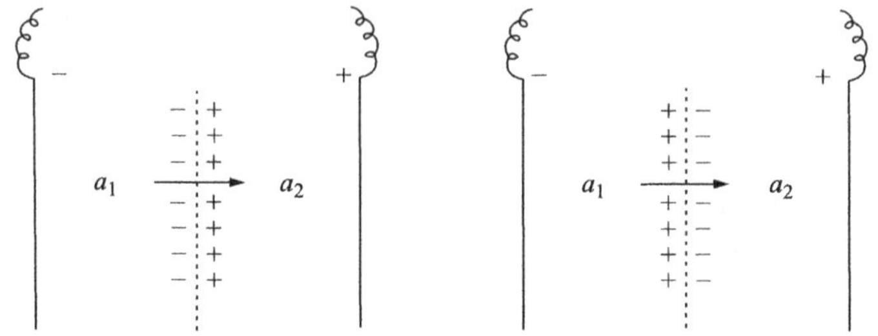


**Fig. 8.21.4 Correlation of the sign of liquid junction potential with the separation of charges at the boundary; (i) $t_+ > t_-$, $E_{lj}$ is positive and $E_{cell(wlj)} > E_{cell(wolj)}$ and (ii) $t_+ < t_-$, $E_{lj}$ is negative and $E_{cell(wlj)} < E_{cell(wolj)}$**

## Generalization of Results

From Eqs (8.21.4) and (8.21.11), it may be observed that the transference number which appears in expression of $E_{cell(wlj)}$ is that of the ionic species with respect to which the electrodes are not reversible. Expressions of $E_{cell(wlj)}$, $E_{cell(wolj)}$ and $E_{lj}$ of both types of cells discussed above may be written together as shown below:

$$E_{cell(wlj)} = \pm\, 2t_{\mp} \frac{RT}{F} \ln \frac{(a_{\pm 2})_{HCl}}{(a_{\pm 1})_{HCl}} \tag{8.21.18}$$

$$E_{cell(wolj)} = \pm \frac{RT}{F} \ln \frac{(a_{\pm 2})_{HCl}}{(a_{\pm 1})_{HCl}} \tag{8.21.19}$$

$$E_{cell(wlj)} = 2t_{\mp}\, E_{cell(wolj)} \tag{8.21.20}$$

$$E_{lj} = \mp (1 - 2t_{\mp}) \frac{RT}{F} \ln \frac{(a_{\pm 2})_{HCl}}{(a_{\pm 1})_{HCl}}$$
$$= (t_- - t_+) \frac{RT}{F} \ln \frac{(a_{\pm 2})_{HCl}}{(a_{\pm 1})_{HCl}} \tag{8.21.21}$$

where the upper and lower signs are meant for cells involving electrodes reversible with respect to cations and anions, respectively.

**Use of KCl (or $NH_4NO_3$) in Salt Bridge**

From Eqs (8.21.9) and (8.21.16), it may be observed that $E_{lj}$ is zero if $t_+ = t_-$. This fact forms the basis of using potassium chloride (or ammonium nitrate) as a bridge in cell measurements involving liquid junction, since the two transport numbers are very nearly identical in this electrolyte.

---

***Example 8.21.1***

Show that the emf of cell

$$M(s) \mid M_xX_y(a_{\pm 1}) \,\vdots\, M_xX_y(a_{\pm 2}) \mid M(s)$$

is given by

$$E = -t_-\left(\frac{x+y}{x}\right)\frac{RT}{yF}\ln\left(\frac{a_{\pm 1}}{a_{\pm 2}}\right)$$

***Solution***

Since the electrode reactions are reversible with respect to cation of charge number $y$ we write, for the sake of simplicity, the cell reaction for the withdrawal of $y$ faraday of electricity. The various reactions are:

(i) *Reaction at left cell*

$$M \rightarrow (M^{y+})_L + ye^- \qquad (1)$$

(ii) *Reaction at right cell*

$$(M^{y+})_R + ye^- \rightarrow M \qquad (2)$$

(iii) *Transfer of cations from the left to right cell*

$$yt_+ \text{ equivalent of } (M^{y+})_L \rightarrow yt_+ \text{ equivalent of } (M^{y+})_R$$

or $t_+$ mole of $(M^{y+})_L \rightarrow t_+$ mole of $(M^{y+})_R$ i.e. $t_+(M^{y+})_L \rightarrow t_+(M^{y+})_R$ (3)

(iv) *Transfer of anions from the right to left cell*

$$yt_- \text{ equivalent of } (X^{x-})_R \rightarrow yt_- \text{ equivalent of } (X^{x-})_L$$

or $\frac{yt_-}{x}$ mole of $(X^{x-})_R \rightarrow \frac{yt_-}{x}$ mole of $(X^{x-})_L$ i.e. $\left(\frac{yt_-}{x}\right)(X^{x-})_R \rightarrow \left(\frac{yt_-}{x}\right)(X^{x-})_L$ (4)

Adding Eqs (1) to (4) and rearranging, we get

$$t_-(M^{y+})_R + \left(\frac{yt_-}{x}\right)(X^{x-})_R \rightarrow t_-(M^{y+})_L + \left(\frac{yt_-}{x}\right)(X^{x-})_L$$

The free energy change is given by

$$\Delta G = \left\{t_-(\mu_{M^{y+}})_L + \left(\frac{yt_-}{x}\right)(\mu_{X^{x-}})_L\right\} - \left\{t_-(\mu_{M^{y+}})_R + \left(\frac{yt_-}{x}\right)(\mu_{X^{x-}})_R\right\}$$

Writing $\mu$ in terms of $\mu^\circ$ and activity and cancelling the $\mu^\circ$ terms, we get

$$\Delta G = t_-RT\ln\frac{(a_{M^{y+}})_L}{(a_{M^{y+}})_R} + \frac{yt_-}{x}RT\ln\frac{(a_{X^{x-}})_L}{(a_{X^{x-}})_R}$$

Multiplying and dividing the first term by $x$ and merging both the terms, we have

$$\Delta G = \frac{t_-}{x} RT \ln \frac{(a_{M^{y+}})_L^x (a_{X^{x-}})_L^y}{(a_{M^{y+}})_R^x (a_{X^{x-}})_R^y}$$

Making use of the expression

$$(a_\pm)_{M_xM_y}^{x+y} = (a_{M^{y+}})^x (a_{X^{x-}})^y$$

we get $$\Delta G = t_-\left(\frac{x+y}{x}\right) RT \ln \frac{a_{\pm 1}}{a_{\pm 2}}$$

Since $y$ faraday of electricity has been withdrawn, we have

$$\Delta G = -\, yFE$$

and thus

$$E = -\, t_-\left(\frac{x+y}{x}\right) \frac{RT}{yF} \ln\left(\frac{a_{\pm 1}}{a_{\pm 2}}\right)$$

---

***Example 8.21.2***

Find the cell reaction and calculate the potential of following cell with transference at 298 K.

$$Pb(s) \mid PbSO_4(s), CuSO_4(0.2 \text{ mol kg}^{-1}, \gamma = 0.110) \;\vdots\; CuSO_4(0.02 \text{ mol kg}^{-1}, \gamma = 0.320), PbSO_4(s) \mid Pb$$

The transport number of $Cu^{2+}$ is 0.370. Is the cell reaction spontaneous, as written? What would be the cell potential if the above cell is converted into one without liquid junction potential?

***Solution***

The cell reaction can be obtained by adding various reactions that take place when one faraday of electricity is withdrawn from the cell. The various reactions are:

(i) *Reaction at left electrode (anode)*

$$\tfrac{1}{2}Pb(s) + \tfrac{1}{2}(SO_4^{2-})_L \rightarrow \tfrac{1}{2}PbSO_4(s) + e^- \qquad \text{(i)}$$

(ii) *Reaction at right electrode (cathode)*

$$\tfrac{1}{2}PbSO_4(s) + e^- \rightarrow \tfrac{1}{2}Pb(s) + \tfrac{1}{2}(SO_4^{2-})_R \qquad \text{(ii)}$$

(iii) *Transference of cations*

$t_+$ equivalent of $(Cu^{2+})_L \rightarrow t_+$ equivalent of $(Cu^{2+})_R$

or $(t_+/2)$ mole of $(Cu^{2+})_L \rightarrow (t_+/2)$ mole of $(Cu^{2+})_R$ (iii)

(iv) *Transference of anions*

$t_-$ equivalent of $(SO_4^{2-})_R \rightarrow t_-$ equivalent of $(SO_4^{2-})_L$

or $(t_-/2)$ mole of $(SO_4^{2-})_R \rightarrow (t_-/2)$ mole of $(SO_4^{2-})_L$ (iv)

Adding Eqs (i) to (iv) and rearranging, we get

$$\frac{t_+}{2}(\mathrm{Cu}^{2+})_\mathrm{L} + \frac{t_+}{2}(\mathrm{SO}_4^{2-})_\mathrm{L} \rightarrow \frac{t_+}{2}(\mathrm{Cu}^{2+})_\mathrm{R} + \frac{t_+}{2}(\mathrm{SO}_4^{2-})_\mathrm{R}$$

The cell potential is given by

$$E_{\mathrm{cell(wlj)}} = -\frac{t_+}{2}\frac{RT}{F}\ln\left\{\frac{(a_{\mathrm{Cu}^{2+}})_\mathrm{R}(a_{\mathrm{SO}_4^{2-}})_\mathrm{R}}{(a_{\mathrm{Cu}^{2+}})_\mathrm{L}(a_{\mathrm{SO}_4^{2-}})_\mathrm{L}}\right\} = -t_+\frac{RT}{F}\ln\left\{\frac{(a_\pm)_\mathrm{R}}{(a_\pm)_\mathrm{L}}\right\}$$

$$= -t_+\frac{RT}{F}\ln\left\{\frac{(m'_\pm)_\mathrm{R}(\gamma_\pm)_\mathrm{R}}{(m'_\pm)_\mathrm{L}(\gamma_\pm)_\mathrm{L}}\right\} = -t_+\frac{RT}{F}\ln\left\{\frac{(m')_\mathrm{R}(\gamma_\pm)_\mathrm{R}}{(m')_\mathrm{L}(\gamma_\pm)_\mathrm{L}}\right\}$$

where $m' = m/m^\circ$. Substituting the given data, we get

$$E_{\mathrm{cell(wlj)}} = -(0.370)\,(0.059\,13\ \mathrm{V})\log\left\{\frac{(0.02)(0.320)}{(0.2)(0.110)}\right\}$$

$$= 0.011\,7\ \mathrm{V}$$

Since $E_{\mathrm{cell(wlj)}}$ positive, the cell reaction will be spontaneous. Now since

$$E_{\mathrm{cell(wlj)}} = 2t_+E_{\mathrm{cell(wolj)}}$$

if follows that

$$E_{\mathrm{cell(wolj)}} = \frac{E_{\mathrm{cell(wlj)}}}{2t_+} = \frac{0.011\,7\ \mathrm{V}}{2\times 0.370} = 0.015\,81\ \mathrm{V}$$

## 8.22 COMMERCIAL CELLS

In this section, we describe some of the electrochemical cells which are commonly employed as sources of electrical energy. The electrochemical cells may be broadly classified into two categories as described below.

### THE PRIMARY CELLS

Such types of cells can be used only so long the active materials are present. Once these are consumed, the cell cannot be recharged by passage of current through it, and hence it has to be discarded.

**Leclanche or Dry Cell**

One of the examples is the Leclanche or dry cell represented as

$$\mathrm{Zn} \mid \mathrm{NH_4Cl(20\%)},\ \mathrm{ZnCl_2} \mid \mathrm{MnO_2} \mid \mathrm{C}$$

The reactions involved are:

*Anode* $\mathrm{Zn(s)} \rightarrow \mathrm{Zn}^{2+} + 2\mathrm{e}^-$

*Cathode* $2\mathrm{MnO_2(s)} + \mathrm{H_2O(1)} + 2\mathrm{e}^- \rightarrow \mathrm{Mn_2O_3(s)} + 2\mathrm{OH}^-$

*Overall* $\mathrm{Zn(s)} + 2\mathrm{MnO_2(s)} + \mathrm{H_2O(1)} \rightarrow \mathrm{Zn}^{2+} + 2\mathrm{OH}^- + \mathrm{Mn_2O_3(s)}$

The $OH^-$ generated in the above electrochemical reaction causes the following secondary reactions:

$$NH_4Cl + OH^- \rightarrow NH_3 + Cl^- + H_2O$$

$$Zn^{2+} + 2NH_3 + 2Cl^- \rightarrow Zn(NH_3)_2Cl_2$$

Since these reactions are not involved directly in the electrode reactions, they do not contribute anything towards the cell emf of 1.5 V. Since the cathode potential is a function of pH, the cell potential falls rapidly on continuous discharge.

**Ruben-Mallory Cell**

Another example of the primary cell which provides more constant voltage is the Ruben-Mallory cell. Here the use of a large excess of hydroxyl ions makes the cell less sensitive to pH change. The cell is represented as

$$Hg \mid HgO \mid KOH \mid Zn(OH)_2 \mid Zn$$

The reactions involved are:

*Anode* $Hg(l) + 2OH^-(aq) \rightarrow HgO(s) + H_2O(l) + 2e^-$

*Cathode* $Zn(OH)_2(s) + 2e^- \rightarrow Zn(s) + 2OH^-(aq)$

*Overall* $Hg(l) + Zn(OH)_2(s) \rightarrow HgO(s) + Zn(s) + H_2O(l)$

## THE SECONDARY CELLS

Such types of cells can be used again and again by recharging the cell.

**Load Storage Cell**

One of the examples is the lead storage cell where lead acts as anode and lead impregnated with lead dioxide acts as cathode. The electrolyte is a solution of approximately 20 per cent sulphuric acid with a specific gravity of about 1.15 at room temperature. The reactions involved are:

*Anode* $Pb(s) + SO_4^{2-}(aq) \rightarrow PbSO_4(s) + 2e^-$

*Cathode* $PbO_2(s) + 4H^+(aq) + SO_4^{2-}(aq) + 2e^- \rightarrow PbSO_4(s) + 2H_2O(l)$

*Overall* $Pb(s) + PbO_2(s) + 2H_2SO_4(aq) \rightarrow 2PbSO_4(s) + 2H_2O(l)$

The emf of the cell depends on the activity of sulphuric acid in solution. At 25 °C, some of the values are 1.90 V at 7.4 per cent $H_2SO_4$, 2.0 V at 21.4 per cent $H_2SO_4$, and 2.14 V at 39.2 per cent $H_2SO_4$.

To recharge the cell, it is connected with a cell of potential higher than that of the cell in such a way that the lead storage cell now acts as an electrolytic cell where lead is deposited on the cathode, lead dioxide is formed at the anode and $H_2SO_4$ is regenerated. The specific gravity of the solution rises due to the increase in $H_2SO_4$ concentration. The recharging is carried out only up to a certain specific gravity value.

**Nicad Cell**

Another example of the secondary cell is the 'nicad' or nickel-cadmium battery. The anode and cathode are Cd and $NiO_2$, respectively, and the electrolytic solution is KOH. The reactions are:

| | |
|---|---|
| *Anode* | $Cd(s) + 2OH^-(aq) \rightarrow Cd(OH_2(s) + 2e^-$ |
| *Cathode* | $NiO_2(s) + 2H_2O(1) + 2e^- \rightarrow Ni(OH)_2(s) + 2OH^-(aq)$ |
| *Overall* | $Cd(s) + NiO_2(s) + 2H_2O(1) \rightarrow Cd(OH)_2(s) + Ni(OH)_2(s)$ |

The emf of the cell is 1.4 V. This cell has comparatively longer life than a lead storage cell and can be packed in a sealed unit. This cell find its much applications in electronic calculators and electronic flash.

### A Few More Examples

A few other examples of secondary cells are:

1. Edison cell (anode—iron, cathode—nickel oxide, electrolyte—KOH solution).
2. Silver-Zinc cell (anode—zinc, cathode—silver oxide, electrolyte—KOH solution).
3. Silver-Cadmium cell (anode—cadmium, cathode—silver oxide, electrolyte—KOH solution).

The term *battery* is used wherever a number of cells are connected in series or in series-parallel.

### Fuel Cell

Fuel cells are another means by which chemical energy may be converted into electrical energy. Here the cathode and anode constituents are continually supplied and thus energy can be withdrawn indefinitely from a fuel cell as long as the outside supply of fuel is maintained. One of the examples is the hydrogen-oxygen fuel cell. The cell consists of three compartments separated from one another by porous electrodes. The hydrogen gas is fed into one compartment and the oxygen gas is fed into another compartment. These gases then diffuse slowly through the electrodes and react with an electrolyte that is in the central compartment. The electrodes are made of a conducting material, such as graphite, with a sprinkling of platinum to act as a catalyst, and the electrolyte is an aqueous solution of a base. The reactions are:

| | |
|---|---|
| *Anode* | $H_2(g) + 2OH^-(aq) \rightarrow 2H_2O(1) + 2e^-$ |
| *Cathode* | $O_2(g) + 2H_2O(1) + 4e^- \rightarrow 4OH^-(aq)$ |
| *Overall* | $2H_2(g) + O_2(g) \rightarrow 2H_2O(1)$ |

One of the advantages of the fuel cells is that the energy is extracted from the reactants under more nearly reversible conditions. This type of cell finds great importance in space vehicles.

## 8.23 SOLVED PROBLEMS

**1.** A half-cell A consisting of a strip of cobalt dipped into a 1 M solution of $Co^{2+}$ and a half-cell B consisting of a strip of chromium dipped into a 1 M solution of $Cr^3$. The magnitudes of the individual half-cell potentials were then determined as

$$Co^{2+} + 2e^- \rightarrow Co; \quad |E^\circ| = 0.277 \text{ V}$$

$$Cr^{3+} + 3e^- \rightarrow Cr; \quad |E^\circ| = 0.744 \text{ V}$$

Answer the following:

(a) When both half-cells were connected separately with the standard hydrogen half-cell, the metallic electrode was found to be negative. What are the correct signs of the electrode potentials?

(b) Will a reaction occur when metallic cobalt is placed in a 1 M solution of $Cr^{3+}$ and when metallic chromium is dipped into a 1 M solution of $Co^{2+}$?
(c) Of the substances, Co, $Co^{2+}$, Cr and $Cr^{3+}$, which one is the strongest oxidant and the strongest reductant?
(d) Cobaltous ion forms complex with ammonia $[Co(NH_3)_6]^{2+}$. If ammonium hydroxide were added to half-cell A, would its electrode potential become more positive, less positive or be unaffected?
(e) If the half-cells A and B were connected together, which electrode would be negative? What would be the cell voltage? Suppose the above cell is connected with an external battery of voltage 1.0 V with (i) negative terminal of the cell attached to the negative terminal of the external battery, and (ii) negative terminal of the cell attached to the positive terminal of the external battery. What would be the cell reaction in the two cases?

***Solution*** (**a**) Since the metallic electrode was found to be negative, we may write the cell as

$$\mathrm{Co} \mid \mathrm{Co}^{2+} \,\vdots\vdots\, \mathrm{H}^+ \mid \mathrm{H}_2 \mid \mathrm{Pt} \qquad \text{(i)}$$

$$\mathrm{Cr} \mid \mathrm{Cr}^{3+} \,\vdots\vdots\, \mathrm{H}^+ \mid \mathrm{H}_2 \mid \mathrm{Pt} \qquad \text{(ii)}$$

The emfs of the above two cells will be positive and are given by

$$E^\circ_{\mathrm{cell(i)}} = E^\circ_{\mathrm{H^+ \mid H_2 \mid Pt}} - E^\circ_{\mathrm{Co^{2+} \mid Co}} = -\,E^\circ_{\mathrm{Co^{2+} \mid Co}}$$

$$E^\circ_{\mathrm{cell(ii)}} = E^\circ_{\mathrm{H^+ \mid H_2 \mid Pt}} - E^\circ_{\mathrm{Cr^{3+} \mid Cr}} = -\,E^\circ_{\mathrm{Cr^{3+} \mid Cr}}$$

Since $E^\circ_{\mathrm{cells}}$ are positive, it is obvious that the potentials $E^\circ_{\mathrm{Co^{2+} \mid Co}}$ and $E^\circ_{\mathrm{Cr^{3+} \mid Cr}}$ must be negative. Hence the values of these potentials are

$$E^\circ_{\mathrm{Co^{2+} \mid Co}} = -\,0.277\ \mathrm{V}$$

$$E^\circ_{\mathrm{Cr^{3+} \mid Cr}} = -\,0.744\ \mathrm{V}$$

(**b**) The expected reactions would be

(i) $3Co + 2Cr^{3+} \rightarrow 3Co^{2+} + 2Cr$

and (ii) $2Cr + 3Co^{2+} \rightarrow 3Co + 2Cr^{3+}$

The cells giving these reactions would be

(i) $\mathrm{Co} \mid \mathrm{Co}^{2+} \,\vdots\vdots\, \mathrm{Cr}^{3+} \mid \mathrm{Cr}$

(ii) $\mathrm{Cr} \mid \mathrm{Cr}^{3+} \,\vdots\vdots\, \mathrm{Co}^{2+} \mid \mathrm{Co}$

and the cell potentials would be

$$E^\circ_{\mathrm{cell(i)}} = E^\circ_{\mathrm{Cr^{3+} \mid Cr}} - E^\circ_{\mathrm{Co^{2+} \mid Co}}$$
$$= -\,0.744\ \mathrm{V} - (-\,0.277\ \mathrm{V}) = -\,0.467\ \mathrm{V}$$

$$E^\circ_{\mathrm{cell(ii)}} = E^\circ_{\mathrm{Co^{2+} \mid Co}} - E^\circ_{\mathrm{Cr^{3+} \mid Cr}}$$
$$= -\,0.277\ \mathrm{V} - (-\,0.744\ \mathrm{V}) = 0.467\ \mathrm{V}$$

Since $E^\circ_{\text{cell(ii)}}$ is positive, it is obvious that Cr can react with $Co^{2+}$ whereas Co cannot react with $Cr^{3+}$.

(**c**) Of the two potentials, $E^\circ_{Co^{2+}|Co}$ is more positive and thus $Co^{2+}$ ions are more easily reducible than $Cr^{3+}$ ions. Hence $Co^{2+}$ ions act as a more powerful oxidizing agent (strongest oxidant). On the other hand, Cr will act as a more powerful reducing agent (strongest reductant).

(**d**) The reaction of $Co^{2+}$ with $NH_3$ is

$$Co^{2+} + 6NH_3 \rightarrow [Co(NH_3)_6]^{2+}$$

which reduces the concentration of free $Co^{2+}$ ions in solution. The Nernst equation for the reaction

$$Co^{2+} + 2e^- \rightarrow Co$$

is given by
$$E_{Co^{2+}|Co} = E^\circ_{Co^{2+}|Co} - \frac{RT}{2F} \ln \frac{1}{a(Co^{2+})}$$

Thus, on decreasing the activity of $Co^{2+}$ ions, the potential $E_{Co^{2+}|Co}$ will become less positive.

(**e**) If the half-cells A and B are connected to give a galvanic cell, then the more negative electrode will serve as the negative electrode, i.e. the chromium electrode. The potential would be

$$E^\circ_{\text{cell}} = E^\circ_{Co^{2+}|Co} - E^\circ_{Cr^{3+}|Cr} = -\,0.277\text{ V} - (-\,0.744\text{ V}) = 0.467\text{ V}$$

If this cell is connected with an external battery of 1.0 V in the two given ways, we would have:

(i) *Negative terminal of the cell connected to the negative terminal of the external battery:* Since the potential of the external battery is larger than the emf of the cell, the electrons will flow from the external battery into the cell. Thus, the given cell will become an electrolytic cell with the following cell reaction:

| | | |
|---|---|---|
| *At cathode* *(negative terminal)* | $Cr^{3+} + 3e^- \rightarrow Cr$ | (i) |
| *At anode* *(positive terminal)* | $Co \rightarrow Co^{2+} + 2e^-$ | (ii) |

Multiplying Eq. (i) by 2 and Eq. (ii) by 3, then adding we get

$$2Cr^{3+} + 3Co \rightarrow 2Cr + 3Co^{2+}$$

(ii) *Negative terminal of the cell connected to the positive terminal of the external battery:* In this case, the electrons flow from the negative terminal of the cell to its positive terminal through the external battery, i.e. the cell continues to serve as the galvanic cell with the cell reaction:

$$2Cr + 3Co^{2+} \rightarrow 2Cr^{3+} + 3Co$$

---

**2.** (a) The standard potentials of the two reactions

$$M^{n+} + ne^- = M$$

$$M^{m+} + me^- = M$$

are $E^\circ_{M^{n+}|M}$ and $E^\circ_{M^{m+}|M}$, respectively. Assuming $n$ to be greater than $m$, show that the standard potential of the reaction

$$M^{n+} + (n - m)e^- \rightarrow M^{m+}$$

is given by

$$E^\circ_{M^{n+}, M^{m+} | Pt} = \frac{nE^\circ_{M^{n+}|M} - mE^\circ_{M^{m+}|M}}{n - m}$$

(b) From the data

(i) $Cr^{3+} + 3e^- \rightarrow Cr$ $\quad E^\circ_{Cr^{3+}|Cr} = -0.74$ V

(ii) $Cr^{3+} + e^- \rightarrow Cr^{2+}$ $\quad E^\circ_{Cr^{3+}, Cr^{2+}|Pt} = -0.40$ V

Compute $E^\circ$ for the reaction

(iii) $Cr^{2+} + 2e^- \rightarrow Cr$

(c) A cell is to be constructed to give a reaction

$$Cr + 2Cr^{3+} \rightarrow 3Cr^{2+}$$

show how it can be constructed by combining the half-cells $Cr^{3+} | Cr$ and $Cr^{3+}, Cr^{2+} | Pt$; $Cr^{3+} | Cr$ and $Cr^{2+} | Cr$; and $Cr^{3+}, Cr^{2+} | Pt$ and $Cr^{2+} | Cr$. Do you expect the same $E^\circ$ value in each case? If not, explain why. Calculate the equilibrium constant for the above reaction at 298 K.

***Solution***

(**a**) We have

| *Electrode reaction* | *Electrode potential* | *Free energy change* |
|---|---|---|
| (i) $M^{n+} + ne^- \rightarrow M$ | $E^\circ_{M^{n+}|M}$ | $\Delta G_1 = -nFE^\circ_{M^{n+}|M}$ |
| (ii) $M^{m+} + me^- \rightarrow M$ | $E^\circ_{M^{m+}|M}$ | $\Delta G_2 = -mFE^\circ_{M^{m+}|M}$ |
| (iii) $M^{n+} + (n - m)e^- \rightarrow M^{m+}$ | $E^\circ_{M^{n+}, M^{m+}|Pt}$ | $\Delta G_3 = -(n-m)FE^\circ_{M^{n+}, M^{m+}|Pt}$ |

Now reaction (iii) can be obtained by subtracting reaction (ii) from reaction (i). Hence, we must have

$$\Delta G^\circ_3 = \Delta G^\circ_1 - \Delta G^\circ_2$$

$$-(n-m)FE^\circ_{M^{n+}, M^{m+}|Pt} = -nFE^\circ_{M^{n+}|M} - (-mFE^\circ_{M^{m+}|M})$$

Thus $\quad E^\circ_{M^{n+}, M^{m+}|Pt} = \dfrac{nE^\circ_{M^{n+}|M} - mE^\circ_{M^{m+}|M}}{(n-m)}$

(**b**) We have:

| *Electrode reaction* | *Electrode potential* | *Free energy change* |
|---|---|---|
| (i) $Cr^{3+} + 3e^- \rightarrow Cr$ | $E^\circ_{Cr^{3+}|Cr} = -0.74$ V | $\Delta G^\circ_1 = -3F(-0.74\text{ V})$ |
| (ii) $Cr^{3+} + e^- \rightarrow Cr^{2+}$ | $E^\circ_{Cr^{3+}, Cr^{2+}|Pt} = -0.40$ V | $\Delta G^\circ_2 = -F(-0.40\text{ V})$ |
| (iii) $Cr^{2+} + 2e^- \rightarrow Cr$ | $E^\circ_{Cr^{2+}|Cr} = ?$ | $\Delta G^\circ_3 = -2FE^\circ_{Cr^{2+}|Cr}$ |

The reaction (iii) can be obtained by subtracting reaction (ii) from reaction (i) Hence, we must have.

$$\Delta G_3^\circ = \Delta G_1^\circ - \Delta G_2^\circ$$

$$-2FE^\circ_{Cr^{2+} | Cr} = 3F(0.74 \text{ V}) - F(0.40 \text{ V})$$

Thus $$E^\circ_{Cr^{2+} | Cr} = -\frac{2.22 \text{ V} - 0.40 \text{ V}}{2} = -\frac{1.82 \text{ V}}{2} = -0.91 \text{ V}$$

(**c**) The given cell reaction is

$$Cr + 2Cr^{3+} \rightarrow 3Cr^{2+}$$

The cell giving the above reaction may be constructed as follows.

(i) *Using half-cells* $Cr^{3+} | Cr$ *and* $Cr^{3+}, Cr^{2+} | Pt$.

The given reaction may be split into the following half-cell reactions:

*Right electrode, Reduction* $Cr^{3+} + e^- \rightarrow Cr^{2+}$

*Left electrode, Oxidation* $Cr \rightarrow Cr^{3+} + 3e^-$

Multiplying the former reaction by 3 and then adding it to the latter reaction, we get the given cell reaction. Thus, the cell would be

$$Cr \,|\, Cr^{3+} \,\vdots\vdots\, Cr^{3+}, Cr^{2+} \,|\, Pt$$

with $$E^\circ_{cell} = E^\circ_{Cr^{3+}, Cr^{2+} | Pt} - E^\circ_{Cr^{3+} | Cr} = -0.40 \text{ V} - (-0.74 \text{ V}) = 0.34 \text{ V}$$

and $$\Delta G^\circ = -nFE^\circ_{cell} = -3F(0.34 \text{ V}) = -F(1.02 \text{ V})$$

(ii) *Using half-cells* $Cr^{3+} | Cr$ *and* $Cr^{2+} | Cr$.

In this, case, we will have

*Right electrode, Reduction* $Cr^{3+} + 3e^- \rightarrow Cr$

*Lift electrode, Oxidation* $Cr \rightarrow Cr^{2+} + 2e^-$

Multiplying the former reaction by 2 and the latter reaction by 3 and then on adding, we get the given cell reaction. Hence, the cell would be

$$Cr \,|\, Cr^{2+} \,\vdots\vdots\, Cr^{3+} \,|\, Cr$$

with $$E^\circ_{cell} = E^\circ_{Cr^{3+} | Cr} - E^\circ_{Cr^{2+} | Cr} = -0.74 \text{ V} - (-0.91 \text{ V}) = 0.17 \text{ V}$$

and $$\Delta G^\circ = -nFE^\circ_{cell} = -6F(0.17 \text{ V}) = -F(1.02 \text{ V})$$

(iii) *Using half-cells* $Cr^{3+}, Cr^{2+} | Pt$ *and* $Cr^{2+} | Cr$.

In this case, we have

*Right electrode, Reduction* $Cr^{3+} + e^- \rightarrow Cr^{2+}$

*Left electrode, Oxidation* $Cr \rightarrow Cr^{2+} + 2e^-$

Multiplying the former reaction by 2 and then adding to it the latter reaction, we get the given cell reaction. Hence, the cell would be

$$\mathrm{Cr \mid Cr^{2+} \,\vdots\vdots\, Cr^{3+}, Cr^{2+} \mid Pt}$$

with $E^\circ_{\text{cell}} = E^\circ_{\mathrm{Cr^{3+}, Cr^{2+} \mid Pt}} - E^\circ_{\mathrm{Cr^{2+} \mid Cr}} = -0.40\ \text{V} - (-0.91\ \text{V}) = 0.51\ \text{V}$

and $\Delta G^\circ = -nFE^\circ_{\text{cell}} = -2F(0.51\ \text{V}) = -F(1.02\ \text{V})$

In all the three cases discussed above, $E^\circ_{\text{cell}}$ is different since it depends on the make up of the cell. It may be observed that the value of $\Delta G^\circ$ is the same in all the three cases as it depends only on the cell reaction.

*Calculation of Standard Equilibrium Constant* This can be calculated using the expression

$$\Delta G^\circ = -RT \ln K^\circ$$

or
$$\ln K^\circ = -\frac{\Delta G^\circ}{RT} = -\frac{-F(1.02\ \text{V})}{RT} = \frac{F(1.02\ \text{V})}{RT}$$

or
$$\log K^\circ = (1.02\ \text{V})\left(\frac{F}{2.303RT}\right) = \frac{1.02}{0.059\ 13} = 17.25$$

Hence $K^\circ = 1.78 \times 10^{17}$

---

**3.** Given the cell

$$\mathrm{Cd \mid Cd(OH)_2(s) \mid NaOH(0.01\ mol\ kg^{-1}) \mid H_2(1\ bar) \mid Pt}$$

with $E_{\text{cell}} = 0.0$ V at 298 K. If $E^\circ_{\mathrm{Cd^{2+} \mid Cd}} = -0.40$ V. Calculate (a) $K_s$ for $Cd(OH)_2$, and (b) $\Delta G$, $\Delta S$ and $\Delta H$ for the cell reaction if $(\partial E/\partial T)_p = 0.002\ \text{V K}^{-1}$.

***Solution***

For the given cell

$$\mathrm{Cd \mid Cd(OH)_2(s) \mid NaOH(0.01\ mol\ kg^{-1}) \mid H_2(1\ bar) \mid Pt}$$

we have:

| *Electrode* | *Reduction reaction* | |
|---|---|---|
| Right | $2H^+ + 2e^- \rightarrow H_2$ | (i) |
| Left | $Cd^{2+} + 2e^- \rightarrow Cd$ | (ii) |

Subtraction of Eq. (ii) from Eq. (i) gives the cell reaction:

$$\mathrm{Cd + 2H^+ \rightarrow Cd^{2+} + H_2}$$

The emf of the cell is given by

$$E_{\text{cell}} = E^\circ - \frac{RT}{2F} \ln \frac{(a_{\mathrm{Cd^{2+}}})(f_{\mathrm{H_2}}/f^\circ)}{(a_{\mathrm{H^+}})^2}$$

Assuming $f_{\mathrm{H_2}} = p(\mathrm{H_2})$ and the fact that $p(\mathrm{H_2}) = 1$ bar, we get

$$E_{\text{cell}} = E^\circ - \frac{RT}{2F} \ln \left(\frac{a_{\mathrm{Cd^{2+}}}}{a^2_{\mathrm{H^+}}}\right) \qquad \text{(iii)}$$

(**a**) Equation (iii) may be written as

$$E_{\text{cell}} = E^\circ - \frac{RT}{2F} \ln \left( \frac{a_{\text{Cd}^{2+}}\, a^2_{\text{OH}^-}}{a^2_{\text{H}^+}\, a^2_{\text{OH}^-}} \right)$$

$$= E^\circ - \frac{RT}{2F} \ln \frac{K_s^\circ}{K_w^{\circ 2}}$$

Substituting the given data, we get

$$0 = \{0 - (-40\text{ V})\} - \frac{0.059\,13\text{ V}}{2} \log \left( \frac{K_s^\circ}{K_w^{\circ 2}} \right)$$

or $$\log \left( \frac{K_s^\circ}{K_w^{\circ 2}} \right) = \frac{(0.40\text{ V})}{(0.029\,57\text{ V})} = 13.527$$

This gives $\dfrac{K_s^\circ}{K_w^{\circ 2}} = 3.36 \times 10^{13}$

i.e. $$\frac{K_s/\text{M}^3}{K_w^2/\text{M}^4} = 3.36 \times 10^{13} \quad \text{or} \quad \frac{K_s}{K_w^2} = 3.36 \times 10^{13}\text{ M}^{-1}$$

Hence $$K_s = (3.36 \times 10^{13}\text{ M}^{-1})\, K_w^2 = (3.36 \times 10^{13}\text{ M}^{-1})\,(10^{-28}\text{ M}^4)$$

$$= 3.36 \times 10^{-15}\text{ M}^3$$

(**b**) We have

$$\Delta G = -nFE = 0$$

$$\Delta S = nF\left(\frac{\partial E}{\partial T}\right)_p = 2(96\,500\text{ C mol}^{-1})(0.002\text{ V K}^{-1})$$

$$= 386\text{ C V K}^{-1}\text{ mol}^{-1}, \quad \text{i.e.} \quad 386\text{ J K}^{-1}\text{ mol}^{-1}$$

$$\Delta H = \Delta G + T\Delta S = T\Delta S = (298\text{ K})\,(386\text{ J K}^{-1}\text{ mol}^{-1})$$

$$= 115\,028\text{ J mol}^{-1} = 115.028\text{ kJ mol}^{-1}$$

---

**4.** When metallic copper is shaken with a solution of a copper salt, the reaction $Cu + Cu^{2+} \rightarrow 2Cu^+$ proceeds. When equilibrium is established at 298 K, $[Cu^{2+}]/[Cu^+]^2 = 1.66 \times 10^6\ (\text{mol dm}^{-3})^{-1}$. If the standard potential of the $Cu^{2+} \mid Cu$ half-cell is + 0.337 V, what is the standard (reduction) potential of $Cu^+ \mid Cu$ half-cell?

***Solution***

The cell producing the given reaction

$$Cu + Cu^{2+} \rightarrow 2Cu^+$$

is $$Cu \mid Cu^+ \,\vdots\vdots\, Cu^+, Cu^{2+} \mid Pt$$

and its standard emf is given by

$$E^\circ_{\text{cell}} = E^\circ_{\text{Cu}^{2+},\,\text{Cu}^+ \mid \text{Pt}} - E^\circ_{\text{Cu}^+ \mid \text{Cu}} \qquad (1)$$

The equilibrium constant of the reaction is

$$K_{eq} = \frac{[Cu^+]^2}{[Cu^{2+}]} = \frac{1}{1.66 \times 10^6 \text{ mol}^{-1} \text{ dm}^3} = 6.024 \times 10^{-7} \text{ mol dm}^{-3}$$

From the relations

$$\Delta G^\circ = -nFE^\circ_{cell} \quad \text{and} \quad \Delta G^\circ = -RT \ln K^\circ_{eq}$$

we can determine $E^\circ_{cell}$ by using the expression

$$E^\circ_{cell} = \frac{RT}{nF} \ln K^\circ_{eq} = \frac{0.059\ 13 \text{ V}}{n} \log K^\circ_{eq}; \qquad \text{(at 298 K)}$$

Substituting the data, we get

$$E^\circ_{cell} = \frac{0.059\ 13 \text{ V}}{1} \log (6.024 \times 10^{-7}) = (0.059\ 13 \text{ V})(-6.22)$$

$$= -0.367\ 8 \text{ V} \qquad (2)$$

Now the standard potential of $Cu^{2+} \mid Cu$ will be related to the potentials of $Cu^+ \mid Cu$ and $Cu^{2+}, Cu^+ \mid Pt$. We have

| *Electrode* | *Reaction* | *Potential* | $\Delta G^\circ$ |
|---|---|---|---|
| (i) $Cu^+ \mid Cu$ | $Cu^+ + e^- \rightarrow Cu$ | $E^\circ_{Cu^+ \mid Cu}$ | $-FE^\circ_{Cu^+ \mid Cu}$ |
| (ii) $Cu^{2+}, Cu^+ \mid Pt$ | $Cu^{2+} + e^- \rightarrow Cu^+$ | $E^\circ_{Cu^{2+}, Cu^+ \mid Pt}$ | $-FE^\circ_{Cu^{2+}, Cu^+ \mid Pt}$ |
| (iii) $Cu^{2+} \mid Cu$ | $Cu^{2+} + 2e^- \rightarrow Cu$ | $E^\circ_{Cu^{2+} \mid Cu}$ | $-2FE^\circ_{Cu^{2+} \mid Cu}$ |

Adding reactions of Eqs (i) and (ii), we get the reaction of Eq. (iii). Therefore

$$-FE^\circ_{Cu^+ \mid Cu} - FE^\circ_{Cu^{2+}, Cu^+ \mid Pt} = -2FE^\circ_{Cu^{2+} \mid Cu}$$

or

$$E^\circ_{Cu^{2+} \mid Cu} = \tfrac{1}{2}[E^\circ_{Cu^+ \mid Cu} + E^\circ_{Cu^{2+}, Cu^+ \mid Pt}] \qquad (3)$$

From Eqs (1), (2) and (3), we have

$$E^\circ_{Cu^{2+}, Cu^+ \mid Pt} - E^\circ_{Cu^+ \mid Cu} = -0.367\ 8 \text{ V}$$

$$E^\circ_{Cu^{2+}, Cu^+ \mid Pt} + E^\circ_{Cu^+ \mid Cu} = 2E^\circ_{Cu^{2+}, Cu} = 2 \times 0.337 \text{ V} = 0.674 \text{ V}$$

Solving for $E^\circ_{Cu^{2+}, Cu^+ \mid Pt}$ and $E^\circ_{Cu^+ \mid Pt}$, we get

$$E^\circ_{Cu^{2+}, Cu^+ \mid Pt} = 0.153 \text{ V} \quad \text{and} \quad E^\circ_{Cu^+ \mid Cu} = 0.521 \text{ V}$$

---

**5.** When silver chloride is dissolved in a large excess of ammonia, practicaly all silver can be assumed to exist as $Ag_m(NH_3)_n^{m+}$. Compute the values of $m$ and $n$ using the following two cells:

Ag | $0.379 \times 10^{-3}$ M AgCl, 1 M $NH_3$ ⁞⁞ $37.9 \times 10^{-3}$ M AgCl, 1 M $NH_3$ | Ag
$E_{cell} = 0.118\ 5$ V at 298 K
Ag | $3.4 \times 10^{-3}$ M AgCl, 1 M $NH_3$ ⁞⁞ $3.4 \times 10^{-3}$ M AgCl, 0.1 M $NH_3$ | Ag
$E_{cell} = 0.126\ 3$ V at 298 K

***Solution***

For the given cells, we have

| *Electrode* | *Reduction reaction* | |
|---|---|---|
| Right | $(\mathrm{Ag}_m(\mathrm{NH}_3)_n^{m+})_\mathrm{R} + me^- \rightarrow m\,\mathrm{Ag} + n(\mathrm{NH}_3)_\mathrm{R}$ | (i) |
| Left | $(\mathrm{Ag}_m(\mathrm{NH}_3)_n^{m+})_\mathrm{L} + me^- \rightarrow m\,\mathrm{Ag} + n(\mathrm{NH}_3)_\mathrm{L}$ | (ii) |

Subtraction of Eq. (ii) from Eq. (i) gives the cell reaction as

$$(\mathrm{Ag}_m(\mathrm{NH}_3)_n^{m+})_\mathrm{R} + n(\mathrm{NH}_3)_\mathrm{L} \rightarrow (\mathrm{Ag}_m(\mathrm{NH}_3)_n^{m+})_\mathrm{L} + n(\mathrm{NH}_3)_\mathrm{R}$$

The emf of the cell is given by

$$E_\mathrm{cell} = -\frac{RT}{mF}\ \ln\left(\frac{[\mathrm{Ag}_m(\mathrm{NH}_3)_n^{m+}]_\mathrm{L}[\mathrm{NH}_3]_\mathrm{R}^n}{[\mathrm{Ag}_m(\mathrm{NH}_3)_n^{m+}]_\mathrm{R}[\mathrm{NH}_3]_\mathrm{L}^n}\right)$$

For the first cell, we have

$$0.118\,5\ \mathrm{V} = -\left(\frac{0.059\,13\ \mathrm{V}}{m}\right)\log\left(\frac{0.379\times10^{-3}}{37.9\times10^{-3}}\right)$$
$$= -\left(\frac{0.059\,13\ \mathrm{V}}{m}\right)(-2)$$

Hence $\quad m = \dfrac{2\times0.059\,13}{0.118\,5} = 1$

For the second cell, we have

$$0.1263\ \mathrm{V} = -\left(\frac{0.059\,13\ \mathrm{V}}{m}\right)\log\left(\frac{0.1}{1}\right)^n$$

Since $m = 1$, we have

$$n = \frac{0.1263}{0.059} \simeq 2$$

Thus, the formula of the complex is $\mathrm{Ag(NH_3)_2^+}$.

---

**6.** The emf of the cell

$$\mathrm{Pt}\mid \mathrm{H_2(g)}(p = 1\ \mathrm{bar})\mid \mathrm{HA_1(0.1\ M)} \,\vdots\vdots\, \mathrm{HA_2(0.1\ M)},\ \mathrm{QH_2},\ \mathrm{Q}\mid \mathrm{Pt}$$

is found to be 0.731 4 V at 298 K, where $\mathrm{HA_1}$ and $\mathrm{HA_2}$ are two monoprotic weak acids. Answer the following:

(a) Which of the two acids is stronger?

(b) How many times is it stronger?

(c) If the dissociation constant of the stronger acid is $2.1 \times 10^{-4}$ mol dm$^{-3}$, calculate the dissociation constant of the other.

(d) What are the values of pH of the two solutions?

***Solution***

(**a**) For the given cell, we have

| *Electrode* | *Reduction reaction* | |
|---|---|---|
| Right | $Q + 2H_R^+ + 2e^- \rightarrow QH_2$ | (i) |
| Left | $2H_L^+ + 2e^- \rightarrow H_2$ | (ii) |

Subtraction of Eq. (ii) from Eq. (i) gives the cell reaction as

$$H_2 + Q + 2H_R^+ \rightarrow 2H_L^+ + QH_2$$

Since $p(H_2) = 1$ bar, the emf of the cell is given by

$$E = E^\circ - \frac{RT}{2F} \ln \left\{ \left( \frac{(a_{H^+})_L}{(a_{H^+})_R} \right)^2 \left( \frac{a_{QH_2}}{a_Q} \right) \right\}$$

Assuming $a_{QH_2} = a_Q$, we get

$$E = E^\circ - \frac{RT}{F} \ln \left( \frac{(a_{H^+})_L}{(a_{H^+})_R} \right)$$

Now $\quad E^\circ = E^\circ_{QH_2,\,Q,\,H^+\,|\,Pt} - E^\circ_{H^+|\,H_2\,|\,Pt} = E^\circ_{QH_2,\,Q,\,H^+\,|\,Pt} = 0.699\ 4\ \text{V}$

With this, the previous expression becomes

$$E = 0.699\ 4\ \text{V} - \frac{RT}{F} \ln \left\{ \frac{(a_{H^+})_L}{(a_{H^+})_R} \right\}$$

At 298 K, we have

$$E = 0.699\ 4\ \text{V} - (0.059\ 13\ \text{V}) \log \left\{ \frac{(a_{H^+})_L}{(a_{H^+})_R} \right\}$$

or

$$-\log \left\{ \frac{(a_{H^+})_L}{(a_{H^+})_R} \right\} = \frac{E - 0.699\ 4\ \text{V}}{0.059\ 13\ \text{V}}$$

Since $E = 0.731\ 4$ V, we get

$$-\log \left\{ \frac{(a_{H^+})_L}{(a_{H^+})_R} \right\} = \frac{0.731\ 4\ \text{V} - 0.699\ 4\ \text{V}}{0.059\ 13\ \text{V}}$$

or

$$\log \frac{(a_{H^+})_R}{(a_{H^+})_L} = 0.541\ 2$$

Hence

$$\frac{(a_{H^+})_R}{(a_{H^+})_L} = 3.477$$

Thus, the acid $HA_2$ is stronger than the acid $HA_1$.

(**b**) The acid $HA_2$ is 3.477 times stronger than the acid $HA_1$.

(**c**) For the two acids, we have

$$K_1 = \frac{(a_{H^+})_L (a_{A_1^-})}{a_{HA_1}} \quad \text{and} \quad K_2 = \frac{(a_{H^+})_R (a_{A_2^-})}{a_{HA_2}}$$

Assuming

$$(a_{H^+})_L = a_{A_1^-}$$

$$(a_{H^+})_R = a_{A_2^-}$$

we get $$K_1 = \frac{(a_{H^+})_L^2}{a_{HA_1}} \quad \text{or} \quad (a_{H^+})_L = (K_1 a_{HA_1})^{1/2}$$

$$K_2 = \frac{(a_{H^+})_R^2}{a_{HA_2}} \quad \text{or} \quad (a_{H^+})_R = (K_2 a_{HA_2})^{1/2}$$

Thus $$\frac{(a_{H^+})_R}{(a_{H^+})_L} = \left(\frac{K_2}{K_1}\frac{a_{HA_2}}{a_{HA_1}}\right)^{1/2}$$

As the acids are weak, we may assume

$$a_{HA_1} = a_{HA_2}$$

Thus $$\frac{(a_{H^+})_R}{(a_{H^+})_L} = \left(\frac{K_2}{K_1}\right)^{1/2}$$

Since $(a_{H^+})_R/(a_{H^+})_L$ is equal to 3.477, we get

$$\left(\frac{K_2}{K_1}\right)^{1/2} = 3.477$$

Now $K_2 = 2.1 \times 10^{-4}$ mol dm$^{-3}$, therefore

$$K_1 = \frac{K_2}{(3.477)^2} = \frac{2.1 \times 10^{-4}\ \text{mol dm}^{-3}}{(3.477)^2} = 1.74 \times 10^{-5}\ \text{mol dm}^{-3}$$

(**d**) Assuming activities equal to concentrations, we have

$$(c_{H^+})_R = (K_2 c_{HA_2})^{1/2} = \{(2.1 \times 10^{-4}\ \text{mol dm}^{-3})\ (0.1\ \text{mol dm}^{-3})\}^{1/2}$$

$$= 4.58 \times 10^{-3}\ \text{mol dm}^{-3}$$

$$(\text{pH})_R = -\log\{(c_{H^+})_R/\text{mol dm}^{-3}\} = 2.34$$

$$(c_{H^+})_L = (K_1 c_{HA_1})^{1/2} = \{(1.74 \times 10^{-5}\ \text{mol dm}^{-3})\ (0.1\ \text{mol dm}^{-3})\}^{1/2}$$

$$= 1.32 \times 10^{-3}\ \text{mol dm}^{-3}$$

$$(\text{pH})_L = -\log\{(c_{H^+})_L/\text{mol dm}^{-3}\} = 2.88$$

---

**7.** Determine the degree of hydrolysis and hydrolysis constant of anilinium hydrochloride in M/32 solution of salt at 298 K from the following cell data at 298 K:

$$\text{Pt} \mid H_2(1\ \text{bar}) \mid H^+(a = 1) \parallel \frac{M}{32}\, C_6H_5NH_2HCl\ \text{soln.} \mid H_2(1\ \text{bar}) \mid \text{Pt}$$

$$E_{cell} = -0.188\ \text{V}.$$

***Solution***

Since $E_{cell} = E_R - E_L$

therefore $E_R = E_{cell} = -\ 0.188\ V$

Now $$E_R = -\frac{RT}{F}\ln\frac{(p_{H_2}/p°)^{1/2}}{([H^+/c°])}$$

$$= \frac{RT}{F}\ln([H^+]/c°); \qquad (\text{as } p(H_2) = 1\ \text{bar})$$

Thus $$\log\{[H^+]/c°\} = E_R\left(\frac{F}{2.303RT}\right)$$

At 298 K, we have

$$\log([H^+]/c°]) = \frac{-\ 0.188\ V}{(0.059\ 13\ V)} = -\ 3.179$$

or $[H^+] = 6.62 \times 10^{-4}\ \text{mol dm}^{-3}$

The hydrolysis reaction is

$$C_6H_5NH_3^+ + H_2O \rightleftharpoons C_6H_5NH_2 + H_3O^+$$

Let $\alpha$ be the degree of hydrolysis. It is obvious that

$$[H_3O^+] = c\alpha$$

where $c$ is the concentration of anilinium hydrochloride solution.

Thus $$\alpha = \frac{[H_3O^+]}{c}$$

Substituting the values, we get

$$\alpha = \frac{6.62 \times 10^{-4}\ M}{(1/32)\ M} = 2.118 \times 10^{-2}$$

Now $$K_h = \frac{[C_6H_5NH_2][H_3O^+]}{[C_6H_5NH_3^+]} = \frac{(c\alpha)(c\alpha)}{c(1-\alpha)} = \frac{c\alpha^2}{1-\alpha}$$

$$= \frac{(1\ \text{mol dm}^{-3}/32)\ (2.118 \times 10^{-2})^2}{(1 - 0.021\ 18)}$$

$$= 1.43 \times 10^{-5}\ \text{mol dm}^{-3}$$

---

**8.** The standard potentials of the two half-cells

$$A^{(x+n)+}, A^{x+} \mid Pt \quad \text{and} \quad B^{(y+m)+}, B^{y+} \mid Pt$$

are $E_1^\circ$ and $E_2^\circ$, respectively. If $E_2^\circ$ is greater than $E_1^\circ$, answer the following.

(i) The cell

$$Pt \mid A^{x+}(m), A^{(x+n)+}(km) \,\vdots\vdots\, B^{(y+m)+}(n), B^{y+}(kn) \mid Pt$$

was constructed, where $k$ has some constant value. The cell was operated reversibly up to its exhaustion point, i.e. to a point where no more reaction in the cell takes place and hence no more current can be withdrawn from the cell. Show that

(a) the half-cell potentials at the exhaustion point are given by the expression

$$E = \frac{nE_L^\circ + mE_R^\circ}{(n+m)}$$

(b) The value of $k$ at the exhaustion point at 298 K is given by the relation

$$\log k_e = \frac{nm(E_R^\circ - E_L^\circ)}{(0.059\ 13\ \text{V})(m+n)}$$

(ii) Which of the following two cells will give a spontaneous cell reaction when $k < k_e$?

$$\text{Pt} \mid A^{(x+n)+}(km),\ A^{x+}(m) \ \vdots\vdots\ B^{(y+m)+}(n),\ B^{y+}(kn) \mid \text{Pt}$$

and

$$\text{Pt} \mid B^{(y+m)+}(n),\ B^{y+}(kn) \ \vdots\vdots\ A^{(x+n)+}(km),\ A^{x+}(m) \mid \text{Pt}$$

Which of the four species is more powerful oxidant and reductant?

***Solution***

(i) The given cell is

$$\text{Pt} \mid A^{x+}(m),\ A^{(x+n)+}(km) \ \vdots\vdots\ B^{y+}(kn),\ B^{(y+m)+}(n) \mid \text{Pt}$$

Its cell reaction may be obtained as follows:

| *Electrode* | *Reduction reaction* | |
|---|---|---|
| Right | $B^{(y+m)+} + me^- \rightarrow B^{y+}$ | (i) |
| Left | $A^{(x+n)+} + ne^- \rightarrow A^{x+}$ | (ii) |

Multiplying Eq. (1) by $n$ and Eq. (2) by $m$ and then subtracting the latter from the former, we get

$$mA^{x+} + nB^{(y+m)+} \rightarrow mA^{(x+n)+} + nB^{y+} \quad (3)$$

The cell potential would be

$$E_{\text{cell}} = E^\circ_{\text{cell}} - \frac{RT}{mnF} \ln \left\{ \frac{[A^{(x+n)+}]^m [B^{y+}]^n}{[A^{x+}]^m [B^{(y+m)+}]^n} \right\} \quad (4)$$

Now as the cell operates, concentrations $A^{x+}$ and $B^{(y+m)+}$ will decrease and those of $A^{(x+n)+}$ and $B^{y+}$ will increase. If the concentration $x$ of $A^{x+}$ disappear, concentrations of various species will be:

$$[A^{x+}] = m - x$$

$$[B^{(y+m)+}] = n - \frac{n}{m}x = n\left(\frac{m-x}{m}\right)$$

$$[A^{(x+n)+}] = km + x$$

$$[B^{y+}] = kn + \frac{n}{m}x = n\left(\frac{km+x}{m}\right)$$

From these concentrations, it follows that at any stage of the reaction, we will have

$$\frac{[A^{x+}]}{[B^{(y+m)+}]} = \frac{[A^{(x+n)+}]}{[B^{y+}]} = \frac{m}{n} \tag{5}$$

(a) The electrode potentials of the two half-cells are given by

$$E_R = E_R^\circ - \frac{RT}{mF} \ln \left( \frac{[B^{y+}]}{[B^{(y+m)+}]} \right) \tag{6}$$

$$E_L = E_L^\circ - \frac{RT}{nF} \ln \left( \frac{[A^{x+}]}{[A^{(x+n)+}]} \right) \tag{7}$$

At the exhaustion point (when no current can be withdrawn from the cell), the two potentials will become identical. The electrode potentials at the exhaustion point can be obtained by multiplying Eq. (6) by $m$, and Eq. (7) by $n$ and then adding the resultant expressions. Thus, we have

$$(mE_R + nE_L) = (mE_R^\circ + nE_L^\circ) - \frac{RT}{F} \ln \left\{ \frac{[B^{y+}][A^{x+}]}{[B^{(y+m)+}][A^{(x+n)+}]} \right\}$$

or $$(m + n)E_e = (mE_R^\circ + nE_L^\circ) - \frac{RT}{F} \ln \left\{ \frac{[A^{x+}][B^{y+}]}{[B^{(y+m)+}][A^{(x+n)+}]} \right\}$$

Making use of Eq. (5), we get

$$(m + n)E_e = mE_R^\circ + nE_L^\circ$$

or $$E_e = \frac{mE_R^\circ + nE_L^\circ}{m + n} \tag{8}$$

(b) At the exhaustion point, $E_{cell}$ will be zero. Thus from Eq. (4), we get

$$E_{cell}^\circ = \frac{RT}{mnF} \ln \left\{ \frac{[A^{(x+n)+}]^m [B^{y+}]^n}{[A^{x+}]^m [B^{(y+m)+}]^n} \right\}$$

Let $k$ at the exhaustion point be represented as $k_e$. Substituting the concentrations of various species in the above expression, we get

$$E_{cell}^\circ = \frac{RT}{mnF} \ln \left\{ \left( \frac{k_e m}{m} \right)^m \left( \frac{k_e n}{n} \right)^n \right\}$$

$$= \frac{RT}{mnF} \ln (k_e)^{m+n}$$

$$= \left( \frac{m + n}{mn} \right) \frac{RT}{F} \ln k_e \tag{9}$$

Hence

$$\ln k_e = \left( \frac{nm}{m + n} \right) \frac{F}{RT} E_{cell}^\circ$$

At 298 K, we have

$$\ln k_e = \frac{nm}{m+n} \frac{1}{(0.059\ 13\ \text{V})} (E_R^\circ - E_L^\circ) \tag{10}$$

(ii) Let us take the cell

$$\text{Pt} \mid A^{x+}(m),\ A^{(x+n)+}(km) \ \vdots\vdots\ B^{(y+m)+}(n),\ B^{y+}(kn) \mid \text{Pt}$$

Its emf as given by Eq. (4) is

$$E_{\text{cell}} = E_{\text{cell}}^\circ - \frac{RT}{mnF} \ln \left\{ \frac{[A^{(x+n)+}]^m [B^{y+}]^n}{[A^{x+}]^m [B^{(y+m)+}]^n} \right\}$$

Substituting the given concentrations, we get

$$E_{\text{cell}} = E_{\text{cell}}^\circ - \frac{RT}{mnF} \ln \left\{ \left(\frac{km}{m}\right)^m \left(\frac{kn}{n}\right)^n \right\} = E_{\text{cell}}^\circ - \frac{RT}{mnF} \ln k^{m+n}$$

$$= E_{\text{cell}}^\circ - \left(\frac{m+n}{mn}\right) \frac{RT}{F} \ln k$$

Making use of Eq. (9), we get

$$E_{\text{cell}} = \left(\frac{m+n}{mn}\right) \frac{RT}{F} \ln k_e - \left(\frac{m+n}{mn}\right) \frac{RT}{F} \ln k$$

$$E_{\text{cell}} = \left(\frac{m+n}{mn}\right) \frac{RT}{F} \ln \frac{k_e}{k} \tag{11}$$

Now since $k < k_e$, it follows that $k_e/k$ will be greater than one. Thus $E_{\text{cell}}$ as given by Eq. (11) will be positive. Hence the cell reaction will be spontaneous. Obviously, the given second cell will not produce the spontaneous cell reaction. From Eq. (3), it follows that $A^{x+}$ is the most powerful reducing agent and $B^{(y+m)+}$ is the most powerful oxidizing agent.

## REVISIONARY PROBLEMS

8.1 (a) What is the difference between a galvanic cell and an electrolytic cell? Set up galvanic cells for the following spontaneous reactions:
(i) $Zn(s) + Cu^{2+}(aq) \rightarrow Zn^{2+}(aq) + Cu(s)$
(ii) $Zn(s) + 2AgCl(s) \rightarrow 2Ag(s) + Zn^{2+}(aq) + 2Cl^-(aq)$
(b) What are the sign conventions of electrodes in galvanic and electrolytic cells? What types of reaction take place at these electrodes?

8.2 Illustrate with suitable examples, the distinction between a reversible and an irreversible cell. State the conditions for a thermodynamically reversible cell.

8.3 (a) Explain, why we cannot use a voltmeter for determining the electromotive force (emf) of a galvanic cell. Describe, briefly the potentiometric method for determining the emf of a cell.
(b) Describe briefly the construction and working of Weston standard saturated cell.

8.4 (a) Illustrate by taking the example of Daniell cell, the conventions which are adopted while writing an electrochemical cell.
(b) An electrochemical cell is written as

$$\text{Pt} \mid H_2(g)(p = 1\ \text{bar}) \mid \text{HCl(soln)}(a_1) \mid \text{AgCl(s)} \mid \text{Ag(s)}$$

Answer the following:
(i) Which electrode serves as the negative terminal and which one as the positive terminal?
(ii) In which direction the electrons flow in the external circuit (not shown)?
(iii) What are the half-cell reactions occurring in the left and right half-cells?
(iv) What is the overall cell reaction ?
Suppose now the cell is written in the reverse direction, i.e.,

$$Ag(s) \mid AgCl(s) \mid HCl(soln)\ (a_1) \mid H_2(g)(p = 1\ bar) \mid Pt$$

What would be the answers to the above four questions? How would you determine experimentally which of the above two cells lead to the spontaneous reaction?

8.5 Show by taking the typical example of the cell

$$Pt \mid H_2(g) \mid H^+(a_1) \ \vdots\vdots\ Zn^{2+}(a_2) \mid Zn$$

that the cell reaction can be generated by subtracting the reduction reaction of the left half-cell (L) from that of the right half-cell (R). Correspondingly, the emf of the cell can be determined by subtracting the reduction potential of the left half-cell from that of the right half-cell.

Suppose that the above cell is written in the reverse direction, i.e.

$$Zn \mid Zn^{2+}(a_2) \ \vdots\vdots\ H^+(a_1) \mid H_2(g) \mid Pt$$

Show that the value of emf of this cell is the same as that of the previous cell but with the opposite sign.

8.6 Derive the relation between the emf of a cell and the change in the free energy of its cell reaction. Hence show that the emf of the cell is an intensive property. How would you predict the nature of a cell reaction (whether spontaneous, or nonspontaneous or at equilibrium) from the sign of its emf value?

8.7 (a) The escaping tendency of a species carrying $z_i$ charge from a phase is represented by the electrochemical potential $\tilde{\mu}$ defined as

$$\tilde{\mu} = \mu + z_i F\phi$$

where $\mu$ is the chemical potential and $\phi$ is the electrical potential of the phase. Derive the above expression.

8.8 Derive the Nernst equation for the following half-cells.
(1) Gas-ion half-cell:
  (i) The hydrogen-hydrogen ion half-cell.
  (ii) The chlorine-chloride ion half-cell.
  (iii) The oxygen-hydroxide ion half-cell.
(2) The metal-metal ion half-cell:
  (i) $Zn\text{-}Zn^{2+}$ half-cell.
  (ii) $M\text{-}M^{n+}$ half-cell where M can be any metal.
(3) Metal-insoluble salt-anion half-cell:
  (i) Silver-silver chloride-chloride half-cell.
  (ii) Mercury-mercuric oxide-hydroxide half-cell.
  (iii) The calomel half-cell.
  (iv) Antimony-antimony oxide-hydroxide half-cell.

8.9 (a) What do you understand by the term standard electrode potential?
(b) Comment upon the statement: The absolute value of an electrode potential cannot be determined. However, its value relative to a reference electrode whose potential is taken to be zero may be determined.
(c) Explain to which electrode this arbitrary zero potential is assigned and how are the values of other electrode potentials determined relative to this electrode?

8.10 Justify the following statements:

(a) If the standard potential $E^{\circ}_{M^{n+}|M}$ is positive, then $M^{n+}$ ions can be reduced to M by hydrogen gas under standard conditions.

(b) If the standard potential $E^{\circ}_{M^{n+}|M}$ is negative, then $M^{n+}$ ions cannot be reduced by hydrogen gas under standard conditions. In fact, reverse of the reaction, occurs, i.e. $H^+$ will be reduced to $H_2$ gas by the metal M.

(c) The compounds of active metals such as Zn, Na and Mg are not reducible by hydrogen gas under standard conditions while those of noble metals such as Cu and Ag are reducible by hydrogen gas.

(d) Sign of the electrode potential determines the nature of the electrode (whether a positive terminal or a negative terminal) of a cell constituted by combining the given electrode with the standard hydrogen-hydrogen ion electrode and having a positive emf.

(e) Comparison of the values of two electrode potentials is equivalent to comparing the relative easiness with which the two reduction reactions can be carried out.

(f) If the two half-cells are combined to make a voltaic cell, the half-cell of a more positive potential forms a positive terminal whereas the half-cell of lesser positive potential forms a negative terminal.

(g) Substance which forms the soluble complex (or precipitate) with the oxidized member of the couple (oxidized form + $ne^-$ = reduced form) reduces the reduction potential of the electrode.

(h) The classification of a metal into noble or active is not absolute but very much dependent upon its environment. For example, ordinarily Ag is a noble metal but in the presence of $I^-$, $S^{2-}$ or $CN^-$ ions; it becomes an active metal.

8.11 What effect the change in the ionic activity has on the value of electrode potential involving a reaction $M^{n+} + ne^- \rightarrow M$? Show how can this change be utilized in determining the charge on the ion?

8.12 (a) Show that the metal-metal ion half-cell potential $E_{Ag^+|Ag}$ is related to the corresponding metal-insoluble salt-anion half-cell potential $E_{X^-|AgX|Ag}$ through the relation

$$E^{\circ}_{X^-|AgX|Ag} = E^{\circ}_{Ag^+|Ag} - \frac{2.303\,RT}{F} pK^{\circ}_s(AgX)$$

where $X^-$ is the halide ion ($Cl^-$ or $Br^-$ or $I^-$).

(b) It is known that

$$K_s(AgCl) > K_s(AgBr) > K_s(AgI)$$

Arrange the potential $E_{X^-|AgX|Ag}$ (where $X^-$ can be $Cl^-$, $Br^-$ and $I^-$) in increasing order of magnitude.

8.13 Justify the statement:

For a given overall reaction, the value of $\Delta G$ is a constant quantity but the emf of the cell which produces this overall reaction depends on the make up of the cell. Thus, various cells which produce the same cell reaction and which differ from each other may have different values of the emf.

Support your answer by constructing cells out of the following half-cells.

(a) $Fe^{2+} + 2e^- \rightarrow Fe$; $E^{\circ}_{Fe^{2+}|Fe} = -0.440$ V

(b) $Fe^{3+} + 3e^- \rightarrow Fe$; $E^{\circ}_{Fe^{3+}|Fe} = -0.036$ V

(c) $Fe^{3+} + e^- \rightarrow Fe^{2+}$; $E^{\circ}_{Fe^{3+}, Fe^{2+}|Pt} = 0.771$ V so as to produce the following cell reaction:

$$Fe + 2Fe^{3+} \rightarrow 3Fe^{2+}$$

8.14 What is a reference half-cell? Outline its merits relative to that of the hydrogen-hydrogen ion half-cell. Describe the calomel half-cell and derive its Nernst equation.

8.15 Starting from the Gibbs-Helmholtz equation and the relation $\Delta G = -nFE$, derive the following expressions of $\Delta H$ and $\Delta S$.

$$\Delta H = -nF\left[E - T\left(\frac{\partial E}{\partial T}\right)_p\right]$$

$$\Delta S = nF\left(\frac{\partial E}{\partial T}\right)_p$$

What are the units of $\Delta G$, $\Delta H$ and $\Delta S$ if $F$ is expressed in coulomb mole$^{-1}$ and $E$ in volt?

8.16 (a) Show that the standard equilibrium constant of the cell reaction at 298 K can be computed from the relation

$$\log K^\circ = \frac{n(E^\circ/\text{volt})}{0.059\ 13}$$

(b) Show that the essential condition for a cell reaction to be at equilibrium is

$$E_R = E_L$$

Show from the above equality, how the equilibrium constant of a reaction can be computed.

8.17 Explain, how can an accurate value of the standard half-cell potential be determined experimentaly?

8.18 Set up an electrochemical cell which can determine accurately the dissociation constant of a given weak acid.

8.19 Set up an electrochemical cell with the help of which it is possible to determine accurately the ionic-product of water.

8.20 Explain by taking an example of AgCl, how solubility product of the latter can be determined from the emf measurement of a suitable electrochemical cell.

8.21 (a) Define the term pH of a solution. Explain, how the pH value of a solution is determined by employing the following half-cells.

(i) The hydrogen-hydrogen ion half-cell
(ii) The quinhydrone half-cell.
(iii) The glass half-cell.
(iv) Antimony-antimony oxide-hydroxide half-cell.

(b) Justfy the statement that the linear relationship between the quinhydrone half-cell potential and pH of the solution, i.e.

$$E_{Q,\,QH_2,\,H^+\,|\,Pt} = E^\circ_{Q,\,QH_2,\,H^+\,|\,Pt} - \frac{2.303\,RT}{F}\text{pH}$$

holds good only up to certain pH value ($\simeq$ up to 8.5). Beyond this pH, the half-cell potential is no longer linear and hence cannot be employed in determining the pH of highly alkaline solutions.

(c) Describe the glass electrode. Explain, how the pH of a solution is determined by using the glass electrode.

8.22 (a) What do you understand by the terms:
(i) Concentration cell without liquid junction?
(ii) Concentration cell with liquid junction?

8.23 Derive the following expressions as applicable to the appropriate concentration cell without transference.

(a) $E_{cell} = \dfrac{RT}{2F} \ln \dfrac{a_1}{a_2}$ for a cell involving amalgam electrodes

(b) $E_{cell} = \dfrac{RT}{2F} \ln \dfrac{p_1}{p_2}$ for a cell involving two gas electrodes operating at different pressures

(c) $E_{cell} = \dfrac{2RT}{F} \ln \dfrac{(a_{\pm 2})}{(a_{\pm 1})}$ for a cell involving different electrolytic concentrations

8.24 (a) Describe the factors which are responsible for the liquid junction potential. Derive the following expressions for a concentration cell with transference involving 1 : 1 electrolyte.

$$E_{wlj} = \pm\, 2t_{\mp} \frac{RT}{F} \ln \frac{(a_{\pm 2})}{(a_{\pm 1})}$$

$$E_{wolj} = \pm \frac{RT}{F} \ln \frac{(a_{\pm 2})}{(a_{\pm 1})}$$

$$E_{lj} = 2t_{\mp} E_{wolj}$$

$$E_{lj} = \mp (1 - 2t_{\mp}) \frac{RT}{F} \ln \frac{(a_{\pm 2})}{(a_{\pm 1})}$$

where the upper and lower signs are meant for cells involving electrodes reversible with cations and anions, respectively, and $a_{\pm 2}$ and $a_{\pm 1}$ are the activities of the electrolytes in the right and left half cells, respectively.
(b) Explain, how is the transport number of either cation or anion determined from the study of cells with and without transport?
(c) Justify the use of KCl (or $NH_4NO_3$) in the salt bridge.

8.25 The standard potentials of the two reactions

$$M^{n+} + ne^- = M$$

and $$M^{m+} + me^- = M$$

are $E^\circ_{M^{n+}|M}$ and $E^\circ_{M^{m+}|M}$, respectively, with $n > m$. Show that for reaction

$$M^{n+} + (n - m)e^- = M^{m+}$$

the standard potential is given by

$$E^\circ_{M^{n+}, M^{m+}|Pt} = \frac{nE^\circ_{M^{n+}|M} - mE^\circ_{M^{m+}|M}}{(n - m)}$$

8.26 Explain the essential features on which the potentiometric titrations are based. Describe them with suitable examples.

8.27 Construct the curves of $E$ versus $V$, $\Delta E/\Delta V$ versus $V$ and $\Delta^2 E/\Delta V^2$ versus $V$ for the following titrations:
(a) 50 cm$^3$ of 0.1 M HCl versus 0.1 M NaOH.

(b) 50 $cm^3$ of 0.1 M $Fe^{2+}$ versus 0.1 M $Ce^{4+}$.
(c) 50 $cm^3$ of an acid solution (pH = 1.0) of 0.1 M $Fe^{2+}$ versus 0.016 67 M (= 0.1 N) $Cr_2O_7^{2-}$.
(d) 50 $cm^3$ of 0.1 M $Cl^-$ versus 0.1 M $Ag^+$.

8.28 What are redox indicators? Describe the working of redox indicators.

8.29 Discuss the role of phosphoric acid in the volumetric titration of $Fe^{2+}$ versus $Cr_2O_7^{2-}$ using diphenylamine as the internal indicator.

8.30 Show that the emf of the cell

$$\mathrm{M} \mid \mathrm{M}_x\mathrm{X}_y(a_{\pm 1}) \mid \mathrm{A}_x\mathrm{X}_n(\mathrm{s}) \mid \mathrm{A} - \mathrm{A} \mid \mathrm{A}_x\mathrm{X}_n(\mathrm{s}) \mid \mathrm{M}_x\mathrm{X}_y(a_{\pm 2}) \mid \mathrm{M}$$

is given by

$$E = -\left(\frac{x+y}{x}\right)\frac{RT}{yF}\ln\left(\frac{a_{\pm 1}}{a_{\pm 2}}\right)$$

8.31 Show that the emf of the cell

$$\mathrm{A} \mid \mathrm{A}_x\mathrm{X}_n(\mathrm{s}) \mid \mathrm{M}_x\mathrm{X}_y(a_{\pm 1}) \mid \mathrm{M}\text{-}\mathrm{M} \mid \mathrm{M}_x\mathrm{X}_y(a_{\pm 2}) \mid \mathrm{A}_x\mathrm{X}_n(\mathrm{s}) \mid \mathrm{A}$$

is given by

$$E = -\left(\frac{x+y}{y}\right)\frac{RT}{xF}\ln\left(\frac{a_{\pm 2}}{a_{\pm 1}}\right)$$

8.32 Show that the emf of the cell

$$\mathrm{M(s)} \mid \mathrm{M}_x\mathrm{X}_y(a_{\pm 1}) \vdots \mathrm{M}_x\mathrm{X}_y(a_{\pm 2}) \mid \mathrm{M(s)}$$

is given by

$$E = -t_-\left(\frac{x+y}{x}\right)\frac{RT}{yF}\ln\left(\frac{a_{\pm 1}}{a_{\pm 2}}\right)$$

8.33 Show that the emf of the cell

$$\mathrm{A(s)} \mid \mathrm{A}_x\mathrm{X}_n(\mathrm{s}) \mid \mathrm{M}_x\mathrm{X}_y(a_{\pm 1}) \vdots \mathrm{M}_x\mathrm{X}_y(a_{\pm 2}) \mid \mathrm{A}_x\mathrm{X}_n(\mathrm{s}) \mid \mathrm{A(s)}$$

is given by

$$E = -t_+\left(\frac{x+y}{x}\right)\frac{RT}{xF}\ln\left(\frac{a_{\pm 2}}{a_{\pm 1}}\right)$$

## TRY YOURSELF PROBLEMS

8.1 From the standard potential values, predict that out of the members of Group 1 of the periodic table, lithium and sodium are, respectively, the best and poorest reducing agents.

8.2 Will the permanganate ion under standard conditions be stable in aqueous medium?

8.3 Compute the potential values given in Example 8.9.6 in an acidic solution having $[H^+]$ = 1 M. From the obtained values, show that:
(i) the reduction potential of the reaction

$$\tfrac{1}{2}\mathrm{Br_2} + \mathrm{e^-} \rightarrow \mathrm{Br^-}$$

remains unchanged.
(ii) The $BrO_3^-$ ion is a better oxidizing agent (more easily reduced) in an acidic solution than in a basic solution.

(iii) $Br_2$ will not disproportionate in acidic solution to $Br^-$ and HOBr or to $Br^-$ and $BrO_3^-$.
(iv) HOBr, will however, disproportionate in the acidic solution to give $Br_2$ and $BrO_3^-$.
The values of potentials are

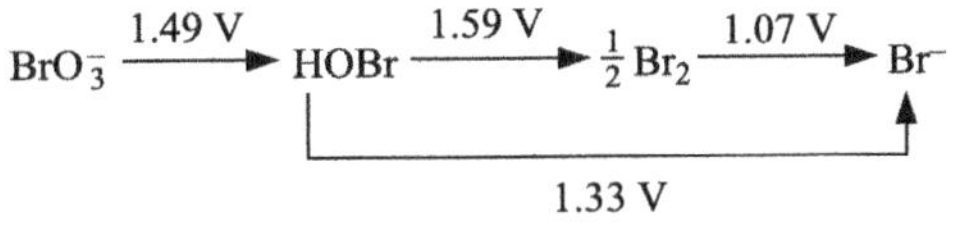


8.4 (a) Given below are the potential (in volts) data on halogens.

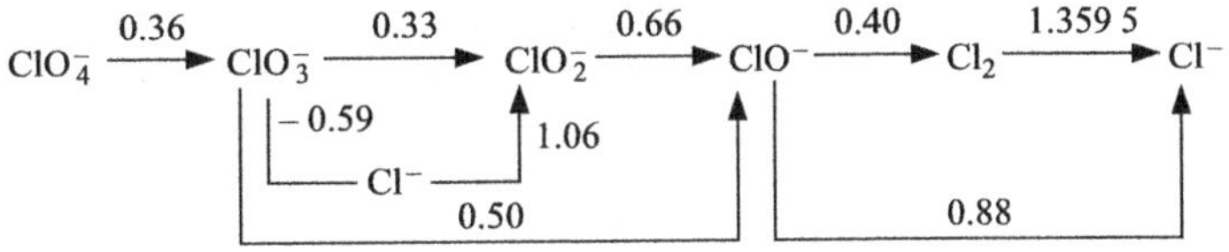


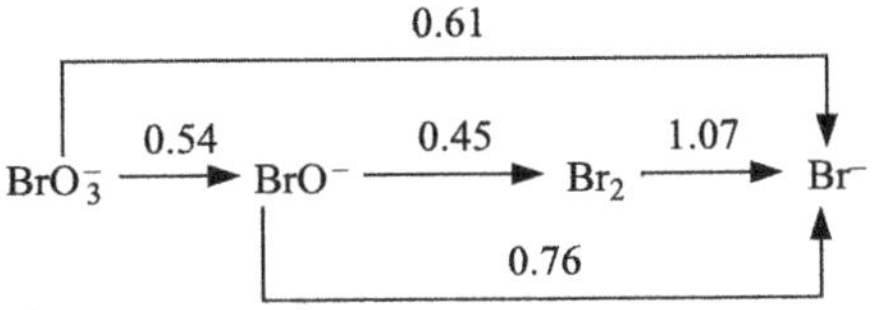


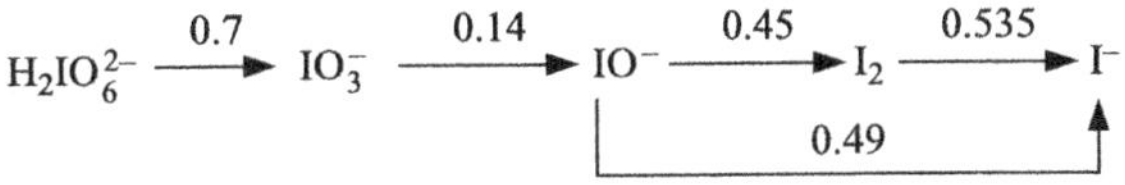


Show that
(i) $Cl_2$, $Br_2$, and $I_2$ disproportionate in basic solution as follows:

$$3X_2 + 6OH^- \rightarrow XO_3^- + 5X^- + 3H_2O.$$

(ii) $Cl_2$, $Br_2$ and $I_2$ are stable in acidic medium.
(iii) $Cl_2$ can displace $Br_2$ from $Br^-$ and $Br_2$ can displace $I_2$ from $I^-$.
(iv) The standard equilibrium constant for the disproportionation of $ClO_3^-$ to $Cl^-$ and $ClO_4^-$ is $10^{20}$.
(v) The $IO_3^-/I^-$ couple oxidizes much better in acid media (pH = 0) then in basic media (pH = 14).
(b) Justify the statement:
Thermodynamic predictions are not always kinetically reliable.

8.5 Is water easier to oxidize or reduce as the pH increases?

8.6 Calculate the equilibrium constants for the following reactions:
(a) $2Cu^+ \rightarrow Cu^{2+} + Cu$
(b) $Cu_2O + H_2O \rightarrow Cu(OH)_2 + Cu$
by using the potential data given below

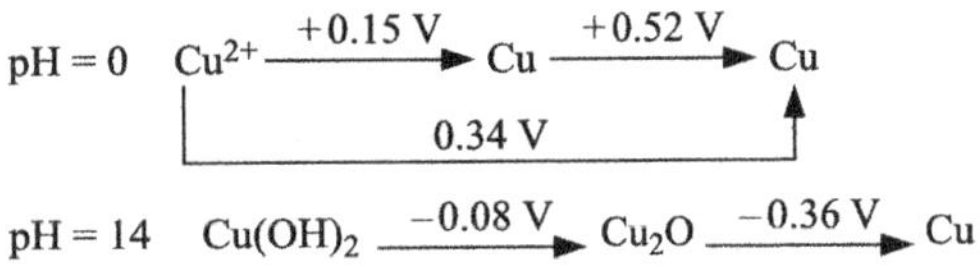

8.7 (i) Calculate the solubility product of the following (the required data are shown along with the reaction):

(a) $Cd(OH)_2$

$$Cd(OH)_2 + 2e^- \rightarrow Cd + 2OH^-; \qquad E^\circ = -0.81 \text{ V at pH} = 14$$
$$E^\circ = -0.40 \text{ V at pH} = 0$$

(b) $Pb(OH)_2$

$$Pb(OH)_2 + 2e^- \rightarrow Pb + 2OH^-; \qquad E^\circ = 0.07 \text{ V at pH} = 14$$
$$E^\circ = 0.09 \text{ V at pH} = 0$$

(ii) Calculate the potential of the $Fe^{2+} \mid Fe$ couple at pH 7.

Given: $E^\circ = 0.036$ V at pH = 0 and $K_s^\circ(Fe(OH)_2) = 6 \times 10^{-38}$.

8.8 (a) The commonly observed formation of rust when iron is exposed to moist air may be explained on the basis of reactions given below, followed by the conversion of $Fe^{2+}$ to reddish brown $Fe_2O_3$.

(i) In acidic medium:

$$Fe \rightarrow Fe^{2+} + 2e^-$$
$$O_2 + 4H^+ + 4e^- \rightarrow 2H_2O(l)$$

(ii) In neutral solutions:

$$Fe \rightarrow Fe^{2+} + 2e^-$$
$$O_2 + 2H_2O + 4e^- \rightarrow 4OH^-$$

Show that under standard conditions, the overall reactions given above will take place spontaneously.

(b) One way to prevent a buried iron pipe from rusting is to connect it with a piece of wire to a magnesium or zinc rod. What is the electrochemical principle for this action?

8.9 Justify the value $-0.828$ V for $E^\circ$ of the following half-cell in basic solution:

$$H_2O(l) + e^- \rightarrow \tfrac{1}{2} H_2(g) + OH^-$$

8.10 What will be the potential of standard hydrogen electrode at 323 K?

8.11 In the metal amalgam-metal ion electrode, if the amalgam is saturated with the solute metal, the electrode is equivalent to the pure metal-metal ion electrode. Explain, why is it so?

8.12 Can you employ potassium chloride salt bridge as a connecting devise to connect silver nitrate solution?

8.13 The emf of a cell is found to be 0.30 V. What would happen if this cell is connected to a 1.5 V battery with its negative terminal connected to the (a) negative terminal of the battery, and (b) positive terminal of the battery?

8.14 In the titration of $Fe^{2+}$ versus $Cr_2O_7^{2-}$, the internal indicator *N*-phenylanthranilic acid is usually used in the presence of larger amount of $H^+$ ions. Discuss, what effect the latter has on the transition potential of the indicator and how far it is helpful in locating the end point.

8.15 Derive the following expressions:

(a) For 1 : 1 electrolyte (e.g., HCl)

$$m_+ = m_- = m_\pm$$

(b) For 1 : 2 electrolyte (e.g., $SnCl_2$)

$$m_+ = \frac{1}{(2)^{2/3}} m_\pm \quad \text{and} \quad m_- = (2)^{1/3} m_\pm$$

(c) For 1 : 3 electrolyte (e.g., $FeCl_3$)

$$m_+ = \frac{1}{(3)^{3/4}} m_\pm \quad \text{and} \quad m_- = (3)^{1/4} m_\pm$$

(d) For 1 : 4 electrolyte (e.g., $SnCl_4$)

$$m_+ = \frac{1}{(4)^{4/5}} m_\pm \quad \text{and} \quad m_- = (4)^{1/5} m_\pm$$

## NUMERICAL PROBLEMS

**Construction of Galvanic Cell**

8.1 Construct the galvanic cell for each of the following reactions. Draw the diagram of the cell indicating clearly the anode, cathode and the direction of flow of electrons in the external circuit.

(i) $\frac{1}{2} Br_2(l) + Ag(s) \rightarrow AgBr(s)$

(ii) $\frac{1}{2} Br_2(l) + I^-(aq) \rightarrow Br^-(aq) + \frac{1}{2} I_2(s)$

(iii) $2Hg(l) + Cl_2(g) \rightarrow Hg_2Cl_2(s)$

(iv) $AgCl(s) + \frac{1}{2} H_2(g) \rightarrow HCl(aq) + Ag(s)$

(v) $H_2(g) + \frac{1}{2} O_2(g) \rightarrow H_2O(l)$

(vi) $HgO(s) + H_2(g) \rightarrow Hg(l) + H_2O(l)$

(vii) $AgCl(s) + I^-(aq) \rightarrow AgI(s) + Cl^-(aq)$

**Cell Reaction and Cell Potential**

8.2 Write the cell reactions for the following cells and determine the emfs of the cells under standard conditions.

(i) $Pb \mid PbSO_4(s) \mid SO_4^{2-}(aq) \,\vdots\vdots\, Cu^{2+}(aq) \mid Cu$

(ii) $Ag \mid AgBr(s) \mid Br^-(aq) \,\vdots\vdots\, Cl^-(aq) \mid AgCl(s) \mid Ag$

8.3 Determine the cell reaction, $E^\circ_{cell}$ and $E_{cell}$ for the following cells at 298 K. Assume activity and fugacity equal to molar concentration and pressure, respectively.

(i) $Cu \mid CuSO_4(0.02\ M) \,\vdots\vdots\, Fe^{2+}(0.2\ M), Fe^{3+}(0.01\ M) \mid Pt$

(ii) $Zn \mid ZnCl_2(0.02\ M) \,\vdots\vdots\, Na_2SO_4(0.1\ M) \mid PbSO_4(s) \mid Pb$

(iii) $Pt \mid Fe^{2+}(a = 1.0), Fe^{3+}(a = 0.1) \,\vdots\vdots\, Cl^-(a = 0.001) \mid AgCl(s) \mid Ag$

(iv) $Pt \mid Cl_2(g, 1.12\ bar) \mid Cl^-(a = 0.526) \,\vdots\vdots\, I^-(a = 0.10) \mid AgI(s) \mid Ag(s)$

8.4 Write the cell reactions and potentials of the following cells:

$Pt \mid H_2(g, 0.4\ bar) \mid HCl(a_\pm = 3.0) \mid AgCl(s) \mid Ag$

$Pt \mid H_2(g, 1\ bar) \mid H_2SO_4(m = 0.05\ mol\ kg^{-1}, \gamma = 0.34) \mid Hg_2SO_4(s) \mid Hg$

**Computation of Cell Potential**

8.5 Given are the following data.

$MnO_4^- + 8H^+ + 5e^- \rightarrow Mn^{2+} + 4H_2O; \quad E^\circ = 1.51\ V$

$Mn^{3+} + e^- \rightarrow Mn^{2+}; \quad E^\circ = 1.51\ V$

Calculate the standard potential for

$MnO_4^- + 8H^+ + 4e^- \rightarrow Mn^{3+} + 4H_2O$ (*Ans.* 1.51 V)

8.6 (a) For a 1 : 1 electrolyte in aqueous solution the extended Debye-Hückel law is given as

$$\log \gamma_{\pm} = -\ 0.509\ m'^{1/2} + Bm'$$

where $B$ is constant and $m' = m/m^{\circ}$ (here $m^{\circ}$ is 1 mol $kg^{-1}$). Show that emf of the cell

$$\text{Pt} \mid H_2(\text{g, 1 bar}) \mid \text{HCl}(m) \mid \text{AgCl} \mid \text{Ag}$$

is given by

$$E_{\text{cell}} + (0.118\ 3\ \text{V}) \log m' - (0.06\ \text{V}) \sqrt{m'} = E^{\circ}_{\text{Cl}^- | \text{AgCl} | \text{Ag}} - 0.118\ 3\ \text{V})\ Bm'$$

(b) If a plot is drawn between $E_{\text{cell}} + (0.118\ 3\ \text{V}) \log m' - (0.06\ \text{V}) \sqrt{m'}$ and $m'$, what will be its slope and intercept?
(c) Given the following data for the cell

$$\text{Pt} \mid H_2(\text{g, 1 bar}) \mid \text{HCl}(m) \mid \text{AgCl(s)} \mid \text{Ag}$$

| | | | | |
|---|---|---|---|---|
| $m$/mol $kg^{-1}$ | 0.009 | 0.014 | 0.025 | 0.055 |
| $E$/V | 0.469 5 | 0.447 8 | 0.419 6 | 0.381 2 |

Compute $E^{\circ}_{\text{Cl}^- | \text{AgCl} | \text{Ag}}$

(d) What will be the mean activity coefficient of 0.02 mol $kg^{-1}$ HCl at 298 K?
(e) The emf of the cell given in part (c) is 0.352 4 V at 298 K when the HCl is at a concentration of 0.1 mol $kg^{-1}$. What is (i) the activity of the HCl, (ii) the mean activity coefficient of the HCl at this concentration, and (iii) the pH of the acid?

(*Ans.* (c) 0.225 V, (d) 0.875, (e) 0.080, 0.795, 1.10)

8.7 The standard free energy change for the reaction

$$H_2(\text{g, 1 atm}) + \tfrac{1}{2} O_2(\text{g, 1 atm}) \rightarrow H_2O(\text{l})$$

is – 237.19 kJ $mol^{-1}$ at 298 K. (i) Write the electrode reactions if the reaction is to occur in a galvanic cell. (ii) Calculate the standard electrode potential for the electrodes $OH^- \mid O_2 \mid$ Pt and $H^+ \mid O_2 \mid$ Pt. The cell potential producing the reaction

$$H_2O(\text{l}) \rightarrow H^+(a = 1) + OH^-(a = 1)$$

is – 0.827 7 V.

(*Ans.* 0.403 V, 1.229 V)

**Computation of Thermodynamic Data**

8.8 For the lead storage cell

$$\text{Pb} \mid PbSO_4 \mid H_2SO_4(m = 1\ \text{mol kg}^{-1}) \mid PbSO_4 \mid PbO_2 \mid \text{Pt}$$

the emf is found to vary

$$E_{\text{cell}}/\text{V} = 1.917\ 31 + 51.1 \times 10^{-6}(t/^{\circ}\text{C}) + 108 \times 10^{-8}(t/^{\circ}\text{C})^2$$

in the temperature range 0 to 60 °C
(a) Calculate $\Delta G$, $\Delta H$ and $\Delta S$ for the cell reaction at 273 K and 298 K.
(b) For the half-cells at 298 K

$$PbO_2(\text{s}) + SO_4^{2-} + 4H^+ + 2e^- \rightarrow PbSO_4(\text{s}) + 2H_2O; \quad E^{\circ} = 1.684\ 9\ \text{V}$$

$$PbSO_4(\text{s}) + 2e^- \rightarrow \text{Pb(s)} + SO_4^{2-}; \quad E^{\circ} = -\ 0.355\ 3\ \text{V}$$

Calculate the mean ionic activity coefficient in 1 molal $H_2SO_4$ at 298 K. Assume that the activity of water is unity.

(*Ans.* (a) at 0 °C: $\Delta G = -$ 370.03 kJ mol$^{-1}$; $\Delta H = -$ 366.80 kJ mol$^{-1}$; $\Delta S$ = 10.83 J K$^{-1}$ mol$^{-1}$
at 25 °C: $\Delta G = -$ 370.43 kJ mol$^{-1}$; $\Delta H = -$ 364.11 kJ mol$^{-1}$; $\Delta S$ = 21.23 kJ mol$^{-1}$
(b) 0.130).

8.9 For the cell

$$\text{Pt} \mid H_2(\text{g, 1 bar}) \mid \text{NaOH(1 M), NaCl(1 mol dm}^{-3}) \mid \text{AgCl(s)} \mid \text{Ag}$$

the potentials at different temperatures are

| | | | |
|---|---|---|---|
| $t$/°C | 20 | 25 | 30 |
| $E$/V | 1.047 68 | 1.051 88 | 1.056 08 |

(a) Calculate the values of p$K_w^o$ at these temperatures.
(b) Calculate the enthalpy and entropy of ionization of water at 298 K.
Given: $E^o_{Cl^- \mid AgCl \mid Ag}$ = 0.222 5 V
(*Ans.* 14.185, 14.018, 13.858, 55.86 kJ mol$^{-1}$, $-$ 81.049 J K$^{-1}$ mol$^{-1}$)

8.10 The emf of the cell

$$\text{Cd} \mid \text{CdCl}_2\cdot 2\tfrac{1}{2}\,H_2O\text{, sat. solution} \mid \text{AgCl} \mid \text{Ag}$$

at 298 K is 0.675 33 V and $(\partial E/\partial T)_p$ is $-$ 6.5 × 10$^{-4}$ V K$^{-1}$. Calculate $\Delta G$, $\Delta H$ and $\Delta S$ for the cell reaction at 298 K.

(*Ans.* $-$ 130.33 kJ mol$^{-1}$, $-$ 167.74 kJ mol$^{-1}$ and $-$ 125.52 J K$^{-1}$ mol$^{-1}$)

8.11 The emf of the cell

$$\text{Ag} \mid \text{AgCl(s)} \mid \text{KCl(soln)} \mid Hg_2Cl_2\text{(s)} \mid \text{Hg}$$

is 0.045 5 V at 298 K and the temperature coefficient is 3.38 × 10$^{-4}$ V K$^{-1}$. What is reaction that takes place in the cell and what are free energy, enthalpy and entropy changes at 298 K?

(*Ans.* $-$ 8.81 kJ mol$^{-1}$, 10.66 kJ mol$^{-1}$ and 63.43 J K$^{-1}$ mol$^{-1}$)

8.12 At 273 K, a calorimetric determination of $\Delta H$ for the reaction

$$\text{Zn} + 2\text{AgCl} \rightarrow \text{ZnCl}_2 + 2\text{Ag}$$

yielded $-$ 217.78 kJ mol$^{-1}$, while the emf of the corresponding cell was 1.015 4 V. What was $(\partial E/\partial T)_p$ of the cell? (*Ans.* $-$4.14 × 10$^{-4}$ V K$^{-1}$)

8.13 Consider the following cell arrangement.

$$\text{Ag(s)} \mid \text{AgCl(s)} \mid \text{NaCl(aq)} \mid Hg_2Cl_2\text{(s)} \mid \text{Hg}$$

(a) Write down the half-cell reactions.
(b) The standard emfs of the cell at various temperatures are as follows:

| | | | | |
|---|---|---|---|---|
| $T$/K | 291 | 298 | 303 | 311 |
| $E°$/mV | 43.0 | 45.4 | 47.1 | 50.1 |

Calculate $\Delta G$, $\Delta S$ and $\Delta H$ for the reaction

$$2\text{Ag(s)} + Hg_2Cl_2\text{(s)} \rightarrow 2\text{AgCl(s)} + 2\text{Hg(l)}$$

8.14 At 298 K the free energy of formation of $H_2O(l)$ is – 237.23 kJ $mol^{-1}$, while that of its ionization to hydrogen and hydroxyl ions is 79.71 kJ $mol^{-1}$. Determine the emf at 298 K of the cell

$Pt \mid H_2(g, 1 \text{ bar}) \mid H^+(1 \text{ M}) \,\vdots\vdots\, OH^-(1 \text{ M}) \mid O_2(g, 1 \text{ bar}) \mid Pt$ (*Ans.* 0.403 V)

**Computation of Equilibrium Constant**

8.15 Evaluate the standard equilibrium constant for the formation of the triiodide ion

$$I_2(aq) + I^- \rightleftharpoons I_3^-$$

from the knowledge that

$I_2(aq) + 2e^- \rightarrow 2I^-$; $E° = 0.619\,7$ V

and $I_3^- + 2e^- \rightarrow 3I^-$; $E° = 0.535\,5$ V (*Ans.* 708)

8.16 Given that

$Ag(CN)_2^- + e^- \rightarrow Ag + 2CN^-$; $E° = -\,0.31$ V

and $Ag^+ + e^- \rightarrow Ag$; $E° = +\,0.80$ V

Calculate the standard equilibrium constant for the reaction

$Ag(CN)_2^- \rightleftharpoons Ag^+ + 2CN^-$ (*Ans.* $1.68 \times 10^{-19}$)

8.17 Given the information that

$Cu^{2+} + 2e^- \rightarrow Cu$; $E° = 0.34$ V

$Cu^+ + e^- \rightarrow Cu$; $E° = 0.52$ V

$Cu^{2+} + Cl^- + e^- \rightarrow CuCl(s)$; $E° = 0.54$ V

Calculate the standard solubility product of CuCl. (*Ans.* $3.6 \times 10^{-7}$)

8.18 Calculate the partial equilibrium pressure of $O_2$ above $H_2O_2$ at 298 K.

(*Ans.* $2.758 \times 10^{-12}$ atm)

8.19 Calculate the standard equilibrium constant for each of the following reactions:

(i) $Fe^{2+} + Cu^{2+} \rightarrow Fe^{3+} + Cu^+$

(ii) $5Fe^{2+} + MnO_4^- + 8H^+ \rightarrow 5Fe^{3+} + Mn^{2+} + 4H_2O$

(iii) $Zn^{2+} + 4CN^- \rightarrow Zn(CN)_4^{2-}$

(iv) $PbSO_4 \rightarrow Pb^{2+} + SO_4^{2-}$

(v) $2Fe^{2+} + Cl_2 \rightarrow 2Cl^- + 2Fe^{3+}$

(vi) $AlF_6^{3-} \rightarrow Al^{3+} + 6F^-$

(*Ans.* (i) $3.6 \times 10^{-11}$, (ii) $8.0 \times 10^{61}$, (iii) $8.0 \times 10^{-16}$, (iv) $1.7 \times 10^{-8}$, (vi) $1.6 \times 10^{-21}$)

8.20 Compute the standard equilibrium constant of the reaction at 298 K

$2Au + Au^{3+} \rightarrow 3Au^+$. Given:

$Au^{3+} + 2e^- \rightarrow Au^+$; $E° = 1.41$ V

$Au^{3+} + 3e^- \rightarrow Au$; $E° = 1.50$ V

(*Ans.* $7.37 \times 10^{-10}$)

8.21 The emf of the cell

$$Ag \mid 0.1\ M\ AgNO_3 \mid sat.\ NH_4NO_3 \mid AgCl\ in\ 0.1\ M\ KCl \mid Ag$$

is – 0.45 V at 298 K. Calculate the standard solubility product and solubility of AgCl in water. Given that 0.1 M KCl is 85% and 0.1 M $AgNO_3$ is 82% dissociated.

(*Ans.* $1.707 \times 10^{-10}$, $1.875 \times 10^{-5}$ mol $dm^{-3}$)

8.22 The standard equilibrium constant for the reaction

$$2Cu^+ \rightleftharpoons Cu^{2+} + Cu(s)$$

at 298 K is $1.646 \times 10^6$. (a) Construct a cell in which the reaction could occur. (b) Calculate the standard emf of the cell. (c) Evaluate the standard potential of $Cu^+ \mid Cu$ electrode.

(*Ans.* (b) 0.368 V, (c) – 0.521 V)

8.23 Calculate the equilibrium constants of the reactions

$$Hg + Hg^{2+} \rightleftharpoons Hg_2^{2+}$$

$$Cu + Cu^{2+} \rightleftharpoons Cu_2^{2+}$$

8.24 How would you use the arrangement

$$Ag \mid AgCl\ sat.\ soln. \mid AgCl(s) \mid Ag$$

to determine the solubility product of silver chloride at 298 K.

8.25 (a) Calculate $K_s^\circ$ for $CrSO_4$ from the following data:

$$Cr + SO_4^{2-} \rightarrow CrSO_4(s) + 2e^- \qquad E^\circ_{298K} = 0.4\ V$$

$$Cr \rightarrow Cr^{3+} + 3e^- \qquad E^\circ = 0.5\ V$$

(b) Given $E^\circ$ is 0.152 V for $Ag + I^- = AgI + e^-$ at 298 K and $E^\circ$ for $Ag \rightarrow Ag^+ + e^-$ is – 0.80 V at 298 K, calculate $K_s^\circ$ for AgI.

(*Ans.* (a) $10^{-6}$, (b) $1.2 \times 10^{-17}$)

8.26 The solubility of AgBr is $2.1 \times 10^{-6}$ mol $kg^{-1}$ at 298 K. Find the emf of the cell

$$Ag \mid AgBr\ sat.\ solution \mid AgBr(s) \mid Ag$$ (*Ans.* – 0.536 V)

8.27 Calculate ideally the free energy change upon diluting 1 mol NaCl from 0.020 mol $kg^{-1}$ to 0.001 mol $kg^{-1}$ at 298 K.

(*Ans.* – 148 40 J)

8.28 Calculate $\Delta G^\circ$ and the standard equilibrium constant for the reaction

$$H_2(g) + O_2(g) \rightleftharpoons H_2O_2(l)$$

at 298 K. At what pressure would $\Delta G$ be equal to zero? Given that:

$$O_2 + 2H^+ + 2e^- = H_2O_2; \quad E^\circ = 0.682\ V$$

(*Ans.* – 131.26 kJ $mol^{-1}$, $1.148 \times 10^{23}$, $5.9 \times 10^{-12}$ atm)

8.29 At 298 K, the potential of the cell

$$Ag \mid AgI \mid KI(1\ M) \,\vdots\vdots\, AgNO_3(0.001\ M) \mid Ag$$

is 0.72 V. The mean ionic activity coefficient of 1 M KI may be taken as 0.65, and of 0.001 M $AgNO_3$ as 0.98. (a) What is the solubility of AgI? (b) What is the solubility of AgI in pure water? (*Ans.* $6.59 \times 10^{-16}$ M, $2.07 \times 10^{-8}$ M)

8.30 If the standard solubility product of $Hg_2Cl_2$ is $4.2 \times 10^{-18}$ for the reaction

$$Hg_2Cl_2(s) \rightleftharpoons Hg_2^{2+} + 2Cl^-$$

and if the decimolar calomel electrode with half-cell reaction

$$\frac{1}{2} Hg_2Cl_2(s) + e^- \rightarrow Hg(l) + Cl^-(0.1 \text{ M})$$

has electrode potential of 0.334 V, find $E^\circ$ of the half reaction

$$\frac{1}{2} Hg_2^{2-} + e^- \rightleftharpoons Hg(l) \qquad (Ans.\ 0.788 \text{ V})$$

8.31 (a) Excess solid silver is placed in a solution of 0.1 M ferric ion. Calculate the equilibrium concentration of $Ag^+$ and $Fe^{3+}$ ions.
(b) Finely divided Ni is added to a solution in which the activity of $Sn^{2+}$ is 0.1. What will be the activities of $Ni^{2+}$ and $Sn^{2+}$ when equilibrium is established.
(*Ans.* (a) $[Ag^+] = 0.08$ M (b) $a_{Ni^{2+}} = 0.1$, $a_{Sn^{2+}} = 5.7 \times 10^{-5}$)

**Computation of pH**

8.32 For the cell

$$Pt \mid H_2(1 \text{ bar}) \mid HCl \mid AgCl \mid Ag$$

$E^\circ = 0.222$ V and $E = 0.332$ V at 298 K. What is the pH of the HCl solution? (*Ans.* 0.93)

8.33 A hydrogen electrode and calomel cell are used to determine the pH of a solution on a mountain where the barometric pressure is 500 Torr. The hydrogen is allowed to bubble out of the electrode at the atmospheric pressure prevailing there. If the pH is calculated to be 4.0, what is the correct pH of the solution?

**Liquid Junction Potential**

8.34 Calculate the emf of the following cell at 298 K.

$$Ag \mid AgCl(s) \mid NaCl(a = 0.01) \,\vdots\, NaCl(a = 0.025) \mid AgCl(s) \mid Ag$$

The transport number of $Na^+$ is 0.39. (*Ans.* – 0.018 4 V)

**Concentration Cell**

8.35 Calculate the emf of the amalgam concentration cell at 333 K:

$$Zn(Hg)(c_1) \mid ZnSO_4(aq) \mid Zn(Hg)(c_2)$$

when $c_1$ and $c_2$ are $6.08 \times 10^{-5}$ and $2.28 \times 10^{-3}$ mol dm$^{-3}$, respectively. The amalgams may be considered ideal. (*Ans.* – 0.046 6 V)

8.36 If activities equal concentrations, what is the potential at 298 K of

$$Pt \mid Na(0.2\% \text{ in Hg}) \mid NaCl(0.15 \text{ mol kg}^{-1}) \mid AgCl(s) \mid Ag \mid AgCl(s) \mid NaCl\ (0.037\,2 \text{ mol kg}^{-1}) \mid Na\ (0.2\% \text{ in Hg}) \mid Pt$$

(*Ans.* – 0.072 V)

8.37 Determine the cell reaction and potential of the cell

$$Zn(s) \mid ZnCl_2(m = 0.02 \text{ mol kg}^{-1}, \gamma = 0.643) \mid AgCl(s) \mid Ag - Ag \mid AgCl(s) \mid ZnCl_2(m = 1.50 \text{ mol kg}^{-1}, \gamma = 0.290) \mid Zn(s)$$

(*Ans.* 0.093 V)

8.38 For the cell with transference

$$Pt \mid H_2(p = 1 \text{ bar}) \mid H_2SO_4(m = 0.005 \text{ mol kg}^{-1}, \gamma = 0.643) \,\vdots\, H_2SO_4(m = 2.0 \text{ mol kg}^{-1}, \gamma = 0.125) \mid H_2(p = 1 \text{ bar}) \mid Pt$$

$E = 0.03$ V at 298 K. Find the cell reaction and the transport number of the hydrogen ions. (*Ans.* 0.82)

8.39 The concentration cell

$$Ag(s) \mid AgCl(s) \mid KCl(0.5\ M) \mid K(Hg) \mid KCl(0.05\ M) \mid AgCl \mid Ag$$

has an emf of – 0.107 4 V at 298 K. (a) Formulate the corresponding cell with transference, (b) Write the equations for the cell reactions for the withdrawal of $2F$ of electricity, (c) The emf of the cell in (a) is 0.053 57 V. Calculate $t(Cl^-)$.

(*Ans.* 0.55)

8.40 The emf at 298 K is – 0.335 V for the cell

$$Pt \mid H_2(1\ bar) \mid HCl(0.1\ mol\ kg^{-1}) \mid AgCl \mid Ag - Ag \mid AgCl \mid HCl(10^{-4}\ mol\ kg^{-1}, \gamma_\pm = 1) \mid H_2(1\ bar) \mid Pt$$

Calculate (a) $\Delta G$ for the cell reaction at 298 K (write down the cell reaction), (b) the mean activity coefficient of 0.1 mol kg$^{-1}$ HCl, and (c) $\Delta G$ for the cell reaction at 273 K, assuming $\Delta H$ to be zero and $\Delta S$ independent of temperature.

(*Ans.* $\Delta G$ = 32 200 J mol$^{-1}$, $\gamma_\pm$ = 0.69, $\Delta G°$(273 K) = 29 600 J mol$^{-1}$)

8.41 Given the cell

$$Ag \mid AgCl(s) \mid KCl(0.01\ mol\ kg^{-1}, \gamma_\pm = 0.90) \vdots KCl(0.1\ mol\ kg^{-1}, \gamma_\pm = 0.80) \mid AgCl(s) \mid Ag$$

(a) Write the cell reaction and (b) Calculate the emf of the cell at 298 K. Given $t_- = 0.60$.

(*Ans.* $t_+K^+(0.01\ mol\ kg^{-1}) + t_+Cl^-(0.01\ mol\ kg^{-1}) \rightarrow t_+K^+(0.1\ mol\ kg^{-1}) + t_+Cl^-(0.1\ mol\ kg^{-1})$; $E$ = – 0.045 V)

# Annexure I

# Concept of Activity

**Definition of Activity** The variation of chemical potential with pressure of an ideal gas is given by

$$\mu_i = \mu_i^\circ + RT \ln\left(\frac{p_i}{p^\circ}\right) \tag{1}$$

where $\mu_i^\circ$ is the standard chemical potential of the $i$th gas at the standard-state pressure ($p^\circ$) of 1 bar. To have the same type of dependence of any substance (gas, liquid, solid), G.N. Lewis modified Eq. (1) as

$$\mu_i = \mu_i^\circ + RT \ln a_i \tag{2}$$

where $a_i$, a dimensionless quantity, is known as activity of the $i$th substance.

The activity of a substance existing in different conditions may be defined as follows.

**For a Substance in the Gaseous Phase** We have

$$\text{Ideal gas} \qquad a_i = \frac{p_i}{p^\circ} \tag{3}$$

$$\text{Real gas} \qquad a_i = \frac{f_i}{f^\circ} \qquad \text{(where } f \text{ is fugacity of the gas)} \tag{4}$$

**For a Substance in Solid or Liquid Phase** The variation of chemical potential of a liquid or solid with pressure is given by

$$\mu_i = \mu_i^\circ + V_{\mathrm{m},\,i}(p - p^\circ) \tag{5}$$

Comparing this with Eq. (2), we get

$$RT \ln a_i = V_{\mathrm{m},\,i}(p - p^\circ) \tag{6}$$

or $\quad a_i = \exp\{V_{\mathrm{m},\,i}(p - p^\circ)/RT\}$

Small variations in pressure do not have a significant effect on the activity.

**For Components in an Ideal Liquid Solution** Since each component of liquid solution is in equilibrium with its vapour, we will have

$$\mu_i(\mathrm{l}) = \mu_i(\mathrm{g})$$

The expressions of $\mu_i(\text{l})$ and $\mu_i(\text{g})$ are

$$\mu_i(\text{l}) = \mu_i^\circ(\text{l}) + RT \ln a_i$$

$$\mu_i(\text{g}) = \mu_i^\circ(\text{g}) + RT \ln\left(\frac{p_i}{p^\circ}\right) \qquad \text{(assuming vapour to behave ideally)}$$

Hence $\quad \mu_i^\circ(\text{l}) + RT \ln a_i = \mu_i^\circ(\text{g}) + RT \ln\left(\frac{p_i}{p^\circ}\right)$

For a pure liquid, $a_i = 1$ and $p_i = p_i^*$. Thus

$$\mu_i^\circ(\text{l}) = \mu_i^\circ(\text{g}) + RT \ln\left(\frac{p_i^*}{p^\circ}\right)$$

Substituting this in previous expression, we get

$$RT \ln a_i = RT \ln\left(\frac{p_i}{p_i^*}\right)$$

or $$a_i = \left(\frac{p_i}{p_i^*}\right) \qquad (7)$$

Since the liquid solution behaves ideally, $p_i = x_i p_i^*$ (Raoult's law), we get

$$a_i = x_i \qquad (8)$$

**For Componetns in Nonideal Liquid Solution**

The components of a nonideal solution exhibit deviations from Raoult's law. To account for the nonideality, the activity is defined as

$$a_i = x_i \gamma_i \qquad (9)$$

where $\gamma_i$, known as activity coefficient, is a measure of deviations from Raoult's law. Since the activity of a component approaches its amount fraction as $x_i$ approaches unity, we will have

$$\gamma_i \rightarrow 1 \quad \text{as} \quad x_i \rightarrow 1 \qquad (10)$$

The components exhibiting positive deviations from Raoult's law, $\gamma_i$ is greater than unity, and those exhibiting negative deviations, $\gamma_i$ is less than unity.

Since $a_i = (p_i/p_i^*)$, we will have

$$x_i \gamma_i = \frac{p_i}{p_i^*}$$

or $$\gamma_i = \frac{p_i}{x_i p_i^*} \qquad (11)$$

**For Components in Dilute Real Solutions**

In dilute real solutions, solute follows Henry's law whereas solvent follows Raoult's law, i.e.

For solvent $\quad p_i = x_i p_i^*$ (Raoult's law)

For solute $\quad p_i = x_i k_H$ (Henry's law)

Proceeding similarly to nonideal solution, it can be shown that

For solvent $$\gamma_i = \frac{p_i}{x_i p_i^*} \tag{12}$$

For solute $$\gamma_i' = \frac{p_i}{x_i k_H} \tag{13}$$

The activity coefficient for solute follows the following characteristics.

(i) $\gamma_i' \to 1$ as $x_i \to 0$
(ii) $\gamma_i' > 1$ if there are positive deviations from Henry's law
(iii) $\gamma_i' < 1$ if there are negative deviations from Henry's law

**For Electrolytes in Solution**

The chemical potential of a completely dissociated electrolyte is given by

$$\mu(M_xX_y) = x\mu(M^{y+}) + y\mu(X^{x-})$$

Since $\mu = \mu^\circ + RT \ln a$, we get

$$\mu(M_xX_y) = x[\mu^\circ(M^{y+}) + RT \ln a_{M^{y+}}] + y[\mu^\circ(X^{x-}) + RT \ln a_{X^{x-}}]$$
$$= x\mu^\circ(M^{y+}) + y\mu^\circ(X^{x-}) + RT \ln \{(a_{M^{y+}})^x (a_{X^{x-}})^y]$$
$$= \mu^\circ(M_xX_y) + RT \ln a_{M_xX_y} \tag{14}$$

where $$a_{M_xX_y} = (a_{M^{y+}})^x (a_{X^{x-}})^y \tag{15}$$

written simply as

$$a = (a_+)^x (a_-)^y \tag{16}$$

**Activity in Terms of Molality**

While dealing with electrolytic solution, it is customary to use the molality. The expressions relating activities of ions with the corresponding molalities are

$$a_+ = \left(\frac{m_+}{m^\circ}\right)\gamma_+ \tag{17}$$

$$a_- = \left(\frac{m_-}{m^\circ}\right)\gamma_- \tag{18}$$

where $m^\circ$ is the standard-state molality and is equal to 1 mol kg$^{-1}$. Here after, Eqs (17) and (18) are simply written as

$$a_+ = m_+'\gamma_+ \quad \text{and} \quad a_- = m_-'\gamma_- \tag{19}$$

where $m_+' = m_+/m^\circ$ and $m_-' = m_-/m^\circ$.

**Expressions of Mean Activity**

Since individual activities cannot be determined, it is more appropriate to express the activity of an ion in terms of mean activity of the electrolyte. The expression is

$$(a_\pm)^\nu = (a_+)^x (a_-)^y \tag{20}$$

where $\nu = x + y$ is the total number of ions produced due to the dissociation of 1 molecule of electrolyte $M_xX_y$. Comparing Eqs (16) and (20), we get

$$a = (a_\pm)^\nu$$

Now since $a_+ = m'_+\gamma_+$ and $a_- = m'_-\gamma_-$, it follows that

$$(a_\pm)^\nu = (m'_+\,\gamma_+)^x (m'_-\,\gamma_-)^y = (m'^x_+\,m'^y_-)(\gamma^x_+\,\gamma^y_-) = m'^\nu_\pm\,\gamma^\nu_\pm$$

or $$a_\pm = m'_\pm\,\gamma_\pm \tag{21}$$

To write mean molality $m'_\pm$ of the electrolyte $M_xX_y$ in terms of its molality, we proceed as follows.

$$m'^{x+y}_\pm = m'^x_+\,m'^y_- = (x\,m')^x(y\,m')^y = x^x y^y m'^{x+y}$$

Thus $$m'_\pm = (x^x y^y)^{1/(x+y)} m' \tag{22}$$

The above relations may be supplemented by the following examples.

$$a(\mathrm{MX}) = a^2_\pm = m'^2_\pm\gamma^2_\pm = (m'_+m'_-)\gamma^2_\pm = m'^2\gamma^2_\pm$$
$$a(\mathrm{MX_2}) = a^3_\pm = m'^3_\pm\gamma^3_\pm = (m'_+m'^2_-)\gamma^3_\pm = (m')(2m')^2\gamma^3_\pm = 4m'^3\gamma^3_\pm$$
$$a(\mathrm{MX_3}) = a^4_\pm = m'^4_\pm\gamma^4_\pm = (m'_+m'^3_-)\gamma^4_\pm = (m')(3m')^3\gamma^4_\pm = 27m'^4\gamma^4_\pm$$
$$a(\mathrm{M_2X_3}) = a^5_\pm = m'^5_\pm\gamma^5_\pm = (m'^2_+m'^3_-)\gamma^5_\pm = (2m')^2(3m')^3\gamma^5_\pm$$
$$= 108m'^5\gamma^5_\pm$$

# Annexure II

## Derivation of Debye-Hückel Law

Because of ionic interactions, many of bulk properties of a solution containing an electrolyte are not related directly with the actual concentration of electrolyte but with the effective concentration or activity of the electrolyte.

Depending on the way the concentration is expressed, the actual concentration $c$ and the activity $a$ are related to each other through the expression

$$a = (c/c^{\circ})\gamma_c$$

$$a = (m/m^{\circ})\gamma_m \tag{1}$$

$$a = x\gamma_x$$

where $\gamma_x$ is known as the rational activity coefficient whereas $\gamma_c$ and $\gamma_m$ are practical activity coefficients.

The individual ion activities and activity coefficients cannot be determined experimentally. Since both the positively- and negatively-charged ions produced from a given electrolyte determine the bulk properties of a system, mean ion activity ($a_{\pm}$) and mean ion activity coefficients ($\gamma_{\pm}$) are used instead of individual ion activities and activity coefficients. These are defined by

$$(a_{\pm})^{\nu} = (a_{+})^{\nu_+}(a_{-})^{\nu_-} \tag{2}$$

and

$$(\gamma_{\pm})^{\nu} = (\gamma_{+})^{\nu_+}(\gamma_{-})^{\nu_-} \tag{3}$$

where $\nu_+$ and $\nu_-$ are the number of cations and anions, respectively, derived from each molecule of the electrolyte and $\nu = \nu_+ + \nu_-$.

The chemical potential $\mu_i$ of an ion $i$ in an ideal solution is given by

$$\mu_i = \mu_i^* + RT \ln x_i \tag{4}$$

For nonideal solution, Eq. (4) is written as

$$\mu_i = \mu_i^* + RT \ln a_i$$

or

$$\mu_i = \mu_i^* + RT \ln x_i + RT \ln \gamma_i \tag{5}$$

**Physical Significance of Activity Coefficient**

The activity coefficient can theoretically be predicted on the basis of the Debye-Hückel theory of ionic atmosphere. According to the latter, each ion in a solution is surrounded by a spherical and symmetrical ionic atmosphere consisting of other ions whose net charge is opposite to that of the central ion. The activity coefficient expression in Eq. (5) may be considered as the contribution of this ionic atmosphere. Since $\mu_i$ represents the change in free energy of the system which would occur if 1 mol of the *i*th ions were added to a large quantity of it, the term $RT \ln \gamma_i$ in Eq. (5) may be regarded as the contribution of ion atmosphere towards the total energy of the ion. The contribution per ion is $kT \ln \gamma_i$ and may be equated to the work which must be performed to give an ion of electrical potential $\phi_i$ (due to its atmosphere) its charge $z_i e$. The work done in this process is given by

$$w = \int_0^{z_i e} \phi_i \, \mathrm{d}(z_i e)$$

Equating this with $kT \ln \gamma_i$, we get

$$kT \ln \gamma_i = \int_0^{z_i e} \phi_i \, \mathrm{d}(z_i e) \tag{6}$$

**Electrical Potential of an Ion Due to Its Ionic Atmosphere**

The electrostatic potential $\phi$ at a given point in the vicinity of a positive (or negative ion is related to the charge density $\rho$ at the given point by the Poisson equation, viz.;

$$\frac{\partial^2 \phi}{\partial x^2} + \frac{\partial^2 \phi}{\partial y^2} + \frac{\partial^2 \phi}{\partial z^2} = -\frac{\rho}{\varepsilon} \tag{7}$$

where $x$, $y$ and $z$ are the rectangular coordinates of the given point and $\varepsilon$ is the permittivity of the solvent. In terms of polar coordinates, Eq. (7) is given by

$$\frac{1}{r^2}\frac{\partial}{\partial r}\left(r^2 \frac{\partial \phi}{\partial r}\right) = -\frac{\rho}{\varepsilon} \tag{8}^{\dagger}$$

**Charge Density at the Given Point**

Let the charge density per unit volume at some point in the vicinity of an ion (positive or negative ion) be required. On the average there will be more ions of opposite charge in the vicinity of a given ion. The concentrations of positive and negative ions ($N_+$, $N_-$) in terms of the total concentrations of these ions ($N_+^0$, $N_-^0$) may be written following the Boltzmann distribution law as

$$N_+ = N_+^0 \exp(-z_+ e\phi / kT) \tag{9a}$$

$$N_- = N_-^0 \exp(-z_- e\phi / kT) \tag{9b}$$

where $\phi$ is the electrical potential at the point under study and $z_+$ and $z_-$ are the charge numbers (including sign) of positive and negative ions, respectively. If it be assumed that $kT$ is much larger than $z_+ e\phi$ or $|z_-| e\phi$, Eqs. (9) may be simplified by expanding the two exponential terms and ignoring all but the first terms in the expansion. Thus, we get

$$N_+ = N_+^0 (1 - z_+ e\phi / kT) \tag{10a}$$

$$N_- = N_-^0 (1 - z_- e\phi / kT) \tag{10b}$$

---

$^{\dagger}$There is no angular dependence in Eq. (8) as the charge density is spherical symmetrical in the vicinity of a positive ion.

The charge density at the given point will be given by

$$\rho = N_+ z_+ e + N_- z_- e$$

where $e$ is the elementary charge (i.e. magnitude of electronic charge). Substituting for $N_+$ and $N_-$ from Eqs. (10), we get

$$\rho = z_+ e N_+^0 (1 - z_+ e\phi/kT) + z_- e N_-^0 (1 - z_- e\phi/kT)$$

Making use of electroneutrality condition

$$N_+^0 z_+ e + N_-^0 z_- e = 0$$

in the above expression, we get

$$\rho = -(N_+^0 z_+^2 e^2 \phi/kT) - (N_-^0 z_-^2 e^2 \phi/kT)$$

that is $$\rho = -\sum_i N_i^0 z_i^2 e^2 \phi/kT \qquad (11)$$

where the summation is to be carried over all the ions.

**Expression of Electrical Potential $\phi$**

Substituting Eq. (11) in Eq. (8), we get

$$\frac{1}{r^2}\frac{\partial}{\partial r}\left(r^2 \frac{\partial \phi}{\partial r}\right) = \frac{1}{\varepsilon}\left\{\sum_i N_i^0 z_i^2 e^2 \phi/kT\right\} \qquad (12a)$$

or $$\frac{1}{r^2}\frac{\partial}{\partial r}\left(r^2 \frac{\partial \phi}{\partial r}\right) = \kappa^2 \phi \qquad (12b)$$

where $$\kappa = \left\{\frac{1}{\varepsilon}\left(\frac{e^2}{kT}\right)(\Sigma N_i^0 z_i^2)\right\}^{1/2} \qquad (13a)$$

The expression of $\kappa$ in terms of molar concentrations of ions is

$$\kappa = \left[\frac{1}{\varepsilon}\left(\frac{e^2}{kT}\right)\{\Sigma(N_A\, c_i) z_i^2\right]^{1/2} = \left[\frac{2}{\varepsilon}\left(\frac{e^2 N_A}{kT}\right)\mu\right]^{1/2} \qquad (13b)$$

where $N_A$ is the Avogadro's constant and $\mu$ is the ionic strength of the solution defined as $\frac{1}{2}\Sigma c_i z_i^2$.

The general solution of Eq. (12b) is given by

$$\phi = \frac{A \exp(-\kappa r)}{r} + \frac{A' \exp(\kappa r)}{r}$$

where $A$ and $A'$ are constants. In order to have finite value of $\phi$ for larger value of $r$ and the fact that $\phi \to 0$ as $r \to \infty$, the constant $A'$ must be set equal to zero. Thus, we have

$$\phi = \frac{A \exp(-\kappa r)}{r} \qquad (14)$$

**Expression of the Constant *A***

The condition of electroneutrality requires that the total charge of the atmosphere around a given charged ion ($z_i e$) must be equal and of opposite sign of the charge of the ion itself, i.e. $-z_i e$. The charge around the ion can be determined by considering the charge carried by a spherical shell of thickness d$r$ at a distance $r$ from the ion and then integrating the expression from the closest distance $a$ that atmosphere and central ion may approach to a distance of infinity. Thus, we have

$$-z_i e = \int_a^\infty \rho(4\pi r^2)\,\mathrm{d}r \tag{15}$$

The dependence of $\rho$ on $r$ can be found by substituting the expression of $\phi$ from Eq. (14) in Eq. (11). Thus, we have

$$\rho = -\sum_i N_i^0 z_i^2 (e^2/kT)\phi$$

$$= -\sum_i N_i^0 z_i^2 (e^2/kT)\left\{\frac{A\exp(-\kappa r)}{r}\right\}$$

which on making use of Eq. (13a) becomes

$$\rho = -A\kappa^2\varepsilon\frac{\exp(-\kappa r)}{r} \tag{16}$$

Substituting Eq. (16) in Eq. (15), we get

$$-z_i e = -\int_a^\infty (4\pi A\kappa^2\varepsilon) r\exp(-\kappa r)\,\mathrm{d}r$$

$$= -(4\pi A\kappa^2\varepsilon)\left[r\frac{\exp(-\kappa r)}{(-\kappa)} - \frac{\exp(-\kappa r)}{\kappa^2}\right]_a^\infty$$

$$= -(4\pi A\kappa^2\varepsilon)\left(\frac{\kappa a + 1}{\kappa^2}\right)\exp(-\kappa a)$$

$$= -4\pi A\varepsilon(\kappa a + 1)\exp(-\kappa a)$$

or
$$A = \frac{z_i e}{4\pi\varepsilon}\,\frac{\exp(\kappa a)}{\kappa a + 1} \tag{17}$$

Substituting Eq. (17) in Eq. (14), we get

$$\phi = \frac{z_i e}{4\pi\varepsilon}\,\frac{\exp(\kappa a)}{1 + \kappa a}\,\frac{\exp(-\kappa r)}{r} \tag{18}$$

For the distance $a$ of closest approach to the given ion, Eq. (18) becomes

$$\phi = \frac{z_i e}{4\pi\varepsilon a}\frac{1}{1+\kappa a} \quad \text{or} \quad \phi = \frac{z_i e}{4\pi\varepsilon a} - \frac{z_i e}{4\pi\varepsilon}\frac{\kappa}{1+\kappa a} \tag{19}$$

Equation (19) indicates that the electrical potential is composed of two terms; the first term represents the potential due to the given ion itself and the second term represents the potential on the ion due to its ionic atmosphere, i.e.

$$\phi = \phi(\text{ion}) + \phi(\text{atmosphere}) \tag{20}$$

where $$\phi(\text{ion}) = \frac{z_i e}{4\pi\varepsilon a} \tag{21}$$

$$\phi(\text{atmosphere}) = -\frac{z_i e}{4\pi\varepsilon}\frac{\kappa}{1+\kappa a} \tag{22}$$

**Relation between $\gamma_i$ and $\mu$**

Substituting the expression of $\phi$ from Eq. (22) in Eq. (6), we get

$$kT \ln \gamma_i = -\int_0^{z_i e}\frac{(z_i e)}{4\pi\varepsilon}\frac{\kappa}{1+\kappa a}\,\mathrm{d}(z_i e)$$

or $$\ln \gamma_i = -\frac{z_i^2 e^2}{2kT(4\pi\varepsilon)}\frac{\kappa}{1+\kappa a} \tag{23a}$$

or $$\log \gamma_i = -\frac{z_i^2 e^2}{(2.303)(2kT)(4\pi\varepsilon)}\frac{\kappa}{1+\kappa a}$$

Substituting the expression of $\kappa$ (Eq. 13b), we get

$$\log \gamma_i = -\frac{z_i^2 e^2}{(2.303)(2kT)(4\pi\varepsilon)}\frac{\{(2e^2 N_A/\varepsilon kT)\mu\}^{1/2}}{1+\{(2e^2 N_A/\varepsilon kT)\mu\}^{1/2}a}$$

$$= -\frac{z_i^2 A\sqrt{\mu}}{1+Ba\sqrt{\mu}} \tag{24}$$

where $$A = \frac{e^2}{(2.303)(2kT)(4\pi\varepsilon)}\left(\frac{2e^2 N_A}{\varepsilon kT}\right)^{1/2} \tag{25}$$

$$B = \left(\frac{2e^2 N_A}{\varepsilon kT}\right)^{1/2} \tag{26}$$

**Expression of Mean Ionic Activity Coefficient**

Since $\gamma_\pm^{\nu} = \gamma_+^{\nu_+}\gamma_-^{\nu_-}$, we get

$$\nu \log \gamma_\pm = \nu_+ \log \gamma_+ + \nu_- \log \gamma_-$$

Substituting log $\gamma_+$ and log $\gamma_-$ from Eq. (24), we get

$$\nu \log \gamma_\pm = -\frac{(\nu_+ z_+^2 + \nu_- z_-^2)A\sqrt{\mu}}{1 + Ba\sqrt{\mu}} \tag{27}$$

Using the condition of electroneutrality $\nu_+ z_+ = -\nu_- z_-$, we get

$$\nu_+ z_+^2 + \nu_- z_-^2 = \nu_+\left(-\frac{\nu_- z_-}{\nu_+}\right)^2 + \nu_- z_-^2 = \nu_- z_-^2\left(\frac{\nu_-}{\nu_+} + 1\right)$$

$$= \left(\frac{\nu_-}{\nu_+}\right) z_-^2(\nu_+ + \nu_-) = -z_+ z_-(\nu_+ + \nu_-)$$

With this Eq. (27) becomes

$$\log \gamma_\pm = \frac{(z_+ z_-)A\sqrt{\mu}}{1 + Ba\sqrt{\mu}} \qquad (\text{since } \nu = \nu_+ + \nu_-) \tag{28}$$

### Numerical Values of the Constants *A* and *B*

The values of *A* and *B* for water at 298 K are as follows.

$$A = \left[\frac{(1.6 \times 10^{-19}\ \mathrm{C})^2}{(2.303)(2)(1.38 \times 10^{-23}\ \mathrm{J\ K^{-1}})(298\ \mathrm{K})(4)(3.14)(78 \times 8.854 \times 10^{-12}\ \mathrm{C^2\ s^2\ kg^{-1}\ m^{-3}})}\right]$$

$$\times \left[\frac{2(1.6 \times 10^{-19}\mathrm{C})^2(6.022 \times 10^{23}\ \mathrm{mol^{-1}})}{(78 \times 8.854 \times 10^{-12}\ \mathrm{C^2\ s^2\ kg^{-1}\ m^{-3}})(1.38 \times 10^{-23}\ \mathrm{J\ K^{-1}})(298\ \mathrm{K})}\right]^{1/2}$$

$$= (1.558 \times 10^{-10}\ \mathrm{m})(1.042 \times 10^8\ \mathrm{m^{1/2}\ mol^{1/2}})$$

$$= 0.016\,24\ \mathrm{m^{3/2}\ mol^{-1/2}} = (0.016\,24)(10^{3/2}\ \mathrm{dm^{3/2}\ mol^{-1/2}})$$

$$= 0.514\ \mathrm{mol^{-1/2}\ dm^{3/2}}$$

$$B = \left[\frac{2(1.6 \times 10^{-19}\ \mathrm{C})^2(6.022 \times 10^{23}\ \mathrm{mol^{-1}})}{(78 \times 8.854 \times 10^{-12}\ \mathrm{C^2\ s^2\ kg^{-1}\ m^{-3}})(1.38 \times 10^{-23}\ \mathrm{J\ K^{-1}})(298\ \mathrm{K})}\right]^{1/2}$$

$$= 1.042 \times 10^8\ \mathrm{m^{1/2}\ mol^{-1/2}} = 1.042 \times 10^8 (10\ \mathrm{dm})^{1/2}\ \mathrm{mol^{-1/2}}$$

$$= 3.30 \times 10^8\ \mathrm{dm^{1/2}\ mol^{-1/2}}$$

### Limiting and Extended Forms of the Debye-Hückel Equation

For a very dilute solution $Ba\sqrt{\mu} << 1$, hence Eq. (28) simplies to

$$\log \gamma_\pm = -z_+|z_-|A\sqrt{\mu} \tag{29}$$

Equation (29) is known as the limiting form of the Debye-Hückel equation. According to Eq. (29), log $\gamma_\pm$ decreases with increase in the ionic strength of the solution. In actual practice, activity coefficients first decreases and show a turning point at some value of $\mu$, after which they progressively increase. Equation (28) may be modified by adding a term which increases with increase in ionic strength such as

$$\log \gamma_{\pm} = -\frac{A\ z_{+}\ |z_{-}|\ \sqrt{\mu}}{1 + Ba\sqrt{\mu}} + b\mu \tag{30}$$

Equation (30) is usually referred to as extended Debye-Hückel equation.

# Annexure III
## Operation of Definition of pH

The defining equation pH = – log $a_H$ (or commonly used definition pH = – log $\{[H^+]/\text{mol dm}^{-3}\}$ in dilute solution) is replaced by the following operational definition.

$$\text{pH(X)} = \text{pH(S)} - \frac{E_S - E_X}{(RT \ln 10)/F}$$

where $E_S$ is the emf of the cell

$$\begin{matrix}\text{Reference} \\ \text{electrode}\end{matrix} \left|\right| \text{KCl (aq, } m > 3.5 \text{ mol kg}^{-1}) \,\vdots\vdots\, \begin{matrix}\text{Solution S} \\ \text{of known pH}\end{matrix} \left|\right. H_2(g) \left|\right. \text{Pt}$$

and $E_X$ is the emf of the cell in which the solution S in the above cell is replaced by the solution X of unknown pH.

The various solutions S at different temperatures are available. For example, the following aqueous solutions at 25 °C may be used

| | | |
|---|---|---|
| (i) | KH tartrate (saturated solution) | pH = 3.557 |
| (ii) | KH phthalate ($m$ = 0.05 mol kg$^{-1}$) | pH = 4.008 |
| (iii) | $KH_2PO_4$ ($m$ = 0.025 mol kg$^{-1}$) + $Na_2HPO_4$ ($m$ = 0.25 mol kg$^{-1}$) | pH = 6.865 |
| (iv) | $KH_2PO_4$ ($m$ = 0.008 695 mol kg$^{-1}$) + $Na_2HPO_4$ ($m$ = 0.030 43 mol kg$^{-1}$) | pH = 7.413 |
| (v) | $Na_2B_4O_7$ ($m$ = 0.01 mol kg$^{-1}$) | pH = 9.180 |

The reference electrode is usually a calomel electrode, silver|silver chloride electrode or a thallium amalgam|thallous chloride electrode. In practice, a glass electrode is used in place of hydrogen electrode.

In the restricted range of dilute aqueous solution (concentration < 0.1 mol dm$^{-3}$) which is neither strongly acidic or alkaline (2 < pH < 12), the above procedure yields pH which lies in the range

$$\text{pH} = -\log\{([H^+]/\text{mol dm}^{-3})\,\gamma_\pm\} \pm 0.02$$

where $\gamma_\pm$ denotes the mean ionic activity coefficient of a typical unit-univalent electrolyte in the solution.

# Appendix I

## Units and Conversion Factors

**CGS Units vis-à-vis SI Units**

| *Physical quantity* | *CGS units* | | *SI units* | |
|---|---|---|---|---|
| | *Name* | *Symbol* | *Name* | *Symbol* |
| length | centimetre | cm | metre | m |
| | Angstrom ($10^{-8}$ cm) | Å | | |
| mass | gram | g | kilogram | kg |
| time | second | s | second | s |
| temperature | celsius<br>kelvin | °C<br>°K | kelvin | K |
| energy | calorie | cal | joule | J |
| | kilocalorie | kcal | kilojoule | kJ |
| | litre-atmosphere | lit-atm | | |
| | ergs | erg | | |
| electric current | ampere | A | ampere | A |

**Units Derived From the Basic SI Units**

| *Physical quantity* | *Name of unit* | *Symbol and definition* |
|---|---|---|
| force | newton | $N = kg\ m\ s^{-2}$ or $J\ m^{-1}$ |
| energy | joule | $J = kg\ m^2\ s^{-2}$ or N m |
| electric charge | coulomb | C = A s |
| potential difference | volt | $V = kg\ m^2\ s^{-3}\ A^{-1}$ or $J\ A^{-1}\ s^{-1}$ |
| resistance | ohm | $\Omega = kg\ m^2\ s^{-3}\ A^{-2}$ or $V\ A^{-1}$ |
| frequency | hertz = cycle per second | Hz |
| area | square metre | $m^2$ |
| volume | cubic metre | $m^3$ |
| density | kilogramme per cubic metre | $kg\ m^{-3}$ |
| velocity | metre per second | $m\ s^{-1}$ |
| angular velocity | radian per second | $rad\ s^{-1}$ |
| acceleration | metre per square second | $m\ s^{-2}$ |
| pressure | newton per square metre or pascal | $N\ m^{-2}$ or Pa |

**Conversion of CGS Units to SI Units**

| *Quantity* | *Unit* | *Equivalent** |
|---|---|---|
| length | Å | $10^{-10}$ m = $10^{-1}$ nm = $10^{2}$ pm |
| volume | litre | $10^{-3}$ m$^3$ = dm$^3$ |
| force | dyne | $10^{-5}$ N |
| energy | erg | $10^{-7}$ J |
| | cal | 4.184 J |
| | eV | 1.602 1 × $10^{-19}$ J |
| | eV/mole | 96.5 kJ mol$^{-1}$ |
| pressure | atmosphere | 101.325 kN m$^{-2}$ |
| | mmHg (or Torr) | 133.322 N m$^{-2}$ |
| | bar ($10^6$ dynes/cm$^2$) | $10^5$ N m$^{-2}$ |
| viscosity | poise | $10^{-1}$ kg m$^{-1}$ s$^{-1}$ |

* Symbols used for fractions and multiples are as follows.

| *Fraction* | *Prefix* | *Symbol* | *Multiples* | *Prefix* | *Symbol* |
|---|---|---|---|---|---|
| $10^{-1}$ | deci | d | 10 | deca | da |
| $10^{-2}$ | centi | c | $10^{2}$ | hecto | h |
| $10^{-3}$ | milli | m | $10^{3}$ | kilo | k |
| $10^{-6}$ | micro | μ | $10^{6}$ | mega | M |
| $10^{-9}$ | nano | n | $10^{9}$ | giga | G |
| $10^{-12}$ | pico | p | $10^{12}$ | tera | T |
| $10^{-15}$ | femto | f | $10^{15}$ | peta | P |

**Values of Some Physico-Chemical Constants**

| *Constant* | *SI units* |
|---|---|
| Acceleration of gravity, $g$ | 9.806 65 m s$^{-2}$ |
| Avogadro's constant, $N_A$ | 6.022 05 × $10^{23}$ mol$^{-1}$ |
| Bohr magneton, $\mu_B$ | 9.274 08 × $10^{-24}$ A m$^2$ |
| Bohr radius, $a_0$ | 5.2971 77 × $10^{-11}$ m |
| Boltzmann constant, $k$ | 1.380 66 × $10^{-23}$ J K$^{-1}$ |
| Elementary charge, $e$ | 1.602 19 × $10^{-19}$ C |
| Electronic rest mass, $m_e$ | 9.109 53 × $10^{-31}$ kg |
| Faraday constant, $F$ | 9.648 46 × $10^{4}$ C mol$^{-1}$ |
| Gas constant, $R$ | 8.314 J K$^{-1}$ mol$^{-1}$ |
| Molar volume of ideal gas at 0 °C and 1 atm, $V_m$ | 2.241 4 × $10^{-2}$ m$^3$ mol$^{-1}$ |
| Planck's constant, $h$ | 6.626 18 × $10^{-34}$ J s |
| Proton rest mass, $m_p$ | 1.672 65 × $10^{-27}$ kg |
| Vacuum speed of light, $c_0$ | 2.997 925 × $10^{8}$ m s$^{-1}$ |
| Standard atmospheric pressure | 101.325 kPa |
| Permittivity of vacuum, $\varepsilon_0$ | 8.854 188 × $10^{-12}$ C$^2$ s$^2$ kg$^{-1}$ m$^{-2}$ |
| Relative permittivity of water at 25 °C | 78 |

# Index

Made in the USA
Monee, IL
08 November 2026

68007853R00326